Cognitive Psychology

COGNITIVE PSYCHOLOGY

Gabriel Radvansky
Mark Ashcraft
Antonia Ypsilanti
Lambros Lazuras

Pearson

Harlow, England • London • New York • Boston • San Francisco • Toronto • Sydney
Dubai • Singapore • Hong Kong • Tokyo • Seoul • Taipei • New Delhi
Cape Town • São Paulo • Mexico City • Madrid • Amsterdam • Munich • Paris • Milan

PEARSON EDUCATION LIMITED
KAO Two
KAO Park
Harlow
CM17 9NA
United Kingdom
Tel: +44 (0)1279 623623
Web: www.pearson.com/uk

First edition published 2025 (print and electronic)

ISBN: 978-1-292-73015-8 (print)
978-1-292-46197-7 (eTextbook)
978-1-292-73016-5 (ePub)

British Library Cataloguing-in-Publication Data
A catalogue record for the print edition is available from the British Library

Library of Congress Cataloging-in-Publication Data

10 9 8 7 6 5 4 3 2 1
29 28 27 26 25

Cover design by Kelly Miller
Front cover image Jeronimo Ramos / Adobe

Print edition typeset in Sabon MT Std 10/12 by Straive
Printed and bound by CPI UK

NOTE THAT ANY PAGE CROSS REFERENCES REFER TO THE PRINT EDITION

While we work hard to present unbiased, fully accessible content, we want to hear from you about any concerns or needs with this Pearson product so that we can investigate and address them:

- Please contact us with concerns about any potential bias at https://www.pearson.com/report-bias.html
- For accessibility-related issues, such as using assistive technology with Pearson products, alternative text requests, or accessibility documentation, email the Pearson Disability Support team at support@pearson.com

Brief contents

Preface xi

1 **Understanding cognitive psychology** 1
2 **The neural basis of cognition** 33
3 **Sensation and perception** 56
4 **Attention** 101
5 **Short-term and working memory** 134
6 **Learning and remembering** 172
7 **Memory and forgetting** 213
8 **The building blocks of language** 246
9 **Language in context** 297
10 **Reasoning and decision-making** 336
11 **Problem-solving** 386
12 **Social cognition** 425
13 **Cognition and emotion** 445
14 **Research methods in human cognition** 464

Answers to end of chapter quizzes 485
References 498
Glossary 555
Index 569

Contents

Preface xi

1 Understanding cognitive psychology 1

Learning objectives 1
1.1 Thinking about thinking 3
1.2 Memory and cognition defined 6
1.3 An introductory history of cognitive psychology 8
1.4 Cognitive psychology and information processing 17
1.5 Measuring information processes 18
1.6 The standard theory and cognitive science 24
1.7 Themes of human cognition 29
Summary: Cognitive psychology 31
Chapter 1 quiz 32

2 The neural basis of cognition 33

Learning objectives 33
2.1 Brain and cognition 34
2.2 Understanding how neurons work 35
2.3 Brain structures and their function 42
2.4 Cognitive neuropsychology 51
2.5 Connectionism 52
Summary: Neural basis of cognition 53
Chapter 2 quiz 54

3 Sensation and perception 56

Learning objectives 56
3.1 Psychophysics 57
3.2 Visual sensation and perception 61
3.3 Pattern recognition 75
3.4 Top-down processing 81
3.5 Object recognition and agnosia 87
3.6 Auditory sensation and perception 93
Summary: Sensation and perception 98
Chapter 3 quiz 99

4 **Attention** 101

Learning objectives 101
4.1 Understanding attention 102
4.2 Automatic attentional processes 104
4.3 Controlled, voluntary attention 118
4.4 Attention as a mental resource 125
Summary: Attention 132
Chapter 4 quiz 133

5 **Short-term and working memory** 134

Learning objectives 134
5.1 A limited-capacity bottleneck 135
5.2 Short-term memory retrieval 141
5.3 Working memory 148
5.4 Assessing working memory 158
5.5 Working memory and cognition 166
Summary: Short-term working memory 170
Chapter 5 quiz 171

6 **Learning and remembering** 172

Learning objectives 172
6.1 Preliminary issues 174
6.2 Storing information in episodic memory 180
6.3 Boosting episodic memory 187
6.4 Context 193
6.5 Facts and situation models 195
6.6 Autobiographical memories 198
6.7 Memory for the future 202
6.8 Semantic networks and connectionist models 203
Summary: Learning and remembering 210
Chapter 6 quiz 212

7 **Memory and forgetting** 213

Learning objectives 213
7.1 The seven sins of memory 214
7.2 Forgetting through decay and interference 215
7.3 False memories, eyewitness memory and 'forgotten memories' 224
7.4 Amnesia and implicit memory 236
Summary: Memory and forgetting 244
Chapter 7 quiz 245

8 The building blocks of language 246

Learning objectives 246
8.1 Linguistic universals and functions 248
8.2 Phonology 258
8.3 Syntax 267
8.4 Lexical factors 276
8.5 Semantics 280
8.6 Brain and language 287
Summary: Language 294
Chapter 8 quiz 295

9 Language in context 297

Learning objectives 297
9.1 Conceptual and rule knowledge 298
9.2 Reading 305
9.3 Reference, situation models and events 311
9.4 Conversation and gesture 323
Summary: Comprehension 333
Chapter 9 quiz 334

10 Reasoning and decision-making 336

Learning objectives 336
10.1 Formal logic and reasoning 338
10.2 Decision-making 351
10.3 Heuristics and biases in decision-making 352
10.4 Framing and risky decisions 365
10.5 Adaptive thinking and 'fast and frugal' heuristics 367
10.6 Other explanations 372
10.7 Limitations in reasoning 376
Summary: Reasoning and decision-making 382
Chapter 10 quiz 383
Appendix: Algorithms for coin tosses and hospital births 384

11 Problem-solving 386

Learning objectives 386
11.1 Understanding problem-solving 386
11.2 Basics of problem-solving 388
11.3 Gestalt psychology and problem-solving 395
11.4 Insight and analogy 401
11.5 Means–end analysis 412
11.6 Improving your problem-solving 418
Summary: Problem-solving 423
Chapter 11 quiz 424

12 Social cognition 425

Learning objectives 425
12.1 The neural basis of social cognition 427
12.2 Understanding the self 429
12.3 Understanding others 433
12.4 Responding to adverse social signals 438
Summary: Social cognition 443
Chapter 12 quiz 444

13 Cognition and emotion 445

Learning objectives 445
13.1 What is emotion? 446
13.2 Emotion and perception 448
13.3 Emotion and memory 452
13.4 Emotion and language 455
13.5 Emotion and decision making 457
13.6 Emotion regulation 460
Summary: Emotion 462
Chapter 13 quiz 463

14 Research methods in human cognition 464

Learning objectives 464
14.1 The purpose of research methods in cognitive psychology 464
14.2 Reaction time-based tasks 465
14.3 Visual search tasks 472
14.4 Neurophysiological methods 474
14.5 Neuroimaging and brain stimulation methods 478
14.6 Research integrity and ethics 480
Summary: Research methods in human cognition 483
Chapter 14 quiz 483

Answers to end of chapter quizzes 485
References 498
Glossary 555
Index 569

Lecturer Resources

For password-protected online resources tailored to support the use of this textbook in teaching, please visit **http://www.pearson.com**

ON THE WEBSITE

Preface

Cognition is an indispensable part of human living. The study of cognitive processes helps us understand how we perceive and make sense of ourselves and the world around us. This book is about how we think, how we remember, how we communicate and how we make sense of other people. *Cognitive Psychology* offers you unique insights into basic cognitive processes, such as attention, perception and memory, as well as more complex ones, like decision-making and social cognition. It will also teach you how to use cutting-edge methodologies and related technology to assess cognitive processes, from electric impulses between neurons in your brain, to how your eye gaze patterns predict what products you're likely to buy while browsing through a supermarket aisle.

The unique features of *Cognitive Psychology* include:

- Cutting-edge research in social cognition and human emotions, covering topics such as theory of mind, empathy, loneliness and social rejection.
- Insights into research ethics and integrity and the 'reproducibility crisis' in psychology – a timely and important topic that will sharpen your critical thinking skills and help you distinguish disinformation and fake news from valid research data.
- Globally relevant, timely and inclusive research examples, showcasing the relevance of cognitive psychology to everyday life.
- Key figures in cognitive psychology are presented at the beginning of each chapter (except Chapter 1) to familiarise readers with their work and celebrate the diversity of researchers in cognitive psychology from across the globe.

How to use this book

This book is primarily targeted to students who study for a psychology (or behavioural science) degree, as well as students from related disciplines who undertake modules relevant to cognitive psychology or who are independently studying human cognition. We want to inspire you to broaden your understanding of human behaviour by looking into different cognitive systems and processes. We have purposefully included a chapter on social cognition – a unique feature of this textbook – to provide you with a well-rounded perspective about how cognitive psychology can explain much of everyday social interactions (or the lack thereof, as in the case of loneliness and social rejection).

Being seasoned academics with over 20 years of experience in higher education teaching and a track record of research funding and publications, we also wrote this book with fellow academics in mind. We provide you with resources and interactive features that will complement and support your teaching, and help you improve the student experience. Importantly, we present updated empirical evidence and include key figures in every chapter (except in the Introduction) to help you showcase the diversity of researchers in cognitive psychology.

1 Understanding cognitive psychology

Learning objectives

1.1 Analyse the mental processes underlying our thoughts

1.2 Differentiate memory and cognition

1.3 Summarise the history of cognitive psychology

1.4 Interpret how planning guides behaviours involved in problem-solving

1.5 Compare human information processing to the operations of a computer program

1.6 Explain the mental processes that take place while doing a task

1.7 Describe the themes of human cognition

Introduction

How do you know what you know? How can you tell if you have control over your actions? Are you certain that your experiences are real and not merely the product of computer coding, as the 1999 film *The Matrix* depicted? These are some scary questions bordering on existentialism. Let's consider some simpler, less threatening, everyday questions: when was the last time you forgot something? Or the last time you went to the grocery store having forgotten your shopping list? All these questions have something in common: to answer them we need a good understanding of human cognition.

cognitive psychology
Concerned with the scientific study of diverse mental processes and their relationship with behaviour.

This book is focused on human cognition and is a primer of **cognitive psychology**: the scientific discipline that examines mental processes and their relation to behaviour. Initially thought to be focused on information processing, cognitive psychology has expanded to address a wide range of topics, from attention and language processing to memory, problem-solving and decision-making. We discuss each of these topics in detail in the following chapters. In doing so, we emphasise two main aspects: theoretical underpinnings (i.e. key models and theories explaining mental processes) and methodologies used in empirical research. The methodologies used by cognitive psychologists to study mental processes are robust and have a long history and diverse origins. We present these methodologies throughout the book when describing relevant studies (e.g. in a Research Box), but we offer a more detailed discussion in Chapter 14.

The Matrix depicted that 'reality' is more than meets the eye, and invited us to think more critically about different cognitive processes including (but not limited to) sensation, perception and decision-making.
Source: A-R-T Shutterstock

How do people think?

One of the central features of modern cognitive psychology is its allegiance to objective, empirical methods of investigation. We are experimentalists, and you will read about this approach in this chapter. Although we present many studies, we also try to make connections with our everyday experiences and how they are relevant to the discussion of pertinent issues.

Within the boundaries of science, cognitive psychology asks a wide range of fascinating questions. With the advance of machine learning and artificial intelligence, there has been an explosion of interest in cognition both inside and outside psychology. Questions that were too complex to answer for too long are now active areas of research. For example: How do we read? How do we use language? How do we make decisions? These are now relevant to understanding both human and machine processing. The pent-up interest in these questions, unleashed during the **cognitive revolution** of the late 1950s and early 1960s, has yielded tremendous progress. Furthermore, we now acknowledge, seek and sometimes participate in important disciplines such as linguistics, computer science, anthropology and neuroscience. This interdisciplinary approach is called **cognitive science** and is concerned with the scientific study of the mind.

This book aims to share what has been discovered about human cognition and the insights those discoveries provide about human behaviour. It also aims to describe how cognitive psychology has made these discoveries. You will appreciate this information more if you also understand how one conducts research and acquires knowledge. Regardless of whether you choose to become a cognitive psychologist (or, more broadly, a cognitive scientist) an understanding of cognitive psychology can only benefit your inquiry into the human mind and behaviour. Because the cognitive approach influences many areas of psychology, your understanding of cognitive psychology will enhance your mastery of psychology as a whole.

Finally, this book will also illustrate the pervasiveness of cognitive psychology and its impact on fields outside psychology. Cognitive science is a multidisciplinary field. This fusion and cross-pollination of ideas stem from the conviction that researchers in linguistics, artificial intelligence, neurosciences, economics, biology, anthropology and even consumer behaviour can contribute important ideas to psychology and vice versa. Psychology has a long tradition of influencing educational practice, and it is important that it continues to do so. Even fields as diverse as medicine, law, and business use findings from cognitive psychology. For example, a cognitive psychologist named Daniel Kahneman won the Nobel Prize in Economic Sciences in 2002 for his work on decision-making – a topic we discuss at great lengths in Chapter 11. But it should not surprise you that cognitive psychology is relevant to so many other fields. After all, what human activity doesn't involve mental processes?

1.1 Thinking about thinking

LEARNING OBJECTIVE
Analyse the mental processes underlying our thoughts

What is going on when we are thinking? What are the cognitive processes that shape our thoughts? Cognitive psychology attempts to study not only what we are thinking but also why and how we are thinking it. Memory, perception, emotions, beliefs, reasoning, imagination and how we acquire knowledge all factor into cognitive processes.

Let's begin to develop our understanding of cognitive psychology by considering three examples. For all three, you should read and answer the question, but more importantly, try to be as aware as possible of the thoughts that cross your mind as you consider the question.

1.1.1 The first question is easy: How many hands did Aristotle have?

Here we are not particularly interested in the correct answer: two. We are more interested in the thoughts you had as you considered the question. Most students report a train of thought something like this: 'Silly question. Of course he had two hands. Wait a minute, why would a professor ask such an obvious question? Maybe Aristotle had only one hand. Nah, I would have heard of it if he had had only one hand – he must have had two.'

An informal analysis will uncover some of the thoughts you had. These are tracked in Table 1.1. Bear in mind that this illustrates intuitive analysis and is not a full description of these processes.

First, perceptual processes were used for the written words of the question to focus your eyes on the printed line, then you moved your focus across the line bit by bit, registering the material into one of your memory systems. Smoothly and

Table 1.1 Summary of intuitive cognitive analysis

Processes	Topic
Sensory and perceptual	
Focus eyes on visual stimuli (e.g. picture or words)	Visual perception, sensory memory
Encode and recognise visual stimuli	Pattern recognition, reading
Memory and retrieval	
Look up and identify words in memory	Memory retrieval
Retrieve word meanings	Semantic retrieval
Comprehension	
Combine word meanings to yield sentence meaning	Semantic retrieval, comprehension
Evaluate sentence meaning, consider alternative meanings	Comprehension
Judgement and decision	
Retrieve answer to question	Semantic retrieval
Determine reasonableness of question	Comprehension, conversation
Judge speaker's intent and knowledge	Decision-making and reasoning
Computational (Question 2)	
Retrieve factual knowledge	Semantic retrieval
Retrieve knowledge of how to divide and execute procedure	Procedural knowledge

rapidly, other processes took the material into memory to identify the letters and words. Of course, few readers consciously attend to the nuts and bolts of perceiving and identifying words unless the vocabulary is unfamiliar (or the print is bad!). Yet your lack of awareness does not mean that these processes did not happen. Ask any primary school teacher about the difficulties children have identifying letters and putting them together into words.

We have encountered two important lessons already. First, mental processes, such as reading, can occur with little conscious awareness, especially if they are highly practised. Second, even though these processes can operate very quickly, they are complex. Their complexity makes it even more amazing how efficient, rapid and seemingly automatic they are.

As you identified the words in the question, you were looking up their meanings and fitting them together to understand the question. Surely, you were not consciously aware of looking up the meaning of *hands* in a mental dictionary. But just as surely, you did find that entry, along with your general knowledge about the human body.

Now we are getting to the most fundamental aspects. With little effort, we retrieve information from memory that *Aristotle* refers to a human being, a historical figure from the past. Many people know little about Aristotle beyond the fact that he was a Greek philosopher of the ancient time. Yet this seems to be enough, combined with what we know about people in general, to determine that Aristotle was probably just like everyone else: he had two hands.

At a final (for now) stage, people report thoughts about the reasonableness of the question. In general, people do not ask obvious questions, at least not of other adults. If they do, it is often for another reason – a trick question, maybe, or sarcasm. So, students report that for a time they decided that maybe the question was not so obvious after all. In other words, they returned to memory to see whether there was some special knowledge about Aristotle that pertains to his hands. The next step is truly fascinating. Most students claim to think to themselves, 'No, I would have known about it if he had had only one hand,' and decide that it was an obvious question after all. This lack-of-knowledge reasoning is fascinating because so much everyday reasoning is done without the benefit of complete knowledge. In an interesting variation, if students are asked, 'How many hands did Beethoven have?' their knowledge of Beethoven's musical fame typically leads to the following inference: 'Because he was a musician, he played the piano, and he could not possibly have been successful at it with only one hand. Therefore, he must have had two.' An occasional student goes even further with, 'Two, but he did go deaf before he died.' A rock music aficionado could further extend this thought: 'Wait, two hands are not a prerequisite for playing an instrument. Rick Allen, the drummer of the Sheffield-based rock band Def Leppard, was amputated and carried on playing the drums and enjoyed a successful musical career with only one arm.'

Now that's interesting! Someone found a connection between the disability implied by the question 'How many hands?' and related ideas in memory, Beethoven's hard of hearing. Such an answer shows how people can also consider implications, inferences and other unstated connections as they reason: The thinking process can consider a great deal of knowledge, and this illustrates the role of prior knowledge in reasoning, where richer knowledge about Beethoven or Def Leppard can lead to an inference.

One other thing to note from this example is that there are different cognitive processes that are all operating at the same time or similar times – perception, attention, memory, language comprehension, and so forth. These processes are also providing input and influencing one another. In essence, cognition is a complex and interactive thing, and it is going to take a lot of time and effort to tease it all apart and understand how it works.

Def Leppard's Rick Allen continued playing the drums following an arm amputation in the mid-1980s, and contributed to the band's commercial success.

Source: Davide Sciaky/Alamy Stock Photo

1.1.2 The second question: What is 723 divided by 6?

This question uses your knowledge of arithmetic. Just as with the first question, many of your mental processes happened more or less automatically: identifying the digits, accessing knowledge of arithmetic procedures, and so on. Yet you may be aware of the steps in doing long division: Divide 6 into 7, subtract 6 from 7 to get the first remainder, bring down the 2, then divide 12 by 6, and so on. These steps are mentioned at the bottom of Table 1.1 (see 'Computational', which includes your knowledge of how to do long division). Cognitive psychology is also interested in your mental processing of arithmetic problems and knowledge you acquired in school, not just the kind of reasoning you used in the Aristotle question.

1.1.3 The third question: Does a robin have wings?

Most adults have little to say about their train of thought when answering this question. Many people insist, 'I just knew the answer was yes.' The informal analysis for Question 1 showed how much of cognition occurs below awareness. The assertion that 'I just knew it' is not useful, however certain you are that no other thoughts occurred. You had to read the words, find their meanings in memory, check the relevant facts and make your decision as in the previous examples. Each of these steps is a mental act, the very substance of cognitive psychology. Furthermore, each step takes some amount of time to complete.

Question 3 takes adults about one second to answer. However, the question 'Does a robin have feet?' takes a little longer, around 1.2 to 1.3 seconds. Even small time differences can give us a wealth of information about cognition and memory. The difference in Question 3 is that most of the mental processes do not require much conscious activity; the question seems to be processed automatically. Because such automatic processes are so pervasive, we are particularly interested in understanding them.

1.2 Memory and cognition defined

LEARNING OBJECTIVE
Differentiate memory and cognition

To better understand the topic of this title, we need to be more explicit about the terms we use. Just what do we mean when we use the terms **memory** and **cognition**? In this section, we will address these two concepts.

Understanding the terms 'memory' and cognition

Now that you have an idea of the topics under cognitive psychology, we need more formal definitions of the terms *memory* and *cognition*.

Memory

Most of us have a good idea of what the term memory means, something like 'being able to remember or recall some information' or 'the act of recalling previously learned facts or events'. Note that both of these definitions are hopelessly circular; memory is 'being able to remember' or 'the act of recalling'.

However, the definitions do point to several critical ideas:

1. The information recalled from memory is from the past. The past could be a childhood memory from years ago or something that happened only moments ago.
2. Memory is a process of storing information or recovering it for current use. Note that **retrieval** includes both the conscious, intentional recalling to mind and the more automatic (or even unaware) retrieval of the earlier examples.
3. Memory is also a place where all the knowledge of a lifetime is stored. This is evident in theories of cognition that rely on divisions such as short-term and long-term memory. Although there is some physical location in your brain for storage, 'location' is often taken metaphorically; regardless of where it happens, there is some memory system that holds information. With the advent of neuroimaging, we are making progress in understanding where functions and processes occur in the brain.

A formal definition of memory captures these essential ingredients. Consider memory to be the mental processes of acquiring and retaining information for later retrieval and the mental storage system that enables these processes. Memory is demonstrated whenever the processes of retention and retrieval influence behaviour or performance in some way, even if we are unaware of it. Furthermore, this definition includes retention across not only hours, weeks or years, but also even very brief spans of time, in any situation in which the original event is no longer present.

Note that memory refers to three kinds of mental activities: acquisition (also called learning or encoding), retention and retrieval (Melton, 1963[1]). Because all three are needed to demonstrate remembering, we include them in our broader definition.

Cognition

The term cognition is a much richer one. In Ulrich Neisser's landmark book *Cognitive Psychology* (1967),[2] he stated that cognition 'refers to all the processes by which the sensory input is transformed, reduced, elaborated, stored, recovered, and used . . . [including] such terms as sensation, perception, imagery, retention, recall, problem solving, and thinking' (p. 4).

For the present, we use the following definition: *cognition* is the collection of mental processes and activities used in perceiving, remembering, thinking and understanding, as well as the act of using those processes.

Cognitive psychology is largely, though not exclusively, interested in what might be everyday, ordinary mental processes. These processes are entirely commonplace – not simple, by any means, but certainly routine. Our definition should not include only 'normal' mental activities, however. Although cognitive psychology generally does not deal with psychologically 'abnormal' states, such as schizophrenia, such 'non-normal' processes, although unusual or rare, may enrich our science.

Most cognitive research deals with the sense modalities of vision and hearing and focuses heavily on language. Some people may be concerned that the reliance on seemingly sterile experimental techniques and methods, techniques that ask simple questions, may yield overly simple-minded views about cognition. This reflects a concern that cognitive research lacks **ecological validity**, or generalisability to the real-world situations in which people think and act (e.g. Holleman et al., 2020[3]). To some, this criticism is sensible, meaning that the findings derived from work in cognitive psychology should, in some way, find value and applicability in the real world, even if that value may be several steps removed from the original study. A primary reason that cognitive psychologists often do not try to do studies that have an immediate and direct implication for real-world activities is the glaring fact that cognition is complex, even when using artificially simple tasks. At our current level of sophistication, we would be quickly overwhelmed if tasks were very complex or if we tried to investigate the full range of a behaviour in all its detail and nuance. In this stage of investigation, it is reasonable for scientists to take an approach called **reductionism**, attempting to understand complex events by breaking them down into their components. An artificially simple situation can reveal an otherwise obscure process. Once the basic processes and components of cognition are understood, then better accounts of how they work together can be put forward. The greater goal is for scientists to eventually put the pieces back together and deal with the larger events as wholes.

1.3 An introductory history of cognitive psychology

LEARNING OBJECTIVE
Summarise the history of cognitive psychology

Let's now turn to cognitive psychology's history and development (for an excellent history of cognitive psychology, see Mandler, 2007[4]). Figure 1.1 summarises the main patterns of influence that produced cognitive psychology and cognitive science, along with approximate dates.

To a remarkable extent, the bulk of the scientific work on memory and cognition is quite recent, although some elements, and many experimental tasks, appeared even in the earliest years of psychology. However, interest in memory and thinking is as old as recorded history. Aristotle, born in 384 BC, considered the basic principles of memory and proposed a theory in his treatise *De Memoria* (*Concerning Memory*; see Hothersall, 1984[5]). Even a casual reading of ancient works such as Homer's *Iliad* or *Odyssey* reveals that people have always wondered how the mind

René Descartes proclaimed *'cogito ergo sum'*, placing cognition at the very core of our conscious experience of existence.
Source: Georgios Kollidas/Shutterstock

works and how to improve it (in Plato's *Phaedrus*, Socrates fretted that the invention of written language would weaken reliance on memory and understanding, just as modern parents worry about social media). Philosophers of every age have considered the nature of thought. Descartes even decided that the proof of human existence is our awareness of our own thought: *Cogito ergo sum*, 'I think, therefore I am' (Descartes, 1637/1972, p. 52[6]).

The critical events at the founding of psychology, in the mid-to-late 1800s, converged most strongly on one man, Wilhelm Wundt, and on one place, Leipzig, Germany. In 1879, Wundt established the first laboratory for psychological experiments that had a lasting impact, at the University of Leipzig. Yet Wundt's was not the first psychology laboratory. For example, Ferdinand Ueberwasser founded a psychology laboratory in 1783. However, for various reasons, it did not have a lasting or widespread impact (Schwarz & Pfister, 2016[7]). Also, several people had already

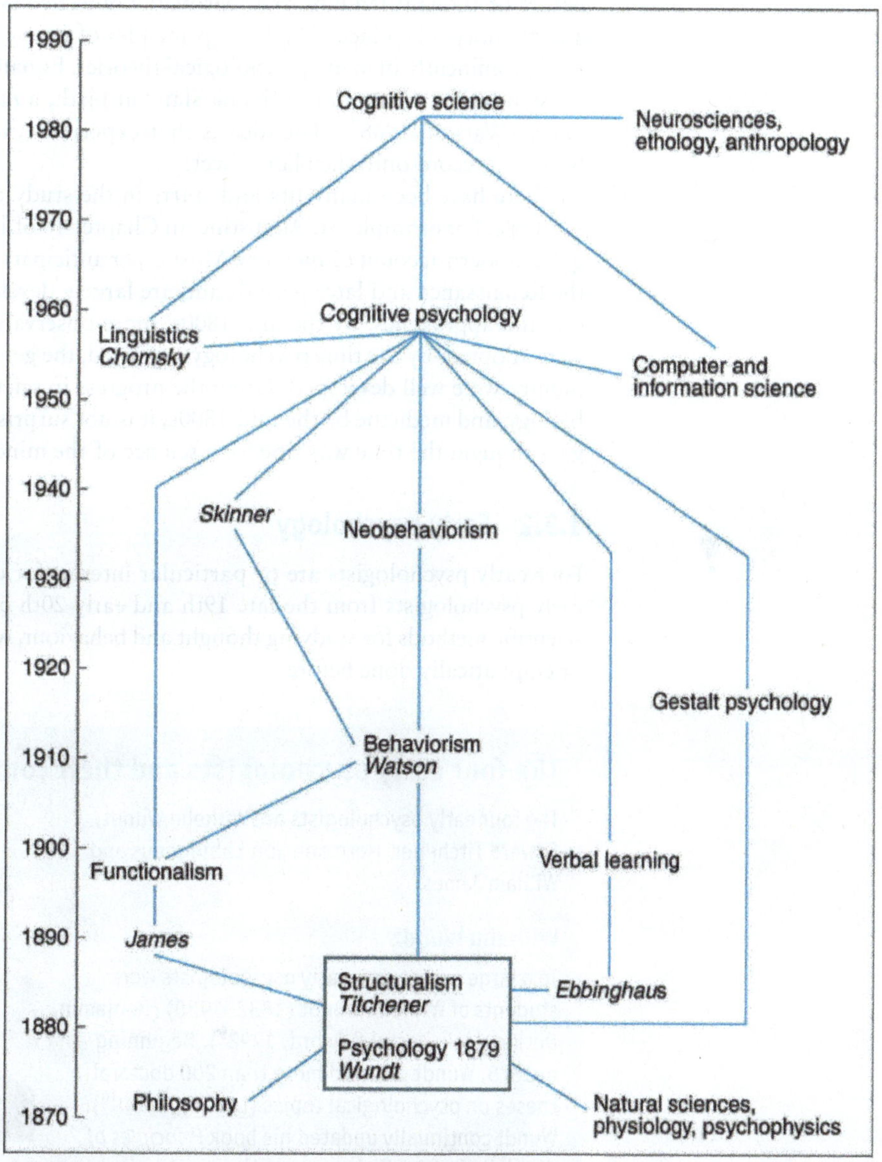

Figure 1.1 The main patterns of influence that produced cognitive psychology and cognitive science

been doing psychological research, but did not fully identify themselves as psychologists, but more as physiologists and the like (e.g. Weber's and Fechner's work in psychophysics, Helmholtz's studies of the speed of neural impulses, and Broca's and Wernicke's identification of linguistic brain regions). American psychologist William James even established an early laboratory in 1875, although apparently he used it more for classroom demonstrations than for genuine experiments. Still, the consensus is that 1879 is the beginning of the discipline of psychology, separate from philosophy and physiology.

1.3.1 Anticipations of psychology

Aristotle, for two reasons, is one of the first historical figures to advocate an empirically based, natural science approach. First, although he was certainly not the only great thinker to insist on observation as the basis for all science, he was the first to express this – a position known as **empiricism**. Second, Aristotle's inquiry into the nature of thought led him to a reasonably objective explanation of how learning and memory take place. The basic principles of association he identified have figured prominently in many psychological theories. Equally important was Aristotle's insistence that the mind is a 'blank slate' at birth, a ***tabula rasa***, or clean sheet of paper (Watson, 1968[8]). The idea is that experience, rather than inborn factors, 'writes' a record onto the blank sheet.

There have been many fits and starts in the study of memory over time since Aristotle. For example, St. Augustine, in Chapter 10 of his *Confessions*, presented a quite modern account of memory. Most other anticipations of psychology date from the Renaissance and later periods and are largely developments in scientific methods and approaches. By the mid-1800s, more observational or empirical methods were adopted. By the time psychology appeared, the general procedures of scientific inquiry were well developed. Given the progress in scientific fields such as physics, biology, and medicine by the mid-1800s, it is not surprising that the early psychologists thought the time was ripe for a science of the mind.

1.3.2 Early psychology

Four early psychologists are of particular interest for cognitive psychology. These early psychologists from the late 19th and early 20th centuries worked to develop scientific methods for studying thought and behaviour, which had not been explicitly or emphatically done before.

The four early psychologists and their contributions

The four early psychologists are Wilhelm Wundt, Edward Titchener, Hermann von Ebbinghaus and William James.

Wilhelm Wundt

To a large extent, the early psychologists were students of Wilhelm Wundt (1832–1920) (Benjamin, Durkin, Link, Vestal & Acord, 1992[9]). Beginning in 1875, Wundt directed more than 200 doctoral theses on psychological topics (Leahey, 2000[10]). Wundt continually updated his book *Principles of Physiological Psychology*, reporting new results from his laboratory. He also founded the first

Wilhelm Wundt
Source: INTERFOTO/Alamy Stock Photo

psychology journal, *Philosophical Studies* (neither of these titles matches its modern connotations). Unfortunately, Wundt's later interests went largely unrecognised until recently (Leahey, 2000[11]). His work on language, child psychology and other applied topics foreshadowed some modern insights but was rejected or ignored at the time.

In terms of psychology, Wundt believed that its study was 'of conscious processes and immediate experience' – what today we consider areas of sensation, perception and attention. To study these, in addition to extensive use of response time measures, Wundt used the method of *Selbst-Beobachtung*. Translated literally as 'self-observation', this generally is known as **introspection**, a method in which one looks carefully inward, reporting on inner sensations and experiences. Wundt intended this to be a careful, reliable and scientific method in which the observers (who were also the participants) needed a great deal of training to report only the elements of experience that were immediate and conscious. Reports in which memory intruded were to be excluded.

Edward Titchener

For American psychology in Wundt's tradition, the most important figure was Edward Titchener, an Englishman who came to Cornell University in 1892. Working with Wundt convinced Titchener that psychology's progress depended critically on introspection. Topics such as mental illness and educational and social psychology (including Wundt's broader interests) were 'impure' because they could not be studied this way. Titchener insisted on rigorous training for his introspectors, who had to avoid 'the stimulus error' of describing the physical stimulus rather than the mental experience of it. Moreover, Titchener made himself the final authority on whether introspection reports were correct or not. By these means, Titchener attempted to study the structure of the conscious mind: the sensations, images and feelings that, for Titchener, were the very elements of the mind's structure. He called this **structuralism**, an early movement or school of psychological thought (see Figure 1.1). Such a system was destined for difficulties. For example, it is unscientific for one person, Titchener, to be the ultimate authority to validate observations. As other researchers used introspective methods, differences and contradictory results began to crop up, producing disputes that hastened the decline of Titchener's once-powerful structuralism.

Hermann von Ebbinghaus

Hermann Von Ebbinghaus
Source: Darling Archive/Alamy Stock Photo

In contrast to Titchener's structuralism, there was the theoretically modest but eventually more influential work of Hermann von Ebbinghaus. Ebbinghaus was a contemporary of Wundt's in Germany, although he never studied with Wundt in person. Ebbinghaus's achievements in studying memory and forgetting are all the more impressive because he worked outside the establishment of the time. Historical accounts suggest that Ebbinghaus read Wundt's book, decided that a study of the mind by objective methods was possible, and set about the task of figuring out how to do it.

Lacking a formal laboratory and in an academic position with an absence of sufficiently like-minded colleagues, Ebbinghaus had to rely on his own resources, even to the extent that he alone served as a subject in his research. Ebbinghaus's aim was to study memory in a 'pure' form. To do this, he needed materials that had no pre-existing associations, so he constructed lists of *nonsense syllables*, consonant–vowel–consonant (CVC) trigrams that, by definition, had no meaning. Ebbinghaus would learn a list (e.g. of 16 items) to a criterion of mastery (e.g. two perfect recitations), then set the list aside. Later, he would relearn the same list, noting how many fewer trials he needed to relearn it. His measure of learning was the 'savings score', the number (or proportion) of trials that had been saved in memory between the first and second sessions. His savings measure of memory is based on the idea that if information is stored in memory, even in a form that is not strongly consciously available, it can still ease the relearning of that material. This is in contrast to modern memory researchers who place a greater emphasis on methods such as recall and recognition, and miss some of the potential advantages of the savings method. Using his savings method, Ebbinghaus was able to study retention and forgetting of memories as a function of time, degree of learning or overlearning, and even the effect of nonsense versus meaningful material (he compared forgetting curves for nonsense syllables and meaningful poetry).

Ebbinghaus's work, described in his 1885 book, gained wide acclaim as a model of scientific inquiry into memory. For instance, Titchener praised Ebbinghaus's work as the most significant progress since Aristotle (cited in Hall, 1971[12]). It is difficult to point to another psychologist of his day whose contributions or methods continue to be used. The field of verbal learning owes a great deal to Ebbinghaus. The Ebbinghaus tradition, depicted in Figure 1.1, is one of the strongest influences on cognitive psychology.

William James

William James
Source: Chronicle/Alamy Stock Photo

American philosopher and psychologist William James, a contemporary of Wundt, Titchener and Ebbinghaus, provided at Harvard an alternative to Titchener's rigid system. His approach, influenced by the writings of Darwin, was **functionalism**, in which the functions of consciousness, rather than its structure, were of interest. Thus, James asked questions such as 'How does the mind function?' and 'How does it adapt to new circumstances?'

James's informal analyses led to some useful observations. For example, he suggested that memory consists of two parts: an immediately available memory that we are currently aware of and a larger memory that is the repository for past experience. The idea of memory being divided into parts, based on different functions, is popular today. Indeed, the first serious models of human cognition included the two kinds of memory James discussed in 1890.

Probably because of his personal distaste for experimentation and his broad interests, James did not do much actual research. However, his far-reaching ideas were more influential than any of Titchener's work, as evidenced by his classic 1890 book *Principles of Psychology*. James's influence on the psychology of memory and cognition was delayed, however, for it was John B. Watson, in 1913, who solidified a new direction in American psychology away from both the structuralist and functionalist approaches. This new direction was behaviourism.

1.3.3 Behaviourism

Not all of American psychology from 1910 through the 1950s was behaviourist. The fields of clinical, educational and social psychology, to name a few, continued in their own development in parallel to behaviourism. Furthermore, there were changes within behaviourism that smoothed the transition to cognitive psychology. This was a kind of neobehaviourism with some unobservable, mediating variables. Nonetheless, it was still a behaviourist environment.

Most people who take introductory psychology know of John B. Watson, the early behaviourist who stated in his 1913 'manifesto' that observable, quantifiable behaviour was the proper topic of psychology, not the fuzzy and unscientific concepts of thought, mind and consciousness. Attempts to understand the 'un-observables' of the mind were inherently unscientific, in his view, and he pointed to the unresolved debates in structuralism as evidence. Thus, psychology was redefined as the scientific study of observable behaviour, the programme of **behaviourism**. There was no room for mental processes because they were not observable behaviours.

Why did such a radical redefinition of psychology's interests have such broad appeal? Part of this was a result of the work that Pavlov and others were doing on conditioning and learning. Here was a scientific approach that was going somewhere, as opposed to the endless debates in structuralism. Furthermore, the measurement and quantification of behaviourism mirrored successful sciences such as physics. Modelling psychology on the methods of these sciences might help it become more scientific (Leahey, 2000, calls this mentality 'physics envy'[13]). One of behaviourism's greatest legacies is the emphasis on methodological rigour and observables, traditions that continue to be in force to this day.

During the behaviourist era, there were a few psychologists who pursued cognitive topics – Bartlett of Great Britain, for example – but most American experimental psychology focused on observable, learned behaviours, especially in animals (but see Dewsbury, 2000, for a history of research on animal cognition during the behaviourist era[14]). Even the strongly cognitive approach of Tolman – whose article 'Cognitive maps in rats and men' (1948[15]), a molar (as opposed to molecular) approach to behaviourism, is still worth reading – included much of the behaviourist tradition:

- concern with the learning of new behaviours
- animal studies
- interpretation based closely on observable stimuli.

Gestalt psychology, which immigrated to the USA in the 1930s (Mandler & Mandler, 1969[16]), always maintained an interest in human perception, thought and problem-solving but never captured the imaginations of many American experimentalists.

Thus, the behaviourist view dominated American experimental psychology until the 1940s, when B. F. Skinner emerged as a vocal, even extreme, advocate. In keeping with Watson's earlier sentiments, Skinner also argued that mental events such as thinking have no place in the science of psychology – not that they are not real, but that they are unobservable and hence unnecessary to a scientific explanation of behaviour.

1.3.4 Emerging cognition

It is often difficult to determine precisely when historical change takes place. Still, many psychologists favour the idea that a cognitive revolution occurred in the mid-to-late 1950s, with a relatively abrupt change in research activities, interests, scientific beliefs and a definitive break from behaviourism (Baars, 1986[17]). Because of the nature and scope of these changes, some see the current approach as a revolution that rejected behaviourism and replaced it with cognitive psychology. However, some historians claim that this was not a true scientific revolution but merely 'rapid, evolutionary change' (see Leahey, 1992[18]). In either case, the years from 1945 to 1960 were a period of rapid reform in experimental psychology. The challenges to neobehaviourism came both from within its own ranks and from outside, prodding psychologists to move in a new direction.

World War II

Lachman, Lachman and Butterfield (1979)[19] made a point about the growing dissatisfaction among the neobehaviourists. They noted that many academic psychologists were involved with the US war effort during World War II. Psychologists accustomed to studying animal learning in the laboratory were 'put to work on the practical problems of making war . . . trying to understand problems of perception, judgment, thinking, and decision making' (p. 56). Many of these problems arose because of soldiers' difficulties with sophisticated technical devices: skilled pilots who crashed their aircraft, radar and sonar operators who failed to detect or misidentified enemy blips, and so on.

Lachman et al. (1979) were very direct in their description of this situation:

> Where could psychologists turn for concepts and methods to help them solve such problems? Certainly not to the academic laboratories of the day. The behaviour of animals in mazes and Skinner boxes shed little light on the performance of airplane pilots and sonar operators. The kind of learning studied with nonsense syllables contributed little to psychologists trying to teach people how to operate complex machines accurately. In fact, learning was not the central problem during the war. Most problems arose after the tasks had already been learned, when normally skilful performance broke down. The focus was on performance rather than learning; and this left academic psychologists poorly prepared. (pp. 56–57)[20]

Tasks, such as the vigilance needed for air traffic control, require cognitive processes at a fundamental level.

Source: Burben/Shutterstock

As Bruner, Goodnow and Austin (1956)[21] put it, the 'impeccable peripheralism' of stimulus–response (S–R) behaviourism became painfully obvious in the face of such practical concerns.

To deal with practical concerns, wartime psychologists were forced to think about human behaviour very differently from how they had been up until that point. The concepts of attention and vigilance, for instance, were important to understand sonar operators' performance. Experiments on the practical and theoretical aspects of vigilance began (see especially Broadbent, 1958[22]). Decision-making was a necessary part of this performance too, and from this came such developments as signal detection theory. These wartime psychologists rubbed shoulders with professionals from different fields – those in communications engineering, for instance – from whom they gained new outlooks and perspectives on human behaviour. Thus, these psychologists returned to their laboratories after the war, determined to broaden their own research interests and those of psychology as well.

Verbal learning

Verbal learning was the branch of experimental psychology that dealt with humans as they learned verbal material composed of letters, nonsense syllables or words. The groundbreaking research by Ebbinghaus started the verbal learning tradition, which derives its name from the behaviourist context in which it found itself. Thus, verbal learning was defined as the use of verbal materials in various learning paradigms. Throughout the 1920s and 1930s there was a large body of verbal learning research, with well-established methods and procedures. Tasks such as serial learning, paired-associate learning and, to an extent, free recall were the accepted methods.

Proponents of verbal learning were similar to the behaviourists. For example, they agreed on the need to use objective methods. There also was widespread acceptance of the central role of learning, conceived as a process of forming new associations, much like the learning of new associations by a rat in a Skinner box. From this perspective, a theoretical framework was built that used a number of concepts that are accepted today. For example, a great deal of verbal learning was oriented around accounts of interference among related but competing newly learned items.

The more moderate view in verbal learning circles made it easy for people to accept cognitive psychology in the 1950s and 1960s: There were many indications that an adequate psychology of learning and memory needed more than just observable behaviours. For instance, the presence of meaningfulness in 'nonsense' syllables had been acknowledged early on: Glaze (1928) titled his paper 'The association value of nonsense syllables'[23] (and apparently did so with a straight face). At first, such irksome associations were controlled for in experiments to avoid contamination of the results. Later, it became apparent that the memory processes that yielded those associations were more interesting.

In this tradition, Bousfield (1953; Bousfield & Sedgewick, 1944[24]) reported that, with free recall, words that were associated with one another (e.g. *car* and *truck*) tended to cluster together, even though they were arranged randomly in a study list. There were clear implications that existing memory associations led to the reorganisation. Such evidence of processes occurring between the stimulus and the response – in other words, mental processes – led proponents of verbal learning to propose a variety of mental operations such as rehearsal, organisation, storage and retrieval.

The verbal learning tradition led to the derivation and refinement of laboratory tasks for learning and memory. Its advocates borrowed from Ebbinghaus's example of careful attention to rigorous methodology to develop tasks that measured the outcomes of mental processes in valid and useful ways. Some of these tasks were more closely associated with behaviourism, such as the paired-associate learning task that lent itself to tests of S–R associations in direct ways. Nonetheless, verbal learning gave cognitive psychology an objective, reliable way to study mental processes – research that was built on later (e.g. Stroop, 1935[25]) – and a set of inferred processes such as storage and retrieval to investigate. The influence of verbal learning on cognitive psychology, as shown in Figure 1.1, was almost entirely positive.

Linguistics

The changes in verbal learning were a gradual shifting of interests and interpretations that blended almost seamlessly into cognitive psychology. By contrast, 1959 saw the publication of an explicit, defiant challenge to behaviourism. Watson's 1913 article was a behaviourist manifesto, crystallising the view against introspective methods. To an equal degree, Noam Chomsky's 1959 article was a cognitive manifesto, a rejection of a purely behaviourist explanation of the most human of all behaviours: language.

Chomsky's views on human behaviour and language

Noam Chomsky provided a new framework for explaining language that has had profound implications for studying the mental processes that support language use in humans.

In 1957, B. F. Skinner published a book titled *Verbal Behavior*, a treatment of human language from the radical behaviourist standpoint of reinforcement, stimulus–response associations, extinction, and so on. His central point was that the psychology of learning – that is, the conditioning of new behaviour by means of reinforcement – provided a useful and scientific account of human language. In oversimplified terms, Skinner's basic idea was that human language, 'verbal behaviour', followed the same laws that had been discovered in the animal learning laboratory: a reinforced response increased in frequency, a non-reinforced response should extinguish, a response conditioned to a stimulus should be emitted to the same stimulus in the future, and so on. In principle, then, it is possible to explain human language, a learned behaviour, by the same mechanisms given knowledge of the current reinforcement contingencies and past reinforcement history of the individual.

Noam Chomsky, a linguist at the Massachusetts Institute of Technology, reviewed Skinner's book in the journal *Language* in 1959. The first sentence of his review noted that many linguists and philosophers of language 'expressed the hope that their studies might ultimately be embedded in a framework provided by behaviourist psychology' and therefore were interested in what Skinner had to say. Chomsky alluded to Skinner's optimism that the problem of verbal behaviour would yield to behavioural analysis because the principles discovered in the animal laboratory 'are now fairly well understood . . . [and] can be extended to human behavior without serious modification' (Skinner, 1957, cited in Chomsky, 1959, p. 26[26]).

But by the third page of his review, Chomsky stated that 'the insights that have been achieved in the laboratories of the reinforcement theorist, though quite genuine, can be applied to complex human behavior only in the most gross and superficial way. . . . The magnitude of the failure of [Skinner's] attempt to account for verbal behavior serves as a kind of measure of the importance of the factors omitted from consideration' (p. 28, emphasis added). The fighting words continued. Chomsky asserted that if the terms stimulus, response, reinforcement, and so on are used in their technical, animal laboratory sense, then 'the book covers almost no aspect of linguistic behavior' (p. 31) of interest.

To Chomsky, Skinner's account used the technical terms in a non-technical, metaphorical way, which 'creates the illusion of a rigorous scientific theory [but] is no more scientific than the traditional approaches to this subject matter, and rarely as clear and careful' (pp. 30–31).

To illustrate his criticism, Chomsky noted the operational definitions that Skinner provided in the animal laboratory, such as for the term reinforcement. But, unlike the distinct and observable pellet of food in the Skinner box, Skinner claimed that the person exhibiting the behaviour could even administer reinforcement for their verbal behaviour; that is, self-reinforcement. In some cases, Skinner continued, reinforcement could be delayed for indefinite periods or never be delivered at all, as in the case of a writer who anticipates that their work may gain them fame for centuries to come.

When an explicit and immediate reinforcer in the laboratory, along with its effect on behaviour, is generalised to include non-explicit and non-immediate (and even non-existent) reinforcers in the real world, it seems that Skinner had brought along the vocabulary of scientific explanation but left the substance behind. As Chomsky bluntly put it, 'A mere terminological revision, in which a term borrowed from the laboratory is used with the full vagueness of the ordinary vocabulary, is of no conceivable interest' (p. 38). Chomsky's own view of language emphasised its novelty and the internal rules for its use. Language was an important behaviour – and a learned one at that – for psychology to understand. An approach that offered no help in understanding this was useless.

To a significant number of people, Chomsky's arguments summarised the dissatisfactions with behaviourism that had become so apparent. The irrelevance of behaviourism to the study of language and, by extension, any significant human behaviour was now painfully obvious. In combination with the other developments – the wartime fling with mental processes, the expansion of the catalogue of such processes by verbal learning, and the disarray within behaviourism itself – it was clear that the new direction for psychology would take hold.

1.4 Cognitive psychology and information processing

LEARNING OBJECTIVE
Interpret how planning guides behaviours involved in problem-solving

If we had to pick a date to mark the beginning of cognitive psychology, we might pick 1960. This is not to say that significant developments were not present before this date, for they were. It is also not to say that most experimental psychologists who studied humans became cognitive psychologists that year, for they did not. As with any major change, it takes a while for the new approach to catch on and for people to decide that the new direction is worth following. However, several significant events clustered around 1960 that were significant departures from what came before. Just as 1879 is considered the formal beginning of psychology, and 1913 the beginning of behaviourism, so 1960 approximates the beginning of cognitive psychology.

In his 1959 review, Chomsky made a forceful argument against a purely behaviourist position. He argued that the truly interesting part of language was exactly what Skinner had omitted: mental processes and cognition. Language users follow rules when they generate language, rules that are stored in memory and operated on by mental processes. In Chomsky's view, it was exactly there, in the organism, that the key to understanding language would be found.

Researchers in verbal learning and other fields were making the same claim. As noted, Bousfield (1953)[27] found that people cluster or group words together based on the associations among them. Memory and a tendency to reorganise clearly were involved. Where were these associations? Where was this memory? And where was this tendency to reorganise? They were in the person, in memory and mental processes.

During the 1950s, certain reports on attention, first from British researchers such as Colin Cherry and Donald Broadbent, pertained to the wartime concerns of attention and vigilance. Again, mental processes were being isolated and investigated. No one could deny their existence any longer, even though they were unseen mental processes. A classic paper, Sperling's monograph on visual sensory memory, appeared in 1960. [MacLeod (1991) noted an increase around 1960 of citations to the rediscovered Stroop (1935) task.[28]]

Another startling development of this period was the invention of the modern digital computer. At some point in the 1950s, certain psychologists realised the relevance of computing to psychology. In some interesting and possibly useful ways, computers behave like people (not surprising, according to Norman, 1986, p. 534, because 'the architecture of the modern digital computer . . . was heavily influenced by people's naive view of how the mind operated'[29]). They take in information, do something with it internally, and then produce some observable product. The product gives clues to what went on internally. The operations done by the computer were not unknowable because they were internal and unobservable. They were under the control of the computer program, the instructions given to the machine to tell it what to do.

The realisation that human mental activity might be understood by analogy to this machine was a breakthrough. The computer was an existence proof for the idea that unobservable processes could be studied and understood. Especially important was the idea of symbols and their internal manipulation. A computer is a symbol-manipulating machine. The human mind might also be a symbol-manipulating system, an idea attributed to Allen Newell and Herb Simon. According to Lachman et al. (1979),[30] their conference in 1958 had a tremendous impact on those who attended. Newell and Simon presented an explicit analogy between information processing in the computer and that in humans. This important work was the basis for the Nobel Prize awarded to Simon in 1978 (see Leahey, 2003, for a full account of Simon's contributions[31]).

Among the indirect results of this conference was the 1960 publication of a book by Miller, Galanter and Pribram called *Plans and the Structure of Behavior*.[32] The book suggested that human problem-solving could be understood as a kind of planning in which mental strategies or plans guide behaviour toward its goal. The mentalistic plans, goals and strategies in the book were not just unobservable, hypothetical ideas. Instead, they were ideas that in principle could be specified in a program running on a lawful, physical device: the computer.

1.5 Measuring information processes

LEARNING OBJECTIVE
Compare human information processing to the operations of a computer program

The aim of cognitive psychology is to reverse-engineer the brain in much the same way that engineers reverse-engineer devices that they cannot get into. Putting it simply, we want to know:

- What happens in there?

- What happens in the mind – or in the brain, if you prefer – when we perceive, remember, reason and solve problems?
- How can we peer into the mind to get a glimpse of the cognition that operates so invisibly?
- What methods can we use to obtain some scientific evidence on mental processes?

Guiding analogies

With the development of cognitive psychology, the seemingly unrelated fields of communications engineering and computer science supplied psychology with some intriguing ideas and useful analogies that were central to developing the cognitive science.

Channel capacity

To highlight one, psychologists found the concept of **channel capacity** from communications engineering useful (a similar, more popular term would be bandwidth). In the design of telephone systems, for instance, one of the built-in limitations is that any channel – any physical device that transmits messages or information – has a limited capacity. In simple terms, one wire can carry just so many messages at a time, and it loses information if capacity is exceeded. Naturally, engineers tried to design equipment and techniques to get around these limitations to increase overall capacity.

Psychologists noticed that, in several ways, humans are limited-capacity channels, too. There is a limit on how many things you can do, or think about, at a time. This insight lent a fresh perspective to human experimental psychology. It makes sense to ask questions such as 'How many sources of information can people pay attention to at a time?', 'What information is lost if we overload the system?' and 'Where is the limitation, and can we overcome it?'

The computer analogy

Even more influential than communications engineering was computer science. Computer science developed a machine that reflected the essence of the human mind. Because such things are unseen when both computers and humans do them, there was good reason for drawing the computer analogy to cognition. Basically, this analogy held that human information processing might be similar to the steps and operations in a computer program, similar to the flow of information from input to output. If so, then thinking about how a computer does various tasks gives some insights into how people process information.

1.5.1 Interpreting graphs

If you are good at interpreting data in graphs, go ahead and just study the figures. Some students struggle with graphs, not understanding what is being shown. Because you will encounter many graphs in this text, you need to understand what you are looking at and what it means.

Take a moment to go through these graphs to see how they present data and to what you should pay attention.

Interpreting data in graphs

Here is a series of graphs that contrast college and fourth-grade-student vocal response times to multiplication problems.

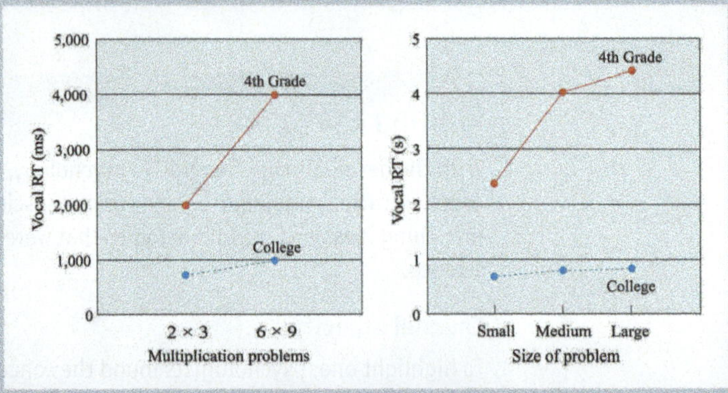

Vocal response times (RTs) to multiplication problems.

The above graph presents response time data, the time it takes to respond to an item. We abbreviate response time as RT, and it is usually measured in milliseconds (ms), thousandths of a second (because thought occurs so fast). In the figure, the label on the *y*-axis is 'Vocal RT'; these people were making vocal responses (speaking), and we measured the time between the onset of a multiplication problem and the vocal response. The numbers on the *y*-axis show you the range of RTs that were observed. The dependent variable is always the measure of performance we collected – here it is vocal RT – and it always goes on the *y*-axis.

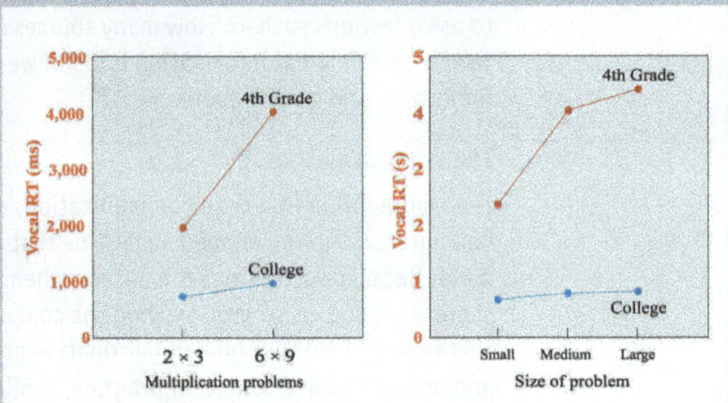

The *x*-axis label in the left panel is 'Multiplication problems', and we have plotted two problems: 2 × 3 and 6 × 9. It is customary to show a more general variable than this on the x-axis, as shown in the right panel.

There you see a point for a whole set of:

- small multiplication problems: 2 × 3 up to 4 × 5
- medium-size problems: 2 × 7 and 8 × 3'
- large problems: 6 × 8 and 9 × 7

So the *x*-axis label in the right panel is 'Size of problem'. Notice that the *y*-axis is now in whole seconds, to save some space.

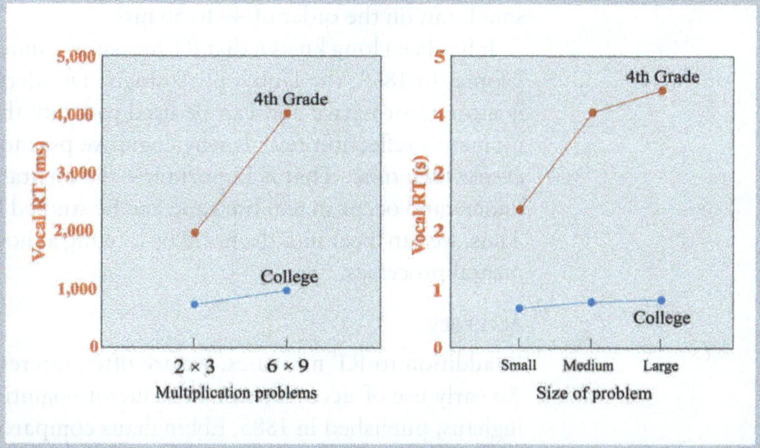

Now the data. The points in the graph are often a mean or average of the dependent variable, RT in this case. Both panels show two curves or lines each, one for college students and one for fourth-grade students (Campbell & Graham, 1985[33]), for multiplication problems. Notice that the curves for fourth-graders are higher. Looking at the *y*-axis in the left panel, the average fourth-grader took 1,940 ms to answer '6' to the problem 2 × 3, compared with 737 ms for the average college student. In the right panel, the average fourth-grader took about 2,400 ms to respond to small problems, 4,100 ms to medium, and 4,550 ms to large. Compare this much greater increase in RT as the problems get larger with the pattern for college students: There was still an increase, but only from 730 ms to about 900 ms.

Why did Campbell and Graham find that fourth-graders were slower? No doubt this is because college students have had more practice in doing multiplication problems than fourth-graders. In other words, college students know multiplication better, have the facts stored more strongly in memory, and so can retrieve them more rapidly. It is a sensible cognitive effect that the strength of information in memory influences the speed of retrieval. And it is easily grasped by looking at and understanding the graphed results.

Source: Data from Campbell and Graham (1985).[34]

1.5.2 Time and accuracy measures

How we peer into the mind to study cognition depends on using acceptable measurement tools to assess otherwise unseen, unobservable mental events. Other than Wundt's method of introspection, what can we use? There are many measures, but two of the most prominent behavioural measures are:

1. The *time* it takes to do some task.
2. The *accuracy* of that performance.

Because these measures are so pervasive, it is important to discuss them at the outset. We provide more information about how reaction time tasks are used in cognitive psychology in Chapter 14.

Response time

Many research programs in cognitive psychology rely heavily on **response time (RT)**, a measure of the time elapsed between some stimulus and the person's response to the stimulus (RT is typically measured in milliseconds, abbreviated *ms*; a millisecond is one thousandth of a second).

Why is this so important, especially when the actual time differences can seem so small, say, on the order of 40 to 50 ms?

It has been long known that RT measures can reveal individual differences among people. In 1868, the Dutch physiologist Donders (1868/1969)[35] observed that RT is more informative and can be used to study the 'speed of mental processes'. A moment's reflection reveals why cognitive psychology uses response times: *mental events take time*. That is important – the mental processes and events we want to understand occur in real time and can be studied by measuring how long they take. Thus, we can 'peer into the head' by looking at how long it takes to complete certain mental processes.

Accuracy

In addition to RT measures, we are often interested in **accuracy**, broadly defined. An early use of accuracy as a measure of cognition was the seminal work by Ebbinghaus, published in 1885. Ebbinghaus compared correct recall of information in a second learning session with recall of the same material during original learning as a way of measuring how much material had been saved in memory.

Figure 1.2 is a classic serial position graph, showing the percentage of items correctly recalled. The *x*-axis indicates each item's original position in the list.

In this experiment (Glanzer & Cunitz, 1966[36]), the list items were shown one at a time, and people had to wait 0, 10 or 30 seconds before they could recall the items. Making it even more difficult, the retention interval was filled with counting backward by 3s. Here, it is clear that memory was influenced by an item's position in the list – recall was better for early items than for those in the middle. Also notice the big effect that delaying recall with backward counting had at late positions. So, we cannot conclude that early list items *always* had an advantage over late list items – look how accurately the very last items were recalled when there was no

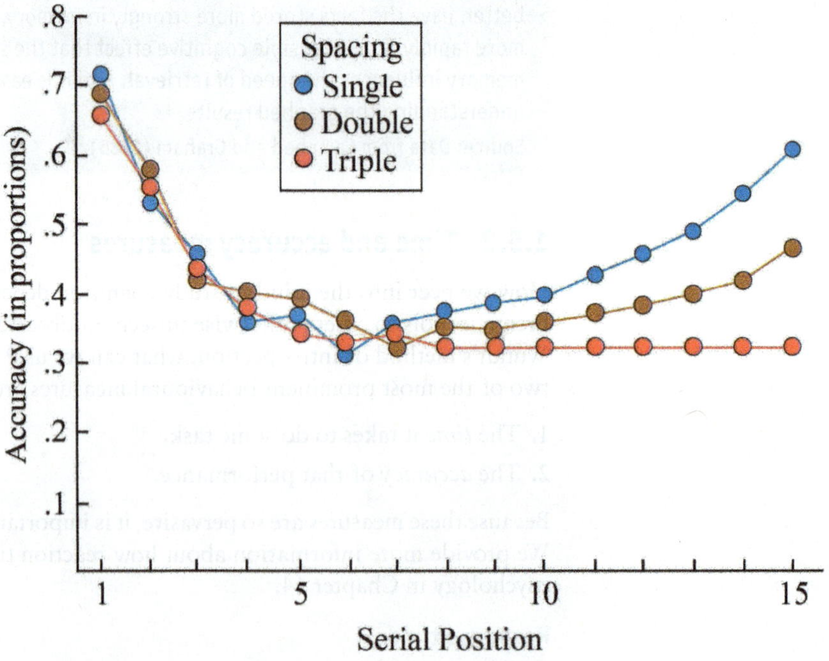

Figure 1.2 Hypothetical serial position curves, showing the decrease in accuracy at the end of the list when 0, 10 or 30 seconds of backward counting intervenes between study and recall

delay. Instead, the overall bowed shape of the graph tells us something complex and diagnostic about memory: recalling the items from the end of the list may depend on a different kind of memory than recalling the early words, and it may be that memory can be disrupted by activity-filled delays.

More modern variations on simple list-learning tasks look not only at proportion correct on a list but also at incorrect responses, such as any recalled words that were not on the studied list (called intrusions). Did the person remember a related word such as *apple* rather than the word that was studied, *pear*? Was an item recalled because it resembles the target in some other way, such as remembering *G* instead of *D* from a string of letters? This approach is similar to the Piagetian tradition of examining children's errors in reasoning, such as failure to conserve quantity or number, to examine their cognitive processes.

Replication

Although the various measures for studying the mind that are available to cognitive psychologists are all well and good, it is important to remember that any one study is often insufficient to provide the amount of evidence needed to adequately support a given model or theory. The results from any one study, considered in isolation, can often be interpreted in multiple ways and be consistent with multiple theories. To provide solid support for a particular theory might require multiple experiments to rule out various alternative explanations and triangulate on a theory that best captures the truth about how cognition works.

Foremost in this search for additional evidence is the need for findings – particularly surprising and important findings – to be replicated. Psychological research has taken some heat recently for a large number of published findings that are difficult, if not impossible, to replicate (Open Science Collaboration, 2015[37]). A finding is only scientifically useful and meaningful if it can be replicated. The simplest type of replication would be for the researchers who first found a result to try to replicate it themselves. Better yet is if the replication can be done by other researchers at other institutions. This would promote confidence that the result is not due to some, typically unintentional, implicit bias or anomaly in the first lab that might be producing the result. Replications are even more convincing if the basic pattern is found even when various aspects of the original study are changed, such as the specific materials, the modality of the presentation, the precise instructions used, and so on. The persistence of a finding under a wide variety of conditions and materials would be the hallmark of a robust finding. Robust findings can regularly be replicated.

Another way of dealing with the issue of replicability in psychology is the increasing reliance on measures of effect size. Traditional inferential statistics in psychology, such as *t*-tests, analysis of variance, correlation, regression, and the like, have been interpreted using some measure of statistical significance, often abbreviated as p. By convention, many effects are considered statistically significant when $p < .05$ *(that is, there is an estimated less than 5% probability that an observed effect in a study is due to chance). The p-value* tells the researcher only whether the effect is statistically significant or not. In comparison, a measure of effect size (e.g. Cohen, 1988[38]) provides the researcher with an index of how extensive the effect is. For example, a study with a large number of observations may find a statistically significant difference between two conditions, although the effect size itself may be very small. This may lead researchers to question the utility and theoretical importance of the findings of their studies. Typically, it is more likely to easily replicate study findings of a difference between different conditions when the p-values are quite small and the effect sizes are quite large.

1.6 The standard theory and cognitive science

LEARNING OBJECTIVE
Explain the mental processes that take place while doing a task

Here we present a standard theory of human cognition along with major outlines that are widely accepted. Although its details are inaccurate, this theory is generally accurate enough to provide a useful heuristic or guide to thinking about human memory and cognition. Because many researchers make wide use of it, it is generally known as the standard model or **modal model of memory**.

1.6.1 The standard theory

The basic structure of the standard model is useful because it is relatively simple, direct and is easy to think about and use when discussing ideas about cognition.

The basic system includes three components:

1. **Sensory memory**
2. **Short-term memory**
3. **Long-term memory**

At the input end, environmental stimuli enter the system, with each sense modality having its own sensory register or memory. Some of this information is selected and forwarded to short-term memory, a temporary working memory system with several **control processes** at its disposal. The short-term store can transmit information to and retrieve information from long-term memory. It is also the component responsible for response output, for communicating with the outside world. If consciousness is anywhere in the system, it is here.

Let's use a multiplication example.

You read '2 × 3 = ? ' and encode the visual stimulus into a visual sensory register. **Encoding** is the act of taking in information and converting it to a usable mental form. Because you are paying attention, the encoded item is passed to short-term memory (STM). This STM is a working memory system where the information you are aware of is held and manipulated. For this example, the system may search long-term memory (LTM) for the answer. A control process in working memory initiates this search, while others maintain the problem until processing is completed. After the memory search, LTM 'sends' the answer, 6, to STM, where the final response is prepared and output, say by speech.

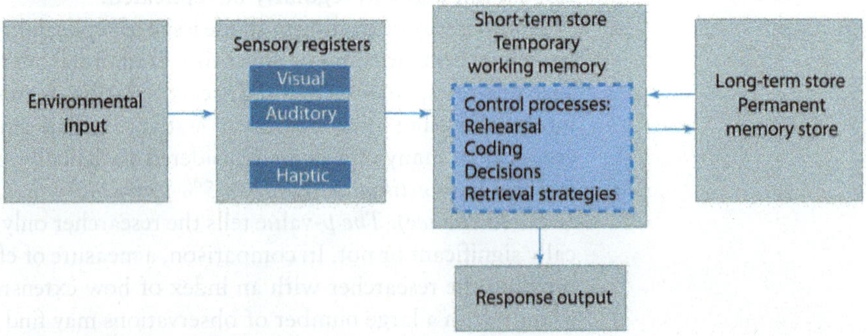

Figure 1.3 Information flow through the memory system in the Atkinson and Shiffrin (1968, 1971[39]) model, the original standard theory in the information-processing approach

Each step in this sequence consumes some amount of time. By comparing these times, we start to get an idea of the underlying mental processing. As you saw in Figure 1.3, a problem such as 2 × 3 takes about 700 ms to answer, compared with more than 1,000 ms for 6 × 9 (values are taken from Campbell & Graham, 1985[40]). The additional 300 ms might be due to long-term memory retrieval, say because of differences in how easily the problems can be located in LTM.

1.6.2 A process model

Although the modal model provides a useful summary, we often need something more focused to explain our results. A common technique is to conceptualise performance in terms of a **process model**, a small-scale model that delineates the mental steps involved in a task and makes testable predictions. Formally, a process model is a hypothesis about the specific mental processes that take place when a particular task is performed.

A process model for lexical decision

A task that is often used in research in cognitive psychology to explore process models is the **lexical decision task**, a timed task in which people decide whether letter strings are or are not English words (see Meyer, Schvaneveldt & Ruddy, 1975[41]). In this task people are shown a series of letter strings. The task is to decide on each trial whether they form a word. So, the letter string might be a word, such as MOTOR, or a non-word, such as MANTY. People are asked to respond rapidly but accurately, and response time is the main dependent measure.

Logically, what sequence of events must happen in this task?

In the process model shown in Figure 1.3, the first stage involves *encoding*, taking in the visually presented letter string and transferring it to working memory. Working memory polls long-term memory to assess whether the letter string is stored there. Some kind of *search* through long-term memory takes place. The outcome of the search is returned to working memory and forms the basis for a decision: either 'yes, it's a word', or 'no, it's not'. If the decision is yes, then one set of motor responses is prepared and executed, say, pressing the button on the left; the alternative set of responses is prepared and executed for pressing the other button.

Lexical decision and word frequency

Say that our results revealed a relationship between RT and the frequency of the words (word frequency is almost always an influence on RT in lexical decision).

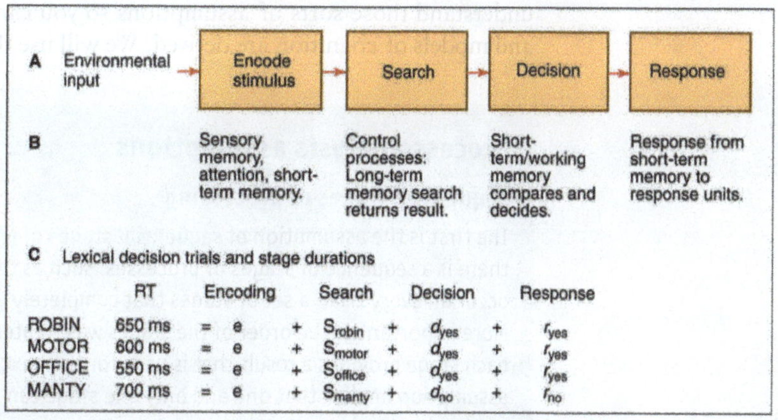

Figure 1.4 A: A general process model, adapted from Sternberg (1969)[42]. B: A list of the memory components and processes that operate during separate stages of the process model. C: A process analysis of the lexical decision task, where RT to each letter string is the sum of the durations of the separate stages

Note that for the three-word trials, the only systematic difference arises from the search stage; encoding, decision and response times should be the same for all three-word trials, according to the logic of process models and the assumptions of sequential and independent stages of processing.

We might test words at low, medium and high levels of frequency in the language:

- ROBIN occurs infrequently, about twice per million words.
- MOTOR is of moderate frequency, occurring 56 times per million.
- OFFICE is of high frequency, occurring 255 times per million (Kucera & Francis, 1967[43]; the most frequent printed word in English is THE, occurring 69,971 times per million).

It takes longer to judge words of lower frequency than higher-frequency words (Allen & Madden, 1990[44]; Whaley, 1978[45]). This is the **word frequency effect**. Other variables also affect response times, but word frequency is enough for our example.

For the sake of argument, say that average responses to low-frequency words, such as ROBIN, took 650 ms; those to medium-frequency words, such as MOTOR, took 600 ms; and those to high-frequency words, such as OFFICE, took 550 ms. What does the process model in Figure 1.5 tell us about such a result?

Logically, we would not expect that word encoding would be influenced by frequency, with high-frequency words being easier to see. So, we assume that encoding is unaffected by word frequency and is relatively constant.

Likewise, all three cases will net a successful search. So, we would not expect time differences in the decision stage because the decision is the same (yes). And finally, 'yes' responses should all take about the same amount of time for any word. Thus, the encoding, decision and response stage times are constants, regardless of word frequency.

The only stage left is the search stage. On reflection, this seems likely to be influenced by word frequency. For instance, it could easily be that words used more frequently are stored more strongly in memory, or stored repeatedly (e.g. Logan, 1988[46]); either possibility could yield shorter search times. Thus, we can tentatively conclude that word frequency has an effect on the search stage. Any factor that affects long-term memory search should influence this stage and should produce a time or accuracy difference. Using the numbers from earlier, the search process would take an extra 50 ms for each change from high to medium to low word frequency.

1.6.3 Revealing assumptions

Several assumptions are made when doing a process analysis. It is important to understand those sorts of assumptions so you can better appreciate how theories and models of cognition are derived. We will use the forgoing example as a guide.

Process analysis assumptions

Sequential stages of processing

The first is the assumption of **sequential stages of processing**. It was assumed that there is a sequence of stages or processes, such as those depicted in Figure 1.5, that occur on every trial, a set of stages that completely accounts for mental processing. More importantly, the order of the stages was treated as fixed on the grounds that each stage provides a result that is used for the next one. More to the point, this assumption implies that one and only one stage can be done at a time, which may not be the case in reality.

The influence of the computer analogy is very clear here. Computers have achieved high speeds of operation, but use **serial processing**: They do operations one by one, in a sequential order. And yet there is no a priori reason to expect that human cognition has this quality in all situations. It may well be that some operations are done in parallel, rather than sequentially.

Independent and non-overlapping

The second assumption is that the stages were **Independent and non-overlapping stages.** That is, any single stage was assumed to finish its operation before the next stage could begin, and the duration of any single stage had no bearing or influence on the others. Thus, at the beginning of a trial, encoding starts, completes its operations and passes its result to the search stage. Then and only then could the search stage begin, followed after its completion by the decision and response stages. With these assumptions, one could interpret the total time for a trial as the sum of the durations for each stage.

Because mental processes take time and each stage is separate, one could view the total time as the sum of the times for all the individual stages. In our earlier example, then, the 50-ms differences between ROBIN, MOTOR and OFFICE were attributed to the search stage. It may well be that different stages of process begin using partial information as each emerges from a prior stage.

Parallel processing

As research has been done, evidence has accumulated that casts doubt on the assumptions of serial, non-overlapping stages of processing. Instead, some evidence exists that multiploe mental processes can operate *simultaneously* – which is termed **parallel processing**. One example involves typing. Salthouse (1984)[47] did a study of how skilled typists type and how performance changes with age. His data argued for a four-process model. The input stage encoded the to-be-typed material, a parsing stage broke large reading units (words) into separate characters, a translation stage transformed the characters into finger movements, and an execution stage triggered the keystrokes. Significantly, his evidence indicated that these stages operate in parallel: while one letter is typed, another is translated into a finger movement, and the input stage is encoding upcoming letters, even as many as eight characters in advance of the one being typed. Moreover, older adults counteracted the tendency towards slower finger movements by increasing their 'look ahead' span at the upcoming letters (see Townsend & Wenger, 2004, for a thorough discussion of serial versus parallel processing[48]).

In moving away from the simpler computer analogy, the cognitive science approach embraced the idea that we need to understand cognition with some reference to the brain. An important lesson we have learned from neuroscience is that the brain shows countless ways in which different cognitive components and processes operate simultaneously, in parallel. Furthermore, there is now ample neurological evidence that different regions of the brain are more specialised for different processing tasks, such as encoding, responding, memory retrieval and controlling the stream of thought (Anderson, Qin, Jung & Carter, 2007[49]).

Context effects

A second difficulty with the early assumptions of sequential stages and non-overlapping processes arose when context effects were taken into account. A simple example of this is the speeding up in deciding – that you are faster to decide MOTOR is a word if you have seen MOTOR recently. A more compelling demonstration comes from work on lexical ambiguity – the fact that many words have more than one meaning. As an example, Simpson (1981)[50] had people do a modified lexical decision task, judging letter strings such as DUKE or MONEY (or MANTY or ZOOPLE) after they had read a context sentence. When the letter string and sentence were related – for instance, 'The vampire was disguised as a handsome count,' followed by DUKE – the lexical decision on DUKE was faster than normal. The reason involved **priming**, the idea that concepts in memory become activated and hence easier to process. In this case, because the context sentence primed the royalty sense of the word *count*, the response time to DUKE was speeded up.

This was an issue for earlier cognitive models because there was no mechanism to account for priming. Look again at Figure 1.5. Is there any component that allows a context sentence to influence the speed of the processes? No, you need a meaning-based component to keep track of recently activated meanings that would speed up the search process when meanings matched but not when they were unrelated.

Let's look more deeply at the influence of context on cognition. Information that is active in long-term memory, for example, can easily have an effect right now on *sensory memory*, the input stage for external stimuli. Here is a simple example:

> As you read a sentence or paragraph, you begin to develop a feel for its meaning. Often you understand well enough that you can then skim through the rest of the material, possibly reading so rapidly that lower-level processes such as proofreading and noticing typograpical errors may not function as accurately as they usually do. Did you see the mistake?

What? Mistake? If you fell for it, you failed to notice the missing *h* in the word *typographical*, possibly because you were skimming but probably because the word *typographical* was expected, predictable based on meaning. You may have even 'seen' the missing *h* in a sense. Why? Because of your understanding of the passage, its meaningfulness to you may have been strong enough that the missing *h* was supplied by your long-term memory.

We call such influences **top-down or conceptually driven processing** when existing context or knowledge influences earlier or simpler forms of mental processes. It is one of the recurring themes in cognition. For another example, adapted from Reed (1992),[51] read the following sentence:

> FINISHED FILES ARE THE RESULT OF YEARS OF SCIENTIFIC STUDY COMBINED WITH THE EXPERIENCE OF MANY YEARS.

Now, read it a second time, counting the number of times the letter *F* occurs.

If you counted fewer than six, try again – and again, if necessary. Why is this difficult? Because you know that function words such as OF carry very little meaning, your perceptual input processes are prompted to pay attention only to the content words. Ignoring function words, and consequently failing to see the letter *F* in a word such as OF, is a clear-cut example of conceptually driven processing (for an explanation of the 'missing letter effect', see Greenberg, Healy, Koriat & Kreiner, 2004[52]).

Other issues

Another issue with these assumptions involves other, often slower mental processes of interest to cognitive psychology. Some process models are aimed at accuracy-based investigations – percentage correct or the nature of one's errors in recall. In a similar vein, many cognitive processes are slower and more complex. Studies of decision-making and problem-solving often involve processing that takes much longer than most RT tasks; for example, some cryptarithmetic problems (e.g. substitute digits for letters in the problem SEND + MORE) can take 15 to 20 minutes! A more meaningful measure of these mental processes involves a verbal report or **verbal protocol** procedure, in which people verbalise their thoughts as they solve the problems. This type of measure in cognitive research is less widely used than time and accuracy measures, but is important nonetheless (see Ericsson & Simon, 1980[53], for the usefulness of verbal protocols).

1.6.4 Cognitive science

Cognitive psychology is now firmly embedded in a larger multidisciplinary effort focused on the study of mind. This broader perspective is cognitive science. As noted earlier, *cognitive science* draws from, and influences, a variety of disciplines such as computer science, linguistics, and neuroscience, and even such far-flung fields as law and anthropology (e.g. Spellman & Busey, 2010[54]). It is a true study of the mind, in a broad sense, as illustrated in Figure 1.5.

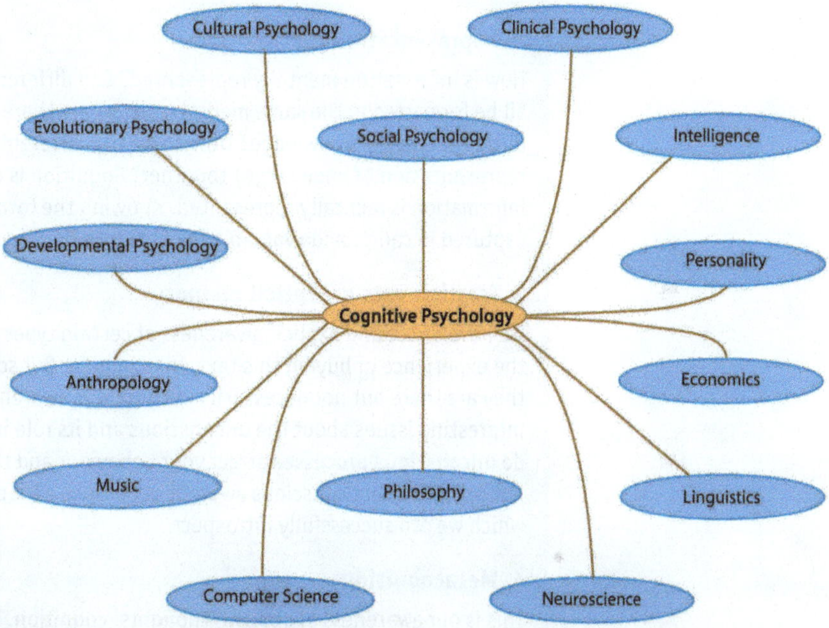

Figure 1.5 Cognitive psychology and the various fields of psychology and other disciplines to which it is related

In general, cognitive science is the study of thought, using available scientific techniques and including all relevant scientific disciplines for exploring and investigating cognition. One of the strongest contributions to this expanded body of evidence has been the consideration of the neurological bases and processes that underlie thought.

1.7 Themes of human cognition

LEARNING OBJECTIVE
Describe the themes of human cognition

Across the various topics in cognition, a number of themes appear repeatedly. You will not necessarily find sections in this text labelled with them. Instead, they crop up across several areas of cognitive science, in different contexts. If you can read a section and identify and discuss the themes that pertain to it, then you probably have a good understanding of the material.

Seven themes of cognition

Here is a list and brief description of seven important themes that occur throughout the text.

1. Data-driven versus conceptually driven processing

Some processes rely heavily on information from the environment (data-driven or bottom-up processing). Others rely heavily on our existing knowledge (conceptually driven or top-down processing). Conceptually driven processing can be so powerful that we often make errors, from mistakes in perception up through mistakes in reasoning. But could we function without it? As you will see, much of cognition involves a combination of both data-driven and conceptually driven processes, although their balance may vary depending on the circumstances.

2. Representation

How is information mentally represented? Can different kinds of knowledge all be formatted in the same mental code, or are there separate codes for the different types of knowledge? How do we use different types of representation (or representation of knowledge) together? Cognition is critically dependent on how information is mentally represented. Knowing the format that the information is captured in can provide valuable insights into how it can be used.

3. Implicit versus explicit memory

We have direct and explicit awareness of certain types of memories; you remember the experience of buying this text, for example. But some processes are implicit; they are there but not necessarily in conscious awareness. This raises all sorts of interesting issues about the unconscious and its role in cognition; for instance, how do unconscious processes affect your behaviour and thinking? Much of cognition is not open to direct conscious awareness because a lot of it happens below the level at which we can successfully introspect.

4. Metacognition

This is our awareness of our own thoughts, cognition, knowledge and insight into how the system works. Such awareness prompts us to write reminders to ourselves to avoid forgetting. How accurate is this awareness and knowledge? Does it sometimes mislead us? Metacognition is important because our insights into our memory and cognition guide many of our daily decisions, such as deciding whether we have understood or learned class material.

5. Brain

Far more than the cognitive psychology of the past, brain–cognition relationships and questions are a primary concern. It is important to understand how our neural hardware works to produce the kinds of thinking that we are capable of, and what its limitations are. Our cognition is a direct function of our neural states and how they function well or poorly, depending on the circumstances.

6. Embodiment

Cognitive psychology is not just about the way we think about and represent information. It also reflects the fact that we need to interact physically with the world – it is embodiment. How do we capture the world in our mental life? How do the ways that our bodies interact with the world influence our thinking? How do we incorporate and take into account physical realities in how we think about and process information? Cognition is grounded in our actions and interactions with the world.

7. Future orientation

To do well in the world, cognition needs to be future-oriented. We must be able to predict what will happen next or sometime soon in the future. We also need to be able to plan our actions and anticipate the likely outcomes. What aspects of the world are regularly predicted by cognition? How well do people anticipate what they will need to do? Memory and cognition are often discussed as if they were focused on the past, but keep in mind that much of what they are doing is aimed at the future.

Summary: Cognitive psychology

1.1 Thinking about thinking

- Cognitive psychology is the scientific study of human mental processes. This includes perceiving, remembering, using language, reasoning and solving problems.
- Intuitive analysis of examples such as 'How many hands did Aristotle have?' and 'Does a robin have wings?' indicates that many mental processes occur automatically (very rapidly and below the level of conscious awareness).

1.2 Memory and cognition defined

- Memory is composed of the mental processes of acquiring and retaining information for later use (encoding), the mental retention system (storage), and then using that information (retrieval).
- Cognition is the collection of mental processes and activities used in perceiving, remembering, thinking and understanding, as well as the act of using those processes.

1.3 An introductory history of cognitive psychology

- The modern history of cognitive psychology began in 1879 with Wundt and the beginnings of experimental psychology as a science.
- The behaviourist movement rejected the use of introspection and substituted the study of observable behaviour.
- Modern cognitive psychology, which dates from approximately 1960, rejected much of the behaviourist position but accepted its methodological rigour. Many diverse viewpoints, assumptions and methods converged to help form cognitive psychology. This was at least a rapid, evolutionary change in interests, if not a true scientific revolution.

1.4 Cognitive psychology and information processing

- Cognitive psychology began as a separate field around 1960 with the decline of behaviourism; the developments in linguistic and verbal learning; and important papers by researchers on attention, visual processing and memory.
- The advent of the modern digital computer played a key role in the development of cognitive psychology and served as the basic metaphor for how the mind processes knowledge. From this, models of the mind as a means of processing information were developed.

1.5 Measuring information processes

- Although channel capacity was an early, useful analogy in studying information processing, a more influential analogy was later drawn between humans and computers: that human mental processing might be analogous to the sequence of steps and operations in a computer program. Computers still provide an important tool for theorising about cognitive processes.
- Measuring information processes, the mental processes of cognition, has relied heavily on time and accuracy measures. Differences in response time (RT) can yield interpretations about the speed or difficulty of mental processes, leading to inferences about cognitive processes and events. Accuracy of performance, whether it measures correct recall of a list or accurate paraphrasing of text, also offers evidence about underlying mental processes. The findings of individual studies can be interesting and informative. However, in order for a finding to be truly valuable, it is best if it has been replicated to show that it is stable and robust. Good science is rooted in being able to accurately predict outcomes.

1.6 The standard theory and cognitive science

- The modal model of memory suggests that mental processing can be understood as a sequence of independent processing stages, such as the sensory, short-term and long-term memory stages.
- Process models are appropriate for fairly simple, rapid tasks that are measured by response times, such as the lexical decision task.

- There is substantial evidence to suggest that cognition involves parallel processing and is influenced by context; for example, research on skilled typing shows a high degree of parallel processing. Also, slower, more complex mental processes, such as those in the study of decision-making and problem-solving, may be studied using verbal protocols.
- Cognitive psychology is better understood as residing within the context of a broader cognitive science. This approach describes cognition as the coordinated, often parallel operation of mental processes within a multicomponent system. The approach is deliberately multidisciplinary, accepting evidence from all the sciences interested in cognition.

1.7 Themes of cognition

- A number of themes running throughout the study of cognition cut across many of the subdomains. These include data-driven versus conceptually driven processes, representation, implicit versus explicit memory, metacognition, the brain, embodiment and a future orientation to cognition.

Chapter 1 quiz

Question 1

You are in a lecture and are asked whether the phrase 'the sky is green with blue clouds' is true or false. What is the correct order of the first three cognitive processes involved in thinking about and answering this question?

A. Perceptual, memory and comprehension.
B. Comprehension, memory and retrieval.
C. Computational, comprehension and sensory.
D. Perceptual, judgement, computational.

Question 2

Ebbinghaus conducted an investigation whereby he created a list of nonsense syllables, learnt the list and then recalled the nonsense syllables to see how many were stored in memory. Given the early work on memory with the use of no laboratory, what was a criticism of the research from Ebbinghaus?

A. There was no clear theory developed from Ebbinghaus.
B. The ideas from Ebbinghaus in relation to the savings measure of memory have not been useful in the area of memory.
C. There was a lack of control over other variables now associated with memory.
D. The investigations from Ebbinghaus had shown little progress within the early cognition research.

Question 3

The behaviourist position in cognitive psychology posits the idea that all behaviours are acquired through some form of conditioning, and conditioning occurs through the interaction with the environment. Why did other researchers such as Chomsky, disagree with this position?

A. Researchers suggested that the environment was not the concept that helped to develop language.
B. In the application of language, it was found that other cognitive processes such as attention, recognition and computation were involved in language development.
C. Behaviourism was not seen to use mental processes and there was a wide range of evidence to support this in terms of studies looking at infant language development.
D. Skinner developed a clear theoretical framework and he suggested that this could not be disproved due to the application to verbal learning.

Question 4

You have been asked to bake a cake using a written recipe and this is a task you have not completed before. You have only seen online recordings of someone baking a cake. Which process is the first process in helping you bake a cake?

A. Perception
B. Encoding
C. Decision making
D. Long-term memory storage

Question 5

How does top-down processing differ from bottom-up processing?

A. Top-down processing involves existing knowledge rather than environmental information.
B. Top-down processing requires memorisation instead of intuitive understanding.
C. Top-down processing happens nearly instantaneously rather than gradually.
D. Top-down processing is a standard model concept instead of a process-driven concept.

2 The neural basis of cognition

Learning objectives

2.1 **Explain the relevance of neuroscience to cognitive psychology**

2.2 **Describe the structure and function of neurons**

2.3 **Identify key brain areas and neural networks, including their function**

2.4 **Understand cognitive processes in patients with brain damage**

2.5 **Summarise connectionism as a computer-based method in cognitive science**

Key figures

Professor Geraint Rees

Professor Geraint Rees is considered one of the most influential neuroscientists and the author of the book *Neurobiology of Attention*. He is a Professor of Cognitive Neurology at University College London and has held numerous leadership positions in academia.

Professor Michael Gazzaniga

Professor Michael Gazzaniga is a leading researcher in cognitive neuroscience at the University of California in Santa Barbara and some of his most influential work is on split-brain patients, showing that language comprehension could be executed by the right hemisphere.

Introduction

An important aspect of modern cognitive science is an understanding of neural processes and how they relate to thought. In this chapter, we provide a basic overview of the fundamental principles of neural processing and various brain structures that are important for a wide range of cognitive processes. We begin with an exposition of the characteristics of individual neurons, how they communicate information, and the influence of some basic classes of neurotransmitters. After this, we look at various structures of the brain, outline the major regions of the cortex, and cover some of the major methods of assessing how neural processing corresponds to cognitive processing. Finally, we discuss a computational approach, called neural net or connectionist modelling, which tries to capture the basic principles of neural processing in computer-based mathematical models of thought.

2.1 Brain and cognition

LEARNING OBJECTIVE
Explain the relevance of neuroscience to cognitive psychology

We start with a brief description of a cognitive disruption to understand more about **cognitive neuroscience**. Tulving (1989)[1] described a patient he called K.C., a young man who had sustained brain damage in a motorcycle accident. Some nine years after the accident, he still showed pervasive disruption of long-term memory. The fascinating thing about his memory impairment was that it was selective: K.C. remained competent with language, his intelligence was normal, and he was able to converse on a number of topics. But K.C. explained that he could not remember any experience from his own past – in the sense of bringing back to conscious awareness, a single thing that he had ever done or experienced in the past. For example, even though he could remember how to play chess, he could not remember ever having played it before. He knew his family had a holiday house on a lake, but he did not have any recollection of having been there. K.C.'s brain damage seemed to destroy his ability to access what we will call **episodic memory**, his own autobiographical knowledge, while leaving his general knowledge system – his **semantic memory** – intact. This pattern is called a **dissociation**, a disruption in one component of mental functioning but no impairment of the other. Can these two forms of long-term memory, episodic and semantic memory, be the same, given K.C.'s dissociation between the two? Probably not. Better understanding of how cognitive systems and processes are organised at a neural level can help us to explain dissociations, such as the one observed in K.C.

Although the evidence of brain damage is important, we also need other kinds of evidence – for example, information about the neurochemical and neurobiological activities that support normal remembering, learning and thinking, or the changes in the brain that accompany ageing. Therefore, we are interested in contributions from different neuroscience disciplines – neurochemistry, neurobiology, neuroanatomy and so on – as they relate to human cognition.

Although it is important to understand impairments in cognitive processes and the value of rehabilitation and retraining for patients with brain damage, our interest in cognitive science extends to the study of normal cognition from the standpoint of the human brain. We need to learn about normal cognition through any relevant means available. Thus, more and more investigators have been examining the behavioural and cognitive effects of brain damage (e.g. McCloskey, 1992; Fagerholm et al., 2015[2]), using those observations to develop and refine theories of normal cognition (de Haan et al., 2020; Martin, 2000[3]) and to inform rehabilitation interventions and treatment approaches (e.g. Maggio et al., 2019[4]). As you will see throughout this course, the great irony of brain damage is that it sometimes leads to a clearer understanding of cognitive processes in healthy people.

2.1.1 Dissociations and double dissociations

The concept of dissociation – the opposite of association – is important, so we should learn more about it.

Dissociations and the lack of dissociation

Consider two mental processes that 'go together' in some cognitive tasks, called process A and process B. By looking at these processes as they may be disrupted in brain damage, we can determine how separable the processes are.

Complete separation is a **double dissociation**. Evidence of a double dissociation requires at least two patients, with 'opposite' or reciprocal deficits. For example, assume that Patient X has a brain lesion that has disrupted process A. His performance on tasks that use process B is intact, not disrupted at all.

Patient Y, on the other hand, has a lesion that has damaged process B, but tasks that use process A are normal, not disrupted by the damage.

In a simple dissociation, process A could be damaged while process B remains intact, yet no other known patient has the reciprocal pattern. For example, *semantic retrieval* (retrieving the meaning of a concept) could be intact, while *lexical retrieval* (finding the name for the concept) could be disrupted: this is called **anomia**. In this situation, lexical retrieval is dissociated from semantic retrieval, but it is probably impossible to observe the opposite pattern. How can you name a concept if you cannot retrieve the concept in the first place?

In a full or complete association (lack of dissociation), disruption of one of the processes always accompanies disruption in the other process. This pattern implies that process A and process B rely on the same region or brain mechanism, such as recognising objects and recognising pictures of those objects.

Likewise, theoretical models and methods used in neuroscience can inform theories of cognitive processes and, accordingly, indicate areas of intervention. For instance, using non-invasive and non-pharmacological methods, such as positron emission tomography (PET) and functional magnetic resonance imaging (fMRI) scans, as well as transcranial magnetic stimulation (TMS or TDMS), we can localise regions of activity during different kinds of cognitive processing, and better understand the functioning of healthy brains and brains that have undergone injury or any other kind of damage, such as those observed in brain tumours or neurodegenerative diseases such as dementia (e.g. Coffey et al., 2021; Smalle et al., 2015[5]).

The transfer of knowledge and insights from cognitive psychology to neuroscience is bidirectional, as both fields inform each other and advance. What has become clear is that cognition and neuroscience are highly interlinked. This is particularly evident when you consider how different cognitive functions depend on neurological accounts of perceptual and motor process and how they influence even seemingly abstract tasks, such as using semantic memory or even dreaming (Fernandino et al., 2022; Simor et al., 2022[6]). In this chapter, we focus only on a few of the many examples in the intersection between cognitive psychology and neuroscience.

2.2 Understanding how neurons work

LEARNING OBJECTIVE
Describe the structure and function of neurons

At birth, the human brain weighs approximately 400 g (about 14 oz). It grows to an average of 1,450 g in adults, slightly more than 3 lb and is roughly the size of a ripe grapefruit.

The basic building block of the brain (indeed of the entire nervous system) is the **neuron**, a specialised cell that receives and transmits neural impulses. How many neurons are there in the brain? Available estimates vary. Herculano-Houzel (2012)[7] reported that there are about 100 billion neurons in the brain, although humans will lose about 10 per cent of this mass over the course of a lifetime. To put that figure in perspective, consider that the Milky Way galaxy has about 100 billion stars. Moreover, there is a very high degree of interconnectivity in the brain among these cells, with these neurons making about 100 trillion connections. Talk about a complex mental machine! In addition to neurons, there are many other types of cells in the brain, including glial cells, and connective and circulatory tissue that also abound.

2.2.1 The structure of a neuron

Neurons are the nerve cells that form nerve tracts throughout the body and in all structures of the central nervous system (CNS), that is, the brain and the spinal cord. In this section, we examine neuronal structure in more detail.

At one end of the neuron, many small branchlike fingers, called **dendrites**, gather neural impulses from neighbouring neurons into the neuron itself. In somewhat more familiar terms, the dendrites are the *input* structures of the neuron, receiving the neural impulses that are being passed along in a particular neural tract. The dendrites receive either excitatory or inhibitory neural impulses, respectively known as excitatory post-synaptic potentials (EPSPs) and inhibitory post-synaptic potentials (IPSPs) – the *synapse* is a crucial component of neuronal communication, and it is discussed in more detail later in Section 2.2.2.

The central portion of each neuron is the cell body, or **soma** (Greek for 'body'), where the biological activity of the cell is regulated. Extending from the cell body is a longish extension or tube, the **axon**, which ends in another set of branch-like structures called *axon terminals* or sometimes *terminal buttons* or *arborisations* – the latter term derives from the tree-like form of these structures. The axon terminals represent the *output* end of the neuron, the place where the neural impulse ends within the neuron itself, before it is communicated on to the next neuron along a neural tract. If the dendrites represent the *post-synaptic* area of neurons, the axon terminals represent the *pre-synaptic* area.

Some neurons also have a fatty coating along the axon. This coating is called the **myelin sheath** and serves as an insulator for the neuronal axon. The myelin sheath is typically not continuous, but has gaps along the way, which are known as the **nodes of Ranvier**.

An important job of the myelin sheath is to make the propagation of neural impulses along the axon more efficient. It does this by allowing neural impulses to 'jump' between Ranvier nodes, rather than travelling down the entire length of the axon, which would require more time and energy. This functionally shortens the

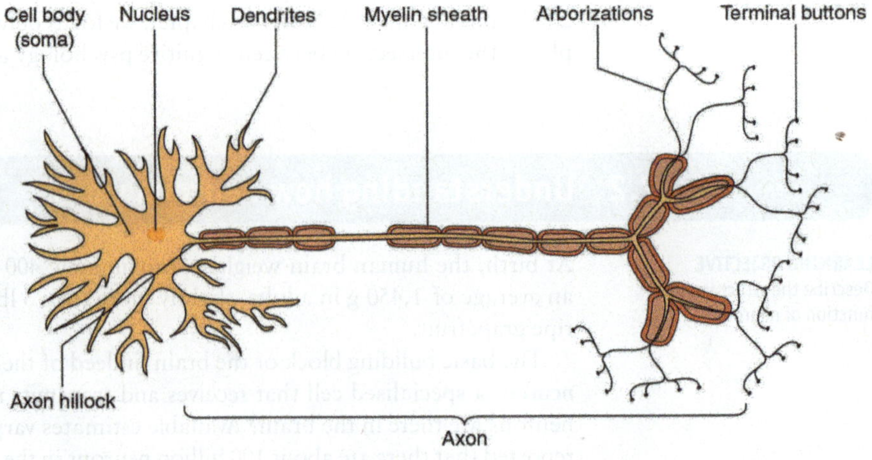

Figure 2.1 Different parts of a neuron

distance the neural signal travels, thereby speeding up the propagation of neural impulses and the process of neuronal communication. Note that certain health conditions known as *demyelinating diseases*, such as multiple sclerosis, are characterised by damage in the myelin sheath of neurons.

Not all neurons are myelinated. Often, myelin is found in neurons that need to transmit signals at a relatively long distance, such as out in the periphery of the nervous system. Most of the cortical neurons from which cognition arises are not myelinated. You can easily see differences between bundles of myelinated and unmyelinated neurons even without the aid of a microscope. Neurons that have myelin sheaths appear to be white, because the myelin is made of fat. However, neurons that do not have a myelin sheath appear to be grey. So, when people talk about 'white matter' and 'grey matter' in a person's brain, they are talking about bundles of myelinated and unmyelinated neurons, respectively. Also, neurons themselves do not produce the myelin. Instead, myelin is produced by a special group of nerve cells in the nervous system, known as the *glial cells*.

Receptor cells react to the physical stimulus and trigger a pattern of firing down *sensory neurons*. These neuron tracts pass the message along into the spinal cord. For a simple reflex, the message loops quickly through the spinal cord and goes back out to the arm muscles through a tract of motor neurons that terminate at *effector cells*, which connect directly to the muscle fibres and cause the muscles to pull your arm away. For example, nociceptors (sensory neurons responsive to noxious stimuli) are responsible for the sensation of pain. Nociceptors respond to different types of noxious stimulation (e.g. tissue damage, pressure and extreme heat or cold) and transfer such sensory messages to the brain, which then sends information to motor neurons and muscle fibres in order to prevent further damage. René Descartes' dualism presents a rudimentary, yet not entirely accurate, approach to understanding the physical pathways involved in the neural sensation and response to painful stimuli.

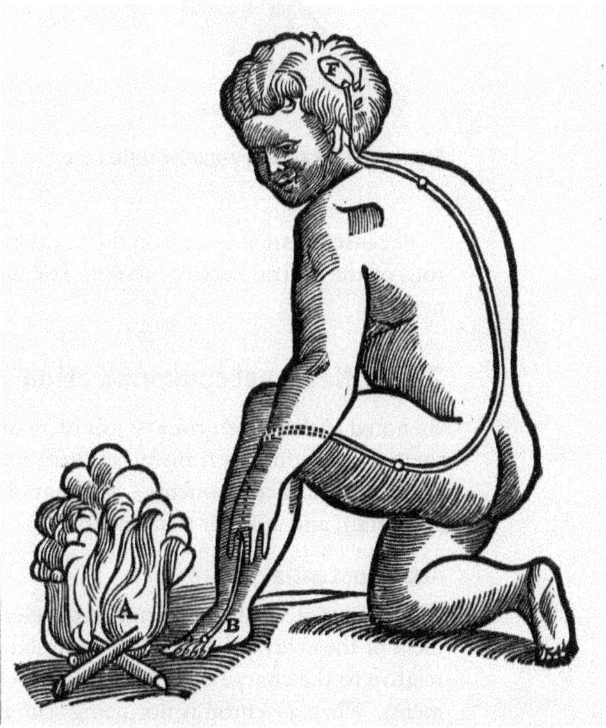

The Descartes reflex, showing how the stimulus (heat emitted from fire, A) is received by the skin (B), which sends the message to the brain (F) to initiate action.

Source: World History Archive/Alamy Stock Photo

As the reflex triggers the quick return of a message out to the muscles, it simultaneously routes a message up the spinal cord and into the brain. Thus, the second route involves only the central nervous system (CNS). There is only one kind of neuron in the CNS, called an *interneuron* or *association neuron*.

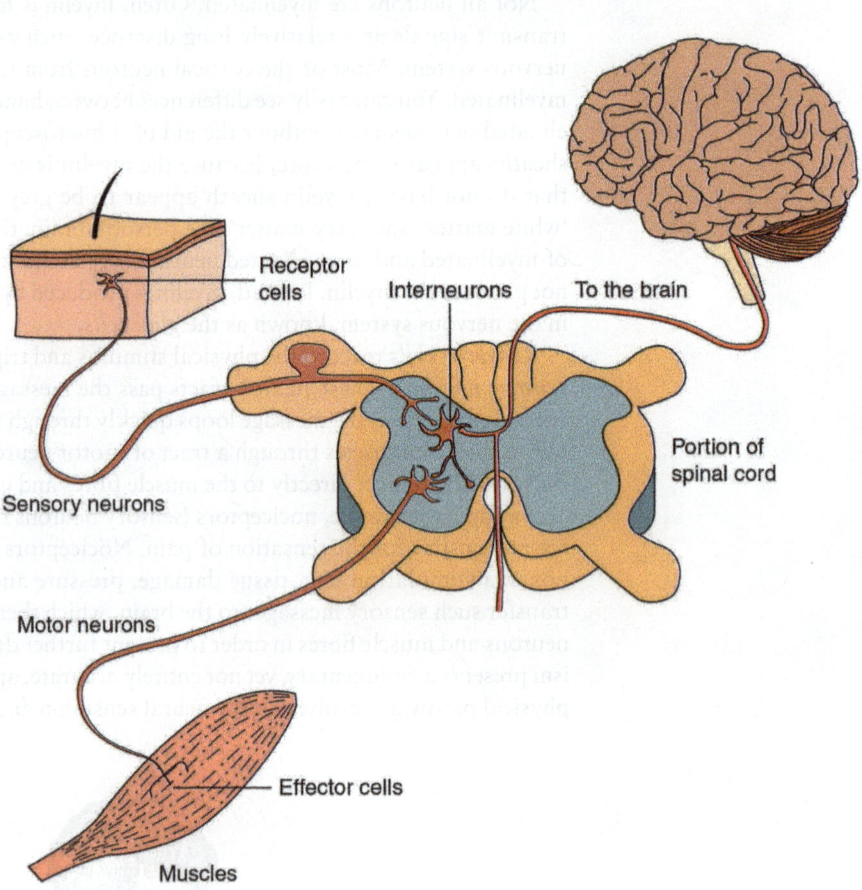

Figure 2.2 A sensory-motor reflex arc

Because we are focusing on the brain here, we are interested only in the interneurons of the central nervous system. For simplicity, we will just refer to them here as neurons.

2.2.2 Neuronal communication

As noted earlier, the primary job of neurons is to communicate information, via the propagation and transmission of neural impulses. This is done by two basic electrochemical components – one that occurs within a single neuron (called action potential), and another that occurs between neurons (called synapse).

Action potential

The **action potential** is a change in the electrical charge of the intracellular environment of the neuron (i.e. inside the cellular membrane of the neuron's body or soma) relative to the charge outside of the cellular membrane (i.e. extracellular environment). When a neuron is not being stimulated, it has a resting electrical charge of −70 mV (millivolts). When a neuron is stimulated, there is a depolarisation of its electrical potential (i.e. the electrical charge of the neuron gradually becomes more positive), and the neuron is said to 'fire' when its electrical charge reaches +40 mV

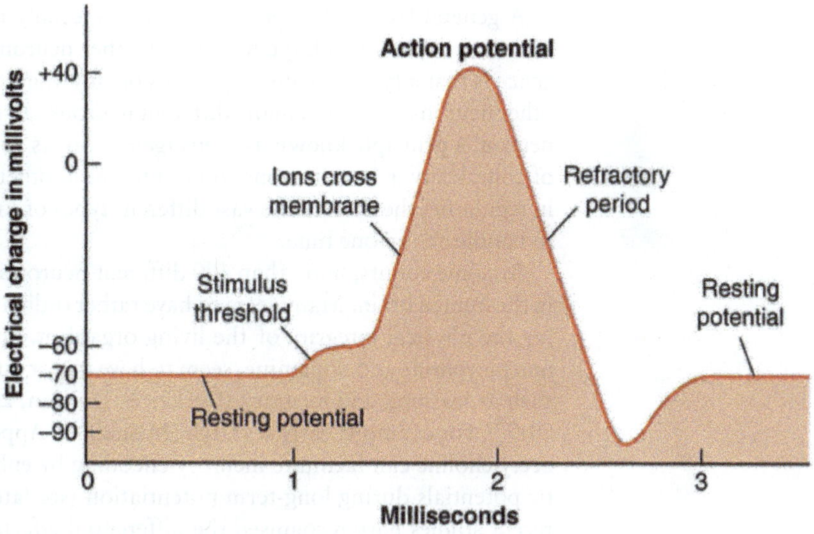

Figure 2.3 Action potential

(see Figure 2.3). Action potentials follow the **all-or-none principle**. That is, there is either an action potential, which is always the same anywhere in the nervous system, or there is no action potential. It is *not* the case that some action potentials are larger or smaller than others.

Note that the entire neuron does not change its electrical charge for the action potential to be triggered. Instead, the action potential is **propagated** from the dendrites, through the soma and down the axon, almost like a wave of electrical charge. After the neuron has fired, there is a recovery period in which the neuron resets itself and is ready to fire again.

The propagation of the action potential down the axon is facilitated by the exchange of positively charged sodium (Na^+) ions entering the interior of the cell, while positively charged potassium (K^+) ions are forced out of the cell. The influx of positively charged Na^+ ions is responsible for the gradual depolarisation of the intracellular environment of the neuron, which then triggers the action potential. The pushing out of positively charged K^+ ions renders the base charge of the neuron negative again. The equilibrium between Na^+ and K^+ ions inside the cell is regulated by the **sodium–potassium pump**, a specialised structure along the neuronal axon that controls how many K^+ ions are exchanged for every Na^+ ion.

Synapse

There may be relatively few or many axon terminals emanating from a single neuron. In either case, these terminals in the brain are adjacent to dendrites from many other neurons. Thus, the impulse of the action potential within a neuron terminates at the axon terminals and is taken up by the dendrites of the next neurons in the pathway. These are the neurons whose dendrites are adjacent to the axon terminals. The region where the axon terminals of one neuron and the dendrites of another come together is the **synapse**. Accordingly, the neuron that transmits the neural impulse is the pre-synaptic neuron, whereas the neurons receiving the neural impulse are the post-synaptic neurons. For the most part, the neurons do not actually touch one another (some regions of the brain contradict this rule). The synapses in the nervous system are extremely small physical gaps or clefts between the neurons. As such, the space between the pre-synaptic axon terminals and the post-synaptic dendrites is called the **synaptic cleft**. Each synapse is 100–200 angstroms wide, with an angstrom being 1/10,000th of a millimetre. Note that the word *synapse* is also used as a verb: a neuron is said to synapse on another, meaning that it passes its message on to that other neuron.

A general law of the nervous system, especially in the brain, is that any single neuron synapses on a large number of other neurons. The evidence for this *divergence* is that a typical neuron synapses on anywhere from 100 to as many as 15,000 other neurons. Likewise, many different neurons can synapse on a single destination neuron, a principle known as *convergence*. So, as you can see, there is a great deal of complexity in how neurons connect to one another. This complexity is important in regulating the flow of the vast different types of knowledge that our brains need to handle at any one time.

By some counts, more than 100 different neurotransmitters have been identified in the human brain. Many seem to have rather ordinary functions, such as maintaining the physical integrity of the living organism. Others, especially **acetylcholine**, **norepinephrine** and dopamine, seem to have major influences on cognitive processes such as learning and memory (Decker & Duncan, 2020; Roozendaal & Hermans, 2017[8]). For example, early research (Mishkin & Appenzellar, 1987[9]) suggested that acetylcholine can facilitate memory encoding by enhancing the strength of synaptic potentials during long-term potentiation (see later discussion). However, more recent studies have recognised the differential effects of acetylcholine on multiple memory processes, such as memory consolidation and retrieval (Decker & Duncan, 2020; Micheau & Marighetto, 2011[10]).

Another important neurotransmitter for human cognition is **glutamate**, which is an excitatory neurotransmitter. This is important for creating or strengthening the connections between neurons. This is how we learn, by changing how neurons are connected to one another, and glutamate is an important neurotransmitter for doing this. A crucial neurotransmitter that is derived from glutamate is *gamma-aminobutyric acid*. However, rather than remembering such a long name, most neuroscientists simply refer to this neurotransmitter as **GABA**. GABA is an inhibitory neurotransmitter that works by weakening the connections between neurons and, in some sense, it is the opposite of glutamate.

2.2.3 Neurons and learning

So, what exactly do these neurotransmitters change when learning occurs? Although it is not completely understood, one of the basic processes involved is something called **long-term potentiation (LTP)**, first discovered in 1973 (Lømo, 2018; Nicoll, 2017[11]).

Long-term potentiation is the process by which connections between neurons are strengthened. This strengthening essentially changes the ease with which two connected neurons will fire. More specifically, how strongly the pre-synaptic neuron excites the post-synaptic neuron to fire is a function of LTP.

Let's look at the processes that bring about LTP during learning in a bit more detail, following work done by Pittenger and Kandel (2003)[12]. The process described here is illustrated in Figure 2.4.

When a neuron is stimulated, it can release glutamate into the synapse. This is taken up by the next neuron. Some of these glutamate neurons are taken up by what are known as *AMPA receptors*. These are fast-acting receptor sites that will result in a depolarisation of the neuron at that moment and may cause it to fire. What is more interesting here is that some of the glutamate will also be taken up by what are known as *NMDA receptors*. The consequence of these is a series of chemical and genetic reactions that ultimately result in the creation of new AMPA receptors. What these new AMPA receptors do is to make it more likely that the neuron will take up glutamate that is released in the future, making it more likely that the neuron will fire. In other words, a simple way of thinking about this is that the AMPA receptors cause the neuron to fire, but the NMDA receptors lead to a strengthening

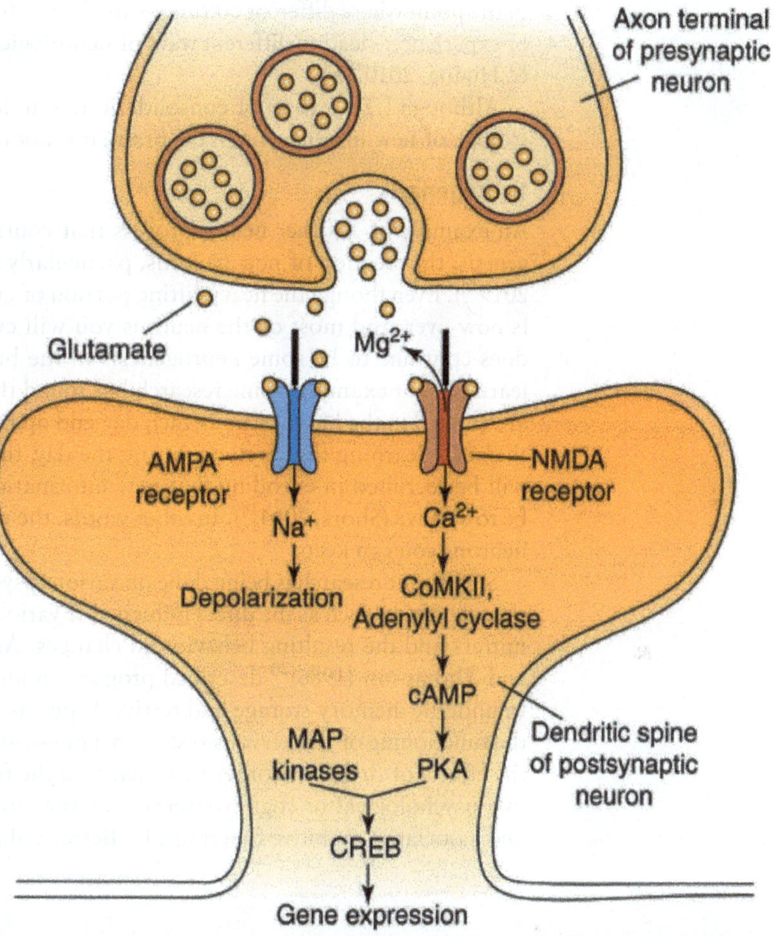

Figure 2.4 Short- and long-term effects of glutamate on a post-synaptic neuron, which can result in LTP

of the connection between the neurons. This idea of changing the strength of the connection between neurons will be important when we discuss the development of connectionist models of cognition later in the module.

Often LTP can last for days or weeks.

Memory consolidation

We need to be able to remember things for longer periods of time. This is accomplished by a process known as **consolidation** (Abraham, 2006[13]), which takes place over long spans of time lasting days, weeks and years. Consolidation makes memories more and more permanent over time. Interestingly, one of the processes that occur when we sleep can aid this consolidation process. People have been shown to remember more information if they had slept during the delay between study and test than if they had not (Hu, Stylos-Allan & Walker, 2006[14]), particularly if they experienced slow-wave (stages 3 and 4) sleep, as compared to REM (dreaming) sleep (Rasch & Born, 2008[15]).

At a small scale, LTP captures changes in individual neural connections. At a larger scale, consolidation captures changes in assemblies of neurons over long periods of time. At an even larger scale, changes in brain connections from experience lead to different ways of processing information over a person's lifetime, even to the point where different cultures – and hence different collections and complexes of experience – lead to different ways of neurologically processing information (Park & Huang, 2010[16]).

Although LTP, a form of consolidation, is an important process involved in the storage of new information in the brain, it is not the only process that is operating.

Neurogenesis

An example of another neural process that contributes to new learning is **neurogenesis**, the creation of new neurons, particularly in the hippocampus (Toda et al., 2019[17]). Even though the heavy-lifting portion of creating new neurons in your brain is now over, and most of the neurons you will ever have are already made, there does continue to be some neurogenesis in the brain, and this can influence new learning. For example, some research has found that most of the new neurons that are created in the hippocampus each day end up not making it – they die. However, the more learning that is done during the day, the more likely these new neurons will be recruited in encoding that new information, and the more likely they will be to survive (Shors, 2014[18]). In other words, the more you learn each day, the more neurons you can keep.

Significant research is being done on various psycho-biochemical properties of the neural system, such as the direct influence of various chemical agents on neurotransmitters and the resulting behavioural changes. As an example, Abraham (2006)[19] and Thompson (1986)[20] described progress in identifying neural changes believed to underlie memory storage and retrieval. Just as various psychoactive drugs affect the functioning of the nervous system in a physical sense, research is now identifying the effects of drugs and other treatments on the functioning of the nervous system in a psychological or cognitive sense (e.g. the effect of alcohol abuse on the brain and associated cognitive functions; La Berre et al., 2017[21]).

2.3 Brain structures and their function

LEARNING OBJECTIVE
Identify key brain areas and neural networks, including their function

Ignoring many levels of intermediate neural functioning and complexity, we now take a tremendous leap from the level of single neurons to the level of the entire brain. To account for all human behaviour, including bodily functions that occur involuntarily (e.g. digestion), entails an extensive discussion of the central (CNS) and peripheral nervous systems (PNS). But to explore cognitive neuroscience, we can limit ourselves to just the CNS. In fact, our discussion of the different cognitive functions throughout this book even omits much of the CNS, save for the **neocortex (or cerebral cortex)**, which sits at the top of the human brain and subcortical structures, such as the limbic system.

2.3.1 Subcortical brain structures

The physically lower brain structures are collectively called the old brain or brain stem. The subcortical brain includes nuclei (white and grey matter) of the thalamus, hypothalamus, corpus callosum, hippocampus, putamen and amygdala – all of which play key roles in cognition (see Figure 2.5).

All these subcortical (meaning 'below the cortex') structures play a significant role in supporting cognitive functions. Let's consider them individually.

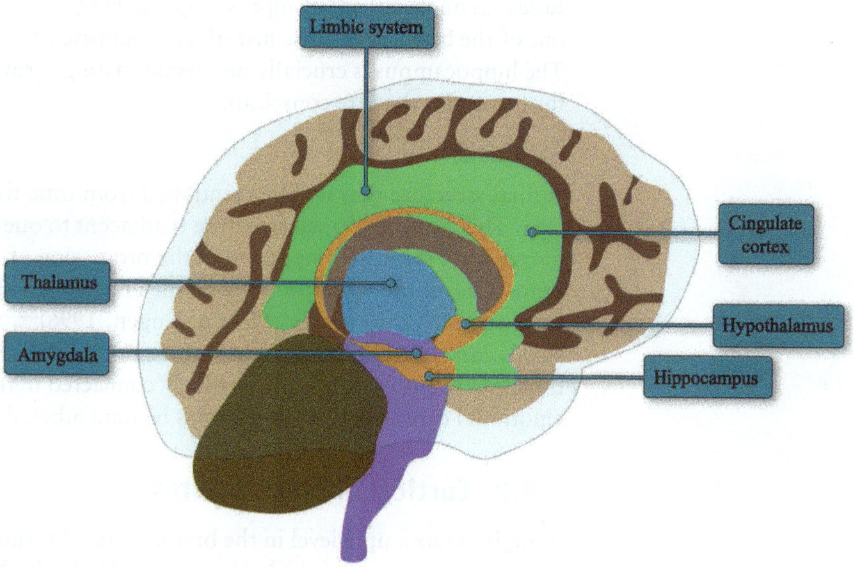

Figure 2.5 Lower brain structures

Source: miha de/Shutterstock

Thalamus

First, deep inside the brain is the **thalamus**, meaning 'inner room' or 'chamber'. It is often called the gateway to the cortex because almost all messages entering the cortex come through the thalamus (a portion of the olfactory sense of smell is one of the very few exceptions). In other words, the thalamus is the major relay station from the sensory systems of the body into the neocortex Made up of several nuclei, each linked to different sensory systems (e.g. visual, auditory), the thalamus filters information that goes to the cortex, With links to other subcortical structures, such as the hippocampus and the mammillary bodies (both part of the limbic system), it supports learning and episodic memory. The thalamus also plays an important role in sleep and wakefulness.

Hypothalamus

Under the thalamus sits the hypothalamus (*hypo* means 'under' in Greek), an important structure in maintaining homeostasis by controlling the endocrine and autonomic nervous system. With links to the limbic system with three zones (supraoptic region, tuberal region and mammillary region) the hypothalamus is implicated in regulating stress responses via the hypothalamo–pituitary–adrenocortical (HPA) axis.

Corpus callosum

Just above the thalamus is a broad band of nerve fibres called the **corpus callosum**. As described later, the corpus callosum ('callous body') is the primary bridge across which messages pass between the left and right halves – the hemispheres – of the cortex. This is the broadband connection of the brain that allows the two halves to communicate and pass information between each other.

Hippocampus

Another structure is the **hippocampus**, from the Latin word for 'sea horse', referring to its curved shape. The hippocampus is located immediately interior to the temporal lobes; that is, underneath the temporal lobes but in the same horizontal plane.

Research on the effects of hippocampal damage is described in Chapter 7, including one of the best-known case histories in cognitive neuroscience, that of patient H.M. The hippocampus is crucially important in long-term memory processes, especially for memories that are conscious.

Amygdala

A final structure that will be mentioned from time to time in this text is the **amygdala**. This almond-shaped structure is adjacent to one end of the hippocampus. The amygdala is critically important in the processing of emotional information in the brain. One of the unique aspects of the amygdala is that it gets nearly direct inputs from the olfactory nerves (Herz & Engen, 1996)[22], which is why how something smells can be strongly associated with emotional responses (e.g. love, relaxation or disgust). The amygdala is also strongly connected to the hippocampus, which is why emotional experience can sometimes be remembered very well.

2.3.2 Cortical brain structures

Now, let's move up a level in the brain. Figure 2.6 shows the neocortex, or cerebral cortex, the top layer of the brain, responsible for higher-level mental processes.

The cortex is a wrinkled, convoluted structure that nearly surrounds the old brain. The two halves or hemispheres cover about 2,500 cm^2 and are from about 1.5 to 3 mm thick. The wrinkling comes about by trying to get such a large surface area in a small volume. It compares to trying to get a piece of paper into a cup. To get the paper in, you wrinkle it up. The neocortex is the most recent structure to have evolved in the human brain (*neo* means 'new') and is much larger in humans than in other animals. Compare the average weight of it in humans (1,450 g) with that in the great apes (400 g). In addition, because the neocortex is primarily responsible for higher mental processes such as language and thought, it is not surprising that

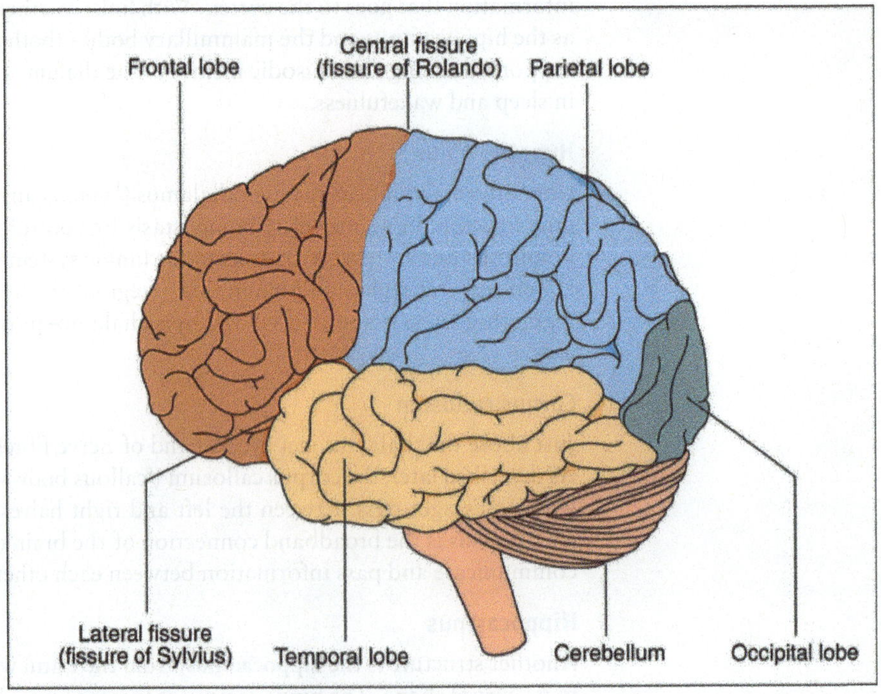

Figure 2.6 The four lobes of the neocortex

Source: miha de/Shutterstock

it is so large relative to the rest of the brain – about three-quarters of the neurons in the human brain are in the neocortex.

The side, or lateral, view (*lateral* simply means 'to the side') in Figure 2.6 reveals the four general regions, or **Lobes (of the brain)** Frontal, parietal, occipital and temporal.

Clockwise from the front, these are the **frontal lobe**, **parietal lobe**, **occipital lobe** and **temporal lobe**, named after the skull bones on top of them (e.g. the temporal lobes lie beneath your temples). Note that these lobes are not separate from one another in the brain. Instead, each hemisphere of the neocortex is a single sheet of neural tissue. The lobes are formed by the larger folds and convolutions of the cortex, with the names used as convenient reference terms for the regions. As an example, the central fissure, or fissure of Rolando, shown in the figure is merely one of the deeper folds in the brain, serving as a convenient landmark between the frontal and parietal lobes.

Note that the lobes of the cortex are generally specialised for processing different kinds of information. For example:

- The occipital lobe is used for visual processing.
- The temporal lobe is used for auditory, linguistic and memory processing.
- The parietal lobe is used for spatial and sensory processing.
- The frontal lobe is used for cognitive control.

Throughout this book, we may refer to mental functions critically involving different lobes of the cortex. However, we know that a more fine-grained consideration of the localisation of different brain activities than at the level of just the four lobes is needed. Another way of labelling areas in the brain is to assign numbers to them called **Brodmann's areas**. There are maps of the outside and inside of the cortex laid out with the different numbered areas to help orient you to different locations in the brain. So, for example, if you were to read that attention can involve activity in the anterior cingulate cortex (B.A. 24), you would know that you could click on that B.A. number to reveal the Brodmann map with that area highlighted to know just where this is. Through the book we may refer to brain regions using the Broadmann map, but we will always use the names of the structures to ensure consistency in terminology.

Also, as different topics are covered in this chapter, various smaller areas of the cortex will be identified as being critically involved for doing specific tasks. For example, the fusiform face area (B.A. 37) is important for the processing of human faces, and damage in this area can lead to a deficit known as prosopagnosia (an inability to identify faces). Similarly, the parahippocampal place area (also B.A. 37) is important when processing spatial locations, and damage there can lead to an inability to process location information (Epstein & Kanwisher, 1998)[23]. Thus, different parts of the brain are critically involved in different types of cognitive processes. This does not mean that those processes occur only in those parts of the brain, but that those parts of the brain are important and critically involved in that type of thinking. An analogy would be with a car. The accelerator pedal is critically involved in the speed of the car, but by itself it does not make the car go (you need an engine, wheels and many other components for that).

2.3.3 Principles of functioning

Two important principles of functioning in the neocortex are described here. This is necessary background knowledge for understanding the effects of brain function on cognitive processes – that is, how the mind relates to the brain. These principles involve the ideas of **contralaterality** and hemispheric specialisation.

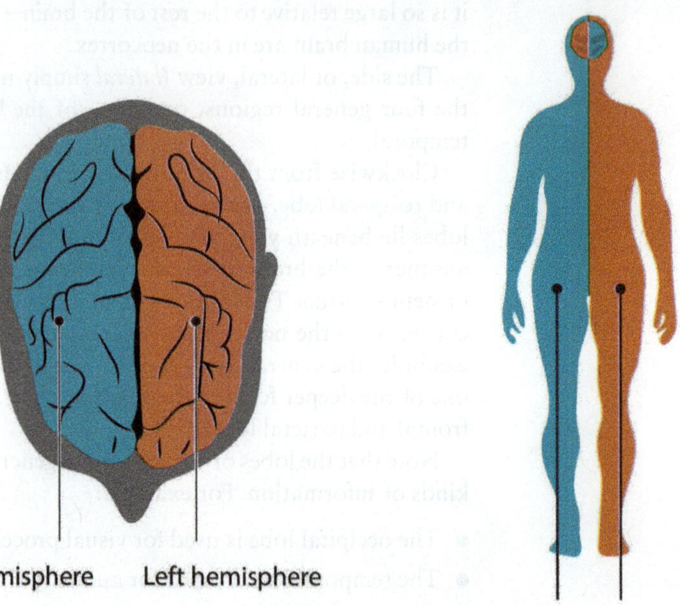

Figure 2.7 **Contralaterality – the right hemisphere receives input from the left side of the body and controls the left side, and vice versa**

Contralaterality

When viewed from the top, the neocortex is divided into two mirror-image halves – the left and right **cerebral hemispheres**. This follows a general law of anatomy that, except for internal organs such as the heart, the body is basically bilaterally symmetrical. What is somewhat surprising, however, is that the receptive and control centres for one side of the body are in the opposite hemisphere of the brain. This is contralaterality (*contra* means 'against' or 'opposite'). In other words, for evolutionary reasons that will probably remain obscure, the right hemisphere of the brain receives its input from the left side of the body and controls the left side. Likewise, the left hemisphere receives input from and controls output to the right side of the body. As an example, people who have a stroke in the left hemisphere will often have some paralysis in the right half of the body. There are a few exceptions, such as the olfactory nerves, in which there are ipsilateral (same side) connections.

Hemispheric specialisation

The second principle concerning lateralisation involves different specialisations of the two cerebral hemispheres, referred to as **hemispheric specialisation**. Despite their mirror-image appearance, the two hemispheres are not functionally identical. Instead, each hemisphere tends to specialise in different abilities and tends to process different types of information. **cerebral lateralisation** is the functional dominance of each hemisphere to carry out specific functions. This is not to say that a process or function takes place only in one hemisphere. It merely says that there is often a tendency, sometimes strong, for one or the other hemisphere to be especially **dominant** in different processes or functions. Information flows from one hemisphere to the other via a bundle of neurons called the corpus callosum.

dominant
Hemisphere of the brain that is more involved in certain cognitive functions. For example, in a right-handed individual the left hemisphere is dominant for language.

Many people have heard of 'left brain versus right brain' functions, often from the popular press. Such descriptions are notorious for exaggerating and oversimplifying laterality and specialisation. For instance, in these descriptions, the left hemisphere ends up with the rational, logical and symbolic abilities – the boring ones – whereas the right hemisphere gets the holistic, creative and intuitive processes – the interesting ones! Corballis (1989, p. 501)[24] noted that the right hemisphere achieves 'a certain cult status' in some approaches.

But even ignoring that oversimplification, it is far too easy to misunderstand the principles of lateralisation and specialisation; too easy to say 'process X happens in *this* hemisphere, process Y in *that* one'. Even the simplest act of cognition, say naming a picture, involves multiple components, distributed widely across both hemispheres, and complex coordination of the components. Disruption of any one of those could disrupt picture naming. Thus, several different patients, each with dramatically different localised brain damage, could show an inability to name a picture, each for a different reason relating to different lateralised processes.

Nonetheless, there is a striking division of labour in the neocortex, in which the left cerebral hemisphere is specialised for certain language functions such as syntax, pronunciation and speech movements (Ferstl et al., 2009[25]). There are also theoretical suggestions that lateralisation is closely linked to handedness. For example, up to 90 per cent of the right-handed population is thought to have left hemisphere dominance for language. However, Annett's (1998)[26] right shift theory suggested that handedness is a continuous variable and that approximately one-third of the population are mixed-handers (i.e. use both hands), a concept that has been consistently supported using self-report measures and performance tasks. Left-handedness is more common in males (Papadatou-Pastou et al., 2008[27]) but there is no clear justification as to why this is the case (McManus, 2019[28]). Several theories have been put forward including X-linked genetic models (e.g. Jones & Martin, 2010[29]) but these have not been strongly supported by empirical evidence. Similarly, a large meta-analysis showed that there are no sex differences in language processing therefore, the increased prevalence of left-handedness among males remains a mystery.

2.3.4 Split-brain research and lateralisation

Despite the exaggerated claims you often read, there has been a good deal of careful work on the topic of lateralisation and specialisation of different regions in the two hemispheres. Among the best known is the research on **split-brain** patients.

Although we have discussed some localisation of function in this chapter, implying that different hemispheres, lobes or parts of the brain are largely responsible for different types of cognition (also referred to as 'dominance'), it should be kept in mind that there is some individual variability in this. Some people are different. For example, some people have the distribution of processes associated with the left and right hemisphere reversed (and it is not just being left-handed that does this). Also, there can be influences of culture on cognitive neurological processing (Ambday & Bharucha, 2009[30]). For example, Tang et al. (2006)[31] found that whereas English speakers tended to use more language areas when processing Arabic numerals, Chinese speakers tended to use more visual/spatial areas of the brain, suggesting that different cultures can lead people to use different parts of the brain to process the same kinds of information.

2.3.5 Cortical specialisation

Not only are the different hemispheres of the cortex differentially able to do different things, but also the different parts of the cortex are tuned for different kinds of processes. As with lateralisation of brain function, the fact that a part of the brain is implicated in a certain kind of mental process, such as remembering, does not mean that that part of the brain does only that process, nor does it mean that other parts of the brain are not involved in that activity. The localisation or specialisation of function in the cortex means that that part of the brain is critically important for a certain kind of cognition. There will be numerous examples of the localisation of function throughout this chapter for different types of cognition. To give you a better feel for this, we will cover three types of cortical specialisation and the implications they have for human thought.

Sensory and motor cortices

First, consider a kind of cortical specialisation in which a section of cortex is specialised for a certain type of processing. As you know, your brain is connected to the rest of your body. When you sense things touching your body or move part of your body, this involves specific parts of your cortex. These are called the sensory and motor cortices, respectively, and are shown below in Figure 2.8.

The sensory **cortex** is a band of cortex at the front of the parietal lobes, just behind the central sulcus (B.A.s 1, 2 and 3), and is responsible for processing sensory information from throughout the body. In comparison, the **motor cortex** is a band of cortex at the back of the frontal lobe, just in front of the central sulcus (B.A. 4), and is responsible for controlling all of your voluntary muscle movements. But it is not the case that the whole of the sensory or motor cortices is responsible for any and all sensations and voluntary movements you make. Instead, there are different places along these cortices that are specialised for different parts of the body, such as the fingers, leg or mouth. So, not only is there a localisation of sensory and motor processes in the brain, but there is even a localisation of different body parts within each of those parts of the cortex.

Mirror neurons

A second kind of specialisation in the cortex does not involve a region of the brain, but instead involves certain kinds of neurons that seem to be used to do a specific kind of mental processing. Here, this specialisation is with neurons found in association with processing in or near the motor cortex. These neurons are active when a person is performing an action or, more interestingly, watching another person do

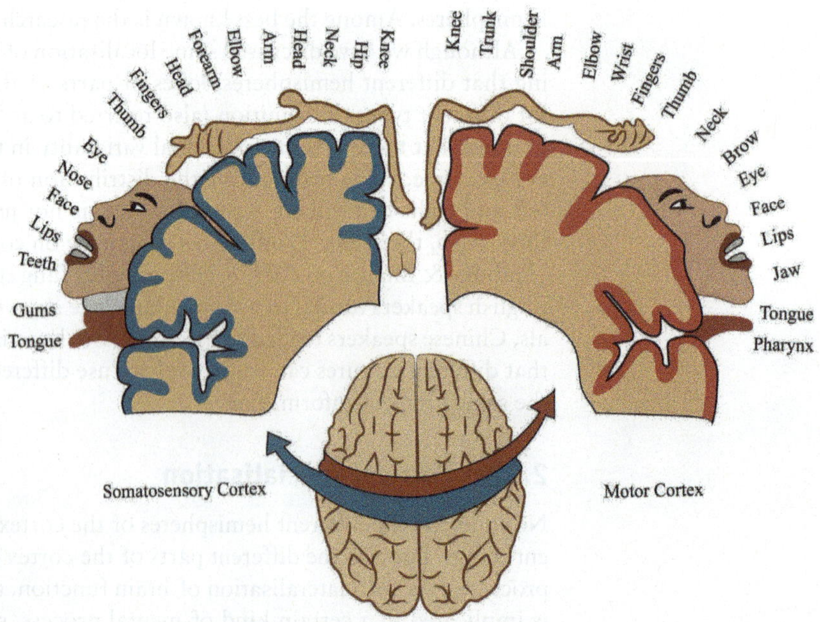

Figure 2.8 The locations of the sensory and motor cortices
Source: Science History Images/Alamy Stock Photo

something (see Glenberg, 2011[32]). These were originally discovered in work involving monkeys (Di Pellegrino, Fadiga, Fogassi, Galese & Rizzolatti, 1992[33]). Because it was initially thought that these neurons are involved when people mentally imitate or mirror another person's actions, they were called **mirror neurons**. These neurons appear in several places in the cortex, such as the inferior frontal cortex (B.A. 44 and 45). Although the exact nature and function of mirror neurons is not yet completely understood (see the July 2011 Mirror Neuron Forum issue of *Perspectives on Psychological Science* for a lively debate on the issue), they are involved in some aspects of action perception and understanding, speech perception, and more so in facilitating imitative behaviour (Heyes & Catmur, 2022[34]).

Dorsal and ventral pathways

A third type of specialisation of function in the cortex does not involve certain regions of the cortex, or certain types of cells, but a chain or sequence of cells and areas that are all working towards a processing goal. A good example of this involves the visual pathways in the cortex and the processing of different types of visual information. Many initial, early visual processes occur at the back of the brain in the occipital lobe, a region called V1 (for 'visual 1') (B.A. 17). The visual pathways that emerge from the early visual processing done in this and subsequent areas diverge into two pathways, depending on the type of visual information being processed (Mishkin, Ungerleider & Macko, 1983; Ungerleider & Haxby, 1994[35]), as shown in Figure 2.9.

One of these pathways is the **dorsal pathway**, which moves along the top of the cortex across the parietal lobe. The dorsal pathway is primarily involved in determining *where* things are in space. As such, it is also called the 'where' system in vision. The other of these is the **ventral pathway**, which moves along the bottom of the cortex into the temporal lobe. The ventral pathway is primarily involved in determining *what* things are. As such, it is also called the 'what' system in the visual system. Although it seems odd when you first come across this fact, the visual system

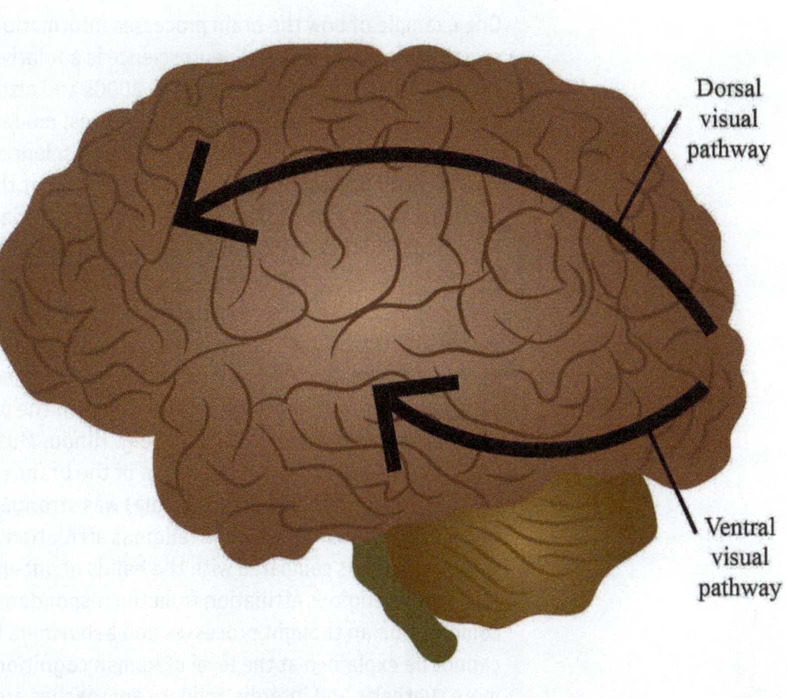

Figure 2.9 The dorsal and ventral pathways

Source: miha de/Shutterstock

does process where a thing is and what it is differently, although the system does bring this information together later on.

2.3.6 Levels of explanation and social neuroscience

Although understanding how neurons communicate with each other is important for a deeper comprehension of human cognition, it is not necessarily the case that all our understanding of thought, learning, memory and other cognitive functions and processes can be reduced to understanding how neurons do their job. Instead, assemblies of neurons working together can produce cognitive processes that are **emergent properties**: they are present when neurons work together but are absent in the individual neurons, as noted by early cognition scholars (Minsky, 1986[36]). A relevant example is empathy, our capacity to understand and relate to the pain, discomfort and misfortune of other people around us. Empathy is not the result of a single neuron firing at a specific pace, but rather emerges as a result of the activation of different neural pathways, involving both cortical and subcortical structures (Decety, 2015[37]).

Because of this, researchers of human cognition often use different levels of explanation to capture how a specific cognitive process operates and how it relates to behaviour. Thus, we can sometimes understand how problem-solving may be progressing without knowing what the individual neurons are doing (let alone the individual neurotransmitters flowing in the synaptic cleft or the Na^+ and K^+ ions inside and outside the cellular membrane!). That said, cognitive scientists understand that there are important bridges across these different levels of explanation, and that understanding what is happening at one level can help to explain and give an understanding of what is happening at another level.

Social neuroscience

One example of how the brain processes information comes from research on social neuroscience. Social neuroscience is a relatively 'young' interdisciplinary field that was developed in the early 2000s and attempts to explain complex human social behaviour by merging theories, models and methods from social psychology, cognitive psychology and neuroscience (Cacioppo & Decety, 2011[38]). One of the premises of social neuroscience is that the ways in which we perceive, evaluate and respond to information from our social environment are reflected in neural functioning. An illustrative example comes from the study of Vaughn et al. (2018)[39] who recorded, with the use of brain imaging methods (see Chapter 14), the brain responses of human participants who observed other people experiencing pain. Participants were put in a brain scanner while watching pictures of other people being stabbed in the hand with a syringe needle. Interestingly, researchers manipulated social information: the hands on the pictures were tagged with religious group membership labels (e.g. Hindu, Muslim, atheist, Christian). The researchers found that the activation of the brain's 'empathic neural network' (e.g. anterior cingulate cortex, insula) was stronger when the hands of in-group members (i.e. having the same religious affiliation as respondents) were stabbed with a needle, as compared with the hands of out-group members (i.e. having a different religious affiliation from the respondents). Such studies tell us that complex human thought processes and behaviours that shape our social relations cannot be explained at the level of human cognition or neuroscience alone. Rather, more synthetic and interdisciplinary approaches are needed.

2.4 Cognitive neuropsychology

LEARNING OBJECTIVE
Understand cognitive processes in patients with brain damage

Cognitive neuropsychology is a branch of cognitive psychology that tries to understand and develop models of normal brain functioning through the observation and assessment of patients with localised brain damage. It seeks to learn more about cognitive processes involved in language, memory, object perception and face recognition by examining *acquired* or *developmental* disorders in cognition. Let's assume that a theory about normal cognition posits that animate and inanimate objects (e.g. a table and a cat) are processed is a similar way, that is, there is a module in the brain that activates when we see a table and cat. If one patient with brain damage (patient A) shows selective impairment in recognising one of the two objects (can recognise a table but not a cat), and another (patient B) shows the opposite impairment (can recognise a cat but not a table) then we can assume that these modules are separate in the brain (also referred to as double dissociation). Modularity is the first assumption of cognitive neuropsychology, which refers to the architectural organisation of cognitive functions in 'modules'. This does not necessarily assume anatomical modularity. These modules may not be supported by a small specific area of the brain; rather, modules can be supported by widely distributed neural networks in the brain. Moreover, it is assumed that this functional and structural architecture is common among people and, therefore, that evidence from a single patient can inform models of normal cognitive functioning (generalisation).

One of the most common methods used in cognitive neuropsychology are case studies or multiple case studies of patients with selective brain damage. The cause of brain damage is not relevant to the development of theories; however, they can inform neuropsychology in understanding more about the brain architecture. There are three types of impaired patterns explored in case studies: associations, dissociations and double dissociations. Triple associations have also been recorded but are quite rare, and therefore we will not discuss them here. In associations, a patient is impaired in function A and function B, and therefore it is assumed that these functions are associated in some way. For example, a patient is impaired in understanding printed words (naming words) and understanding spoken words – therefore, these functions are associated. In dissociations, a patient is impaired in function A but not in function B; for example, there is impairment in printed words but not spoken words, so therefore these functions must be dissociated. Finally, in double dissociations, two patients present opposite patterns of impairment, supporting the idea that these functions are separate modules. Double dissociations can also be inferred from a single patient (see the following case study).

Case study of double dissociation in prosopagnosia: Patient J.R. (Vuilleumier et al., 2003)[40]

Face recognition is right hemisphere-dominant, and therefore damage to this region can cause the inability to recognise faces. Here is an example of a female right-handed patient with damage to the left lateral temporo-occipital region following thrombosis (blockage) in the left sinuses. The patient had no prior neurological history and was 21 years old studying history at university. J.R. showed selective impairment in over-identifying unknown faces as familiar. This is opposite to prosopagnosia caused by right hemisphere damage, which refers to the inability to recognise familiar faces.

2.5 Connectionism

LEARNING OBJECTIVE
Summarise connectionism as a computer-based method in cognitive science

To better understand how the nervous system codes and processes information, it is helpful to have scientific models that capture these complex processes. To this end, we conclude the chapter with a brief presentation on **connectionism or computational cognitive modelling**, an important computer-based method in cognitive science.

Research in this field aims to uncover the nature of cognition and various cognitive functions by developing process-based theoretical models using algorithms. Given the complexity of the human mind and its behavioural adaptability, process-based computational models are likely essential to explain these processes and to predict future outcomes. Computational models use algorithmic functions to specify the sequence of events and predict behaviour precisely. The interdisciplinary nature of these models suggests that in the future they will play a crucial role in understanding artificial intelligence in the context of cognition.

connectionist models are also called neural net models or parallel distributed processing (PDP) models; for our purposes the three terms are treated as synonymous. They refer to a computer-based technique for modelling complex systems that is inspired by the structure of the nervous system. A fundamental principle in connectionist models is that the simple nodes or units that make up the system are interconnected. Knowledge, all the way from the simplest to the most complex, is represented in these models as simple interconnected units. The connections between units can be either excitatory or inhibitory. In other words, the connections can have positive or negative weights. The basic units receive positive and negative activation from other units; depending on these patterns, they in turn transmit activation to yet other units. Furthermore, the interconnectedness of these basic units is usually described as 'massive' because there is no particular restriction on the number of interconnections any unit can have. In principle, any bit of knowledge or information can be connected or related to an almost limitless number of other units.

Figure 2.10 illustrates an early connectionist model by McClelland and Rumelhart (1981)[41], a model that dealt with word recognition.

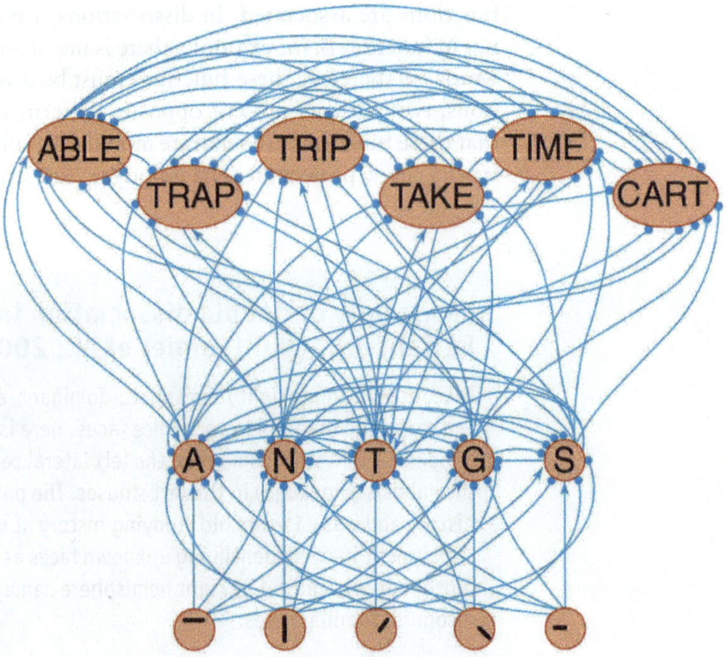

Figure 2.10 An illustration of part of McClelland and Rumelhart's (1981)[42] PDP model of feature, letter and word recognition

The bottom row of nodes or units represents simple features, such as a horizontal line and a vertical line, each connected to letters at the next higher level, which in turn are connected to words at the top level. For simplicity, look at the feature on the far left, the horizontal line. The connection directly up from that to the capital letter A would be a positive, excitatory connection because the letter A has a horizontal line. The connection from this feature up to the letter N, however, would be a negative, inhibitory one: if the feature detection system detects a horizontal line, this works against recognition of the letter N. In the same fashion, the capital A would have a positive connection up to the word ABLE because A is in the first position there, but it would have a negative, inhibitory connection to TRAP because TRAP does not begin with the letter A.

Referring to such models by the term *parallel distributed processing* highlights a different facet of the brain and the computer system. Mental processes operate in a thoroughly parallel fashion and are widely distributed across the brain. Likewise, processing in a PDP model is thoroughly parallel and distributed across multiple levels of knowledge. As an example, even as the feature detectors at the bottom of the model are being matched to an incoming stimulus, word units at the top of the model may already be activated. Thus, activation from higher levels may influence processing at lower levels, even as the lower levels affect activation at higher levels.

Consider how the system would recognise the word TRAP in the sentence 'After the bear attacked the visiting tourists, hunters went into the forest to set a trap.' Even as the feature and letter detector units would be working on the T, then the R, and so forth, word units at the top would have already received activation based on the meaning and context of the sentence. With bears attacking tourists and hunters going into the forest, the word TRAP is highly predictable from the context, but CART would not be. Given this context, TRAP would be more easily recognised, perhaps because the features within the letters would have been activated by the context already. In this fashion, the comprehension and word recognition systems would be operating in parallel with the feature and letter detection systems, each system making continuing, simultaneous contributions to the other systems and to overall mental processing.

The similarity of the connectionist scheme to the functioning of the brain is obvious and vitally important – it is widely believed that connectionist models operate on the same, or at least very similar, basic principles as the brain (Thomas & McClelland, 2008[43]). In other words, the connectionist framework may give us an excellent way of modelling and simulating intact cognitive processes (Monaghan & Pollmann, 2003[44]). New research also suggests that connectionist models may also allow us to simulate the effects of neural damage on cognition (Guest, Caso & Cooper, 2020[45]).

Summary: Neural basis of cognition

2.1 The brain and cognition together

- By understanding the structure and function of the nervous system, we can obtain a better insight into the structure and function of cognition. Patients, such as K.C., have brain damage that reflects differences in various components of cognition.
- The best evidence for a brain–cognition link is if there is a double dissociation, with one function being affected but not a second: one with one type of damage, and the reverse with another type of brain damage.

2.2 Basic neural functions

- Neurons are the basic building blocks of the nervous system. Understanding how these elements work can provide a better understanding of the machinery that underlies cognition.
- Neurons communicate information via an electrical component and a chemical component. The electrical component is the action potential by which, obeying the all-or-none principle, a signal is eventually sent down the axon to communicate with other neurons.

- The chemical component involves neurotransmitters released at the synapse. There is a wide variety of neurotransmitters, each involved in a different type of neural process. Some of these neurotransmitters are excitatory, stimulating the post-synaptic neuron, and some are inhibitory, suppressing the post-synaptic neuron.
- Information is stored in the nervous system via a pattern of interconnections among neurons and how strong these interconnections are. Early on in memory formation, adjustments to this pattern are affected by processes such as long-term potentiation, but later, memory solidification is affected by a process of consolidation. The storage of new memories may also be affected by the processes involved in neurogenesis.

2.3 Important brain structures and function

- Many subcortical structures make a meaningful contribution to cognition, including the thalamus, hypothalamus, the corpus callosum, the hippocampus and the amygdala.
- The cerebral cortex is the largest brain structure in humans and is the seat of many cognitive functions. The cortex is divided into four lobes and two hemispheres, with some localisation of function in these various brain regions, such as the occipital lobe being primarily involved in visual processing. Damage to certain areas of the brain can cause specific cognitive deficits, such as prosopagnosia.
- There are contralateral connections of the brain with the body, as well as some lateralisation of cognitive function in the two hemispheres. This separation of different mental tasks by the two halves of the cortex is most clearly illustrated in split-brain patients who have had the corpus callosum cut, thereby separating the two hemispheres.
- Although there are different levels of analysis and explanation, a good understanding of neurophysiology can be very helpful in comprehending human cognition.

2.4 Cognitive neuropsychology

- Cognitive neuropsychology tries to explain normal cognitive functioning by obesrving deficits in brain-damaged patients.
- Functional modularity is one of the key assumptions in cognitive neuropsychology.
- Associations, dissociations and double dissociations are patterns of impairment that contribute the understanding of cognitive functions.
- Case studies are important in understanding deficits in cognition.

2.5 Connectionism

- The notion that human cognition is analogous to processing in a computer system has largely been abandoned at the detailed level, especially because of evidence of widespread parallel processing in humans. Connectionist (neural net, PDP) models can simulate such parallel processes and therefore may be excellent ways of modelling human cognitive processes. These models use the nervous system, and how it represents and processes information, as the inspiration for how they are structured and developed.

Chapter 2 quiz

Question 1

A researcher in cognitive psychology would like to investigate how an individual knows how to ride a bike even when they have not experienced riding a bike themselves. They ask a group of participants to remember the steps taken of how to ride a bike and to write down the list of steps. Which type of memory will be used?

A. Episodic memory
B. Semantic memory
C. Short-term memory
D. Declarative memory

Question 2

Which of the following best describes the function of the dendrites within a neuron?

A. Dendrites contain the biological information for the cell, regulating the information within the cell.
B. Dendrites are cables, thinner than a human hair, where electrical impulses travel away to be received by other neurons.
C. Dendrites gather electrical and chemical information that is received from another neuron
D. Dendrites are the fatty-like structures that wrap around the axon of a cell. They protect the cell from damage.

Question 3

Trisha is on holiday and is swimming in the sea. She has a phobia of fish and has been hoping to avoid them; however, she finds a large fish not far from where she is swimming. Which part of the brain is activated during the response to seeing the fish?

A. Hypothalamus
B. Corpus callosum
C. Hippocampus
D. Amygdala

Question 4

A patient is admitted to hospital with difficulties in language processing and issues with their vision. They cannot process visual images accurately as they are having difficulties seeing them (e.g. blurred vision). The patient has issues with naming objects even when someone else states the name of them first. For example, when they look at a chair, they cannot say that this is a chair. Which two brain regions could be damaged in this case?

A. The parietal lobe and the frontal lobe.
B. The occipital lobe and temporal lobe.
C. The frontal lobe and temporal lobe.
D. The frontal lobe and the occipital lobe

Question 5

Which of the following best describes the feature detection mechanism of a connectionist model?

A. Nodes representing individual symbols are matched against collections of symbols.
B. Collections of symbols are shuffled around until they produce a coherent semantic meaning.
C. Complex symbols are encoded and then broken down into constituent symbols.
D. Basic units match elementary shapes against more complex symbols.

3 Sensation and perception

Learning objectives

3.1 Explain the process of relating mental experience to physical reality

3.2 Distinguish the processes of visual sensation and perception

3.3 Evaluate the process by which we identify patterns and objects

3.4 Explore ideas about how what we know influences what we see

3.5 Summarise the science of object recognition and agnosia

3.6 Describe the science of hearing

Key figures

Professor Doris Tsao

Professor Tsao is a prominent vision scientist, and her research has provided insights into how the primate and the human brain construct and represent the visual world. Her research has focused on object and face recognition, among other topics.

Professor Semir Zeki

Professor Zeki has made major contributions to understanding the organisation and function of the visual cortex in the human brain. His research has revealed how our brains encode and process visual information, including motion, colour and form.

Introduction

Unlike other species, humans have a unique creative potential. Over thousands of years, we have created music that ranges from the western African tribal music of Dogon to the psychedelic rock of Pink Floyd and the earworms of K-pop. Our creations appeal to other senses too. From the absorbing ceiling of the Sistine Chapel to the mysterious Banksy murals, we have found ways to tantalise visual perception. We are also quick to detect creaky sounds in the middle of the night – a remnant of our ancestors' capacity to detect threats in the environment and seek cover. Come to think of it, however, it is truly a wonder we can see or hear anything at all given the implausible, even backward, structure of the eye and the indirect, unlikely structure of the ear. Yet, early theorists noted that the human eye could detect the flame of a candle from as far as 30 miles away – although more recent accounts give more conservative, but still impressive, estimates. Likewise, our ears can detect sounds as low as 20 Hz and, as you will notice in the next chapter, our brains can readily pick up personally relevant auditory information (e.g. someone mentioning our name) even in a noisy crowd. Yet the complexity of the mental processing of perception exceeds these sensitivities. Because we 'understand' what we have seen so quickly, with seemingly little effort, 'we can be deceived into thinking that vision should therefore be fairly simple to perform' (Hildreth & Ullman, 1989[1]). As this book repeatedly emphasises, a rapid and unconscious process is neither necessarily simple nor simple to investigate. If anything, just the opposite is probably true.

This chapter presents a basic study of visual and auditory sensation and perception. We focus on the visual and auditory sensory registers because they are our most prominent intersections with the world (some animals make more use of other senses, such as the sense of smell in dogs, or hearing in elephants). We will discuss several theories, including an elaboration of the connectionist model. In a later section, we consider the neural basis of visual perception, as well as brain-related disruptions in perception. Lastly, we will review what perceptual deficits tell us about the normal processes we take for granted.

3.1 Psychophysics

LEARNING OBJECTIVE
Explain the process of relating mental experience to physical reality

How do you know what exists out in the world and what does not? In other words, how does mental experience compare with physical reality? In what ways do our sensations and perceptions capture the world as it is, and in what ways do they fall short? The relationship of mental experience and physical reality is a specialised sub-domain of research on sensation and perception known as **psychophysics**.

In this section, we cover two issues of interest to psychophysicists: the detection and the discrimination of sensory experiences. Detection studies indicate the properties of physical stimuli that can be sensed and perceived, either within or outside one's conscious awareness. In comparison, studies of discrimination tell us how much a stimulus needs to change to make a noticeable difference to a (human) observer.

3.1.1 Detection and absolute thresholds

One basic question that psychophysicists ask is 'How much physical energy (i.e. light, heat, sound, etc.) is needed for a person to detect that something is present in the environment?' The minimal amount of energy needed to detect a stimulus – whether it is visual, auditory, tactile or any other type – is referred to as a sensory threshold. Conventionally, a sensory threshold is defined as the amount of physical energy needed for a person to detect the presence of a stimulus 50 per cent of the time over many trials. For example, imagine that you are in a completely dark environment. How bright would a light have to be for you to notice it had been turned on? The brightness at which you can perceive the light half of the time is said to be at your threshold. Stimuli that are detected more than half the time are said to be supraliminal, and stimuli that are so weak that they are detected less than half the time are said to be subliminal.

3.1.2 Discrimination

just noticeable difference, or JND
The smallest amount of physical change in a stimulus a person can detect.

Another basic question that psychophysicists ask is 'How much does the world need to change before a person notices that it is different?' The smallest amount of physical change that a person can detect is also known as **just noticeable difference, or JND**.

Although psychophysics primarily refers to the comparison of sensory magnitudes, it is possible to extend its basic principles to making any cognitive comparisons, including those between two mental symbols. One basic finding is that the greater the distance between two items, the easier it is to make a discrimination. This distance effect also holds for both physical and symbolic differences.

Briefly, the *symbolic distance effect* is that we judge differences between symbols more rapidly when they differ considerably on some symbolic dimension, such as value.

Consider the stimuli in panels A and B in this image. Which dot is higher?

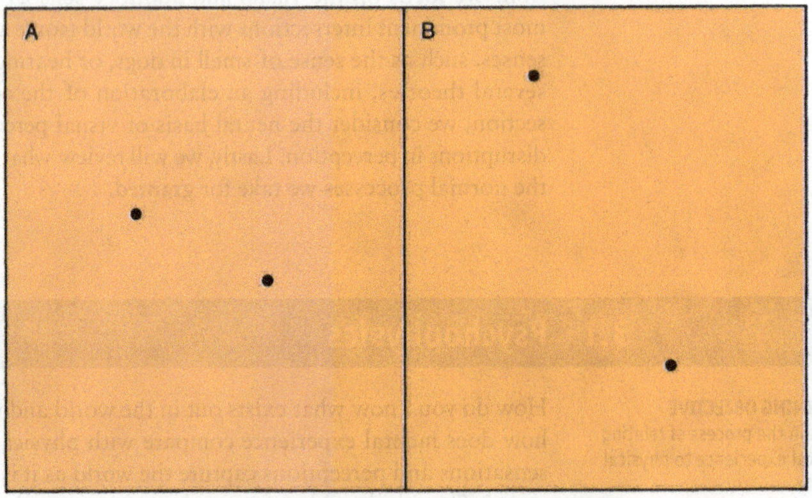

Despite its simple nature, it takes some amount of time to make this decision. The time it takes to decide which dot is higher depends on the separation of the dots; the greater the separation, the faster the response. This is the symbolic distance effect: two stimuli can be discriminated more quickly when they differ more (Moyer & Bayer, 1976[2]). Psychophysicists were already investigating such effects as far back as late 1930s (Johnson, 1939[3]). Contemporary research indicated that the symbolic distance effect could be extended to animals (e.g. monkeys and pigeons) for stimuli that were symbolically represented, such as pairs of numbers (i.e. numerical magnitudes) or other stimuli (Scarf, 2020[4]).

For example, consider the two illustrations below. For panel C, which point is higher? For panel D, which one is lower?

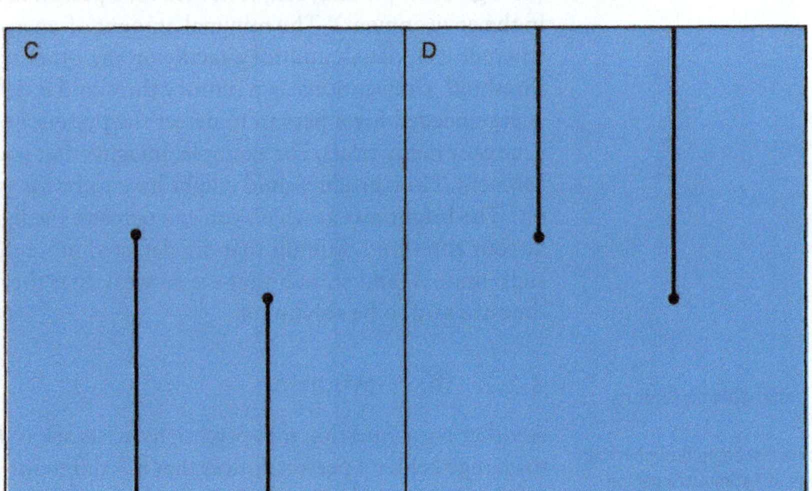

It is probably not obvious to you at a conscious level, but when the question is 'Which balloon is higher?' people's responses depend not only on the discriminability of the two heights but also on the semantic dimension needed (Banks, Clark & Lucy, 1975[5]). In other words, semantic knowledge that balloons are held at the bottom by strings, float up in the air, and are therefore oriented in terms of highness is a significant influence on the decision times; describing the illustrations as balloons led people to treat the pictures symbolically rather than as merely physical stimuli. When the same pictured display was accompanied by the question 'Which balloon is lower?', responses were much slower. And, as you would expect, the situation was reversed when people judged items such as those in panel D, the yo-yos. 'Which yo-yo is lower?' yielded faster responses than 'Which yo-yo is higher?' because knowledge about yo-yos is that they hang down from their strings.

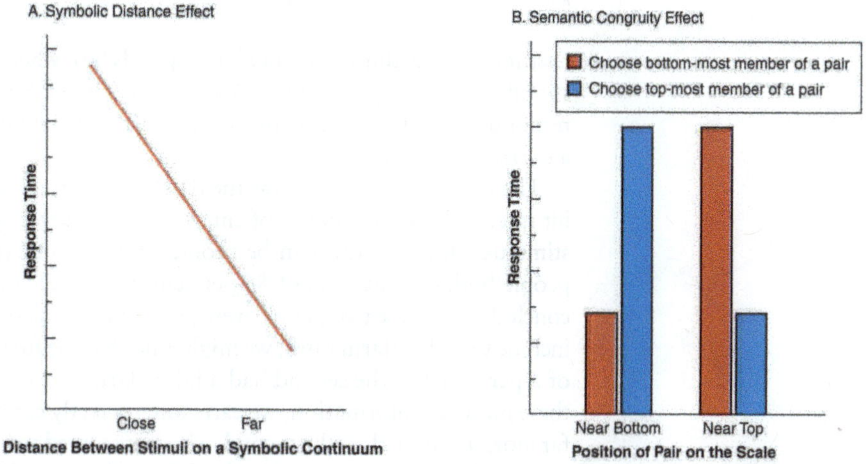

The name for this is the *semantic congruity effect* (Banks, 1977; Banks et al., 1976[6]) in which a person's decision is faster when the dimension being judged matches or is congruent with the implied dimension. In other words, the implied dimension in the balloon illustration is height because balloons float up. When asked to judge 'how high' some 'high' object is, the judgement is faster because 'height' is congruent with 'high.' Likewise, 'lowness' is implied in the yo-yo display, so judging which of two 'low' things is lower is also a congruent decision. The figure above displays the general form of the symbolic distance effect and the semantic congruity effect (Banks, 1977[7]).

3.1.3 Signal detection theory

To better understand how we detect changes between stimuli, there is a fundamental method of cognitive psychology developed through research in psychophysics which is referred to as *signal detection theory*, originally developed in the field of communication theory (Peterson, Birdsall & Fox, 1954[8]) and then applied in psychology (for an overview of signal detection theory and its applications in psychological science, see Wixted, 2020[9]).

When people engage in a psychophysical task, they are often making yes–no judgements, such as assessing whether a stimulus is present or not, whether a stimulus has changed or not, and so on. In these situations, there are four possible types of responses, as shown in Figure 3.1.

The first is if a person responds 'yes' in cases when that is appropriate (e.g. the stimulus is actually present). This is called a *hit*. The second is if a person responds 'yes' when that is inappropriate (e.g. the stimulus is not there). This is called a

	Perceiver says yes	Perceiver says no
Stimulus absent	False alarm	Correct rejection
Stimulus present	Hit	Miss

Figure 3.1 The influence of observed stimuli according to signal detection theory
According to signal detection theory, there are four possible outcomes determined by the presence (or absence) of changes in observed stimuli.

false alarm. The third is if a person responds 'no' when that is inappropriate (e.g. the stimulus is actually present). This is called a *miss*. Finally, the fourth is if a person responds 'no' when that is appropriate (e.g. the stimulus is not there). This is called a *correct rejection*.

The intuitive way to analyse the data from a study might be to look simply at the hit rate, such as the number of times a person correctly identified the presence of a stimulus. However, this can be problematic. As an extreme example, suppose two people both had hit rates of 86 per cent. If we were to use only hit rates, we might conclude that these two people were performing similarly. However, if we were also to include the false alarm rates, we might find that the first person had a false alarm rate of 5 per cent and the second had a false alarm rate of 82 per cent. By using both of these pieces of information, we can more correctly conclude that the first person was far more accurate than the second, who just seemed to have a bias toward saying 'yes'.

Using both the hit and false alarm rates, it is possible to calculate two *independent* indexes of how accurately people are using the hit and false alarm rates. This exposition will be easier to understand if you consult Figure 3.2.

Although psychophysicists doing sensation and perception research primarily developed these measures, or independent indexes, in psychology, their use has been extended to other domains of cognition, such as research on memory (e.g. Kellen et al., 2021[10]), as well as other applied psychology fields, such as decision-making and behavioural economics (Lynn & Barrett, 2014[11]). Consider the application of the principles of signal detection theory in aviation psychology, and more specifically in the ability of air traffic controllers to distinguish between stimuli when visually analysing information in a control panel – such an ability is life-saving, ensuring the safety of passengers on a daily basis! Experimental paradigms derived from Signal Detection Theory could be used for personnel selection and training in this field (Donald & Gould, 2020[12]).

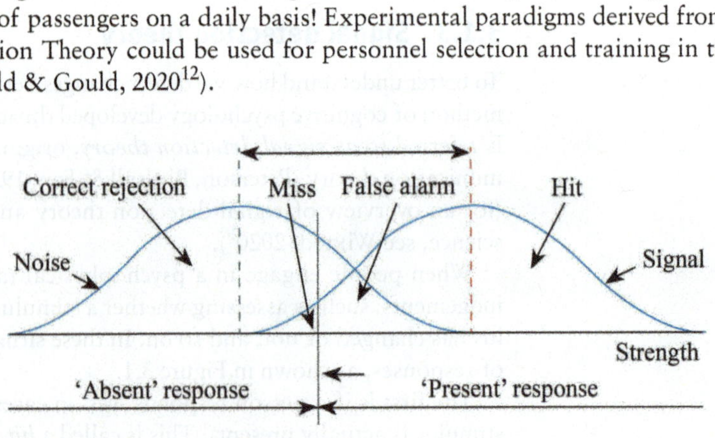

Figure 3.2 Signal and noise distributions

3.2 Visual sensation and perception

LEARNING OBJECTIVE
Distinguish the processes of visual sensation and perception

The human eye is a complex structure with several elements working together to allow vision to happen. Each of its structures plays an important role in the transduction of the proximal stimulus (the light waves entering the eye and being focused on the retina) into the neural impulses that result in sensation. Farther along the visual pathway, perceptual processes interpret these sensory signals.

Figure 3.3 illustrates the basic sensory equipment of human vision. Light waves enter the eye, are focused and inverted by the lens, and are projected onto the retina. The retina has three layers of neurons: rods and cones, bipolar cells and ganglion cells.

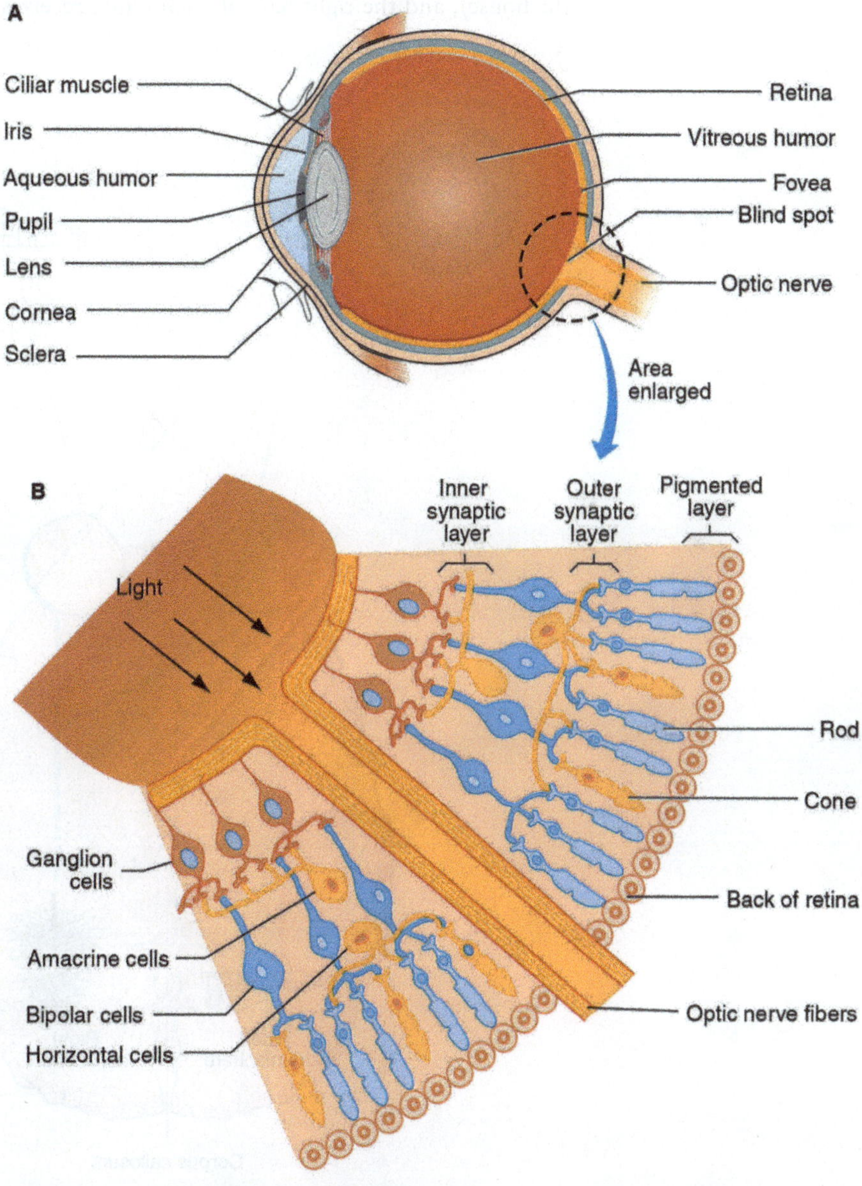

Figure 3.3 The structure of the human eye

The rods and cones are neurons stimulated by light, beginning the process of vision. While rods are specialised for vision in dim light, cones are responsible for colour vision and contain light-sensitive photopigments that process different light wavelengths Patterns of neural firing from these cells pass on to a second layer, the bipolar cells, which then collect the messages and move them along to a third layer, the ganglion cells. The axons of the ganglion cells converge at the rear of the eye, forming the bundle of fibres that make up the optic nerve. The optic nerve signal exits the eye and continues through various structures, eventually projecting to the visual cortex of the occipital lobe in the lower rear portion of the brain.

A brief explanation is in order about how the eyes transmit information to the brain. The contralaterality principle is not as simple as 'left eye to right hemisphere'. Instead, each eye transmits to both hemispheres. In this sense, each retina gathers information from both the right and the left visual fields. As shown in Figure 3.4, where you are looking is your fixation point.

The left half of the retina in each eye receives images from the right visual field (the house), and the right half of each retina receives images from the left visual

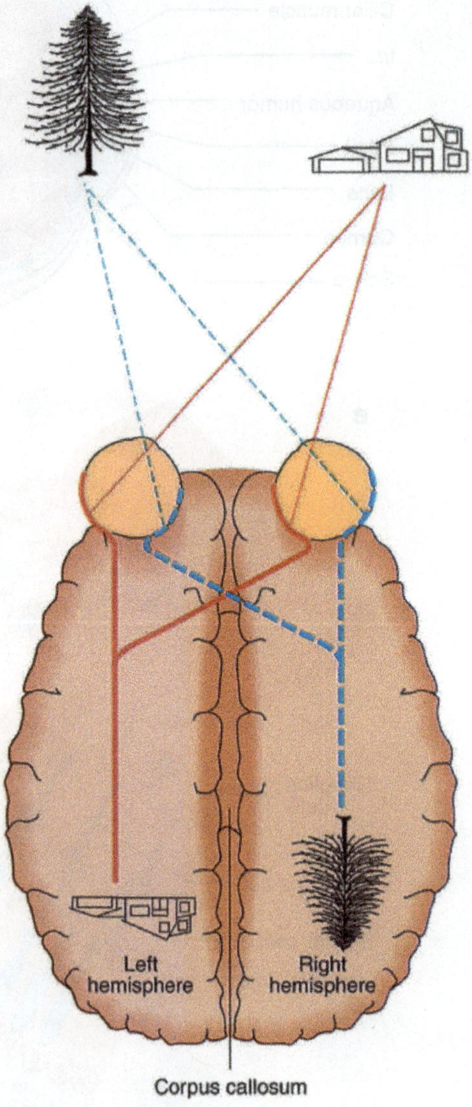

Figure 3.4 The contralaterality principle

field (the tree). Thus, stimuli in the right visual field – the solid lines in the figure – project to the left half of the retina in both eyes, and this is then transmitted to the left hemisphere. Similarly, stimuli in the left visual field – the dotted lines – project to the right half of both retinas and are then sent to the right hemisphere.

At each step – from rods and cones to bipolar cells, then from bipolar cells to ganglion cells – there is some loss of information. Because there is a great deal of compression of information in the early stages of vision, the message that finally reaches the visual cortex is an already processed and summarised record of the original stimulus (Grill-Spector & Malach, 2004[13]). There are about 120 million rods and about 7 million cones on each retina. Most of the cones are in a small area known as the **fovea**, which provides our most accurate, precise vision. Some of the cones in the fovea have 'private' bipolar cells for relaying impulses: one cone synapses with one bipolar cell. In contrast, in peripheral vision, tens to hundreds of rods converge on a single bipolar cell. Such convergence results in a loss of information because a bipolar cell cannot 'know' which of its many rods triggered it. Finally, about 1 million ganglion cells combine to form the optic nerve. So, essentially, vision is compressed from 120 million bits of information down to 1 million. Despite this compression, human vision is still amazingly sensitive and acute. Like all good summaries, the visual system preserves the most useful information – the edges, the contours and any kind of change – and omits the less useful, steady-state information.

fovea
The highly sensitive region of the retina responsible for precise, focused vision, composed largely of cones.

Although much of the coverage here is on the visual processing of individual objects, we must keep in mind that sensory and perceptual processes even influence cognition based on general visual properties or impressions of a scene as a whole. For example, some evidence suggests that people experience and perceive rooms with lighter-coloured ceilings and/or walls (but not floors) as being higher than rooms with darker ceilings or walls (Oberfeld, Hecht & Gamer, 2010[14]). Thus, sensation

Ocular dominance: Nobel Prize-winning research

In early 1980s, neurophysiologists David Hubel and Torsten Wiesel won the Nobel Prize for Physiology or Medicine, for their pioneering research on the visual system. Studying the visual development of small kittens, as a model for the human visual system, Hubel and Wiesel discovered that a group of neurons in the visual cortex was responsive to information that came only from a single eye – a phenomenon they defined as *ocular dominance*. Using experimental methods that may have raised eyebrows in today's research ethics review boards, Hubel and Wiesel blocked visual input from one eye in small kittens, and then observed visual development after the damage was restored. In a series of experiments, the researchers demonstrated that neural responses to visual stimuli differed between the normal eye and the eye that was deprived of visual information. Hubel and Wiesel concluded that the neural cells responsible for visual information processing of the deprived eye repurposed their function in favour of the normal one that was not experimentally deprived. This was a major breakthrough at a time when the visual system was poorly understood.

David Hubel and Torsten Wiesel
Source: Associated Press/Alamy Stock Photo

and perception are not just about bits and pieces of a scene; ultimately, the vision system integrates all the information into a greater whole that influences cognition on a broad scale.

With regard to early visual processing that takes place in the retina and optic nerve, we are talking about *sensation*. All sensory systems (sight-visual, taste-gustatory, touch-tactile, hearing-auditory and smell-olfactory) have receptors that convert physical energy into neural impulses that can be interpreted by the brain. Perception is the process of interpreting and understanding sensory information. As such, we need to explore the stages of visual perception. We begin with how the eye gathers information from the world, and then turn to the memory that retains that information: visual sensory memory.

3.2.1 Gathering visual information

To start out, let us eliminate an apparently common misunderstanding about vision. Winer, Cottrell, Gregg, Fournier and Bica (2002)[15] asked college students, 'How does vision work?' using several variations of the task. In every variation, they found a substantial percentage of college students exhibited *extramission*, the belief that vision involves some kind of ray or wave going out from the eyes to the object being perceived (think of the rays emanating from Superman's eyes). For instance, 69 per cent of the students drew outward-pointing arrows on diagrams. Another 33 per cent gave extramission responses when asked about looking at a shining lightbulb, when the correct answer should be obvious (the bulb emits light, which goes to the eyes). We correct this common misunderstanding here: vision is not the result of anything coming out from our eyes toward what we are looking at. Instead, vision is triggered when the reflection of light from objects in our environment is perceived by our visual system.

Another interesting question is how does the eye see colour? Although painters, visual artists and graphic designers entertain our eyes with a seemingly endless palette of colour, in reality, colour perception in humans is very narrow (Figure 3.5)!

One of the most popular theories of colour perception is the **trichromacy theory** which suggests that there are three different types of cone receptor that respond to

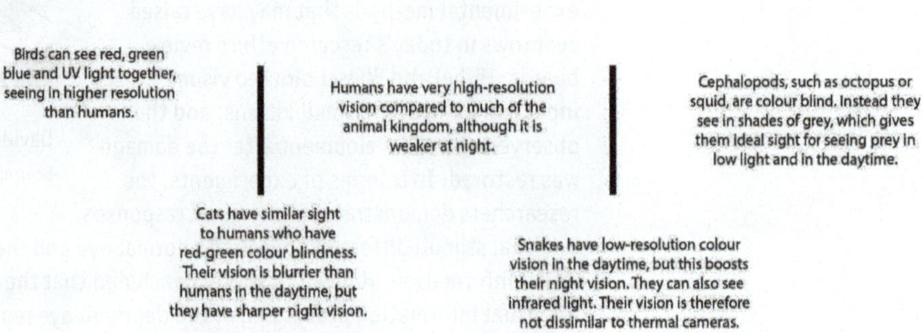

Figure 3.5 The visible light spectrum, indicating the range of visible light the human eye can detect.

short, medium and long wavelengths, which correspond to blue, yellow-green and red colours, respectively. We experience colours, through the activation of at least two of these cones – hence the blend of colours. The theory is supported by studies on colour deficiency, which is the condition where one of the cones is missing (dichromacy), leading to difficulty in recognising specific colours (e.g. protanopia, where someone is missing 'red' cones). However, trichromacy theory cannot account for **negative afterimages**, which are post-vision effects that occur when we look at a coloured square for a long time. These afterimages are generated when we shift our gaze away from the coloured item, in the complementary colour (e.g. a green square produces a red afterimage). This observation led to Hering's (1878) **colour opponent theory**, which suggests that there are three opponent visual channels, red-green, blue-yellow and black-white (achromatic channel), in the cones that respond to colours in opposite ways. The assertions of colour opponent theory are supported by animal studies and some types of colour deficiencies where individuals are lacking short or long wavelengths and they cannot perceive blue-yellow and red-green colours, a condition known as colour blindness. Based on the colour-opponent theory, various tests, known as the pseudoisochromatic plate tests, have been developed to diagnose colour blindness, including the American Optical Hardy-Rand-Rittler test, the Dvorine Colour Blind test and the Ishihara test shown in Figure 3.6.

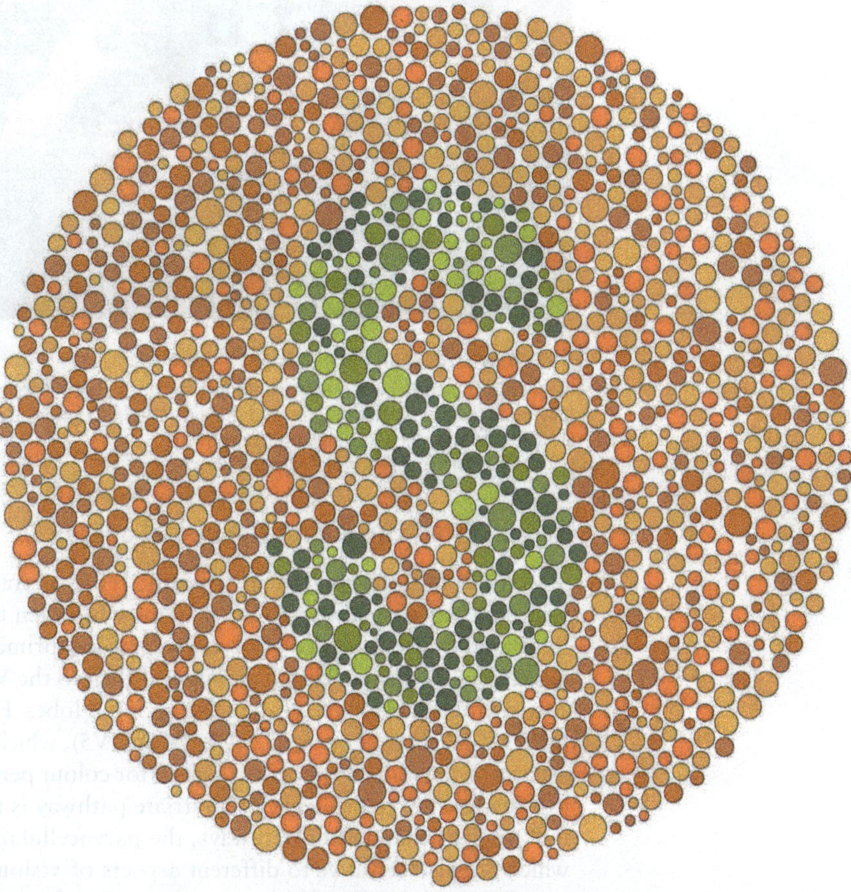

Figure 3.6 An image used in the Ishihara test to diagnose colour blindness
Source: CHEN WEI SHIN/123RF

A subtler misbelief is that we take in visual information in a continuous fashion whenever our eyes are open. After all, our visual experience is of a connected, coherent visual scene that we scan and examine at will. However, this is largely an illusion. If you watch people's eyes as they read, you will see that their eyes do not sweep smoothly across the page. Instead, they jerk, bit by bit, with pauses between successive movements.

Several studies have shown that we sometimes fail to see an object we are looking at directly because our attention is directed elsewhere (Mack, 2003[16]); this inattention blindness is due, in some sense, to our lack of attention to an object. In a dramatic demonstration of this, Haines (1991)[17] tested experienced pilots in flight simulators. A few of them proceeded to land the simulator, paying attention to the gauges and dials on the instrument panel yet failing to notice that another aeroplane was blocking the runway (see Mack & Rock, 19987[18]).

Figure 3.7 An example of a flight simulator
Research showed how experienced pilots in flight simulators proceeded to land the simulator, paying attention to the gauges and dials on the instrument panel yet failing to notice that another aeroplane was blocking the runway.
Source: Franklyn Mosher/Alamy Stock Photo

3.2.2 From the eye to the visual cortex

The visual information that reaches the retina is transferred through the **retina–geniculate–striate pathway** to the thalamus and then to the primary visual cortex. Retinal activation via the optic nerve projects primarily to the **lateral geniculate nucleus** (LGN) of the hypothalamus and then to the V1 area of the primary visual cortex at the back of the head in the occipital lobes. From there, signals are sent to other visual cortical areas (V2, V3, V4 and V5), which are responsible for different aspects of the visual experience (e.g. V4 for colour perception). To complicate matters further, the retina–geniculate–striate pathway is then divided into two largely independent channels or pathways, the parvocellular (P) and magnocellular (M), which are also sensitive to different aspects of vision. The P pathway is sensitive to colour and detail, and hence receives input from cones, and the M pathway is responsive to movement and receives input from rods, which are naturally referred to as the 'what' (ventral) and 'where' (dorsal) systems (see Figure 3.8).

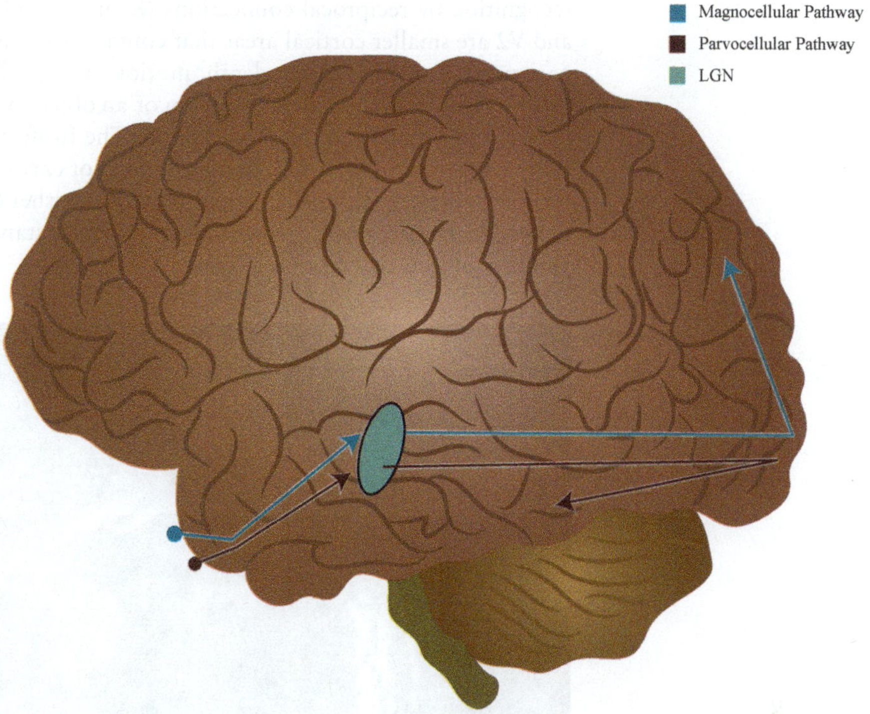

Figure 3.8 The retina–geniculate–striate pathway
Source: miha de/Shutterstock

Although independent, the two pathways are interconnected and there is an additional pathway (the koniocellular K-pathway) that processes p-retinal ganglion cells which aid the colour and shape discrimination (see Figure 3.9).

Information from the LGN is sent to the fourth layer of V1 with magnocellular cells projecting to layer 4Ca and parvocellular cells to 4Cb. The primary (V1) and secondary (V2) visual cortices are responsible for rapid initial visual detection and

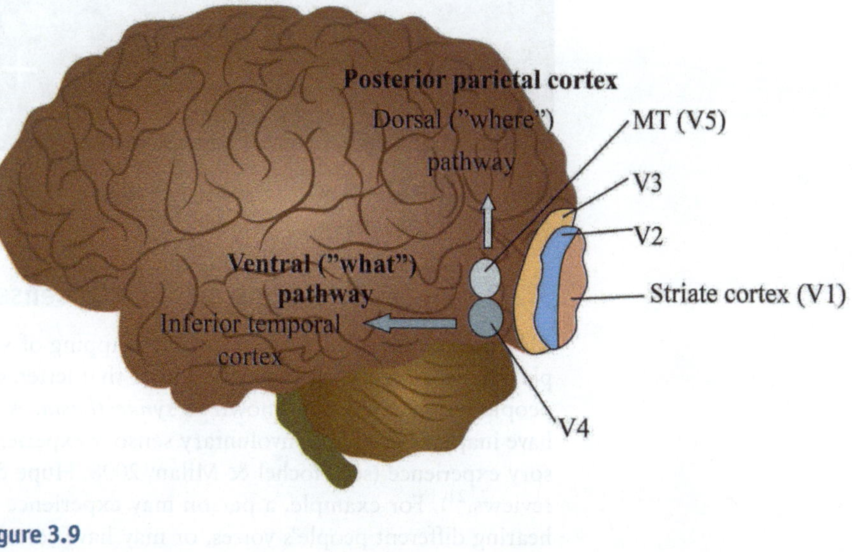

Figure 3.9
Source: miha de/Shutterstock

recognition by reciprocal connections (Kennedy and Bullier, 1985[19]). Beyond V1 and V2 are smaller cortical areas that contain neurons that respond to differential parameters, such as colour, depth, motion etc. Specifically, V3 is responsible for processing information about the form of an object, V4 about its colour and orientation, and V5 about motion perception. The fusiform face area (or FFA) is especially responsive to faces (e.g. human, animal or cartoon faces) and is located in the fusiform gyrus, as depicted in Figure 3.10 (Kanwisher & Yovel, 2006[20]). Combined, these visual areas provide us with a unique understanding of the world around us and how objects and motion interact

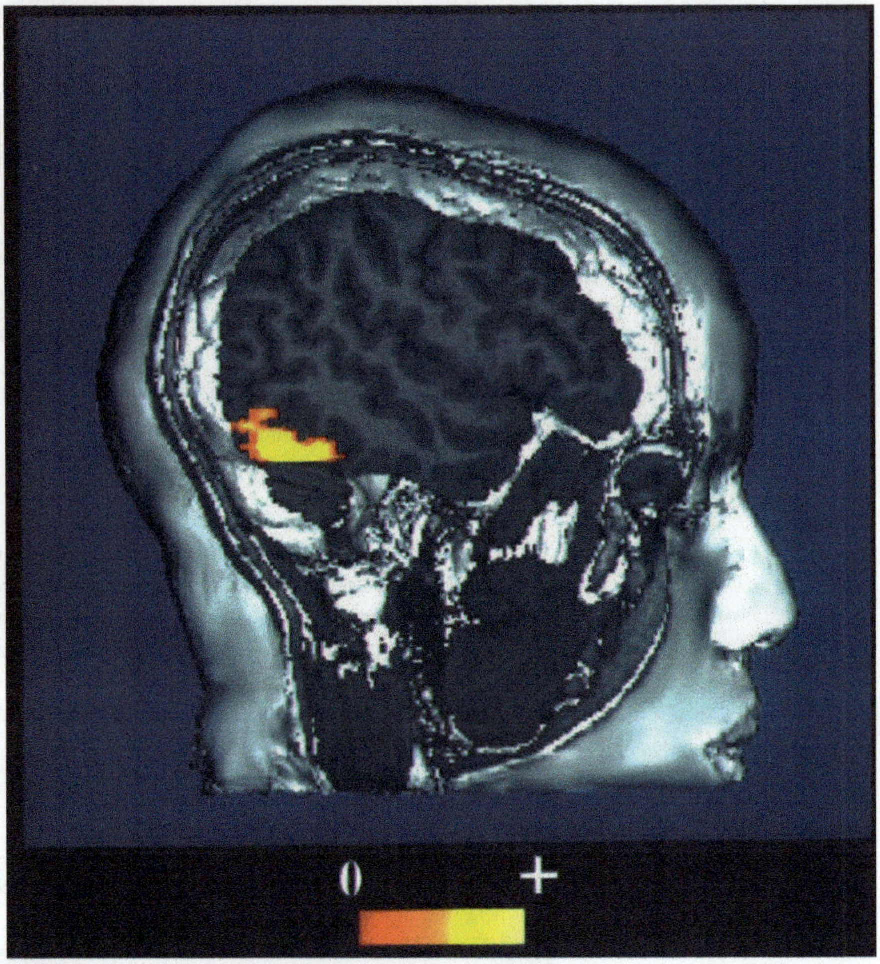

Figure 3.10 From the eye to the visual cortex
Source: Yogi Black/Alamy Stock Photo

3.2.3 Synaesthesia: mixing up your senses

Most people experience a fairly stable mapping of sensory experiences. For example, seeing a letter elicits a perception of that letter, and that is all. That said, some people have a condition known as *synaesthesia*. A person with synaesthesia will have inappropriate and involuntary sensory experiences in addition to typical sensory experience (see Hochel & Milán, 2008; Hupe & Dojat, 2015; Ward, 2013 for reviews.[21]). For example, a person may experience seeing different colours when hearing different people's voices, or may have certain taste experiences when feeling different textures. For some people with synaesthesia (called projectors), their additional sensory qualia are experienced as if they were out there in the world; whereas for others (called associators), their additional experiences are in their mind, but not necessarily out there in the world (Dixon, Smilek & Merikle, 2004[22]). Perhaps the most common form of synaesthesia is when a person reports experiencing colours (photisms) when reading words or numbers. This is called grapheme–colour synaesthesia.

Most people with synaesthesia (called synaesthetes) either are born this way or develop the condition early on. If you are not a synaesthete, you might think of this as a bewildering and confusing way to experience the world. However, synaesthetes do not experience such distraction or confusion. Keep in mind that their additional sensory experiences of the world have always been there! Many synaesthetes do not even realise that they have synaesthesia until secondary school or college when they figure out that most other people are *not* having those additional experiences. An analogy for what synaesthesia may be like is to imagine that you are a person with colour vision in a world in which most people are colour blind. You would be receiving additional sensory and perceptual information that most others do not. Although you would not find it at all confusing, and might even find it helpful, your accounts of your experiences would baffle others.

How does the condition of synaesthesia come about? There are two primary explanations for this phenomenon. First, there is the idea that synaesthesia is caused by a lower ability to suppress inappropriate feedback loops in perceptual processing (Grossenbacher & Lovelace, 2001[23]). That is, sensory signals that would normally be dampened are allowed to spread, resulting in additional sensory experiences under certain circumstances. The second explanation is that there is an incomplete pruning of extra cortical connections during development (Maurer, 1997[24]). The normal pruning of inappropriate and unnecessary neural connections as infants mature does not happen in some synaesthetes.

In addition to the differences in sensory and perceptual experiences, synaesthesia may influence other aspects of cognition. For example, some work with grapheme–colour synaesthetes has shown that memory for materials, such as lists of words, that elicit the synaesthetic experience is superior to that of normal controls, either through subjective reports or through experimental verification (Ward, 2013[25]). Thus, the heightened sensory-perceptual experiences of synaesthesia can actually improve some kinds of memory. Some researchers have posited that grapheme–colour synaesthesia may reflect a memory and associative learning mechanism, whereby specific mnemonic associations are automatically retrieved when relevant stimuli are encountered (e.g. the letter A evokes the colour red; Witthoft & Winawer, 2013[26]).

3.2.4 Visual sensory memory

We turn our attention now to visual sensory memory, also known as *iconic memory* (Rensink, 2014[27]). Because this memory system has such a short duration (less than 500 milliseconds), we have few useful intuitions about its operation. Although our primary concern here is with the normal operation of visual sensory memory, unusual circumstances can give us some clues.

Everyone has seen a flash of lightning. Think about that for a moment, and guess what the duration of a bolt of lightning is. Most people guess that the flash of light lasts a little more than half a second or so, or maybe closer to a whole second. If your estimate was in this neighbourhood, then it is reasonable – but wrong. The bolt of lightning is actually three or four separate bolts. Each one lasts about 1 millisecond (ms), and there is a separation of about 50 ms between bolts. Thus, the entire lightning strike lasts no more than about 2/10 of a second, or 200 ms, and is composed of several individual flashes (Trigg & Lerner, 1981[28]).

Figure 3.11 **Our perception of lightning is a mental event that reflects visual persistence**
Source: Fesus Robert/Shutterstock

Given that an estimate of 0.5–1 second was so off, what was reasonable about it? It was the perception of a flash of light that extended in time. This phenomenon is visual persistence, the apparent persistence of a visual stimulus beyond its physical duration. The neural activity on the retina that the lightning flash causes does not outlast the flash itself. The eye does not continue to send 'lightning' messages after the flash is over. Your perception of the lightning is a mental event that reflects visual persistence in that you perceive a lighted scene that then quickly fades away. Because any persistence of information beyond its physical duration defines the term *memory*, the processes of visual perception (as opposed to sensation) must begin with memory, a temporary visual buffer that holds visual information for brief periods.

Amount and duration of storage

Sperling and his co-workers (Averbach & Sperling, 1961; Sperling, 1960[29]) reported the characteristics and processes of visual sensory memory in their classic research. See Figure 3.12 for a schematic diagram of a typical trial.

Sperling's classic study of iconic memory

Sperling presented a visual stimulus for a controlled period, usually in milliseconds, to study 'the information available in brief visual presentations', the title of his paper in 1960.

A typical iconic memory experiment presented arrays of letters and digits to people for brief durations. In all cases, their task was to report what they could remember from the display. For example, people were shown a series of trials, each with a 3-by-4 array of letters (3 rows, 4 letters per row). The array was shown for 50 ms and was followed by a blank postexposure field. Finally, a signal was given to report the letters.

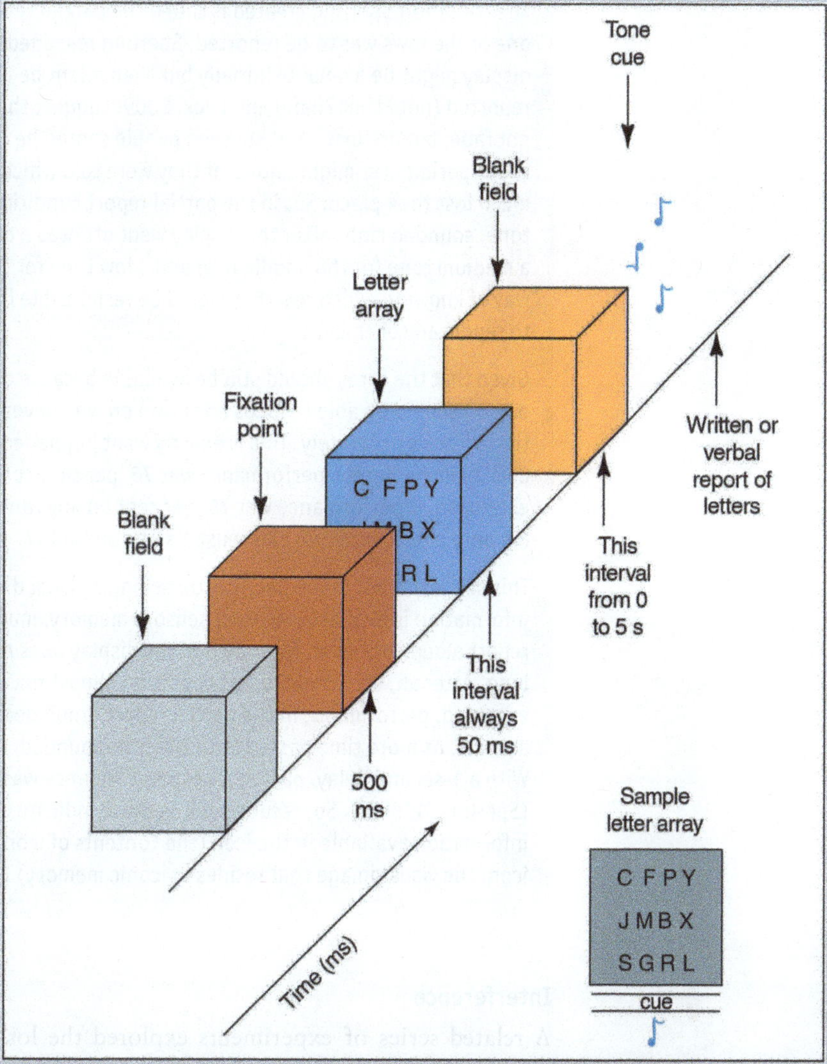

Figure 3.12 A schematic diagram of a typical trial in Sperling's (1960)[30] experiments

After a fixation point appears for 500 ms, the letter array is displayed. The visual field after the display is blank. The tone cue can occur at the same time as the postexposure field, or it can be delayed up to 5 seconds (s). Data from Sperling (1960).

Sperling found that people generally reported no more than 4 or 5 items correctly. Furthermore, this level of accuracy is essentially the same for exposures as long as 500 ms and as short as 5 ms (Sperling, 1963)[31]. It seemed that an average of 4.5 items correct reflected a default strategy. People said they could not remember all 12 letters because the display seemed to fade from view too rapidly. Their level of performance, about 4 or 5 items, was what is expected based on the **span of apprehension**, the number of items recallable after any short display (also known as the **span of attention** or the span of immediate memory).

What distinguished Sperling's research was a condition he developed to contrast with these results. The condition just described, in which people are to report any letters they can, is called the **whole report condition** because the whole display is to be reported.

The condition Sperling created is called the **partial report condition**, in which only one of the rows was to be reported. Sperling reasoned that all the letters of the display might be available initially but then might be lost faster than they could be reported (note that Zhang and Luck, 2009, suggest that these representations do not fade, but are lost[32]). If so, then people should be accurate on any of the rows the experimenter might choose if they were told which row to report before too much loss took place. So, in the partial report condition, he had a signal: a high tone, sounded right after the display went off, was a cue for reporting the top row; a medium tone for the middle row; and a low tone for the bottom row. People had no way of knowing which row they would be responsible for, so they had to be prepared to report any of them.

Given that the array should still be available because of visual persistence, the person should be able to focus attention on whichever row was cued, and read out those letters accurately. This is exactly what happened. When the tone followed the display immediately, performance was 76 per cent correct, regardless of row. By extension, if performance was 76 per cent on any randomly selected row, then visual memory of the entire display must also be around 76 per cent.

This suggests that immediately after seeing a visual display, a great deal of information is available in visual sensory memory, much more than people can report aloud. However, this much of the display does not remain available for long. After all, the whole report condition almost never exceeded 4 or 5 items. As expected, performance in the partial report group declines as the iconic memories are lost. As more time passed until the tone sounded, performance dwindled further. With a 1-second delay, partial report performance was back down to 36 per cent (Sperling, 1960[33]). So, results such as these indicate that there is a large amount of information available in the icon (the contents of iconic memory often are called the **icon**, the visual image that resides in iconic memory) but that it is lost quickly.

Interference

A related series of experiments explored the loss of information from iconic memory. The original research suggested that forgetting was a passive process like fading or decay; that is, the mere passage of time degraded the icon. However, in normal vision we look around continuously, shifting our gaze from one thing to another. What happens to iconic memory when one visual scene immediately follows another? The answer, in short, is interference, forgetting caused by the effects of intervening stimulation or mental processing.

A study by Averbach and Coriell on interference

One study on interference was by Averbach and Coriell (1961; reprinted in Coltheart, 1973[34]).

Averbach and Coriell presented a display of two rows of letters, eight letters per row, for 50 ms. A blank white postexposure field, varying in duration, followed the display and then was followed by a partial report cue. Unlike Sperling, Averbach and Coriell used a visual cue, either a vertical bar marker or a circle marker. The bar marker was just above (or below) the position of the to-be-reported letter, and the circle marker surrounded the position where the to-be-reported letter had just been.

Result

With the bar marker, the results were similar to those obtained by Sperling, such as higher performance with short delays of the cues. But the results with the circle marker were quite different. With the circle marker, accuracy was lower than with the bar marker, especially when the target letter resembled the circle marker (for instance, when the letter c had been shown at that position).

Analysis

These results suggest that the circle marker in some way interfered with or disrupted the memory trace for the letter in that position much more than the bar marker did. In other words, a later visual stimulus interfered with the memory for the prior stimulus at that location. This is backward **masking** or **erasure**. The masking stimulus, if it occurs soon enough after the display, interferes with the perception of the earlier image at the same spot. In some backward masking studies, people claim that they had seen only the mask, even though their performance indicates that sensation did indeed register the first stimulus (data on this go back as far as Werner, 1935; see Kahneman, 1968, for a review[35]).

3.2.5 The early parts of a fixation

The evidence reported by Sperling, Averbach and Coriell and many others, led cognitive psychology to propose that iconic memory existed and that it was an early phase in visual perception (Neisser, 1967[36]). However, later evidence shows results that are even more fascinating. A study by Rayner, Inhoff, Morrison, Slowiaczek and Bertera (1981)[37] examined performance during reading, using an eye tracker for precise timing measurements. After people had fixated a word for a mere 50 ms, the word was replaced with an irrelevant stimulus, which remained in view for another 175 ms to fill up the rest of the fixation time. Surprisingly, this did not affect reading at all – people often did not notice that the word had even changed!

This is very important, so stop and think about it for a moment. Despite what it feels like, as you read these words on this page, you are not viewing them continuously. You are seeing them in brief bursts, extracting information quickly and then devoting mental energy to processing them further, unaware of your 'downtime' during the fixation and of your blindness during the following saccade.

Several investigators have also collected evidence on what might be called 'dynamic icons', that is, iconic images that contain movement (see Finke & Freyd, 1985; Irwin, 1991, 1992; Loftus & Hanna, 1989[38]). Treisman, Russell and Green (1975)[39] presented a brief (100 ms) display of six moving dots and asked people to report the direction of movement. Partial report performance was superior to whole report performance; it is more accurate. It starts out higher and declines over time. In short, the moving images of the dots were decaying just as the static letter grid had in Sperling's procedures. Visual perception, therefore, is not a process of flipping through successive snapshots, with three or four snapshots per second. Instead, it is a process of focusing on the visually attended elements of successive fixations, where each fixation encodes a dynamic segment of the visual environment.

In fact, integration across brief intervals of time can occur even without eye movements. Loftus and Irwin (1998)[40] showed that temporal integration – perceiving two separate events as if they had occurred at the same time – transpires seamlessly when visual events occur within about 20 ms of each other. This happens without any conscious awareness that the two separate events have occurred. Events separated by 40 ms or more, or separate events that last for 40 ms or more, tend not to be integrated as completely, however. With these longer durations, people can more easily detect that two separate events happened rather than just one.

3.2.6 Visual attention

How do all these different results make sense: the wholesale input of visual stimulation; the persistence, decay, interference and integration of information; and the concept of visual attention? The entire sequence of encoding visual information – selecting part of it for further processing, planning subsequent eye movements, and so on – is very active and rapid. The visual continuity we experience – our feeling that we see continuously – is due to the constant updating of visual sensory memory and to our focus on attended information. As we attend to a visual stimulus, we are examining the readout from iconic memory. In the meantime, a new visual scene is being registered in iconic memory. Our mental processes then pick up the thread of the newly registered scene, providing a smooth transition from one display to the next.

Focal attention was Neisser's (1967)[41] term for this mental process of visual attention, such as the mental redirection of attention when the partial report cue is presented. It seems that focal attention, or simply visual attention, helps bridge between successive scenes in visual sensory memory. This prevents us from sensing the blank space of time occupied by the eye's saccades by directing focal attention to elements of the icon. Although we sense a great deal of visual information, what we perceive is the part of a visual scene selected for focal, visual attention. What you perceive right now is not the printed screen in front of you, but the processed and attended portions that were registered in sensory memory, your iconic trace, as processed by visual attention – this is important for understanding the processes underlying visual sensory memory and the visual experiences we create.

3.2.7 Trans-saccadic memory

To build up an understanding of the visual world, we need to move our eyes, head and body, gathering visual information across each successive fixation (see Higgins & Rayner, 2015, for a review[42]). First, it is important to point out that vision is not continuously processing input from the eyes. Instead, during a saccadic eye movement, the processing of visual information is suppressed, which is why you do not see a blur from your eyes moving. This is saccadic suppression.

Next, we turn to the question of how we put the information from all of these fixations together. This is done using iconic memory in a way known as trans-saccadic memory (e.g. Irwin, 1996[43]), the memory that is used across a series of eye movements. How does iconic memory track information to figure out how to put together information from different fixations? It does not use retinal coordinates (where the images fall on the eyes) or spatial coordinates (where things are in space) to do this (Irwin, Yantis & Jonides, 1983[44]). Instead, trans-saccadic memory works by using object files (Kahneman, Triesman & Gibbs, 1992[45]), which are iconic representations of individual objects used to track what is going on in the world. Evidence for their use comes from studies that ask people to detect changes in objects after a saccadic eye movement (Henderson & Anes, 1994[46]) – for example, detecting whether a letter changed to a plus sign in a display. In general, people's responses are fairly accurate in noticing changes in objects they focused on. This does not occur for all objects in a visual scene, but only those to which people attend.

To accomplish trans-saccadic memory and integration effectively, our brain assumes that everything that we are not attending to is more or less stable, which is why we may miss those changes. Moreover, consistent with the theme that some of cognition is future-oriented, some evidence suggests that trans-saccadic memory is predictive. That is, memory is, to some degree, predicting that the world is likely to be stable as a default assumption (Higgins & Rayner, 2015[47]). What is noticed and processed further are those circumstances that violate this prediction of stability; hence, the world is different from what it had been before, which may result in visual attention capture.

3.3 Pattern recognition

LEARNING OBJECTIVE
Evaluate the process by which we identify patterns and objects

We turn now to one of the most intriguing and debated topics in visual perception, the identification of patterns and objects. As you will see, pattern recognition does not occur instantly, although it does happen automatically and spontaneously. Perceptual pattern recognition is, in many ways, a problem-solving process, with much of the mental work occurring subconsciously and very rapidly. Essentially, during perception a person needs to identify the nature of the distal objects in the world based on the proximal images reaching the retina. Often these images are compromised in some way, such as being occluded by another object or being against a complex visual background. Vision parses the visual image in a number of ways to extract information about the objects that are actually present and it follows a number of perceptual principles in doing so.

3.3.1 Gestalt grouping principles

Perhaps the best known and established of these perceptual principles are the Gestalt grouping principles laid out by the Gestalt psychologists in the early to mid-20th century. Although the roots of Gestalt psychology are in the 20th century, it still has important implications today (for a companion paper review, see Wagemans, Elder, et al., 2012; Wagemans, Feldman, et al., 2012[48]). These principles identify those characteristics of perception in which ambiguities in a stimulus are resolved to help determine which objects are present. They are also aimed more at processing information about the whole of an object, rather than simply, and only, building up a mental representation from more basic elements. In fact, obtaining a Gestalt for a perceptual whole of an object may actually disrupt the perception of the parts of an object (Poljac, de-Wit & Wagemans, 2012[49]). fMRI neuroimaging work has supported the Gestalt-based processing of objects (Kubilius, Wagemans & Op de Beeck, 2011[50]). This neuroimaging has shown that some parts of the brain, namely the lateral occipital lobe (B.A. 19) and the posterior fusiform gyrus (B.A. 37), are involved in processing whole objects, apart from the occipital brain areas involved in processing individual elements (see the section on feature detection later).

Figure-ground principle

One of these Gestalt principles is the figure-ground principle. When viewing an image, part of the image is treated as the figure or foreground (the object identified), which is segregated from the visual information upon which it is set (the background).

Classic examples illustrating difficulties in determining figure-ground are reversible figures, in which a person shifts back and forth between what is the foregrounded object and what is the background. See the logo of the Olympic Games 2024 in Paris in Figure 3.13. At one moment it might be a white flame on a golden background, at the next it can be the face of a woman.

Closure and other principles

The aim of some Gestalt grouping principles is to provide a more complete percept from an incoming image that may be fragmentary or incomplete. They follow the principle of **closure**, in which a person 'closes up' an image that has gaps or parts missing, perhaps because they are being occluded (blocked) by some other object (golden hair and lips). An example of closure can be seen in Figure 3.13.

Figure 3.13 An illustration of closure
Source: GK Images/Alamy Stock Photo

In the Gestalt principle of **proximity**, elements that are near to one another tend to be grouped together. This is shown in Figure 3.14. Because of this principle, you see a flock of geese in flight as forming a classic V. Similarly, you may perceive groups of dots moving together as a person walking (Johansson, 1973[51]).

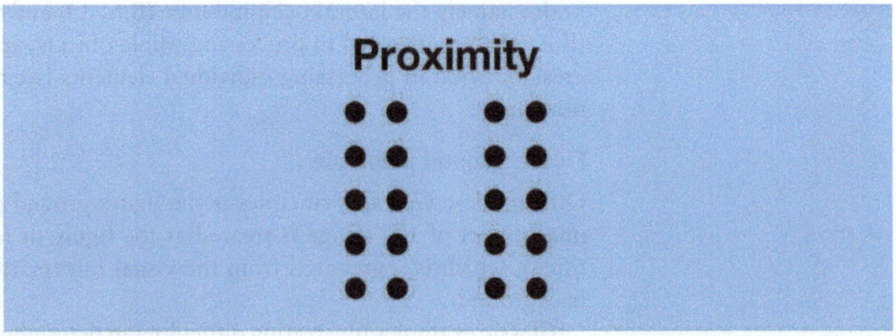

Figure 3.14 An illustration of proximity

Another Gestalt principle is **similarity**, in which elements that are visually similar in some way, such as having a similar color or texture, tend to be grouped together. This is shown in Figure 3.15.

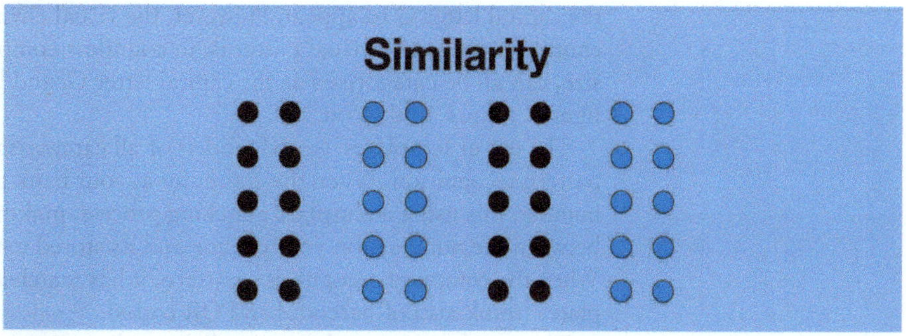

Figure 3.15 An illustration of similarity

Similarity is seeing the individual dots on a television or computer screen as being part of the same object if they have a similar color or visual texture. Certain Gestalt grouping principles take into account some form of trajectory. In some cases, the trajectory is the edge of an object. The principle of **good continuation** assumes that when an edge is interrupted, people assume that it continues along in a regular fashion. In the example shown in Figure 3.16 people tend to organise this as lines from A to B and from C to D, not A to D, B to C, and so on.

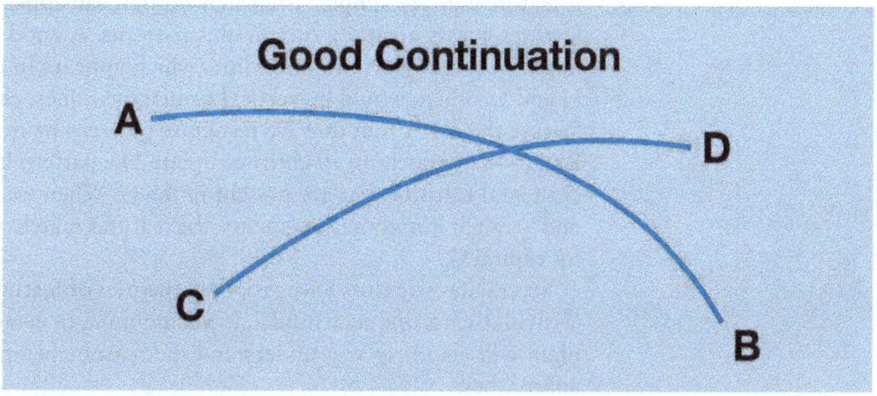

Figure 3.16 An illustration of good continuation

Finally, with **common fate**, entities that move together are grouped together. For example, when an animal moves in the forest, it is easier to spot than if it remains motionless. The movement allows perception to group those points together because they are moving together.

Although we use Gestalt grouping principles to some degree, we need to go deeper to understand how to recognise a visual stimulus as a familiar pattern. How does cognition manage to input a visual stimulus such as G or a tree and end up recognising it as familiar and meaningful? How do we recognise patterns of handwriting, or different printed fonts, despite incredible variability? The following sections present some ideas about how this occurs, by looking at the case of written language.

3.3.2 The template approach

As Neisser (1967)[52] pointed out, pattern recognition would be simplified, though still thorny, if all the equivalent patterns we saw were identical. That is, it would be easier to explain pattern recognition if there were one and only one way for the capital letter G to appear. However, the visual environment is not like this. An enormous variety of visual patterns, in countless combinations of orientation and size, can all be categorised as the capital letter G, and likewise for all other letters, figures, shapes, and so on.

The use of templates, stored models of all categorisable patterns, may help with pattern recognition. When the computer at your bank reads your checking account number, it is using a template matching process, making physical identity matches between the numbers on your cheque and its stored templates for the digits 0 to 9. When the computer recognises a pattern, it has matched it to one of its stored templates (think also of barcodes and QR codes).

The template approach has simplicity and economy on its side. However, beyond this simplicity and economy, the template approach has little else to recommend it, and it has some serious flaws. We have already covered the primary reason, the enormous variability in the patterns that we can recognise. Also, just think of how long it would take you to learn the infinite number of possible patterns (for all of the objects in the world, the different orientations they can be in, the various distances they can be from you, etc.) and then search through those patterns in memory. Would you have time left for anything else?

3.3.3 Visual feature detection

An improvement over the template approach is feature analysis or feature detection. A feature is a very simple visual element that can appear in combination with other features across a variety of stimulus patterns. A good example of a visual feature might be a straight horizontal line, which appears in capital letters such as A, G, H and L; others would be vertical or diagonal lines, curves, and so on. In general, feature theories claim that we recognise patterns by first identifying their features. Rather than matching an entire template-like pattern for capital G, we identify the elemental features that are present in the G. When we detect 'circle opening right' and 'straight horizontal' segments, the features match with those stored in memory for capital G.

Several investigators have proposed theories of feature-based pattern recognition. We will discuss one such model, Pandemonium, in detail. Understanding Pandemonium will also help you understand the reasons behind interactive, connectionist approaches.

Pandemonium

Selfridge (1959),[53] an early advocate of feature detection, described a model of pattern recognition called Pandemonium; an illustration of the model is shown in Figure 3.17. This model has many similarities to the connectionist models that will be discussed later.

In Selfridge's imaginative description, the mechanisms involved are little mental demons that shout aloud as they attempt to identify patterns. As the figure shows, a set of data or image demons encodes the pattern. Next, the computational demons begin to act. They are the feature analysers in Selfridge's model; each one has a single simple feature it is trying to match in the stimulus pattern. For instance, one demon might be trying to match a simple horizontal line, another a vertical line, another a curve opening to the right, and so on. When a computational demon matches a stimulus feature, it begins to shout excitedly.

At the next level up, listening to all this shouting is a set of cognitive demons. These demons represent the different letters of the alphabet, one for each letter. Each one is listening for a particular combination of demons: For instance, the G demon is listening for the 'open curve' and the 'horizontal bar' feature analysers or demons to shout. Any evidence from the computational demons that suggests a match with

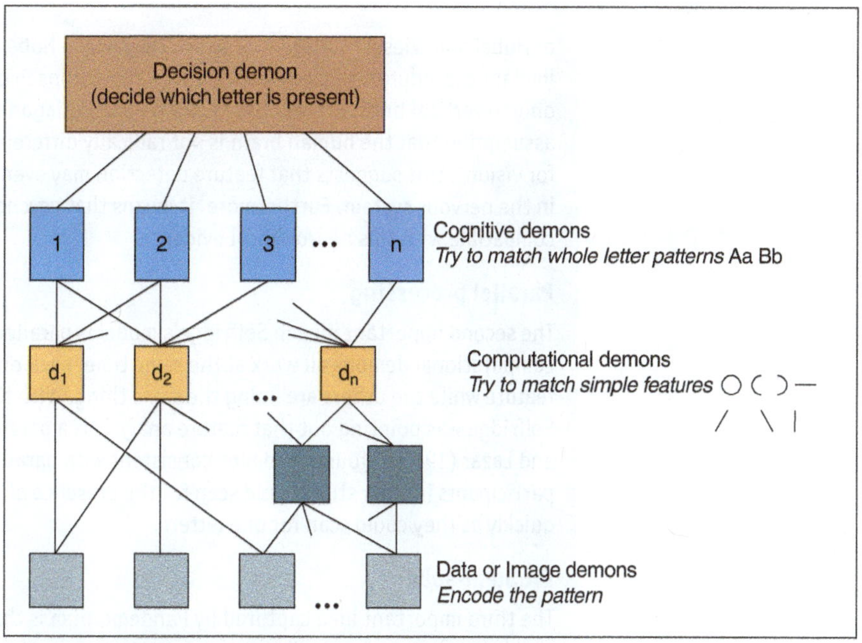

Figure 3.17 Selfridge's Pandemonium model

the stimulus also causes a cognitive demon to begin shouting: Based on the feature analysis evidence, it thinks that it is the matching pattern. Several of the cognitive demons will be shouting at once because several letters usually share some features (e.g. C and G). Thus, the one that shouts the loudest is the one whose pattern most nearly matches the stimulus. The decision demon hears the loudest cognitive demon and has the final say in recognising and categorising the pattern.

Three important ideas

The utility of our thinking about perception as a process that captures some of the stages of cognition highlights a number of important points. Three of these ideas are feature detection, parallel processing and problem solving.

Selfridge's model for pattern recognition

Aside from the vividness of the model's description of scores of shouting demons producing a noisy Pandemonium, Selfridge's model incorporates several important ideas about pattern recognition.

Feature detection

First, it is a feature detection model. The demons detect and report elementary, simple features. There are several related lines of evidence for feature detection in visual pattern recognition (e.g. Pritchard, 1961[54]). Especially convincing are neurophysiological studies showing that special visual cortex cells exist for simple visual features.

These are found in area V1 (B.A. 17) of the occipital lobe (the 'V' stands for 'visual'). The most widely known evidence of this kind comes from the pioneering research of Hubel and Wiesel (1962),[55] for which they won a Nobel Prize. Using electrode implant procedures, these researchers found neurons in cats' brains that respond only to vertical lines, others that respond only to diagonals, and so on. On the assumption that the human brain is not radically different from a cat's brain (at least for vision), this suggests that feature detection may even have a physiological status in the nervous system. Furthermore, it means that psychological theories must be compatible with this neurological evidence.

Parallel processing

The second important idea in Selfridge's model is parallel processing: the computational demons all work at the same time, each one trying to match its own feature while the others are doing the same thing. With this aspect of his model, Selfridge was pointing out that feature analysis is a parallel process. Neisser, Novick, and Lazar (1963)[56] found evidence consistent with parallel feature processing when participants in their study could scan for the presence of 10 different letters just as quickly as they could scan for one letter.

Problem-solving

The third important idea captured by Pandemonium is that perception is a problem-solving process. The world presents the visual system with bits and pieces of features and relations. The visual system must put these together to accurately identify the objects that are out there in the world. Often the system gets this correct. However, occasionally errors do occur, as with visual illusions or when you mistakenly identify one object as something else; for example, when you are driving and swerve to avoid hitting a hedgehog, only to discover a moment later it is only a dead leaf.

Thinking about how we interact with the world further emphasises the inadequacy of template approaches, and the advantage of feature-based approaches, for visual feature detection. Specifically, humans are often in motion – walking, running, driving, and so on. Moreover, the things we are looking at also are often moving. Thus, our views of objects can be constantly shifting. The area of the brain that leads us to see objects as constant and stable across such movement and different viewing angles is area MT (B.A. 19), part of the dorsal stream of visual processing, at the conjunction of the occipital, parietal and temporal lobes. This area is also involved in the perception of apparent movement (Larsen & Bundesen, 2009[57]), as well as real movement, when cognition must make inferences about what goes on in-between. For example, watching a film or TV show involves a series of still pictures presented to you in rapid fashion. The pictures change faster than your iconic memory can decay, replacing one another. Your brain fills in any jumps in position, producing the illusion of motion. This mental perception of illusory motion is **beta movement** (Wertheimer, 1912[58]). A related perceptual illusion occurs when you see lights moving or flowing around on a movie marquee or chasing Christmas lights. This is the **phi phenomenon** (Wertheimer, 1912[59]). Essentially, when iconic memory receives visual images in relatively close proximity in space and time, it will infer virtual movement. Beta movement occurs when making inferences from one picture to the next, as in a film, but phi movement involves illusory tracking of an object in space.

3.4 Top-down processing

LEARNING OBJECTIVE
Explore ideas about how what we know influences what we see

The typical way that most people think about vision is that there is stuff out there in the world and we see what it is. This is the bottom-up, data-driven view of visual processing. The information in the world, the data, drives what we perceive as being out there. Although this is true, it is only half the story. Because the information in the world that reaches us may be fragmentary or incomplete, cognition needs to fill in what is missing. The topic of this section is the filling in of that information.

3.4.1 Conceptually driven pattern recognition

Even Selfridge (1959)[60] knew that Pandemonium was missing an important ingredient. Basically, Pandemonium is a completely **bottom-up, data-driven processing** system in which processing is driven by the stimulus pattern, the incoming data. The patterns to be recognised came in to the image demons at the bottom and then were processed at higher and higher levels until the top-level demon finally recognised the pattern. Yet, Selfridge presented examples like those shown in Figure 3.18, showing that **context** influences pattern recognition.

How adequate is the bottom-up approach as a sole explanation of visual pattern recognition? Did you 'see' the words *went* and *event* despite the fact that they are written identically in the two sentences? So what was the missing ingredient in Selfridge's model?

The missing ingredient was context and a person's expectations. Such effects are called top-down or conceptually driven effects, in which context and higher-level knowledge influence lower-level processes (remember 'typograpical'?). In Figure 3.18 your knowledge of English words and your understanding of the meaning of sentences lead you to perceive what is written as either *went* or *event*, depending on the context.

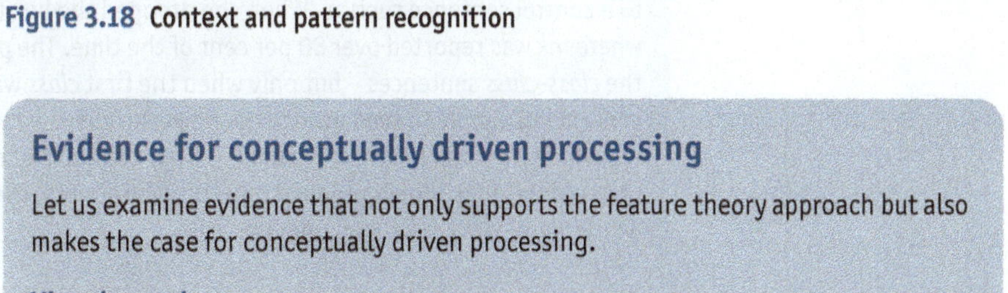

Jack and Jill went up the hill.
The pole vault was the last event.

Figure 3.18 Context and pattern recognition

Evidence for conceptually driven processing

Let us examine evidence that not only supports the feature theory approach but also makes the case for conceptually driven processing.

Visual search

In Neisser's (1964)[61] classic work on visual search, people saw pages of characters, 50 lines of printed letters, with four to six letters per line. One task was to scan the page as rapidly as possible to find the single occurrence of a pre-specified letter (in another task, people had to find the line without a certain letter).

Pattern recognition

Pattern recognition starts by processing the incoming pattern, a bottom-up process. Nonetheless, this bottom-up emphasis slights the contribution of top-down processes. It misses the effect of *context*, the surrounding information and your own knowledge. The influence of knowledge and expertise on perception is fairly common.

Think of the ability of people who are experts at reading X-rays, tasting wines or identifying fingerprints (e.g. Busey & Parada, 2010[62]) to distinguish subtle variations and detect things that most people find very hard to notice. Still, you have some expertise that is revealed in your ability to read. As an example, we often identify a pattern that is not in original stimulus at all, such as the word *the* in the last clause. You misread that part of the sentence, didn't you? And now you know where the missing *the* came from:

Here is a set of sentences (from Morris & Harris, 2002[63]) that illustrate top-down processing:

- When she spilled the ink there was ink all over.
- She ran her best time yet in the rice last week.
- I broke a wine class in my class yesterday.

In the first sentence, the word *ink* repeats. When the sentence is presented word by word at a rapid rate (called the RSVP method, for rapid serial visual presentation), a surprising result is **repetition blindness**. People often fail to report the second occurrence of ink; even though it makes the sentence ungrammatical, people report having seen 'When she spilled the ink there was all over' (see also Kanwisher, 1987, 1991[64]). *Repetition blindness* is the tendency not to perceive a pattern, whether a word, a picture, or any other visual stimulus, when it is quickly repeated. The basic explanation is that cognition has just identified the stimulus, so it does not expect to see the same thing again.

Misreading effect

Another interesting effect noted by Morris and Harris (2002)[65] is a misreading effect, a tendency to read a word that should be in the sentence, based on context, such as *race* instead of *rice*, and *glass* instead of *class* in the second two sentences above. Morris and Harris found that repetition blindness and misreading can combine. For example, in *ink-ink* trials, a strong repetition blindness effect occurred – the second *ink* was reported less than 50 per cent of the time, compared to a control sentence such as 'When she dropped the box there was ink all over,' where *ink* was reported over 80 per cent of the time. The percentages were similar in the *class-class* sentences – but only when the first *class* was read correctly. When the context led people to read *wine glass*, however, there was no repetition blindness – reports of *class* were correct in about 70 per cent of the trials. So, even in cases (*class-class*) that should produce repetition blindness, the critical variable was what word (*glass*) the person had understood.

Note that top-down, conceptually driven processing can be influenced not only by things such as knowledge of a language or the context of a sentence. Even one's culture can influence perception to some degree. For example, evidence suggests that when Westerners view a scene, they are more likely to focus perceptual processes on a foreground object and its visual properties. Eastern Asians, by comparison, are more likely to focus on the context and the background of the scene, taking into account the relations among the various elements (e.g. Nisbett & Masuda, 2003[66]). Evidence of different levels of neural activity in perceptual processing regions of the brain supports this conclusion (Gutchess, Welsh, Boduroglu & Park, 2006[67]). Contextual effects of top-down processing are not limited to linguistic materials. For example, Gartus, Klemer and Leder (2015)[68] reported that artistic judgements of modern art were higher when the art was embedded in a museum context than in a street context (although judgements of graffiti were similar in both contexts).

Conceptually driven and data-driven processes work in combination in most situations involving pattern recognition. An excellent way to model this, to explore how this combination works, is within the connectionist model. Think of this model as Pandemonium Plus, a bottom-up model like Selfridge's with an added top-down processing effect.

3.4.2 Connectionist modelling

Connectionist modelling is a computational approach often used in cognitive science. Connectionist models involve a massive number of mathematical computations. Essentially, each unit in a connectionist layer connects with many or all of the units in the next layer. The impact of each experience on each of these connections needs to be computed. Even if the number of units in a layer is fairly small, the number of separate computations in a single run of the model is staggeringly large because of the very large number of connections among the units.

To flesh out the word recognition model (McClelland & Rumelhart, 1981; Rumelhart & McClelland, 1986[69]), we will use a model that recognises four-letter words, such as *tree*.

Basic statement of parallel distributed processing principles

We start with certain basics of connectionist modelling, including some of the vocabulary. Complex mental operations are the combined effects of the massively parallel processing that characterises a network. The processing is distributed across all the levels of the network (hence, the term *parallel distributed processing*, or PDP).

A network is often composed of three levels of units: the input level, the hidden level and the output level. The internal 'hidden' layer is invisible to an outsider. Units in each of these levels are interconnected (hence the term *connectionism*). Weights of the connections are either positive or negative.

Positive connection weights pass excitatory activation to the connected unit; negative connection weights pass inhibition. A unit transmits its activation to connected units if it has received enough positive activation to reach threshold. The analogue here is excitatory and inhibitory neurotransmitters, which play a similar role in the nervous system.

Connection weights vary as a function of training, in which feedback as to correctness or incorrectness leads to a mathematical adjustment of weights. When a network uses this procedure and the weights have stabilised, the network has been trained up. Back propagation is a commonly used training method, although others exist.

The similarities between PDP models and the neurological structures and activities in the brain are intentional: connections are sometimes called synapses; excitation and inhibition parallel those processes in the cortex; and the approach is known as *neural net* modelling.

The connectionist approach to modelling and understanding cognition differs in a number of ways from more traditional and psychological approaches. Connectionism comes more from a computer science approach to issues of the mind and thinking. Because of this, certain sets of terms may be unfamiliar, but you will need to learn them to understand this approach to perceptual processes.

Top-down and bottom-up influences in a connectionist model

It is important to distinguish between top-down and bottom-up effects in the connectionist model. In this section, we will look at the connectionist modelling of four-letter words to better understand top-down and bottom-up influences.

Connectionist modelling of four-letter words

Let us build the simple model for recognising four-letter words using connectionist modelling piece by piece. Here there are three levels of units. A portion of the PDP network for recognising four-letter words. The bulk of the illustration involves identifying the first letter of the word.

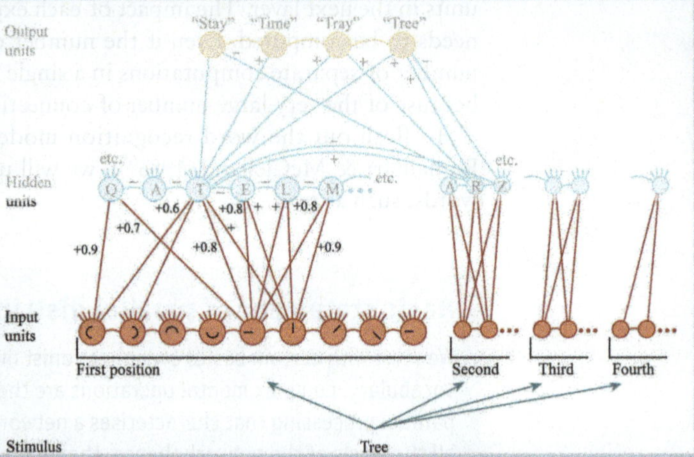

Figure 3.19 Connectionist modelling of four-letter words

Input units

At the bottom are the **input units**. These elementary 'cells' receive input from the environment. Our example is visual word recognition, so our input units are simple visual detectors. That is, we have nine input units, each of which responds to different visual features of the letters of the alphabet. To build on what you already understand, consider the input units to have the same function as the data and computational demons in Pandemonium shown in the Figure 3.19. Our input units encode and respond to simple visual features in letters of the alphabet. Thus, the input unit level in this illustration is the feature-detector level.

Hidden units

When a stimulus presents to the input level, one or more of the input units match the features in the stimulus. When this happens, each unit that matches activates a set of connected units in the middle or *hidden unit* level; hidden here simply means that this level is internal, always one step removed from either input or output. Note that the activation is sent across the connections that link the units together; these are the connections in connectionism.

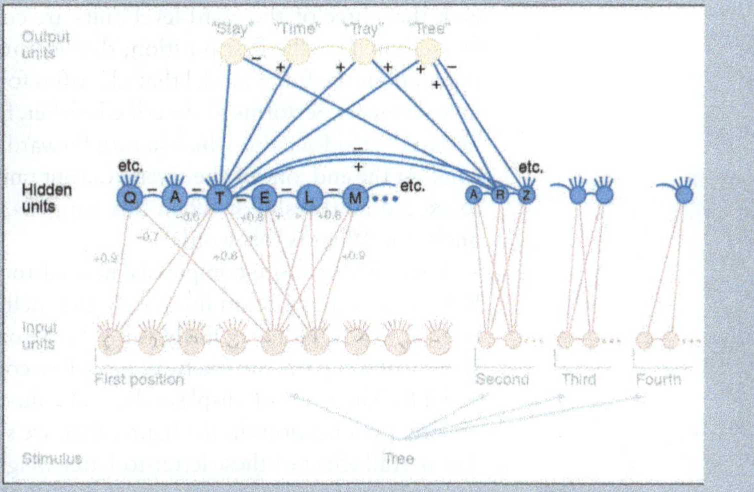

Figure 3.20 Connectionist modelling of four-letter words

The connections have a weight that represents the relationship between the linked units. Some weights are positive, and some are negative. So, in Figure 3.20 the horizontal straight bar feature has positive weights connecting it to the letters T, E and L because they all contain that feature (to minimise confusion, many of the connections have not been drawn fully, and only a few numerical weights are given). Likewise, all the curved features at the input level have positive weights to curved letters and negative weights to angular letters. Eventually, after factoring all the weights into the computational formulae, activations at the output level come into play.

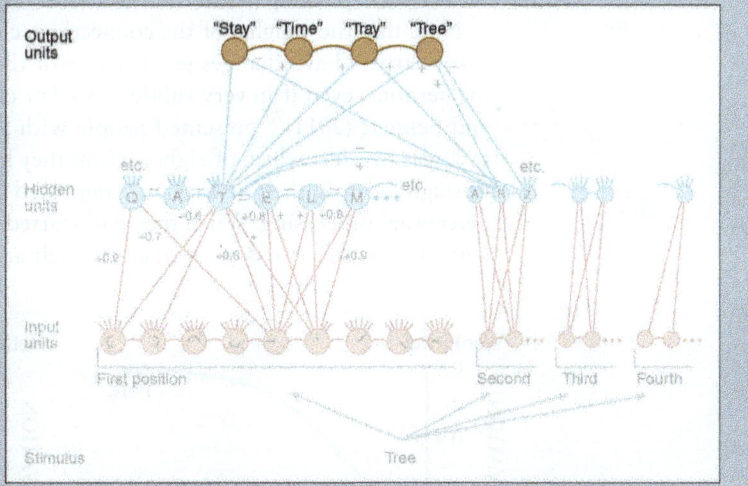

Figure 3.21 Connectionist modelling of four-letter words

Output units

What is all of this getting us? Imagine that you were building a machine (programmeming a computer) that could identify visually presented words. What you see in Figure 3.21 is the connectionist network for the first position in four-letter words. Three more sets of connections, shown in reduced form at the right of the figure, duplicate the same connections again, once for each position. Given these additional positions, we can now talk about the output units.

For simplicity, the figure shows only a handful of four-letter words. Note, however, that three of the word-level units are consistent with the letter detection performed on T in the first position; that is, three of the words begin with a T. Now think about the fuller model that identifies four-letter words. Each of the four input unit segments performs as described earlier, forwarding both activation and inhibition to the **hidden units**, which in turn forward activation and inhibition to the **output units**. At the end, one of the several output units will have received enough activation to exceed its threshold. When this happens, that unit responds by answering the question, 'What is this word?'

There is one more complexity needed to get top-down effects into the model. Reflect for a moment on how likely the spelling pattern TZ is at the beginning of English words. Not very likely, is it? On the other hand, TA, TE, TI, and other consonant–vowel pairs are likely, as are a few consonant–consonant pairs such as TH and TR. The network displays these likelihoods; to distinguish them visually from the other connections in the figure, they are shown with curve-shaped connections. The overall effect of these letter-to-letter weights is that the activations in the system can make up for missing features at the perceptual level.

Figure 3.22 shows the final levels of activation for three possible words, given the partially obscured stimulus pattern at the bottom.

This shows an important feature of connectionist models: Enough knowledge is represented in the system, in the weights for letter-to-letter sequences, that the model identifies the word *fork* even when the last letter could also be an R.

This is important because it illustrates the general theme of top-down, conceptually driven processing. If you saw the partially obscured pattern in Figure 3.22, you would identify the word as *fork*, based on your knowledge that *forr* is not an English word. Your knowledge of English assists your perception. This is what happens in the connectionist model; higher-level knowledge, coded as weighted connections in the network, participates in the lower-level task of identifying letters

Note that the weights of the connections change with experience. Every experience that you have changes you because of the changes in the connections among the neurons, even if in very subtle ways. For example, a study by Hussain, Sekuler and Bennett (2011)[70] presented people with long series of faces or patterns. The task was to select which of eight options they had seen before. As people progressed through the task, their performance improved – they gained expertise in this kind of perceptual processing. When they had started, their accuracy was around 20–40% correct, but after two days of practice, their accuracy increased to around 60–80%.

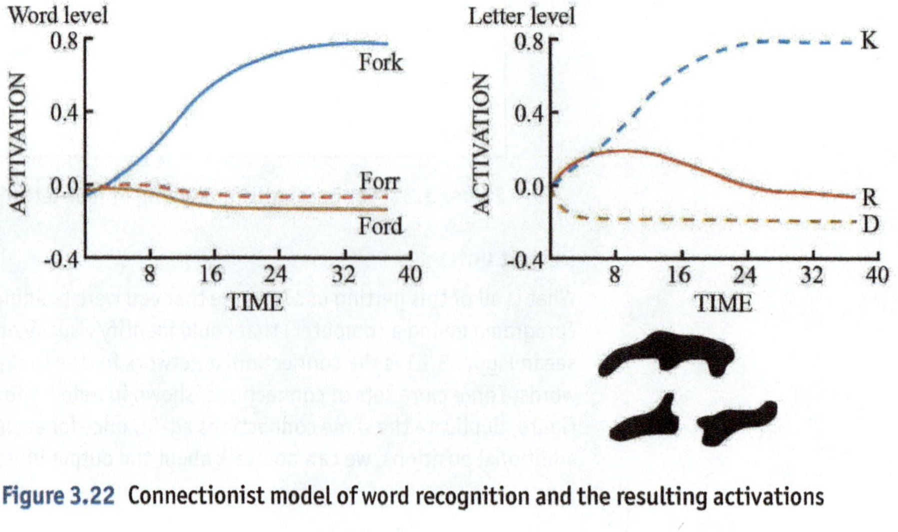

Figure 3.22 Connectionist model of word recognition and the resulting activations

Importantly, this change was not temporary. When the researchers asked the same people to do the task over a year later, their performance only dropped 10–20% for items they had seen before, and a bit more for new items. The point is that they did not have to start back at the beginning. The experience of being in this study changed the connections between their neurons that persisted for over a year (and probably longer).

Such connectionist models satisfy a difficulty you read about earlier: the need for top-down processing. In Figures 3.19–3.21 the top-down effect is prominent in the curved connections, which represent mutual excitation and inhibition. As you will read at several points in this text, connectionist accounts of a whole range of processes can provide new insights into ways of modelling and understanding human cognition. Indeed, connectionist models are finding applications in a stunningly large number of fields (e.g. see Corder, 2004, on a neural net application to landing a crippled airliner[71]).

3.5 Object recognition and agnosia

LEARNING OBJECTIVE
Summarise the science of object recognition and agnosia

How does this approach to identifying letters and words extend to other objects, such as recognising a tree, a briefcase, a human face or a knife hidden in a carry-on bag going through airport security (McCarley, Kramer, Wickens, Vidoni & Boot, 2004; Smith, Redford, Washburn & Taglialatela, 2005[72])? Some of the most significant work on the topic of object recognition involves a process very similar to the feature detection ideas you have been studying.

3.5.1 Marr's computational theory of object recognition

One of the most influential theories of object recognition was provided by David Marr in 1982 using a computational model that set the foundations for future research in the field. He suggested that we recognise objects in stages, starting from the construction of basic 2D entities that determine edges, contours and shades to full 3D shapes. He describes three processing stages:

1. Primal sketch: a rough 2D outline of an object which contains basic information about the main contours, edges and light intensity of the object.
2. $2^1/_2$-D sketch: this adds to the primal sketch information about the depth and orientation of the object including shading, and motion from the perspective of the observer.
3. 3D model: the final stage of object recognition that goes beyond the observer's point of view and allows the full object (with all its details) to be recognised and represented (see Figure 3.23).

Marr's theory has been criticised in that it relies heavily on bottom-up processing and doesn't take into account how previous knowledge can shape the way we recognise objects in our environment (top-down processing).

3.5.2 Recognition by components

The idea in Biederman's (1987, 1990)[73] recognition by components (RBC) theory is that we recognise objects by breaking them down into their parts, and then look up this combination in memory to see which object matches it. Here, pattern recognition has a small number of basic 'primitives', simple three-dimensional geometric forms like those shown in Figure 3.24.

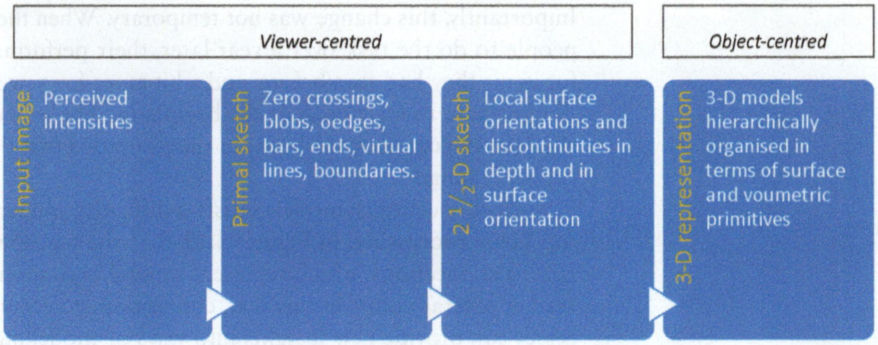

Figure 3.23 Marr's theory of object recognition describes different processing stages

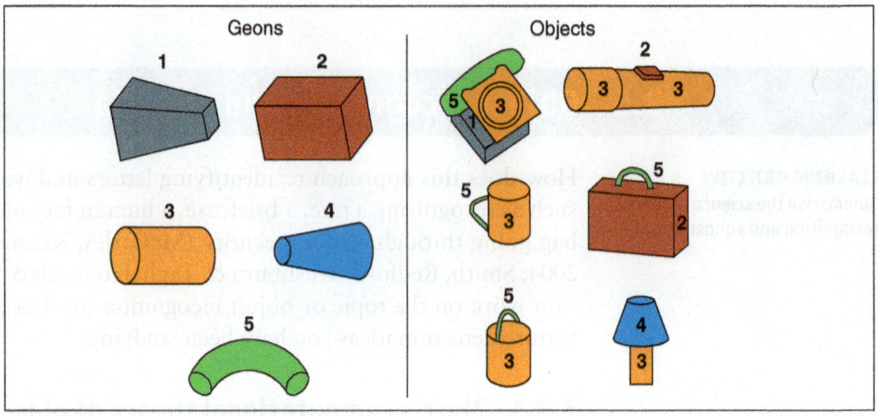

Figure 3.24 Primitives in pattern recognition

These forms, called geons, are a combined form of geometric ions (remember ions from chemistry?). Recognising a briefcase, for example, involves analysing the object into its two geons, the rectangular box (geon 2 in the figure) and the curved cylinder (geon 5). By itself, the rectangular box geon would match the memory for brick or box. But when that component and the curved part on top are detected, the combination matches the memory for briefcase or suitcase.

Biederman (1987)[74] argued that mental representations of three-dimensional objects are composed of geons, much as written language is composed of letters, combined and recombined in different ways. Thus, when we recognise objects, we break them down (*parse* is the technical term) into their components and note where the components join together. We match this pattern to information stored in memory to yield recognition. Two aspects of these patterns are particularly important. First, we find the edges of objects. This enables us to determine which edges maintain the same relationships to one another regardless of viewing orientation. However you look at a brick, the two long edges that are visible remain parallel to one another.

Second, we scan regions of the pattern where the lines intersect (vertices), usually places that form deep concave angles. Look at the deep concave angles on the briefcase in Figure 3.24 where the curved component joins the rectangle. Examining the edges and the areas of intersection enables us to determine which components are present in the pattern: rectangular solid joined on the upper surface by a curved segment. Then we compare this description with stored descriptions in memory, something like 'briefcase: rectangular solid joined on the upper surface by a curved segment'. When we find a match between the identified components and the stored representation, we recognise the pattern.

Evidence for RBC

In his investigations of the RBC model, Biederman (1987[75], 1990[76]) discovered several facts about object recognition. For one, the emphasis on the importance of vertices turns out to be critical.

Figure 3.25 shows several drawings for which people either cannot recover from the deletions or take longer before recognising the objects. Look at these carefully and try to figure out what the objects are. It is difficult because of the deletion of the vertices.

Now look at Figure 3.26.

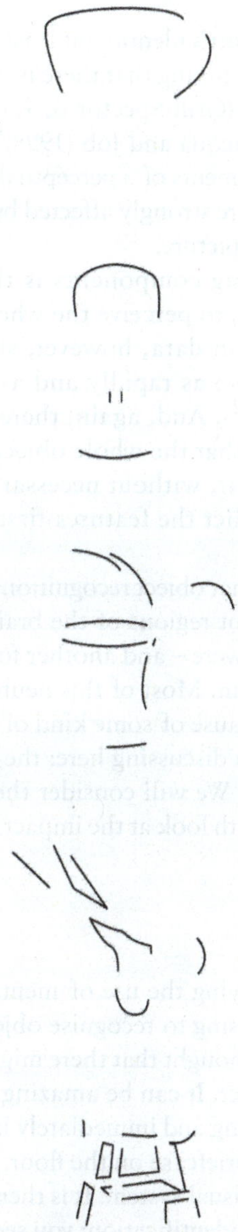

Figure 3.25 Non-recoverable objects

Figure 3.26 Recoverable objects

In Biederman's data (Biederman & Blickle, 1985[77]), people never made more than 30 per cent errors in identifying recoverable patterns, even when 65 per cent of the continuous line contours were deleted and the pattern was shown for only 100 ms. But when the same percentage of the junctions or intersections were deleted as in Figure 3.25, people made errors in the 100-ms condition almost 55 per cent of the time.

Shortcomings of RBC and embodied perception

As useful as it is, RBC is incomplete. First, one difficulty is its ties to bottom-up processing, and object recognition is strongly influenced by context and prior knowledge (e.g. Biederman, Glass & Stacy, 1973; Palmer, 1975[78]). For example, Tanaka and Curran (2001)[79] tested people with some expertise, 'bird experts' and 'dog experts' who had more than 20 years of experience in local bird and dog organisations. These people showed neurological evidence of enhanced early recognition in their areas of expertise, compared to how they recognised objects outside of those areas (e.g. plants).

There is also evidence that retrieval of an object's identity (at least in terms of the category it belongs to) occurs as quickly as identifying that there is even something there – that is, the presentation of a stimulus (Grill-Spector & Kanwisher, 2005; but see Bowers & Jones, 2008[80]). Indeed, Dell'acqua and Job (1998)[81] claimed that object recognition is automatic, given that judgements of a perceptual feature (is this picture elongated horizontally or vertically?) were strongly affected by the top-down knowledge of the identity of the object in that picture.

Second, the model suggests that perceiving components is the first major step in object recognition, thus claiming that, to perceive the whole, it is necessary to first identify the components. Certain data, however, show that people can perceive the overall shape of an object as rapidly and as accurately as the components (e.g. Cave & Kosslyn, 1989[82]). And, again, there are even data from Dell'acqua and Job (1998)[83] indicating that the whole object is recognised automatically using stored knowledge about it, without necessarily identifying components or features. All of these contradict the features-first aspect of the RBC model.

Finally, neuropsychological evidence shows that object recognition is a joint effort between two mental processes and two different regions of the brain, one for features and components – 'bits and pieces', as it were – and another for overall shape and global patterns – the Gestalt or overall form. Most of this neuropsychological evidence comes from studying people who, because of some kind of brain damage, have lost the fundamental ability we have been discussing here: the ability to look at something and rapidly recognise what it is. We will consider the challenges of these people in detail after taking a more in-depth look at the impact of context and embodied perception on perceptual processes.

3.5.3 Agnosia

You have been reading about perception, studying the use of mental mechanisms such as feature detection and top-down processing to recognise objects around us. But we have not questioned that it happens or thought that there might be problems in actually recognising a simple, ordinary object. It can be amazing to learn that a person may lose the ability to glance at something and immediately know what it is. Say there is a cup on the kitchen counter or a briefcase on the floor. We encode the stimulus, the set of features or geons, into the visual system. It is then an automatic, seemingly instantaneous step from encoding to identification: you see the thing, and you immediately know that it is a cup or a briefcase – right?

Wrong. You will now read about a deficit, caused by brain damage, whereby people can no longer do the seemingly instantaneous mental steps of pattern recognition. There are certainly types of brain damage that can disrupt the recognition of printed language, letters and words, which we will discuss in a later module. But for now, we look at a different kind of disruption of recognition, when the recognition of objects – real-world things – is disrupted. This is agnosia, defined as a failure or deficit in recognising objects. This happens either because the person cannot synthesise the pattern of features into a whole or because the person cannot then connect the whole pattern to meaning (from the prefix *a*, meaning 'not,' and the Greek root *gnostic*, meaning 'to know'). Often this is associated with damage to the left occipital and/or temporal lobes (B.A. 37).

When this disruption affects a person's recognition of faces, sometimes while leaving object recognition intact, it is prosopagnosia, a disruption of face recognition. This is often a result of damage to the fusiform gyrus (also B.A. 37), also along the ventral stream. The fact that there are separate conditions involving objects and faces is important because it shows that perception is complex, and that different brain areas emphasise different qualities of information. Perception is not a one-size-fits-all system, but rather a number of speciality systems that typically work in seamless harmony.

Bear in mind that when we talk about agnosia and agnosics (patients with agnosia), we are not talking about people whose basic sensory systems are damaged. In testing an agnosic patient, the person can see and detect visual stimuli; this is not blindness. Instead, it is a cognitive mental loss. The agnosic can input the basic visual stimulus, but then cannot do anything with that encoded information.

When it's difficult to recognise faces

In August 2022, Brad Pitt spoke publicly about his struggle with prosopagnosia, and how this condition makes him come across as self-absorbed or egotistical.

Brad Pitt

Source: Pictorial Press Ltd/ Alamy Stock Photo

A famous case of agnosia – and prosopagnosia, too – is related in the novel *The Man Who Mistook His Wife for a Hat* (Sacks, 1970[84]), about an elderly music professor (called Dr P.) who had lost his ability to recognise objects and faces. At the end of a session with his doctor, he reached over and grasped his wife's head as if reaching to pick up his hat. In another meeting with the doctor, he was able to describe the components or elementary features of an object yet was unable to identify the object he was looking at:

'About six inches in length,' he commented. 'A convoluted red form with a linear green attachment.'

'Yes,' I said encouragingly, 'and what do you think it is, Dr P.?'

'Not easy to say.' . . .

'Smell it,' I suggested.

'Beautiful!' he exclaimed. 'An early rose. What a heavenly smell!'

(Sacks, 1970, pp. 13–14)

Dr P. mistook the grandfather clock in the hall for a person and started to greet it with an outstretched hand. Although he could describe the parts of an object (there were five 'out-pouchings,' and so forth), he could not identify a glove that the doctor held in front of him. Dr P. had serious and pervasive visual agnosia, a profound loss in the ability to visually recognise things.

Although agnosia is not limited to vision – there can be auditory agnosias, for example – an agnosia is modality-specific. That is, a person with visual agnosia has disrupted recognition of objects presented visually but no disruption of hearing, touch or other sensory systems (Dr P. recognised the rose by smelling it).

3.5.4 Implications for cognitive science

What do these neurological disruptions mean for our understanding of normal perception? How does evidence like this advance our understanding of cognition?

Start with the deficit of apperceptive agnosia, where the *a* prefix to *perceptive* denotes some kind of perceptual failure. Here we have a serious disruption in an early stage of perception. It is a disorder of feature detection, a malfunction in the process of extracting features from visual stimuli. Biederman's (1990)[85] geons, for instance, are not being identified or at least not processed much beyond noticing small segments or junction points. Furthermore, it may be important that **apperceptive agnosia** seems to result from damage in the right hemisphere, in the parietal region; growing evidence exists that the right hemisphere is more involved in global processing to include forming global patterns, and that the left hemisphere plays more of a role in local processing (i.e. processing small components and features). If this is so, then it seems reasonable to talk about a disrupted mechanism for forming a Gestalt from the features, where this disrupted mechanism would correspond to the symptoms of apperceptive agnosia.

Associative agnosia is a deeper dysfunction: although the Gestalt or pattern has been formed, it seems to have lost the pathway to the meaning and name of the object. The damaged regions in associative agnosia are lower, more toward the temporal lobe, and in both hemispheres. This pathway, from the vision centres in the occipital lobe forward and down toward the temporal lobe, is the 'what' pathway, which is activated when you look at something to decide what it is. The temporal lobes are particularly associated with language and word meaning. The ability to connect a perceived pattern to its meaning and name is the impairment in associative agnosia.

In conclusion, the varieties of agnosia tell us at least three important things about the perception:

1. Detecting the features in a visual stimulus is a separate (and later) process from sensation. The basic features – whether horizontal lines in a capital A, geons, or something else – must be extracted from the sensory signal.
2. Detecting visual features is critical in constructing a perceived pattern, a percept. If the person cannot extract the features, then he or she cannot 'get' the Gestalt, cannot form an overall pattern or percept.

3. There is a separate step for hooking up the pattern with its meaning and name in memory. This is different from knowing the meaning and name of an object in verbal form. Indeed, given that P. T. only later realised what his pantomime meant, the visual association path can be isolated from all of the other ways of knowing about objects and patterns.

In short, simple, 'immediate' recognition of objects – the cup, the briefcase – is neither simple nor immediate. The disruptions of agnosia, whether caused by difficulties in feature detection or in associating patterns with meaning, provide additional evidence of the complexity of perception.

3.6 Auditory sensation and perception

LEARNING OBJECTIVE
Describe the science of hearing

Auditory stimuli consist of sound waves moving the air. Human hearing responds to these stimuli using an awkward combination of components, a Rube Goldberg-type mechanism that translates the sound waves into a neural message. (Google Rube Goldberg if you do not know about the contraptions he drew.) First, the sound waves funnel into the ear, causing the tympanic membrane, or eardrum, to vibrate. This in turn causes the bones of the middle ear to move, which then sets in motion the fluid in the ear's inner cavity. The moving fluid then moves the tiny hair cells along the basilar membrane, generating the neural message, which is sent along the auditory nerve into the cerebral cortex (e.g. Forgus & Melamed, 1976[86]). Thus, from the unpromising elements of funnels, moving bones and fluid, and the like (Figure 3.27) comes our sense of hearing, or audition.

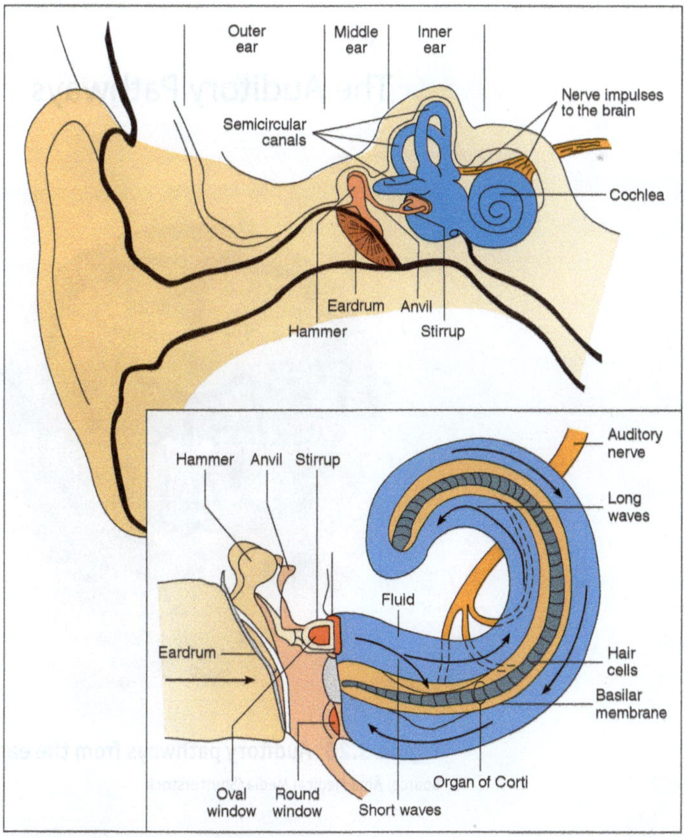

Figure 3.27 The human ear

Interestingly, both ears project auditory information to both cerebral hemispheres, although the majority of the input obeys the principle of contralaterality.

The sensitivity of our hearing defines our auditory world. A pure tone, such as that generated by a tuning fork, is a travelling sound wave with a regular frequency, a smooth pattern of up-and-down cycles per unit of time. Generally, humans are sensitive to patterns as low as 20 cycles per second (cps) and as high as 20,000 cps – although the upper limits decline with age, which is why older adults cannot hear very high pitches. Most of the sound patterns we are interested in, such as those in spoken speech or music, are very complex, combining dozens of different frequencies that vary widely in intensity or loudness. In terms of the sound wave patterns, these different frequencies are superimposed and can be summarised in a spectrum.

In one sense, human hearing is not that impressive: Dogs, for instance, are sensitive to higher frequencies than humans are. Yet in a quite different sense, our hearing is very complex. For instance, we can accurately discriminate between highly similar sounds even from birth: Newborns notice the slight difference between the sounds *pah* and *bah* (Eimas, 1975[87]). Also, we routinely convert the continuous stream of sounds in speech into a meaningful message with little or no effort, at a rate of about two or three words per second. How does audition work? How does it coordinate with our knowledge of language to work so rapidly?

3.6.1 From the ear to the brain

From the ear, specialised cell receptors (hair cells) transform vibrations to electrical impulses that travel along the auditory nerve to the primary auditory cortex, which is conveniently located in the left temporal lobe near language centres (Figure 3.28).

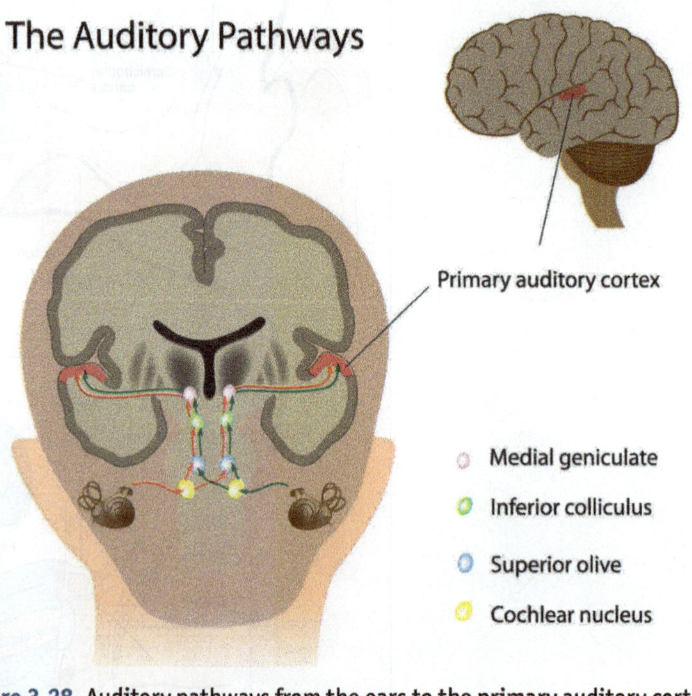

Figure 3.28 Auditory pathways from the ears to the primary auditory cortex
Source: Alila Medical Media/Shutterstock

There are two pathways, one that integrates auditory information with vision (ventral pathway) and one with motor functions (dorsal pathway). The distinction occurs at the auditory nerve which has two branches: the anterior and the posterior. The anterior branch is responsible for sound localisation, and the posterior for complex sound analysis.

3.6.2 Auditory sensory memory

Auditory sensory memory is also called echoic memory (Neisser, 1967[88]). Both terms refer to a brief memory system that receives auditory stimuli and preserves them for some amount of time. Neisser's argument for the existence of echoic memory is still airtight:

> Perhaps the most fundamental fact about hearing is that sound is an intrinsically temporal event. Auditory information is always spread out in time; no single millisecond contains enough information to be very useful. If information were discarded as soon as it arrived, hearing would be all but impossible. Therefore, we must assume that some 'buffer,' some medium for temporary storage, is available in the auditory cognitive system (pp. 199–200).

The function of echoic memory is to encode sensory stimulation into memory and hold it just long enough for the rest of the mental system to gain access to it.

Amount and duration of storage

What is the duration of echoic memory? Darwin, Turvey and Crowder (1972; see also Moray, Bates & Barnett, 1965[89]) answered this question using an auditory analogue of Sperling's work.

They used a task that presents auditory stimuli briefly, in different locations, so that we can cue selected parts for partial report.

The 'three-eared man' procedure

Darwin et al. (1972)[90] used what they called the three-eared man procedure, in which three different spoken messages come from three distinct locations.

Procedure

People heard recorded letters and digits through stereo headphones, with one message played to the left ear, one message to the right ear, and the final message to both ears. The message played into both ears seemed to be localised in the middle of the head, at the 'third ear.' Each of the messages had three stimuli, say, T 7 C on the left ear, 4 B 9 on the right ear, and so on. Each sequence lasted 1 second on the recording, and all three sequences were played simultaneously. Thus, in the space of 1 second, three different sequences of letter and digit combinations were played, for a total of nine separate stimuli.

Result

After the presentation of the messages, people in the whole report condition reported as many of the nine items as they could remember. Their performance was about four items correct, as shown in the above figure.

People in the partial report condition were shown a visual cue, prompting recall of the left, right or centre message. When the cue was given immediately after the stimuli had been heard, performance on the cued location was well above 50 per cent, suggesting that nearly five items out of the original nine were still available.

The advantage of partial report over whole report was maintained even with a 4-second delay in presenting the cue, although performance did decline during that period (also shown in Figure 3.29). The decline in accuracy suggested a decrease in the useful contents of auditory sensory memory, presumably because of a passive loss of information across longer delays.

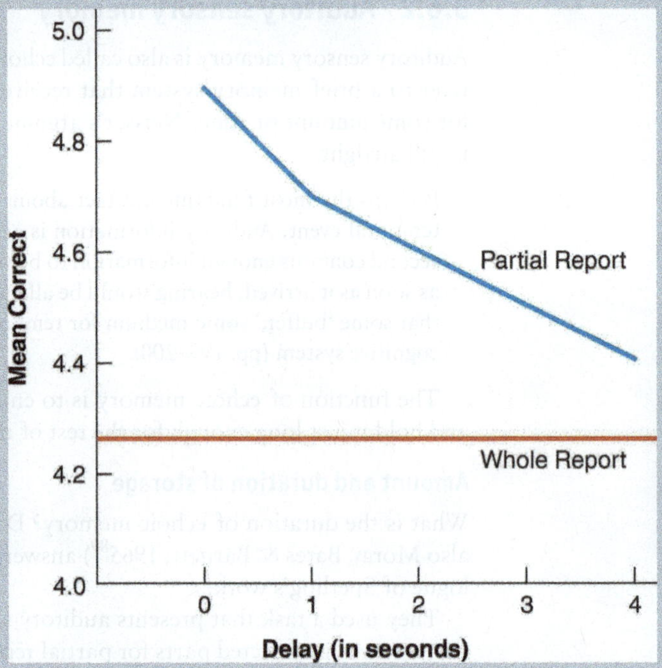

Figure 3.29 Partial report results in the 'three-eared man' procedure
The graph shows the average number of items recalled correctly for the first, second and third items in the three lists, across varying delays in the presentation of the partial report cue. Based on Darwin et al. (1972)[91].

Persistence and interference of auditory information

Without the process of redirected attention, the echoic memory degrades with the passage of time, similar to the loss in iconic memory. However, remember that there is also another kind of forgetting in iconic memory due to interference by subsequent stimuli. Is there any evidence of this in auditory sensory memory? In a word, yes, although a straightforward parallel with iconic persistence and interference may be misleading. We consider the original evidence, and then discuss the controversy over the current status and understanding of auditory sensory memory.

Auditory persistence and interference

The best-known evidence on auditory persistence was done by Crowder and Morton (1969[92]; also Crowder, 1970, 1972[93]).

In their work, a list of nine digits was presented visually, at the fairly rapid rate of two items per second. In the 'silent vocalisation' condition, people saw the nine numbers and read them silently as they appeared. In the 'active vocalisation' condition, people not only saw the list but also named the digits aloud. In the 'passive vocalisation' condition, people heard an accompanying recording that named the digits for them.

For recall, Crowder and Morton found hardly any errors on the last item in the active and passive vocalisation groups. Errors here, as shown in Figure 3.30, were below 10 per cent.

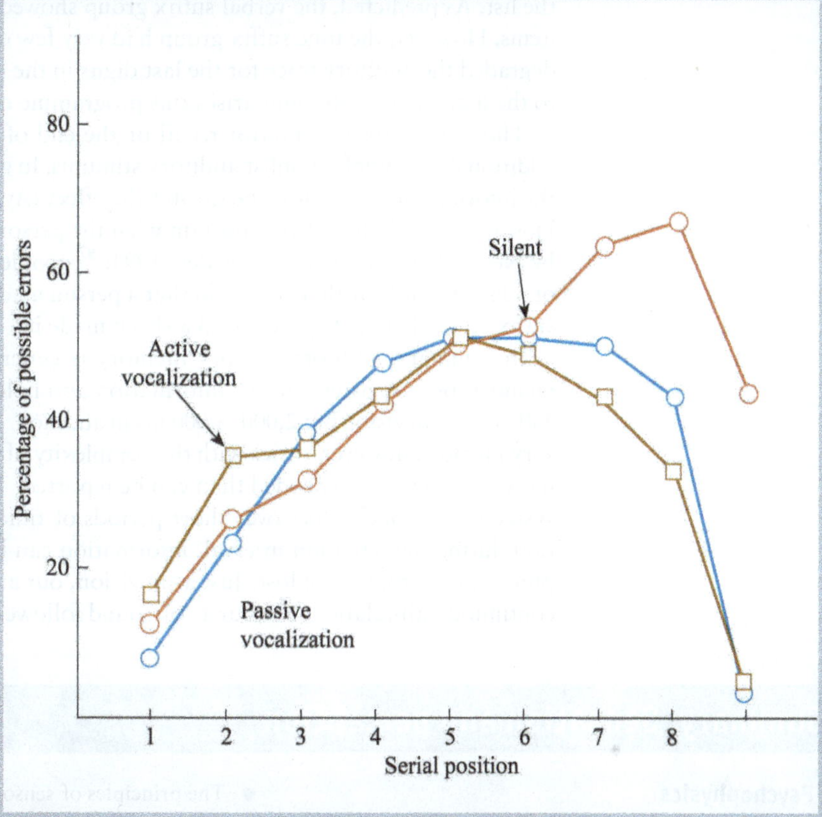

Figure 3.30 Auditory persistence and interference
The number of errors in recall as a function of position in the list of items to be recalled. Note that the active and passive vocalisation groups show almost no errors on the last item, compared with the silent group, which had no auditory trace of the list.

These people had actually heard that last item, so they could simply read it out of echoic memory (in fact, the last three positions showed this advantage). In other words, there was a lingering trace for the last sounds that had been heard. However, the silent group had many errors for the last items, around 50 per cent, because there was no auditory memory trace. (Recall for the earlier positions presumably resulted from a combination of short- and long-term memory – rehearsal of some sort – and so is not of interest here.)

Crowder suggested that still-present traces in echoic memory assisted the vocalisation groups' recall for the last items. This is the **modality effect**, superior recall of the end of the list when heard instead of seen. Crowder (1972)[94] argued that this supports two ideas:

1. The existence of auditory sensory memory.
2. The persistence of auditory traces across a short interval of time.

Interference of auditory information

Having established the persistence of auditory traces, Crowder (1972)[95] went on to investigate auditory interference. After people had heard the items in the list, people in the three suffix groups (auditory, verbal, tone) then heard an additional auditory stimulus: the word *zero* or a simple tone. The groups were told that this final item was merely a cue to begin recalling the list. In reality, the intent of the auditory suffix was to interfere with the lingering auditory trace for the last items in the list. As predicted, the verbal suffix group showed a higher error rate on the last items. However, the tone suffix group had very few errors. The auditory suffix had degraded the auditory trace for the last digits in the list when the suffix was similar to the list. Figure 3.30 summarises this programme of research.

This suffix effect is inferior recall of the end of the list in the presence of an additional meaningful nonlist auditory stimulus. In general, the more a suffix is like the information on the list, the greater the effect (Ayers, Jonides, Reitman, Egan & Howard, 1979[96]). It is also important what the person thinks the suffix is. In a study by Neath, Surprenant and Crowder (1993),[97] people heard a *baa* sound at the end of a list of words. If they were told that a person made the sound, there was a larger suffix effect than if they were told a sheep made it.

In summary, auditory sensory memory is generally similar to visual sensory memory. Both register sensory information and hold it for a brief period of time: 250–500 ms in vision but 2,000–4,000 ms in audition. This duration for auditory sensory memory, however, varies with the complexity of stored information. Generally, more information is encoded than can be reported. The items held in both sensory systems are prone to loss over short periods of time. Finally, by redirecting attention during the retention interval, information can be sent to short-term memory, preventing it from being lost. Just as in vision, our auditory world is usually one of continuous stimulation, not bursts of sound followed by empty intervals.

Summary: Sensation and perception

3.1 Psychophysics

- The mental world depends to a great extent on physical reality. However, the two do not always directly correspond. The realm of psychophysics tries to assess the relationship between our mental experience and the realities of the world we inhabit.
- The most basic process that psychophysics investigates is the sensory threshold, or the minimal amount of something that we can detect. That is, how can we tell whether something is there or not? This is operationalised as the ability to detect the presence of a low-energy stimulus half the time over a large number of trials. Anything above this threshold is said to be supraliminal, and anything below it is said to be subliminal.
- Psychophysics is also concerned with the ability of a person to detect when something has been changed from what it was before. This is operationalised as the just noticeable difference, or JND. In other words, how much does the world have to change before a person notices that it is now no longer the same as it was? Several formulae have been devised to capture this ability.
- The principles of sensory processing that have been developed in the arena of psychophysics can be and have been extended to other domains of cognition, such as making decisions about physical and symbolic, or mental, differences. This also includes the use of signal detection measures, which account for measures of both discrimination and response bias.

3.2 Visual sensation and perception

- The eye sweeps across the visual field in short, jerky movements called saccades, taking in information during brief fixations. The information encoded in these fixations is stored in visual sensory memory for no more than about 250 ms. This iconic image, which may include movement, is lost rapidly or can be interfered with by subsequent visual stimulation. More information is stored in visual sensory memory than can be reported immediately. Information that is reported has been selected by focal attention.
- We do not continuously extract information from the visual scene around us, but instead extract most of the information we need within the first 50 ms of fixation.

Thus, visual sensory memory is a fast-acting and rapidly adapting system, ideally suited for processing information in real time in a continuously dynamic world.

- To build up a complete mental representation of the world, we use trans-saccadic memory. This integration tracks the various entities in the world using object files of what they are doing and how they might be changing. However, this also requires that a person be actively attending to those objects.

3.3 Pattern recognition

- Recognising visual patterns follows principles that have been known for quite some time. The most familiar of these are the Gestalt grouping principles, including figure-ground segregation, closure, proximity, similarity and good continuation.
- Recognition of visual patterns is not a process of matching stored templates to a visual stimulus. Feature detection is a much more convincing account. The features detected are elementary patterns that can be combined to form more complex visual stimuli.

3.4 Top-down processing

- A feature detection account of pattern recognition, such as Pandemonium, must be augmented by conceptually driven processes to account for the effects of context in visual recognition. Current models of this sort include the connectionist approach.

3.5 Object recognition and agnosia

- The recognition by components (RBC) theory claims that we recognise objects by extracting or detecting three-dimensional components, geons, from visual stimuli and then accessing memory to determine what real-world objects contain these components. The most informative parts of objects tend to be where the components join together; people have difficulty recognising objects when these intersections are degraded.
- Studies of patients with visual agnosia demonstrate the complexity of perception. Patients with apperceptive agnosia are sometimes unable to detect even elementary features and therefore have difficulty in perceiving a whole pattern or Gestalt. Those with associative agnosia can perceive the whole but still cannot associate the pattern with stored knowledge.

3.6 Auditory sensation and perception

- Auditory stimuli consist of waves moving from the outer ear to the inner ear, causing the tympanic membrane to vibrate. This in turn causes the movement of the cochlear fluid that sends signals to the auditory cortex which makes sense of this information.

Chapter 3 quiz

Question 1

How do studies of detection differ from studies of discrimination?

A. Detection studies determine what sensory data can be detected, while discrimination studies calculate how much change is required to detect a difference.
B. Detection studies determine what information is outside a person's awareness, while discrimination studies differentiate between types of physical sensations.
C. Detection studies determine how the brain processes physical stimuli, while discrimination studies postulate how the brain translates perception into meaning.
D. Detection studies determine how visual processing operates, while discrimination studies measure the changes in stimulus levels that meet a sensory threshold.

Question 2

When we sense visual information, why does every detail not end with being perceived?

A. The rods and cones can only process a certain amount of light.
B. As information is passed to the different hemispheres of the brain, information is lost.
C. At each step of visual perception (from the eye to the brain), there is some loss of information.
D. Random information is omitted within visual perception.

Question 3

Rajit has been having some perceptual difficulties and has visited the GP due to his concerns. When Rajit sees the letter A, he tastes apples and when he reads the letter S he can taste strawberries. What type of psychological phenomenon is this referring to?

A. Prosopagnosia
B. Synaesthesia
C. Agnosia
D. Apraxia

Question 4

What does the figure-ground principle refer to within Gestalt psychology?

A. A person 'closes up' an image that has gaps or parts missing, perhaps because they are being occluded (blocked) by some other object (golden hair and lips).
B. Elements that are near to one another tend to be grouped together.
C. Elements that are visually similar in some way, such as having a similar colour or texture, tend to be grouped together
D. When viewing an image, part of the image is treated as the figure or foreground (the object identified), which is segregated from the visual information upon which it is set (the background).

Question 5

An individual is driving a car to work. During that time, they are also listening to the radio and singing along. The individual knows when to stop the car for pedestrians and is also aware of when the sat nav changes direction. What type of processing is occurring while the individual is driving?

A. Parallel processing
B. Serial processing
C. Serial processing
D. Feature detection

4 Attention

Learning objectives

4.1 Identify different types of attention

4.2 Explain the processes of attention allocation

4.3 Contrast controlled attention with voluntary attention

4.4 Compare automatic processing with conscious (non-automatic) processing

Key figures

Professor Michael Posner

Professor Michael Posner is one of the most influential researchers in attention. He developed the Posner cueing task that assesses one's ability to shift attention. He has received numerous awards from prestigious bodies such as the APA and the Cognitive Science Society.

Professor Elaine Fox

Professor Elaine Fox has published over 200 studies on attention, cognitive flexibility and emotional resilience and founded the Oxford Centre for Emotions and Affective Neuroscience. She currently works on psychological resilience in young people.

Introduction

Attention – one of cognitive psychology's most important topics and one of our oldest puzzles. What does it mean to pay attention to something? To direct your attention to something? To be unable to pay attention because of boredom, lack of interest or fatigue? What types of stimuli grab or capture our attention? How much control do we have over our attention? (Cognitive science uses 'attend to', meaning 'pay attention', even though some dictionaries cite this is an archaic usage.) We have to work at paying attention to some things (in a lecture, for example). But for other topics, it seems effortless: a good spy novel can rivet your attention.

4.1 Understanding attention

LEARNING OBJECTIVE
Identify different types of attention

According to the *Dictionary of Psychology* of the American Psychological Association, attention is defined as 'a state in which cognitive resources are focused on certain aspects of the environment rather than on others and the central nervous system is in a state of readiness to respond to stimuli'. No matter how straightforward this definition seems to be, attention is one of the thorniest topics in cognitive psychology, possibly because the definition of attention means so many different things. Which cognitive resources are involved? Which aspects of the environment are attended to, and how are other aspects treated? How is the central nervous system involved in attentional processes? This has led some scholars to speculate that the word 'attention' is so broad as to be almost meaningless; that attention is not a 'thing' in itself; and that it would be better to talk of attentional effects instead of attention as a process in and of itself (Anderson, 2011[1]).

Here, we do not take such an extreme position, but use the term *attention* to describe a wide range of phenomena, from the basic idea of arousal and alertness all the way up to consciousness and awareness. Some attention processes are extremely rapid (automatic), so we are aware only of their outcomes, whereas other processes are slow enough that we seem to be aware of them – and able to control them – throughout.

In some cases, attention is reflexive. Even when we deliberately concentrate on something, that concentration can be disrupted and redirected by an unexpected, attention-grabbing event, such as the sudden loud noise in the otherwise quiet library. In other cases, we are frustrated that our deliberate attempts to focus our attention on some task are so easily disrupted by another train of thought or external stimuli. For example, we try very hard to pay attention to a lecture, only to find ourselves daydreaming about last weekend's party.

For organisational purposes, this chapter is structured around three different main types of attention: selective (or focused) attention; divided attention; and lack of attention (or inattention). The final type of attention is nearly synonymous with short-term or working memory, so it is not discussed until later. Although it is possible to organise and present the types of attention in other ways, this approach should help you to develop an understanding of attention and the different attentional processes and modalities and to see how some topics flow into others.

Throughout the chapter we confront four fundamental and interrelated premises about attention:

1. We are constantly presented with more information than we can attend to.
2. There are serious limits in how much we can attend to at once.
3. We can respond to some information and perform some tasks with little, if any, attention.
4. With sufficient practice and knowledge, some tasks become less and less demanding of our attention.

Table 4.1 Different types of attention and examples from everyday life

Types of attention	Everyday example
Selective (or focused) attention	When studying for an important exam your attention is (supposedly) focused on the task at hand (reading).
Divided attention	When driving using a satnav, your attention is divided between finding your way and listening to the satnav's instructions.
Lack of attention (or inattention)	When you read a captivating book and you don't pay attention to the wind blowing outside.

4.1.1 Attention as a mental process

Attention can be thought of as the mental process of concentrating effort on a stimulus or a mental event. By this, we mean that attention is an activity that occurs within cognition.

This activity focuses on a mental resource – effort – either on an external stimulus or on an internal thought. When it refers to an external stimulus, attention is the mental mechanism by which we actively process information in the sensory registers pertaining to that entity. Attention can be conceptualised as an automatic or controlled process. This distinction (also known as dual-process model) of cognitive processes has been the subject of research for many decades (Evans, 2019[2]). Table 4.2 presents the different 'properties' of automatic and controlled mental processes.

Table 4.2 Properties of automatic and controlled mental processes

Automatic	Controlled
Fast	Slow
Parallel	Serial
Implicit	Explicit
Impulsive	Reflective
Effortless	Effortful

4.1.2 Attention as a limited mental resource

Now consider attention as a mental resource, a kind of mental fuel. In this sense, attention is the limited mental energy or resource that powers cognition. It is a mental commodity, the stuff that gets used when we pay attention. According to this metaphor, attention is the all-important mental resource needed to run cognition.

A fundamentally important idea here is that of limitations: attention is limited, finite. We usually state this by talking about the limited capacity of attention. Countless experiments, to say nothing of everyday experiences, reveal the limits of our attention (the capacity to attend to stimuli, to remember events that just happened, to remember things we are supposed to do). In short, there is a limit to how many different things we can attend to and do at once.

It does not take long to think of daily situations that reveal these limitations. You can easily drive on an uncrowded motorway in daylight while carrying on a conversation. You can easily listen to the news on the radio under normal driving conditions. However, in the middle of a heavy rainstorm, you cannot talk to the person sitting in the passenger seat; in rush hour traffic, you cannot (and should not try to) do business on the mobile phone. Under such demanding circumstances, the radio and the conversations are annoyances or irritating – and dangerous – distractions. You must turn down the volume or turn off the phone.

Attention to places and objects

So, what do we pay attention to? Do we attend to locations in space, such as what is happening on the road in front of us instead of in the cow pasture to the side? Or do we attend to objects, such as what that other car is doing as it moves down the road?

Well, the answer is that we do a bit of both. Although these seem to be different kinds of attention, they interact intimately (Leonard, Balestreri & Luck, 2015[3]). Some attention is clearly oriented towards specific locations in space. This will be clear later on when we talk about attention as a spotlight. However, other kinds of attention are clearly object-based.

Figure 4.1 Global or local processing

Not only can attention be directed to either location or objects, but also the breadth of attention to either of these can vary under different circumstances. One attends to an area of space that can be either large or small. For example, we can either attend to a large amount of the video game display, looking for anything moving that might be of importance, or focus on just one part of it to solve a puzzle. Similarly, when attending to objects, we can attend to either the object whole or its parts: the forest or the trees. This is often referred to as the **global–local distinction** (Förster, 2012; Navon, 1977[4]), and an example of it is shown in Figure 4.1.

Not attending

At this point, let's talk about when we are not concentrating our attention. What is our mind doing at that time? Surely, it is not shut off.

Instead, a portion of the brain is operating, processing information in a different way than when attention is being concentrated. This is not a single structure in the brain. Like the network of brain areas that make up the attention network, the portions of the brain that are functioning when a person is not doing anything in particular form a network of allied brain structures called the **default mode network, or DMN** (Buckner, Andrews-Hanna & Schacter, 2008[5]).

Activity in the DMN is negatively correlated with activity in the attention network (Andrews-Hanna, 2012[6]). That is, when one of these brain networks is active, the other is not, and vice versa. The DMN is often active when people are daydreaming, their minds are wandering, or they are thinking about past events in their lives (spontaneous autobiographical remembering). The DMN is also active when people are doing routine activities that do not require intense concentration, such as watching a television show or film (Hasson, Furman, Clark, Dudai & Davachi, 2008; Lerner, Honey, Silbert & Hasson, 2011; Regev, Honey, Simony & Hasson, 2013[7]). This may be the part of the brain we use in our everyday comprehension of stories and events.

4.2 Automatic attentional processes

LEARNING OBJECTIVE
Explain the processes of input attention

Returning to the main topic of attention, we will start with a section on the more basic types of attention listed in Table 4.1, those occurring early in the stream of processing. These processes seem either reflexive or **automatic**, are low-level in terms of informational content, and occur rapidly. They are especially involved in getting sensory information into the cognitive system, so they can generally be called forms of **automatic attention**.

4.2.1 Alertness and vigilance

It almost seems axiomatic to say that part of what we mean by attention involves the basic capacity to respond to the environment. This most basic sense refers to **alertness** as a necessary state of the nervous system: The nervous system must be awake, responsive and able to interact with the environment. We cannot attend to something while we are unconscious. Certain things can impinge on us and rouse us to a conscious state (e.g. alarm clocks, smoke detectors or other loud noises).

Despite the importance of consciousness, there also needs to be some element of alertness. We need to monitor the environment for new, interesting and important events. Sometimes this can be difficult, especially when this alertness must be strung out over a long period during which nothing much happens. The maintenance of attention for infrequent events over long periods of time is known as **vigilance or sustained attention**. The study of vigilance began on British radar operators during World War II (Mackworth, 1948[8]). However, vigilance is important in many other domains, including air traffic control, sonar detection and nuclear power plant operations (Warm, 1984[9]). Even quality inspections in a factory involve some degree of vigilance as workers constantly monitor for important but relatively infrequent flaws in products (Wiener, 1984[10]). The maintenance of attention during vigilance is neurologically complicated, involving a network of brain structures, primarily localised in the right hemisphere of the brain (Langner & Eickhoff, 2013[11]). The need for this sustained coordination of many brain regions may explain why sustained attention can be difficult at times.

Several fundamental vigilance phenomena have been observed over the years (see See, Howe, Warm & Dember, 1995, for a review[12]). For instance, there is a decline in performance as time on the task wears on, showing that people have difficulty maintaining attention on a single task over long periods of time. This decline takes place after about 20 to 35 minutes. Interestingly, the problems that occur with a decline in vigilance do not appear to involve people failing to notice the signal in the task they are doing. Instead, people have difficulty making the decision to respond that they have detected something, a shift in response bias (such as being more or less willing to say that they have seen something). Vigilance is also affected by the neurological and physiological state of a person, such being too hot or cold, the level of arousal, or drug use (Warm, 1984[13]).

Finally, several aspects of the task can influence how effective people are, such as how long the signal is (longer is better), how often there is a signal (more frequent is better), and how busy the background is (less busy is better) (Warm & Jerison, 1984[14]). Performance on vigilance tasks can be improved by giving people rest breaks or the opportunity to do some other task for a brief period before returning to the primary vigilance task (Helton & Russell, 2015[15]). There is also some evidence that meditation training can improve certain aspects of vigilance performance (MacLean et al., 2010[16]).

Explicit and implicit processing

Although nobody disputes that arousal and alertness are a necessary precondition for most cognitive processes, this may overemphasise a kind of thinking known as **explicit processing**. Explicit processes are those involving conscious awareness that a task is being actively done, and usually conscious awareness of the outcome. The opposite is known as **implicit processing**, processing with no necessary involvement of conscious awareness (Schacter, 1989, 1996[17]). As you will see, the distinction between implicit and explicit is often in terms of memory performance, especially long-term memory. When you are asked to learn a list of words and then recite them back, that is a more explicit cognitive task: you are consciously aware of being tested and aware that you are remembering words you just studied. By contrast, you can also demonstrate memory for information without awareness, which is a more implicit cognitive task. For example, you can reread a text more rapidly than you read it the first time, even if you have no recollection of ever having read it before (Masson, 1984[18]). Evidence shows that some mental processing can be done with only minimal attention. Much of this is discussed later. For now, consider a study by Bonebakker et al. (1996; see Andrade, 1995, for a review of learning under anaesthesia[19]) in which they played recorded lists of words to surgery patients. One list was played just before and another during surgery, then a patient's memory was tested up to 24 hours later. Despite the fact that all the patients were given general anaesthesia and were unconscious during the surgery itself, they showed memory for words they had heard. However, they remembered only 6–9 per cent more words compared to a control condition of new words. They certainly did not learn any complex ideas. So, you do need to pay attention to learn well. It is just that certain small amounts of learning can sometimes occur unconsciously.

A powerful part of the study was that performance was based on an implicit memory task, the **word stem completion task**. Patients were given word stems and told to complete them with the first word they thought of. To ensure the task was measuring implicit memory, patients were further asked to exclude any words they explicitly remembered hearing, such as those they remembered hearing before receiving the anaesthesia. For example, say that they heard *BOARD* before surgery and *LIGHT* during surgery. When tested 24 hours after surgery, the patients completed the word stems (e.g. *LI___*) with words they had heard during surgery (*LIGHT*) more frequently than they did with pre-surgery words (*BO___*) or with control words that had never been presented. In other words, they remembered hearing *BOARD* and excluded it on the word stem task. Because they did not explicitly remember *LIGHT*, they finished *LI___* with *GHT*, presumably because their memory of *LIGHT* was implicit. The results demonstrated that the patients had implicit memory of the words they had heard while under the anaesthesia.

Implicit memory versus explicit memory

Look at the list of words below. In the following screens, answer questions associated with this list.

Schlep Schappe Shoal School Shock Scholar Schlep Scheme Schilling Shelf Shuttlecock Shrewd Shipment Shammer

Review your responses below and read the feedback that follows:

1. Recollect the words you just saw and relying on explicit memory complete the following stem: *SCH___*.
2. Now, exclude words you explicitly remember to complete the following stem: *SCH___*.
3. Let's try another example. Complete the following word stems with the first word that comes to your mind: *PAP__*, *GRE__*

Relying on explicit memory, you would have probably completed the stem *SCH___* with *SCHOOL*. But when I asked you to exclude words you explicitly remembered, you would find another way of completing that stem, say *SCHEME*.

In the case of the second example, *PAPER* is a pretty common completion for the first one, probably because paper is a fairly common word (it has not appeared in this module yet). But if you completed the second word stem as *GREEN* without explicitly remembering that you had read about 'green ideas' earlier, then that probably was an implicit memory effect.

4.2.2 Orienting reflex and attention capture

Now consider another kind of attention, the kind caused by a reflexive response in the nervous system.

In a quiet room, an unexpected noise grabs your attention away from what you were doing and may involve a reflexive turning of your head towards the source of the sound. In vision, you move your eyes and head towards the unexpected stimulus, the flash of light or sudden movement in your peripheral vision. This is the **orienting reflex**, the reflexive redirection of attention toward the unexpected stimulus. This response is found at all levels of the animal kingdom and is present very early in life.

Although a host of physiological changes accompany the orienting reflex, including changes in heart rate and respiration (Bridgeman, 1988)[20], we focus on its more mental aspects. The cognitive manifestation of all of this is a redirection of attention towards something, even if the eyes and body do not actually move towards the source. We refer to this process as **attention capture**, which is the spontaneous redirection of attention to stimuli in the world based on physical characteristics.

The orienting reflex is a location-finding response of the nervous system. An unexpected stimulus, a noise or a flash of light, triggers the reflex so that you can locate the stimulus – find where it is in space. This allows you to protect yourself against danger, in the reflexive, survival sense. After all, what if the unexpected movement is from a rock thrown at you or some other threat (e.g. Öhman, Flykt & Esteves, 2001[21])? Note that this system also allows you to monitor for more positive survival stimuli, such as noticing a baby's face (Brosch, Sander, Pourtois & Scherer, 2008[22]). In general, people are more likely to have their attention captured by something important to them in some way (Anderson & Yantis, 2013[23]).

However, people may miss things that are important to them, such as pedestrians or cyclists in a street scene, if several other things capture their attention, such as other cars and trucks. This is especially true if they are near the more vulnerable road users (Sanocki, Islam, Doyon & Lee, 2015[24]). Thus, our attention can be helpful and guide us in many situations, but it is not perfect and may let us down in critical circumstances.

The 'where' pathway projects from the visual cortex to upper (superior) rearward (dorsal) regions of the parietal lobe (and the 'what' pathway is also called the ventral pathway).

Social cues

social cues
Verbal or non-verbal indicators (e.g. body language, tone of voice) that convey information about the intentions of another person and guide conversations.

Attention not only is directed by objects and entities in the environment but also is directed by **social cues**. Perhaps the main cue is noticing where other people are looking (Birmingham, Bischof & Kingstone, 2008[25]; Kingstone, Smilek, Ristic, Friesen & Eastwood, 2003[26]). It has even been suggested that our face and eyes have evolved in such a way to communicate this sort of attention-directing information (Emery, 2000[27]). Even our language can influence how we direct attention. For example, a study by Estes, Verges and Barsalou (2008)[28] showed that attention can be directed based on the meanings of words activated in long-term memory. In this study, people saw a cue word in the middle of the screen, which was soon followed by either an

X or an O at either the top or the bottom of the screen. The task was to indicate which of these two letters was seen by pressing one of two buttons. Researchers found that people were faster to respond to a letter probe if the meaning of the cue word signified a direction consistent with the location of the letter. So, if the cue word was *hat*, people would respond to the letters faster if they were on the top of the screen rather than the bottom. Similarly, if the word was *boot*, the opposite was true (see Figure 4.2).

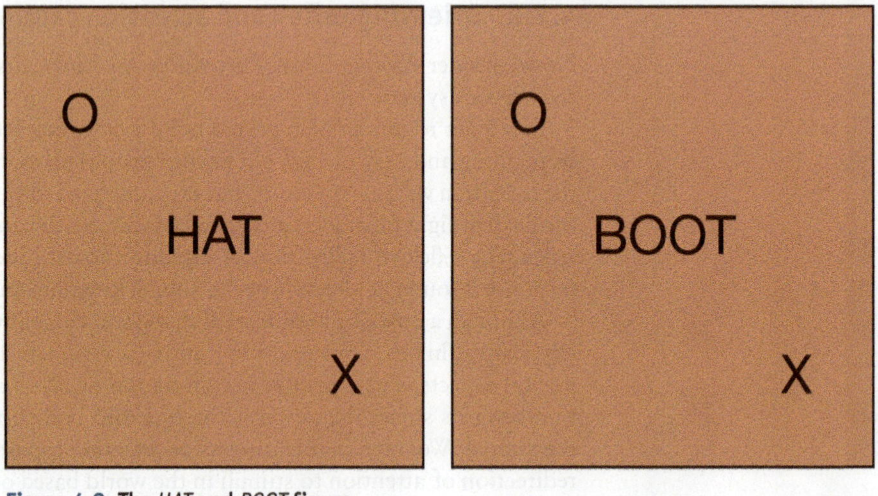

Figure 4.2 The *HAT* and *BOOT* figure

Modulating attention capture

We also orient toward things when something unexpected occurs: the unexpected sound in the quiet library, sudden and startling movement (Abrams & Christ, 2003; Franconeri & Simons, 2003[29]), the abrupt onset of a new object (Davoli, Suszko & Abrams, 2007; Yantis & Jonides, 1984[30]), a change in the colour of an object (Lu & Zhou, 2005[31]), an animate object moving (Pratt, Radulescu, Guo & Abrams, 2010)[32], the change in pitch in a professor's voice during a lecture, or maybe the word *different* in italics in a textbook paragraph. Notice that what seems to capture attention is the occurrence of something *unexpected*, not just something *new* (Vachon, Hughes & Jones, 2012[33]). Thus, not all visual changes capture attention equally. Moreover, a visual offset, when something suddenly disappears from a scene, is much less likely to capture attention than a visual onset, something suddenly appearing (e.g. Cole & Kuhn, 2010[34]).

Orienting focuses us so we can devote deliberate attention to the stimulus if warranted; Cowan (1995)[35] called these *voluntary attentive processes*. In this sense, orienting is a future-oriented preparatory response, one that prepares the system for further voluntary processing. In **visual attention**, fMRI (functional magnetic resonance imaging) neurological scanning has shown that the attention capture process itself seems to involve retinotopic (specific places on the retina in your eye) portions of the occipital lobe, the part of the brain dedicated to vision. This contrasts with more controlled aspects of attention that involve portions of the dorsal parietal and frontal cortex, farther down the stream of neural brain activity (Serences et al., 2005[36]; Yantis, 2008[37]).

However, if the stimulus that triggers an orienting reflex occurs repeatedly, it is no longer novel or different; it has become part of the normal, unchanging background. The process of **habituation** begins to take over, a gradual reduction of the orienting response back to baseline. For example, if the unexpected noise in the quiet library is the ventilation fan coming on, you first notice it but then grow accustomed to it. You have oriented to it, and then that response habituates to the point that you will probably orient again when the fan stops running. When the constant noise stops, that is a change that triggers the orienting response.

4.2.3 Visual search

The last sense of attention to be considered among the input attentional processes is a kind of visual attention. It is related to perceptual space – the spatial arrangement of stimuli in your visual field and the way you search that space. It is different from the orienting response in that there is no necessary movement of the eyes or head, although there is a strong correlation with eye movements (researchers often exploit this relationship to have a general idea of where attention is directed, using eye-tracking devices). Instead, there is a mental shift of attentional focus, as if a spotlight beam were focused on a region of visual space, enabling you to pick up information in that space more easily (think of a 'Superman beam').

Numerous studies that have been done on this kind of visual attention include work that has found regions of the brain that seem to be involved in focused, visual attention.

Posner's spatial cuing task

Consider Figure 4.3, which depicts three kinds of displays in Posner's spatial cuing task (Posner, Nissen & Ogden, 1978; Posner, Snyder & Davidson, 1980[38]).

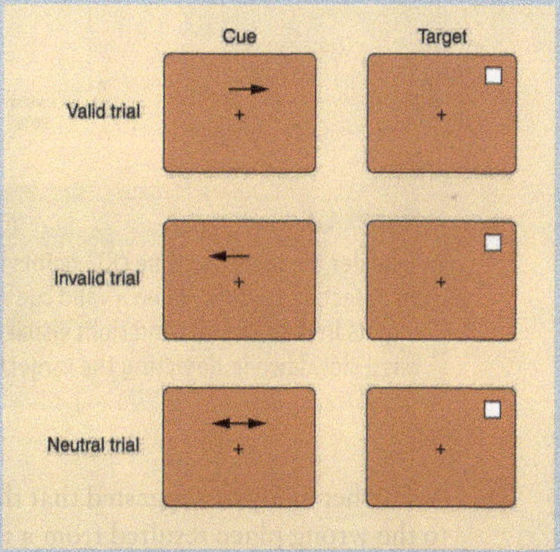

Figure 4.3 Posner's spatial cuing task

The person fixates on the plus sign in the centre of the screen, then sees an arrow pointing left or right or a two-headed arrow. For the targets shown in the figure, with a target appearing on the right, the right-pointing arrow is a valid cue, the left-pointing arrow an invalid cue, and the two-headed arrow a neutral cue. In this experiment, one-headed arrow cues were valid on 80 per cent of the trials.

People in this task are first asked to fixate the centred plus sign on the display, are then shown a directional cue, and finally see a simple target (the thing they are supposed to respond to). For Posner, the directional cue was an arrow. However, attention can be directed in many other different ways, such as by eye gaze or language (e.g. 'left'), although these have more minor influences on attention (Gibson & Sztybel, 2014[39]). The task had people press a button when they detected the target.

For 80 per cent of the cued trials, the arrow correctly pointed to the direction where the target actually did appear 1 second later. In the remaining 20 per cent of the cued trials, the cue was invalid: tt pointed to the wrong side. Neutral trials provided an uninformative cue, a two-headed arrow indicating that the target would appear equally often on the left or right. Throughout the task, people were required to maintain fixation on the plus sign. They could shift only their mental attention to the space where they thought the target might appear, but were not allowed to move their eyes.

The results, shown in Figure 4.4, were very clear. When people shifted their visual attention to the correct area (the valid 80% point in the figure), response time (RT) to detect the target was significantly faster than the neutral, uncued condition. This speed-up is known as a **benefit** or facilitation, a faster-than-baseline response resulting from useful advance information. Yet when the target appeared in the unexpected location, there was a significant **cost**: a response slower than baseline because of a misleading cue.

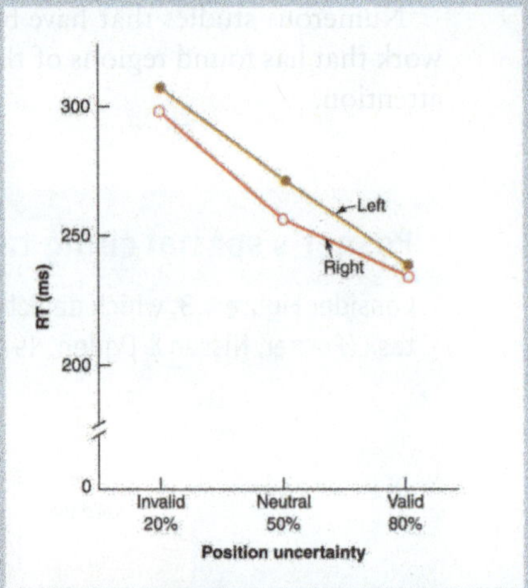

Figure 4.4

Consider the response time (RT) points in the neutral condition to be baseline performance on detecting targets. When a valid cue was presented, there was a reduction in RT for targets in both the left and right visual fields ('valid 80%'). When the cue was invalid, there was a slowdown in detecting the target in both visual fields ('invalid 20%').

Further analysis suggested that the cost of people having directed their attention to the wrong place resulted from a three-part process:

1. Disengaging attention from its current focus.
2. Moving the attentional spotlight to the target's true location.
3. Engaging attention at that new location.

Posner et al. (1980)[40] concluded from this and other related experiments that the attentional focus being switched was a cognitive phenomenon. It was not tied to eye movements but to an internal, mental mechanism. They suggested that attention is like a spotlight that highlights the objects and events it shines on. Thus, **spotlight attention** is the mental attention-focusing mechanism that prepares you to encode stimulus information. Furthermore, Posner et al. (1980) suggested that this shift in attention is essentially the same as the redirection of attention in the orienting reflex, with one big difference: it is voluntary. Therefore, it can happen before a stimulus occurs and can be triggered by cognitive factors such as expectations.

Spotlight metaphor in visual attention

spotlight metaphor
Attention is thought to resemble a spotlight that is directed to a specific location and preferentially attended.

As Cave and Bichot (1999)[41] pointed out, countless studies of visual attention, many of them inspired by Posner's work, have adopted the **spotlight metaphor**. Much of that work has explored the characteristics and limits of visual attention, attempting to evaluate the usefulness of the metaphor. The evidence suggests that the mental spotlight does not sweep, enhancing the intermediate locations along the way. Instead, it jumps (much as the saccade does). Conversely, some evidence supports the similarity between a real spotlight and spotlight attention. For example, the size of the spotlight beam can be altered, depending on circumstances (see Cave & Bichot, 1999, for an extensive review[42]).

Visual search

In the typical result (Treisman & Gelade, 1980, Experiment 1[43]), people could search rapidly for an item identified by the presence of a unique feature. It made little or no difference whether they searched through a small or a large display. For instance, people were able to search through as few as 5 items or as many as 30 in about the same amount of time, approximately 500 ms. The target object just seemed to pop out of the display. Thus, this is called a **pop-out effect**. Because there was no increase in RT across the display sizes, Treisman and Gelade concluded that visual search for a dimension such as shape or colour occurs in parallel across the entire region of visual attention. Such a search must be largely automatic and must represent very early visual processing. In the results shown in Figure 4.5, this is the flat, low function of the graph. (See Finlayson & Grove, 2015[44], for an extension of visual search research into three dimensions, whereby things like distance from the person becomes a factor.)

Visual search: demonstration and analysis

A series of studies by Treisman and her associates (Treisman, 1982, 1988, 1991; Treisman & Gelade, 1980[45]) examined **visual search**. Typically, people were told to search a visual display for either of two simple features (e.g. a letter S or a blue letter) or a conjunction of two features (e.g. a green *T*). The search for a simple feature was called a feature search: people responded 'yes' when they detected the presence of either of the specified features, either a letter *S* or a blue letter. In the conjunction search condition, they had to search for the combination of two features: *T* and the colour green.

The task

Look at the following figures and do the quick demonstrations.

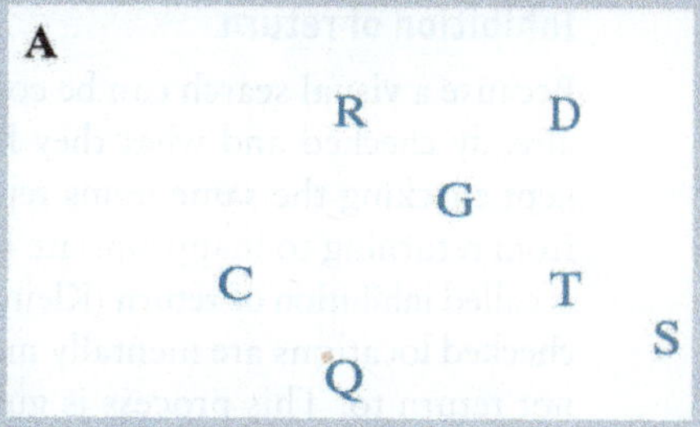

Search for either a letter *T* or a boldfaced letter.

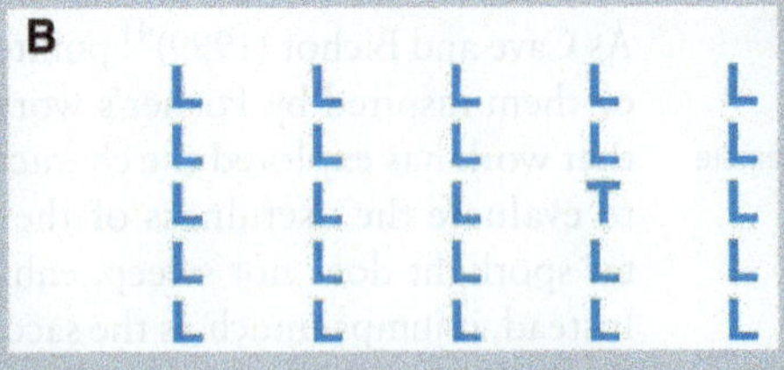

Search for a boldfaced *T*.

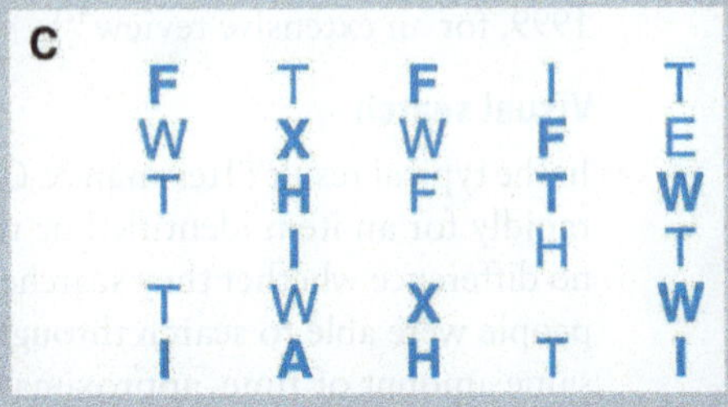

Search for a boldfaced *T*.

Analysis

As you did these visual searches, you surely noticed that searching for *T* in the first figure was stunningly simple. It hardly seemed like a search, did it? Instead, didn't the *T* just 'pop out' at you? In contrast, searching for *T* in the middle figure probably was a slow process, and finding it in the last figure probably took even longer.

A classic, everyday example of difficult visual search can be found in the *Where's Waldo?* children's books in which a person needs to search for Waldo among images of many other objects that share many different kinds of features with Waldo, making it challenging to find him.

In the searches you did, the first were feature searches, and the last figure illustrated a conjunction search (the target had to be both boldfaced and a *T*).

But when people had to do a conjunction search, such as a green *T*, they took more time, up to 2,400 ms, as more and more distractors filled the display (distractors for both conditions were brown *T*s and green *X*s). Such a conjunction search seems a more serial, one-by-one process and a far more conscious, deliberate act. This is the steeply increasing function in Figure 4.5.

Inhibition of return

Because a visual search can be complex, people need a way to track what they have already checked and what they have not. A big problem would occur if a person kept checking the same items repeatedly without checking others. To help people from returning to inappropriate locations, there is a special attention process. This is called **inhibition of return** (Klein, 2000; Posner & Cohen, 1984[46]), in which recently checked locations are mentally marked by attention as places that the search would not return to. This process is guided by the operations of the superior colliculus (which is part of the midbrain structure known as the tectum) and the parietal lobe (Klein, 2000; Vivas, Humphreys & Fuentes, 2006[47]). This is consistent with the idea that inhibition of return is an important visual process (involving the superior colliculus) as well as knowledge of where things are in space, the 'where' neural pathway (involving the parietal lobe). When people are not searching for something, but are simply scanning or memorising a picture, the opposite is shown, with people being more likely to return to a previously fixated location – a *facilitation of return* (Dodd, Van der Stigchel & Hollingworth, 2009[48]). Thus, attention operates in different ways depending on what a person's goals are at the time.

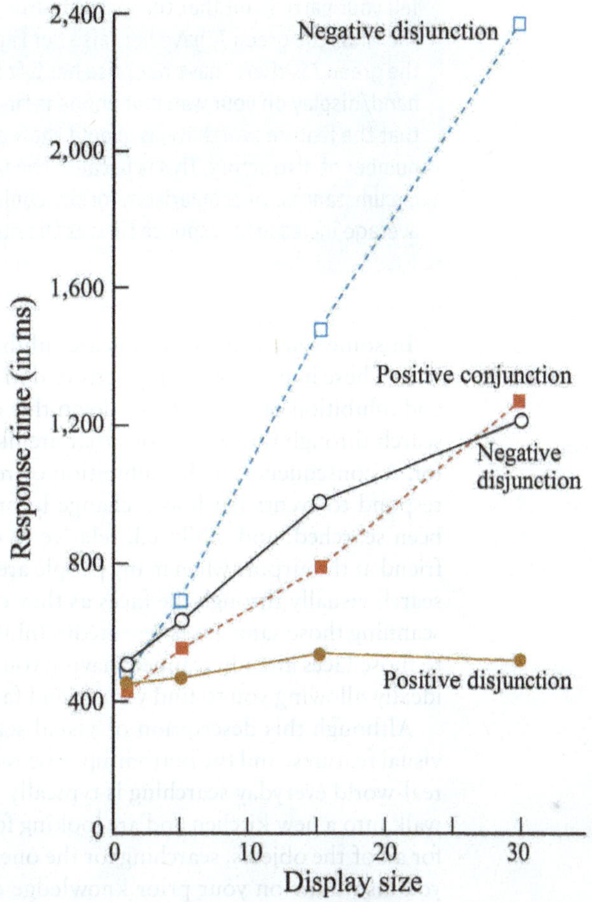

Figure 4.5 Search times when targets were of a specified colour or shape

The dashed lines are for the disjunction search conditions (e.g. search for either a capital *T* or a boldfaced letter). The solid lines show search times for the conjunction condition (e.g. search for a boldfaced *T*). The important result is that disjunction search times did not increase as the display size grew larger, but the conjunction search times did.

The visual search task

Here we have a simple task to do with a friend and a stopwatch (perhaps as an app on your phone) that illustrates the difference between feature and conjunction searches in visual attention. Using two different colours of marker pens, make up a few sheets of paper, or some 4 : 6 index cards, on which you draw letters in two distinct colours such as red and green. For simplicity's sake, let's restrict ourselves to *X*s and *T*s. Have your participant search for a *green T*.

For the *feature* search trials, you will have several red *X*s and *T*s all over the paper. On the 'yes' sheets, you will put a green *T* in one spot on the paper; on the 'no' sheets, you will just have red *X*s and *T*s. Make a separate sheet, one for 'yes' trials and one for 'no' trials, with 4 *X*s and *T*s, then do the same for 6 *X*s and *T*s, then 8, 10 and 12 (do not forget to go back and put the green *T* in for your 'yes' trials). For the *conjunction* search trials, in addition to the green *T* for the 'yes' sheets, you will put green *X*s and red *T*s on the sheets. As before, you will also have 'no' sheets that only have green *X*s and red *T*s.

Tell your participant that the task is to find the green *T* as quickly as possible. When she finds the green *T*, have her raise her right hand. However, if she does not think the green *T* is there, have her raise her left hand. Time the participant (the second hand/display on your watch or phone is fine) each time. The standard result is that the feature search items should show a constant search rate regardless of the number of distractors. This is because the target letter should pop out under these circumstances. In comparison, for the conjunction search items, there should be an average increase in response time as the number of distractors increases.

In some sense, the locations are inhibited from or kept out of the search pattern. These items were highly activated in cognition because they were attended to, and inhibition of return turns down this activation level. So, you only continue to search through those locations that are likely to still have the item you are looking for. A consequence of this inhibition of return process is that people are slower to respond to events (such as a change in brightness) in locations that have recently been searched, and inhibited, relative to other locations. It is like searching for a friend at the airport when many people are arriving from many different flights. You search visually through the faces as they come out of security, but you do not keep scanning those same faces repeatedly. Inhibition of return keeps you from returning to those faces already scanned, having your visual search move on to other faces and ideally allowing you to find your friend faster.

Although this description of visual search heavily emphasises the influence of visual features, and the bottom-up processes that can use such features, most of our real-world everyday searching is typically not this difficult. For example, when you walk into a new kitchen and are looking for the sink, you do not just scan the room for all of the objects, searching for the one with the most sink-like features. Instead, you also draw on your prior knowledge of how things are typically organised in space (Huang & Grossberg, 2010[49]). The sink is more likely to be on an outside wall, perhaps by a window, than somewhere else. Thus, knowledge about the object exerts a top-down influence on the tracking system involved in the visual search.

Also, attention can be influenced by embodied characteristics. For example, people find it easier to spot a target during visual search when their hands are placed near the display as compared to when they are farther away (Davoli, Brockmole & Goujon, 2012[50]). The idea is that hand location corresponds to locations in space that can be easily manipulated, so we are more likely to direct our attention to those locations. Directing attention in this way facilitates even tasks that do not require manipulating the environment, such as visual search. Furthermore, it is important not only *where* your hands are but also *what* they are holding. Biggs, Brockmole and Witt (2013)[51] found that if people are holding a weapon, such as a gun, they are more likely to direct attention to people than objects. Thus, a person's action capabilities can influence where he or she directs attention.

4.2.4 Contrasting input and controlled attention

Treisman's two conditions provide clear evidence of both a very quick, automatic attentional process – essentially the capture of attention due to 'pop-out' – and a much slower, more deliberate attention, the type used for the conjunction search.

In line with Johnston, McCann and Remington's (1995)[52] suggestion, we use the term *input attention* for the fast, automatic process of attention. The slower one, in Johnston et al.'s terms, is **controlled attention**.

The geometric shapes in the figure below refer to different regions of the brain that are involved in attention.

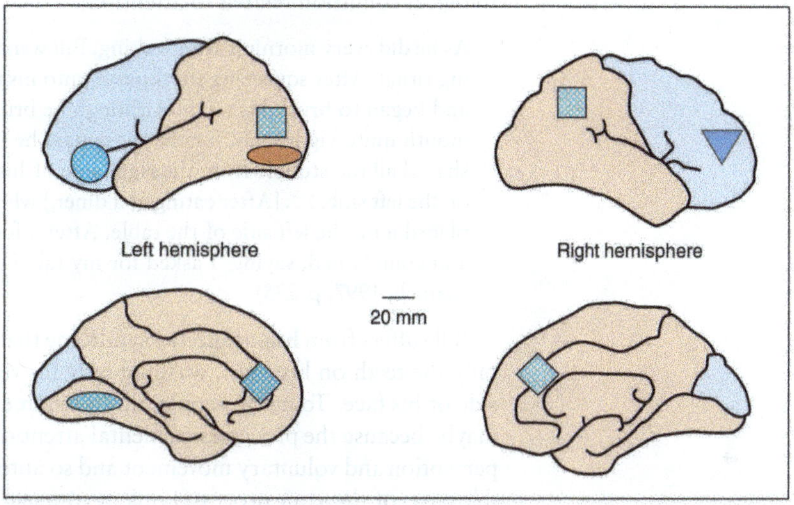

Figure 4.6 Lateral and medial views of the left and right hemispheres of the brain

4.2.5 Video games as mechanisms for improving attention

Much of what we have learned from research in cognitive psychology tells about the human mind's limits and capabilities. It would be ideal to use this scientific knowledge to find ways to improve how we think. Some attempts at improving attention have focused on the use of action video games (such as first-person shooter games). These are of interest because they require a person to use attention in atypical ways, yet place demands on *how* to use that attention. Moreover, some people find them engaging and fun. So, can playing certain kinds of video games actually improve cognition?

Some work on this topic shows improved attention processing in people who regularly play video games compared with people who do not (e.g. Chisholm & Kingstone, 2015[53]). Other studies suggest that people who do not typically play such video games can exhibit improved attention processing after they spend some time playing such games. For example, people who play action video games may have a larger attentional spotlight (Feng, Spence & Pratt, 2007[54]) or make their eye movements (and possibly move attention) faster (Heimler, Pavani, Donk & van Zoest, 2014[55]). Outside of video game playing, evidence exists that training in the visual arts can influence visual attention, such as whether focus is on a local or global level of processing (Chamberlain & Wagemans, 2015[56]).

With that said, there is also evidence that playing such video games does not have a strong and lasting influence on attention (e.g. Gobet et al., 2015[57]). Part of the argument against studies that have found differences between players and non-players is their methodological shortcomings, such as using a limited number of cognitive tasks to compare players and non-players (Boot, Blakely & Simons, 2011; Latham, Patston & Tippett, 2013[58]).

4.2.6 Hemineglect

In many cases, cognitive science has gained insight into a process when there has been some disruption to the system. The study of attention is no exception. For example, under the influence of alcohol, the operation of attention is compromised. Specifically, people attend to a narrow range of information after consuming alcohol (Harvey, Kneller & Campbell, 2013[59]).

Here is a quotation from Banich (1997)[60] about Bill, who has an interesting neurological condition related to attention:

> As he did every morning after waking, Bill went into the bathroom to begin his morning ritual. After squeezing toothpaste onto his toothbrush, he looked into the mirror and began to brush his teeth. Although he brushed the teeth on the right side of his mouth quite vigorously, for the most part he ignored those on the left side. . . . He shaved all the stubble from the right side of his face impeccably but did a spotty job on the left side. . . . [After eating at a diner,] when Bill asked for the check, the waitress placed it on the left side of the table. After a few minutes, he waved the waitress over and complained, saying 'I asked for my tab 5 minutes ago. What is taking so long?' (Banich, 1997, p. 235)

Bill suffers from **hemineglect**, a syndrome that leads to behaviour such as brushing only the teeth on his right, washing only his right arm, and shaving only the right side of his face. To many people, this phenomenon is almost too bizarre to believe, maybe because the processes of mental attention have always been so closely tied to perception and voluntary movement and so automatic that we think they are indivisible parts of the same process. Look at yourself in a mirror, then look at the left side of your face. No problem: you merely move your eyes, shift your direction of gaze, and look at it. If I ask you to stare straight ahead and then attend to something in your left field of vision, say the letter *X* on a computer screen, your normal response is to shift your eyes toward the left and focus on the target. You simply look at the *X* and pay attention to it. You can even shift your mental attention to the left without moving your eyes.

The syndrome known as *hemineglect* (or hemi-inattention) is a disruption in the ability to refocus your attention to one side of your face or the other, say to the X on the left of the computer screen. It is a disruption or decreased ability to attend to something in the (often) left field of vision. *Hemi* means 'half,' and *neglect* mean 'to ignore' or 'to fail to perceive'. Thus, hemineglect is a disorder of attention in which one half of the perceptual world is neglected to some degree and cannot be attended to as completely or accurately as normal. Some form of hemineglect is often observed in stroke victims, even if it is in a more limited and temporary form. Very often, the neglect is of the left visual field, for stimuli to the left of the current fixation, the current focus of attention. And because of the principle of contralaterality, it is not surprising that the brain damage leading to hemineglect is often in the right hemisphere, in particular, certain regions of the right parietal lobe (see Intriligator & Cavanagh, 2001, for evidence that localises selective attention in the parietal lobe[61]).

Here are the facts (see Banich, 1997, or Rafal, 1997, for complete treatments[62]): a patient with hemineglect cannot voluntarily direct attention to half of the perceptual world, whether the to-be-perceived stimulus is visual, auditory or any other type of sensation. In some cases, the neglect is nearly total, as if half of the perceptual world has simply vanished, is simply not there in any normal sense of the word. In other cases, the neglect is partial, so for such people it is more accurate to say that they are less able to redirect their attention than are normal people. Either way, there is a disruption in the ability to control attention. Note that this is not a case of sensory damage like blindness or deafness. The patient with hemineglect receives input from both sides of the body and can make voluntary muscle movements on both sides. And in careful testing situations, such patients can also respond to stimuli in the neglected field. But, somehow, the deliberate devotion of controlled attention to one side is deficient.

Disruptions in patients with hemineglect

Bisiach and Luzzatti (1978)[63] present a compelling description of hemineglect. The afflicted individuals were from Milan, Italy, which they were quite familiar with before their brain damage. This study focused on the main piazza in town, a broad, open square with buildings and shops along the sides and a large cathedral at one end. These patients were asked to imagine themselves standing at one end of the piazza, facing the cathedral, and to describe what they could see. They uniformly described only the buildings and shops on the right side of the piazza. When asked to imagine themselves standing on the steps of the cathedral, facing back the opposite way, they once again described what was on their right side. From this second view, they described exactly what they had omitted from their earlier descriptions. Likewise, they then omitted what they had described earlier.

Critically important here is the observation that these reports, based on memory, were exactly the kind of reports patients with hemineglect give when actually viewing a scene. If these patients had been taken to the piazza, they probably would have seen and described it the same way as they did from memory. (For a similar account, see 'Eyes Right!' in Sacks, 1970; the patient there eats the right half of everything on her dinner plate, then complains about not getting enough food.) Figure 4.7 shows some drawings made by a patient with hemineglect.

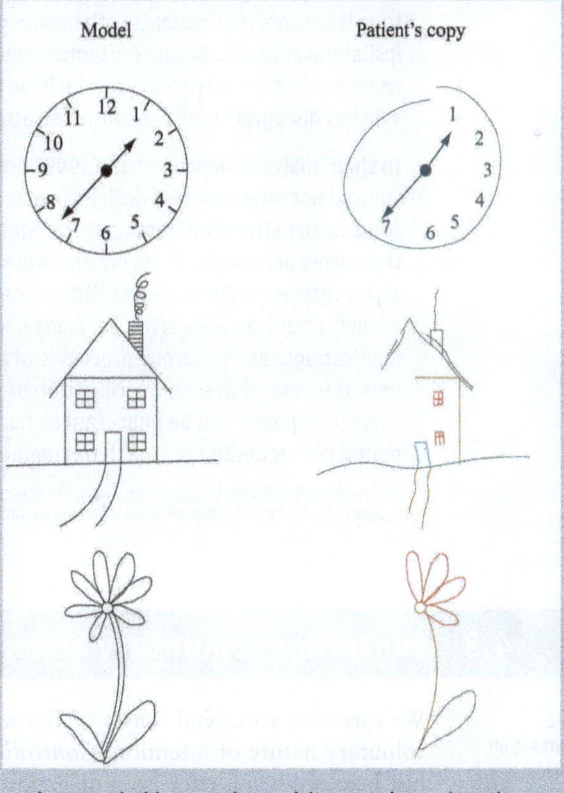

Figure 4.7 Drawings copied by a patient with contralateral neglect

Here, patients were asked to copy drawings or to draw from memory, but the nature of their drawings was no different in either case. These drawings show a dramatic neglect for the left-hand sides of objects: a flower with no petals on the left, a house with the left half missing, a clock face with no numbers on the left. In the standard line bisection task ('draw a slash through the middle of a horizontal line'), patients with hemineglect position the slash too far to the right, as if bisecting only the right half of the entire horizontal line.

Duncan et al. (1999)[64] provided a careful analysis of the disruptions seen in patients with hemineglect in the context of the 'theory of visual attention' (Bundesen, 1990[65]). Duncan et al. noted that several important advances in understanding hemineglect have been made, especially when the patients are tested with some standardised cognitive tasks such as Posner's spatial cueing task, which you read about earlier. For example, it turns out that patients with hemineglect often can attend to stimuli in the neglected field, but only if nothing else is displayed that might attract their attention. This tendency to ignore the contralateral field when a competing stimulus is presented in the ipsilateral field is called *extinction*. It appears to be caused by something like attention capture. When a right-side (ipsilateral) stimulus is presented, it captures the person's attention and prevents attention from being devoted to the left (contralateral) side. In a sense, patients with hemineglect may neglect one side only because there is usually something on the other side that captures their attention.

Hemineglect seems to disrupt both input attention and controlled attention. First, input attention is devoted largely or exclusively to a stimulus in the 'good' or preserved field, the ipsilateral field (the term ipsilesional is also used, meaning 'same side as the brain lesion'). The stimulus in this field captures the patient's input attention. But then it appears that patients with hemineglect cannot disengage attention from that ipsilateral stimulus. Because attention towards the right cannot be disengaged, they cannot shift their attention voluntarily to the left. Thus, capture of attention on one side has disrupted a shift of controlled attention towards the other side.

In their analysis, Duncan et al. (1999) noted that their patients with hemineglect showed not only standard deficits in attention to the contralateral side but also some rather strong bilateral deficits related to attentional capacity. In other words, there were accuracy deficits on the neglected side but capacity difficulties on both sides. Interestingly, there was little evidence that the conceptually driven aspects of their attention were affected. It may be some time before such results and their implications for the normal processes of attention are fully understood. But even now, it is clear that such fractionation of performance – some abilities preserved, some disrupted – will be important in our further understanding of attention (for a neural net modelling approach to hemineglect, see Monaghan & Shillcock, 2004)[66].

Source: From F. E. Bloom & A. Lazerson. *Brain, Mind, and Behavior*, 2nd ed. New York: W. H. Freeman and Co., p. 300.

4.3 Controlled, voluntary attention

LEARNING OBJECTIVE
Contrast controlled attention with voluntary attention

We turn now to several senses of the term *attention* that point to the controlled, voluntary nature of attention. *Controlled attention*, in contrast to what you have just been studying, refers to a deliberate, voluntary allocation of mental effort or concentration. You decide to pay attention to this stimulus and ignore others. Paying attention this way involves effort.

Cognitive psychology has always been intrigued by the fact that at any moment, scores of different sensory messages are impinging on us. We cannot attend to all of them (we would be overwhelmed instantly), nor can we afford for our attention to be captured by one, then another, then another of the multiple sensory inputs (we would lose all coherence, all continuity). Therefore, it makes sense to ask questions about **selective attention**, the ability to attend to one source of information while ignoring other ongoing stimuli around us. How do we do this? How do we screen out the surrounding noises to focus on just one? How can we listen covertly to the person on our right, who is gossiping about someone we know, while overtly pretending to listen to a conversational partner on our left? (And how did we notice that the person on our right was gossiping?) Somewhat the converse of selective attention is the topic of divided attention: how do we divide or share our attentional capacity across more than one source of information at a time, and how much information are we picking up from the several sources?

An example of how these questions involve real-world problems is the issue of whether we really can talk (or text message) on a mobile phone and drive at the same time, dividing our attention between two demanding tasks (Kunar, Carter, Cohen & Horowitz, 2008; Spence & Read, 2003; Strayer & Johnston, 2001[67]). Driving itself can already be taxing on attention with the need to monitor multiple vehicles and road conditions at one time (Lochner & Trick, 2014[68]). Can attention be divided between driving and a mobile phone conversation? In short, the general answer is, 'No, it really can't.' Talking on the mobile phone can lead to *inattention blindness*, in which people fail to attend to or process information about traffic, even if they are looking directly at it. This is equally true for both handheld and hands-free mobile phone conversations, but not for listening to the radio or music (Strayer & Drews, 2007[69]). This is because we are actively involved in mobile phone conversations, but not in what is going on over the radio.

The invisible gorilla

Simons and Chabris (1999)[70] demonstrated the effects of inattentional blindness in the famous 'invisible gorilla' experiment. In their study, participants were asked to observe short videos of different teams of players who were passing and dribbling a basketball, and to count how many passes the players made (e.g. from Player 1 to Player 2, from Player 2 to Player 3, and so on). Each video lasted approximately 1.5 minutes and, after 48 seconds of the video, an unexpected event occurred which lasted 5 seconds: either a woman with an open umbrella or a woman dressed in a gorilla suit walked through the scene while players were passing the ball and dribbling. After watching the videos, participants were requested to write down the counts of passes they observed and were also asked the question, 'Did you see a woman dressed as a gorilla (or a woman with an open umbrella) walk through the scene?' Across all the conditions, almost half of the participants (46 per cent) did not notice the unexpected event, demonstrating a power inattentional blindness effect.[71] You can find out more about the 'invisible gorilla' experiment by visiting this website The Invisible Gorilla: How Our Intuitions Deceive Us[72] and navigate to the video.

In these studies, people drive a simulator while having their eye movements monitored (so the experimenter knows what they are looking at and for how long). Under these circumstances, people who are having phone conversations are less likely to recognise road signs or other important traffic events, even when they look directly at them. This is even revealed in electroencephalographic (EEG) recordings of drivers' brains, not just what they consciously report. When you are actively involved and interacting in a conversation, your limited-capacity attention is drawn away from your immediate environment. You have less attention to devote to driving; consequently, your driving suffers and becomes more dangerous. Moreover, people are often unaware of how impaired their driving becomes when they are talking on a mobile phone (Sanbonmatsu, Strayer, Biondi, Behrends & Moore, 2016[73]). In an interesting twist, it has even been discovered that people find listening to mobile phone conversations more distracting to attention than listening to conversations in which they hear both sides (Galván, Vessal & Golley, 2013[74]). At a more general level, the question is, when do we start reaching the limits of our attentional capacity?

4.3.1 Selective attention and the cocktail party effect

When there are many stimuli or events around you, you may try to focus on just one. The ones you are trying to ignore are distractions that must be excluded. The mental process of eliminating those distractions is called **filtering or selecting**. Some aspect of attention seems to filter out unwanted, extraneous sources of information so we can select the one source we want to attend to.

The process of selective attention seems straightforward in vision: you move your eyes, thereby selecting what you attend to. However, attention is separate from eye movements: you can shift your attention even without eye movements. But in hearing, attention has no outward, behavioural component analogous to eye movements, so cognitive psychology has always realised that selective attention in hearing is thoroughly cognitive. This accounts for the heavy investment in filter theories of auditory perception. If we cannot avoid hearing something, we then must select among the stimuli by some mental process, filtering out the unimportant and attending to the important.

Dual-task or dual-message methods

A general characteristic of many attention experiments involves the procedure of *overload*. We can overload a sensory system by presenting more information than it can handle at once and then test accuracy for some part of the information. This has usually involved a **dual-task method**. Two tasks are presented such that one task captures attention as completely as possible. Because attentional resources are consumed by the primary task, few, if any, resources are left over for attention to the other tasks. By varying the characteristics or content of the messages, we can make the listener's job easier or harder. For instance, paying attention to a message spoken in one ear while trying to ignore the other ear's message is more difficult when both messages are spoken by the same person.

Going a step further, when we examine performance to the attended task, we can ask about the accuracy with which a message is perceived and about the degree of interference caused by a second message. We can also look at accuracy for information that was not in the primary message, the unattended message in the other ear. If there is any evidence of remembering the unattended message, or even some of its features, we can discuss how unattended information is processed and registered in memory.

Shadowing experiments

Some of the earliest cognitive research on auditory selective attention was done by E. Colin Cherry (1953; Cherry & Taylor, 1954[75]). Cherry was interested in speech recognition and attention. Cherry characterised his research procedures, and the question he was asking, as the **cocktail party effect**: how do we pay attention to and recognise what one person is saying when we are surrounded by other spoken messages?

4.3.2 Selection models

It appears that a physical difference between the messages permits people to distinguish between them and eases the job of selectively attending to the target task (Johnston & Heinz, 1978[76]). Investigators routinely call this **early selection**. This refers to some of the earliest phases of perception, an acoustic analysis based on physical features of the message. The evidence is that people can select a message based on sensory information, such as loudness, location of the sound source, or pitch (Egan, Carterette & Thwing, 1954; Spieth, Curtis & Webster, 1954; Wood & Cowan, 1995a[77]). These attentional processes are controlled by specific brain regions, such as the frontal lobes, that have unique EEG signals (e.g. Cavanagh & Frank, 2014[78]).

Early selection theory

This evidence, indicating that people could somehow tune their attention to one message over the other, prompted Donald Broadbent (1958)[79] to propose an early selection theory of attention. In this view, attention acts as a selective filter, as shown in the following example.

Broadbent's early selection theory of attention

Attention can be made selective by focusing in on physical features of a stimulus, as is the case with early filter theories of attention.

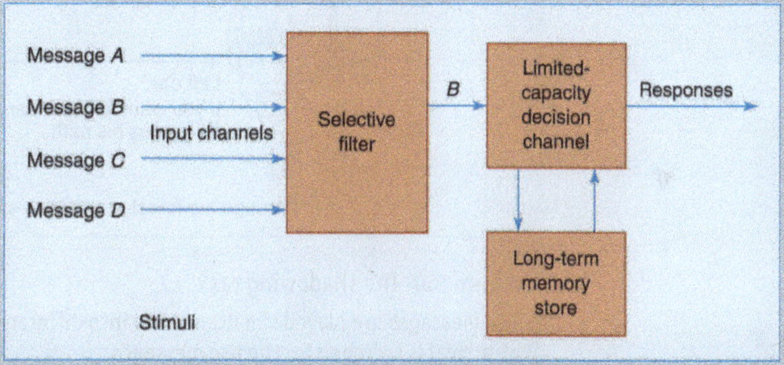

Figure 4.8 Broadbent's filter theory of selective attention

Regardless of how many competing channels or messages are coming in, the filter can be tuned, or switched, to any one of them, based on characteristics such as loudness or pitch. Note that only one message can pass through the filter at a time. In other words, despite the many incoming signals, only one message can be sent through the filter into the 'limited-capacity decision channel' (essentially short-term memory). Only the information on the attended, 'passed along' message affects performance, in Broadbent's view, because only it gets past the filtering mechanism.

It was soon realised that the filter idea had serious problems. For one, intuition tells us that we often notice information from a message we are not attending, as when we hear our name in a crowded, noisy place. Moray (1959)[80] found evidence for this:

Although people did not recall a word presented 35 times to the unattended ear, a third of the people heard their own name (see Wood & Cowan, 1995b[81], for a recent replication of this effect). Not everyone detects his or her name easily. Conway, Cowan, and Bunting (2001)[82] found that people with less cognitive capacity (a.k.a. working memory) were more likely to detect their name. In other words, people who were less able to focus on a task, such as remembering a list of letters, appear to be more prone to distraction and are more likely to process and detect information that they are supposed to ignore, such as information on an unattended channel in this task.

Generally, these findings have implications regarding the nature of attention. If Broadbent's early filter theory were correct, then only the attended and passed-along information should be available where physical cues direct attention. Yet there is clear evidence that unattended information can somehow slip past the filter (see Lachter, Forster & Ruthruff, 2004, who argue that some small amount of attention had been devoted to the 'unattended' stimuli[83]).

Treisman's late selection theory of attention

To show the power of late selection, Treisman did a study now considered a classic (1960).[84]

Right ear:
While Bill was walking through the forest/a bank can lend you the money.
Left ear:
If you want to buy a car/a tree fell across his path.
"through the forest, a tree fell—uh . . . ."

Figure 4.9 The shadowing task

Two messages are played simultaneously into different ears; then, at the slash, the ear-of-arrival is switched for the two messages.

Treisman arranged the recording so that the coherent message being shadowed was unexpectedly shifted to the unattended channel. That is, the sentence that was being said switched from the right to the left ear. Despite the high degree of practice and concentration needed, people routinely switched to the unattended message, although they did not continue to shadow the 'wrong' ear for long. Clearly, there must be some processing of the unattended message. Semantic elements of the unattended channel must be receiving some analysis or there would be no basis for preferring it when the sentences switched ears.

Based on such results, Treisman claimed that all incoming messages receive some amount of low-level analysis. When the unattended messages yield no useful or important information, they are attenuated. They are reduced, not in their volume or physical characteristics, but in their informational importance to ongoing processing. In the process of shadowing, we arrive at an identification of the words and phrases on the attended message.

Treisman (1965) felt that it was during this process of semantic analysis that we make our selection among messages, selection at a 'late' stage. This permits attention to be affected by the semantic aspects of the message – a top-down effect. A more extreme view, by Deutsch and Deutsch (1963), claimed that selection takes place only after all messages have received full acoustic and semantic analysis (i.e., just before the response stage).

So evidence indicates that quite a bit of information is getting into cognition: the meaning of the words on the unattended channel, for example, in Treisman's 1960 study. (See also Lewis, 1970. Carr, McCauley, Sperber & Parmalee, 1982, found comparable results for visual stimuli.[85])

Intrusion of the word tree into the shadow, as shown in Figure 4.9, makes sense only if tree has been recognised as related to the forest theme of the shadowed message. This effect implies some rapid process of accessing the meanings of words. More recent work has shown that information that is not actively attended, and subject to inattention blindness, is processed if it is consistent with our goals and intentions (Koivisto & Revonsuo, 2007; Marsh, Cook, Meeks, Clark-Foos & Hicks, 2007; Most, Scholl, Clifford, & Simons, 2005[86]). In all the cases in which unattended information was processed, it was consistent with some enduring or temporary goal a person had, such as hearing his or her own name, completing the idea conveyed by a sentence, or whatever may be satisfying a person's goals at the time. That is why when you are hungry, images of food are so hard to ignore, even if you are trying to concentrate on something else.

Similarly, even something that is intrinsically not very meaningful, such as a series of tones, can capture and redirect attention if it becomes meaningful, as with a cell phone alert (Stothart, Mitchum & Yehnert, 2015[87]).

So, it seems that people use attention by means of an early filter based on physical qualities, such as spatial location or loudness. On the other hand, it appears that people use attention by means of a later filter that relies on semantic meaning. Which of these is correct? There is evidence for both. Some researchers, such as Lavie (2010),[88] have suggested that an early attention filter is more likely to be used when there is a lot of perceptual information that needs to be sorted through, as in most shadowing situations. This environment leads attention to select information based on physical features and qualities, if possible. However, a later filter is more likely to be used when attention is not being overly taxed, leading it to select information based on meaning.

Source: Adapted from Lindsay & Norman (1977).

Late selection theory

Treisman (1960, 1964[89]) did a series of studies to explore this slippage more closely. She used the standard shadowing task but varied the nature of the unattended message across a subtler range of differences. She first replicated Cherry's findings that selective attention was easy when physical differences existed. Then she turned to the situation in which physical differences were absent – both messages were recorded by the same speaker. Because the same pitch, intonation, stress, and so on were in both messages, early selection should not be possible. Yet she found that people could shadow accurately. The basis for the selection was *message content*, what the message was about rather than what it sounded like. In this situation, the grammatical and semantic features are the basis for selection (*semantic* refers to meaning).

Because attentional selection occurs after all the initial processing of the message is done, this is called late selection. It is later in the stream of processing than early selection based on sensory features, yet before the moment of having to respond aloud with the shadowed speech.

Inhibition and negative priming

Most of the discussion of attention to this point has focused on what gets attended to and activated in cognition, as well as some filter to keep irrelevant information from entering the stream of processing. At this point, we would like to discuss a proposed cognitive mechanism that goes beyond the idea of a filter keeping out irrelevant information and only allowing selected information to be processed further. This is the cognitive attention mechanism of inhibition. Inhibition is thought to actively suppress mental representations of salient but irrelevant information so the information's activation level is reduced, perhaps below the resting baseline level. You already encountered an example of this when we discussed inhibition of return. Here, we look further at how inhibition may be operating to help people select relevant information and filter out irrelevant information.

For inhibition to operate, there needs to be a salient source of interfering and irrelevant information – the irrelevant information needs to be strong and wrong. Under such circumstances, inhibition will be involved. A study by Tipper (1985)[90] provided a classic demonstration of this.

In this study, people were presented with a series of pairs of line drawings of objects, with one object presented in green and the other in red. The task was to name the red object as quickly as possible. The important condition here involved target trials on which the red object had appeared in green on the previous trial (called the *prime* trial). It was observed that people were slower to respond to the target trials (red object) when the trials had been preceded by the to-be-ignored distractor primes (same object in green) compared to control trials where the ignored object on the prime trial was some other object. This response time slowdown is called negative priming Neill, 1977; Tipper, 1985; for a review, see Frings, Schneider & Fox, 2015[91]).

This is because when people are looking at the display, in addition to their processing of the red object, there is also some activation and processing of the green object because people are looking directly at it. The mental representation of the object becomes activated. However, because the identity of this object is irrelevant to the task – it was green, so it did not need to be named – attention actively inhibited and suppressed the object's representation. Then, when the person needed this information on the next trial (because it was now the red object), it took longer to activate and use because it had been inhibited. So the inhibition process slowed down the person's response time. However, see Mayr and Buchner, 2006, and Neill, Valdes, and Terry, 1995, for alternative accounts of negative priming that do not involve an active inhibitory mechanism.[92]

The idea that inhibition is an important part of attention has been extended to many other areas of psychology, particularly those dealing with individual differences. For example, in developmental psychology, attentional inhibition is thought to develop slowly with age (Diamond & Gilbert, 1989), meaning that it is difficult, especially for young children, to maintain focus. In contrast, in older adults there is increased difficulty suppressing irrelevant information (Hasher & Zacks, 1988[93]). Inhibitory problems are also thought be present in schizophrenia (Beech, Powell, McWilliams & Claridge, 1989[94]), with schizophrenics having trouble keeping unwanted thoughts out of consciousness. In less extreme cases, people who are depressed also have trouble inhibiting irrelevant information (MacQueen, Tipper, Young, Joffe & Levitt, 2000[95]), leading to trouble focusing on the task at hand.

4.4 Attention as a mental resource

LEARNING OBJECTIVE
Compare automatic processing with conscious (non-automatic) processing

An important and far-reaching meaning of the term *attention* – this one may be closer to our everyday meaning – treats attention as mental effort, as a mental resource that fuels cognitive activity. If we selectively attend to one particular message, we are deliberately focusing mental energy on that message, concentrating on it to the exclusion of other messages. This sense involves the idea that attention is a limited resource, that there is only so much mental fuel to be devoted here or there at any one time. Kahneman (1973)[96] also suggested that capacity might be elastic, in that increasing the task load might also increase a person's arousal, thus making additional resources available. Approaches that emphasise this meaning of attention are called resource theories.

A corollary to this idea of limited capacity is that attention, loosely speaking, is the same as consciousness or awareness. After all, if you can be consciously aware of only one thing at a time, doesn't that illustrate the limited capacity of attention? Even on a smaller scale, when we process very simple stimuli, there is evidence of this limit to attention. If you are asked to respond to a stimulus and then immediately to a second one, your second response is delayed a bit. This is the **psychological refractory period** or **attentional blink**, which is a brief slowdown in mental processing due to having processed another very recent event (e.g. Barnard, Scott, Taylor, May & Knightley, 2004; Pashler & Johnson, 1998[97]). The implication is that allocating attention to a first stimulus momentarily deprives you of the attention needed for a second stimulus. However, this blink can be overcome if the target item that occurs during the period of the blink is particularly important, such as the future thinking involved in predicting what to do on the next trial (Livesey, Harris & Harris, 2009[98]). The blink may also be overcome if people need to make eye movements (saccades) between the two stimuli. This possibly occurs because the eye movement creates an event boundary so that the two stimuli are processed as part of different events. In such a situation, the blink may not occur (Kamienkowski, Navajas & Sigman, 2012[99]).

In addition, the strain on attention can be reduced when people view nature scenes (which only subtly demand attention) as compared to urban scenes (which have many elements that grab our attention), thereby freeing up attention for other tasks (Berman, Jonides & Kaplan, 2008; Joye, Pals, Steg & Evans, 2013[100]).

A related idea is that this kind of attention is deliberate, wilful, intended – *controlled attention. You* decide to pay attention to a signal, or you decide *not* to attend to it. *You* decide to pay attention to the lecture instead of your memory of last night's date, and when you realise your attention has wandered, *you* redirect it to the lecture, determined *not* to daydream about last night until class is over.

An interesting insight that William James had about attention is the idea that we may do more than one thing at a time if the other processes are habitual. Yet when processes are less automatic, then attention must oscillate among them if they are done simultaneously, with no consequent gain of time. The key point is the idea of automatic processes – that some mental events can happen without draining the pool of attentional resources. Putting it simply, the germ of James's idea, **automaticity**, has become central to cognitive psychology's views on attention, pattern recognition, and a host of other topics. Cognitive science has devoted a huge effort to recasting James's ideas about automaticity and attention into more formal, quantifiable concepts.

4.4.1 Automatic and controlled processing

In place of the former approach, the limited-capacity attentional mechanism and the need for filtering in selective attention, the current view is that a variety of cognitive processes can be done automatically, with little or no conscious involvement necessary. Two such theories of automaticity have been proposed, one by Posner and Snyder (1975)[101] and one by Shiffrin and Schneider (1977;[102] Schneider & Shiffrin, 1977[103]). These theories differ in some of their details but are similar in their overall message (see also Logan & Etherton, 1994; for discussions that oppose the idea of mental resources, see Navon, 1984, and Pashler, 1994[104]).

Automatic and controlled processes of attention vary along a continuum

What differentiates automatic from controlled processing? One way of assessing this is the ease with which a process can be interrupted or stopped. If it continues even when a person wishes to be doing something else, it is more automatic.

Automatic processing

Posner and Snyder described four diagnostic criteria for an automatic process. First, an automatic process occurs without intention. In other words, you cannot prevent it from happening, and once it does start, you cannot stop it. A compelling example of this is the Stroop effect (named after the task described in Stroop, 1935)[105]. Words such as *RED GREEN BLUE YELLOW* were presented visually, written in mismatching colours of ink (e.g. RED printed in green ink). When people have to name the ink colour, they must ignore the printed words themselves. This leads to tremendous interference, a slowing of the ink colour naming, caused by the mismatching information and the contradictory impulses to name the word and the ink colour (this is an extremely easy demonstration to do). This is another case in which inhibition is operating during attention, as the highly salient, but irrelevant, semantic meaning of the word must be suppressed. Note that this requires that a person can automatically read. People who are illiterate would not show a Stroop effect. That said, it is also the case that poor readers can show larger Stroop effects than good readers (Protopapas, Archonti & Skaloumbakas, 2007[106]), perhaps because, under certain circumstances, better readers have greater executive control over their attentional resources. In Posner and Snyder's terms, accessing the meaning of the written symbol RED is automatic: it requires no intention; it happens regardless of whether you want it to. In the research that demonstrates automatic access to word meaning, the term we use is priming. A word activates or primes its meaning in memory and, thus, primes or activates meanings closely associated with it. This priming makes related meanings easier to access: because of priming, they are boosted up, or given an extra advantage or head start (just as well water is pumped more easily when you prime the pump). See also Dunbar and MacLeod, 1984, and MacLeod, 1991, for an explanation of Stroop interference based on priming.[107]

Second, an automatic process does not reveal itself to conscious awareness. You cannot describe the mental processes of looking up the word *RED* in memory.

Third, an automatic process consumes few, if any, resources. Such a process should not interfere with other tasks, certainly not those that rely on conscious resources. As an example, walking is so automatic for adults that it does not interfere with other processes; we can walk and talk at the same time.

Fourth, automatic processes tend to be fast. As a rule, a response taking no more than one second is heavily automatic. (For evidence of slow automatic processing in a person with brain damage, see Wingfield, Goodglass & Lindfield, 1997.[108])

Controlled processing

Let's contrast these diagnostic criteria for automaticity with those for conscious processing or controlled processing. Controlled processes occur only with intention. They are optional and can be deliberately performed or not. Conscious processes are open to awareness. We know they are going on, and, within limits, we know what they consist of. Of greatest importance, conscious processes use attention. They consume some of the limited attentional resources we have. A demanding controlled process should leave few resources available for a second task that also uses controlled processing. Driving during a hard rainstorm consumes too many resources for you to listen simultaneously to the news on the radio. Alternatively, you may stop walking if you are thinking about something that requires intense thought. Of course, if the second task can be done fairly automatically, then both tasks may proceed without interference. For example, you can easily walk and carry on a casual conversation at the same time.

Integration with conceptually driven processes

We can go one step further, integrating this explanation into the idea of conceptually driven processing. Think back to the shadowing research you read about.

Attending to one of two incoming messages and shadowing that message aloud demands controlled attention. Such a process is under direct control. The person is aware of doing the process, and it consumes most of a person's available mental resources.

Presumably, no other conscious process can be done simultaneously with the shadowing task without affecting performance in one or the other task (or both). When the messages are acoustically similar, then people must use differences of content to keep them separate. But by tracking the meaning of a passage, the person's conceptually driven processes come into play in an obvious way. Just as people 'restore' the missing sound in 'the *eel was on the axle' (Warren & Warren, 1970[109]), the person in the shadowing task 'supplies' information about the message from long-term memory. Once you have begun to understand the content of the shadowed message, then your conceptually driven processes assist you by narrowing down the possible alternatives and suggesting what might come next.

Saying that conceptually driven processes 'suggest what might come next' is an informal way of referring to priming. You shadow, 'While Bill was walking through the forest . . . ' Your semantic analysis primes related information and thereby suggests the likely content of the next clause in the sentence. It is likely to be about trees, and it is unlikely to be about banks and cars. At this instant, your 'forest' knowledge is primed or activated in memory. It is ready (indeed, almost eager) to be perceived because it is so likely to be contained in the rest of the sentence. Then *tree* occurs on the unattended channel. Because we access the meanings of words in an automatic fashion, the extra boost given to *tree* by the priming process pushes it over into the conscious attention mechanism. Suddenly, you are saying 'a tree fell across' rather than sticking with the right-ear message. Automatic priming of long-term memory has exerted a top-down influence on the earliest of your cognitive processes, auditory pattern recognition and attention.

The role of practice and memory

If accessing word meaning is automatic, then you might wonder about some of the shadowing research described earlier in which people failed to detect a word presented 35 times, the reversed speech, and so on. If word meaning access is automatic, why didn't these people recognise the words on the unattended channel? A plausible explanation is practice. It seems likely that the inability to be influenced by the unattended message was caused by a lack of practice on the shadowing task.

4.4.2 The role of practice in automaticity

Attention, in its usual everyday sense, is equivalent to conscious mental capacity. We can devote attention to only one demanding task at a time or to two somewhat less demanding tasks simultaneously, so long as they do not exceed the total capacity available. This devotion of resources means that few, if any, additional resources are available for other demanding tasks.

Alternatively, if a second task is performed largely at the automatic level, then it can occur simultaneously with the first because it does not draw from the conscious

The role of practice in automaticity.

Source: Associated Press/Lennox McLenden/Alamy Stock Photo

resource pool (or, to change the metaphor, the automatic process has achieved a high level of skill; see Hirst & Kalmar, 1987[110]). The more automatically a task can be done, the more resources are available for other tasks.

The route to automaticity is practice and memory. With repetition and overlearning comes the ability to perform automatically what formerly needed conscious processing. A particularly dramatic illustration of the power of practice was done by Spelke, Hirst and Neisser (1976).[111] With extensive practice, two people could read stories at normal rates and with high comprehension, while they simultaneously copied words at dictation or even categorised the dictated words according to meaning. Significantly, once practice has yielded automatic performance, it seems especially difficult to undo the practice, to overcome what has now become an automatic and, in a sense, autonomous process (Zbrodoff & Logan, 1986[112]).

4.4.3 Disadvantages of automaticity

We have been talking as if automaticity is always a positive, desirable characteristic; as if anything that reduces the drain on the limited available mental capacity is a good thing. This is not entirely true.

Sometimes we *should* be consciously aware of information or processes that have become too routine and automatic. Barshi and Healy (1993)[113] provided an excellent example, using a proofreading procedure that mimics how we use checklists. People in their study scanned pages of simple multiplication problems. Five mistakes such as '7 : 8 = 63' were embedded in the pages of problems. People saw the same sets of 10 problems over and over. But in the fixed-order condition, the problems were in the same order each time. In the varied-order condition, the problems were in a different order each time. Those tested in the fixed-order condition missed more of the embedded mistakes than those in the varied-order condition; an average of 23 per cent missed in fixed order, but only 9 per cent missed in varied order. Figure 4.10 shows the result across the five embedded errors.

The demands on attention and memory in flying a jet aeroplane are enormous. The pilot must simultaneously pay conscious attention to multiple sources of information while relying on highly practised, automatic processes and overlearned actions to respond to others.

Source: Palm Beach Flight Training

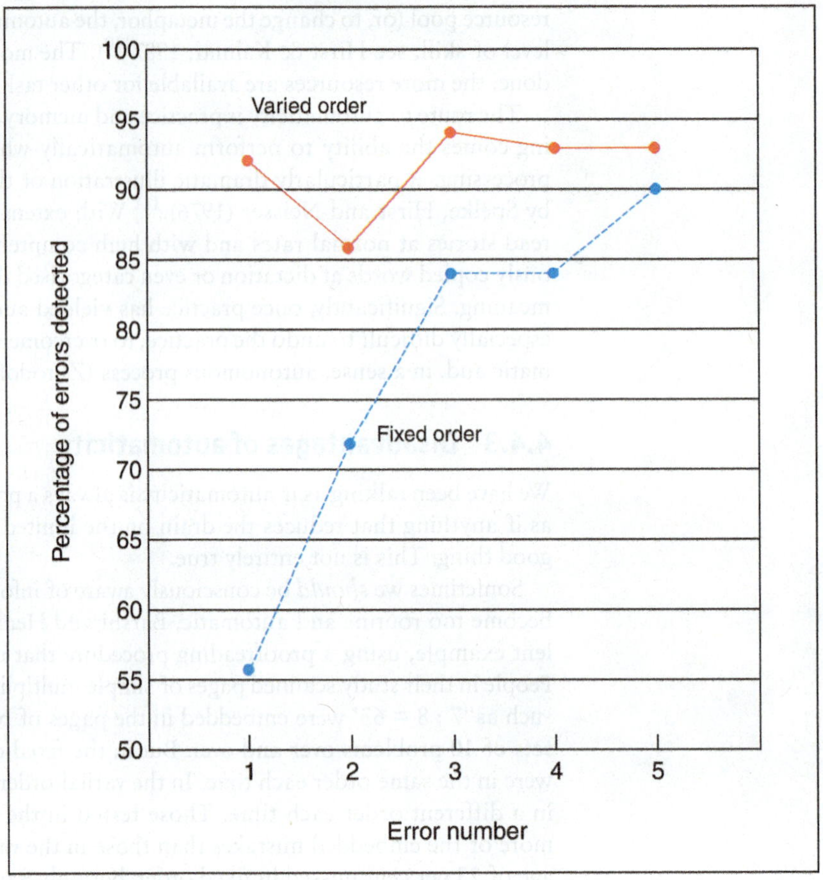

Figure 4.10 Results of Barshi and Healy's (1993)[114] experiment

The results show the percentage of participants detecting the five embedded errors in proofreading multiplication problems. Problems were presented in fixed or varied order. Data are from Barshi and Healy (1993).

Performance did improve in the fixed-order condition, as more and more of the mistakes were encountered. But the first multiplication error was detected only 55 per cent of the time, compared with the 90 per cent detection rate for the varied-order group.

The fixed order of problems encouraged automatic proofreading, which disrupted accuracy in detecting errors. In fact, it took either an earlier error that was detected or a specific alerting signal (Experiment 3) to overcome the effects of routine, automatic proofreading.

The implications of this kind of result should be clear. To ensure safety, pilots are required to go through checklist procedures for take-off, landing, and so forth. Yet because the items on the checklist are in a fixed order, repeated use of the list probably leads to a degree of automaticity and a tendency to miss errors. This is exactly what happened in March 1983, when a plane landed in Casper, Wyoming, without its landing gear down, even though the flight crew had gone through its standard checklist procedure and had 'verified' that the wheels were down. In Barshi and Healy's words, this incident 'reminded the crew and the rest of the aviation community that the countless repetition of the same procedure can lead to a dangerous automatization' (1993, p. 496).[115] It is interesting to wonder which is worse: too much automatisation of procedures, as exemplified by the Barshi and Healy study, or too much attention paid to the procedures.

Mind wandering

Perhaps one of the most obvious and ubiquitous examples of not being able to use our attention the way that we want to is when our minds drift from the task we are supposed to be focusing on to some other, irrelevant idea. **Mind wandering** is the situation in which a person's attention and thoughts wander from the current task to some other, inappropriate line of thought (for a review, see Randall, Oswald & Beier, 2014[116]).

Why the mind wanders

Have you ever daydreamed about a significant other when sitting at a traffic light, or go to the bottom of a page and realised you had no idea what you just read?

In these cases, we have decoupled our attention from the environment to focus more exclusively on our own internal thoughts, often without an awareness that our mind is wandering until we catch ourselves (Smallwood, McSpadden & Schooler, 2007, 2008[117]). This idea is now supported by ERP (*event-related potential*) recordings (Barron, Riby, Greer & Smallwood, 2011[118]). During reading, we can now detect when a person's mind is wandering because his or her pupils become more dilated (Franklin, Broadway, Mrazek, Smallwood & Schooler, 2013[119]), eye movements become more erratic, eyeblinks increase, and eye movements are less tied to characteristics that should influence reading patterns (such as word frequency) (Reichle, Reineberg & Schooler, 2010; Smilek, Carriere & Cheyne, 2010[120]).

Sometimes when we are trying to do one task, attention disengages and our minds wander.
Source: Ariwasabi/Shutterstock

As you know, your mind is more likely to wander when you are bored than when you are engaged and absorbed in something. Essentially, when you are concentrating, all your attention is engaged in the task on which you are focused, and it is difficult for distractions to lure it away. However, when your intended primary task is not taking all your attention, other ideas can break through and take attention away (Smallwood & Schooler, 2006)[121]. Under these circumstances, some control of attention is lost (Kane & McVay, 2012)[122], and it will drift from what you are supposed to be thinking about to something else. This drifting leaves your memory for what you are supposed to be doing much poorer. The surprising prediction here is that people with *more* working memory capacity are *more* likely to mind wander. *Why?*

This is because they will be more likely to have capacity available over and above what is required by the current task. These people get distracted by things that are in the environment, but are not part of what they are supposed to be attending. Such a proneness to mind wander impedes cognition, such as the ability to learn things in school (Smallwood, Fishman & Schooler, 2007)[123]. Because mind wandering involves a loss of attentional control, in some sense, it is not surprising that mind wandering is more likely when a person is under the influence of alcohol (Sayette, Reichle & Schooler, 2009[124]) or craving a cigarette (Sayette, Schooler & Reichle, 2010[125]).

It is important to note that your mind does not wander randomly. Instead, you are more likely to disengage from the current task and mind wander by thinking about things that are current concerns for you, or are relevant to your long-term goals. If you think about those cases when your mind wanders, you will find that you are often daydreaming about things that are important to you in one way or another. Often, there may be something in the environment, such as a person you see, a word that you hear or read, or a smell of perfume, that directs your attention away from what you need to be doing toward something else. In education settings, mind wandering, such as during a class or lecture, can be attenuated if people are quizzed during the lecture about what is being discussed (Szpunar, Moulton & Schacter, 2013)[126]. Such quizzing works because it breaks up the stream of thought, allowing a student to avoid slipping into a mind-wandering state.

Summary: Attention

4.1 Multiple meanings of attention

Attention is a pervasive and complex topic, with meanings and connotations ranging from alertness and arousal up through the notions of automatic and conscious processing. Attention can be thought of as a mental process or mechanism or as a limited mental resource.

Controlled directed attention depends on specific neural networks to function well, and disruptions of these can lead to neurological disorders. However, it should be noted that when people are not attending in this way, the flow of the stream of thought is governed by a collection of neurological structures known as the default mode network.

4.2 Basic input attentional processes

Three basic senses of the term *attention* refer to alertness and arousal, the orienting reflex and the spotlight attention. These correspond to input attention, a fast process involved in encoding environmental stimuli into the mental system. Interestingly, in vision, the mental spotlight attention can be shifted without any movement of the eyes, confirming the mental rather than perceptual nature of attention.

Modern life has resulted in an increase in video game play, often involving action games that require players to redirect their attention in atypical ways. According to some studies, video gaming influences the cognitive process of attention, allowing it to be used more effectively by giving people a larger attentional spotlight. With that said, there is also evidence suggesting that such effects may be limited in scope or even nonexistent.

A disorder known as hemineglect shows how attention can be affected by brain damage, thus informing us about normal attention. In hemineglect, a patient is unable to direct attention voluntarily to one side of space, so he or she neglects stimuli presented on that side. The evidence suggests that this arises from an inability to disengage attention from a stimulus on the non-neglected side, hence disrupting the process of shifting attention to the opposite side.

4.3 Controlled, voluntary attention

Controlled or conscious attention is slower and more voluntary. Selective attention, the ability to focus on one incoming message while ignoring other incoming stimuli, is a complex ability, one investigated since the beginnings of modern cognitive science. The evidence shows that we can select one message, and reject others, based on physical characteristics or on more semantic characteristics. The later the process of selection acts, the more demanding it is of the limited capacity of the attention mechanism.

We are able to use the attention mechanism of inhibition to keep information that would otherwise be highly active, but is irrelevant, out of the current stream of processing. Keeping inappropriate information out helps us to focus on whatever it is that we want to be processing.

4.4 Attention as a mental resource

When attention is viewed as a limited mental resource, issues of task complexity become concerned with how automatic or controlled different mental processes are. Automatic processes are rapid, are not dependent on intent, are unavailable to conscious awareness, and place few, if any, demands on limited attentional resources. Conscious or controlled processes are the opposite: rather slow, requiring intention, open to conscious awareness, and heavily demanding of attentional resources.

Mental processes become more automatic as a function of practice and overlearning. One disadvantage of automaticity is that it is difficult to reverse the effects of practice in an automated task. Automaticity can also lead to errors of inattention, including action slips. When our attention is not fully engaged, our minds can wander off topic. Mind wandering is more likely to occur when there is mental capacity left over and available. Moreover, when our minds wander, the things that we allow our attention to drift to are typically things that we have enduring concerns about, such as things we are anxious or excited about.

Chapter 4 quiz

Question 1

A student is revising for an exam in the library alongside other students. They are reading lecture notes, listening to lecture recordings, chatting with other students about what they have planned for the weekend. They are successfully making an updated set of notes to help with their revision. What type of attention does this refer to?

- **A.** Selective attention
- **B.** Divided attention
- **C.** Lack of attention
- **D.** Sustained attention

Question 2

What is the key difference between explicit and implicit processing?

- **A.** Explicit processing is the type of processing you use on a visual and verbal basis only whereas implicit processing uses spatial and visual information only.
- **B.** Explicit processing is the type of processing that uses memory whereas implicit processing uses no memory at all.
- **C.** Explicit processing is the type of processing that you are consciously aware of whereas implicit processing is the processing you are not consciously aware of.
- **D.** Explicit processing is the type of processing that uses the word stem completion task whereas implicit processing does not use this task.

Question 3

Which of the following is an example of Hemineglect?

- **A.** A person is at a party and is speaking to a friend who is also at the party.
- **B.** A person looks in the mirror and only cleans makeup off the left side of their face.
- **C.** A person is reading a fictional book and is listening to the television at the same time.
- **D.** A person is driving a car and looks at the person crossing the road on the left hand side of the car, following them as they cross over to the right hand side of the car.

Question 4

Which of the following best describes a weakness of the early selection theory of attention?

- **A.** Personally relevant information can bypass the selective filter.
- **B.** The attended object can shift between input channels spontaneously.
- **C.** People can distribute aural attention between more than one stream.
- **D.** People with lower cognitive capacities cannot demonstrate filtering.

Question 5

Which of these can show a disadvantage of automaticity?

- **A.** Proofreading a cognitive psychology dissertation and spotting every single spelling mistake and grammatical error.
- **B.** A train driver using a checklist to ensure that everything has been completed.
- **C.** Completing a memory task and completing it in a longer time each time you take part in the task.
- **D.** Driving an automatic car and when using a manual car, not putting the car into the correct gear.

5 Short-term and working memory

Learning objectives

5.1 **Explain short-term memory encoding and storage**

5.2 **Summarise the process through which short-term memory is recalled**

5.3 **Explain how working memory functions**

5.4 **Identify ways to assess working memory**

5.5 **Evaluate the centrality of working memory in the overall cognitive functions**

Key figures

Professor Alan Baddeley
Professor Alan Baddeley has advanced research on human memory, leading the field by developing one of the most influential theories in cognitive psychology about working memory.

Professors Akira Miyake and Naomi Friedman
Their influential paper in 2012, which was cited over 2,000 times, provided a new framework for examining three executive functions (updating, shifting and switching) taking into account individual differences.

Introduction

Primary memory, elementary memory, immediate memory, short-term memory (STM), short-term store (STS) and working memory (WM) are all terms that refer to memory where the present moment is held in consciousness. It is the seat of conscious attention. In this chapter, we will look at what are short-term memory and working memory, and how they differ.

Research on short-term and working memory came hard on the heels of selective attention studies of the mid-1950s. George Miller's (1956)[1] classic article is an excellent example of this upsurge in interest. A common observation, that we can remember only a small number of isolated items presented rapidly, began to take on greater significance as psychology groped towards a new approach to human memory. Miller's insightful remarks were followed shortly by the Brown (1958)[2] and Peterson and Peterson (1959)[3] reports. Amazingly, simple three-letter stimuli, such as *MHA*, *GPR*, or *KCD*, were forgotten almost completely within 15 seconds if a person's attention was diverted by a distractor task of counting backward by 3s. Such reports were convincing evidence that the limited capacity of memory was finally being pinned down and given an appropriate name: short-term memory.

5.1 A limited-capacity bottleneck

LEARNING OBJECTIVE
Explain short-term memory encoding and storage

If you hear a string of 10 digits, read at a rapid rate, and are asked to reproduce those digits in order, generally you cannot recall more than about 7 (+–2). The same result is found with unrelated words. This is roughly the amount you can say aloud in about 2 seconds (Baddeley, Thomson & Buchanan, 1975[4]) or the amount you can recall in 4 to 6 seconds (Dosher & Ma, 1998; see also Cowan et al., 1998[5]). This limit has been recognised for so long, it was included in the earliest intelligence tests (e.g. Binet's 1905 test; see Sattler, 1982[6]). Young children and people of subnormal intelligence generally have a shorter span of apprehension, or memory span. In the field of intelligence testing, it is almost unthinkable to devise a test *without* a short-term memory-span component. Note that this is a general aspect of short-term memory, not something special about spoken words or digits. For example, a similar finding is observed with letters in American Sign Language (ASL) (Wilson & Emmorey, 2006; Wilson & Fox, 2007[7]), which clearly is more visual/motor relative to spoken English.

The difference between short-term memory and working memory

As we proceed, we shift from the term short-term memory to working memory. Why the two terms? Basically, they have different connotations.

Short-term memory

Short-term memory conveys a simpler idea. It is the label we use to focus on the input and storage of new information. For example, when a rapidly presented series of digits is tested for immediate recall, we generally refer to short-term memory. Likewise, when we focus on the role of rehearsal, we are examining the short-term memory maintenance of new information. Short-term memory is observed whenever short retention is tested – no more than 15 or 20 seconds. It is also observed when little, if any, transfer of new information to long-term memory is involved.

Working memory

Working memory, by comparison, has the connotation of a mental workbench, a place where mental effort is applied (Baddeley, 1992a, 1992b; Baddeley & Hitch, 1974[8]). Thus, when word meanings are retrieved from long-term memory and put together to understand a sentence, working memory is where this happens. Traditional immediate memory tasks are a subset of working memory research but usually are only secondary to reasoning, comprehension, or retrieval processes. For example, the short-term memory responsible for digit span performance is but a single component of the more elaborate working memory system.

5.1.1 Short-term memory capacity

For our purposes, the importance of this limitation is that it reveals something fundamental about human memory. Our immediate memory cannot encode a vast quantity of new information and hold it accurately. Miller stated this limit aptly in the title of his 1956 paper: 'The magical number seven, plus or minus two: Some limits on our capacity for processing information' (for a historical consideration of the paper, see Cowan, 2015[9]). However, more recent work suggests that the situation is worse than this – that people can maintain only 4 plus or minus 1 **units** of information (Cowan, 2010[10]), and that 7 plus or minus 2 is actually a result of some chunking. We process large amounts of information in the sensory memories, and we can hold vast quantities of knowledge in permanent long-term memory. Yet, short-term memory is the narrow end of a funnel, the four-lane bridge with only one open tollgate. It is the bottleneck in our information-processing system.

Overcoming the bottleneck

And so this limitation remains unless what we are trying to remember is grouped in some way, as in the 3–3–4 grouping of a mobile telephone number. In Miller's terms, this is called a chunk of information. By chunking items together into groups, we can overcome this limitation.

Example of bottleneck in car traffic when lanes merge.
Source: egd/Shutterstock

The following is a simple example of the power of chunking:

BYGROUPINGITEMSINTOUNITSWEREMEMBERBETTER

No one can easily remember 40 letters correctly if they are treated as 40 separate, unrelated items. But by chunking the letters into groups, we can retain more information. You can more easily remember the eight words because they are familiar ones that combine grammatically to form a coherent thought. You can remember a National Health Service or National Insurance number more easily by grouping the digits into the 3–2–4 pattern. And you can remember a telephone number more easily if you group the last four digits into two two-digit numbers.

The term for this process of grouping items together, then remembering the newly formed groups, is recoding. By recoding, people hear not the isolated dots and dashes of Morse code but whole letters, words, and so on. The principle behind recoding is straightforward: recoding reduces the number of units held in short-term memory by increasing the richness, the information content, of each unit.

Two conditions are important for recoding:

1. We can recode if there is sufficient time or resources to use a recoding scheme.
2. We can recode if the scheme is well learned, as Morse code becomes with practice.

In a dramatic demonstration of this, one person in a study by Chase and Ericsson (1982),[11] over the period of a few months, could recall 82 digits in order by applying a highly practised recoding scheme he had invented for himself. But what about situations in which an automatic recoding scheme is not available? What is the fate of items in short-term memory? Can we merely hold the usual small number of items?

5.1.2 Forgetting from short-term memory

Short-term memory capacity is quite small (7 plus or minus 2 items) and thus information quickly decays. Numerous experiments have been conducted to understand why information in short-term memory is quickly lost. Some have explored the role of interference tasks, while other researchers have focused on the position of information in an array.

In the research box below, we provide an example of an early distraction task study that shows the influence of rehearsal suppression on subsequent memory recall. More recently, Endress & Szabó (2017)[12] discussed how interference can influence short-term memory recall, either by occupying a 'slot' in short-term memory or by exhausting memory resources, challenging the notion of a single effect on memory performance that was once believed to be true.

Causes of forgetting

Research by Brown (1958)[13] and Peterson and Peterson (1959)[14] provides a compelling demonstration of how the passage of time causes forgetting. The central idea is that forgetting might be caused simply by the passage of time before testing – in other words, forgetting caused by decay.

In their experiments, a simple three-letter trigram (e.g. MHA) was presented to people, followed by a three-digit number (e.g. 728). People were told to attend to the letters, then to begin counting backward by 3s from the number they were given. The counting was done aloud, in rhythm with a metronome clicking twice per second. The essential ingredient here is the distractor task of backward counting. This requires a great deal of attention (if you doubt this, try it yourself, making sure to count twice per second). Furthermore, it prevents rehearsal of the three letters because rehearsal uses the same cognitive mechanism as the backward counting.

At the end of a variable brief time, the people reported the trigram. The results were so unexpected, and the number of researchers eager to replicate them so large, that the task acquired a name it is still known by: the Brown–Peterson task.

The surprising result was that memory of the 3-letter trigram was only slightly better than 50 per cent after 3 seconds of counting; accuracy dwindled to about 5 per cent after 18 seconds (see Figure 5.1).

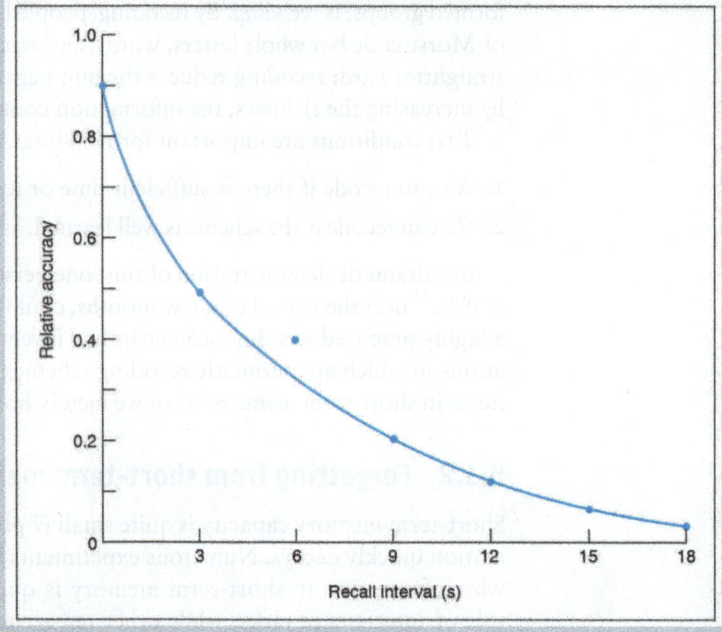

Figure 5.1 The Brown–Peterson task

Relative accuracy of recall in the Brown–Peterson task across a delay interval from 0 to 18 seconds. People had to perform backward counting by 3s during the interval.

The letters were forgotten so quickly even though short-term memory was not overloaded, with a 50 per cent loss after only 3 seconds (assuming perfect recall with a zero-second delay). On the face of it, this seems evidence of a simple decay function: with an increasing time span, less and less information remains in the short-term memory.

Later research, especially by Waugh and Norman (1965),[15] questioned some of the assumptions made. Waugh and Norman thought that the distractor task itself might be a source of interference. If the numbers spoken during backward counting interfered with the short-term memory trace, then longer counting intervals would have created more interference.

Waugh and Norman's reanalysis of several studies confirmed their suspicion. Especially convincing were the results of their own probe digit task (Figure 5.2).

People heard a list of 16 digits, read at a rate of either 1 or 4 per second. The final item in each list was a repeat of an earlier item, and it was the probe or cue to recall the digit that had followed the probe in the original list.

For instance, if the sequence 7 4 6 9 had been presented, then the probe 4 would have cued recall of the 6. The important part of their study was the time it took to present the 16 digits. This took 16 seconds for one group (a long time) but only 4 seconds (a short time) for the other group. If forgetting were caused by decay (a time-based process), then the groups should have differed markedly in their recall because so much more time had elapsed in the 16 seconds group. Yet, as Figure 5.2 shows, the two groups differed little.

This suggests that forgetting was influenced by the number of intervening items, not simply the passage of time. In other words, forgetting in short-term memory was caused by interference, not decay (for cross-species evidence of interference,

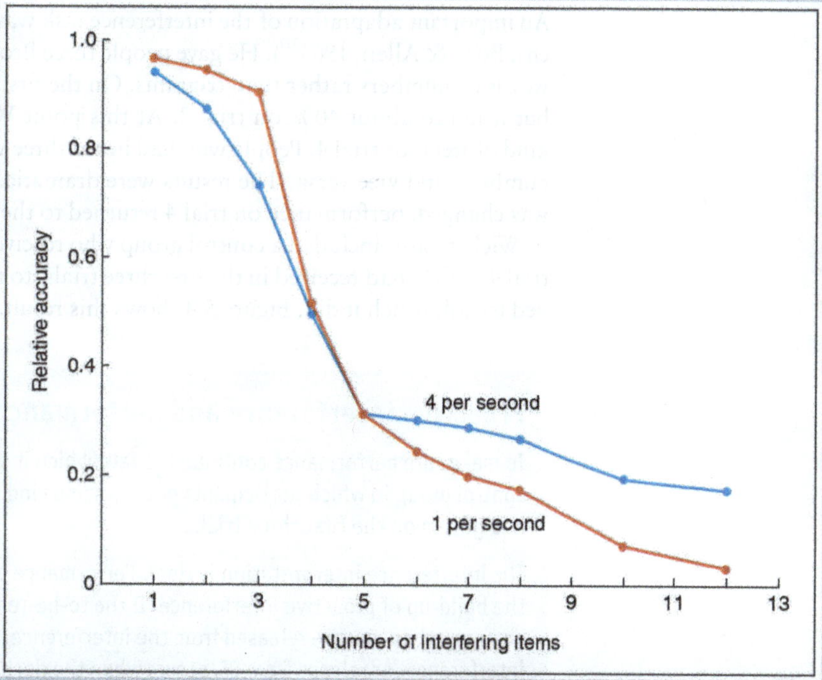

Figure 5.2 The Waugh and Norman (1965) probe digit experiment
Relative accuracy in the Waugh and Norman (1965) probe digit experiment as a function of the number of interfering items spoken between the target item and the cue to recall; rate of presentation was either 1 or 4 digits per second.

see Wright & Roediger, 2003[16]). Although it has been decades since the original work, the issue of decay versus interference explanations for forgetting in short-term memory continues to be of interest to cognitive psychologists (e.g. Unsworth, Heitz & Parks, 2008[17]). Some interesting recent work suggests that whereas the bulk of the forgetting is due to interference, there is still a small amount that can be attributed to a decay process as well (Altmann & Gray, 2002; Altmann & Schunn, 2012; Berman, Jonides & Lewis, 2009[18]).

Source: Peterson & Peterson (1959).

Proactive and retroactive interference (pi and ri)

Shortly after the Peterson and Peterson report, Keppel and Underwood (1962)[19] challenged the decay explanation for forgetting in short-term memory. They found that people forgot at a dramatic rate only after several trials. On the first trial, memory for the trigram was almost perfect.

Keppel and Underwood's explanation was that as you experience more and more trials in the Brown–Peterson task, recalling the trigram becomes more difficult because the previous trials generate interference. This is called **proactive interference (PI)**, in which older material interferes forward in time with your recollection of the current stimulus. This is the opposite of **retroactive interference (RI)**, in which newer material interferes backward in time with your recollection of older items.

The loss of information in the Brown–Peterson task was caused by proactive interference.

Release from pi

An important adaptation of the interference task was done by Wickens (1972; Wickens, Born & Allen, 1963[20]). He gave people three Brown–Peterson trials, using three words or numbers rather than trigrams. On the first trial, accuracy was near 90%, but it fell to about 40% on trial 3. At this point Wickens changed to a different kind of item for trial 4. People who had heard three words per trial were given three numbers, and vice versa. The results were dramatic. When the nature of the items was changed, performance on trial 4 returned to the 90 per cent level of accuracy.

Wickens also included a control group who received the same kind of stimulus in trial 4 as they had received in the first three trials, to make sure performance continued to fall, which it did. Figure 5.3 shows this result.

Proactive interference and performance

To make sure performance continued to fall (which it did), Wickens included a control group in which participants got the same kind of stimulus on trial 4 as they had gotten on the first three trials.

The interference interpretation is clear. Performance deteriorates because of the build-up of proactive interference. If the to-be-remembered information changes, then you are released from the interference. Thus, release from proactive interference, or release from PI, occurs when the decline in performance caused by proactive interference is reversed because of a switch in the to-be-remembered stimuli, as shown in Figure 5.3.

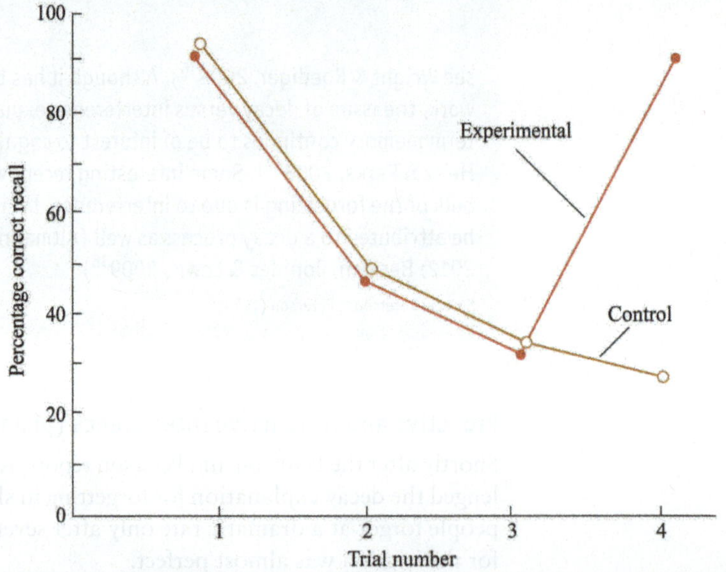

Figure 5.3 Recall accuracy in a release from pi experiment by Wickens, Born, and Allen (1963)[21]

Triads of letters are presented on the first three trials, and proactive interference begins to depress recall accuracy. On trial 4, the control group gets another triad of letters; the experimental group gets a triad of digits and shows an increase in accuracy, known as release from PI.

Release from PI also occurs when the change is semantic, or meaning-based, as when the lists switched from one semantic category to another (see Figure 5.4). However, note that the more related the items on the fourth list were to the original category, the less release from PI was experienced. Thus, short-term memory, to some degree, uses semantic information.

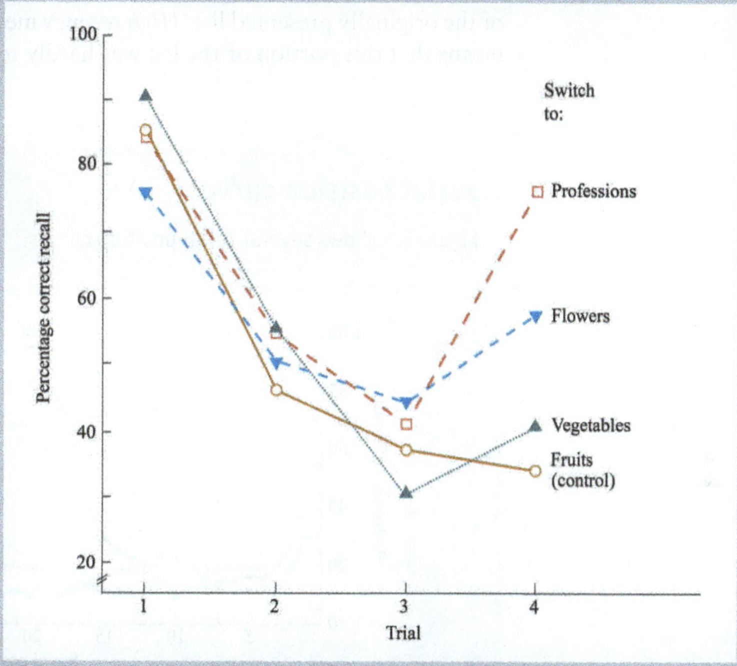

Figure 5.4 Recall accuracy in a release from pi experiment by Wickens and Morisano (reported in Wickens, 1972)[22]

All participants received word triads from the fruits category in trial 4. In trials 1 to 3, different groups received triads from the categories of fruits (control condition), vegetables, flowers and professions.

5.2 Short-term memory retrieval

LEARNING OBJECTIVE
Summarise the process through which short-term memory is recalled

Here, we consider retrieval from short-term memory, which refers to the act of bringing knowledge to the foreground of thinking and perhaps reporting it. Our focus is on two aspects of retrieval: the serial position curve and studies of the retrieval process itself.

5.2.1 Serial position effects

A **serial position curve** is a graph of item-by-item accuracy on a recall task. Serial position simply refers to the original position an item had in a study list.

We will first cover the two tasks used to test people: free recall and serial recall. In **free recall**, people are free to recall the list items in any order, whereas in **serial recall**, people recall the list items in their original order of presentation. Not surprisingly, serial recall is more difficult. Recalling items in order requires people to rehearse them as they are shown, trying to hold on to not only the information itself but also its position in the list. The more items there are, the harder the task becomes. By comparison, with free recall, people can recall the items in any order.

The early list positions are called the primacy portion of a serial position curve. Primacy here has its usual connotation of 'first': It is the first part of the list that was studied. **Primacy effect**, then, refers to the accuracy of recall for the early list positions. A strong primacy effect means good, accurate recall of the early list items, usually because of rehearsal. The final portion of the serial position curve is the recency portion. **Recency effect** refers to the level of correct recall for the final items of the originally presented list. *High recency* means 'high accuracy,' and *low recency* means that this portion of the list was hardly recallable at all.

Serial position curve

Figure 5.5 shows several serial position curves.

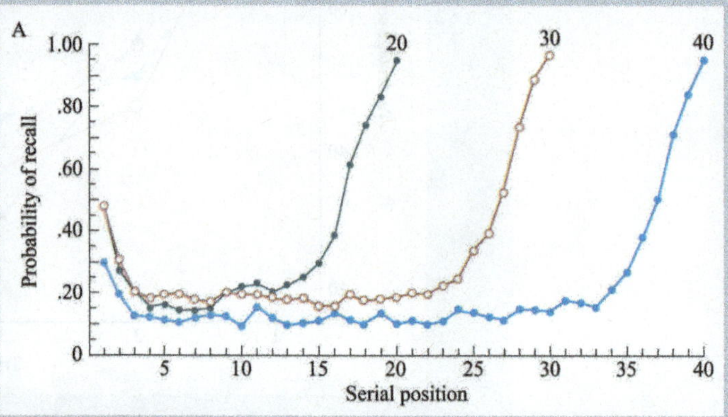

Figure 5.5a

Serial position curves showing recall accuracy across the original positions in the learned list. Rate of presentation was one item per second.

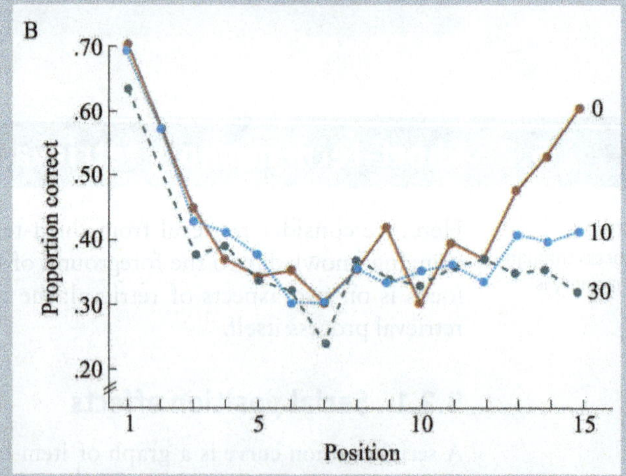

Figure 5.5b

Serial position curves showing the decrease in recency when 10 or 30 seconds of backward counting is interpolated between study and recall.

As Figure 5.5a shows, a strong recency effect is found across a range of list lengths; Murdock (1962)[23] presented 20-, 30- and 40-item lists at a rate of one item per second. Note that there is a slight primacy effect for each list length, but that the middle part had low recall accuracy. Apparently, the first few items were rehearsed enough to transfer them to long-term memory, but not enough time was available for rehearsing items in the middle of the list. For all lists, though, the strong recency effect can be attributed to recall from short-term memory.

The way to eliminate the recency effect should be no surprise. Glanzer and Cunitz (1966)[24] showed people 15-item lists. For some people, after a list, they needed to do an attention-consuming counting task for either 10 or 30 seconds before recalling the items.

In contrast to people who gave immediate recall (0 seconds delay), the people who had to do a counting task showed very low recency (see Figure 5.5b). However, the primacy portion of the list was unaffected. In other words, the early list items were more permanently stored in long-term memory to endure 30 seconds of counting. The most recent items in short-term memory were susceptible to interference.

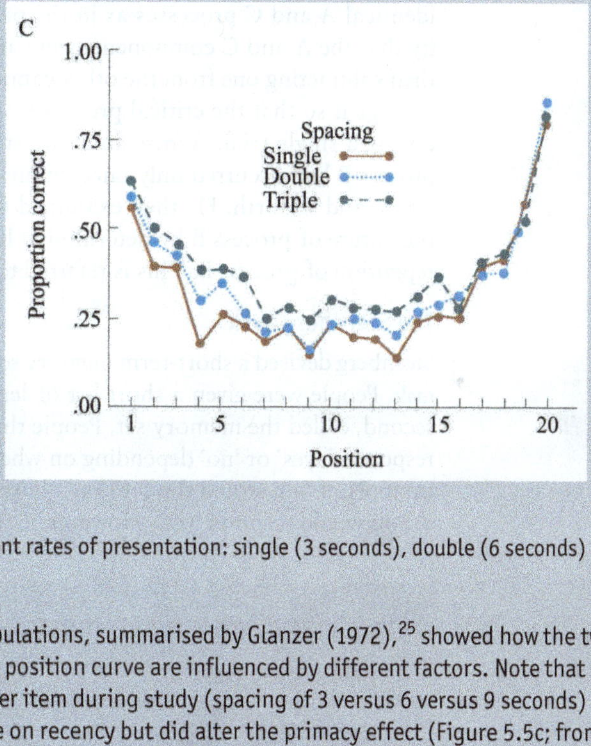

Figure 5.5c

Three different rates of presentation: single (3 seconds), double (6 seconds) and triple (9 seconds).

Other manipulations, summarised by Glanzer (1972),[25] showed how the two parts of the serial position curve are influenced by different factors. Note that providing more time per item during study (spacing of 3 versus 6 versus 9 seconds) had almost no influence on recency but did alter the primacy effect (Figure 5.5c; from Glanzer & Cunitz, 1966[26]). Additional time for rehearsal allowed people to store the early items more strongly in long-term memory. Moreover, additional time did not help the immediate recall of the most recent items. These items were held in short-term memory and recalled before interference could take place.

5.2.2 Short-term memory scanning

We turn now to a different question: How do we access or retrieve the information from short-term memory? To answer this question, we turn to another memory task: recognition.

Essentially, a **recognition task** is one in which people are presented with items and are asked to indicate whether the items were part of what had been studied before. People would select 'yes' to indicate 'Yes, I recognise that as being studied.' Similarly, they would select 'no' to indicate 'no, I didn't study that item.' Making these decisions requires people to access stored knowledge, then compare the items to that knowledge. The important angle is that we can time people as they make their 'yes/no' recognition decisions, and infer the underlying mental processes used on the basis of how long they took. Saul Sternberg used this procedure in addressing the question of how we access information in short-term memory.

Sternberg (1966, 1969, 1975[27]) began by noting that the use of *response time (RT)* to infer mental processes had a venerable history, dating back at least to Donders in the 1800s. Donders proposed a subtractive factors method for determining the time for simple mental events. For example, if your primary task involves processes A, B and C, you devise a comparison task that has only processes A and C in it. After giving both tasks, you subtract the A + C time from the A + B + C time. The difference should be a measure of the duration of process B because it is the process that was subtracted from the primary task.

Sternberg pointed out a major difficulty with Donders' subtractive method. It is virtually impossible to make sure that the comparison task, the A + C task, contains identical A and C processes as in the primary task. There is always the possibility that the A and C components were altered when you removed process B. If so, then subtracting one from the other cannot be justified. Sternberg's solution was to arrange it so that the critical process would have to *repeat* some number of times during a single trial. Across an entire study there would be many trials in which process B had occurred only once, many in which it had occurred two times, three times, and so forth. He then examined the RTs for these conditions, and inferred the nature of process B by determining how much time was *added* to RTs for each repetition of process B. This is referred to as the additive factors method.

The Sternberg task

Sternberg devised a short-term memory scanning task, now simply called a **Sternberg task**. People were given a short list of letters, one at a time, at the rate of one per second, called the memory set. People then saw a single letter, the probe item, and responded 'yes' or 'no' depending on whether the probe was in the memory set. For example, if you stored the set *l r d c* in short-term memory and then saw the letter *d*, you would respond 'yes'. However, if the probe were *m*, you would respond 'no'.

Simple and working memory spans

Here are some suggestions.

Simple memory span

Make several lists, being sure that the items do not form unintended patterns. Use digits, letters or unrelated words. Read the items at a constant and rapid rate (no slower than one item per second) and have the participant name them back in order. Your main dependent variable will be the number or percentage correct.

Sample lists

Digits

- 8 7 0 3 1 4
- 7 1 5 0 5 4 3 6
- 2 8 4 3 6 1 2 9 7 5

Words

- leaf gift car fish rock
- paper seat tire horse film beach forest brush
- bag key book wire box wheel banana floor bar pad block radio boy

Try a few of these variations:

- To illustrate the importance of interference, have people do an interference task on half of the trials. On an interference trial, give them a number such as 437 and have them count backward by 3s, aloud, for 15 seconds, before recalling the list items.
- Keeping list length constant, give different retention intervals before asking for recall (e.g. 5 seconds, 10 seconds, 20 seconds), either with or without backward counting.
- Vary the presentation rate (e.g. one word per second versus one word per 3 seconds) to see how the additional time for rehearsal influences recall.

Working memory span

Follow the examples given in the text to construct a working memory span test – for example, from one to six unrelated sentences, each followed by an unrelated word, whereby the participant must process the sentence and then, at the end of the set, recall the unrelated words that appeared. Span size will be the number of words recalled correctly, assuming the sentences were comprehended.

Probes
The target stimuli that need to be detected among study items (memory sets).

Memory sets were from one to six letters or digits long, within the span of short-term memory, and were changed on every trial. **Probes** also changed on every trial and were selected so the correct response was 'yes' on half the trials and 'no' on the other half. This is illustrated by trials 3 and 4 in Table 5.1. Take a moment to try several of these trials, covering the probe until you have the memory set in short-term memory, then covering the memory set and uncovering the probe, then making your 'yes/no' judgement. For a better demonstration, have someone read the memory sets and probe to you aloud.

In a typical experiment, people saw several hundred trials, each consisting of these two parts, memory set then probe, as shown in Table 5.1.

Table 5.1 Sample Sternberg task

Trial	Memory set items	Probe items	Correct response
1	R	R	Yes
2	LG	L	Yes
3	SN	N	Yes
4	BKVJ	M	No
5	LSCY	C	Yes

Figure 5.6 illustrates the *process model* that Sternberg (1969)[28] proposed, simply a flowchart of the four separate mental processes that occurred during the timed portion of every trial. At the point marked 'Timer starts running here,' the person encodes the probe. Then, the search or scan through short-term memory begins and the mentally encoded probe is compared with items in memory to see whether there was a match. A simple 'yes' or 'no' decision could then be made by pressing one of two buttons.

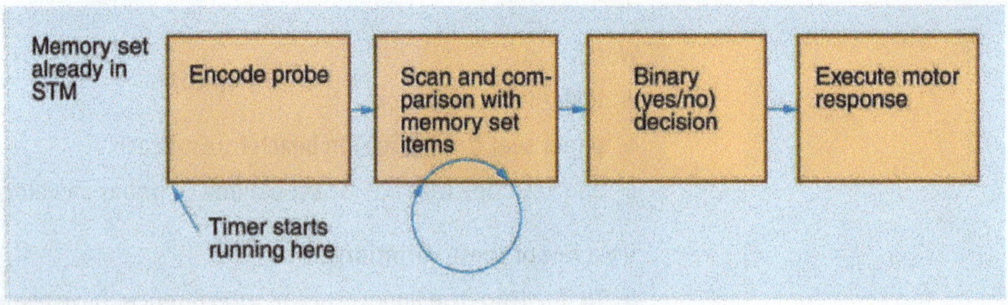

Figure 5.6

In Sternberg's task, it was the search process of the contents of short-term memory that was of interest. Notice – this is critical – that it was this process that was repeated a different number of times, depending on how many items were in the memory set. Thus, by manipulating memory set size, Sternberg influenced the number of cycles through the search process. And by examining the slope of the RT results, he could determine how much time was needed for each cycle.

Sternberg's results

Figure 5.7 shows Sternberg's (1969)[29] results. There was a linear increase in RT as the memory set got larger, and this increase was nearly the same for both 'yes' and 'no' trials. The equation at the top of the figure shows that the *y*-intercept of this RT function was 397.2 ms. Hypothetically, if there had been zero items in short-term memory, the *y*-intercept would be the combined time for the encoding, decision and response stages. More importantly, the slope of the equation was 37.9 ms; for each additional item in the memory set, the mental scanning process took 37.9 ms. In other words, the search through short-term memory is approximately 38 ms per item – very fast.

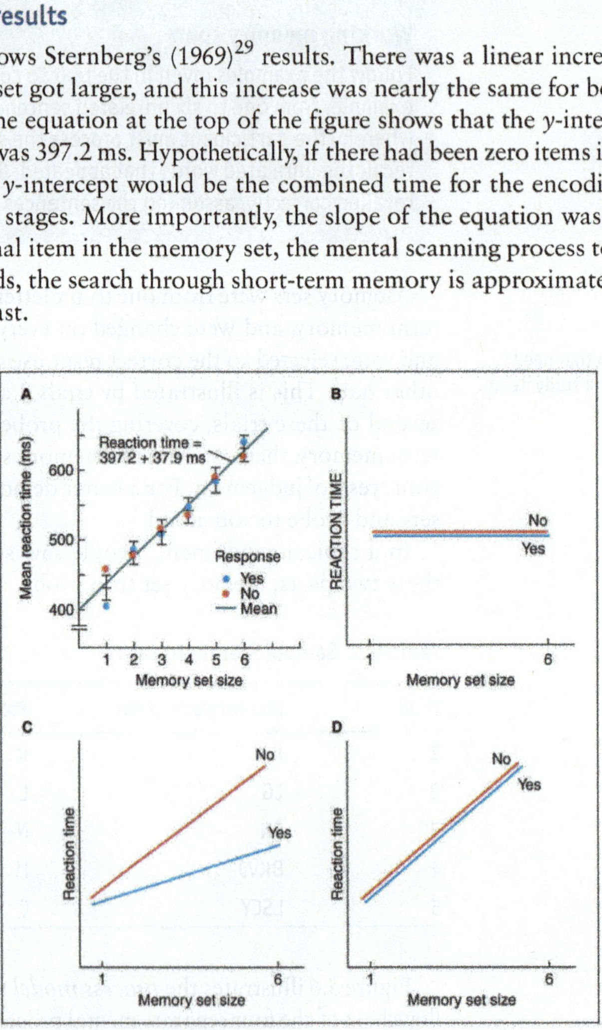

Figure 5.7
Reaction time in the short-term memory scanning task, for 'yes' (blue) and 'no' (red) responses. Reaction time increases linearly at a rate of 37.9 ms per additional item in the memory set.

Three possibilities of mental search

Sternberg considered three possibilities: a serial self-terminating search, a parallel search and a serial exhaustive search.

1. Serial self-terminating search

The most intuitively appealing was a serial self-terminating search in which the positions in short-term memory are scanned one by one, and the scan stops when a match is found; this is how you search for a lost object, say your car keys. On average, the slope of the RT trials for 'yes' responses should be smaller than the slope for 'no' responses. On the 'no' trials, all positions have to be searched before you can decide that the probe was not in the set. But on 'yes' trials, people would encounter matches at all positions in the memory set, sometimes early, sometimes late, with equal frequencies at all positions. The slopes of the positive curves ('yes, I found the target') were always smaller than those of the negative curves ('no, the target was not in the display'). However reasonable such a search appears, Sternberg's data did not match the prediction – he found the same slope for both kinds of trials.

2. Parallel search

The second possibility was a parallel search, in which each position in the memory set is scanned simultaneously. If short-term memory is scanned in parallel, then there should be no increase in RT – if all the positions are scanned simultaneously, it should not take longer to scan six items than three, for example. But again, the data did not match this prediction.

3. Serial exhaustive search

Sternberg inferred that short-term memory is searched in a fashion called a serial exhaustive search. That is, the memory set is scanned one item at a time (serial), and the entire set is scanned on every trial regardless of whether a match is found (exhaustive). Notice that exhaustive search must be correct for 'no' trials because the positions must be scanned exhaustively before you can confidently and accurately make a 'no' decision. Because of the similarity of the 'yes' and 'no' curves, Sternberg argued strongly that both reflect the same mental process, serial exhaustive search (Sternberg, 1969, 1975[30]).

Limitations to Sternberg's conclusions

Across the years, it has been suggested that increasing RTs could be the product of a parallel search in which each additional item slows down the rate of scanning for all items. This is much like a battery, which can run several motors at once, but each runs more slowly when more motors are connected (see Baddeley, 1976, for a review[31]). Others have objected to the assumption that the several stages or processes are sequential and that one must be completed before the next one begins.

For instance, McClelland (1979)[32] proposed that the mental stages might overlap partially in cascade fashion. Still, Sternberg's work pushed the field towards more useful ways of studying cognition. Most research based on RT tasks (e.g. visual search tasks and many long-term memory tasks) owes credit, even if only indirectly, to Sternberg's groundbreaking and insightful work.

5.3 Working memory

LEARNING OBJECTIVE
Explain how working memory functions

Working memory can be viewed as an augmentation of short-term memory. By the mid-1970s, all sorts of functions were being attributed to short-term memory in tasks of problem solving, comprehension, reasoning, and the like. Yet, as Baddeley pointed out, remarkably little work had demonstrated those functions in STM (Baddeley, 1976; Baddeley & Hitch, 1974; Baddeley & Lieberman, 1980[33]).

Going beyond intuitive examples, Baddeley and Hitch (1974)[34] documented their position by describing a dramatic case study, originally reported by Warrington and Shallice (1969; also Shallice & Warrington, 1970; Warrington & Weiskrantz, 1970[35]):

> . . . of a patient who, by all normal standards, had a severely defective short-term memory span (STS). He had a digit span of *only two items* and showed significantly impaired performance on the Peterson short-term forgetting task. If STS does indeed function as a central working memory, then one would expect this patient to exhibit impaired learning, memory and comprehension. No such evidence of general impairment was found either in this case or in subsequent cases of a similar type (Baddeley & Hitch, 1974, pp. 48–49, emphasis added; also Baddeley & Wilson, 1988; Vallar & Baddeley, 1984[36]*)

In a similar vein, McCarthy and Warrington (1984)[37] reported on a patient who had a memory span of only one word, but could nonetheless report back six- and seven-word sentences with about 85 per cent accuracy. Despite both types of lists relying on STM, performance on one type was seriously affected by the brain damage, and the other only minimally.

Baddeley and Hitch reasoned that the problem lies with a simple maintenance theory of STM. They suggested that STM is but one component of a larger, more elaborate system, **working memory**. Because *Baddeley's working memory model* is the most well-known model, we will give extensive coverage to it here. After this, we will describe Engle's model of working memory to provide an alternative idea about the role and processing of working memory.

5.3.1 The components of working memory

A description of Baddeley's working memory model provides a useful context for the studies described later (see Baddeley, 2000a; Baddeley & Hitch, 1974; Salame & Baddeley, 1982[38]).

Baddeley's model, shown in Figure 5.8, has four major components. The main part is the **central executive** (or sometimes executive control), assisted by two auxiliary systems: the **phonological loop** and the **visuo-spatial sketch pad**. Both auxiliary systems had specific sets of responsibilities, assisting the central executive by doing some of the lower-level processing. Thus, in the arithmetic problem, the central executive would be responsible for retrieving values from memory ($4 + 5, 9 \times 2$) and applying the rules of arithmetic. A subsystem, the phonological loop, would then hold the intermediate value 18 in a rehearsal-like buffer until it was needed again. A third auxiliary system, the **episodic buffer** (Baddeley, 2000a[39]), is used to integrate information already in working memory with information retrieved from long-term memory. It is where different types of information are bound together to form a complete memory, such as storing together the sound of someone's voice with an image of his or her face.

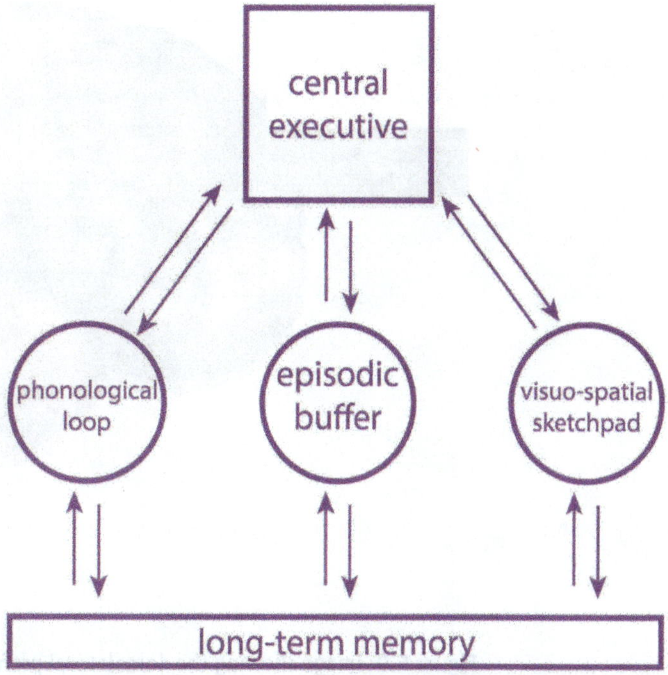

Figure 5.8 Working memory model originally developed by Baddeley and Hitch (1974)[40]

The idea of working memory being divided into components is supported by neurological evidence. Smith and Jonides (1999; also Smith, 2000[41]) reviewed several brain imaging studies to identify regions of heightened activity in various working memory tasks. In general, the thinking is that those brain regions involved in perception are also recruited by working memory for the storage of information, regions toward the posterior (back) of the brain, and that the rehearsal and processing of information are controlled by those aspects of the brain involved in motor control and attention (Jonides, Lacey & Nee, 2005[42]). For the Sternberg task, the scanning evidence showed strong activations in a left hemisphere parietal region and three frontal sites, Broca's area and the left supplementary motor area (SMA) and premotor area (see Chapter 2).

Broca's area is important in the production of language, so finding increased activity here was not surprising. Alternatively, tasks that emphasise executive control, such as switching from one task to another, tend to show strong activity in the **dorsolateral prefrontal cortex (DLPFC)**. (For an argument that task switching does not involve executive control, see Logan, 2003.[43]) This area is central to understanding executive attention (Kane & Engle, 2002[44]). The neurological basis for executive control is also supported by work showing that executive functions in cognition may have a significant genetic basis (Friedman et al., 2008[45]).

dorsolateral prefrontal cortex (DLPFC)
An area in the prefrontal cortex that plays a functional role in supporting executive functions and cognitive flexibility.

Other studies have shown specific brain regions involved in visuo-spatial working memory. In one (Jonides et al., 1993[46]), people saw a pattern of three random dots; the dots were then removed for 3 seconds, and a circle outline appeared; the task was to decide whether the circle surrounded a position where one of the dots had appeared earlier. In a control condition, the dots remained visible while the circle was shown, thus eliminating the need to remember the locations. **Positron emission tomography (PET)** scans revealed that three right hemisphere regions showed heightened activity and so were involved in spatial working memory. They were a portion of the occipital cortex, a posterior parietal lobe region, and the premotor

Positron emission tomography (PET)
Non-invasive imaging method that enables researchers and clinicians to observe the metabolic or biochemical function of organs in the body.

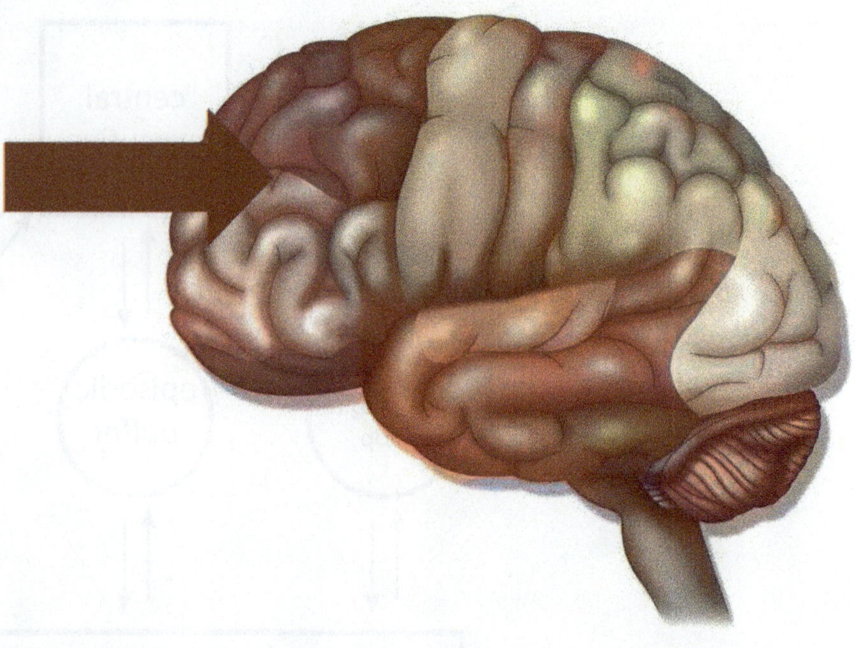

Figure 5.9 Image showing the dorsolateral prefrontal cortex which supports executive functions
Source: BSIP SA/Alamy Stock Photo

and DLPFC region of the frontal lobe (see also Courtney, Petit, Maisog, Ungerleider & Haxby, 1998[47]). In related work, when the task required spatial information for responding, it was the premotor region that was more active; when the task required object rather than spatial location information, the DLPFC was more active (Jonides et al., 1993; see also Miyake et al., 2000[48], for a review of various executive functions attributed to working memory[49]).

5.3.2 The central executive

The *central executive* is the heart of working memory. It is like a large corporation in which the chief executive controls the tasks of planning, initiating activities and making decisions. Likewise, in working memory, the central executive controls the planning of future actions, initiating retrieval and decision processes as necessary, and integrating information coming into the system. Some views of the central executive assume that it is composed of multiple elements that are responsible for different aspects of the control of processing (Vandierendonck, 2016[50]).

Let's take an example to illustrate some of this. To continue with the arithmetic example, the central executive triggers the retrieval of facts such as '4 + 5 = 9' and invokes the problem-solving rules such as 'how to multiply and divide'. Furthermore, the central executive also 'realises' that the intermediate value 18 must be held momentarily while further processing occurs. Accordingly, it activates the phonological loop, sending it the value 18 to rehearse for a few moments until that value is needed again.

Each of the subsystems has its own pool of attentional resources, but the pools are limited. Give any of the subsystems an undemanding task, and it can proceed without disrupting activities occurring elsewhere in working memory. However, if a subsystem is given a particularly difficult task, then either it falters or it must drain additional resources from the central executive. To some degree, working memory resources are shared across processing domains, such as verbal or visual (Vergauwe, Barrouillet & Camos, 2010[51]), which is why closing your eyes can help you think by reducing distractions from the external environment (Vredeveldt, Hitch & Baddeley, 2011[52]).

The central executive has its own pool of resources that can be depleted if overtaxed. For example, people who do something that places a strain on the central executive, such as ignoring distracting information as it scrolls across the bottom of a television screen or exaggerating their emotional expressions, have greater difficulty with central executive processing immediately thereafter (Schmeichel, 2007[53]). Moreover, damage to portions of the frontal lobes, such as the medial frontal lobes and the polar regions, can disrupt working memory's executive function (Banich, 2009)[54]. This produces **dysexecutive syndrome**, in which patients continue to pursue goals that are no longer relevant, and experience heightened distractibility when they do not have clear goals (Gevins, Smith, McEvoy & Yu, 1997)[55].

dysexecutive syndrome
A set of characteristic symptoms, usually following brain damage, that affect executive functions and influence emotional processing and behaviour.

5.3.3 The phonological loop

The *phonological loop* is the speech- and sound-related component responsible for rehearsal of verbal information and phonological processing. This component recycles information for immediate recall, including articulating the information in auditory rehearsal. For a debate on the articulatory versus phonological basis of this subsystem, see Baddeley (2000b); Jones, Macken and Nicholls (2004); and Mueller, Seymour, Kieras and Meyer (2003).[56]

There are two components of the phonological loop: the phonological store and the articulatory loop. The **phonological store** is essentially a passive store component of the phonological loop. This is the part that holds on to verbal information. However, information in the phonological store is forgotten unless it is actively rehearsed and refreshed. Thus, rehearsal is the role of the **articulatory loop**, the part of the phonological loop involved in the active refreshing of information in the phonological store. One way of thinking about these two components of the phonological loop is that the phonological store is like your inner ear – you can hear yourself talk to yourself, or imagine hearing music. Similarly, the articulatory loop is like your inner voice, when you mentally say things to yourself.

Effects of the phonological loop

Researchers have found several effects that provide insights into how the phonological loop works. We cover two of them here:

1. The **articulatory suppression effect** is the finding that people have poorer memory for a set of words if they are asked to say something while trying to remember the words (Murray, 1967). This effect is not complicated; it occurs even when you say something simple, like repeating the word *the* over and over again. What happens here is that the act of speaking consumes resources in the articulatory loop. As a result, words in the phonological store cannot be refreshed and are lost. A related phenomenon is the irrelevant speech effect (Colle & Welsh, 1976[57]). It is hard to keep information in the phonological loop when there is irrelevant speech in the environment. This irrelevant speech intrudes on the phonological loop, consuming resources and causing you to forget verbal information. Note that it is not necessary to even actually say anything. As another illustration of the embodiment of cognition, phonological loop processing can be disrupted if people are using the same muscles as they do when they are speaking, such as by chewing gum (Kozlov, Hughes & Jones, 2012[58]). This is also why it is so difficult to read (and then remember what you read) when you are in a room with other people talking. (So, try to study somewhere quiet.)

2. The **phonological similarity effect** is the finding that memory is poorer when people need to remember a set of words that are phonologically similar, compared to a set of words that are phonologically dissimilar (Baddeley, 1966; Conrad & Hull, 1964[59]). For example, it is harder to remember the set *boat, bowl, bone* and *bore*, compared to the set *stick, pear, friend* and *cake*. This is because words that sound similar can become confused in the phonological store. One thing that happens is that, because the words sound similar, it is hard to keep track of what was rehearsed and what was not. As a consequence, some words may not get rehearsed and are forgotten (Li, Schweickert & Gandour, 2000[60]). In addition, as bits and pieces of words become forgotten or lost, people need to reconstruct them. As a result, people are more likely to make a mistake by misremembering a word that sounded like it should have been in the set, but was not – for example, recalling the word *bold* in the first set of words. In general, when people misremember words in working memory, those are the words that tend to sound similar, rather than having a similar meaning. This suggests that this aspect of working memory relies primarily on phonological rather than semantic information.

Although we have spent a great deal of time covering the spoken/heard language aspects of the phonological loop, evidence exists that there are broader aspects of this part of working memory. For example, it has been found that memory for musical pitches shows similar characteristics to language, in that working memory has a limited capacity for what can be remembered, and people become confused by similar pitches, much like the phonological similarity effect (Williamson, Baddeley & Hitch, 2010[61]). Also, in a very clever study, Shand (1982)[62] tested people who were congenitally deaf and skilled at American Sign Language (ASL). They were given five-item lists for serial recall, presented as either written English words or ASL signs. One list contained English words that were phonologically similar (*shoe, through, new*) though not similar in terms of the ASL signs. Another list contained words that similar in the hand movements necessary for forming the sign (e.g. wrist rotation in the vicinity of the signer's face), although they did not rhyme in English. Recall memory showed confusions based on the hand movement relatedness. In other words, the deaf people were recoding the written words into an ASL-based code and holding *that* in working memory. Their errors naturally reflected the physical movements of that code rather than verbal or auditory features of the words.

Articulatory suppression task

The basic idea behind the articulatory suppression task is that repeated talking consumes the resources of the articulatory loop, making it difficult to maintain other information.

For this task, people are asked to repeat words aloud over and over again while trying to remember another set of verbal/linguistic information. On the face of it, the articulatory suppression task sounds very easy; that it should not be too difficult.

However, actually doing it is a humbling experience that shows how limited our working memory capacity is, and how poor our ability is to do more than one thing at a time when the same part of working memory needs to be used.

Across	Figure
Result	Action
Center	Mother
Reason	Became
Effect	Making
Period	Really
Behind	Either
Having	Office
Cannot	Common
Future	Moment

To illustrate the powerful influence of articulatory suppression, give above are two lists of 10 words each. Copy them down onto a set of note cards. These are just examples, and you can make more lists if you want. Then find a few people to be your participants.

Have them read the cards for each list by allowing them to see each one for 1 second before moving on to the next. When the end of the list is reached, the people should write down as many words as they can remember. Then for one list, have the people simply read the words. However, for the other list, the articulatory suppression list, have them say the word the over and over again (the, the, the . . .) while reading the words. Have them keep repeating the until the end of the list is reached. What you should find is that their performance is worse under articulatory suppression than when they can read in peace and quiet.

5.3.4 The visuo-spatial sketch pad

The *visuo-spatial sketch pad* is a system for visual and spatial information, maintaining that kind of information in a short-duration buffer. If you must generate and hold a visual image for further processing, the visuo-spatial sketch pad is at work. In general, support exists for the idea that the visuo-spatial sketch pad shares some of the same neural processes when manipulating mental images as are used during active perception (Broggin, Savazzi & Marzi, 2012; Kosslyn et al., 1993[63]).

The operation of the visuo-spatial sketch pad can be illustrated by a study by Brooks (1968).[64] People were asked to hold a visual image in working memory (a large block capital *F*), then to scan that image clockwise, beginning at the lower left corner. In one condition, people said 'yes' aloud if the corner they reached while scanning was at the extreme top or bottom of the figure and 'no' otherwise. This was the 'image plus verbal' condition. The other condition was an 'image plus visual' search condition: while people scanned the mental image, they also had to search through a printed page, locating the column that listed the 'yes' or 'no' decisions in the correct order. Thus, two different secondary tasks were combined with the primary task of image scanning. All the tasks used the visuo-spatial sketch pad of working memory. The result was that making verbal responses – saying 'yes' or 'no' – was easy and yielded few errors. However, visual scanning of printed columns was more difficult and yielded substantial errors. This is because scanning the response columns forced the visuo-spatial sketch pad to divide its resources between two tasks, and performance suffered.

Effects of the visuo-spatial sketch pad

A number of effects have been observed that illustrate basic qualities of the visuo-spatial sketch pad. Two of the overarching principles of this aspect of working memory are the influence of embodied cognition and the focus on the future. As you will see, processing in the visuo-spatial sketch pad acts as if people were actively interacting with objects in the world.

The visuo-spatial sketch pad in action

The visuo-spatial sketch pad is not an abstract code; it is a dynamic system that allows people to predict what would happen next if they were actually involved in a situation.

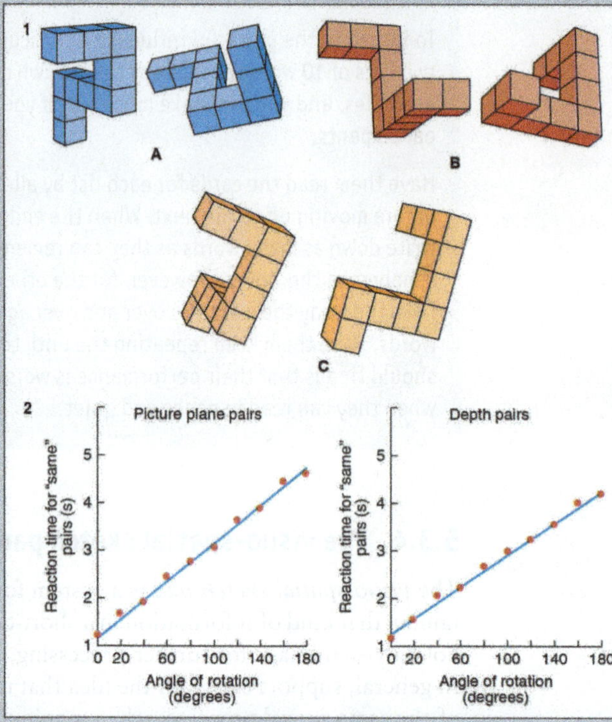

Figure 5.10 Example of stimuli measuring visuo-spatial memory
1. Three pairs of drawings are shown. For each, rotate the second drawing and decide whether it is the same figure as the first drawing. The A pair differs by an 80-degree rotation in the picture plane, and the B pair differs by 80 degrees in depth; the patterns in C do not match. 2. The response times to judge 'same' are shown as a function of the degrees of rotation necessary to bring the second pattern into the same orientation as the first. Response time is a linear function of rotation.

Mental rotation

The most dramatic evidence for the visuo-spatial sketch pad comes from work on mental rotation (Cooper & Shepard, 1973; Shepard & Metzler, 1971[65]). Mental rotation involves people mentally turning, spinning, or rotating objects in the visuo-spatialsketch pad of working memory. In one study, people were shown drawings of pairs of three-dimensional objects and they had to judge whether they were the same shape. The critical factor was the degree to which the second drawing was 'rotated' from the orientation of the first. To make accurate judgements, people had to mentally transform one of the objects, mentally rotating it into the same orientation as the other so they could judge it 'same' or 'different' (see Figure 5.11).

The overall result of the articulatory suppression task was that people took longer to make their judgements as the angular rotation increased. In other words, a figure that needed to be rotated 120 degrees took longer to judge than one needing only 60 degrees of rotation, much as what would be found if a person were to manually turn the objects.

In fact, performance can be enhanced if people are given tactile feedback (by holding an object in their hands) when the object is the same shape and moves in the same way (Wraga, Swaby & Flynn, 2008[66]), consistent with an embodied interpretation. Also consistent with an embodied influence, people find it easier to mentally rotate pictures of easily manipulated objects compared to ones that are hard to physically manipulate (Flusberg & Boroditsky, 2011[67]).

In the Cooper and Shepard (1973) report, people were shown the first figure and were told how much rotation to expect in the second figure. This advance information on the degree of rotation permitted people to do the mental rotation ahead of time. Interestingly, the mental processes seem much the same if you ask people to retrieve an image from long-term memory, then hold it in working memory while performing mental rotation on that image. Researchers have found regular time-based effects of rotation, and activation in the visual (parietal) lobes, when people are asked to retrieve an image from long-term memory and rotate it mentally in working memory (Just, Carpenter, Maguire, Diwadkar & McMains, 2001[68]).

Boundary extension

Another illustration of properties of the visuo-spatial sketch pad is boundary extension, in which people tend to misremember more of a scene than they actually viewed, as if the boundaries of an image were extended farther out (Intraub & Richardson, 1989). (For a review, see Hubbard, Hutchison & Courtney, 2010.)[69] In boundary extension studies, people see a series of pictures. Later, memory is tested for what was seen in the pictures. This can be done by having people either draw what they remember or identify the image they saw earlier. What is typically found is that people tend to misremember having viewed the picture from farther back than was the case.

That is, people misremember information from beyond the bounds of the actual picture.

Figure 5.11 A pair of pictures: if people saw the one on the left, they were more likely to select the one on the right as having been seen before

Source: kurhan/123RF

Overall, it should be clear that there is a lot of active cognition in the visuo-spatial sketch pad. This part of working memory is doing a lot of work, even if you are not consciously aware of much of it. Moreover, this work is oriented toward capturing physical aspects of the world (accurately or not) to help predict what objects will do next so you can better interact with them, such as intercepting or avoiding them, with a minimum of conscious cognitive mental effort. (For example, as can be seen in Figure 5.11, the finding that people will remember more of an image than was seen previously, beyond the boundaries of the actual image). Here, visuo-spatial working memory adds knowledge of what is beyond the picture boundary, based on previous world knowledge of what is likely to be there.

This is then stored in long-term memory. So, when you think back to a show you have seen on television or at the movies, you tend to remember the events as if you were there, with no edge to the world. You do not typically remember the image as it appeared on the screen (or even remember sitting there watching the show).

Representational momentum

The last of the visuo-spatial phenomena considered here (and there are many more) is representational momentum, which is the phenomenon of misremembering the movement of an object farther along its path of travel than where it actually was when it was last seen (Hubbard, 1995[70], 2005[71]). In a typical representational momentum study, people see an object moving along a computer screen. At some point the object disappears. The task is to indicate the point on the screen where the object was last seen.

The typical results show a bias to misremember the object as being farther along its path of travel than it actually was (Freyd & Finke, 1984; Hubbard, 1990[72]). It is as if visuo-spatial working memory is simulating the movement as if it were happening in the world, predicting where that object will be next in the near future. This prediction then enters the decision process and people place the object farther along its path.

Representational momentum can be influenced by other embodied aspects of the situation as well. For example, there is also a bias to remember objects as being farther down than they actually were, as if they were being drawn down by gravity (Hubbard, 1990[73]). There is also evidence that visuo-spatial working memory takes into account friction (Hubbard, 1990[74]), centripetal and impetus forces (Hubbard, 1996[75]), even if physics has shown these ideas to be wrong, as in the case of impetus. Finally, if an object is moving in an oscillating motion, back and forth like a pendulum, and it disappears just before it is about to swing back, people will misremember it as having started its backswing (Verfaille & Y'dewalle, 1991[76]).

5.3.5 The episodic buffer

episodic buffer
The portion of working memory whereby information from different modalities and sources are bound together to form new episodic memories.

As mentioned earlier, the episodic buffer is the portion of working memory whereby information from different modalities and sources is bound together to form new episodic memories (Baddeley, 2000a[77]). This is the part of working memory where the all-important chunking process occurs, but it also includes perceptual processes, such as the integration of colour with shape in visual memory (Allen, Baddeley & Hitch, 2006[78]).

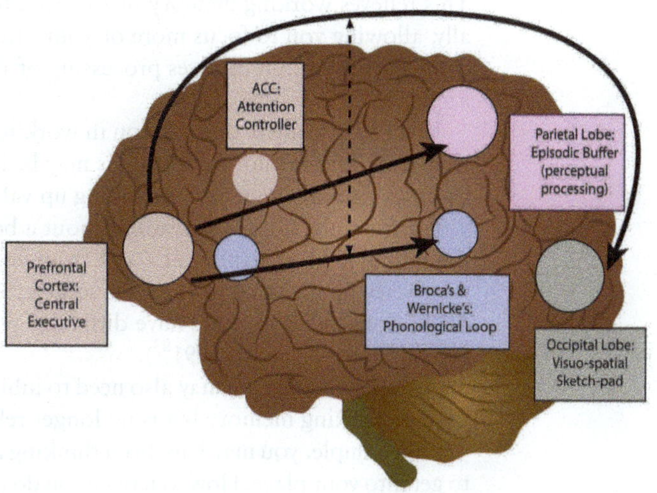

Figure 5.12 An overview of the working memory model and relevant brain areas involved in various components

Source: BSIP SA/Alamy Stock Photo

5.3.6 Engle's controlled attention model

Although Baddeley's multicomponent model is a dominant theory of working memory in research in cognitive psychology, other ideas exist about what working memory is and what it does. One such alternative perspective is **Engle and Kane's controlled attention theory of working memory** (Engle & Kane, 2004[79]).

Unlike Baddeley's theory, Engle's model does not segregate working memory into different bins and components that do different things. Instead, for this view, working memory is not a separate memory system, per se. It is merely that knowledge in long-term memory is currently active and being thought about. Earlier in this course, we pointed out that attention can be thought of as being like a spotlight. Similarly, for this view we can think of working memory as being those ideas and concepts that this mental spotlight is shining on in terms of our own knowledge.

What is particularly important to consider for Engle's view of working memory is how effective working memory is at attending to and processing information. How much control a person can exercise over his or her own train of thought reflects the amount of control a person has over his or her own attention processes. Working memory involves the information that is currently activated in working memory, and the information that is irrelevant and kept out of or suppressed from working memory. Part of the problem people have with working memory processing is not that they do not have enough information active in working memory, but that they have too much of the wrong information in the current stream of thought.

In Engle's model of working memory, two aspects of attention control are particularly important:

1. The scope of attention (i.e. how much information can be attended to at once).
2. The control of attention (i.e. how good people are at directing what is and is not attended to in working memory; Shipstead, Harrison & Engle, 2015[80]).

This control of attention can be improved by giving yourself less to think about to allow yourself to focus more on information that is relevant at that moment.

For example, if you are thinking about something difficult, you may be able to improve working memory processing by closing your eyes (Vredevelt et al., 2011[81]).

This relieves working memory of the need to process information coming in visually, allowing you to focus more of your attention on your internal thoughts. This control of attention involves processing of the frontal lobes of the brain (Kane & Engle, 2002[82]).

Part of the control of attention in working memory is the suppression or inhibition of irrelevant information. This may be information that needs to be kept from entering working memory and taking up valuable working memory capacity.

For example, if you are reading about a boy fishing on a bank, you want to keep the money-bank meaning of *bank* from entering working memory, and to allow the riverbank meaning in. There is some evidence that people who do more poorly on working memory tests have difficulty with this kind of inhibitory processing (Gernsbacher & Faust, 1991[83]).

Alternatively, people may also need to inhibit information that was recently being used in working memory but is no longer relevant.

For example, you may have been thinking about how you need to unlock the door to get into your place. However, once you do that, you need to purge this information from working memory. There is good evidence that information that was recently used is actively suppressed in working memory (Johnson-Laird, 2013[84]).

5.4 Assessing working memory

LEARNING OBJECTIVE
Identify ways to assess working memory

In general, there are two ways to assess working memory, the dual-task method and measures of working memory span:

1. In the dual-task method, performance is examined by having a person do a secondary task, one that consumes working memory resources, at the same time as some primary task. It is often used to see how disruptive the secondary task is.
2. By comparison, for working memory span tests, we get a measure of a person's working memory capacity. By comparing the span scores across a range of people and abilities to performance on other tasks, we see what relationships emerge.

Let's go over each of these methods in turn.

5.4.1 Dual-task method

For the *dual-task method*, one of the tasks done by a person is identified as the primary task we are most interested in. The other is a secondary task that is done simultaneously with the first. Both tasks must rely to some significant degree on working memory. In general, we assess how well the two tasks can be done together and whether there is any competition or interference between them. Any two tasks that are done simultaneously may show complete independence, complete dependence or some intermediate level of dependency. If neither task influences the other, then we infer that they rely on separate mental mechanisms or resources. If one task always disrupts the other, then they presumably use the same mental resources.

Finally, if the two tasks interfere with each other in some circumstances but not others, then there is evidence for a partial sharing of mental resources. Usually such interference is found when the difficulty of the tasks reaches some critical point at which the combination of the two becomes too demanding. Researchers manipulate the difficulty of the two tasks just as you would adjust the volume controls on a stereo, changing the left and right knobs independently until the combination hits some ideal setting. In research, we vary the difficulty of each task separately – we

One-man band, nicely indicating how we can do different things simultaneously at the expense of sharing mental resources.

crank up the 'difficulty knobs' on the two tasks, so to speak – and observe the critical point at which performance starts to suffer.

An important aspect of working memory that the dual-task method highlights is that information processed in one component may not interfere with processing in another. For example, active thinking that uses central executive resources will be relatively unaffected by processing that consumes resources in one of the subsystems.

A dual-task example

In one experiment (Baddeley & Hitch, 1974, Experiment 3[85]), people were asked to do a reasoning task. They were shown an item such as AB and were timed as they read and responded 'yes' or 'no' to sentences about it. A simple sentence here would be 'A precedes B', an active affirmative sentence. An equivalent meaning is expressed by 'B is preceded by A', but it is more difficult to verify because of the passive construction. There were also negative sentences, such as 'B does not precede A' and 'A is not preceded by B' (as well as false sentences; e.g. 'B precedes A'). The sentence difficulty was a way of manipulating the extent of the need for the central executive. While doing the reasoning task, people also had to do one of three secondary tasks:

- Articulatory suppression
- Repeating the numbers 1 to 6
- Repeating a random sequence of digits (the sequence was changed on every trial).

Note how the amount of articulation in the three tasks was about the same (a speaking rate of four to five words per second was enforced), but the demands on the central executive steadily increased. There was also a control condition in which there was no concurrent articulation.

Figure 5.13 shows the reasoning times for these four conditions. The control condition showed that even when reasoning was done alone, it took more time to respond to the difficult sentences. Adding articulatory suppression or repeated counting added more time to reasoning but did not change the pattern of times to any great degree; the curves for *the the the* and *one two three . . .* in the figure have roughly the same slope as the control group. This is because these tasks do not strongly consume working memory resources. However, the random digit condition yielded a different pattern. As the sentences grew more difficult, the added burden of reciting a random sequence of digits took its toll. In fact, for the hardest sentences in the reasoning task, correct judgements took nearly 6 seconds in the random digit condition, compared with only 3 seconds in the control condition. When the secondary task is difficult, the articulatory loop must drain or borrow some of the central executive's resources, thereby slowing down or sacrificing accuracy.

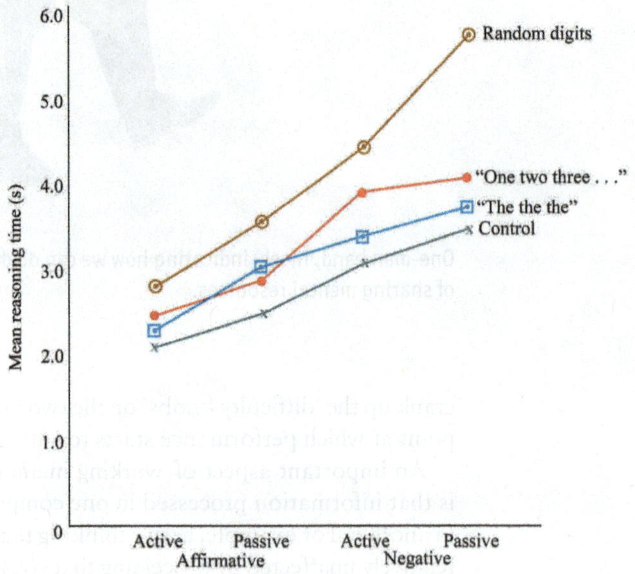

Figure 5.13

This dual-task interference has been shown in a variety of tasks, including showing that dividing attention during driving, such as talking on a cell phone, disrupts the ability to make important judgements, such as when to brake. In general, dual-task processing leads to much slower braking (Levy, Pashler & Boer, 2006[86]). In other words, when you tax working memory resources, it compromises the ability of the central executive to effectively process information.

Another dual-task example

Similar research has also been reported comparing the visuo-spatial sketch pad and the phonological loop.

In the visual memory span task, people saw a grid of squares on the computer screen, with a random half filled in. After a moment, the grid disappeared and was followed by an altered grid in which one of the previously filled squares was now empty. People had to point to the square that was changed, using their visuo-spatial memory of the earlier pattern. In contrast, the letter memory span task, the other primary task, should use the phonological loop. For the secondary tasks, Logie et al. used a mental addition task thought to be irrelevant to the visuo-spatial processing and an imaging task thought to be irrelevant to the phonological loop. The results are shown in Figure 5.14.

Logie, Zucco and Baddeley's (1990) experiment

As one example, Logie, Zucco and Baddeley (1990)[87] selected two different primary tasks: a visual memory span task and a letter span task. (See Morey & Cowan, 2004, for a contrasting view.[88])

These were paired with two secondary tasks, one involving mental addition and the other visual imagery.

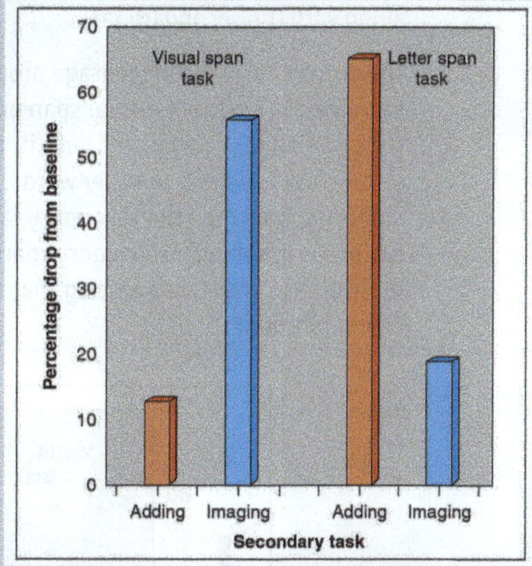

Figure 5.14 Results From Logie, Zucco and Baddeley's (1990)[89] experiment on the visuo-spatial sketch pad

Two secondary tasks, adding and imaging, were combined with two primary tasks, a visual span and a letter span. The results are shown in terms of the percentage drop in performance measured from baseline; the larger the drop, the more disruption there was from the secondary task.

In the visual memory span task, people saw a grid of squares on the computer screen, with a random half filled in. After a moment, the grid disappeared and was followed by an altered grid in which one of the previously filled squares was now empty. People had to point to the square that was changed, using their visuo-spatial

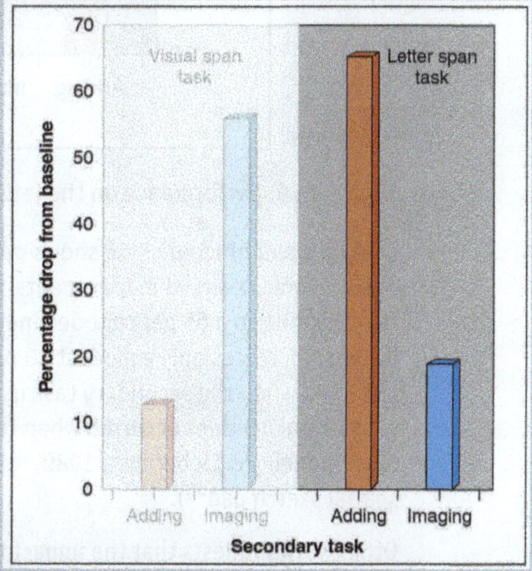

Figure 5.15 Performance on the visual span (grid pattern) task

memory of the earlier pattern. In contrast, the letter memory span task, the other primary task, should use the phonological loop.

For the secondary tasks, Logie et al. used a mental addition task thought to be irrelevant to the visuo-spatial processing and an imaging task thought to be irrelevant to the phonological loop. The results are shown in Figure 5.16.

First, look at the left half of the graph, which reports the results of the visual span (grid pattern) task. Each person did the span task alone, to determine baseline, then along with the secondary tasks.

The graph shows the percentage drop in dual-task performance as compared to baseline. For instance, visual span performance dropped about 15 per cent when the addition task was paired with it. So, dual-task performance was at 85% of the single-task baseline. In other words, mental addition disrupted visual memory only modestly. But when the secondary task was visual imagery, as shown by the second bar in the graph, visual memory span dropped about 55 per cent. This is a large interference effect, suggesting that the visuo-spatial sketch pad was stretched beyond its limits.

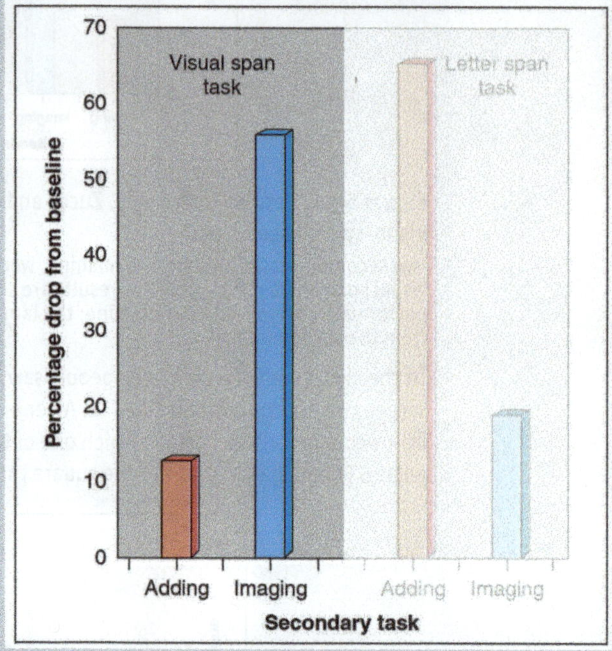

Figure 5.16 Performance on the letter span task

The right half of Figure 5.16 shows performance on the letter span task. Here, the outcome was reversed in that mental addition was very disruptive to the letter span task, leading to a 65 per cent decline. On the other hand, the imaging task depressed letter span scores only a modest 20 per cent. Thus, only minor declines were observed when the secondary task used a different part of working memory. But substantial declines occurred when the two tasks used the same pool of resources (see Baddeley & Lieberman, 1980, for some of the original research on the visuo-spatial sketch pad[90]).

Other work suggests that the impact that dual tasks such as these are having is on the encoding aspect of a task, rather than on the retention of information in working memory per se (Cowan & Morey, 2007[91]).

5.4.2 Working memory span

A different way to study working memory is an individual differences approach. As in any area of psychology, when we speak of individual differences, we are talking about characteristics of individuals – anything from height to intelligence – that differ from one person to the next and can be measured and related to other factors.

Individual differences in working memory are related to various cognitive processes. In this research, people are first given a test to assess their **working memory spans**. They are then given standard cognitive tasks. Consider a program of research by Engle and his coworkers (Engle, 2001; Rosen & Engle, 1997[92]). See Engle (2002)[93] for an excellent introduction.

First, people are given a working memory span task: the task requires simultaneous mental processing and storage of information in working memory. For example, a person might see arithmetic statements along with a word, one at a time (from Turner & Engle, 1989[94]):

(6 × 2) − 2 = 10? SPOT
(5 × 3) − 2 = 12? TRAIL
(6 × 2) − 2 = 10? BAND

People first read the problem aloud, indicated whether the answer was correct, and then said the capitalised word, followed by the second problem and word, then the third. At that point, people tried to recall the three capitalised words, showing that they had stored them in working memory. Scores on this span task are based on the number of capitalised words recalled. Thus, someone who recalled *SPOT TRAIL BAND* and answered the arithmetic questions correctly would have a memory span of 3. In another version of the working memory span task, we use sentences instead of arithmetic problems. But both involve *processing* and *storage*: processing the problem or sentence for meaning, and storing the word for recall.

Many investigators have used span tasks to measure working memory capacity. The original work that used this method (Daneman & Carpenter, 1980[95]) examined reading comprehension as a function of span. There were significant correlations between span scores and performance on the comprehension tasks. One of the most striking correlations was between span and verbal scholastic aptitude test (SAT) scores; it was .59. However, simple span scores seldom correlated significantly with SATs. (Simple span tasks, such as remembering a string of digits, test only the storage of items, whereas working memory span tasks involve both storage and processing.) This strong correlation means that there is an important underlying relationship between one's working memory span and the verbal processing measured by the SAT.

The strongest correlation in Daneman and Carpenter's work was a .90 correlation between memory span and performance on a pronoun reference test. Here, people read sentences one by one and at some point confronted a pronoun that referred back to a previous noun. In the hardest condition, the noun had occurred up to six or seven sentences earlier. The results are shown in Figure 5.17.

Here, people with the highest working memory span of 5 scored 100 per cent correct on the pronoun test, even in the 'seven sentences ago' condition. People with the lowest spans (of 2) got 0 per cent correct. Thus, people with higher working memory spans were able to keep more relevant information active in working memory as they comprehended the sentences.

Research since then has extended these findings. Basically, if a task relies on a need to control attention, scores on the task correlate strongly with working memory span. In fact, Engle (e.g. 2002[96]) argued that working memory capacity is executive attention and offers the equation 'WM = STM + controlled attention' – working memory is the combination of traditional short-term memory plus our controlled

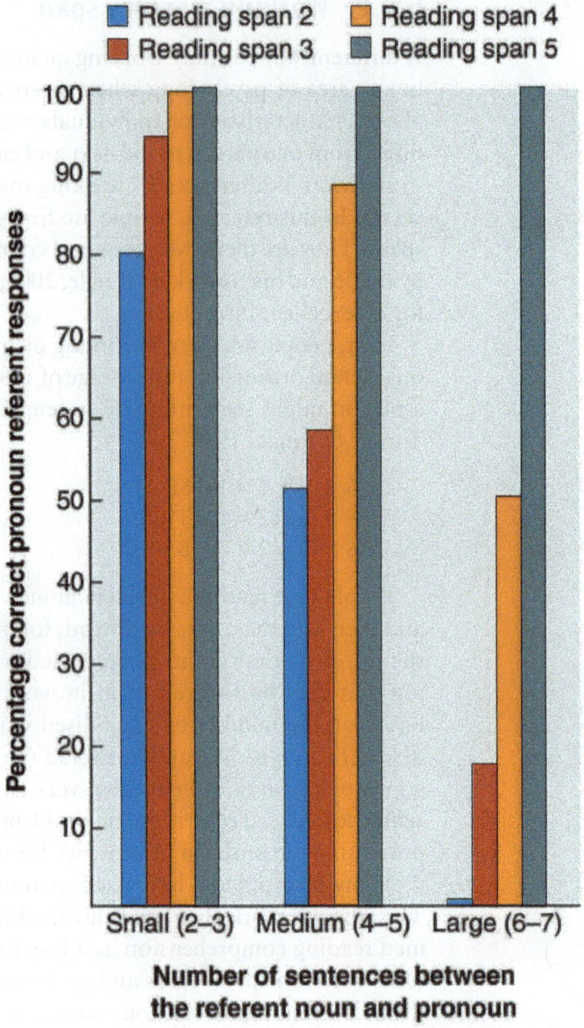

Figure 5.17 The percentage of correct responses to the pronoun reference task when the antecedent noun occurred a small, medium or large number of sentences before the pronoun, as a function of participants' working memory (reading) span

attention mechanism (Kane & Engle, 2003; see also Daneman & Merikle, 1996; Engle, 2002; Miyake & Shah, 1999; Unsworth & Spillers, 2010[97]). This attentional control of the flow of thought involves both the maintenance of information in the short term as well as the ability to access needed information in long-term memory (Unsworth & Engle, 2007[98]).

There is also some evidence that working memory abilities can change with practice. For example, women tend to perform less well than men on visuo-spatial tasks. However, with some training, such as 10 hours of experience playing action-based video games, the performance of females can reach the levels of males (Feng, Spence & Pratt, 2007[99]). Such experience can also boost performance on visuo-spatially based scientific thinking, such as understanding plate tectonics (Sanchez, 2012[100]). It has also been suggested that musical training can boost verbal intelligence (Moreno et al., 2011[101]). Finally, there is some evidence that certain kinds of Buddhist meditation can also improve mental imagery abilities (Kozhevnikov, Louchakova, Josipovic & Motes, 2009[102]). Overall, this work, along with the research by Chase and Ericsson (1982)[103] described earlier (as well as others, e.g. Verhaeghen, Cerella & Basak,

2004[104]), suggests that people can develop strategies to more efficiently and effectively use their working memories over and above any base level of capacity they may have.

That said, memory span scores do not provide insight into all aspects of cognitive abilities. For example, a study by Copeland and Radvansky (2004b)[105] gave people a variety of memory span tasks and assessed performance at more complex levels of comprehension, such as remembering event descriptions, drawing inferences about causes and effects, and detecting inconsistencies in a text. The results showed little evidence of a relation between working memory span and performance. Thus, although memory span highlights important cognitive abilities, it is not the complete story. There is individual variation that can be attributed to other factors as well.

Executive functions and frontal lobe tasks

The diversity in measuring cognitive functions has led to a series of frameworks that aim to provide relevant assessment tasks for each component of executive functions. Miyake et al. (2000, 2012)[106] focused on three executive functions – updating, shifting and inhibition – and suggested that there are specific tasks that measure them:

1. Updating – letter naming task (name the last three letters similar to a digit span).
2. Shifting – categorise objects using different rules (e.g. by shape or by colour).
3. Inhibition – go/no-go task (press a 'go' button for all target except one 'no-go').

Other tasks for each executive function are presented in Figure 5.18.

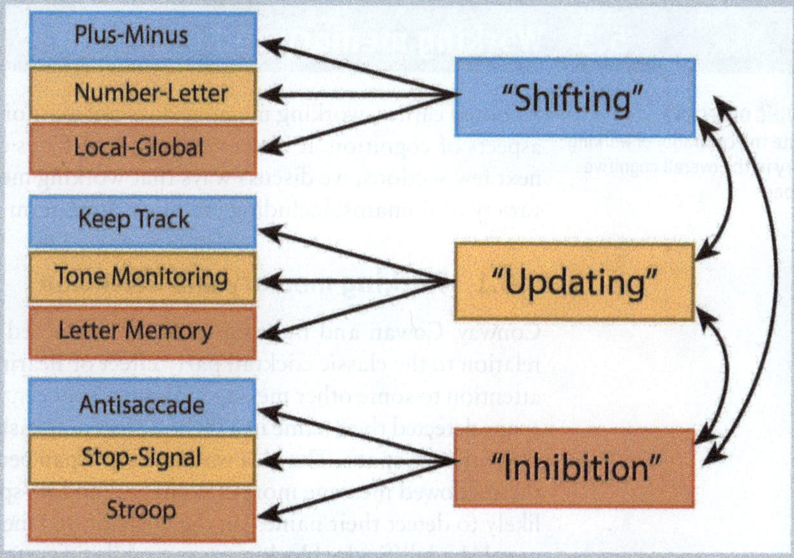

Figure 5.18 The three components of the Miyake et al. model (shifting, updating and inhibition) and examples of relevant tasks used to measure them

5.4.3 Improving working memory

A line of work that has been gathering a great deal of attention and exploration recently is the idea that it might be possible to have people undergo working memory training and improve their overall working memory ability. This idea is similar to that of improving attention processing by having people play action video games. The issue of improving working memory ability is of such great interest because the effectiveness of working memory is strongly related to improved cognitive performance on a wide range of other mental processes, as outlined in the next section of the module. Most salient is the finding that working memory performance is often strongly correlated with measures of fluid intelligence. So, the thinking goes, if you can boost a person's working memory abilities, you should also boost his or her intelligence. And who wouldn't want that?

This question has gained so much attention and interest because certain studies appear to show that working memory training can at least modestly improve working memory performance, often within weeks, as well as performance on intelligence tests (Au et al., 2015[107]). That said, not all the news is so rosy. Some researchers have suggested that, although working memory training regimens certainly improve a person's ability to take working memory tests, it is not clear that such working memory task improvement carries over to improvements in other kinds of cognition or to general intelligence (e.g. Chooi & Thompson, 2012; Harrison et al., 2013; Waris, Soveri & Laine, 2015[108]). There have also been some suggestions that studies showing that working memory training improves cognition may suffer from methodological problems (Bogg & Lasecki, 2015[109]). This does not detract from other research showing that certain kinds of activities are related to superior intelligence, such as musical instrument training (Benz, Sellaro, Hommel & Colzato, 2015; Gordon, Fehd & McCandliss, 2015[110]).

5.5 Working memory and cognition

LEARNING OBJECTIVE
Evaluate the centrality of working memory in the overall cognitive functions

As noted earlier, working memory does not exist or operate independently of other aspects of cognition. It is the vital centre of focus of a great deal of activity. In the next few sections, we discuss ways that working memory influences processing in a variety of domains, including attention, long-term memory and reasoning.

5.5.1 Working memory and attention

Conway, Cowan and Bunting (2001)[111] examined working memory span and its relation to the classic cocktail party effect of hearing one's own name while paying attention to some other message. About 65 per cent of the people with low memory spans detected their name in a dichotic listening task, versus only 20 per cent of people with high spans. The idea was that high-span people were selectively attending to the shadowed message more effectively than low-span people – so they were not as likely to detect their names on the unshadowed message. In contrast, the low-span people had difficulty blocking out or inhibiting attention to the distracting information in the unattended message – so they were more likely to hear their own names.

In a similar demonstration, Kane and Engle (2003)[112] used the classic *Stroop task* – naming the ink colour the word is printed in, when the ink color mismatches. There was a strong Stroop effect, of course – approximately a 100-ms slowdown on mismatching items. More to the point, there was no difference in the Stroop effect for high- and low-span groups when the words were always in a mismatching colour (GREEN printed in red ink) or when half of the words were presented that way.

Everyone remembered the task goal – ignore the word – in these conditions. But when only 20 per cent of the words were in mismatching colours, low-span people made nearly twice as many errors as high-span people. Because mismatching trials were rarer, the low-span people seemed less able to maintain the task goal in working memory. High-span individuals had less difficulty maintaining that goal.

In a more everyday example, Sanchez and Wiley (2006)[113] tested people with different memory spans, giving them texts to read that included illustrations. These illustrations were often irrelevant to the main points of the text, such as having a picture of snow in a passage about ice ages – the snow is related to the topic, but does not provide or support any new information. As such, performance is better if working memory capacity were to focus on the relevant details in the text. What was found was that people with lower memory spans were more likely to be 'seduced' by the irrelevant details in the pictures. They had more difficulty controlling the contents of their current stream of thought and were more apt to be led astray by attractive, but unhelpful, sources of knowledge that served as a distraction.

5.5.2 Working memory and long-term memory

Long-term memory function can also depend on working memory. Rosen and Engle (1997),[114] for instance, had high- and low-span people do a verbal fluency task: generate members of the animal category as rapidly as possible for up to 15 minutes. High-span people outperformed their low-span counterparts, a difference noticeable even 1 minute into the task. Intriguingly, in a second experiment, both groups were tested in the fluency task alone and in a dual-task setting. While naming animals, people had to simultaneously monitor the digits that showed up, one by one, on the computer monitor and press a key whenever three odd digits appeared in the sequence. This task reduced performance on the animal naming task, but only for the high-span people, as shown in Figure 5.19.

Rosen and Engle suggested that the normal, automatic long-term memory search for animal names was equivalent in both groups. But high-span people were able to

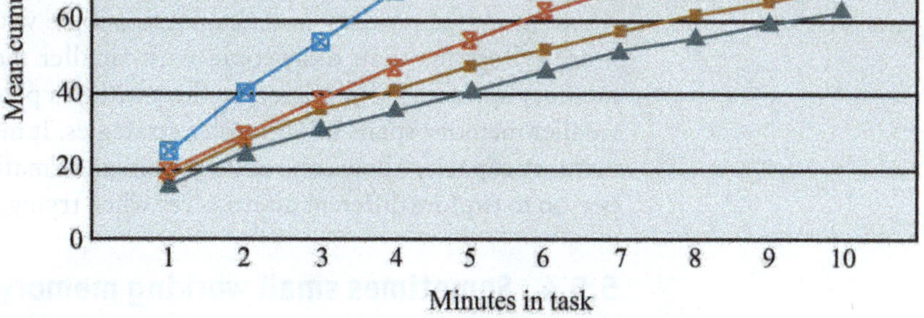

Figure 5.19 The cumulative number of animal names generated by participants of high (open points) or low (filled points) working memory span

Dashed lines indicate performance when participants performed the secondary task of monitoring a stream of digits while generating animal names. Low-span people showed no decrease in performance.

augment this with a conscious, controlled strategic search; in other words, along with regular retrieval, the high-span people could deliberately ferret out additional hard-to-find animal names using a controlled attentional process. This additional 'ferreting' process used working memory. Consequently, the added digit monitoring task used up the working memory resources that had been devoted to the strategic search. This made the high-span group perform more like the low-span group when they had to perform the dual task.

Other studies have also shown the importance of working memory. For instance, Kane and Engle (2000)[115] found that low-span people experience more proactive interference (PI) in the Brown–Peterson task than do high-span people. High-span people used their controlled attentional processes to combat PI, so they showed an increase in PI when they had to do a simultaneous secondary task. (See Bunting, Conway & Heitz, 2004; Cantor & Engle, 1993; and Radvansky & Copeland, 2006a for an exploration of the role of working memory span in managing associative interference during retrieval.[116]) Generally, low-span people appear to search a wider range of knowledge, making them more prone to having irrelevant information intrude on retrieval (Unsworth, 2007[117]). In Hambrick and Engle (2002; see also Hambrick & Meinz, 2011[118]), high-span people had better performance than low-span people on a long-term memory retrieval task, even when both groups were equated for the rather specialised domain knowledge being tested. (In the experiment, people listened to simulated radio broadcasts of baseball games.)

5.5.3 Working memory and reasoning

The idea that working memory involves controlled attention can also be tied to general issues of cognitive and behavioural control, such as those needed in problem-solving. People with lower memory span scores may be less effective at controlling their thought processes. One example of this is a study by Moore, Clark and Kane (2008)[119] that looked at working memory span and choices on moral reasoning problems. So, suppose there is a runaway trolley car. If you let it go, it will kill four unaware people a bit down the track. Alternatively, you could push a very large person next to you in front of the trolley; it will kill him but will derail the trolley and save the other four people. So, how morally acceptable is each of these choices? Moore et al. found that moral reasoning of this type was mediated by a person's working memory capacity, with those with a high working memory capacity making choices on a more consistent (i.e. principled) basis.

The influence of memory span is also seen on more traditional sorts of mental reasoning, such as solving logic problems like categorical syllogisms. In a study by Copeland and Radvansky (2004a[120]; see also Markovits & Doyon, 2004[121]), people with various working memory spans were asked to solve a set of logic problems. There were two primary findings. First, people with greater memory spans solved more syllogisms than did people with smaller memory spans. Second, working memory span was also related to the strategies people used to reason. People with smaller memory spans used simpler strategies. It may be that having more working memory capacity allows one to keep more information active in memory, allowing a person to explore different alternatives when trying to reason and draw conclusions.

5.5.4 Sometimes small working memory spans are better

Intuitively, it would seem that people with more working memory capacity, or those who engage working memory resources more effectively, are more likely to succeed, and this is generally true. However, there are some interesting exceptions. One of these is illustrated in a study by Beilock and DeCaro (2007)[122] in which high- and low-span people were given math problems to solve. Under normal conditions, high-span people tended to do better. However, people were then placed in a high-pressure situation. They were told that they were being timed, that their performance would be videotaped so math experts could evaluate them, that they would be paid for improving their performance, and so forth. In this high-pressure condition, working memory capacity was consumed with task-irrelevant, anxiety-induced thoughts, and performance in both the high- and low-span groups was equivalent (and lower). Thus, when people have their working memory capacity consumed by irrelevant thoughts, they are more likely to use simpler, less effective strategies. This shift to simpler strategies levelled the playing field by causing the high-span people to solve the problems more like the low-span people, who had been using simpler strategies in the first place.

Of particular interest, in a second experiment, people were asked to do a series of word problems that required a complex series of steps (i.e. $B - A - 2 \times C$). Then under low- or high-pressure conditions, people were given a series of new problems, some of which required a simpler solution (i.e. $A - C$). Beilock and DeCaro (2007)[123] found that the low-span people were actually more likely to use the simpler correct solution than the high-span people. The explanation was that low-span people are less likely to derive rule-based strategies for solving problems (because they have less capacity to do so) and are more likely to draw from previous similar experiences. Thus, when given problems with the simpler solutions, the low-span people were less dependent on a complex, rule-based strategy they had derived earlier, and so were more likely to use the more appropriate, simpler strategy. (For other examples of better performance by people with smaller memory spans, see Cokely, Kelley & Gilchrist, 2006; Colflesh & Conway, 2007.[124])

5.5.5 Working memory overview

The general conclusion from all these studies is that *working memory* is a more suitable name for the attention-limited workbench system. Working memory is responsible for the active mental effort of regulating attention, for transferring information into and from long-term memory.

More importantly, there is a limit to the mental resources available to working memory. When extra resources are drained by the subsystems, the central executive suffers along with insufficient resources for its own work. Naturally, as processes become more automatic, fewer resources are tied down (e.g. working memory is unrelated to counting when there are only two or three things to count but is influenced for larger quantities; Tuholski, Engle & Baylis, 2001[125]).

Engle's view of working memory emphasises the general nature of working memory capacity as a measure of executive attention and de-emphasises the multicomponent working memory approach advocated by Baddeley. Part of the reason for this is the generality of the working memory effects. Working memory span predicts performance on a variety of tasks. Of particular importance, working memory span routinely correlates with measures of intelligence, especially so-called fluid intelligence (Fukuda, Vogel, Mayr & Awh, 2010; Kane et al., 2004; Salthouse & Pink, 2008; Shelton, Elliot, Matthews, Hill & Gouvier, 2010[126]), and more so if metacognitive control is taken into account (McCabe, 2010[127]). The relationship between working memory and intelligence is that people who have greater cognitive control and can manage sources of interference better, score better on intelligence tests. This is even supported by neuroimaging data in which people with higher working memory span and intelligence scores show better interference control in cortical areas tied to these processes, such as the dorsolateral prefrontal cortex and portions of the parietal cortex (Burgess, Gray, Conway & Braver, 2011[128]).

Common to both the Baddeley and the Engle approaches, however, is a central set of principles. Working memory is intimately related to executive control, to the deliberate allocation of attention to a task, and to the maintenance of efficient, effective cognitive processing and behavior. However, there is a limitation in the amount of attention available at any one time. Furthermore, the ability to deliberately focus and allocate attention, and to suppress or inhibit attention to extraneous factors, is key to higher-order cognitive processing.

Summary: Short-term working memory

5.1 A limited-capacity bottleneck

- Short-term or working memory is an intermediate system between the sensory and long-term memories. Its capacity for holding information is limited, by some accounts, to only 7 +/− 2 units of information, although other accounts suggest that it may be able to hold only 4 +/− 1 chunks of information. The processes of chunking and recoding, grouping more information into a single unit, are ways of overcoming this limit or bottleneck.
- Whereas a decay explanation of forgetting from short-term memory is possible, most research implicates interference as the primary reason for forgetting. The research suggests two kinds of interference: retroactive and proactive interference.

5.2 Short-term memory retrieval

- Serial position curves reveal two kinds of memory. Early positions in a list are sensitive to deliberate rehearsal that transfers information into long-term memory, called the primacy effect, whereas later positions tend to be recalled with high accuracy in the free recall task. This latter effect, called the recency effect, is due to the strategy of recalling the most recent items first. Asking people to do a distractor task before recall usually eliminates the recency effect because the distractor task prevents them from maintaining the most recent items in short-term memory.
- Sternberg's paradigm, short-term memory scanning, provided a way to investigate how we search through short-term memory. Sternberg's results indicated that this search is a serial exhaustive process occurring at a rate of about 38 ms per item to be searched.

5.3 Working memory

- Working memory consists of a central executive and three major subsystems: the phonological loop for verbal and auditory information, the visuo-spatial sketch pad for visual and spatial information, and the episodic buffer for integrating or binding information from different parts of working memory and/or long-term memory.
- The various components of working memory are thought to operate relatively independently of one another, perhaps by using different neural substrates, although there can be some overlap for demanding tasks.
- An alternative view is that working memory is critically involved in the control of attention and thought. From this perspective, working memory is not a separate system, but the portion of long-term memory that is currently activated.
- There are capacity limits in the system.

5.4 Assessing working memory

- One common method for assessing working memory is to use dual-task methodologies. In these tasks, people are asked to simultaneously do at least two tasks. Researchers then assess how performance on the primary task is affected by the addition of the secondary task, and the theoretical relationship between the two.
- Dual-task methods can be used to study strains on individual components or on the overall capacity of working memory. For example, the subsystems may drain extra needed capacity from the central executive in situations of high working memory demands.
- An alternative research strategy is to test working memory span, then examine differences in performance as a function of span scores. This has revealed several tasks that show a relationship between span and performance. The implication is that working memory span assesses controlled attentional processes, which are significant aspects of performance.
- There have been several efforts at working memory training that aim to boost working memory performance and, hence, general intelligence. Although there has been some reported success on this score, there is also cause for concern as research continues to sort this issue out.

5.5 Working memory and cognition

- Working memory abilities and performance are critical to many tasks assessed by cognitive psychologists. For example, working memory capacity is strongly related to the ability to engage attention. It has also been shown to be strongly related to the efficiency with which simple facts can be retrieved from long-term memory.

- Although larger working memory capacity is associated with superior cognitive performance, there are cases in which circumstances favour smaller working memory capacity. These are typically circumstances in which it is better not to devote too much attention to a task.
- Although there is no clear view on exactly what working memory is, as evidenced by the Baddeley multicomponent model and the Engle attentional control view, there are a number of agreed-on characteristics of what working memory is able to do. These include its limited capacity, the ability to simultaneously handle certain types of non-interfering forms of information, the fact that people differ in their working memory capacities and abilities, and that these individual differences are related to performance on a variety of tasks.

Chapter 5 quiz

Question 1

Which of the following best demonstrates the process of recoding?

A. Being given a list of random numbers to remember and seeing how many can be remembered straight away
B. Repeating the names of 3 animals over again at 3 time points during the day.
C. Remembering the colours of the rainbow
D. Reading a list of random digits in a period of 20 seconds and remembering them a few hours later

Question 2

You have been asked to provide a phone number to a friend. As you have no paper, you are trying to remember all of the digits of the phone number. When you get to your friend, they talk to you about how they are struggling with their maths homework, and this means that you cannot remember the phone number you were going to give them. What is this an example of?

A. Proactive interference
B. Decay
C. Retroactive interference
D. Forgetting

Question 3

Sternberg (1969) developed the process model which provided details of the four processes involved when a person is in the timed portion of a memory task. What is the correct order of the stages in the model?

A. Encode the probe, Scan and comparison with the memory set, Binary (yes/no) decision, Executive motor response.
B. Scan and comparison with the memory set, Binary (yes/no) decision, Executive motor response, Encode the probe.
C. Scan and comparison with the memory set, Binary (yes/no) decision, Encode the probe, Executive motor response.
D. Executive motor response, Encode the probe, Scan and comparison with the memory set, Binary (yes/no) decision.

Question 4

The dual task method assesses the properties of working memory by:

A. Determining whether verbal and spatial tasks are processed simultaneously.
B. Investigating whether several unrelated tasks drain the same pool of resources.
C. Testing the nature and accuracy of recall of lists of related or unrelated information.
D. Calculating how much information can be attended to before bandwidth is exhausted.

Question 5

Working memory training has been shown to improve general working memory ability. What is a key criticism of working memory training?

A. Working memory training can only be used for individuals with cognitive conditions such as Dementia or Alzheimer's disease.
B. Working memory training may be too generic in the resources they target.
C. Working memory training can cause a decrease in working memory as it will strengthen other cognitive functions instead.
D. We are unsure of the effectiveness of working memory training as working memory has associations with other cognitive constructs such as intelligence and attention.

6 Learning and remembering

Learning objectives

6.1 Describe the key components of long-term memory

6.2 Explain the process by which information is stored in long-term memory

6.3 Describe ways to increase the capacity of long-term memory

6.4 Explain the role of context in the storage and retrieval of long-term memories

6.5 Identify the influence of mental representations on long-term memory

6.6 Understand and evaluate research on autobiographical memory

6.7 Describe how memory can be oriented towards the future

6.8 Explain the connectionism approach to understanding semantic memory and human memory more widely

Key figures

Professors Richard Atkinson and Richard Shiffrin

Professors Richard Atkinson and Richard Shiffrin are two of the most famous researchers in the field of memory with their model on multi-store memory (1968). In their model they described the different memory systems, including the sensory store, the short-term memory store and the long-term memory store.

Professor Ivan Izquierdo

Professor Izquierdo worked in the National Institute of Translational Neuroscience of Brazil and published over 650 works on the neuroscience of memory. He was the Head of the Memory Centre, Brain Institute in Porto Alegre.

Introduction

This is the first of two chapters devoted to long-term memory, the storage vault for a lifetime's worth of knowledge and experience. Why two chapters?

1. Long-term memory is fundamental to nearly every mental process, to almost every act of cognition. You cannot understand cognition unless you understand long-term memory.
2. Long-term memory is an enormous area of research because there is a lot to know about it.

3. People are curious about their own memories. Who has not complained at some time about forgetfulness or the unreliability of memory?

Long-term memory is divided and subdivided in various ways. Look at Figure 6.1, a taxonomy suggested by Squire (1986, 1993[1]).

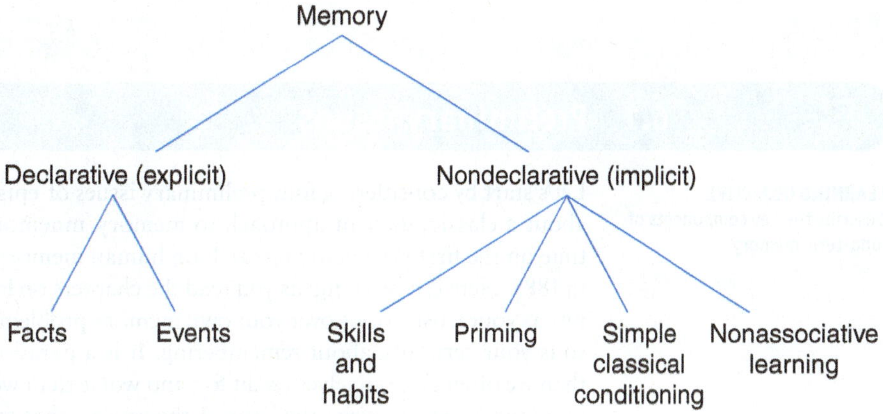

Figure 6.1 A taxonomy of long-term memories

As the figure shows, a distinction can be made between declarative memory (or explicit memory) and non-declarative memory (or implicit memory). Here, **declarative** or **explicit memory** is long-term memory knowledge that is retrieved and reflected on consciously. In contrast, **non-declarative** or **implicit memory** is knowledge that influences thought and behaviour without any necessary involvement of consciousness. The key to this distinction is conscious awareness: either one has it or one does not.

Here is a brief example to clarify the distinction between episodic and semantic memory. If you were asked what happened when you took your driver's licence test, you would retrieve knowledge from *episodic memory* – memory of the personally experienced events. When you retrieve that information, you are conscious of it, you can talk about it, and so on. Episodic memory enables you to record your personal interactions with the world. Alternatively, if you were asked what a driver's licence is, you would retrieve information from *semantic memory*, which is your general world knowledge. You retrieve the concept of a driver's licence, and it is now in your conscious awareness. Whereas episodic memory is your mental slide show, semantic memory is your mental encyclopaedia. Both involve knowledge that you can be consciously aware of (Tulving, 1972, 1983, 1993[2]).

But there is more going on than what rises to the level of conscious awareness. As you read this sentence, you encounter the term *licence* again. Although you are not conscious of it, you are faster at reading that term than you were the first time. The re-reading speed-up is called **repetition priming**. This happens at the non-declarative, implicit level. Likewise, if we gave you some word stems an hour from now and asked you to fill in the blanks with the first word that comes to mind, you would be more likely than chance to complete *LIC___* as *LICENCE* because you had encountered it recently. This happens even if you do not consciously recall having seen the word *licence* (but teasing apart conscious and unconscious influences is no simple matter; see Buchner & Wippich, 2000; Jacoby, 1991; Jacoby, Toth & Yonelinas, 1993[3]).

An important aspect of episodic memories is that they are integrated mental representations. Different bits and pieces of information from different parts of our conscious and unconscious mental worlds are woven together. The *episodic buffer* is the part of working memory that integrates different types of knowledge to form episodic memories. Although many of the examples given here use linguistic materials, such as word lists, episodic memories integrate various types of information including sensory, motor, spatial, language, emotional and narrative information, as well as various other encoding and retrieval processes (Rubin, 2007[4]). Even this is not an exhaustive list. So, as you can see, episodic memory uses a rich variety of information about a broad range of human experience.

6.1 Preliminary issues

LEARNING OBJECTIVE
Describe the key components of long-term memory

Let's start by considering four preliminary issues of episodic memory. First, we talk about a classic, ancient approach to memory, mnemonic devices. We then spend time on the first systematic research on human memory, by Ebbinghaus, published in 1885. Here is a warning: as you read the chapters on long-term memory, note that just as your frustrations over your own memory problems probably are exaggerated, so is your certainty about remembering. It is a paradox; our memories are better than we often give ourselves credit for, and worse than we are often willing to believe or admit. Next, we cover the issue of the process that makes memories permanent: consolidation. This topic was mentioned earlier in the cognitive neuroscience chapter and is discussed further here because this process is vital in making long-term memories long term. Finally, we consider other issues about people's awareness of their own memory content and processes, what is known as *metamemory*.

6.1.1 Mnemonics

The term *mnemonic* (pronounced 'ne-MAHN-ick') means 'to help the memory'; it comes from the same Indo-European base word as *remember*, *mind* and *think*. A **mnemonic device** is an active, strategic learning device or method. Formal mnemonics use pre-established sets of aids and require considerable practice. The strengths of mnemonics include the following:

mnemonic device
Any mental device or strategy that provides a useful rehearsal strategy for storing and remembering difficult material; see *method of loci*, for instance.

1. The material to be remembered is practised repeatedly.
2. The material is integrated into an existing memory framework.
3. The mnemonic provides a way to retrieve the material.

Classic mnemonics

The first historical mention of mnemonics is from the first century B.C., in Cicero's *De oratore*, a treatise on rhetoric. *Rhetoric* is the art of public speaking, which in those days meant speaking from memory.

Techniques of mnemonics

The power of mnemonics is tremendous. Among other things, mnemonics enabled Greek orators to memorise and recite epics such as *The Iliad* and *The Odyssey*.

Cicero describes a technique, based on visual imagery and memorised locations, called the **method of loci** (*loci* is the plural of *locus*, meaning 'a place'; pronounced 'LOW-sigh'). There are two steps to this method:

1. Choose a known set of locations that can be recalled easily and in order, such as locations you encounter in a walk across campus or as you arrive home.

Table 6.1 The method of loci

Set of loci	Words to be remembered	Grocery list and images
Driveway	Grapefruit	Grapefruit instead of rocks alongside driveway
Garage door	Tomatoes	Tomatoes splattered on garage door
Front door of the house	Lettuce	Lettuce leaves hanging over door instead of awning
Coat closet	Oatmeal	Oatmeal oozing out the door when I hang up my coat
Fireplace	Milk	Fire got out of control, so spray milk instead of water
Easy chair	Sugar	Throw pillow is a 2–3 kg bag of sugar
Television	Coffee	Mrs Olson advertising coffee
Dining-room table	Carrots	Legs of table are made of carrots

2. Form a mental image of the first thing you want to remember and mentally place it into the first location, the second item in the second location, and so on (see Table 6.1 for examples). When it is time to recall the items, you mentally stroll through your set of locations, 'looking' at the places and 'seeing' the items you have placed there.

Table 6.2 The peg word mnemonic

Numbered pegs	Word to be learned	Image
One is a bun	Cup	Hamburger bun with smashed cup
Two is a shoe	Flag	Running shoes with flag
Three is a tree	Horse	Horse stranded in top of tree
Four is a door	Dollar	Dollar bill tacked to front door
Five is a hive	Brush	Queen bee brushing her hair
Six is sticks	Pan	Boiling a pan full of cinnamon sticks
Seven is Heaven	Clock	St Peter checking the clock at the gates of Heaven
Eight is a gate	Pen	A picket fence gate with ballpoint pens as pickets
Nine is a vine	Paper	Honeysuckle vine with newspapers instead of blossoms
Ten is a hen	Shirt	A baked hen on the platter wearing a flannel shirt

Principles of mnemonic effectiveness

Mnemonic effectiveness involves three principles.

There are several mnemonic techniques that are commonly used to improve long-term memory. One of the most frequently used is mental imagery (Roediger, 1980[6]). In its simplest form, you can create an image for the items to be remembered (e.g. a shopping list) but each image should prompt the recall of another item too. If done effectively the recall of one item should cue recall of the next (e.g. recall of milk could prompt yogurt). The obvious drawback is that if you forget the first item it's unlikely you will recall its pair.

Three principles of mnemonic effectiveness

First principle

It provides a structure for acquiring the information. The structure may be elaborate, like a set of 40 loci, or simple, like rhyming peg words. It can even be arbitrary if the material is not extensive. (The mnemonic *HOMES* for the names of the five Great Lakes – Huron, Ontario, Michigan, Erie and Superior – is not related to the to-be-remembered material, but it is simple.)

Second principle

Using visual images, rhymes or other kinds of associations and the effort and rehearsal to form them helps to create durable and distinctive memories (What is sticking out of your running shoes?). So, a mnemonic helps safeguard against forgetting (for less optimistic long-term benefits of mnemonics, see Thomas & Wang, 1996[5]).

Third principle

The mnemonic guides you through retrieval by providing cues for recalling information. This function is important because much of forgetting is often a case of retrieval difficulty.

Another one is the **story mnemonic**, which provides an excellent illustration of these features of mnemonics. Creating a narrative or story for a set of information will improve your memory. This can be seen in Figure 6.2, from a study by Bower and Clark (1969[7]).

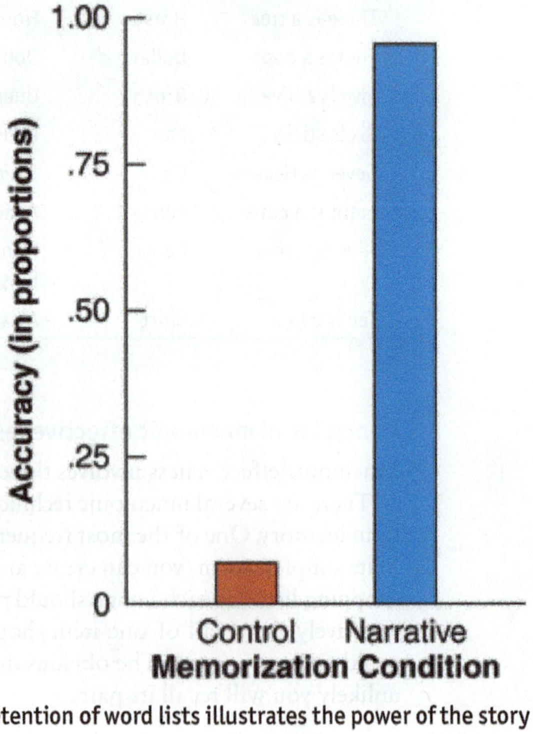

Figure 6.2 The retention of word lists illustrates the power of the story mnemonic

In this study, people were given 12 lists of 10 words each and asked to remember them in order. As can be clearly seen, people who formed a story from the words had much better memory of them than people who simply tried to memorise the lists as best they could. The stories people created for themselves provided structure, imagery and effort, and served as excellent sources of **retrieval cues**. In fact, when Chao Lu set the world record in 2005 for memorising digits of pi (67,890 of them), he created a story out of the digits to help him remember (Hu, Ericsson, Yang & Lu, 2009[8]). Overall, it cannot be stressed enough how the active use of retrieval cues and then practicing retrieval are important for successful remembering. Extended practice was the key in Chaffin and Imreh's (2002)[9] study of how a pianist learned, remembered and performed a challenging piece.

This three-step sequence may sound familiar. It is the sequence we talk about every time we consider learning and memory: the encoding of new information, its retention over time, and the retrieval of the information (Melton, 1963[10]). Performance in any situation that involves memory depends on all three steps. A fault along any one of the three might account for poor performance. A good mnemonic device, including those you invent for yourself (e.g. Wenger & Payne, 1995[11]), ensures success at each of the three stages. Incidentally, do not count on a magic bullet to enhance your memory. Research has found little, if any, evidence that ginkgo biloba, or any other so-called memory enhancer, has any real effect on memory (Gold, Cahill & Wenk, 2002, 2003; Greenwald, Spangenberg, Pratkanis & Eskenazi, 1991; McDaniel, Maier & Einstein, 2002[12]).

6.1.2 The Ebbinghaus tradition

We turn now to the first systematic research on human memory done by the German psychologist Hermann von Ebbinghaus. The Ebbinghaus tradition began with the publication of *Über das Gedächtnis* (1885). The 1964 English translation is titled *Memory: A Contribution to Experimental Psychology*. Ebbinghaus used himself as the only participant in his studies. He also had to invent his own memory task, his own stimuli, and his own procedures for testing and data analysis. Few could do as well today. In devising how to analyse his results, he even came close to inventing a within-groups *t*-test (Ebbinghaus, 1885/1913, footnote 1, p. 67[13]).

Ebbinghaus's research on human memory

It is helpful to consider why Ebbinghaus felt compelled to invent and use nonsense syllables. His rationale was that he wanted to study the properties of memory and forgetting apart from the influence of prior knowledge. As such, words would complicate his results. If he had used words, it would be less clear whether performance reflected the simple use of memory or the influence of prior knowledge. Putting it simply, *learning* implies acquiring new information. Yet, words are not new. And a control factor he adopted, to reduce the possible intrusion of mnemonic factors, was the rapid presentation rate of 2.5 items per second.

The relearning task

Ebbinghaus devised the relearning task (or savings task), in which a list is originally learned, set aside for a period of time, then later relearned to the same criterion of accuracy. In most cases, this was one perfect recitation of the list, without hesitations.

After relearning the list, Ebbinghaus computed a savings score as the measure of learning. The **savings score** was the reduction in the number of trials (or the time) necessary for relearning compared to original learning. Thus, if it took 10 trials to originally learn a list but only 6 trials for relearning, there was a 40 per cent savings (4 fewer trials on relearning divided by the 10 original trials). By this method, *any* information that was left over in memory from original learning could have an influence, conscious or not (see Nelson, 1978, 1985; Schacter, 1987[14]). Work by MacLeod (1988)[15] indicates that relearning seems to help retrieve information that was stored in memory yet was not recallable.

The forgetting curve

Figure 6.3 presents Ebbinghaus's *forgetting curve (*actually a retention curve), showing the reduction in savings as a function of time until relearning. As Slamecka (1985)[16] points out, for the data in Figure 6.3, Ebbinghaus used more than 1,200 lists of nonsense syllables.

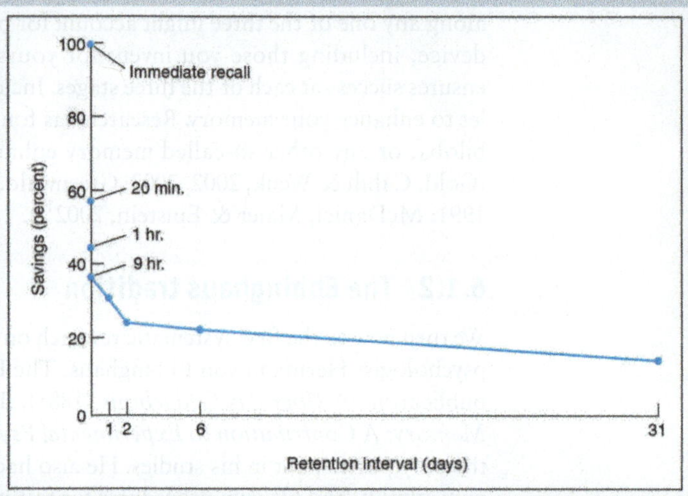

Figure 6.3 The classic forgetting curve from Ebbinghaus (1885/1913)
The figure shows the reduction in savings across increasing retention intervals.

Ebbinghaus relearned the lists after one of seven intervals: 20 minutes, 1 hour, 9 hours, 1 day, 2 days, 6 days or 31 days. As is clear from the figure, the most dramatic forgetting occurs early after original learning. This is followed by a decrease in the rate of forgetting (technically, a negatively accelerating power function; Wixted & Ebbesen, 1991[17]), which is also found for a variety of other memory tests. That said, as outlined in a nice review by Erdelyi (2010),[18] although there tends to be a loss of memories over time, it is possible for items that were previously forgotten to be remembered later (*reminiscence*). There are even cases when the rate of reminiscence may be greater than the rate of forgetting, and a person remembers more over time, a phenomenon called *hypermnesia*.

The effect of repetitions

Other fundamentals Ebbinghaus found were impressive not because they were surprising, but because he demonstrated them first. For example, he studied the effects of repetitions, studying one list 32 times and another 64 times. Upon relearning, the more frequently repeated list showed about twice the savings of the less frequently repeated list. In other words, **overlearning** yields a stronger record in memory. Longer lists took more trials to learn than shorter lists but showed higher savings upon relearning. So, although it is harder to learn a long list originally, the longer list is remembered better, because there was more opportunity to overlearn it (there were more trials in learning before mastery of the whole list). The connection between learning difficulty and memory has been confirmed repeatedly (e.g. Schneider, Healy, & Bourne, 2002[19]).

Finally, in one study, Ebbinghaus continued to relearn the same set of lists across a 5-day period. The savings scores showed no forgetting at all. Ebbinghaus also reported his results on relearning passages of poetry (kept at 80 syllables). After the fourth day, the savings was 100 per cent.

Massed practice versus distributed practice

Beyond the idea of overlearning, there are issues involved with how people repeatedly study to learn something. Specifically, how quickly and effectively something is learned varies as a function of whether study sessions are grouped together or spread out over time. If study time is grouped together into one long session, this is called **massed practice**, although many students know this better as *cramming*. In comparison, if study time is spread out over many shorter sessions, this is called **distributed practice**. The distinction between these two types of studying is important because memory is much better with distributed practice than with massed practice (Glenberg & Lehmann, 1980[20]), although many students incorrectly believe the opposite (Zechmeister & Shaughnessy, 1980[21]). Students often report that they feel like they are working harder when they cram than when they study consistently throughout the term. And in general, that is true – people are working harder with massed practice.

The problem is that people are not learning much. So, you can save yourself a lot of time, and do better in your coursework, if you spread your studying out across the term than if you cram (note that cramming also denies you of the beneficial effects of sleep on memory improvement). These benefits can even be seen a year later (Pashler, Rohrer, Cepeda & Carpenter, 2007[22]).

6.1.3 Memory consolidation

Just because you think about something and understand it does not mean that you will remember it later. Think of how many things you have heard in a lecture or read in a textbook, and then had no useful memory of later when it came time to take an exam. It takes some time for memories to become relatively permanent and reliable. This is the process of *consolidation*, described earlier in this course. Consolidation does not happen all at once, but is stretched out over time and perhaps goes through two phases:

- The first is a fast *synaptic consolidation* phase in which memories may be stored for up to 2 weeks, perhaps in the hippocampus.
- The second is a slower *systems consolidation* phase in which memories are stored for up to a lifetime across the cortex.

Moreover, neurological processes that occur during sleep can aid consolidation (Fenn & Hambrick, 2013; Stickgold & Walker, 2005[23]). Generally, the older a memory is, the more likely it is to have been consolidated, and so it is resistant to forgetting. This consolidation process begins almost immediately as you think of something and can extend for decades into the future. The reason that we do not consolidate some memories is because of factors that can lead to forgetting.

6.1.4 Metamemory

Think about the three issues of mnemonics, Ebbinghaus's work and consolidation from a broader perspective: they involve factors about memory, what makes remembering easier or harder. This self-awareness about memory is **metamemory**, knowledge about (*meta*) one's own memory and how it works or fails to work.

With regard to metacognitive awareness, several studies have focused on metamemory, such as people's 'judgements-of-learning' and 'feeling-of-knowing' estimates (Leonesio & Nelson, 1990; Nelson, 1988[24]). Part of a person's behaviour in a learning task involves self-monitoring, assessing how well one is doing and adjusting study strategies (e.g. Son, 2004[25]). For example, these **metacognitions** guide people to know when to change their answers on multiple-choice exams (Higham & Garrard, 2005[26]). However, metamemory can occasionally mislead us, leading to either over- or underconfidence that we have learned something (e.g. Koriat, Sheffer & Ma'ayan, 2002[27]).

Overall, although we do have some awareness of our memories, we are unaware of many aspects of memory. For example, many college students are unaware that they can improve their memories through various strategies such as dual coding, testing, distributed practice and the generation effect (McCabe, 2011[28]).

6.2 Storing information in episodic memory

LEARNING OBJECTIVE
Explain the process by which information is stored in long-term memory

How do people store information in episodic memory? And how can we measure this storage?

Theories on storing information in episodic memory

Ebbinghaus's theory

Ebbinghaus investigated one storage variable (repetition) and one memory task (relearning). He found that increasing the number of repetitions led to a stronger memory, a trace of the information in memory that could be relearned faster. Thus, frequency has a fundamental influence on memory: information presented more frequently is stored more strongly.

Hasher and Zacks' theory

A corollary of this is that people are good at remembering how frequently something has occurred (e.g. how many films they have seen in the past month). Hasher and Zacks (1984)[29] summarise a large body of research on how sensitive people are to event frequency. Because people's estimates of frequency are generally accurate, Hasher and Zacks propose that frequency information is automatically encoded into memory with no deliberate effort or intent. Although there is dispute about just how automatic this is (Greene, 1986; Hanson & Hirst, 1988; Jonides & Jones, 1992[30]), there is no doubt that event frequency has an impact on long-term memory.

von Restorff's theory

The flip side of frequency is distinctiveness. It is easier to remember unusual, unexpected or atypical events. Technically called the isolation effect, it is more commonly known as the von Restorff effect, named after the woman who did the first study (von Restorff, 1933)[31]. The effect is simply better memory for information that is distinct from the information around it, such as printing one word in a list in red ink or changing its size (Cooper & Pantle, 1967; Kelley & Nairne, 2001[32]). The isolation effect relies on memory for the distinctive item to be noticed as distinctive. Generally, the occurrence of unexpected and distinctive items can produce increased processing in the hippocampus (e.g. Axmacher et al., 2010[33]). Thus, damage to the hippocampus should reduce or eliminate the von Restorff effect.

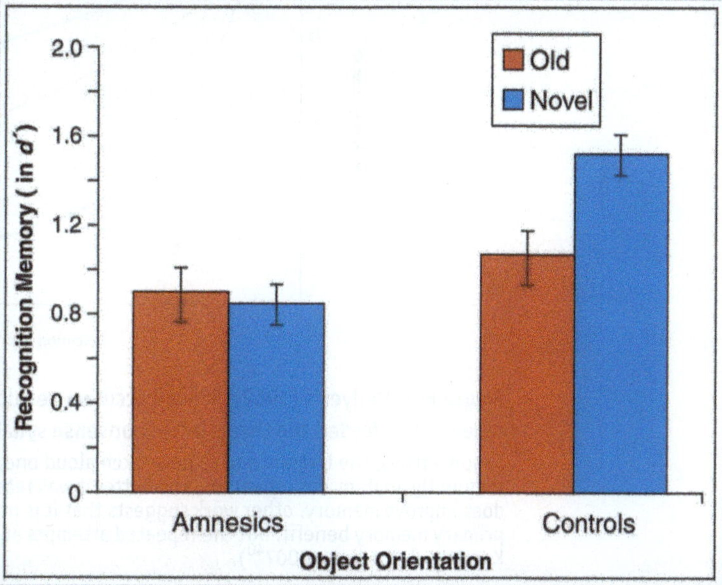

Figure 6.4 **Accuracy data from Kishiyama, Yonelinas and Lazzara (2004)[34] (using the signal detection measure of *d'*)**

The figure shows memory for a novel distinctive item (the von Restorff effect) for normal controls on the left and amnesics on the right. Notice that performance is better for the novel information for the normal controls, but not for the amnesics.

How do you learn and remember something new? There are three important steps to doing so:

1. Rehearsal
2. Organisation
3. Imagery.

A summary of these will then lead us to the topic of retrieval and a discussion of forgetting.

6.2.1 Rehearsal

In Atkinson and Shiffrin's (1968)[35] important model of human memory, information in short-term memory, was subject to rehearsal, a deliberate recycling or practicing of information in the short-term store. They proposed two effects of rehearsal. First, rehearsal maintains information in the short-term store. Second, the longer an item is held in short-term memory, the more likely it will be stored in long-term memory, with the strength of the item's long-term memory trace depending on the amount of rehearsal it received. In short, rehearsal transfers information into long-term memory (see also Waugh & Norman, 1965[36]).

Frequency of rehearsal

What evidence is there of this effect of rehearsal? Aside from Ebbinghaus, many experiments have shown that rehearsal leads to better long-term retention. For example, Hellyer (1962)[37] used the Brown–Peterson task, with CVC trigrams and with an arithmetic task between study and recall (see Figure 6.5).

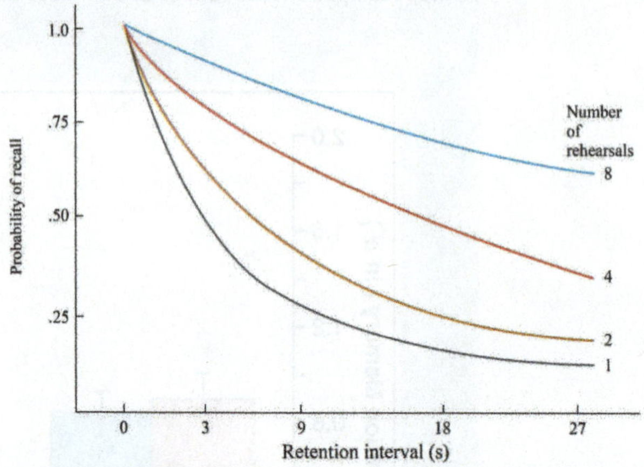

Figure 6.5 Hellyer's (1962)[38] recall accuracy results as a function of the number of rehearsals afforded the three-letter nonsense syllable and the retention interval

In some trials, the trigram had to be spoken aloud one, two, four, or eight times. The more frequently an item was rehearsed, the better it was retained. However, although rehearsal does improve memory, other work suggests that it is not repeated study that produces the primary memory benefit, but the repeated attempts at trying to remember (Karpicke, 2012[39]; Karpicke & Roediger, 2007[40]).

Rehearsal and serial position

Further evidence on the effects of rehearsal was provided in a series of studies by Rundus (1971; Rundus & Atkinson, 1970[41]). In these experiments, Rundus had people learn 20-item lists of words, presenting them at a rate of 5 seconds per word. People were told to rehearse aloud as they studied the lists, repeating whatever words they cared to during each 5-second presentation. Figure 6.6 shows Rundus's most telling results.

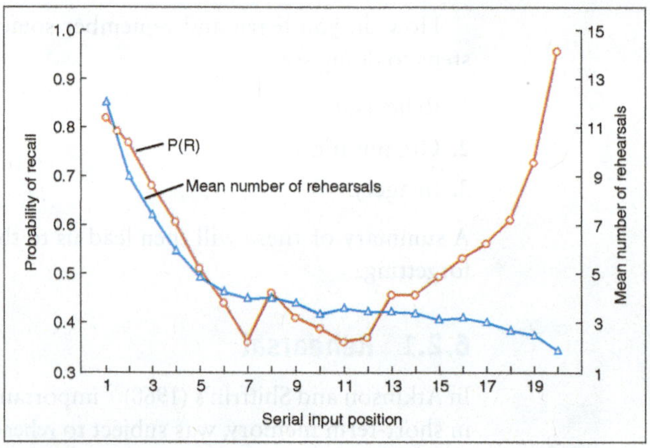

Figure 6.6 Probability of recall

The probability of recall, P(R), is plotted against the left axis, and the number of rehearsals afforded an item during storage is plotted against the right axis. The similar pattern of these two functions across the primacy portion of the list indicates that rehearsal is the factor responsible for primacy effects. Rundus then tabulated the number of times each word was rehearsed and compared this to the likelihood of later recalling the word.

In the early primacy portion of the *serial position curve*, there was a positive relationship between the frequency of rehearsal and the rate of recall. In other words, the *primacy effect* depended on rehearsal. The early items are rehearsed more frequently and so are recalled better. High recall of the late positions, the *recency effect*, was viewed as recall from short-term memory, which is why they were recalled so well despite being rehearsed so little.

Similar curves are observed in long-term memory. That is, given an event of a certain type, such as going to the cinema, people are likely to remember their first and last experiences better, and not so much those in between (e.g. Sehulster, 1989[42]). This even applies to semantic information, such as knowledge of the presidents of the United States (Roediger & Crowder, 1976[43]). Although some people have argued that the cognitive processes involved in long-term memory differ from those involved in short-term memory (Davelaar, Goshen-Gottstein, Ashkenazi, Haarmann & Usher, 2005[44]), some suggestion exists that the same principles drive the serial position curves in both the short- and long-term memory (e.g. Brown, Neath & Chater, 2007[45]).

Massed versus distributed practice

Rehearsing or practising material will make it more memorable. However, not all kinds of practice are equal. A distinction between two kinds of practice is seen in massed and distributed practice:

1. *Massed practice* is when people memorise information in one long session (remember the earlier discussion about cramming).
2. *Distributed practice* is when rehearsal is spread out across multiple, shorter occasions.

This distinction between different types of practice is important – keeping the amount of study time constant – because memory is better with distributed practice than with massed practice (Gerbier & Toppino, 2015; Glenberg & Lehmann, 1980[46]). In other words, although you can learn if you cram, you learn better if you study a bit every day. This benefit of distributed over massed practice occurs for several reasons: the later rehearsals remind people of the prior encounters (thereby strengthening the memories) (Benjamin & Tullis, 2010[47]), people experience the information in multiple contexts (Glenberg, 1976, 1979[48]), and people's minds wander less when they engage in distributed practice (Metcalfe & Xu, 2016[49]).

Testing effect

It is pretty clear that when you are listening to a lecture, reading a book or studying in some other way, you are learning new information. If you want to remember something better, it may seem intuitive that you should study as much as you can. And, for the most part, this is true.

However, it has also been found that, after some initial studying, memory may actually be better if people take a test as compared to just studying more. This is called the *testing effect* (Roediger & Karpicke, 2006[50]). When you take a test, such as an essay exam, a fill-in-the-blank test or a multiple-choice test, the contents of your memory are being assessed. It seems counterintuitive, but taking a test can actually improve memory more than studying. Every time you encounter information, whether you are studying it or being tested on it, it counts as a learning trial (e.g. Gates, 1917; Roediger & Karpicke, 2006[51]).

Essentially, the testing effect is the finding that the additional experience you get from tests actually helps you remember the information better – better even than studying, especially if you take a recall test (McDaniel, Roediger & McDermott, 2007[52]). This testing benefit applies to recognition (multiple-choice) tests

(Marsh, Roediger, Bjork & Bjork, 2007[53]) and to non-verbal material such as maps (Carpenter & Pashler, 2007[54]). It also transfers to new items that were neither studied nor tested (Butler, 2010[55]).

To make things even better, it has been found that after testing, people learn new information faster (Pastötter & Bäuml, 2014[56]). So, an effective tool to help you study and learn the material for this or any other class is to take practice tests if they are available. If they are not available, your study group could create and take practice tests. It may sound like a lot of work, but your memory for the material will be much better than if you had just spent the same amount of time studying by yourself.

6.2.2 Depth of processing

A more refined idea is that there are two kinds of rehearsal (Craik & Lockhart, 1972[57]).

Craik and Lockhart (1972)[58] proposed a theory of memory different from the stage approach of sensory, short- and long-term memory. They embedded their two kinds of rehearsal into what they called the **levels of processing**, or **depth of processing**, framework. Essentially, information receives some amount of mental processing. Some items that get only incidental attention are processed at a shallow level – for example, hearing the sounds of the words without attending to their meaning while daydreaming during a lecture. Other items get more intentional and meaningful processing that elaborates the memory of that item – for example, by drawing relationships between already-known information and what is being processed.

Several predictions from the depth of processing framework were tested with a fair degree of initial success. For example, if information is shallowly processed, using only maintenance rehearsal, then the information should not be well remembered later. If it is only maintained, then it should not be stored at a deep, meaningful level in long-term memory. This was the kind of result that emerged. As an

Kinds of rehearsal

Maintenance rehearsal

Maintenance rehearsal is a low-level, repetitive information recycling. This is the rehearsal you use to recycle a phone number to yourself until you dial it. Once you stop rehearsing, the information is lost. In Craik and Lockhart's view, maintenance rehearsal holds information at a certain level in memory without storing it permanently. As long as information is maintained, it can be retrieved. However, when rehearsal stops, information will likely vanish.

Elaborative rehearsal

Elaborative rehearsal is a more complex rehearsal that uses the meaning of the information to store and remember it. Information that is elaboratively rehearsed is stored more deeply in memory and is remembered better. You might include imagery or mnemonic elaboration in your elaborative rehearsal. You might try to construct sentences from the words in a list. You might impose an organisation on the list. Or you might even convert nonsense syllables like BEF into more meaningful items like BEEF.

In other words, maintenance rehearsal maintains an item at its current level of storage, whereas elaborative rehearsal moves the item more deeply, and more permanently, into memory.

example, Craik and Watkins (1973)[59] devised a monitoring task in which people heard a long list of words but only had to track the most recent word beginning with, say, a g. In a surprise recall test, people showed no recall differences for 'g words' held a long time versus those maintained only briefly (see also Craik & Tulving, 1975[60]).

You can think of levels of processing as being related to the idea of how much attention you pay to different things. If you pay more attention, you will remember more. This idea is illustrated in a study by Henkel (2014)[61] in which students were asked to go to a museum and either look at objects or take pictures of them. According to the study, people remembered the objects better if they did not take pictures because the act of taking a picture detracts attention from the environment and focuses it on the act of taking a picture. Thus, the level of processing is more shallow. So, if you want to remember experiences from your life, you should spend less time taking pictures and more time in the experience.

6.2.3 Challenges to depth of processing

As research continued on the depth of processing framework, some difficulties emerged. One example was Baddeley's (1978)[62] review paper, 'The trouble with levels.' A major point in this review was the problem of defining levels independently of retention scores (see Glenberg & Adams, 1978; Glenberg, Smith & Green, 1977[63]). Essentially, there is no method for deciding ahead of time whether a certain kind of rehearsal would lead to shallow or deep processing. Instead, we had to wait and see whether it improved recall. If it did, it must have been elaborative rehearsal. If it did not improve recall, it must have been maintenance rehearsal. The **circularity** of this reasoning was a serious problem for depth of processing ideas of memory retention.

Task effects

A second point in Baddeley's (1978)[64] review concerned **task effects** in which difficulty arose with the levels of processing approach when different memory tasks were used. The reason was simply that very different results were obtained using one task or another.

We have known since Ebbinghaus that different memory tasks reveal different things about the variables that affect performance. Ebbinghaus used a relearning task, so even material that was hard to retrieve might still influence memory performance. In a similar vein, a substantial difference is found between performance on recall and recognition tasks. In *recognition tasks*, people are shown items that were originally studied, known as 'old' or target items, as well as items that were not studied, known as 'new', lure or distractor items. They must then decide which are targets and which are distractors (multiple-choice tests are recognition tests, by the way).

Standard memory tasks and terminology

Recognition accuracy usually is higher than recall accuracy. This interactive includes descriptions of these tasks. Furthermore, two different factors influence recognition: *recollection* – the actual remembering of the information – and *familiarity*, the general sense that you have experienced the information before (e.g. Curran, 2000; Yonelinas, 2002[65]). Indeed, studies on false memory often ask people whether they actually 'remember' experiencing an event or whether they just 'know' that it happened.

Relearning task

1. Original learning: learn list items (e.g. list of unrelated words) to some accuracy criterion.
2. Delay after learning the list.
3. Learn the list a second time.

Dependent variables. The main dependent variable is the savings score: how many fewer trials are needed during relearning relative to the number of trials for the original learning. If the original learning took 10 trials and relearning took 6, then relearning took 4 fewer trials. Savings score = 4/10; expressed as a percentage, savings was 40 per cent.

Independent or control variables. Presentation rate, list item types, list length, accuracy criterion.

Paired-associate learning task

1. A list of pairs is shown, one pair at a time. The first member of the pair is the stimulus, and the second member is the response (e.g. for the pair *ice-brush, ice* is the stimulus term and *brush* is the correct response).
2. After one study trial, the stimulus terms are shown, one at a time, and the person tries to name the correct response term for that stimulus.
3. Typically, the task involves several successive attempts at learning, each including first a study trial, then a test trial. The order of the pairs is changed each time. In the anticipation method, there is one continuous stream of trials, each consisting of two parts, presenting the stimulus alone, then presenting the stimulus and response together. Across repetitions, people begin to learn the correct pairings.

Dependent variables. Typically, the number of study test trials to achieve correct responding to all stimulus terms ('trials to criterion') is the dependent variable.

Independent or control variables. Presentation rate, length of list, the types of items in the stimulus and response terms, and the types of connections between them. Commonly, once a list had been mastered, then either the stimulus or response terms are changed, or the item pairings rearranged (e.g. *ice-brush* and *card-floor* in the first list, then *ice-floor* and *card-brush* in the second list).

Recall task

Serial recall task. Learn a list of items, and then recall them in the original order of presentation.

Free recall task. Learn the list information, and then recall the items in any order:

1. Learn list items.
2. Optional delay or distractor task during delay.
3. Recall list items.

Dependent variables. The main dependent variable is the number (or proportion) of items recalled correctly. For serial recall, accuracy is scored as a function of the original position of the items in the list. Occasionally, other dependent variables involve speed, or organisation of recall (e.g. items recalled by category - *apple, pear, banana, orange* – before items from a different category are recalled).

Independent or control variables. Rate of presentation (usually experimenter paced), type of list items, length of list.

Recognition task (episodic)

1. Learn list items.
2. Optional delay or distractor task during delay.
3. Make yes/no decisions about the items in a test list: 'yes' the item was on the list, or 'no' it was not. This often refers to deciding whether an item is 'old' – that is, on the original list – or is 'new' and not on the original list. Old items are also called targets, and new items are also called distractors or lures.

Dependent variables. The dependent variable usually is accuracy, such as the proportion correct. Correct 'yes' responses to old items are called *hits*, and incorrect 'yes' responses to new items are called *false alarms.*

Independent or control variables. Same as with recall tasks.

Recognition is easier than recall, in part because the answer is presented to a person, who then only has to make a new versus old decision. Because more information is stored in memory than can be retrieved easily, recognition shows greater sensitivity to the influence of stored information. The issue of *how much* easier recognition is than recall is difficult to resolve (see research by Craik, Govoni, Naveh-Benjamin & Anderson, 1996, and by Hicks & Marsh, 2000[66]).

The relevance of this to the depth of processing framework is that most of the early research that supported it used recall tasks. When recognition was used, however, maintenance rehearsal had clear effects on long-term memory. A clever set of studies by Glenberg et al. (1977)[67] confirmed this. They used a *Brown–Peterson task*, asking people to remember a four-digit number as the (supposedly) primary task. During retention intervals that varied in duration, people had to repeat either one or three words aloud as a distractor task (don't confuse the distractor task here with distractor items, items in a recognition test that were not shown originally). Because people believed that digit recall was the important task, they devoted only minimal effort to the word repetitions, and so probably used only maintenance rehearsal. After the supposedly 'main' part of the task was done, people were given a surprise recall task. The results showed the standard effect. But when they were given a surprise recognition task, the amount of time spent rehearsing *did* influence performance. Words rehearsed for 18 seconds were recognised better than those rehearsed for shorter intervals. Thus, the generalisation that maintenance rehearsal does not lead to improved memory performance was disconfirmed – it did not apply when memory was tested with a recognition task.

6.3 Boosting episodic memory

LEARNING OBJECTIVE
Describe ways to increase the capacity of long-term memory

The most salient thing that many people want to do with their episodic memories is improve them so that they do not forget as much. That is, they would like to reduce the rate of forgetting that they experience. We have already gone over a few things you can do to improve your episodic memory. Specifically, you can repeatedly practise information, especially if your practice is distributed, and you can engage in deeper processing. An overarching idea in the depth of processing framework is that the more you do with information, the better it is remembered. There are numerous examples of this 'hard work has its rewards' principle. In this section we look at six examples of this – six ways that show how information is processed and how the amount of effort a person puts into encoding it affects performance:

1. The self-reference effect
2. The generation effect
3. The production effect
4. The impact of enactment on memory
5. The benefits of imagery and organisation
6. The consequences of taking a survival-based perspective.

6.3.1 The self-reference effect

The self-reference effect is the finding that memory is better for information that you relate to yourself in some way (e.g. Bellezza, 1992; Gillihan & Farah, 2005; Rogers, Kuiper & Kirker, 1977; Symons & Johnson, 1997[68]). If you think about it, you know a lot about yourself and you tend to be motivated by such information. When you relate something to yourself, say a detail you are trying to remember, you link the new knowledge to the old, yielding a more complex structure, an elaborative encoding (e.g. the locations given earlier for the method of loci – driveway, garage door, etc. – were the memorised locations one of this text's authors used for his own mnemonic device for years, based on his own house). Thus, the elaborated structure becomes more memorable.

6.3.2 Generation, production and enactment

Here we discuss three ways in which episodic memory can be improved: the generation effect, the production effect and the enactment effect.

The generation effect is the finding that information you generate or create yourself is better remembered compared to information you have only heard or read. This was first reported by Slamecka and Graf (1978).[69] In their study, for the *read* condition, people simply read words printed on cards. However, for the *generate* condition, people needed to generate the word on their own. This was done by giving a word and the first letter of the word that was to be recalled, with the instruction that the to-be-generated word had to be related to the first word. For example, a person might see *Long–S______* where the word *short* needed to be generated. The results showed that people remembered words better when they had generated them than when they had just read them.

In their review of work on the generation effect, Bertsch, Peta, Wiscott and McDaniel (2007)[70] reported that this robust finding was more likely to occur with free recall, and that the effect grew larger over longer delays. More importantly, the generation effect applies not only to lists of words but also to textbook material (e.g. deWinstanley & Bjork, 2004[71]). In short, the generation effect is another example that the more effort you put into mentally processing information, the more likely it is that you will remember it later and for a longer time.

A related finding is the *production effect* in which people actually produce information rather than simply reading or hearing it (Fawcett, 2013; MacLeod, Gopie, Hourihan, Neary & Ozubko, 2010[72]). For example, people given a set of materials to learn will remember those materials if they say them aloud rather than if they simply read them. The production effect occurs because actually saying something requires more effort than reading. This extra effort is deeper processing, which leads to better memory. Note that the production effect does not necessarily require that a person say something aloud. It can also be observed when people simply mouth the material, whisper it, write it or type it out (Forrin, MacLeod & Ozbuko, 2012[73]).

Another way to engage in deep encoding is to take advantage of the **enactment effect**, in which there is improved memory for participant-performed tasks relative to those that are not participant-performed. In such studies, actually doing some activity is compared to just watching someone else doing it. For example, people might be told to 'break the match', 'point at the door,' or 'knock on the table' or to watch someone else do those actions. In general, people remember things better if they do them themselves (e.g. Engelkamp & Dehn, 2000; Saltz & Donnenwerth-Nolan, 1981[74]). Essentially, the additional mental effort needed to do the task is another form of deep processing.

The value of enactment can be seen in the practical application of learning lines of dialogue. Evidence exists that even untrained non-actors (i.e. novice actors) learn dialogue, as in the script of a play, better when they rehearse the dialogue and stage movements together (Noice & Noice, 2001; see also Freeman & Ellis, 2003, and Shelton & McNamara, 2001, for other multimodality effects on learning[75]). Physical movement, in other words, can be part of an enhanced mnemonic. Enactment improves memory by helping people better organise and structure information about their actions (Koriat & Pearlman-Avnion, 2003[76]).

6.3.3 Organisation in storage

Another important piece of the storage puzzle involves **organisation**, the structuring of information as it is stored in memory. Well-organised material can be stored and retrieved with impressive levels of accuracy. The earliest work on organisation (or **clustering**) was done by Bousfield. For example, in his earliest study (Bousfield & Sedgewick, 1944[77]), he asked people to name as many birds as they could. The result was that people tended to name the words in subgroups, such as *robin, bluejay, sparrow – chicken, duck, goose – eagle, hawk.* To study this further, Bousfield (1953)[78] gave people a 60-item list to be learned for free recall. Unlike other work at that time, Bousfield used related words for his lists, 15 words each from the categories *animals, personal names, vegetables* and *professions.* Although the words were shown in a random order, people tended to recall them by category; for instance, *dog, cat, cow, pea, bean, John, Bob.*

Where did this organising come from? Obviously, people noticed that several words were drawn from the same categories. They used the strategy of grouping the items together on the basis of category (this has a nice metamemory effect as well). The consequence of this reorganisation was straightforward: the way the material had been stored governed how it was recalled.

6.3.4 Improving memory

Baddeley (1978)[79] was one of the critics of the depth of processing framework, concluding that it was valuable only at a rough, intuitive level but not as a scientific theory. Although that may be true, it is hard to beat Craik and Lockhart's insights if you are looking for a way to improve your own memory.

How can memory be improved?

Think of maintenance versus elaborative rehearsal as simple recycling in short-term memory versus meaningful study and transfer into long-term memory.

Inventing mnemonics

Apply this to your own learning. When you are introduced to someone, do you merely recycle that name for a few seconds, or do you think about it, use it in conversation and try to find mnemonic connections to help you remember it? When you read, do you merely process the words at a simple level of understanding, or do you actively elaborate when you are reading, searching for connections and relationships that make the material more memorable? In other words, use the depth of processing ideas in your own metacognition. Try inventing a mnemonic (this invokes the generation effect), applying elaborative rehearsal principles or actively doing something with the information, such as drawing a diagram (this invokes the enactment effect) to something you may need for this course. Most effective mnemonics organise and structure information in some way.

Organisation using hierarchies of words

Bower, Clark, Lesgold and Winzenz (1969)[81] demonstrated the power of organisation as a mnemonic for improving storage in long-term memory. Four hierarchies of words were presented in the organised condition, arranged as lists with headers (one of the four hierarchies is shown in Figure 6.7).

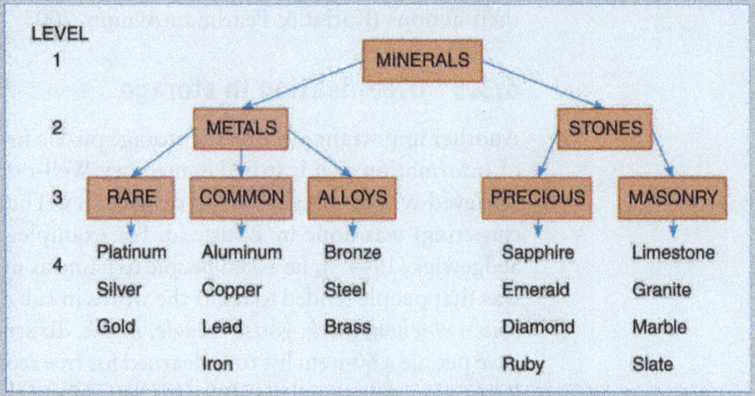

Figure 6.7 One of the hierarchies presented by Bower et al. (1969)[80]

For instance, under *stones* was *precious*, and under that were *sapphire, emerald, diamond* and *ruby*. The control group was shown words in the same physical arrangements, but the words were randomly assigned to their positions. People got 4 trials to learn all 112 words; their performance is shown in Table 6.3 below.

Presenting the words in the organised fashion led to 100 per cent accuracy on trials 3 and 4, an amazing accomplishment given the number of words. In contrast, the control group managed to recall only 70 out of 112 words by trial 4, 62.5 per cent accuracy.

This organisational principle, if used effectively, can lead to astounding feats of memory. People with exceptional memories, such as those who memorise *pi* out to extraordinary numbers of digits, are essentially using some organisational strategy (e.g. Ericsson, Delaney, Weaver & Mahadevan, 2004[82]) along with other basic memory skills, such as imagery (e.g. Takahashi, Shimizu, Saito & Tomoyori, 2006[83]).

Table 6.3 Average percentage of words recalled over four trials as a function of organisation

Conditions	1	2	3	4
Organised	65%	94.7%	100%	100%
Random	18.3	34.7	47.1	62.5

This applies not only to words and numbers but also to complex sets of information. As Anderson (1985)[84] pointed out, a chapter outline can serve the same function as the Bower et al. hierarchies, with obvious implications for students' study strategies (*there's* a hint).

Rehearsing by category

Another reason for clustering studies was that they demonstrated people's strategies for learning in an objective fashion. Many studies used a free recall task, then examined the effects of different degrees of list organisation, different numbers of categories, different numbers of items within categories, and so on. Clustering and organisation were then examined in terms of a creative array of dependent variables, such as order of recall, degree of clustering, speed and patterning of pauses during recall, and rehearsal (see reviews by Johnson, 1970[85]; Mandler, 1967[86], 1972[87]). As an example, Ashcraft, Kellas, and Needham (1975)[88] had people rehearse aloud as they studied clustered or randomised lists. Their results suggested that recalling the words by category was due to reorganisation during rehearsal. That is, people tended to rehearse by category. For instance, when *horse* was presented, this would trigger the rehearsal of *dog, cat, cow, horse* together. When sufficient time is provided, people can reorganise the words as they store them in memory. Furthermore, the number of times a word had been rehearsed during study is predictive of recall order. More frequently rehearsed categories, as well as words within those categories, are recalled earlier than categories and words that were less frequently rehearsed.

Source: Adapted from Bower et al. (1969).[89]

Subjective organisation

Don't misunderstand the previous section: Organisation is *not* limited to sets of items with obvious, known categories. A study by Tulving (1962)[90] showed that people can and do use **subjective organisations** – literally, organisation imposed by the participant (for an update, see Kahana & Wingfield, 2000[91]). Tulving used a multi-trial free recall task, in which the same list of words is presented repeatedly across several trials, wherein each trial had a new reordering of the words. His analysis looked at the regularities that developed in the recall orders. For example, a person might recall the words *dog, apple, lawyer, brush* together on several trials. This consistency suggested that the person had formed a cluster or chunk composed of those four items using some idiosyncratic basis. For example, a person might link the words together in a sentence or story: 'The dog brought an apple to the lawyer, who brushed the dog's hair.' Regardless of how they were formed, the clusters were used repeatedly during recall, serving as a kind of organised unit. Tulving called this *subjective organisation*; that is, organisation developed by a person for structuring and remembering information. In other words, even 'unrelated' items become organised through the mental activity of a person imposing an organisation.

6.3.5 Imagery

As mentioned at the beginning of the chapter one way you can improve memory is to use **visual imagery**, the mental picturing of a stimulus that affects later recall or recognition. We have discussed some imagery effects such as imagery-based mnemonic devices. Now we turn to visual imagery's effect on the storage of information in long-term memory and the boost it gives to material you are trying to learn. In general, visual episodic memory for pictures is very good.

Studies have shown that people can remember an astounding number of pictures after seeing them only once (Brady, Konkle, Alvarez & Oliva, 2008[92]). For example, Standing (1973)[93] showed that people had over 80 per cent memory for 10,000 pictures 2 days later.

Impact of imagery on memory

An early contributor to understanding how imagery impacts memory was Alan Paivio. Paivio (1971)[94] reviewed scores of studies that showed the generally beneficial effects of imagery on memory. These effects are beyond those caused by other variables, such as word- or sentence-based rehearsal or meaningfulness (Bower, 1970; Yuille & Paivio, 1967[95]). One example is a paired-associate learning study by Schnorr and Atkinson (1969[96]; see Table 6.4).

The task

The task is to learn the list so that the correct response item can be reproduced whenever the stimulus item is presented. Thus, if you saw *elephant-book* during study, you would be tested during recall by seeing the term *elephant* and your correct response would be *book*. Schnorr and Atkinson had people study half of a list by forming a visual image of the two terms together. The other half of the list was studied by rote repetition.

On immediate recall, the pairs learned by imagery were recalled at better than 80 per cent accuracy, compared to about 40 per cent for the rote repetition pairs.

Table 6.4 Lists of paired associates

List 1 (A–B)	List 2 (C–D)	List 3 (A–B_1)
tall–bone	safe–fable	plan–bone
plan–leaf	bench–idea	mess–hand
nose–fight	pencil–owe	smoke–leaf
park–flea	wait–blouse	pear–kiss
grew–cook	student–duck	rabbit–fight
rabbit–few	window–cat	tall–crowd
pear–rain	house–news	nose–cook
mess–crowd	card–rest	plan–flea
print–kiss	color–just	grew–flew
smoke–hand	flower–jump	park–few
List 4 (A–B_2)	**List 5 (A–C)**	**List 6 (C–D)**
smoke–arm	tall–bench	smoke–fable
mess–people	plan–pencil	pear–idea
rabbit–several	nose–wait	mess–owe
park–ant	park–student	pear–blouse
plan–tree	grew–window	grew–wait
tall–skeleton	rabbit–house	grew–duck
nose–battle	mess–color	park–rest
grew–chef	pear–print	nose–nest
pear–storm	print–flower	pear–just
print–lips	smoke–safe	tall–jump

The superiority of imagery was found even 1 week later. It is also important to note that the creation of mental images is not automatic. It requires attention and effort, which is part of the reason for the benefit it provides.

The effect

Studies such as this led Paivio to propose the dual-coding hypothesis (Paivio, 1971),[97] which states that words that denote concrete objects, as opposed to abstract words, can be encoded into memory twice: once in terms of their verbal attributes and once in terms of their imaginal attributes. Thus, a word like *book* enjoys an advantage in memory because it can be recorded twice – once as a word and once as a visual image.

There are two ways it can be retrieved from memory – one way for each code. However, a term such as *idea* has only a verbal code because it evokes no obvious image. This is not to say that people cannot create an image for *idea*, such as a lightbulb, but only that the image is much more available and natural for concrete words.

6.3.6 Adaptive memory

For memory to be of value to us, and be better remembered, it needs to give us something useful. It should help us survive in the world. The survival motivation is strong, and knowledge of what can either increase or decrease our survival is important. Thus, if people bring a survival perspective to bear on what they are learning, it can improve performance. This was shown in a study by Nairne, Thompson and Pandeirada (2007; see also Nairne & Pandeirada, 2008; Nairne, Pandeirada & Thompson, 2008; Weinstein, Bugg & Roediger, 2008[98]). In this study, people were given lists of words. During the first part of the study, people rated the words for pleasantness, relevance to moving to a foreign land, personal relevance or survival value (e.g. finding food and water or avoiding predators). Words that were rated high on survival value were more likely to be remembered later. The survival angle has such a strong impact that it can outperform the effects of other well-known memory-enhancing strategies such as imagery, self-reference and generation (Nairne et al., 2008[99]). Part of this benefit comes from thinking about how survival relates to us, as a kind of self-reference effect (Klein, 2012[100]). So, if we think about how information relates to our ability to survive, endure or otherwise be useful, this takes advantage of our fundamental motivations, which we can leverage to improve memory (Wurm, 2007; Wurm & Seaman, 2008[101]).

6.4 Context

LEARNING OBJECTIVE
Explain the role of context in the storage and retrieval of long-term memories

As noted earlier, episodic memory is for memory of personal experiences. These experiences do not occur in a vacuum. Instead, they happen in some sort of setting or *context*. Every experience you have occurs in a particular place, at a particular time, when you are thinking particular thoughts, and when you have encountered particular things. All these and more make up the context of an episodic experience. Context plays a critical role in episodic memory: it can make remembering easier or harder. In addition to the influence of context on our ability to remember, we can even look at memory for the context or source itself.

6.4.1 Encoding specificity

In Tulving and Thompson's (1973; Unsworth, Spillers & Brewer, 2012[102]) view, an important influence on memory is **encoding specificity**. This phrase means that information is encoded into memory *not* as a set of isolated, individual items. Instead, each item is encoded into a richer memory representation that includes the context it was in during encoding. So, when you read *cat* in a list of words, you are likely to store not only the word *cat* but also information about the context you read it in.

In a classic study of encoding specificity, Godden and Baddeley (1975)[103] had people learn a list of words (see Figure 6.8). Half of these people learned the list on land, and the other learned the list underwater (all of these people were scuba divers). They were then given a recall test for the list. The important twist is the context in which they tried to recall the information. Half of the people recalled the items in the same context they had experienced during learning. However, the other half recalled the information in the other context.

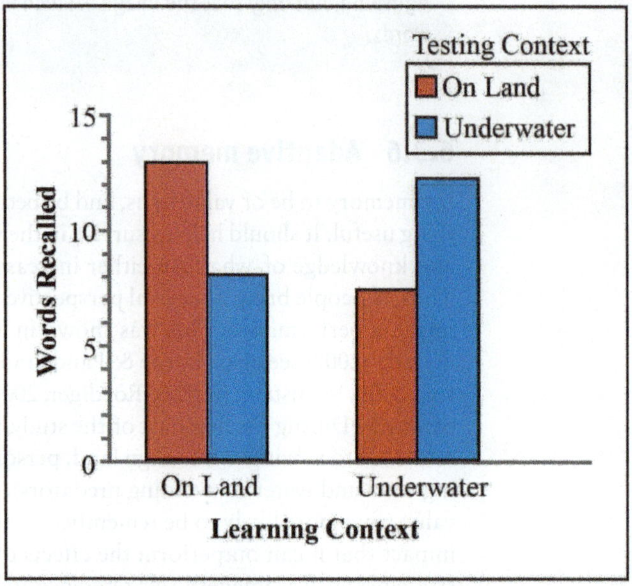

Figure 6.8 The classic encoding specificity result reported by Godden and Baddeley[104]
The figure shows better performance when the encoding context matched the retrieval context. That is, memory for things learned on land was better when tested on land as opposed to underwater, whereas things learned underwater were better remembered when the people were tested underwater as opposed to on land. The interesting finding was that memory was better when the encoding and retrieval contexts were the same, relative to when they were different.

A more everyday example of encoding specificity is the experience of going to a room in your home to do something, but when you arrive, you cannot remember why you are there. However, when you go back to where you started, you remember. So, reinstating the original context helped you remember. This is also why witnesses may return to the scene of a crime. Being there again reinstates the context, helping them remember details they might otherwise forget.

More generally, when your memory is tested, with free recall for instance, you attempt to retrieve the trace left by your original encoding. Encoding the context along with the item allows the context to serve as an excellent *retrieval cue* – a useful prompt or reminder for the information. The original context cues give you the best access to the information during retrieval, and these cues can be verbal, visual, or something else (Schab, 1990[105], for instance, has found that odors are effective contextual cues).

The processing of content (what it is you are to remember), context, and their binding together is more complex than it may seem initially. The brain has separate processing streams for what something is and where it is.

Another variant of the encoding specificity effect is *state-dependent learning*, which is the finding that people are more likely to remember things when their physiological state at retrieval matches that at encoding. For example, Goodwin, Powell, Bremeer, Hoine and Stern (1969)[106] found that people made fewer errors on a memory test when they recalled information when they were drunk (a certain physiological state) if they had learned that information inebriated, than if they tried to recall it when they were sober!

In summary, the storage of information into episodic long-term memory is affected by several factors that can lead to better memory. Moreover, the congruence between study and test contexts can be vital. Relevant rehearsal, including organisational and imaginal elements, improves performance (see Section 6.2 for details on the kinds of rehearsal strategies).

6.4.2 Source monitoring

Context is not only an important cue to access episodic memories, but it can also be used as the target of a memory search – when people seek to remember the context in which information was learned.

In many memory studies, we intentionally lure people into making mistakes. That is, we present meaningful material such as a story or a set of related sentences, then do something that invites mistakes of one sort or another. This is unlike most episodic and semantic memory studies in which people are simply asked to learn and remember material in a straightforward way.

6.5 Facts and situation models

LEARNING OBJECTIVE
Identify the influence of mental representations on long-term memory

In many memory studies, we may intentionally manipulate an experimental design to cause people to make errors. That is, we might present meaningful material such as a story or a set of related sentences, then do something that invites mistakes of one sort or another. This is unusual in most episodic and semantic memory studies in which people are simply asked to learn and remember material in a straightforward way.

6.5.1 The nature of propositions

In this section we discuss the idea that what people remember from meaningful material is the idea or gist – in other words, we do not usually remember superficial details, exact words, or exact phrasings, but we do remember the basic idea. However, psychology needs ways to represent those ideas, needs a scientific way to quantify what the vague term *meaning* means. Also, to do research on content accuracy, we need a way to score recall to see how well people remembered meaningful content. The unit that codes meaning is called a proposition. A **proposition** represents the meaning of a single, simple idea, the smallest unit of knowledge about which you can make true/false judgements.

Remembering propositions

In this section we will explore how we remember propositions by looking at an early study by Sachs (1967)[107] who investigated whether or not people remember *meaning* rather than verbatim information. In her study, participants heard passages of connected text and were then tested on one critical sentence in the passage 0, 80 or 160 syllables after it had been heard. The test was a simple recognition test among four alternatives. One alternative was a verbatim repetition, another represented a change both in surface form and in meaning, and the other two represented changes only in surface form. When recognition was tested immediately, people were very good at recognising the exact repetition. In other words, they rejected changes in superficial structure and changes in meaning. After comprehending the next 80 syllables in the passage, however, performance was accurate only in rejecting the alternative that changed the meaning. So, after the 80-syllable delay, people showed no preference for the repetition over the paraphrases.

Sample passage from Sachs (1967)[108] including multiple-choice recognition test for critical sentence

Read the passage below at a comfortable pace but without looking back. After you have finished reading, your memory of one of the sentences in the paragraph will be tested.

> While most gardeners enjoy tulips, it is unlikely that they would be valued as highly today as they were in Holland during the seventeenth century. For some reason, the tulip became very popular at this time. The tulip was first brought into Europe from Turkey. In 1551, the Viennese ambassador to Turkey wrote of seeing these plants. Years later, in 1562, a cargo of tulip bulbs was sent to Antwerp. The flower then spread through Holland from there. At first, only the rich collected and traded tulips. Eventually, most of Holland was involved in the matter. Almost everyone tried to outdo their neighbors, both in the growing of tulips with rare colours and also in the high prices that were paid for them. A price of 6000 florins was paid for one bulb of the variety Semper Augustus in 1636. At this time, 6000 florins was the price of a house and grounds. Soon, everyone in Holland was working in the tulip trade, and ordinary business was being neglected. People who had been away from Holland and then returned during the craze sometimes made mistakes. A sailor is said to have mistaken a tulip bulb worth several thousand florins for an onion, and cut it up to eat with his herring.

Now, without looking back, decide which of the following sentences occurred in the paragraph. Check to see whether your answer was correct by referring back to the paragraph.

1. The Viennese ambassador to Turkey wrote of seeing these plants in 1551.

No

2. A sighting of these plants was written about in 1551 by the Viennese ambassador in Turkey.

No

3. The ambassador to Turkey from Vienna, in 1551, wrote of seeing these plants.

No

4. In 1551, the Viennese ambassador to Turkey wrote of seeing these plants.

Yes

Sachs concluded that we quickly lose information about the actual, verbatim words that we hear (or read), but we do retain the meaning. We reconstruct what must have been said based on the meaning that is stored in a propositional representation. Only when there is something 'special' about the verbatim words, say in recalling a joke, do we retain surface form as part of our ordinary memory for meaningful discourse (but see Masson, 1984[109]).

Confirmation of this finding was offered by Kintsch and Bates (1977),[110] who gave a surprise recognition test to students either 2 or 5 days after a classroom lecture. Some evidence of verbatim memory was present after 2 days, but very little persisted 5 days afterwards. As expected, verbatim memory for details and extraneous comments was better than verbatim memory for general lecture statements (see also Bates, Masling & Kintsch, 1978[111]). However, even here, reconstructive memory seemed to play a role. Students were better at rejecting items such as jokes that had not been presented than they were at recognising jokes and announcements that had been heard (see also Schmidt, 1994[112], on the effects of humour on sentence memory). For jokes, the wording can be important, so there was better surface form memory because the nature of the information constrains the wording. Similarly, people can have better memory for verbatim information if they are reading poetry, where wording, rather than prose, is important (Tillman & Dowling, 2007[113]).

6.5.2 Situation models

There is a great deal of evidence that basic idea units can influence memory. However, there is more to life than simple ideas. One of the themes of this text is *embodied cognition*, the idea that how we think is influenced by how we act or are otherwise involved with the world. One way that embodied cognition manifests itself is in the idea that people create models (Johnson-Laird, 1983; Zwaan & Radvansky, 1998[114]) of the situations described, and do not just create memories of the simple propositional idea in sentences. It is clear that when people move from one event to another, such as by moving from one spatial location to another, this serves as an event boundary. These distinct events are then stored separately in memory and influence how easy it is for people to retrieve information (Horner, Bisby, Wang, Bogus & Burgess, 2016; Pettijohn, Thompson, Tamplin, Krawietz & Radvansky, 2016[115]). One way of thinking about different types of mental representations and their influence on memory comes from van Dijk and Kintsch's (1983)[116] work on language comprehension.

This division of different kinds of mental representation can be seen in how well people remember information at the different levels over time. For example, in a study by Kintsch, Welsch, Schmalhofer and Zimny (1987),[117] people read a text and then later took a memory test. For the test, people were shown sentences and indicated whether they had read them before. Four types of memory probes were used on the recognition test:

1. Verbatim probes, which were exact sentences that had been read earlier.
2. Paraphrases, which captured the same ideas as those in the text, but with a different wording.
3. Inferences, which were ideas that were likely to be true, but not actually mentioned.
4. 'Wrongs', or incorrect probes that were thematically consistent with the passage but were incorrect if one had read and understood the passage.

Kintsch et al. compared performance on these various types of memory probes to assess the strength of the representations at the various levels. For example, by comparing performance on the verbatim probes and paraphrases, one can estimate memory for the exact wording. This is because both of these probe types refer to ideas that were actually in the text, but only one uses the exact wording. So, the degree to which memory is better in the verbatim condition compared to the paraphrase condition is a measure of surface form memory.

The results of Kintsch et al.'s (1987)[118] study are shown in Figure 6.9.

Although memory for all three levels is reasonably good immediately after reading, there are big differences later on, depending on what is being assessed. First, for the surface form, verbatim information is lost quickly from memory and reaches chance performance by the end of the 4 days. Second, memory was better for the textbase level than for the surface form. So, although people may forget the exact words they read before, they are better at remembering the ideas that were presented (cf. the work by Sachs described earlier in the chapter). However, even memory at this level declines over time. But for the third level, the situation model, performance starts out high and then stays high, with little evidence of forgetting over the 4-day retention interval. Thus, there is something psychologically real about looking at mental representations in this way.

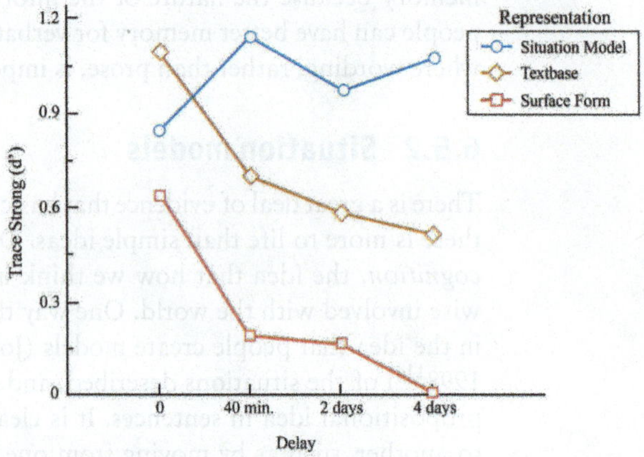

Figure 6.9 Results of Kintsch et al.'s (1987)[119] study

Memory for information at the surface form, textbase and situation model levels over the course of four days. After Kintsch et al. (1987)[120].

A more everyday example of this would be your memory for a newspaper article you might read. Soon after reading the article, your ability to remember verbatim sentences from the article is pretty poor. Furthermore, over time, you start to forget what specific ideas were actually read in the article and what ideas or inferences you may have created when you were trying to understand it. However, you have a relatively good memory over time for the events described in the article, and this memory stays with you for a much longer period of time.

6.6 Autobiographical memories

LEARNING OBJECTIVE
Understand and evaluate research on autobiographical memory

This next section addresses the study of one's lifetime collection or narrative of memories, called **autobiographical memory (ABM)**. These are memories for more natural experiences and information – a self-memory system (Conway & Pleydell-Pearce, 2000[121]).

ABMs are hierarchically organised, with more general events (e.g. I was in Italy for three years) being at the top level, followed by intermediate events (e.g. last year's holiday in Greece) to specific events (e.g. I had lunch with X last week) at the lower level (Anderson & Martin, 1993[122]) The specificity and vividness of the recalled AMBs may be an indicator of the significance or the *centrality* of this memory to one's behaviour. Also, visual distractors (e.g. irrelevant images) can reduce the recollection of autobiographical memories by placing excessive demands on long-term memory retrieval processes (Anderson, Dewhurst & Dean, 2017[123])

The research on autobiographical memory has revealed many phenomena that characterise it. Here, we present three of them to give you a feel for why it is important to consider autobiographical memory beyond what is known about episodic and semantic memory. More specifically, we consider infantile amnesia, the reminiscence bump and involuntary memories.

6.6.1 Psychologists as subjects

Several researchers have tested their own memories in carefully controlled long-term studies following Ebbinghaus (Section 6.1.2). One major difference from Ebbinghaus's procedure was that Linton (1975, 1978),[124] Wagenaar (1986)[125] and Sehulster (1989)[126] tested their memories for naturally occurring events, not artificial laboratory stimuli. For instance, Wagenaar recorded daily events in his own life for more than 6 years, some 2,400 separate events, and tested his recall with combinations of 4 different cue types: what the event was, who was involved, and where and when it happened. Although he found that pleasant events were recalled better than unpleasant ones at shorter retention intervals, his evidence also showed that none of the events could truly be said to have been forgotten (but contrast this with the Bahrick, Hall & Berger, 1996, evidence that bias toward pleasant things affects our memories of high school grades[127]).

6.6.2 Infantile amnesia

Think about the first thing you can remember. The very first thing. What is your earliest memory? Where were you? What was going on? Who were you with? How old were you? For most people, this memory is from the ages of 2 to 4 years old. But, wait a minute! What about all of the things that came before that? Why can't you remember anything earlier than that? Furthermore, if you think about it, your memory for your childhood is quite spotty, getting better only as you got older.

infantile amnesia is the inability to remember early life events and very poor memory for your life at a very young age. One of the first people to discuss this phenomenon was Sigmund Freud (1899/1938).[128] He thought that this was a true amnesia in which there was a catastrophic forgetting of early life events. For him, this was done to protect the ego from threatening psychosexual content. However, there has been little empirical support for this idea across the years, so we do not consider it further.

Instead, it appears more likely that infantile amnesia is not an amnesia at all. That is, it is not a massive forgetting of things that would otherwise be remembered. Instead, this reveals how memory is developing. Keep in mind that humans are born neurologically immature and quite helpless. It takes some time for the nervous system to develop to the point that we can have autobiographical memories. Clearly there is some memory, even before birth. Newborns prefer the sound of their own mother's voice, which they heard *in utero*. Moreover, implicit, procedural learning begins almost immediately as a person learns to do things such as control his or her head, arms and legs, then more complex tasks such as sitting up, using a spoon and dumping cereal on the floor. Children are developing semantic memories as they learn what things are and what they are called. They also have episodic memories in that they are clearly influenced by context. However, young children have a difficult time remembering specific events from their lives, especially as they move away from them.

Infantile amnesia resolves itself as one develops a sense of self and can start organising information around this self-concept (Howe & Courage, 1993[129]). Certain neurological correlates indicate the use of self-referential memories as opposed to more general knowledge (Magno & Allan, 2007[130]). As we have noted earlier, information that you can relate to yourself is often remembered best. In a sense, then, the offset of infantile amnesia marks the beginning of autobiographical memory. Several things contribute to the development of the self-concept. First, this is also the time when there is a tremendous increase in the child's use of expressive language (Nelson, 1993[131]). If you think about your own autobiographical memories, they are heavily influenced by language in terms of their content, their structure and how you think about them. Second, this period of time is when the hippocampus is maturing, allowing more complex memories to be formed (Nadel & Zola-Morgan, 1984[132]). Finally, at this age, a child has started to develop schemas and scripts that are complex enough to begin making sense of the world in a more adult-like fashion, thereby facilitating memory for individual life experiences (Nelson, 1993[133]).

6.6.3 Reminiscence bump

When you talk to people older than the typical college student, you may find that they have a bias to remember events from when they were in college (or that age, thereabouts). Why is that? Are they trying to make a social connection with you? Was that really the best time of their lives? Although we cannot answer all of these questions, we can tell you that your observation is correct and that we can provide some insight into the first question. The **reminiscence bump** is superior memory than would otherwise be expected for life events around the age of 20, between the ages of 15 and 25. This is illustrated in Figure 6.10 (Rubin, Rahhal & Poon, 1998[134]).

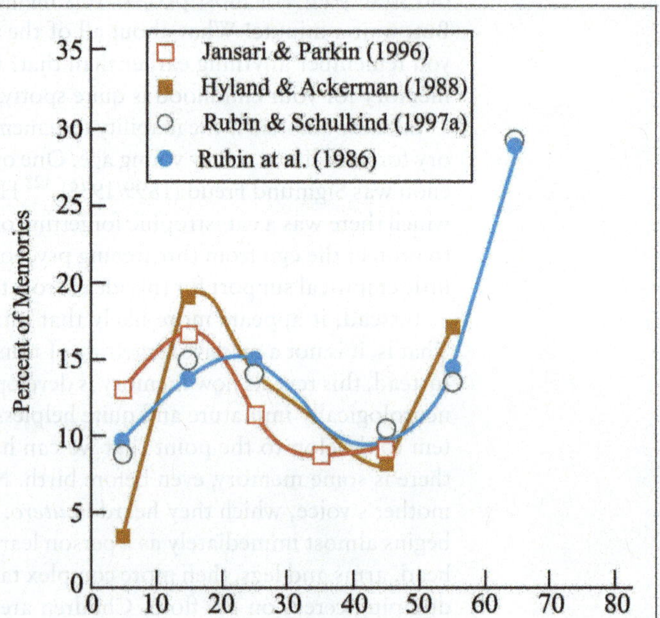

Figure 6.10 Illustration of the reminiscence bump

What causes the reminiscence bump?

There are several ideas about what causes the reminiscence bump.

First-time events

One idea is that memory tends to be better for things that happen for the first time. For example, it is easier to remember your first kiss than your 27th, although you may have just as much fun during each. The early 20s is a period in a person's life when many things are happening for the first time – the first time a person lives alone, drives, votes, gets a real job, gets a speeding ticket, pays taxes, and so on. Because there are so many firsts, it is easier to remember events from this period of life. This idea is supported by the fact that people who move from one country to another, where another language is spoken, show a reminiscence bump for the time of their immigration, regardless of how old they were at the time (Schrauf & Rubin, 1998[135]). In a similar way, evidence shows a relocation bump, a peak in memories surrounding the life transition of changing one's residence, occurring in middle adulthood (Enz, Pillemer & Johnson, 2016[136]). Presumably such moves bring many new experiences, a number of firsts that make those times in life more memorable.

Expectation to remember

Another idea is that we remember so much more from this period of our lives because we expect to. That is, we all carry out cultural life scripts (Berntsen & Rubin, 2004; Rubin & Berntsen, 2003[137]) about when the exciting times in our lives are supposed to be, such as graduating from school, getting married, buying our first home, having our first child, and so on. This reflects the idea that our culture influences many aspects of how we think about our lives and structure our autobiographical memories (Ross & Wang, 2010[138]). There is also a 'life stories' account that is based on the story of our lives that we create over time and as we develop (Gluck & Bluck, 2007[139]).

So, part of what is driving the reminiscence bump is the expectation that people are supposed to remember more from these important points in their lives. In fact, when people recall events from their lives, the intensity that they report for those events, particularly positive events (e.g. weddings, birth of children or graduations), corresponds with the cultural expectation (Collins, Pillemer, Ivcevic & Gooze, 2007[140]). Another source of support for this idea is that people show a reminiscence bump for a character they read about in a novel (Copeland, Radvansky & Goodwin, 2009[141]). People remember events about a story character better when the character was around the age of 20 or when the character went through major life change (for instance, a character's long-time husband dies and she goes out into the working world), as compared to earlier events in the novel (when she was younger) or chapters that were read more recently (when she was older). This seems more likely to occur if people are using life scripts about what should be more memorable to actually guide their memory retrieval.

This figure shows the rate of remembering events as a function of how old the person was at the time (older adults are used here to make the reminiscence bump clearer). These studies often use what is known as the *Galton-Crovitz technique* (Crovitz & Shiffman, 1974; Galton, 1879[142]), in which people are given lists of words and asked to respond with the first autobiographical memory that comes to mind. The first thing to notice is that most of the reported memories come from the recent past, and that the further away in time the event is (the older the event is, hence the younger a person was at the time), the less likely it will be recalled. However, note that around the age of 20 there is a tendency to remember more events than would be expected if this were a normal forgetting curve. So, why does this happen?

6.6.4 Involuntary memory

Autobiographical memories not only are recalled when we actively try to remember but also happen spontaneously and involuntarily, without any clear effort to do so. These types of memories often occur (Berntsen, 2010[143]), and they often refer to single events, rather than general periods of time (Berntsen & Hall, 2004[144]). An example of an **involuntary memory** would be remembering a specific event that happened during a specific chemistry class when you were in high school, compared to just remembering that you took a chemistry class. These involuntary memories are often triggered by some cue – literally a retrieval cue – in the environment. For example, a person may see a highly valued object (an old toy) that then has the power to elicit a strong autobiographical memory (Jones & Martin, 2006[145]). Emotional intensity is commonly a critical aspect of such triggered autobiographical memories (Talarico, LaBar & Rubin, 2004[146]).

Some of the strongest cues to spontaneously bring about autobiographical recollections are odours. Memories elicited by odours produce a strong feeling of being back in time (what it was like to experience the event long ago) (Herz & Schooler, 2002; Willander & Larsson, 2006, 2007[147]). Moreover, odors elicit stronger emotions in the person experiencing the memory.

6.7 Memory for the future

LEARNING OBJECTIVE
Describe how memory can be oriented towards the future

This section of the chapter flips the way in which we typically think about memory. Most of the time when we consider memory, we are thinking about how we remember incidents that happened in the past. And certainly, that is one of the major things for which we use our memories. However, an important role of memory is not only a retrospective one that allows us to go over past information, but also a prospective one that allows us to perform well by anticipating events that have not yet happened. Consistent with one of the themes of this course, memory is just as important for the future as it is for the past. To this end, here we discuss two ways that memory for the future comes about: *prospective memory*, or remembering to do something in the future, and *episodic future thinking*, or imagining future events.

6.7.1 Prospective memory

Prospective memory (Loftus, 1971[148]) is the ability to remember to do something in the future.

For the improvement of daily prospective memory, a study by Trawley, Law and Logie (2011)[149] found that for prospective memory to do a number of tasks, it is best if people make a plan ahead of time and then stick to it. When this plan is interrupted and deviated from, people are more likely to forget something that they are supposed to do. As has been found with other types of memory, sleeping after encoding a prospective memory goal can also help performance (Scullin & McDaniel, 2010[150]).

6.7.2 Episodic future thinking

In addition to remembering to do something in the future, memory is also important for imagining what the future will be like, allowing us to anticipate and plan what we will do. This is *episodic future thinking*: the mental construction of what future events might be like, of the outcome of future situations (Atance & O'Neill, 2001;

Szpunar, 2010; Szpunar & Radvansky, 2016[151]). As an example of this, when you are getting ready to ask someone you are fond of to have coffee or lunch with you, you may imagine how that situation will unfold. You can predict and prepare for how the event will unfold so you can do better when the actual event happens (Klein et al., 2010[152]). Episodic future thinking is actually very common. It has been estimated that this is something that we do about every 15 minutes (D'Argembeau, Renaud & Van der Linden, 2011[153]).

Part of the reason that this type of future thought is called episodic future thinking is because it is believed to use the same memory processes as episodic memory. This is called the *constructive episodic simulation hypothesis* (Schacter & Addis, 2007[154]). When we try to think about a future event, we use our memories of prior, similar events to imagine what the new event might be like. For example, if we decide to fix something in the engine of our car, we may think about other times that we have tried to fix things that are mechanical and remember any problems or troubles we had at those times. To predict how asking someone out for coffee will go, you remember how that went the last time you did it. In general, the closer the event you are trying to imagine conforms to prior events, the easier it is to construct (Szpunar & Schacter, 2013[155]).

Episodic future thinking also exhibits some of the same patterns of behaviour, although in an opposite direction, as episodic remembering. For example, when we think about the past, we have a bias to think about more recent events than older events. Similarly, when we think about the future, we have a bias to think more about future events that are closer to the current point in time than times much further ahead (Addis & Schacter, 2008; Spreng & Levine, 2006[156]).

Although episodic future thinking uses some of the same mental processes and representations as are used in retrospective episodic remembering, differences exist between the two. For instance, our imaginings of the future tend to be more positive and to focus more on events that will be important to our life stories (e.g. marriage or retirement). Also, they tend to be less vivid (Anderson, Dewhurst & Nash, 2012; Berntsen & Bohn, 2010; Grysman, Prabhakar, Anglin & Hudson, 2015; Rasmussen & Berntsen, 2013[157]). Overall, it is clear that memory is important not only for recollecting what we have been through in the past but also for allowing us to do well in the future.

6.8 Semantic networks and connectionist models

LEARNING OBJECTIVE
Explain the connectionism approach to understanding semantic memory and human memory more widely

We conclude this chapter by discussing semantic memory and semantic networks, which forms part of long-term memory and attempts to address issues about how semantic knowledge is stored.

Quillian's representation of the semantic network

The model

His initial model of semantic memory was not a psychological model of memory but a computer program for understanding language. Shortly afterwards, however, Quillian collaborated with Allan Collins, and their psychological model was the first attempt in cognitive psychology to explain the structure and processes of semantic memory.

The Collins and Quillian model of semantic memory (1969, 1970, 1972; Collins & Loftus, 1975[158]) was based on a network metaphor of memory. At the heart of the model were fundamental assumptions about the *structure* and *retrieval processes* of semantic memory.

Figure 6.11 illustrates a portion of the semantic network.

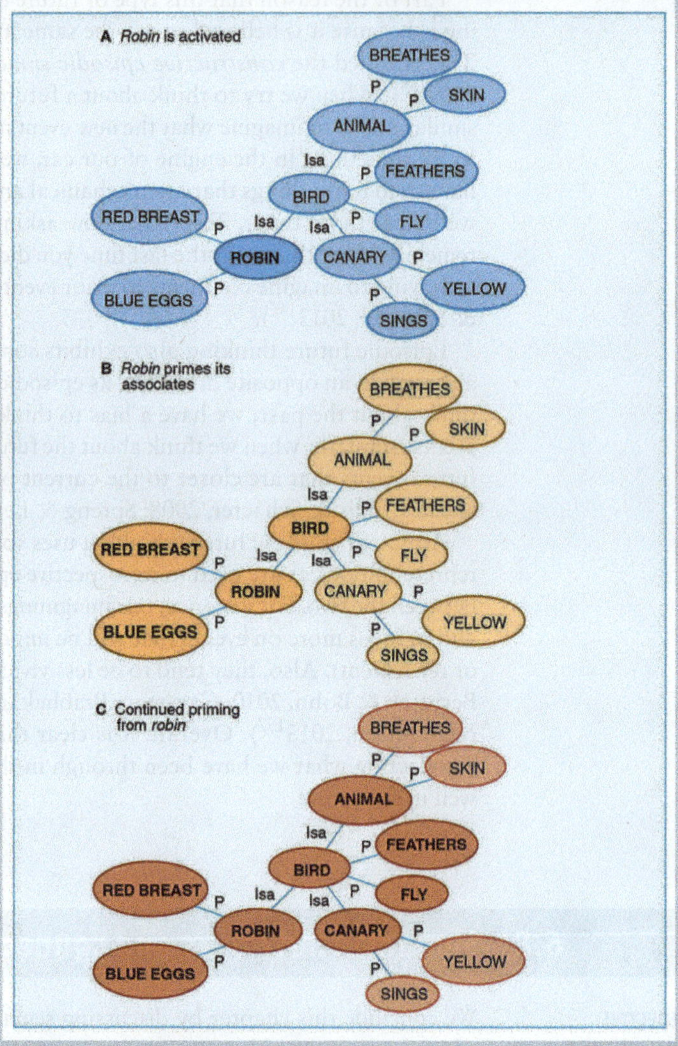

Figure 6.11 **An Illustrated portion of the semantic network**

A. The concept 'ROBIN' has been activated and is shown in boldface.
B. The spreading activation from 'ROBIN' has activated concepts linked to 'ROBIN', such as the boldface 'BIRD', 'RED BREAST' and 'BLUE EGGS'.
C. The continued spread of **activation** that originated from 'ROBIN' is depicted.

The network

Collins and Quillian viewed the concepts of semantic memory as being nodes in a network. In other words, the structure of semantic memory was a network, an interrelated set of concepts. Each concept is represented as a **node**, a point or location in the semantic space.

Furthermore, concept nodes are linked by pathways that serve aslabeled, directional *associations* between concepts. This entire collection – nodes linked via pathways – is the network. Note that every concept is related to every other concept in that some combination of links, however indirect and long, can be traced between any two nodes. The major process that operates on the network is **spreading activation**, which is the retrieval of information from this network.

The concept

Concepts usually are at baseline, in a quiet, unactivated state. For example, as you read this sentence, one of the concepts in your semantic memory that is probably not activated is 'ROMANCE'. But when you read the word, its mental representation is activated. Then 'ROMANCE' is no longer quiet; it has been activated. (Here, words are italicised, and concept names are printed in capital letters.)

Even such a simple network represents a great deal of information. For instance, some of the facts in this network are that 'ROBIN' is a member of the category 'BIRD', that a 'ROBIN' has a 'RED BREAST', that a 'CANARY' is 'YELLOW' and so on.

Each of the links records a simple fact or proposition, a relation between two concepts. Note further that each link specifies a relationship. Thus 'ROBIN is a BIRD' is a member of the category 'BIRD', and 'BIRD has the property FEATHERS'. (The latter statement is called a **property statement**. The *is* a relationship was a bit of jargon from Rumelhart, Lindsay & Norman, 1972, meaning 'is a', as in 'is a member of the category'.[159]) Direction is important because the reversed direction is not true, that is, '*All BIRDS are ROBINS'. (Sentences that are intentionally wrong are marked with an asterisk.) Figure 7.3 shows what happens to this part of a semantic network when the word *robin* is presented. First, the node for *robin* is activated, illustrated by boldface in Figure 7.3A. Then, activation spreads to the concepts it is linked to, like 'BIRD', 'RED BREAST' and 'BLUE EGGS' in Figure 7.3B. Those concepts continue the spread of activation to their associated nodes, as depicted in Figure 7.3C.

The spread of activation is triggered each time a concept is activated. If two concepts are activated, then there are two spreads of activation, one from each node.

You should have found two characteristics of spreading activation. First, activating 'ROBIN' eventually primed a node that was also activated by 'BREATHES'. This explains how information is retrieved from semantic memory. The spreads of activation intersect, and a decision is made to be sure that the retrieved link represents the idea in the sentence. So, although a pathway would be found between 'ANIMAL' and 'RED BREAST', a decision that the idea 'all animals have red breasts' would be invalid. A second characteristic of semantic search is that other concepts also become activated or primed. In our example, many other concepts were primed. There should also be a '1' next to 'RED BREAST' and 'BLUE EGGS' from the first cycle; a '1' next to 'FLY', 'FEATHERS' and 'CANARY' after the second cycle; and so on. Thus, spreading activation activates related concepts. More and more distant concepts, connected by longer pathways, do not receive as much activation. Moreover, these related concepts are not activated forever because activation decays after some amount of time.

6.8.1 Feature comparison models

Semantic networks are not the only way semantic memory has been conceived. Another is as a collection of features, as with the feature comparison model of Smith and his colleagues. In Smith's model (Rips, Shoben & Smith, 1973; Smith,

Rips & Shoben, 1974[160]), semantic memories were captured as feature lists. So, semantic memory is a collection of sets of semantic features: simple one-element properties of the concept. Thus, the concept 'ROBIN' is a list of its features, such as animate, red-breasted and feathered. Furthermore, these features vary in terms of their definingness, with defining features being emphasised more than less defining features. Thus, an essential feature is a defining feature, such as animate for 'BIRD'. Conversely, other features (e.g. 'ROBIN' perches in trees) would be less so but are characteristic features: features that are common but not essential. Thus, characteristic features do not define 'ROBIN': robins may or may not perch in trees. But defining features are essential: if it is a robin, it has to be animate.

Smith's model of feature comparison

The retrieval process in the Smith model is *feature comparison*; follow along with the sequence in Figure 6.12 as you read.

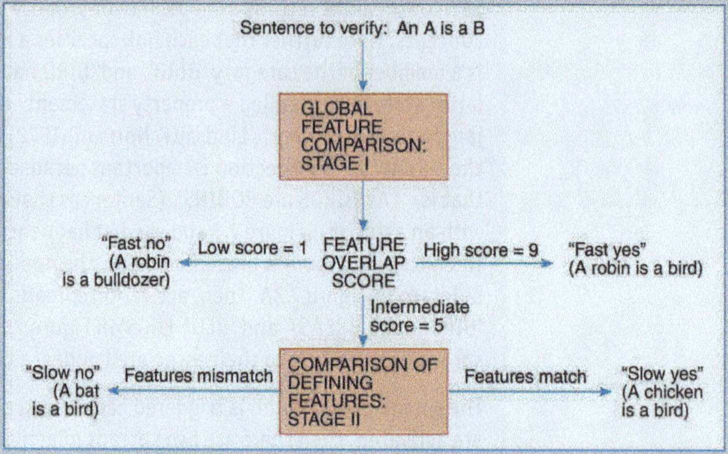

Figure 6.12 **The comparison and decision process in the Smith et al. model (1974)[161], with sample sentences**

Stage I

Suppose you are given the sentence 'A robin is a bird' and have to make a true/false judgement. For this model, you would access the concepts 'ROBIN' and 'BIRD' in semantic memory and then compare the two feature sets. Stage I involves a rapid, global comparison of features, to 'compute' the similarity between the two concepts. For illustration, assume that these similarity scores range from 1 to 10.

For 'A robin is a bird', the feature sets overlap a great deal; there are very few 'ROBIN' features that are not also 'BIRD' features, leading to a high overlap score (e.g. 8 or 9), so high that you confidently respond 'yes' immediately. Conversely, for 'A robin is a bulldozer,' there is so little feature overlap (e.g. 1 or 2) that you can respond 'no' immediately. These are 'fast yes' and 'fast no' stage I responses.

Stage II

What about when the relationship is not so obvious?

First, consider 'A chicken is a bird'. For most people, chickens are less representative of the BIRD category. They do not perch, make nests in trees, fly, and so on. So, the stage I comparison would find only an intermediate degree of overlap, and a second comparison is needed, called the stage II comparison.

Unlike fast stage I comparisons, a stage II comparison is slower, emphasising more defining features. Thus, because it is true that chickens are birds, there is a match on all the features in this stage, yielding a 'slow yes' response. Similarly, for 'A bat is a bird', the overlap score is intermediate, triggering a stage II. Here, however, there are important mismatches. The characteristic features that make bats similar to birds are outweighed by the defining features (mammal, furry, teeth, etc.) mismatch. Thus, the stage II comparison gives evidence that the sentence is false.

6.8.2 Semantic relatedness

Figure 6.13 is a modified network of the BIRD category that incorporates semantic relatedness (Collins & Loftus, 1975[162]).

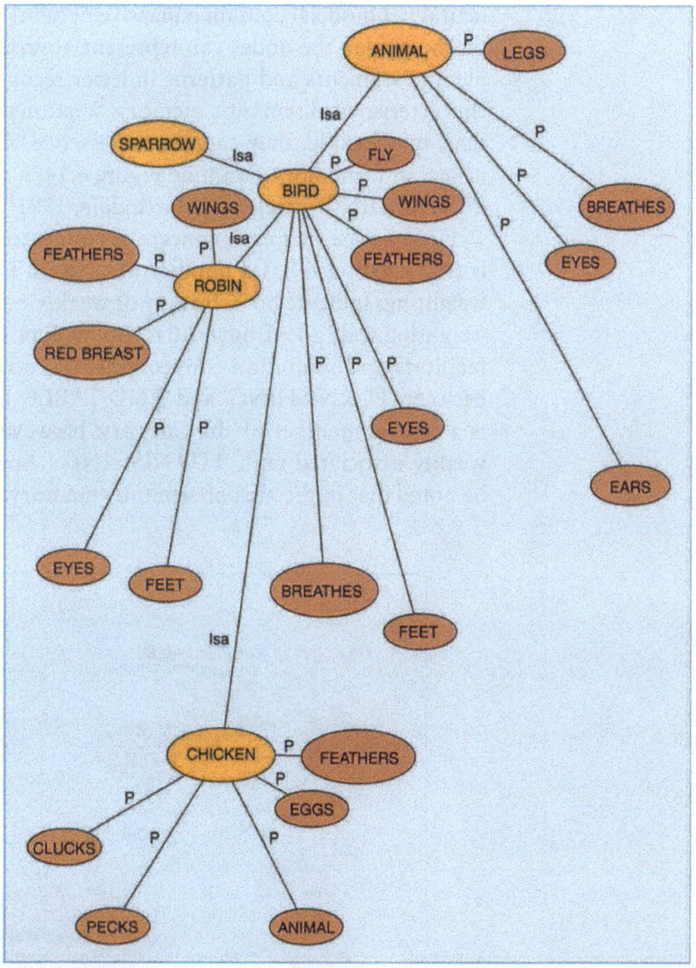

Figure 6.13 An illustration of a portion of the semantic network, taking into account three empirical effects

There is no strict cognitive economy in the hierarchy, so redundant information is stored at several different concepts. Typical members of the category are stored more closely to the category name or prototype member, and properties that are more important are stored more closely to the concept than those of lesser importance.

For example, a strict hierarchy is incorrect. Instead, these links are of different lengths, reflecting different strengths of association, with stronger associations being verified faster than weaker ones. Moreover, for the feature of typicality (elaborated on later), more typical examples of a concept are linked by shorter pathways. Therefore, the link between robin and bird is shorter than that between chicken and bird. It is difficult in a two-dimensional figure to completely capture the associative and strength complexity of semantic memory. Thinking of semantic memory in space with three or more dimensions makes this easier to imagine, but harder to illustrate. Regardless, the idea that performance varies directly as a function of the strength of associations is readily understood. The stronger the relation between concepts, the faster you can retrieve the idea. This is the semantic relatedness effect: concepts that are more highly interrelated are retrieved faster.

6.8.3 Connectionism and the brain

A more complex way to model human memory rather than a semantic network is to use the approach of connectionism. You have had small doses of this already. At the most fundamental level, connectionist models (also known as PDP models and neural net models) contain a massive network of interconnected nodes. Depending on the model, the nodes can represent almost any kind of information, from simple line segments and patterns in letter recognition to more complex features and characteristics in semantic memory. What makes connectionist models attractive is that, in principle, they capture the essence of the neurological processes involved in semantic memory. Examine Figure 6.14, a connectionist network for part of the 'FURNISHING' category (Martindale, 1991[163]).

First, notice that each concept is connected to other nodes. The difference here is that each path has a number next to it. This is its weighting or strength. The weightings indicate how strongly or weakly connected two nodes are. Generally, the weighting scale goes from +1.0 to 1.0, with positive and negative numbers indicating facilitate and inhibition connections, respectively. So, for example, the weighting between 'FURNISHING' and 'END TABLE' is +0.8, indicating that 'END TABLE' is a central member of the category. However, 'COASTER' with its +0.1 is only weakly associated with 'FURNISHING.' Apart from this one example, it should be noted that in the area of semantic memory, there are many connectionist models

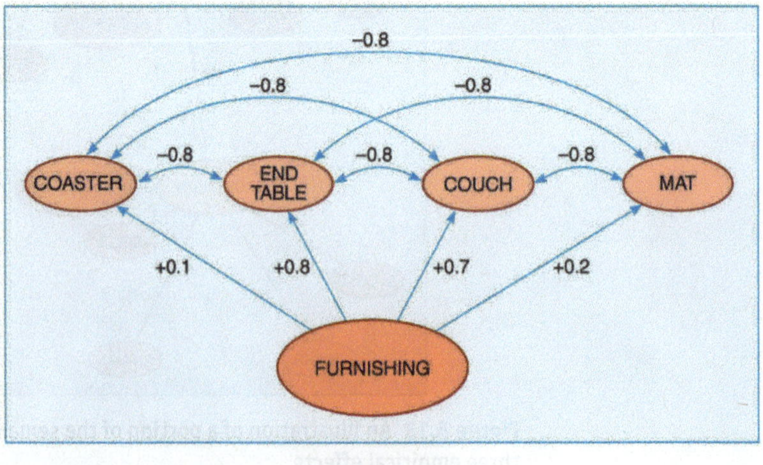

Figure 6.14 A small portion of a connectionist network

Note that the nodes at the same level exert an inhibitory influence on each other and receive different amounts of facilitation from the category name.

Based on Martindale (1991).[164]

that capture the process of semantic memory. These include models that account for things like priming (Masson, 1995[165]) and word meaning (McRae, de Sa & Seidenberg, 1997[166]). The glimpse at connectionism here is very simplified. In many respects connectionist models differ from the older spreading activation models. An important distinction is that in a connectionist mode, a concept is defined as a pattern of activation across units in the network. Priming is explained by the similarity of activation patterns between a prime and a target: The 'FURNISHING' concept has a similar pattern of activation to the pattern for 'END TABLE', so the one serves as an effective prime for the other.

What is exciting about connectionist models is that they give us a tool for understanding the richness of cognition, a working 'machine' that lets us see what happens when multiple layers of knowledge influence even the simplest acts of thought. Particularly compelling (Seidenberg, 1993)[167] are four advantages of connectionist models. First, they are more similar to the network of neurons in the brain. The brain is a massive set of interconnected neurons, just as a connectionist network is a massive set of interconnected units. Second, the units are similar to those in the brain. In the nervous system, a neuron either fires or does not. When it does fire, it affects the neurons it synapses on. This parallels the fire/no-fire nature of connectionist units. Third, the positive and negative weights between units in connectionist models parallel excitatory and inhibitory neural synapses. Fourth, the activity of a connectionist model is massively parallel. Multiple processes are co-occurring in a model at various levels, much as there is parallel processing in the brain (McClelland, Rumelhart & Hinton, 1986; Rumelhart, 1989[168]).

Studies on semantic memory impairment

How could semantic memory be splintered so that a person's categories of animate things are disrupted while categories of non-living things are preserved? More to the point, could semantic memory be structured to reveal these two very broad categories, living and non-living things? Why this distinction, and not some other, such as concrete versus abstract, high versus low frequency, or something else?

Warrington and Shallice's study

Warrington and Shallice (1984)[169] suggested a plausible explanation. Suppose that the bulk of your knowledge about living things is coded in semantic memory in terms of sensory properties: A parrot is a brightly coloured animal that makes a distinctive sound. Likewise, most of what you know about non-living things involves their function: a briefcase is for carrying around papers and books. Warrington and Shallice suggested that this dissociation could be due to a selective disruption to sensory knowledge in semantic memory. If so, that might explain the patients' JBR selective impairment for recalling living things (Warrington & Shallice, 1984).

Farah and McClelland's model

Going a step further, Farah and McClelland (1991)[170] built a connectionist model to evaluate the Warrington and Shallice hypothesis. In their model, semantic memory had two types of features about concepts – visual features (sensory) and functional features – with this knowledge being acquired by visual or verbal input. After constructing and training the model (establishing its memory, in a sense), Farah and McClelland 'lesioned' it. That is, they 'damaged' the visual units in semantic memory by altering the connection weights or disconnecting the visual units. The outcome was strikingly similar to the patients' dissociations. When the visual units were lesioned, the network had poor accuracy in associating names and pictures of living things. This is shown in the left panel of Figure 6.15, as is the modest decline in accuracy for the *non-living things.* Conversely, when the network's functional units were lesioned instead, it was the *non-living* category that suffered (right panel, Figure 6.15; for updates, see Cree & McRae, 2003, and Rogers et al., 2004[171]).

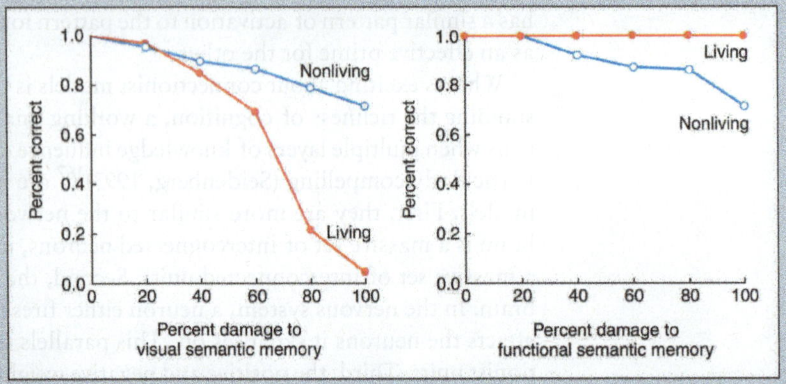

Figure 6.15 Performance of Farah and Mcclelland's model
The performance of the basic model as measured by probability of correctly associating names and pictures for living and non-living things, after different amounts of damage to visual and semantic units.

Does this demonstration prove that impairment of patient J. B. R.'s visual semantic knowledge accounts for the dissociation? No, the model makes the correct prediction, a point in its favor. But this is not proof that the model is correct. Instead, the demonstration essentially asked whether it is possible that impairment of sensory knowledge could produce the dissociation between living and non-living things. An answer is yes, it is possible, because a connectionist model simulates a dissociation. In other words, the model provides a degree of assurance (probably a large degree) that the Warrington and Shallice hypothesis is reasonable and should be pursued further.

Summary: Learning and remembering

6.1 Preliminary issues

- Long-term memory is a divided between declarative memory and non-declarative memory. Declarative memory consists of episodic and semantic memories; non-declarative memory includes priming and procedural or motor learning. Declarative memories can be verbalised, but non-declarative memories cannot. Conscious awareness of the memory is unnecessary for implicit memory tasks but does accompany explicit memory tasks.
- A classic method for improving memory involves mnemonics. Mnemonics, such as the method of loci, use a variety of techniques, especially visual imagery, to improve performance.
- Ebbinghaus was the first person to extensively study memory and forgetting. Working on his own, he invented methods for doing so. The relearning task revealed a sensitivity to the demands of simple recall tasks – that they measure consciously retrievable information but underestimate the amount of information learned and remembered. The classic forgetting curve he obtained, along with his results on practice effects, inspired the tradition of verbal learning and, later, cognitive psychology.
- Memory consolidation is the process that makes memories permanent. This involves two phases: a fast-acting synaptic consolidation and a slow-acting systems consolidation. The older a memory is, the more likely that it will be consolidated.

- Metamemory is one's awareness of the contents and processes of one's own memory. This is an imperfect insight. One way to assess metamemory involves judgements of learning in which people assess whether they have learned something and whether it is in memory. Generally, these assessments are poor immediately after learning, but are better if people wait a short period to make such assessments. How well people assess whether they have learned something influences how they decide to study later. Another way to assess metamemory is examining a person's feeling of knowing whether something that is currently forgotten will be remembered later. A related phenomenon is the tip-of-the-tongue (TOT) effect wherein people cannot remember, but feel that remembering is imminent.

6.2 Storing information in episodic memory

- Important variables in storage are rehearsal and organisation, regardless of whether the information is verbal or perceptual. Maintenance rehearsal and elaborative rehearsal have different functions: the former for mere recycling of information, the latter for more semantically based rehearsal, which was claimed to process the information more deeply into memory. Difficulties in this depth of processing framework involve specification of the idea of depth.
- Generally, the amount of rehearsal is positively related to recall accuracy for the primacy portion of a list. Also, if people rehearse information in a distributed, rather than a massed, manner, later memory is much better. Rehearsal by taking a test can improve memory more than by simply studying alone.
- The more deeply information is processed, as with elaborative rehearsal, the better it is remembered. This elaborative rehearsal may occur not only through organisation, especially by category, but also by subjectively defined chunks, to improve memory because it stores the information securely and provides a useful structure for retrieval. However, the depth of processing framework, while intuitively appealing, is hard to nail down explicitly.

6.3 Boosting episodic memory

- There are several ways of improving what will be remembered from episodic memory. This can be done by referring information to oneself or by taking some action with the information. This action can include the generation of the information, the production of what is heard or written by speaking or writing it, or the enactment of described actions. Memory can also be boosted by structuring and organising the material in some way. This organisation can be objective, such as categories, or can be subjectively imposed on the material by an individual. Forming mental images also improves memory by providing another memory code to allow people to access the knowledge as needed. Finally, memory improvement occurs if people relate the material to be learned to their own survival in some way, taking advantage of evolutionary pressures that have shaped memory.

6.4 Context

- Episodic memory is strongly influenced by context. One type of context is the environment one finds oneself in. According to the encoding specificity principle, memory is better when the encoding and retrieval contexts match than when they are different.
- It is also important to be able to remember the context or source in which something was learned. This can involve distinguishing between what was only imagined and what actually happened, or distinguishing between two external sources. Source monitoring issues have been observed in cases of cryptomnesia, or unconscious plagiarism.

6.5 Facts and situation models

- Comprehending and remembering ideas can involve constructing propositional representations in which meaningful elements are represented as nodes connected by various pathways (e.g. agent, recipient).
- We tend to remember the gist or general meaning of a passage but not the more superficial aspects such as exact wording. We routinely 'recognise' a sentence as having occurred before even if the sentence is a paraphrase.
- Situation models are representations of the described situations. Whereas memories at the propositional level are forgotten rapidly, these memories persist for long periods of time.

6.6 Autobiographical memories

- Studies of autobiographical memory, or memory in real-world settings, show the same kinds of effects as laboratory studies but sometimes more strongly. Autobiographical memory is one area in which we might see modern psychologists studying their own memories. One part of autobiographical memory is the lack of conscious memories of events from when we were infants. This is known as infantile amnesia. Another part is the superior memory people have for the time in their life around the age of 20. This is known as the reminiscence bump. Finally, autobiographical memories are not always deliberately retrieved, but are often retrieved involuntarily.

6.7 Memory for the future

- Memory is important not only for storing and retrieving information encountered but also for allowing us to anticipate and do well in the future. Prospective memory involves remembering to do something when a future event is encountered, or a certain period of time has elapsed.

- Episodic future thinking involves using knowledge that we have in declarative memory to imagine and prepare for situations and events that we think are likely to occur.

6.8 Semantic memory

- Semantic memory contains our long-term memory knowledge of the world. Early studies of the structure and processes of semantic memory generated two kinds of models: network approaches and feature list approaches.
- In network models, concepts are represented as nodes in a semantic network, with connecting pathways between concepts. Memory retrieval involved the process of spreading activation: activation spreads from the originating node to all the nodes connected to it by links.
- Feature comparison models assume that semantic concepts are sets of semantic features. Verification consists of accessing the feature sets and comparing these features.
- Connectionist approaches to memory allow one to simulate these processes in a way that is more analogous to how the brain does this.
- Connectionist approaches even furnish persuasive analyses of semantic disruptions caused by brain damage, so-called category-specific deficits.

Chapter 6 quiz

Question 1

Makena is a teacher and is teaching a class of school children how to remember the order of the planets. She decides to ask the children to create a phrase which begins with each letter of the order of the planets. One of the children decides to use Many Vicious Elephants Make Jam Sandwiches Under Nettle Plants so that they can remember the correct order of the planets. Explain what technique has been used by the child?

A. Mnemonic
B. Retrieval cue
C. Distributed practice
D. Immediate recall

Question 2

Researchers such as Rundus (1971) have studied how rehearsal can link to the serial position curve. Explain this type of link?

A. There is a negative relationship between the frequency of rehearsal and the rate of recall in the first few words of the serial position curve.
B. There is a negative relationship between the items remembered last in the serial position curve and the frequency of rehearsal.
C. There is a positive relationship between the frequency of rehearsal and the rate of recall in the first few words of the serial position curve.
D. There is a positive relationship between the items remembered last in the serial position curve and the frequency of rehearsal.

Question 3

Which person will most likely remember what he or she is trying to commit to memory?

A. Soledad is memorizing her cherry pie recipe, so she can pass it on to her uncle.
B. Nigel is trying to remember the name of an interesting woman he just met at a party.
C. Rob is studying snakebite prevention tips in preparation for a week-long hike in the desert.
D. Bethany is cramming for her final exam in modern art class.

Question 4

Early autobiographical memory research used researchers' own memories. What is a criticism of using this method?

A. The researchers could not investigate infantile amnesia in themselves as they would have been too young at this age to conduct investigations.
B. The researchers could not remain impartial from the data collection and the results could have been biased with the expectation of remembering items from the past.
C. The researchers could not collect enough data by only using themselves.
D. The researchers were unsure about what type of memory they were researching.

Question 5

Rips, Shoben & Smith (1973) and Smith, Rips & Shoben (1974) developed feature comparison models. How does this type of model differ from semantic network models?

A. The semantic network model viewed memory as a computer program whereby networks are used to connect different nodes, like a computer.
B. The semantic network model is the only model which uses meaning as a basis for the development. Everything within this model has some form of meaning behind it.
C. The sematic network model is formed through certain stages of how we remember information.
D. The semantic network model uses a complex view of interconnected nodes, known as connectivism.

7 Memory and forgetting

Learning objectives

7.1 Identify how long-term memory can fail

7.2 Explain how stored information is retrieved from episodic memory

7.3 Describe types of variations in memories

7.4 Differentiate between the types of amnesia

Key figures

Professor Elizabeth Loftus

Professor Elizabeth Loftus focuses on understanding factors that influence the reliability of our memories, particularly in eyewitness testimonials during criminal trials, and how misinformation can be detrimental to the outcome of court decisions.

Professor David Bannerman

Professor David Bannerman is a behavioural neuroscientist investigating the role of AMPA receptors in spatial memory. He is the Head of the Behavioural Neuroscience Unit at the University of Oxford.

Introduction

Having studied remembering from episodic and semantic memory in Chapter 6, we now look at the opposite of remembering: *forgetting*. There are a number of ways that information can be forgotten. In some cases, the forgetting occurs even though the information is in memory. The problem is that a person just cannot get at it. This is more like most of the forgetting that we experience in our lives. In other cases, the forgetting can be more extreme, where the memory no longer exists. So obviously, there can be no way to ever get at this. This extreme forgetting can happen with some cases of amnesia.

The distinction between milder and more severe forms of forgetting reflects a distinction made in cognitive psychology between memory availability and accessibility. The term **availability** refers to whether a memory trace exists somewhere in the memory system so it could be retrieved. Information that no longer exists would not be available in memory. By comparison, the term **accessibility** refers to whether a memory trace that does exist can be found, activated and brought to mind. Forgetting may occur when a memory is available, but not accessible. An analogy that is often used to illustrate this is that of books in a library. Whether a given book is in the library refers to its availability. If the book is in the library, then it is available. The ability to find a book on the library shelves refers to its accessibility. Forgetting in which there is availability, but not accessibility, may be when the book is on a shelf in the library, but the catalogue entry is missing, or the book has been put on the wrong shelf. You would not be able to recover the book unless you happened to stumble upon it.

accuracy
How correct a person is in the responses given; often quantified in terms of both number correct and number incorrect. Separately, or paired with response time, it can give an indication of cognitive functioning.

A major difference between this chapter and the previous two is the greater emphasis on the **accuracy** of remembering versus the *speed* (response time, RT). In many cases of forgetting, the emphasis is on *inaccuracy*. Do people remember something or not? A stunning aspect of many of the results of remembering is inaccurate, error-prone memory such as eyewitness recollection of the details of an event that are just plain wrong. Schacter (1996)[1] writes eloquently of the 'fragile power' of memory, the paradoxical situation in which we are capable of remembering amazing quantities of information yet have a strong tendency to misremember under a variety of circumstances. We focus on the fragile part of this description in this chapter.

7.1 The seven sins of memory

LEARNING OBJECTIVE
Identify how long-term memory can fail

In a very approachable work, Schacter (1999)[2] provides some specifics about the fragile nature of memory by enumerating seven sins of memory, seven ways in which long-term memory lets us down (see Table 7.1). These are ways that the normal operation of memory leads to problems.

Table 7.1 The seven sins of memory

Sin	Description
1. Transience	The tendency to lose access to information across time, whether through forgetting, interference or retrieval failure.
2. Absent-mindedness	Everyday memory failures in remembering information and intended activities, probably caused by insufficient attention or superficial, automatic processing during encoding.
3. Blocking	Temporary retrieval failure or loss of access, such as the tip-of-the-tongue effect, in either episodic or semantic memory.
4. Misattribution	Remembering a fact correctly from past experience but attributing it to an incorrect source or context.
5. Suggestibility	The tendency to incorporate information provided by others into our own recollection and memory representation.
6. Bias	The tendency for knowledge, beliefs and feelings to distort recollection of previous experiences and to affect current and future judgements of memory.
7. Persistence	The tendency to remember facts or events, including traumatic memories, that one would rather forget; that is, failure to forget because of intrusive recollections and rumination.

Considering the seven sins of memory, we can subcategorise them in those that involve forgetting (1–3), those that involve memory distortions (4–6), and one that concerns unwanted recollections from traumatic experiences. There are numerous reasons why we forget information including the fact that we don't always pay attention to events around us (absent-minded). We may also not notice the change in events around us as they unfold over time (change blindness) but whatever the reason, it seems to stem from the fact that most information is processed at a shallow level (i.e. used shallow coding techniques). Unlike the common misperception that memory can be *blocked* if under stress, the most common experience is a tip-of-the-tongue phenomenon, which is when we know that something is in our memory but we cannot recall it. In the following section we will discuss various retrieval processes and why we forget.

7.2 Forgetting through decay and interference

LEARNING OBJECTIVE
Explain how stored information is retrieved from episodic memory

We turn now to the cognitive mechanisms or processes that likely lead to much of the forgetting that we experience every day. Two theories of forgetting have preoccupied cognitive psychology from the very beginning: decay and interference.

Decay

It is a bit unusual for the name of a theory to imply its content as clearly as does the term **decay**. The older a memory trace is, the more likely that it has been forgotten, just as the print on an old newspaper fades into illegibility. The principle dates back to Thorndike (1914),[3] who called it the *law of disuse*. Habits, and by extension the memories that underlie them, are strengthened when they are used repeatedly, and those that are not are weakened through disuse. Thorndike's idea was a beautiful theory, easily understood and straightforward in its predictions.

However, the decay theory of forgetting is problematic because it claims that the passage of time itself causes forgetting. McGeoch (1932)[4] launched a scathing attack on this claim, arguing that it is the *activities* that occur during a period of time that cause forgetting, not time itself. In other words, time alone doesn't cause forgetting – it is what happens during that time that does. Although there are still some arguments for some, at least partial, influence of decay (Schacter, 1999[5]), it is difficult to imagine a study that would provide a clean, uncontaminated demonstration of it. As time passes, there can be any number of opportunities for interference, even if this is due the momentary thoughts you have while your mind wanders. The time interval also gives opportunities for selective remembering and rehearsal, which would boost remembrance of old information.

Interference

Interference
When certain memories prohibit the recall of other memories.

Interference theory was a staple in the experimental diet of verbal learners for at least two reasons. First, the arguments against decay theory and for interference theory were convincing, on both theoretical and empirical grounds. Demonstrations such as the often-cited Jenkins and Dallenbach (1924)[6] study made sense within an interference framework where, after identical time delays, people who had remained awake after learning recalled less than those who had slept (Figure 7.1).

The everyday activities encountered by people who stayed awake interfered with memory. Fewer interfering activities occurred for people who slept, so their memory was better.

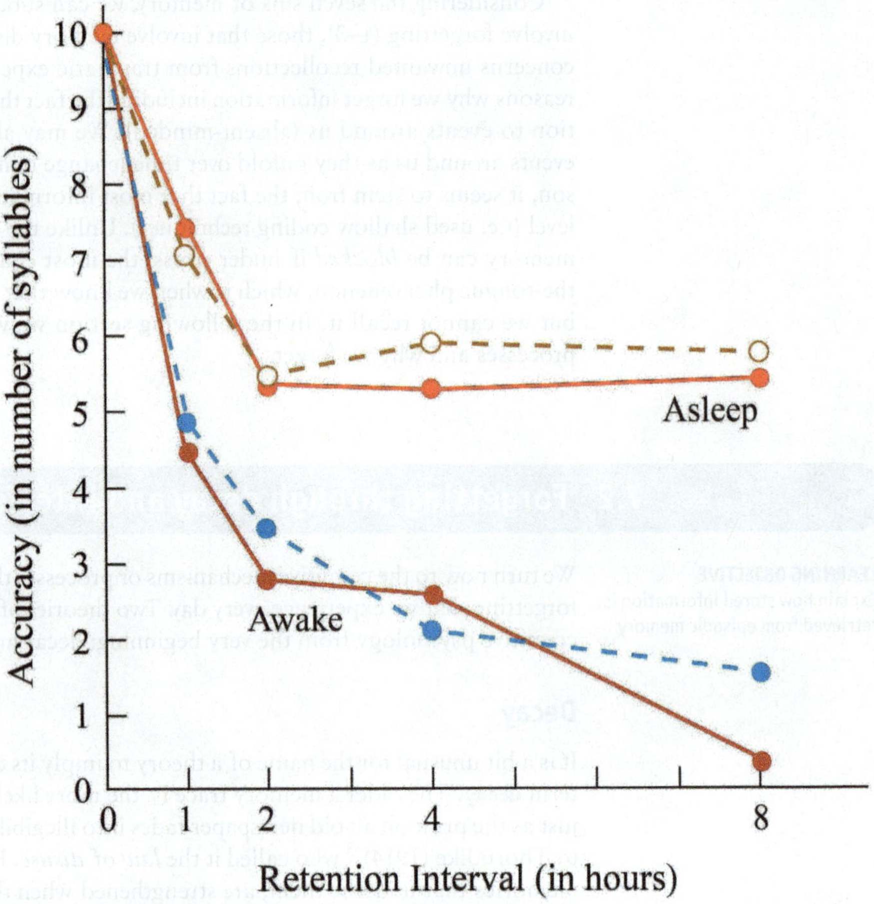

Figure 7.1 The classic Jenkins and Dallenbach (1924)[8] result

Drosopoulos, Schulze, Fischer and Born (2007)[7] replicated this effect. They further concluded that the memory benefit of sleep serves to mitigate the effects of *interference*. However, this was primarily for information that was not strongly encoded (that is, not well learned) to begin with. Although how much interference is experienced is linked to how related the material is, there are other reasons for long-term memory forgetting that are influenced by whether you get some sleep.

Specifically, when you create and store new long-term memories, they do not instantly appear in your brain in a well-established form. Instead, there is a period of time during which memories go through a process of *consolidation*, the more permanent establishment of memories in the neural architecture. Later we see the consequences of a dramatic disruption in consolidation that can result in amnesia. For now, it is important to note that the disruption when people are awake (and not asleep) is interference of the memory consolidation process (Wixted, 2005)[9]. New information is encoded into memory that uses the same neural parts (such as the hippocampus) that were used by the older information. This reuse of same neural networks interferes with memories for the older information, thereby disrupting consolidation and causing forgetting.

This interference is somewhat like writing messages in clay. Imagine writing a message in a bit of clay, then writing even more messages on the clay. Sometimes you will write over messages you had previously written, making the earlier ones harder to recover. This is the type of interference we are talking about. Still, sometimes a message does not get written over, and eventually the clay hardens. In this case, the message consolidates into the clay, making it harder to disrupt. Ultimately, these older memories are more robust and less prone to disruption.

A second reason for the popularity of interference studies is that these effects were easily obtained with a task already in wide use, the paired-associate learning task. This task was a natural for studying interference – and it conformed to the behaviourist Zeitgeist or 'spirit of the times' before the cognitive revolution. Unlike consolidation-disrupting interference, the interference explored using paired-associate tasks was due to *cue overload*. In these cases, the many memories related to a specific cue compete with one another during retrieval.

7.2.1 Paired-associate learning

A few moments studying the *paired-associate learning* task will help you understand interference theory. In other words, how does the relationship between what was learned initially and what is learned later influence how well one remembers something later?

The basic paired-associate learning task is as follows: a list of stimulus terms is paired, item by item, with a list of response terms. After learning, the stimulus terms should prompt the recall of the proper response terms.

Table 7.2 presents several paired-associate lists as a demonstration. Imagine learning List 1 to a criterion of one perfect trial (try it to get a good idea of what the task is like). After that, you switch to the second half of the study, which involves learning another list. The similarity of the first and second lists is critical. If you were

Table 7.2 Lists of paired associates

List 1 (*A–B*)	List 2 (*C–D*)	List 3 (*A–B$_r$*)
tall–bone	safe–fable	plan–bone
plan–leaf	bench–idea	mess–hand
nose–fight	pencil–owe	smoke–leaf
park–flea	wait–blouse	pear–kiss
grew–cook	student–duck	rabbit–flight
rabbit–few	window–cat	tall–crowd
pear–rain	house–news	nose–cook
mess–crowd	card–nest	park–few
print–kiss	colour–just	grew–flea
smoke–hand	flower–jump	print–rain
List 4 (*A–B′*)	**List 5 (*A–C*)**	**List 6 (*A–D*)**
smoke–arm	tall–bench	smoke–fable
mess–people	plan–pencil	print–idea
rabbit–several	nose–wait	mess–owe
park–ant	park–student	pear–blouse
plan–tree	grew–window	rabbit–news
tall–skeleton	rabbit–house	grew–duck
nose–battle	pear–card	park–cat
grew–chef	mess–colour	nose–nest
pear–storm	print–flower	plan–just
print–lips	smoke–safe	tall–jump

switched to List 2, you would experience little or no interference because List 2 has terms that are dissimilar to List 1. In the lingo of interference theory, this was the *A–B, C–D* condition, where the letters *A* to *D* refer to different lists of stimulus or response terms. This condition represented a baseline condition because there is no similarity between the *A–B* and the *C–D* terms (however, you may need fewer trials on the second list because of 'general transfer' effects from List 1, warm-up or learning to learn).

If you shifted to List 3, however, there would have been massive negative transfer. It would have taken you more trials to reach criterion on the second list. This is because the same stimulus and response terms were used again but in new pairings. Thus, your memory of List 1 interfered with the learning of List 3. The term for this was *A–B, A–*B_r, where the subscript *r* stood for 'randomised' or 're-paired' items. Finally, if you switched to List 4 (the *A–B, A–B′* condition), there would have been a great deal of positive transfer; you would need fewer trials to reach criteria on the second list because List 4 (designated *B′*) is related to the earlier one (*B*). For instance, in List 1 you learned *plan–leaf*; in List 4, *plan* went with *tree*.

These are all *proactive interference* effects, showing the effects a prior task has on current learning. Table 7.3 is the general experimental design for a proactive interference (PI) study as well as for a retroactive interference (RI) study. As a reminder, *retroactive interference* occurs when a learning experience interferes with recall of an *earlier* experience; the newer memory interferes backward in time ('retro').

Table 7.3 Designs to study two different kinds of interference

Proactive interference (PI)				
	Learn	**Learn**	**Test**	**Interference effect**
PI group	*A–B*	*A–C*	*A–C*	*A–B* list interferes with *A–C*; e.g. an *A–B* word intrudes into *A–C*
Control group	–	*A–C*	*A–C*	
Retroactive interference (RI)				
	Learn	**Learn**	**Test**	**Interference effect**
RI group	*A–B*	*A–C*	*A–B*	*A–C* list interferes with *A–B*; e.g. an *A–C* word intrudes into *A–B*
Control group	*A–B*	–	*A–B*	

Both proactive and retroactive interference have been examined thoroughly, with complex theories based on the results. Although the literature is extensive, we make no attempt to cover it in depth here (but see standard works, e.g. Postman & Underwood, 1973; Underwood, 1957; Underwood & Schultz, 1960; and Klatzky, 1980, for a very readable summary[10]).

7.2.2 Associative interference

Taking the idea that people can store ideas in memory and that these can be thought of as being organised in a network, we can make some predictions about how this affects performance during remembering. First, a given concept can have multiple associations with it. So, a node in a network can have multiple links to multiple concepts. Second, we can further assume that there is a limit to people's cognitive resources – how much of a network a person can search at once. We have already covered a number of these sorts of mental limits in our discussions of attention and working memory, and we are just extending that logic here. Given these two simple assumptions, we can make some predictions about memory.

In a classic study, Anderson (1974)[11] had people memorise a list of sentences about people in locations, such as 'The hippie is in the park.' The important part of the study was that, across the list, he varied the number of associations with the person and location concepts. That is, a given person in the study list could be described as being in 1, 2 or 3 different locations, and each location could have 1, 2 or 3 different people in it. When we graph a network representation of this type of information, we can see various numbers of links 'fanning' off of a given concept node.

Now, Anderson (1974)[12] further assumed that the amount of activation that could spread along the links of the network was limited. As such, the more links there were fanning off of a concept node in memory, the more widely the activation was distributed, and the longer the processing of the activation along any one of those pathways took. The end result is the prediction that the more links there are off a given concept – the more links fanning off of a concept node – the slower the retrieval process. This, of course, would yield a longer response time, which, as can be seen in Figure 7.2, is exactly what happened in Anderson's (1974) study. After people had memorised the list of sentences, they were given a recognition test in which they had to indicate whether the test sentences were studied on the list or not. The **fan effect** that was found was that when more words associated with a concept, response times were longer.

The fan effect is an **associative interference** effect – the more words associated with a concept, the slower people were to retrieve any one of them. In an interesting extension of this finding, Bunting, Conway and Heitz (2004; see also Radvansky & Copeland, 2006a[13]) looked at the fan effect in terms of the working memory

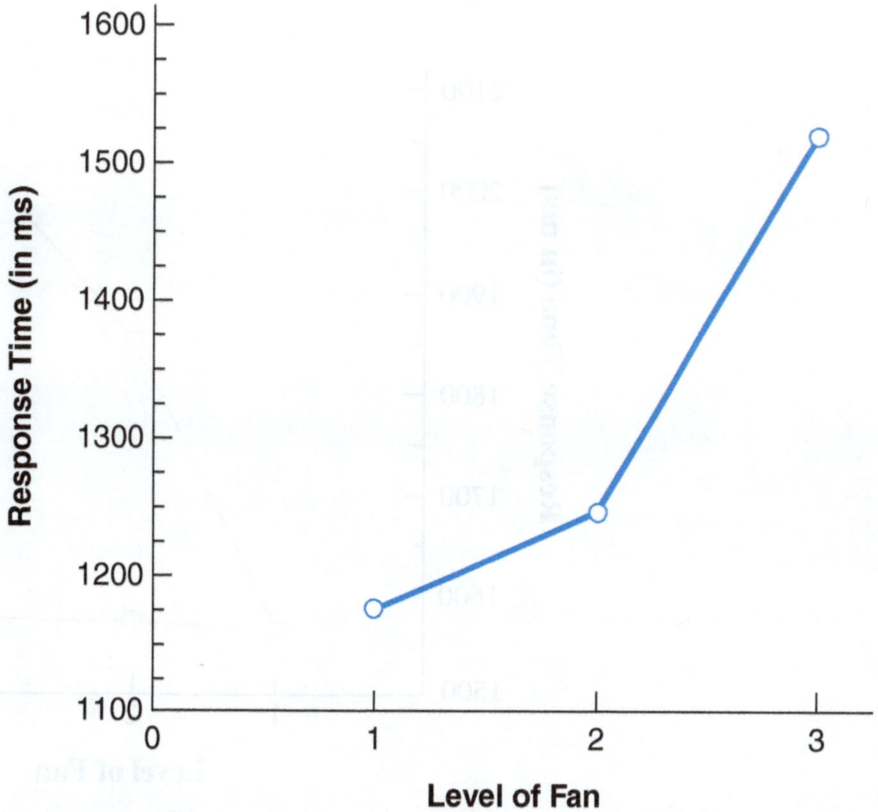

Figure 7.2 The fan effect
An example of data showing a fan effect, showing an increase in response time (RT) with an increase in the number of newly learned associations with a concept (fan).

capacity of their participants. They found that people with lower working memory capacity exhibited greater interference – a larger 'fan effect' – than people with a higher working memory capacity. That is, people with less working memory capacity were further disrupted when their capacity was divided up among many words (larger 'fans'). They were working with fewer working memory resources to begin with, so a cognitive task that places a greater burden on them has a more disruptive effect because they have less extra capacity to compensate for that disruption.

7.2.3 Situation models and interference

Not only are situation model memories (memories for described events) remembered better over the long term, but they also have other benefits, which may contribute to this superior memory.

As you can see in Figure 7.3, Radvansky and Zacks (1991)[14] found a fan effect when there was a single object in multiple locations, but not when there was a single location with multiple objects.

This mental organisation is based on how people think about interaction with the world. In a study by Radvansky, Spieler and Zacks (1993),[15] students learned sentences about people in small locations that typically contain only a single person, such as witness stand, tyre swing or store dressing room. Here, a situation in which multiple people are in one of these locations is unlikely. But, because people can move from place to place, a person-based organisation is plausible – and this is what is observed with a fan effect when a single location was associated with multiple people, but not when a single person was associated with multiple locations.

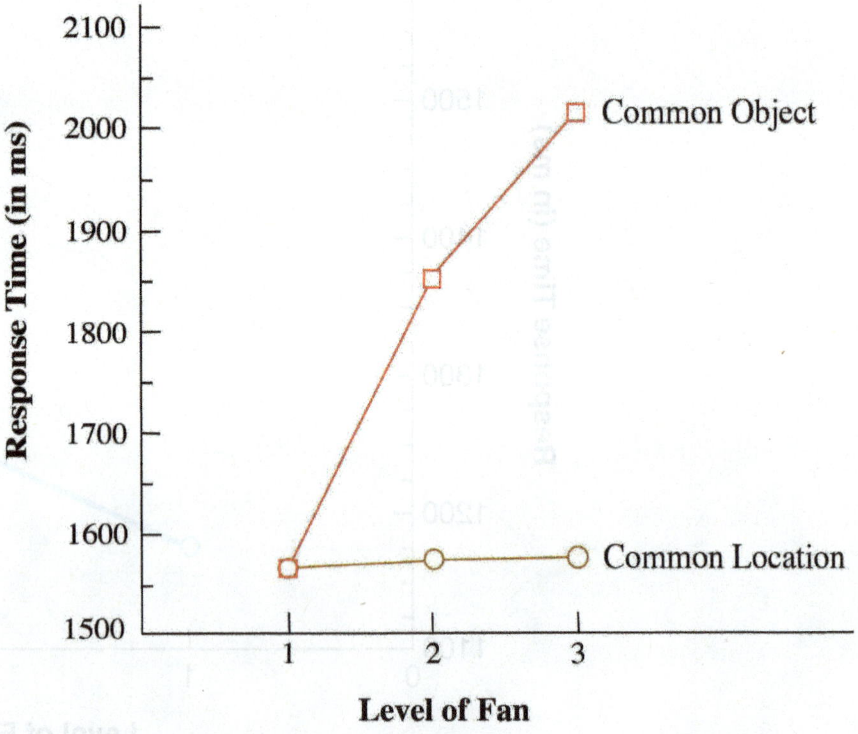

Figure 7.3 Radvansky and Zacks (1991)[17] fan effect

Response times to sentence memory probes as a function of the level of fan (number of associations with a sentence concept). Data are divided based on whether the shared concept is a single location with multiple objects or a single object in multiple locations.

The fan effect

Try and memorise the following sentences:

- The potted palm is in the hotel.
- The potted palm is in the museum.
- The potted palm is in the barber shop.
- The pay phone is in the library.
- The welcome mat is in the library.
- The wastebasket is in the library

Note down the sentences that you can remember

Under these circumstances, how people think about the situations described by the learned sentences influences the ease with which they remember this information later.

In the first three sentences, there is a fan of 3 off of the object concept because the potted palm is described as being in three places. According to *propositional network theories*, the division of the spreading activation causes retrieval to proceed more slowly. Similarly, according to a *situation model view*, because these three sentences are interpreted as referring to different events (that is, different potted palms), people would create a separate situation model for each one. Then during retrieval, these three situation models would each contain a potted palm and would interfere with one another, thereby producing a fan effect.

In comparison, for the last three sentences, there is again a fan of 3, this time off of the location concept, because there are three objects in the library. However, in this case, it is not difficult for people to think of these three sentences as referring to a common event (it is the same library). As such, people can integrate this information into a single situation model and store one representation in memory (e.g. one library, with a pay phone, welcome mat and wastebasket in it). Then, during memory retrieval, because everything is stored together in a single situation model, there should be no retrieval interference and no fan effect. This is exactly what was observed.

For the fan effect, the more things a person knows about a concept, the longer the retrieval time should be. However, this basic principle can shift around somewhat when we start thinking about how the studied information might be organised into *situation models*. For example, in one study, Radvansky and Zacks (1991)[16] had people memorise sentences about objects in locations.

This applies not only to information that people learn from sentences but also to how we interact with the world. Specifically, if a given object is encountered in more than one location, the memory for that object will be worse. For example, in a study by Radvansky and Copeland (2006b; Pettijohn & Radvansky, 2016a, 2016b; Radvansky, Krawietz & Tamplin, 2011; Radvansky, Tamplin & Krawietz, 2010[18]), people navigated a virtual environment in which they picked up objects and moved them either across a room or from one room to another. The researchers found that, compared to moving objects across a room, when people moved objects from one room to another, they had a harder time remembering. That is, walking through doorways caused forgetting. This occurred because people had memories of the object being in more than one event, more than one room. Therefore, when they needed to remember something about these objects, these different memories interfered with one another, and memory was worse. This is much like a fan effect, only with objects and the places where you interact with them.

7.2.4 Overcoming forgetting from interference

Beginning in the mid-1960s, a different theory came to dominate cognitive psychology's view of forgetting. Both the decay and interference theories suggest that information in long-term memory can be lost from memory. This definition of *forgetting* (loss from memory) was implicit in the mechanisms thought to account for it. *Forgetting* is now used without the idea of complete loss from memory to refer to situations in which there is difficulty remembering. In other words, we now focus on cases where the information is available in memory but may not be accessible.

For example, one line of research looks at **retrieval-induced forgetting**, the temporary forgetting of information because of having recently retrieved related information (e.g. Anderson, Bjork & Bjork, 2000; MacLeod & Macrae, 2001[19]). Similarly, Anderson (2003)[20] and Storm (2011)[21] have suggested that forgetting is an active inhibition process, designed to override mistaken retrieval of related information ('activated competitors' in Anderson's terms). Even here, the unwanted information that is causing interference is still in memory – if it weren't, there would be no need to override it. That said, the memories that are more likely to be inhibited are the ones that have recently been processed, whereas older memories are more likely to be facilitated by recently retrieved, related information (Bäuml & Samenieh, 2010[22]).

It is possible, therefore, that there may be no *complete* forgetting from long-term memory aside from loss due to organic or physical factors such as stroke or diseases like Alzheimer's dementia. So, the information is available in memory in some form. Forgetting may be due to retrieval failure or a process of retrieval inhibition, a deliberate (though only partially successful) attempt to forget (e.g. when you try to forget an unpleasant memory or an incorrect fact; Bjork & Bjork, 2003[23]).

7.2.5 Retrieval cues

We have already discussed how access can be increased by reinstating the original learning context. This can be thought of as providing effective retrieval cues. So, let's look at retrieval cues more generally. Any cue that was encoded along with the learned information should increase accessibility. This is perhaps the reason why in the study described in the research box below (Tulving and Pearlstone, 1996)[24], category cues helped participants recall more information than they otherwise would have. Similarly, this is why recognition usually reveals higher performance than recall. In a recognition test, you merely have to pick out which of several alternatives is the correct choice. What better retrieval cue could there be than the very information you are trying to retrieve? Subsequent research has shown the power of retrieval cues in dramatic fashion. (A convincing demonstration is presented in the following sample experiment taken from Bransford & Stein, 1984;[25] do that demonstration now, before reading further.)

More about retrieval failure

An everyday example of retrieval failure

Everyone is familiar with retrieval failure. Students often claim that they knew the information but that they 'blocked' on it during the exam. Sometimes this is an example of retrieval failure. A straightforward experience of this is the classic *tip-of-the-tongue (TOT)* effect. [TOT is pronounced 'tee-oh-tee', not like the word *tot*. Furthermore, it is often used as a verb: 'The subject TOTed (*tee-oh-teed*) seven times on the list of 20 names.'] People are in the TOT state when they are momentarily unable to recall a word, often a person's name, that they know is in long-term memory. Although you may be unable to retrieve a word or name during a TOT state, you usually have access to partial information about it, such as the sound it starts with, its approximate length and the stress or emphasis pattern in pronunciation. (See Brown & McNeill, 1966; Burke, MacKay, Worthley & Wade, 1991; Jones, 1989; Koriat, Levy-Sadot, Edry & de Marcus, 2003; and Meyer & Bock, 1992.[26]) If you want to try this yourself, provide a list of questions that can be used to trigger the TOT state.

However, retrieval failure, like the TOT effect, is not limited to lapses in remembering names or words. As Tulving and Pearlstone (1996) found, it is a fundamental aspect of memory.

Research on retrieval failure

An early demonstration of retrieval failure is a study by Tulving and Pearlstone (1966)[27] in which two groups of people studied the same list of 48 items, four words from each of 12 categories (e.g. animals, fruits, sports). The items were preceded by the appropriate category name, such as 'crimes – treason, theft; professions – engineer, lawyer', and people were told that they had to remember only the items. At retrieval, one group was asked for standard free recall. The other group was given the names of the categories as retrieval cues (a **cued recall** condition).

The study results were both predictable and profound. The free recall group remembered 40 per cent of the items, whereas the cued recall group named 62 per cent. One conclusion we can draw confirms intuitions dating back to Ebbinghaus: recall often underestimates how much information was learned. Recognition scores, not to mention savings scores, usually show higher retention. More importantly, unsuccessful retrieval, say in the absence of cues, might be a critical, possibly major, cause of forgetting.

According to this view, information stored in long-term memory remains there permanently and so is available, just as a book on the library shelf is available. Successful performance depends also on accessibility, the degree to which information can be retrieved from memory. Items that are not accessible are not immediately retrievable, just as the mis-shelved book in the library is difficult to locate or retrieve.

This suggests that information is not lost from memory but is lost *in* memory. This persists until an effective retrieval cue is presented that locates the memory that cannot be retrieved.

Thomson and Tulving (1970)[28] asked people to learn a list of words for later recall. Some of the words were accompanied by cue words printed in lowercase letters; people were told they need not recall the cue words but that the cues might be helpful in learning. Some of the cue words were high associates of the list items, such as *hot–COLD*, and some were low associates, such as *wind–COLD*. During recall, people were tested for their memory of the list in one of three conditions: low-associate cues, high-associate cues or no cues.

The results were that high associates used as retrieval cues benefited recall both when they had been presented during study and when no cue word had been given. When no cue word was given, people spontaneously retrieved the high associate during input and encoded it along with the list item. In contrast, when low associates had been given, only low associates functioned as effective retrieval cues. High-associate retrieval cues were no better than no cues at all. In other words, if you had studied *wind–COLD*, receiving *hot* as a cue for *COLD* was of no value. Thus, retrieval cues can override existing associations during recall.

Demonstrations of the effectiveness of retrieval cues are common. For instance, you hear a 'golden oldie' on the radio, and it reminds you of a particular episode (a special high school dance, with particular classmates, and so forth). This even extends to general context effects. Marian and Neisser's (2000)[29] bilingual participants remembered more experiences from the Russian-speaking period of their lives when they were interviewed in Russian, and more from the English-speaking period when interviewed in English (see also Schrauf & Rubin, 2000[30]). Actors remember their lines better when enacting their stage movements of a performance, even 3 months later (and with intervening acting roles; Noice & Noice, 1999[31]).

7.2.6 Part-set cueing effect

There is no question that, for the most part, retrieval cues help memory. However, there are some notable exceptions. One is the **part-set cueing effect** (Slamecka, 1968[32]) where people cued with a subset of a list have more difficulty recalling the rest of the set than if they had not been cued at all. In other words, cueing people with part of the information impairs memory compared to doing nothing. For example, if someone asked you to name the seven Disney characters, you would have a harder time with the last three if you were told the names of four of them than if you were simply asked to name all seven of them by yourself. One cause of the part-set cueing effect is that when people are provided with part-set cues, these items disrupt the retrieval plan that a person would normally use by imposing a different organisation of the material. Also, part-set cueing involves the use of an active inhibitory mechanism (Aslan, Bäuml & Grundgeiger, 2007[33]), much like what would be occurring in retrieval practice. In short, when a person is given a part-set cue, this causes an implicit retrieval of those items. At that time, the related memory traces serve as competitors and are actively inhibited.

7.3 False memories, eyewitness memory and 'forgotten memories'

LEARNING OBJECTIVE
Describe types of variations in memories

Schacter (1996)[34] spoke of the fragile nature of memory and the seven sins of memory, discussing how our memories can fail us in certain situations. Indeed, we have been discussing a variety of things that show different memory weaknesses. Is this what Schacter meant when he talked about the 'sins' of memory as being the weaknesses in our memory systems? What is the weakness that Schacter pinpointed – what situations did he have in mind? A straightforward answer is that memory fails us in exactly those situations that call for absolutely accurate recall, completely correct recollection of real-world events exactly as they happened. The weakness of memory seems to be that we are often unable to distinguish between what really happened and what our existing knowledge and comprehension processes might have contributed to recollection. We will discuss two research programmes that show incorrect or distorted memory, then tackle the issues raised by these results.

7.3.1 False memories

A simple yet powerful laboratory demonstration of **false memory**, memory of something that did not happen, was reported by Roediger and McDermott (1995),[35] based on a demonstration by Deese (1959).[36] [See Roediger & McDermott, 2000, for an introduction to this work and the *DRM* (Deese–Roediger–McDermott) task as it is often called; see Gallo, 2010, for a review of this work.[37]] Roediger and McDermott had people study 12-item lists made up of words such as *bed rest awake* and *pillow*, words highly associated with the word *sleep*. The word *sleep* was never presented. Instead, it was the **critical lure** word, a word that was highly related to the other words in the list but that never actually appeared. In immediate free recall, 40 pee cent of the people recalled *sleep* from the list and later recognised it with a high degree of confidence. This is a false memory.

In a second study, people studied similar lists, recalling the list either immediately or after a distractor task (doing arithmetic for 2 minutes). Then, everyone was given a recognition task. During free recall, 55 per cent of the people recalled the lure. The recognition results, shown in Figure 7.4, were even more dramatic. Of course, a few people 'recognised' non-studied words that were unrelated to the study list words (e.g. *thief* for the *sleep* list). More important, correct recognition for studied words increased to well above chance for study/arithmetic lists and even higher for study/recall lists. However, false recognition of the critical lure was higher than correct recognition of words actually shown on the list, showing the same pattern

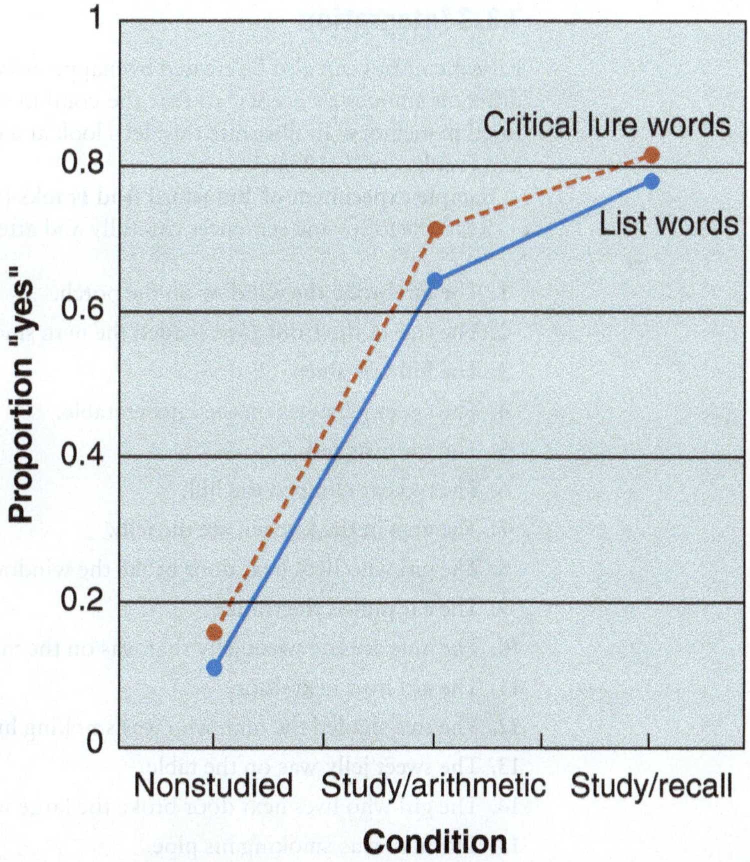

Figure 7.4 Roediger and McDermott's (1995)[38] results showing the occurrence of false memories after hearing lists of related words

of increases across conditions. There was an 81 per cent false alarm rate for critical lures when the lists had been studied and recalled. In other words, falsely remembering the lure during recall strengthened memories of the lure word, leading to a higher false recognition rate. When questioned further, most people claimed to 'remember' the critical lure word rather than merely 'know' it had been on the list.

In terms of *content accuracy* – memory for the ideas – performance is good, exactly what we would expect; you see a list of words such as *bed, rest, awake* and *pillow*, and, because the list is 'about' sleep, you then recall *sleep*. But in terms of *technical accuracy*, memory for the exact experience, performance is poor because people came up with the word *sleep* based on their understanding of the list and then could not distinguish between what had really been there and what was supplied from memory. These sorts of meaning-based false memories are formed quickly, in as little as 4 seconds (Atkins & Reuter-Lorenz, 2008). It appears that the critical process is not the automatic spread of activation, such as that observed in lexical decision priming, but the use of the thematic information in the study lists during efforts to remember what had been heard earlier (Meade, Watson, Balota & Roediger, 2007[39]). Roediger and McDermott's (1995, p. 812)[40] conclusion about this compelling memory illusion summarised the situation aptly:

> All remembering is constructive in nature. The illusion of remembering events that never happened can occur quite readily. Therefore, the fact that people may say they vividly remember details surrounding an event cannot, by itself, be taken as convincing evidence that the event actually occurred.

7.3.2 Integration

False memories can also be created by inappropriately combining information from different sources or events, so that the combined information becomes linked or fused in memory. To illustrate this, let's look at a classic set of studies by Bransford and Franks (1971, 1972).[41]

Sample experiment of Bransford And Franks (1971) – Part 1
Read the following sentences carefully and attempt the activity that follows.

1. The girl broke the window on the porch.
2. The tree in the front yard shaded the man smoking his pipe.
3. The hill was steep.
4. The sweet jelly was on the kitchen table.
5. The tree was tall.
6. The old car climbed the hill.
7. The ants in the kitchen ate the jelly.
8. The girl who lives next door broke the window on the porch.
9. The car pulled the trailer.
10. The ants ate the sweet jelly that was on the table.
11. The girl lives next door.
12. The tree shaded the man who was smoking his pipe.
13. The sweet jelly was on the table.
14. The girl who lives next door broke the large window.
15. The man was smoking his pipe.
16. The old car climbed the steep hill.
17. The large window was on the porch.

18. The tall tree was in the front yard.
19. The car pulling the trailer climbed the steep hill.
20. The jelly was on the table.
21. The tall tree in the front yard shaded the man.
22. The car pulling the trailer climbed the hill.
23. The ants ate the jelly.
24. The window was large.

Bransford and Franks (1971)[42] were interested in the general topic of how people acquire and remember ideas, not merely individual sentences but integrated wholes. They asked people to listen to sentences like those in Part 1 of the sample experiment you just read, one by one, and then (after a short distractor task) answer a simple question about each sentence. After going through this procedure for all 24 sentences and taking a 5-minute break, people were given another test. During this second test, people had to make yes/no recognition judgements, saying 'yes' if they remembered reading the sentence in the original set and 'no' otherwise. They also had to indicate, on a 10-point scale, how confident they were about their judgements: positive ratings (from 1 to 5) meant they were sure they had seen the sentence; negative ratings (from –1 to –5) meant they were sure they had not.

All 28 sentences in the recognition test presented in Part 2 of the experiment were related to the original ideas in Part 1. However, the clever aspect of the recognition test is that only 4 of these 28 sentences had in fact appeared on the original list; the other 24 were new. The separate sentences were all derived from four basic idea groupings, such as 'The ants in the kitchen ate the sweet jelly that was on the table.' Each of the complete idea groupings consisted of four separate simple *propositions*; for example:

The ants were in the kitchen.

The ants ate the jelly.

The jelly was sweet.

The jelly was on the table.

The original set of sentences in Part 1 of the sample experiment of Bransford and Franks presented six sentences from each idea grouping:

- Two of the six were called 'ones': simple, one-idea propositions such as 'The jelly was on the table.'
- Two of the six were 'twos': two merged simple propositions as in 'The ants in the kitchen ate the jelly.'
- Two of the six were 'threes': as in 'The ants ate the sweet jelly that was on the table.'

In Bransford and Franks' experiments, the final recognition test (Part 2) presented ones, twos, threes and the overall four for each idea grouping. Their findings are shown in Figure 7.5.

People were recognising the sentences that expressed the overall idea most thoroughly, even when they had not seen exactly those sentences during study. Such responses are called **false alarms**, saying 'OLD' when the correct response is 'NEW.'

Moreover, people were not confident about having seen old sentences; the only sentences they were sure about were sentences in which ideas from different groupings had been combined. Furthermore, they were fairly confident that they had not

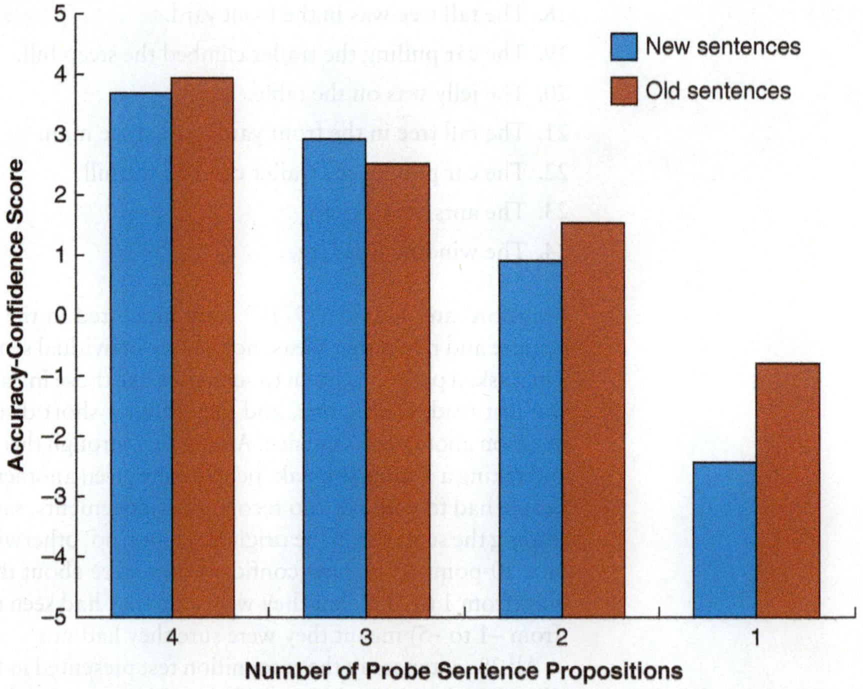

Figure 7.5 Confidence ratings for people's judgements of new and old sentences

Just as your performance probably indicated, people overwhelmingly judged threes and fours as having been on the study list (just as you probably judged question 20, the 4, as old). Furthermore, they were confident in their ratings, as shown in the figure.

Source: Bransford & Franks (1971)[43].

seen ones, even though they had seen several sentences that were that short. Because the shorter sentences did not express the whole idea (which would be captured by the single situation model that captured all of the ideas), people believed that they had not seen them before.

These results (Bransford & Franks, 1971, 1972[44]) suggest that people had acquired a more general idea than any of the individual study sentences had expressed. In essence, people were reporting a *composite* memory in which related ideas were fused together, forming one memory of the whole idea. Therefore, later recognition performance was entirely reasonable. They were matching the combined ideas in the recognition sentences to their composite memory. Rather than verbatim memory, Bransford and Franks found 'memory for meaning,' memory based on the integration of related material.

7.3.3 Leading questions and memory distortions

Another line of research provides a simple yet powerful demonstration of how inaccurate memories can be. This was begun by Elizabeth Loftus and her colleagues on the topic of leading questions and memory distortions (see Loftus, 2003, 2004, for highly readable introductions to this area[45]).

7.3.4 The misleading information effect

Investigators have developed several tasks to test for the effects of misleading information (e.g. Zaragoza, McCloskey & Jamis, 1987[46]). In a typical study, people see the original event in a film or set of pictures (e.g. pictures depicting a car accident, with one showing a stop sign). Later, they are exposed to additional information, such as a narrative about the accident. Some people receive only neutral information, whereas others get a bit of misinformation (the narrative mentions 'the give way sign', for instance). Finally, there is a memory test, often a yes/no recognition task that asks about the critical piece of information: was there a stop sign or a give way sign?

A common result is that some people incorrectly claim to remember the misinformation, the give way sign here. This is the **misinformation effect**. Belli (1989),[47] for instance, found that misled people had more than 20 per cent lower accuracy than did control groups who were not exposed to the misinformation. Furthermore, Loftus, Donders, Hoffman and Schooler (1989)[48] found that misled groups were faster in their incorrect judgements – picking the give way sign, for example – than in their correct decisions. This suggests a high degree of misplaced confidence on the part of the misled people. What is particularly troubling is that misinformation effects can persist even if the information is later corrected. For example, a study by Lewandowsky, Stritzka, Oberauer and Morales (2005)[49] looked at misinformation about the Iraq war that was reported by the press and then later retracted. However, people continued to inappropriately remember the misinformation as true. This was particularly true for Americans who had more support for the war than for Australians or Germans, who had less support. These latter groups better remembered that the misinformation was false and had been retracted.

The misleading information effect does not always result from incorrect information that was encountered after a previously witnessed event. Another way that misinformation can influence memory is if people encounter incorrect information that contradicts knowledge that they previously knew. For example, although most people know that the Pacific Ocean is the world's largest ocean, people can be misled by simply reading a passage that presents incorrect information. In a study by Fazio, Barber, Rajaram, Ornstein and Marsh (2013),[50] if people read a text that (incorrectly) stated that the Atlantic is the largest ocean, they were more likely 2 weeks later to mistakenly report that the Atlantic is the largest ocean (71 per cent correct) compared to people who were not misled (82 per cent correct), even if they had previously known that the Pacific is the largest. Thus, encountering new but incorrect information can also distort our knowledge.

A study by Elizabeth Loftus on leading questions and memory distortions

Loftus started by examining the effects of leading questions; that is, questions that tend to suggest to a person what answer is appropriate. She wondered whether there were long-term consequences of leading questions in terms of what people remember about events they have witnessed.

Part 1

In one study, Loftus and Palmer (1974)[51] showed several short traffic safety films depicting car accidents to college classes. They asked the students to describe each accident after seeing the film and then answer a series of questions about what they had seen. One of the questions asked for an estimate of the car's speed, something people are notoriously poor at. The longer-term importance of this effect gets to the heart of issues about eyewitness testimony and memory distortion. Loftus and Palmer wondered whether the question about speed altered the people's memories of the filmed scene. In other words, if participants are exposed to the implication that the cars had 'smashed' together, would they remember a more severe accident than they had actually seen? This is called **memory impairment**: a genuine change or alteration in memory of an experienced event as a function of some later event.

Part 2

This memory distortion is exactly what Loftus and Palmer found. A week after the original film and questions, people answered another set of questions (but they did not see the film again). One of the questions was 'Did you see any broken glass?' Many of the participants in the 'smashed' group (they had read that the cars smashed together) said 'yes' even though there had been none in the film.

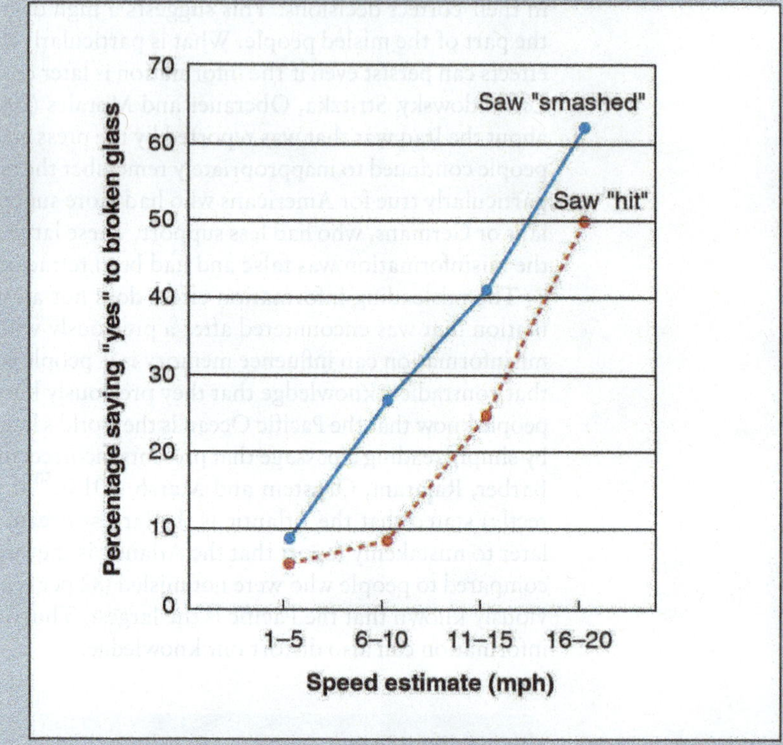

Figure 7.6 Loftus and Palmer's (1974)[52] results

The study results show the increased probability of saying 'yes' to the question of whether there was broken glass as a function of the severity of the word used to describe the event.

The result

In fact, 34 per cent of the 'smashed' group said 'yes', compared with only 14 per cent of the group who had read that one car 'hit' the other. Furthermore, the likelihood of saying 'yes' grew stronger as the speed estimates went up, as shown in the figure above.

At each point in the graph, 'remembering' broken glass was more common for people who had read 'smashed'.

Think about that again.

Although there was no broken glass in the film, people remembered it, partly as a function of their own speed estimate and the verb they had been questioned with earlier. The occurrences *after* the memory was formed altered its nature. The question using 'smashed' was not just a leading question; it was a source of misleading information.

7.3.5 Source misattribution and misinformation acceptance

Several reviews and summaries (Ayers & Reder, 1998; Loftus, 1991; Loftus & Hoffman, 1989; Roediger, 1996[53]) outline the overall message of this research. As Loftus (1991) noted, alteration of the original memory may be only one part of memory distortion. There seem to be three important memory distortion effects:

1. Source misattribution
2. Misinformation acceptance
3. Overconfidence in the accuracy of memory.

Two of the three important memory distortion effects

Source misattribution

Sometimes people come to believe that they remember something that never happened. This is *source misattribution*, the inability to distinguish whether the original event or some later event was the true source of the information. Source misattribution suggests a confusion in which we cannot clearly remember the true origin of a piece of knowledge (Zaragoza & Lane, 1994[54]). Using the stop sign/give way sign example, source misattribution occurs when we cannot correctly distinguish whether memory of the give way sign came from the original film or from another source, possibly the mistaken narrative that was read later or perhaps from prior knowledge and memory (Lindsay, Allen, Chan & Dahl, 2004[55]).

Another example of source misattribution can be seen in studies of the false fame effect. In these studies, people read a list of non-famous names, which increases the familiarity of those names. Later, people are more likely to judge the names as famous, essentially confusing familiarity with fame (Jacoby, Woloshyn & Kelly, 1989[56]). They have lost memory for the source of the feeling of familiarity (that it had been read on a list explicitly labelled as non-famous names), so they make their decisions solely on the basis of how familiar the names were. This confusion is particularly likely when people did not remember reading the original list of names, suggesting that the effect occurred at an implicit level (Kelley & Jacoby, 1996; see also Busey, Tunnicliff, Loftus & Loftus, 2000[57]).

Misinformation acceptance

For Loftus (1991)[58], a second, possibly larger component of memory distortion is *misinformation acceptance*, in which people accept additional information as having been part of an earlier experience without actually remembering that information. For example, a person in a misinformation experiment may not remember seeing a stop sign but is willing to accept that there was a give way sign when the narrative mentions it. Later on, the person reports having seen the give way sign. In short, people accept information presented after the fact and often become certain about these second-hand memories. These tendencies probably grow stronger as more time elapses after the event and the original memory becomes less accessible (Payne, Toglia & Anastasi, 1994[59]).

Misinformation acceptance may be partly related to **reconsolidation**. Essentially, reconsolidation is the idea that when a memory is retrieved, it enters a plastic, malleable state where it can be changed before it is stored in memory again (e.g. Sara, 2000)[60]. It may be that reminding people of the event before the presentation of misleading post-event information makes the memories pliable again, making it more likely that the incorrect information will be incorporated into the memory trace. Consistent with this, it has been found that having a person recall an eyewitnessed event before hearing misleading information actually makes it more likely that a person will misremember the event later (Chan, Thomas & Bulevich, 2009[61]).

Implanted memories

Yet another way to examine the acceptance of misinformation is by creating **implanted memories** of events that never happened.

Studies on acceptance of misinformation and pseudo-events

Not everything we remember actually happened. Sometimes these memories come from incorrect information we hear from other people, yet we often believe and accept these false memories as real.

Early use of this approach (e.g. Hyman, Husband & Billings, 1995; Loftus & Hoffman, 1989[62]) involved telling people childhood stories about themselves that their parents had supplied to the researchers, then questioning them about their memory for the episodes. Unknown to the person, one of the stories was a fictional, although plausible, event (called a *pseudo-event*; for instance, 'when you were 6, you knocked over a punch bowl at a wedding reception'). A large number of people came to accept the bogus story as true and claimed to 'remember' it. For example, none of the Hyman et al. participants claimed to remember the pseudo-event when they were first told about it, but 25 per cent 'recalled' it by their third session of questioning.

More recently, several studies have looked at other ways of conveying misinformation or pseudo-events. For example, Wade, Garry, Read and Lindsay (2002)[63] showed people a photo of themselves as children, riding in a hot-air balloon. There had never been a hot-air balloon ride. However, the photos were digitally altered to include an actual picture of the person as a child. Fully 50 per cent of the (now adult) people later reported memories of the ride in the hot-air balloon. However, this may be too persuasive – if you saw a picture of yourself riding in a hot air balloon at age 6, wouldn't you just decide that it must have happened and that your memory was faulty? As a follow-up, Lindsay, Hagen, Read, Wade and Garry (2004)[64] implanted a memory of a pseudo-event with the aid of a more conventional photograph – they got participants' grade-school class photos from their families and showed those photos as 'memory cues'. The pseudo-event was a prank the person was told he or she and a friend had pulled in grade school, sneaking some Slime (that brightly coloured gelatinous toy, made by Mattel) into the teacher's desk. Three childhood stories, two true events and one pseudo-event, were read to everyone, and half of participants were also shown their grade-school class picture. People were asked to recall whatever they could about the three stories, then to come back in a week for further testing. They were encouraged to spend some time every day 'working at remembering more about that event' during the week.

Figure 7.7 presents the results for the pseudo-event, tabulating in each condition the proportion of people having no memory of the event, the proportion having images

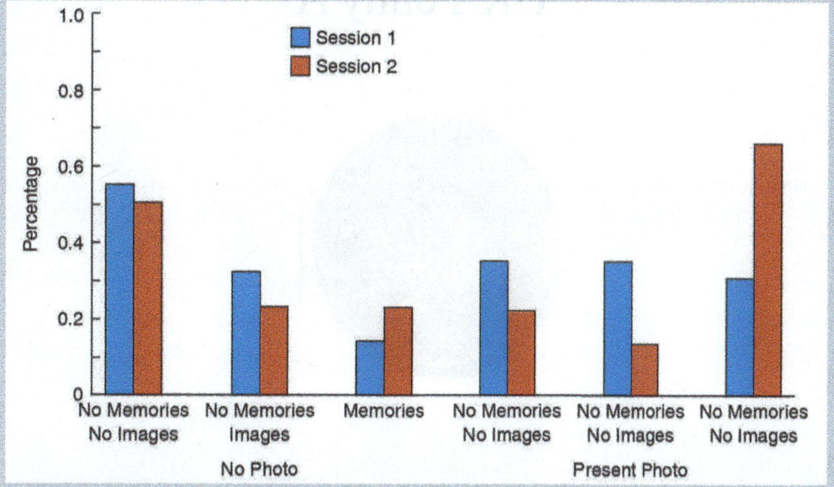

Figure 7.7 The percentages of people who have memories for a false event

but not memories of the event, and the proportion that reported having genuine memories of the pseudo-event. According to the figure, well over half of the people who merely heard the story said they had no memory of the event, a correct memory, and roughly 15 per cent claimed they did remember it, an incorrect memory.

The most stunning result involved how the photo group changed from Session 1 to 2. The percentage of people who incorrectly claimed to remember the event climbed from 30 per cent to nearly 70 per cent in the photo group. Apparently, having a photo that 'took them back' to grade-school years boosted the acceptance of misinformation and seemed to implant the bogus memory more thoroughly and convincingly. Indeed, the researchers reported that after they told the people that the Slime event had never happened, they often expressed surprise (e.g. 'You mean that didn't happen to me?'). Not only can implanted memories come from other people, they can also come from ourselves. People who are asked to invent eyewitnessed events sometimes later remember them as being actual events, even though they made them up themselves (e.g. Chrobak & Zaragoza, 2008[65]). Conversely, people who feign amnesia then find it harder to remember accurately later when they are not trying to fake it (Sun, Punjabi, Greenberg & Seamon, 2009[66]).

Source: From Lindsay, Hagen, et al., 2004, p. 152. American Psychological Society.

Overconfidence in memory

Despite our feeling that we remember events accurately ('I saw it with my own eyes!'), we often misremember experiences. And, as you have just read, we can be induced to form memories for events that never happened on the basis of suggestion, evidence or even just related information (e.g. the class photos).

As if this weren't bad enough, we often become overly confident in the accuracy of our memories (see Wells, Olson & Charman, 2002, for an overview[67]) and surprisingly unaware of how unreliable memory can be (a classic illustration is shown in Figure 7.8).

As you read a moment ago, Roediger and McDermott's (1995)[68] participants not only recalled (falsely) and recognised the critical lure, the majority claimed that they genuinely remembered it and had explicit, 'vivid memory' of hearing the word in

Figure 7.8 Which penny drawing is accurate?

the list. Aside from a basic belief in ourselves, this overconfidence seems to involve two factors:

1. **Source memory**, our memory of the exact source of information.
2. **Processing fluency**, the ease with which something is processed or comes to mind: as if you thought to yourself, 'I remembered "sleep" too easily to have just imagined that it was on the list, so it must have been on the list' (see Kelley & Lindsay, 1993)[69].

7.3.6 Stronger memory distortion effects

Can something as simple as this false memory effect in the laboratory explain real-world inaccuracies in memory? Probably.

It doesn't take much to realise the implications of this work: memory is pliable. Memories of events can be altered and influenced, both by the knowledge people have when the event happens and by what they encounter afterwards. People report that they remember events that did not happen. And in many cases, they become confident about their accuracy for those events. Unfortunately, at this point, because true and false memories overlap so much on so many dimensions, there is no reliable way to assess whether any given memory is true or false without other corroborating evidence (Bernstein & Loftus, 2009[70]).

7.3.7 Repressed and recovered memories

There are broad, disturbing implications of these findings for when people are trying to remember real-world events (Mitchell & Zaragoza, 2001[71]). If we can 'remember' things with a high degree of confidence and conviction, even though they never happened, then how seriously should eyewitness testimony be weighed in court proceedings? Juries are usually heavily influenced by eyewitnesses. Is this justified? Should a person be convicted of a crime based solely on someone's memory of a criminal act? The controversy over recovered memories is an obvious and worrisome arena in which our understanding of human memory is critical.

This sort of work is difficult because there are no reliable indicators of true versus false memories, although some overall patterns distinguish them. For example, people overall show less confidence in false than true memories, tend to recall them later, and provide less detail about false than true memories (Frost, 2000; Heaps & Nash, 2001[72]). Moreover, certain evidence indicates different physiological processing for true versus false memories. For example, it has been shown that true memories are more likely to produce distinct patterns of gamma oscillations in certain regions of the brain (Sederberg et al., 2007[73]). (EEG recordings show regular oscillations at different frequencies, and gamma oscillations occur in the 28–100 Hz range.)

7.3.8 The irony of memory

Memory presents us with great irony. How can this powerful, flexible system be so fragile, so prone to errors? We complain about how poor our memories are, how forgetful we are, how hard it is to learn and remember information. We deal with the difficulties of the transience of our memories, our absent-mindedness, the occasionally embarrassing blocking we experience when trying to remember. Are these accurate assessments of our memories (e.g. Roediger & McDermott, 2000[74])?

Well, they are probably exaggerated. First, as Anderson and Schooler (1991)[75] note, when we complain about memory failures, we neglect the huge stockpile of facts and information that we expect memory to store and to provide immediate access to. We underestimate the complexities, not to mention the sheer volume, of information stored in memory. For example, some have estimated that the typical adult has at least a half-million-word vocabulary, and it is almost impossible to estimate how many people we have known across the years.

We also fall into the trap of equating remembering with recall. When we say we have forgotten something, we probably mean we are unable to recall it at that moment. However, recall is only one way of testing memory. Recognition and relearning are far more forgiving in terms of showing that information has indeed been retained in memory. Finally, we focus on the failures of retrieval, without giving credit for the countless times we remember accurately.

How much cognitive psychology will you remember in a dozen years? A study by Conway, Cohen and Stanhope (1991)[76] examined exactly that: students' memory of the concepts, specific facts, names, and so on from a cognitive psychology course taken up to 12 years earlier (see Figure 7.9). Recall of material dwindled quite a bit across the 12 years, from 60 to 25 per cent for concepts, for example. But recognition for the same material dropped only a bit, from 80 per cent to around 65–70 per cent. Correct recognition for all categories of information remained significantly above chance across all 12 years. Your honest estimate – your metacognitive awareness of having information in storage – can be quite inaccurate.

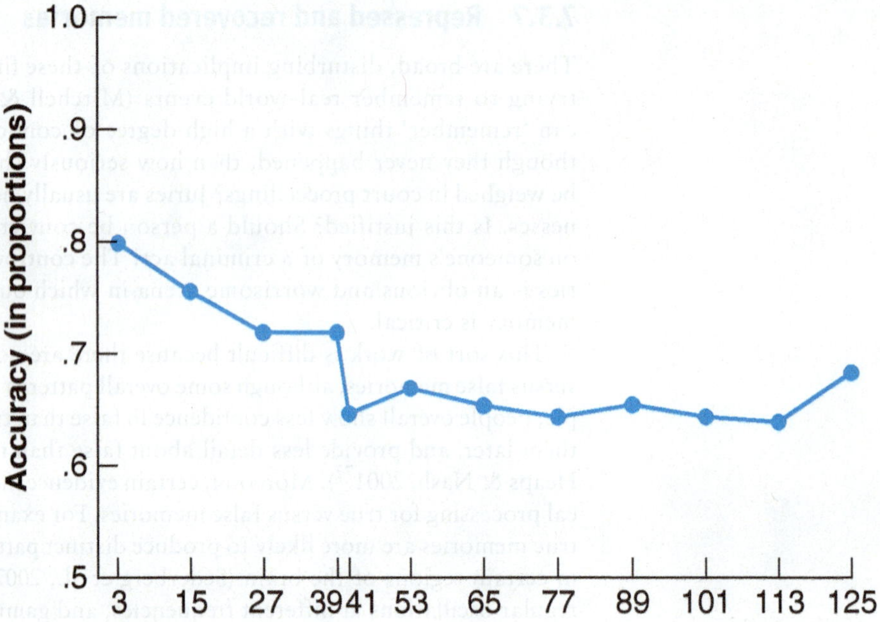

Figure 7.9 Memory for materials learned in a college course on cognitive psychology over many years

7.4 Amnesia and implicit memory

LEARNING OBJECTIVE
Differentiate between the types of amnesia

We study dysfunctions caused by brain damage to understand cognition and its organisation. Sometimes the patterns of disruptions and preserved abilities can tell us a great deal about how cognition works. This has been especially fruitful for understanding long-term memory in cases of amnesia.

amnesia is the catastrophic loss of memories or memory abilities caused by brain damage or disease. Amnesia is one of the oldest and most thoroughly studied mental disruptions caused by brain disorders, as well as a common result of brain injury and damage. Although some amnesias are temporary, due to a blow to the head or even acute emotional or physical stress (e.g. transient global amnesia;

Kinds of amnesia

Many kinds of amnesias have been studied, and we have space to discuss only a few. A few bits of terminology will help you understand the material and alert you to the distinctions in memory that are particularly relevant.

Amnesia related to the time of injury

First, the loss of memory in amnesia is always considered in relation to the time of the injury. If a person suffers loss of memory for events before the brain injury, this is retrograde amnesia. Interestingly, retrograde amnesia commonly shows a temporal gradient – memories that are more distant in time from the injury are less impaired (e.g. Brown, 2002; Wixted, 2004[78]). This temporal gradient is called

Ribot's law (Ribot, 1882[79]). Another form of amnesia is **anterograde amnesia**, disruption in acquiring new memories for events occurring after the brain injury. A person can show both forms of amnesia, although the extent of the memory loss usually is different for events before and after the damage. For example, anterograde amnesia often seems more extensive, simply because it disrupts learning from the time of the brain damage to the present.

The cases we talk about here are extreme in that the memory disruption is extensive, which is uncharacteristic of most cases of amnesia.

Amnesia related to dissociations

Second, we are trying to understand the architecture of memory, how its components are interrelated, whether some are independent of others, and so on. This is an analysis of dissociations, where the term *dissociation* refers to a disruption in one component of cognition, but no impairment of another. If two mental processes – *A* and *B* – are dissociated, then *A* might be disrupted by brain damage while *B* remains normal; patient K.C., described later, displays this kind of pattern. Sometimes another patient is found who has the reverse pattern: *B* is disrupted by the brain damage, but *A* is intact.

When two such complementary patients are found, with reciprocal patterns of cognitive disruption, then the abilities *A* and *B* are *doubly dissociated* (a simple example would be seeing and hearing, either of which can be damaged without affecting the other). A double dissociation implies not only that *A* and *B* are functionally independent, but also that *A* and *B* involve different regions of the brain. A simple dissociation is not as strong. If process *A* is damaged while *B* remains normal, it could be that research has not yet found a patient with the reciprocal pattern. Or it could be that process *A* can be selectively damaged without affecting *B* but that damage to *B* would always disrupt *A*.

The opposite of a dissociation is an *association*: a situation in which *A* and *B* are so completely connected that damage to one would always disrupt the other (e.g. recognising objects and recognising pictures of objects).

Amnesia related to focal brain lesions

Finally, the most useful cases are those of focal brain lesions, in which the damage is to a small, restricted area of the brain. Cases such as that of patient K.C. illuminate the underlying mental processes more clearly because many of his mental processes are intact despite the dysfunction caused by the focal lesion. Unfortunately, the widespread damage and neural deterioration of some injuries or diseases, such as Alzheimer's disease, make it difficult to pin down the neurogenerator of the cognitive functions that are disrupted. So many regions are damaged that no single one can be pinpointed as the region responsible for a particular ability.

Brown, 1998[77]), the amnesias in which we are interested here are relatively permanent, caused by enduring changes in the brain.

7.4.1 Dissociation of episodic and semantic memory

We begin with a case history. Tulving (1989)[80] described patient K.C., who as the result of a motorcycle accident experienced serious brain injury, especially in the frontal regions. As a result of this injury, K.C. showed a seemingly complete loss of episodic memory: he was completely amnesic for his own autobiographical knowledge. K.C. had profound retrograde and anterograde amnesia. He showed great difficulty in both storing and retrieving personal experiences in long-term memory.

Although K.C.'s episodic memory no longer worked, his semantic memory did. He was adept at answering questions about his past by relying on general, semantic knowledge. When asked about his brother's funeral, he responded that the funeral was very sad, not because he remembered attending the funeral (he did not even remember that he had a brother) but because he knew that funerals are sad events.

K.C.'s memory disruption, intact semantic memory yet damaged episodic retrieval, was evidence of a dissociation between episodic and semantic memory. This suggests that episodic and semantic memories are separate systems, enough so that one can be damaged while the other stays intact. In Squire's (1987)[81] taxonomy, K.C. had lost one of the two major components of declarative knowledge, his episodic memory.

Functional neuroimaging evidence

There are limitations on what can be learned about normal cognition from data from brain-damaged patients. Brain-damaged patients may be unique. The data from them may not be generalisable. Because we might worry about the generality of such results – K.C. could have been atypical before his accident – Tulving presented further support for his conclusions, studies of brain functioning among normal individuals (Nyberg, McIntosh & Tulving, 1998[82]).

When brain activity is measured in a scanner, such as is done with a PET scan (as in Figure 7.10), certain areas may become more active, relative to their baseline activity. This is often taken as an indication that those parts of the brain are more involved in the task being done at the time.

Tulving and his colleagues developed what they called the Hemispheric Encoding/Retrieval Asymmetry model, or **HERA model** (Habib, Nyberg & Tulving, 2003; Nyberg, Cabeza & Tulving, 1996[83]). Data from PET studies, such as the Nyberg et al. (1998) study, show that the left frontal lobe is more likely to be involved in the retrieval of semantic memories and the encoding of episodic memories (shown in Figure 7.10 as more activated in yellow and red). This makes sense, because when you encounter a new event, to create a new episodic memory you need to understand the event using your semantic knowledge. By comparison, the right frontal lobe is more likely to be involved in the retrieval of episodic memories. So, based on the HERA model, different parts of the brain are involved in different types of memory processing (see also Buckner, 1996[84]; Shallice, Fletcher & Dolan, 1998[85]). The HERA model is *not* making the claim that episodic memories are stored in the right hemisphere and semantic memories in the left. Instead, it is just that those brain regions are more involved in those kinds of activities. Even researchers who do not agree with the HERA model do agree that the brain processes semantic and episodic memories differently (e.g. Ranganath & Pallar, 1999; Wiggs, Weisberg & Martin, 1999[86]).

7.4.2 Anterograde amnesia

The story of anterograde amnesia begins with a classic case history. A popular theoretical stance in 1950 was that memories are represented throughout the cortex, rather than concentrated in one place. Karl Lashley articulated this position in his famous 1950 paper 'In search of the engram.' Three years later, neurosurgeon

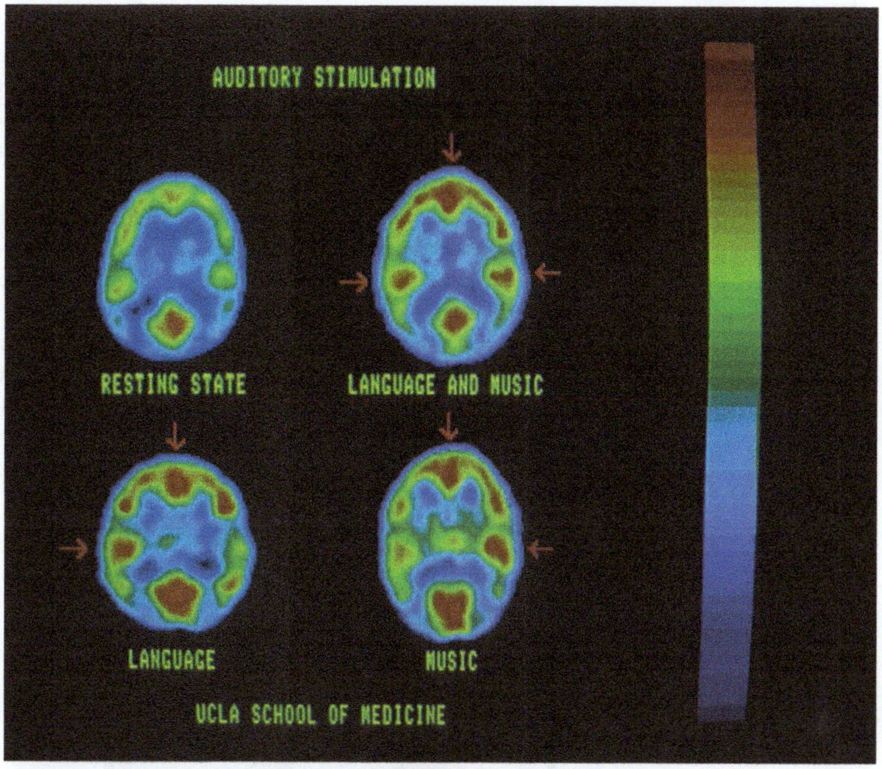

Figure 7.10 Illustration of blood flow as revealed in PET scan images

Source: Dan McCoy/Rainbow/RGB Ventures/SuperStock/Alamy Stock Photo

William Scoville made an accidental discovery. Scoville performed radical surgery on a patient, Henry Molaison (1926–2008), more commonly known as H.M., sectioning (lesioning) H.M.'s hippocampus in both the left and right hemispheres in an attempt to gain control over his severe epilepsy. To Scoville's surprise, the outcome of this surgery was pervasive anterograde amnesia: H.M. was unable to learn and recall anything new. Although his memory of events before the surgery remained intact, as did his overall IQ (118, well above average), he lost the ability to store new information in long-term memory.

Across the years, H.M. served as a participant in hundreds of tasks (e.g. Milner, Corkin & Teuber, 1968[87]) documenting the many facets of his anterograde amnesia. His memory of events prior to the surgery, including his childhood and school days, was quite good, with some gaps. His language comprehension was normal, and his vocabulary was above average. Yet any task that required him to retain information across a delay showed severe impairment, especially if the delay was filled with an interfering task. These impairments applied equally to non-verbal and verbal materials. For instance, after a 2-minute interference task of repeating digits, he was unable to recognise photographs of faces. He was unable to learn sequences of digits that went beyond the typical short-term memory span of seven. In a conversation reported by Cohen (in Banich, 1997[88]), he told about some rifles he had (it was a childhood memory). This reminded him of some guns he had also had, so he told about them. Telling about the guns took long enough, however, that he forgot he had already talked about the rifles, so he launched into the rifle story again, which then reminded him of the guns – and so on until his attention was diverted to some other topic.

H.M.'s implicit memory

Interestingly, the evidence from the H.M. studies also suggests that H.M.'s memory was normal when it involved more unconscious *implicit memory* because he was able to learn a motor skill, mirror drawing. This task requires a person to trace between the lines of a pattern while looking at it and the pencil only in a mirror (Figure 7.11). H.M.'s performance showed a normal learning curve, with very few errors on the third day of practice.

Note, though, that on days 2 and 3, H.M. did not remember having done the task before; he had no explicit memory of ever having done it, despite his normal performance based on implicit memory.

Likewise, H.M. showed systematic learning and improvement on solving the Tower of Hanoi problem in which a stack of discs needs to be moved from one position to another following specific rules. Although he did not remember the task

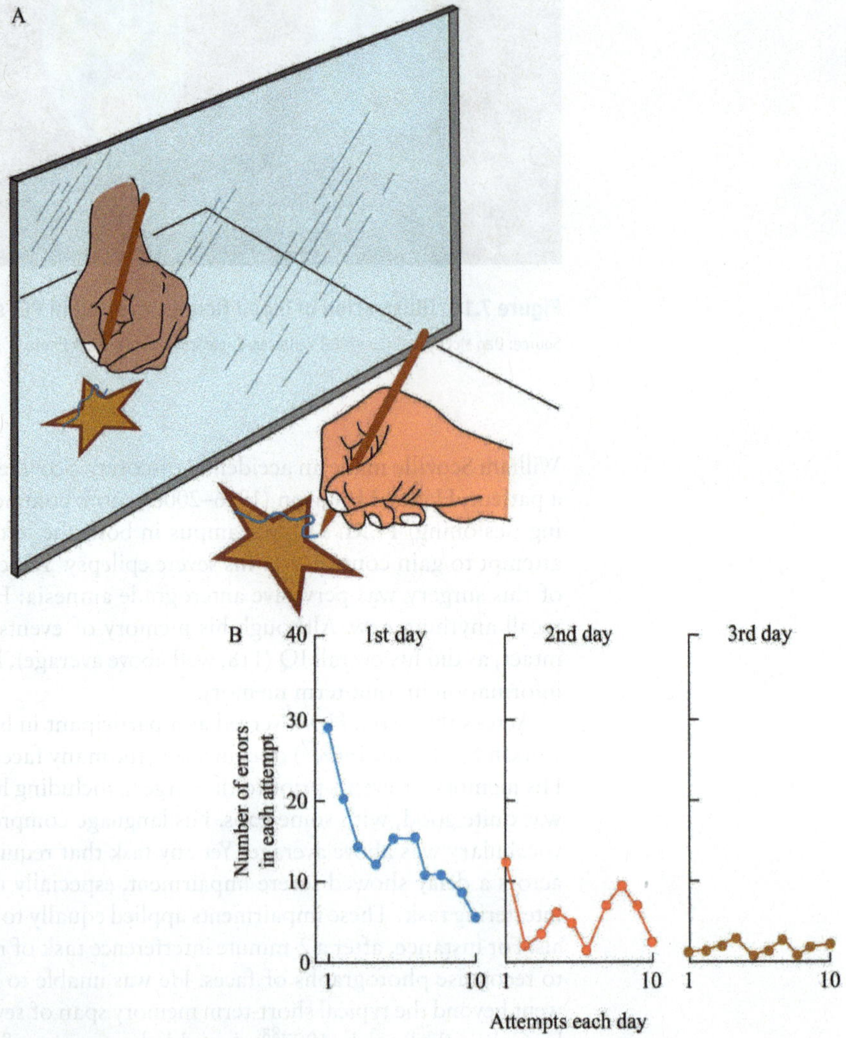

Figure 7.11 H.M. studies

(A) In this test, people trace between the two outlines of the star while viewing their hand in a mirror. The reversing effect of the mirror makes this a difficult task initially. Crossing a line constitutes an error. (B) Patient H.M. shows clear improvement in the motor learning star task, an influence of implicit learning and memory.

After Blakemore (1977)[91].

itself, his performance nonetheless improved across repeated days of practice. Such empirical demonstrations confirm what clinicians working with amnesia patients have known or suspected for a long time: despite profound difficulties in what we normally think of as memory, aspects of patients' behaviour do demonstrate a kind of memory – in other words, implicit memory (see Schacter, 1996)[89]. All the subtypes underneath 'non-declarative (implicit)' memory – skill learning, priming, simple classical conditioning, and non-associative learning – represent different aspects of implicit memory; different forms and types of performance in which implicit memories can be displayed (Squire, 1993; see Gupta & Cohen, 2002, and Roediger, Marsh & Lee, 2002, for reviews[90]).

Implications for memory

What do we know about human memory as a function of H.M.'s disrupted and preserved mental capacities? How much has this person's misfortune told us about memory and cognition?

The most apparent source of H.M.'s amnesia was a disruption in the transfer of information to long-term memory. H.M.'s retrieval of information learned before surgery was intact, indicating that his long-term memory per se, including retrieval, was unaffected. Likewise, his ability to answer questions and do other simple short-term memory tasks indicates that his attention, awareness and working memory functions were also largely intact. But he had a widespread disability in transferring new declarative information into long-term memory. This disability affected most or all of H.M.'s explicit storage of information in long-term memory (Milner et al., 1968[92]), including real-world episodic material.

It is a mistake to conclude from this that H.M.'s memory disruption – say, the process of explicit rehearsal – takes place in the hippocampus. Instead, it seems more likely that the hippocampus is on a critical pathway for successful transfer to long-term memory. Other research on patients with similar lesions (e.g. Penfield & Milner, 1958[93]; Zola-Morgan, Squire & Amalral, 1986[94]) confirms the importance of the hippocampus to this process of storing new information in long-term, explicit memory. In some sense, then, the hippocampus is a gateway into long-term memory. Thus, the hippocampus is essential for declarative or explicit memory. See Eichenbaum and Fortin (2003)[95] for an introduction to the relationship between the hippocampus and episodic memory, and see Barnier (2002)[96] for an extension of these effects to posthypnotic amnesia.

7.4.3 Implicit and explicit memory as revealed by amnesia

To repeat a point made earlier, the operative word in the definitions of explicit and implicit memories is *conscious*. Explicit memories, whether episodic or semantic, come to us with conscious awareness and therefore have an explicit effect on performance, an effect that could be verbalised. For example, name the third letter of the word meaning 'unmarried man'. The very fact that you can say *c* and name the word *bachelor* attests to the fact that this is an explicit memory. In contrast, fill in the following word stems: 'ki__, swe__, fl__.' Even without any involvement of conscious awareness, you may have filled these in with the words *kitchen, sweet, flag* with greater likelihood than would have been expected by chance. The words *kitchen, sweet* and *jelly* occurred in this chapter. Even reading material a second time through is done more rapidly, an effect called *repetition priming*. That is, when information is encountered, it unconsciously activates prior memories of the same thing, making it easier to process. We all demonstrate such implicit effects as repetition priming, amnesic or not (Graf & Schacter, 1987; Kolers & Roediger, 1984[97]).

Repetition priming tests

Repetition priming has been established in several tasks, such as word identification and lexical decisions (Morton, 1979),[98] word and picture naming (Brown, Neblett, Jones, & Mitchell, 1991[99]), and rereading fluency (Masson, 1984[100]). In all these, a prior encounter with the stimulus yields faster performance on a later task, even though you may not consciously remember having seen it before (see Logan, 1990[101], for the connection of repetition priming).

Explicit memory test

In a classic demonstration of repetition priming, Jacoby and Dallas (1981)[102] had people study a list of familiar words, answering a question about each as they went through the list. Sometimes the question asked about the physical form of the word, as in, 'Does it contain the letter L?' Sometimes it asked about the word's sound, as in, 'Does it rhyme with train?' And sometimes, the question asked about a semantic characteristic, as in, 'Is it the centre of the nervous system?' This was a direct manipulation of depth of processing. Asking about the physical form of the word should induce shallow processing, leading to poor memory. Asking about rhymes demands somewhat deeper processing, and asking about semantic characteristics should demand full, elaborative processing of the list words.

At test, explicit memory was assessed by a yes/no recognition task ('Did this word occur in the study phase?'). Here, recognition accuracy was affected by the type of question answered during study. When a question related to the physical form, recognition was at chance, 51 per cent. When it had asked about the sound of the word, performance improved. And when semantic processing had been elicited, recognition accuracy was high, 95 per cent. This a test of explicit memory because people had to say 'yes' or 'no' based on whether they had seen the word earlier. As expected, more elaborative processing led to better explicit memory performance.

Implicit memory test

The other test given, the implicit memory test, was a perceptual test. Here, words were shown one at a time for only 35 ms, followed by a row of asterisks as a mask. People had to report the word they saw. In other words, the perceptual test did not require the people to remember which words they had seen earlier. They just had to identify the briefly presented words. For this test, word identification averaged about 80 per cent, regardless of how the words had been studied, in comparison to only 65 per cent of control words that had not appeared earlier.

This is a typical implicit memory result. Even without conscious recollection of the original event, there is facilitation for a repeated stimulus. Explicit measures of memory, such as recall or recognition, show strong influences of how information was studied. However, implicit measures, such as a perceptual or word stem completion task, show clear priming regardless of how information was studied (see also Roediger, Stadler, Weldon & Riegler, 1992; Thapar & Greene, 1994[103]); for work on forgetting and interference in implicit memory, see Goshen-Gottstein & Kempinsky, 2001, and Lustig & Hasher, 2001, as well as Kinder & Shanks, 2003, for a counterargument[104]).

Implicit memory is also involved in motor tasks, such as knowing how to ride a bicycle, play a musical instrument, play a sport, and so on. Here, implicit memory is often called *procedural memory*. Like other implicit memories, procedural memories are very durable and, once acquired, show a very shallow forgetting curve. Remember the saying that once you learn to ride a bike, you never forget? This is also seen in cases of profound amnesia. These people may lose a great deal of declarative knowledge, but their procedural knowledge or skills remain largely intact. They can even acquire new skills, as shown by H.M.'s performance on the mirror drawing task.

Just because a memory is implicit does not mean it has no influence on conscious experience. For example, implicit memory may be involved in causing the *déjà vu* experience (Brown, 2004; Cleary, 2008[105]). A new place may seem familiar to you, even though you have never been there before – not because of some psychic connection, but because the place is similar enough to other places you have been to. As a result, the new place elicits a feeling of familiarity. However, you are not consciously aware of being reminded of these other places. The result is this eerie feeling of familiarity when you enter a place you know you have never been to before.

In addition, cognitive science learned an important lesson from patients such as H.M. and K.C. If we had stuck to laboratory-based experiments alone, and never paid attention to patients with amnesia, we would have failed to realise the importance of that second, less obvious kind of long-term memory, the kind not dependent on conscious recollection. We would have missed implicit memory.

Accelerated long-term forgetting

As demonstrated in the cases of K. C. and H. M., acquired brain damage led to significant deficits in acquiring new information. Even in such cases, once new information is acquired it tends to be retained and, when forgetting occurs, it follows the normal pace observed in healthy individuals (Baddeley, Atkinson, Hitch & Allen, 2021[106]). Nevertheless, over the last three decades, a growing body of research has indicated a special case of forgetting known as Accelerated Long-term Forgetting or ALF. In ALF, new information that is normally learned and retained becomes rapidly forgotten within a short period of time, such as days or weeks. ALF was initially attributed to seizure activity in patients with Temporal Lobe Epilepsy (TLE). Specifically, it was theorised that epileptic seizures disrupted the normal function of memory consolidation processes, but subsequent studies have challenged this view on the grounds of methodological flaws and limitations (Elliott et al., 2014[107]). Alternative explanations purport that ALF in TLE may be due to undetected deficits in learning (i.e., information acquisition) that only become observed at later stages, or due to individual differences in memory retention deficits in people suffering from TLE (Cassel & Kopelman, 2019[108]). Other studies have explored the possible role of ALF in dementia. Specifically, ALF has been observed people who have a higher genetic risk for Alzheimer's Disease (AD), suggesting that ALF may serve as an early marker of AD (Tort-Merino et al., 2021; Rodini et al., 2022[109]). A lot of questions remain unanswered and further research is needed to improve our understanding of the role of ALF in neurological and neurodegenerative disorders, including AD, and to indicate novel methodologies to reliably assess ALF in both research and clinical practice.

Summary: Memory and forgetting

7.1 The seven sins of memory

- Forgetting may occur because the knowledge no longer exists in memory. This is forgetting due to a lack of availability. Alternatively, forgetting may occur because the knowledge is in memory, but cannot be successfully retrieved. This is forgetting due to a lack of accessibility.
- Schacter has defined seven 'sins' in which memory performance is compromised. Three of these are sins of omission, namely transience, absent-mindedness and blocking. Three more are sins of commission, namely, misattribution, suggestibility and bias. Finally, the sin of persistence is when we wish to forget knowledge, but it persists in memory.

7.2 Forgetting through decay and interference

- Although there has been some suggestion that forgetting occurs through a decay process associated with the passage of time, most normal forgetting appears to be due to a process of interference. Using paired-associate lists, memory researchers have found that this can be proactive interference, in which older memories impair the ability to access newer memories, or retroactive interference, in which newer memories impair the ability to access older memories.
- Associative interference, or a fan effect, can be observed as more things are learned about a concept. This is brought about by activation being divided among the various associations. This type of interference can be attenuated if information can be integrated in some way, as with a situation model. This is the cognitive process leading to such phenomena as the finding that walking through doorways causes forgetting.
- The loss of memories through forgetting can also be reduced if people have the appropriate retrieval cues available. These allow people to access those memories more directly than if people were left to their own devices. Note that memory retrieval cues can sometimes backfire, as with the part-set cueing effect.

7.3 False memories, eyewitness memory and 'forgotten memories'

- Several paradigms give clear evidence of false memories, such as the Roediger and McDermott (1995)[110] list presentation studies and eyewitness memory research by Loftus and others. 'Remembering' in such situations is affected by source misattribution, the acceptance of misinformation and bias. People tend to be overconfident about their memories, regardless of the distortions that might be involved.
- In eyewitness memory and testimony, any new information about an event is integrated with relevant existing knowledge. Thus, we are less than accurate when we attempt to retrieve such knowledge because we are often unable to discriminate between new and original information.
- Cases of 'forgotten' and 'recovered' memories are particularly difficult to assess because of the fragile, reconstructive nature of memories. It is a concern that therapeutic techniques used to assist in 'recovering' memory of trauma are so similar to variables such as repetition and repeated questioning, variables that increase the false memory effect.

7.4 Amnesia and implicit memory

- Studies of people with amnesia caused by brain damage have taught us a great deal about long-term memory. Patient K.C. showed total amnesia for episodic information, although his semantic memory was unimpaired, suggesting a dissociation between episodic and semantic memories. Patients like H.M., a person with anterograde amnesia, are typically unable to acquire new explicit memories, but have intact implicit memory. The medial temporal area and especially the hippocampus are very important for the formation of new explicit memories, but different brain structures underlie implicit learning.
- Recognition memory for information acquired across an extended period is remarkably accurate across many years, whereas recall performance begins to decline within months.

Chapter 7 quiz

Question 1

Rachel is with a group of friends and is talking about how she attended a festival over the summer with her parents. The festival was a classical festival with music from classical orchestras and artists. During the conversation her friend Yuri talks about how she really enjoys rock music and explains how she has seen lots of rock music artists. Rachel then begins to talk about how the classical festival included rock music. Upon reflection, Rachel had not told the truth and there was no rock music at the festival. What is this an example of?

A. Transience
B. Bias
C. Absent-mindedness
D. Suggestibility

Question 2

Why are retrieval cues important in memory?

A. Research has shown that they can dramatically increase memory performance.
B. Research has suggested that they can distinguish clearly between visual and verbal memory resources.
C. Research has shown that visual retrieval cues are more effective than spatial retrieval cues.
D. Research has shown that retrieval cues help us to remember the meaning of items (semantic memory) rather than their experiences (episodic memory).

Question 3

An eyewitness is in court being questioned as part of a case against someone who committed a crime. The eyewitness is asked to recall the scene very carefully and is then asked some questions by a solicitor. The solicitor is trying to trick the eyewitness into giving incorrect information. What strategy could they use?

A. Source misattribution
B. Misinformation acceptance
C. Leading questions
D. Implanted memories

Question 4

Patient H.M (Henry Molaison) is a classic case study within memory research. H.M. had a bilateral medial temporal lobectomy in an attempt to cure his epilepsy. After the surgery, H.M displayed normal language comprehension. He had some gaps in his memory of childhood. He was unable to retain new information and was more susceptible to interference from verbal and non-verbal sources. Why has patient H.M. developed anterograde amnesia?

A. Retrograde amnesia impacts verbal memories whereas anterograde amnesia impacts visual and spatial memories.
B. Anterograde amnesia is only formed after a purposeful intervention such as brain surgery whereas retrograde amnesia is from accidental causes.
C. Anterograde amnesia is specifically related to implicit memories whereas retrograde amnesia is not.
D. Anterograde amnesia impacts the formation of new memories whereas retrograde amnesia impacts memories from the past.

Question 5

Which of the following is the best definition of priming?

A. The process of organising individual pieces of information into a meaningful whole
B. The activation of related concepts and meanings.
C. The use of generalised knowledge structures
D. The accessing and comparison of features within feature sets

8 The building blocks of language

Learning objectives

- **8.1 Describe how language is defined**
- **8.2 Explain how phonology influences spoken language**
- **8.3 Summarise the rules governing syntax**
- **8.4 Analyse lexical factors to understand the meaning of language**
- **8.5 Compare semantics and syntax to understanding intended meaning**
- **8.6 Describe how the brain comprehends language**

Key figures

Professor Steven Pinker

Professor Pinker is a cognitive psychologist who developed his own theory of language acquisition and conducted pioneering research on the neural basis of language learning in children.

Professor Elizabeth Bates

Professor Bates was a prominent cognitive scientist, studying the neural basis of language learning in infants and children, post-stroke language affect, and advancing the field of psycholinguistics. Her theory of language acquisition posited that language emerges through social interaction and challenged traditional notions of innate language capacity.

Introduction

Language, along with music, is one of the most common and universal features of human society. Language pervades every facet of our lives, from our most public behaviour to our most private thoughts. We might imagine a society that has no interest in biology, or even one with no formal system of numbers and arithmetic. But a society without language is inconceivable. Every culture, no matter how primitive or isolated, has language. Every person, unless deprived by nature or accident, develops skill in the use of language. Human use of language is astounding because we are typically able to process approximately three words per second, drawing on a vocabulary of about 75,000 words.

In this chapter, we will delve into the basics of language, including characteristics, functions, structure and form.

How do humans develop language?

Towards the end of the 1950s the dominant view of behaviourism as an explanation for all human behaviour, including language, would start to decline. The cognitive revolution would shift the focus away from observable, overt behaviours, towards previously unseen mental processes that guided much of human action – the very topic of this textbook! Noam Chomsky, a pioneering linguist, challenged the then dominant behaviourist approach that viewed language as the outcome of learning processes, such as imitation, operant conditioning and reinforcement. Chomsky argued that language acquisition and development was an innate capacity. It would be impossible for children to master the complexities of human language only based on the limited input from their social environment (e.g. social learning/imitation and reinforcement). Instead, children have an innate capacity to learn language which Chomsky labelled a 'language acquisition device' (or LAD). By being exposed to spoken language, specialised neural areas that support the function of the LAD enable young children to quickly develop their linguistic skills and, effectively, learn how to communicate verbally with others. The LAD is not language-specific but, as Chomsky noted, allowed children to acquire and gradually master certain universal grammatical rules and principles that govern most, if not all, spoken languages.

Chomsky argued that language acquisition and development was an innate capacity.

Source: Jeff Titcomb/Alamy Stock Photo

Steven Pinker, and experimental psychologist with a particular interest in language acquisition in children and the underlying neural processes, further extended Chomsky's views of innate language by arguing that language is instinctual.

Steven Pinker

Source: ZUMA Press, Inc/ Alamy Stock Photo

Far from being a skill learned only through language and grammar classes in schools, Pinker suggests that all people are born with a 'language instinct'. However, unlike Chomsky who dismissed evolutionary explanations, Pinker argues that language developed in humans as an adaptation, through natural selection processes. The language acquisition debate is far from over. Psycholinguistics is a continuously developing field.

Linguistics is the study of language and has had a profound influence on cognitive psychology. It was a major turning point when Chomsky rejected behaviourism's explanation of language. Because approaches such as Chomsky's seemed likely to yield new insights and understanding, psychology renewed its interest in language in the late 1950s and early 1960s, borrowing heavily from linguistic theory. Elizabeth Bates' theory of language acquisition challenged the dominant nativism views supported by Chomsky (and later by Pinker) and purported that language abilities are distributed across different brain regions, and that we acquire language through the interaction of our brains with our social environments.

Regardless of which theory has received more empirical (and popular!) support, one thing became clear: language is a purposeful activity. It is there to do something: to communicate, to express thoughts and ideas, to make things happen. However, linguistics focused on language itself as a formal, almost disembodied system so that the *use* of language by humans was seen as less interesting, tangential or even irrelevant. This view denied a fundamental interest of psychology–behaviour associations. Thus, a branch of cognitive psychology evolved called **psycholinguistics**, which is the study of language as it is learned and used by people. In this chapter, we present psycholinguistics but mainly focus on the cognitive aspects of language and verbal communication, as well as language and communication impairment.

8.1 Linguistic universals and functions

LEARNING OBJECTIVE
Describe how language is defined

One might define *language* as 'the use of words and how they are spoken in various combinations so a message can be understood by other people who also speak that language'. That is not a bad start. For example, one critical idea in the definition is that meaning and understanding are *attributed* to the words and their pronunciations, rather than being part of those words. As an illustration, the difference in sound between the words *car* and *cars* is the *s* sound, denoting plural in English. But this meaning is not inherent in the *s* sound, any more than the word *chalk* necessarily refers to the white stuff used on blackboards. This is an important idea: language is based on arbitrary connections between linguistic elements, such as sounds (pronunciations) and the meanings they denote.

8.1.1 Defining language

This definition of *language* is a bit confining and restricts language to human speech. By this rule, writing would not be language, nor would sign language for people with deafness. It is true that writing is a recent development, dating back only about 5,000 years, compared to the development of articulate speech, thought to have occurred some 100,000 years ago (e.g. Corballis, 2004[1]). It is equally true that the development of writing depends critically on a spoken language. Thus, the spoken, auditory form of a language is more basic than the written version. Is there any doubt that children would fail to acquire language if they were only exposed to books instead of speech? Nonetheless, we include written language because reading and writing are major forms of communication.

8.1.2 Language universals

subject-verb-object (SVO) language
Place the subject before the verb and followed by the object, as in 'Tom ate the ice cream'.

There are a large number of differences between languages. Take word order, for example. English is largely a **subject-verb-object (SVO) language** in which the sentence subject comes first, followed by the verb and then the object of the sentence. Other

subject-object-verb (SOV)
Place the subject and the object before the verb, as in 'Tom, the ice cream, ate'.

languages have other structures. Japanese, for instance, is a **subject-object-verb (SOV)** language in which the verb typically comes at the end. Despite the many differences among human languages, they all share some universal properties.

Language as a shared symbolic system for communication

Let's offer a definition that is more suitable here. **Language** is a shared symbolic system for communication.

1. **Language is symbolic.**

   It consists of units (e.g. sounds that form words) that symbolise or stand for the referent of the word. The referent, the thing referred to by the final *s*, in words such as *cars*, for example, is the meaning 'plural'.

2. **The symbol system is shared by all users of a language culture.**

   Language users all learned the same set of arbitrary connections between symbols and meaning, and they also share a common rule system that translates the symbol-to-meaning connections.

3. **The system enables communication.**

   The user translates from the thought into a public message, according to the shared rule system. This enables the receiver to retranslate the message back into the underlying thought or meaning.

Hockett (1960a, 1960b, 1966)[2] proposed a list of 13 **linguistic universals**, features or characteristics that are common to all languages. To distinguish human language from animal communication, Hockett proposed that only human language contains all 13 features. Several of the universals he identified, such as the vocal–auditory requirement, are not essential characteristics of human language, although they were probably essential to its evolution. Other features are critically important to our analysis.

Hockett's linguistic universals

Here are 12 of Hockett's linguistic universals, along with short explanations. We limit our discussion to four of these, plus two others implied, but absent from the list.

Vocal–auditory channel

The channel or means of transmission for all linguistic communications is vocal–auditory.

Hockett excluded written language by this universal because it is a recent invention and not found in all language cultures.

Broadcast transmission and directional reception

Linguistic transmissions are broadcast in all directions from the source and can be received by any hearer within range. Therefore, the transmission is public. By virtue of binaural hearing, the direction or location of the transmission is conveyed by the transmission itself.

Rapid fading

Linguistic transmission is transitory and has to be received at an exact time or it will fade. This implies that the hearer must record the message on paper or store it in memory.

Interchangeability

'Any speaker of a human language is capable, in theory, of saying anything he can understand when someone else says it. For language, humans are what engineers call "transceivers": units freely usable for either transmission or reception' (Hockett, 1960a[3]).

This means that because I can understand a sentence you say to me, I can say that sentence back at you. I can both receive and transmit any message. By contrast, there are certain animal systems in which males and females produce different calls or messages that cannot be interchanged.

Total feedback

The human speaker has total auditory feedback for the transmitted message, simultaneous with the listener's reception of the message. This feedback is used for moment-to-moment adjustments of sound production.

Specialisation

The sounds of language – versus non-language sounds – are specialised to convey meaning or linguistic intent. Consider a jogger saying, 'I'm exhausted', when the speech act conveys a specific meaning. Contrast this with a jogger panting loudly at the end of a run, when the sounds being produced have no necessary linguistic function, although a hearer may infer from the panting that the jogger is exhausted.

Semanticity

Linguistic utterances, whether simple phrases or complete sentences, convey meaning by means of the symbols we use to form the utterance.

Discreteness

Although sound patterns can vary continuously across several dimensions (e.g. duration of sound, loudness of sound), language uses only a small number of discrete ranges on those dimensions to convey meaning. Thus, languages do not rely on continuous variation of vowel duration, for example, to signal changes in meaning.

Displacement

Linguistic messages are not tied to communication in time and space. This implicates an elaborate memory system within the speaker or hearer to recall the past and manipulate the future.

Productivity

Language is novel, consisting of utterances that have never been uttered or comprehended before. New messages, including words, can be coined freely by means of rules and agreement among members of the language culture.

Duality of patterning

A small set of sounds, or *phonemes*, can be combined and recombined into an infinitely large set of sentences or meanings. The sounds have no inherent meaning, but the combinations do.

Cultural or traditional transmission

Language is acquired by exposure to the culture and the language of the surrounding people. Contrast this to various nonhuman animal courtship and mating communications in which genetics govern specific messages.

Semanticity

An important aspect of language is **semanticity**: that it conveys meaning. For example, the sounds of human language carry meaning, whereas other sounds, such as coughing or whistling, are not part of our language because they do not convey meaning in the usual sense. We are ignoring here a situation such as a roomful of students coughing in unison at a professor's boastful remark to indicate their collective opinion. The coughing sound in this case is *paralinguistic* and functions much as rising vocal pitch does to indicate anger. Then again, it could just be a roomful of coughing students.

Arbitrariness

Arbitrariness means there is no inherent connection between the units (sounds, words) used in a language and their meanings. There are a few exceptions, such as **onomatopoeias** like *buzz*, *hum* and *zoom*. But as Pinker (1994)[4] notes, some units we consider onomatopoetic, such as a pig's oink, are not because they are different sounds in other languages. In Japanese, for example, a pig's sound is 'boo-boo'. But far more common, the language symbol bears no relationship to the thing itself. The word *dog* has no inherent correspondence to the four-legged furry creature, just as the spoken symbol *silence* does not resemble its referent, true silence. Hockett's example drives the point home: '*Whale* is a small symbol for a very big thing, and *microorganism* is a big symbol for an extremely small thing.'

Because there are no built-in connections between symbols and their referents, knowledge of language must involve learning and remembering the arbitrary connections. Thus, we refer to language as a shared system. We all have learned essentially the same connections, the same set of word-to-referent associations, and stored them in memory as part of our knowledge of language. By convention – by agreement with the language culture – we all know that *dog* refers to one particular kind of physical object. Obviously, we have to know which word goes with a particular referent because there is no way to look at an object and decide what its name must be.

Two important consequences of the arbitrariness of language deserve special attention, partly because they help to distinguish human language from animal communication and partly because they tell us about the human language user. These two consequences concern flexibility and the principle of naming. Hockett listed neither of these, although they are derived from his point about arbitrariness.

Flexibility of symbols

Note that arbitrariness makes language symbolic. *Desk* and *pupitre* are the English and French symbols for a particular object. Were it not for the history of our language, we might call that object a *zoople* or a *manty*. A consequence of this *symbolic* aspect of language is that the system demonstrates tremendous **flexibility**. Because the connection between symbol and meaning is arbitrary, we can change those connections and invent new ones. We routinely shift our terms for the things around us, however slowly such change takes place.

Contrast this flexibility with the characteristics of the opposite of a symbolic system, called an iconic system. In an *iconic system*, each unit has a physical resemblance to its referent, just as a map is physically similar to the terrain it depicts. In such a system there is no flexibility because changing the symbol for a referent would make the connection arbitrary.

Although our names for things such as dog seem so strongly tied to the object, the word is actually arbitrary and can vary from language to language. For example, instead of dog, the same things would be called *chien, gǒu* and *canis* in French, Chinese and Latin.
Source: Ladanifer/Alamy Stock Photo

Naming

A corollary to arbitrariness and flexibility involves **naming** (Glass & Holyoak, 1986[5]). We assign names to all the objects in our environment, to all the feelings and emotions we experience, to all the ideas and concepts we conceive of. So, wherever it is you are sitting right now as you read this text, each object in the room has a name. In an unfamiliar or unusual place (an airport control tower or a car repair shop), you may not know the name of something, but it rarely, if ever, occurs to you that the thing might not have a name.

Furthermore, we do not stop by naming just the physical objects around us. We have a vocabulary for referring to unseen characteristics, privately experienced feelings, and other intangibles and abstractions. Terms such as *perception, mental process spreading activation* and *knowledge* have no necessary physical referent, nor do words such as *justice, cause, truth, likewise* and *however* refer to concrete objects. Indeed, we even have words such as *abstractions* and *intangibles* that refer to the *idea* of being abstract. Going one step further, we generate or invent names for new objects, ideas, activities, and so forth. For instance, think of the new vocabulary that had to be invented and mastered to describe the various actions and operations for using the Internet and modern technology. Because we need and want to talk about new things, new ideas and new concepts, we invent new terms. (See Kaschak & Glenberg, 2000, on how we invent new verbs from already known words; e.g. *to crutch* or *to google*.)[6]

Displacement

One of the most powerful language tools is the ability to talk about something other than the present moment, a feature called **displacement**. By conjugating verbs to form past tense, future tense, and so on, we can communicate about objects, events, and ideas that are not present but are remembered or anticipated. When we use constructions such as *'If I go to the library tomorrow, then I will be able to . . .'* we demonstrate a particularly powerful aspect of displacement: we can communicate about something that has never happened and indeed might never happen, while anticipating future consequences of that never-performed action. To illustrate the power and importance of displacement, try speaking only in the present tense for about 5 minutes. You will discover how incredibly limiting it would be if we were 'stuck in the present'.

Productivity

By most accounts, the principle of **productivity** (*also called generativity*) is important because it gives language a notable characteristic – novelty. Indeed, the novelty of language, and the productivity that novelty implies, formed the basis of Chomsky's (1959)[7] critique of Skinner's book and the foundation for Chomsky's own theory of language (1957, 1965[8]).

Importance of productivity

It is an absolute article of faith in both linguistics and psycholinguistics that the key to understanding language and language behaviour lies in an understanding of novelty, an understanding of the productive nature of language.

Consider the following: aside from trite phrases, customary greetings, and so on, hardly any of our routine language is standardised or repetitive. Instead, the bulk of what we say is novel. Our utterances are not memorised, are not repeated, but are new.

This is the principle of *productivity*, that language is a productive and inherently novel activity, that we generate utterances rather than repeat them. We (the authors of your text) lecture on the principle of productivity every time we teach our memory and cognition classes, each time uttering a stream of sounds, a sequence of words and sentences, that is novel, new, literally invented on the spot – the ideas we talk about may be the same semester after semester, but the sentences are new each time. Even in somewhat stylised situations, as in telling a joke, the language is largely new. Only if the punchline requires a specific wording do we try to remember the exact wording of a previously used sentence.

What does this mean?

It means that language is *creative*, not just repetitive. We do not recycle sentences. Instead, we create them on the spur of the moment, now in the active voice, now in the passive, with a prepositional phrase sometimes at the beginning, sometimes at the end, and so on. In a very real sense then, applying our productive rules of language to the words in our vocabulary permits us to generate an infinite number of utterances.

How can we understand any and all of the infinite set of sentences? What does it mean for a theory of language that speakers and listeners can generate and comprehend any one of this numberless set?

In brief, it means that language users must have some flexible basis for generating novel utterances, for coming up with the different sequences of symbols that can be comprehended. And, likewise, comprehenders must have the same flexible basis to hear the sequence of words and recover from them what the intended meaning is. By most accounts, the basis for such productivity is a set of rules. To anticipate later sections of the module, rules form the basis for each level of language we discuss, from our phonological system up through the highest level of analysis, the conceptual and belief systems we hold as we comprehend language.

8.1.3 Animal communication

animal communication
The ways animals use different channels and modalities to communicate with each other.

In contrast to flexible and productive human language, **animal communication** is neither. Animal communication is seen in a wide range of circumstances, from insects to primates. For example, bees communicate the location of honey (a form of displacement) through a waggle dance (Dyer, 2002; Sherman & Visscher, 2002; von Frisch, 1967[9]). Essentially, they orient themselves within the hive to the relative position of the sun, and then act out a dance that conveys how the flight will progress to get to the source of the nectar. This is even more impressive given that it is fairly dark in a beehive, and the dance is performed on a vertical surface.

Compression in the human language

Source: Ernie Janes/Alamy Stock Photo

Recall Treebeard, the Ent in the *Lord of the Rings: The Two Towers*. Treebeard was a tree-like creature that saved hobbits Merry and Pippin from Orcs, and later fought against Saruman. What was characteristic about Treebeard and his fellow Ents, was not just their sheer strength and size, but also their language, *Entish* – it takes a really long time to say anything in Old Entish. In our world, however, *brevity* or the ability to compress as much information as possible into smaller language units is of paramount importance for survival. Human language follows the principles of compression. Zipf's law states that shorter words tend to be used more frequently than longer ones in human verbal communication. Risueno-Segovia et al. (2023)[10] studied whether the principles of compression observed in humans can also be found in primates. To find out, they trained marmoset monkeys (*Callithrix jacchus*), like the ones shown in the picture above, to vocalise utterances when presented with a certain set of visual stimuli, and they rewarded them after every utterance. The monkeys could freely choose among vocal utterances of either shorter (e.g. chirps and tsiks) or longer lengths (e.g. phees and trills). The results supported Zipf's law: the vocalisations that were more frequently used tended to be shorter, thus, supporting the idea that the origins of human language communication patterns (e.g. brevity) can be traced back in our ancestral primates. This idea challenges Chomsky's theory of how humans develop language or the notion that language is a uniquely human capacity.

Closer to humans, consider the signalling system of vervet monkeys (Marler, 1967[11]). This consists of several distress and warning calls, alerting an entire troupe to imminent danger. These monkeys produce a guttural 'rraup' sound to warn of eagles, one of the monkey's natural predators; they 'chutter' to warn of snakes and 'chirp' to warn of leopards. The system thus exhibits *semanticity*, an important characteristic of language. That is, each signal in the system has a different, specific referent (eagle, snake and leopard). Furthermore, these seem to be arbitrary connections: 'Rraup' doesn't resemble eagles in any physical way.

But as Glass and Holyoak (1986)[12] note, the troupe of monkeys cannot get together and decide to change the meaning of 'rraup' from *eagle* to *snake*. The arbitrary connections to meaning are completely inflexible. (This inflexibility results at least in part from genetic influence; compare this with Hockett's last universal, cultural transmission.) Furthermore, there is a vast difference between naming in human languages and in animal communication. There seem to be no words in the monkey system for other important objects and concepts in their environment, such as 'tree'. And as for displacement and productivity, consider the following quotation from Glass and Holyoak (1986, p. 448[13]): 'The monkey has no way of saying "I don't see an eagle", or "Thank heavens that wasn't an eagle", or "That was some huge eagle I saw yesterday".' Even human infants can use displacement by pointing to refer to things that are not immediately present, whereas chimpanzees cannot (Liszkowski, Schäfer, Carpenter & Tomasello, 2009[14]), suggesting that non-human animals lack even the basic cognitive abilities required by language.

Although there are no true languages among the animal communication systems, this is not to say that nothing can be learned about language from studying animals. As one illustration, work by Hopkins, Russell, and Cantalupo (2007)[15] used magnetic resonance imaging (MRI) with chimpanzees to show that there was a lateralisation of function as a consequence of tool use. Moreover, those regions of the brain that were more affected corresponded to Broca's and Wernicke's areas in humans, which correspond to critical areas of human language production and comprehension (as you will see later in the chapter). This suggests that our development of language may be tied, to some extent, to the development of tool use by our ancestors. Similarly, there is evidence that some great apes and non-verbal human infants can think about and signal absent objects (Bohn, Call & Tamasello, 2015[16]).

Miller (1973)[17] proposed that language is organised on five levels (see Table 8.1). In addition to the three traditional levels of phonology, syntax and lexical or semantic knowledge, Miller suggested two higher levels as well. He called these the *conceptual knowledge* and *belief* levels. For organisational purposes, we focus primarily on the first three of the levels in this chapter.

Table 8.1

Level of language analysis	Explanation
Phonology	Analysis of the sounds of language
Syntax	Analysis of word order and grammar
Lexical/semantic knowledge	Analysis of word meanings and their integration in sentences and phrases
Conceptual knowledge	Analysis of sentence and phrase meaning with reference to semantic memory knowledge
Belief	Analysis of sentence and discourse meaning with reference to one's own beliefs and one's beliefs about a speaker's intentionality

In short, beyond a level of arbitrariness, animal communication does not exhibit the characteristics that appear to be universally true of human language. There are no genuine languages in animals, although there may be genuine precursors to human language among various apes. In human cultures, genuine language is the rule. (For a more up-to-date discussion of animal cognition, see Bekoff, Allen & Burghardt, 2002.[18]) Note that there has been some suggestion that our close relatives, the Neanderthals, also possessed speech (D'Anastasio et al., 2013; Dediu & Levinson, 2013[19]), although this view is controversial (Berwick, Hauser & Tattersall, 2013[20]).

8.1.4 Levels of analysis

The traditional view of language from linguistics is that it is the set of all acceptable, well-formed utterances. In this scheme, the set of rules used to generate the utterances is called a **grammar**. In other words, the grammar of a language is the complete set of rules that will generate all the acceptable utterances and will not generate any unacceptable, ill-formed ones. According to most linguists (e.g. Chomsky, 1965), such a grammar operates at three levels: **Phonology** of language deals with the sounds of language; **syntax** deals with word order and grammaticality; and **semantics** deals with accessing and combining the separate word meanings into a sensible, meaningful whole.

A critical distinction between competence and performance

Chomsky (1957, 1965)[21] insisted that there is an important distinction in any investigation of language, the distinction between competence and performance. **Competence** is the internalised knowledge of language and its rules that fully fluent speakers of a language have. It is an ideal knowledge, to an extent, in that it represents a person's complete knowledge of how to generate and comprehend language. **Performance** is the actual language behaviour a speaker generates, the string of sounds and words that the speaker utters.

When we produce language, we are not only revealing our internalised knowledge of language, our competence; we are also passing that knowledge through the cognitive system. So, it is not surprising that our performance sometimes reveals errors. Speakers may lose their train of thought as they proceed through a sentence, and so may be forced to stop and begin again. We pause, repeat ourselves, stall by saying 'ummm,' and so on. We can attribute all these **dysfluencies**, these irregularities or errors in otherwise fluent speech, to the language user. Lapses of memory, momentary distractions, intrusions of new thoughts, 'hiccups' in the linguistic system – all of these are imperfections in the language user rather than in the user's basic competence or knowledge of the language. Chomsky, as a linguist, was not particularly interested in these performance-related aspects of language. Psychology, on the other hand, views them as rich sources of evidence for understanding language and language users.

The strong and weaker versions of Sapir–Whorf hypothesis

In its strongest version, the hypothesis claims that language controls both thought and perception to a large degree, so you cannot think about ideas or concepts your language does not name. In its weaker version, the hypothesis claims that your language influences and shapes your thought, making it merely more difficult, rather than impossible, to think about ideas without having a name for them.

The strong Sapir–Whorf hypothesis

In a series of studies assessing the Sapir–Whorf hypothesis, Eleanor Rosch tested members of the Dani tribe in New Guinea on a perceptual and memory test (Rosch-Heider, 1972[22]). She administered both short- and long-term memory tasks, using chips of different colours as the stimuli. She found that the Dani learned and remembered more accurately when the chips were 'focal' colours rather than 'non-focal' colours, as when the learning trial presented a 'really red' red as opposed to a 'sort-of-red' red. In other words, the central, perceptually salient, 'good' red was a better aid to accuracy than the non-focal 'off-red.' The compelling aspect of these studies involved the language of the Dani people, which contains only two colour terms, one for 'dark' and one for 'light'. Nothing in their language expresses meanings such as 'true red' or 'off-red', and yet their performance was influenced by the centrality of focal versus non-focal colours.

Thus, in this example, a person's language could have affected cognition (the language had very few colour terms), and yet did not. Moreover, others have suggested that some reported linguistic relativity effects are not reliable (Wright, Davies & Franklin, 2015[23]).

That said, there is evidence that language may influence the shape of perceptual colour spaces (Regier, Kay & Khetarpal, 2009[24]). Other research shows that different ways of referring to objects – such as the distinction between count and mass nouns in English, which does not occur in Japanese and Chinese – do not influence object perception (Barner, Li & Snedeker, 2010[25]). Results such as these seem to disconfirm the strong Sapir–Whorf hypothesis.

The weaker Sapir–Whorf hypothesis

However, current research has found increasing support for the weaker form of the Sapir–Whorf hypothesis, the hypothesis that language does influence our thoughts, sometimes to a surprising degree (e.g. Boroditsky, 2001, but see Chen, 2007; January & Kako, 2007[26]). Here are two examples: The first involves number, and the second involves intentionality.

For the first, 'English speakers have no difficulty expressing the idea that, if there are 49 men and 37 pairs of shoes, some men will have to go without shoes. There are non-literate societies where this would be a difficult situation to describe, because the language may have number terms only for ;one-two –many' (Greenberg, 1978; Hunt & Agnoli, 1991, p. 385; see also Roberson, Davies & Davidoff, 2000; Malt, Sloman & Gennari, 2003, discuss how one's linguistic and cultural history influence perception and naming[27]). In other words, one person's language and culture can make it very difficult to think and talk about this situation of 12 men going barefoot, whereas another person's language and culture support that kind of thinking and expression.

Second, consider intentionality: whether someone did something on purpose or by accident. In English, we would say 'Sue broke the vase' both when she intended to break the vase and when she accidentally knocked over the vase. But in Spanish and Japanese, the agent is less likely to be mentioned in the case of an accident; the appropriate sentence is much more like 'the vase broke.' In a test of this language effect, Fausey and Boroditsky (2011)[28] showed English and Spanish speakers several videos of the two kinds of scenarios, intentional and accidental acts, and then gave them a surprise memory test. Both groups remembered the agent (Sue) well when the act was intentional, but the Spanish speakers tended not to remember the agent as well when the act was accidental. After all, their language-based way of thinking about the accident de-emphasises the agent and focuses on the breaking itself. (See also Dolscheid, Shayan, Majid & Casasanto, 2013, for an account of differences in pitch perception in Dutch and Farsi speakers and their references to pitch as being either high-low or thin-thick.[29])

Thus, the participants' language clearly affected their interpretation of the event and their later memory for the event (see Boroditsky, 2011, for a highly readable summary of similar language effects[30]).

The Sapir–Whorf hypothesis

We tend to think of mental processes, including those related to language, as universal, as being equally true of all languages. Even slight familiarity with another language, however, reveals at least some of our beliefs to be misconceptions.

An organising issue in studies of cultural influences on language and thought is how one's language affects one's thinking. This topic is called the **Sapir–Whorf hypothesis**, or more formally the **linguistic relativity hypothesis** by Whorf (1956).[31] This idea comes out of work by Edward Sapir, a linguist and anthropologist, and his student Benjamin Whorf. The basic idea is that the language you know shapes the way you think about events in the world around you.

8.2 Phonology

LEARNING OBJECTIVE
Explain how phonology influences spoken language

In any language interaction, the task of a speaker is to communicate an idea by translating that idea into spoken sounds. The hearer goes in the opposite direction, translating from sound to intended meaning. Essentially, a person is transferring the contents of his or her mind to another person (a lot like ESP, only in a plausible – spoken – way). Among the many sources of information available in the spoken message, the most obvious and concrete one is the sound of the language itself, the stream of speech signals that must be decoded. Other sources such as the gestures and facial expressions of the speaker are also available, but we focus on the speech sounds here. Thus, our study of the grammar of language begins at this basic level of *phonology*, the sounds of language and the rule system for combining them.

8.2.1 Sounds in isolation

To state an obvious point, different languages sound different: they are composed of different sets of sounds. The basic sounds that compose a language are called *phonemes*. If we were to conduct a survey, we would find around 200 different phonemes across all known spoken languages. However, no single language uses even half that many phonemes. English uses about 46 phonemes (experts disagree on whether some sounds are separate phonemes or blends of two phonemes; the disagreement centres on diphthong vowel sounds, as in *few*, seemingly a combination of 'ee' and 'oo'). Hawaiian uses only about 15 phonemes (Palermo, 1978[32]). There is little significance to the total tally of phonemes in a language; no language is superior to another because it has more (or fewer) phonemes.

Table 8.2 shows the typology of the phonemes of English based on the characteristics of their pronunciation.

Table 8.2 English consonants and vowels

English consonants

Manner of articulation		Bilabial	*Labiodental*	Dental	Alveolar	Palatal	Velar	Glottal
(oral) Stops	Voiceless	P (*p*ut)			t (*t*uck)		k (*c*ap)	
	Voiced	b (*b*ut)			d (*d*ug)		g (*g*ot)	
Nasal (stop)		m (*m*ap)			n (*n*ap)		η (so*ng*)	
Afffricatives	Voiceless	m (map)				č (*c*hurn)		
	Voiced					ǐ (*j*ump)		
Fricatives	Voiceless		f (*f*it)	Q (*t*hink)	s (*s*ad)	š (fish)		h (*h*ad)
	Voiced		v (*v*ote)	∂ (them)	z (*z*ip)	ž (*a*zure)		
Glides		w (won)				y (*y*es)		
Liquids					l (*l*ame)	r (*r*age)		

English vowels

	Front	Center	Back
	i (beef)		u (boom)
High			U (book)
	i (b*i*t)		
		I (b*i*rd)	o (bowl)
Middle	e (b*a*be)	(sof*a*)	
	ε (bed)		(bought)
	æ (b*a*d)	(b*u*s)	
Low			
			a (p*a*lm)

Source: Based on Glucksberg & Danks (1975)[33].

For consonants, three variables are relevant: place of articulation, manner of articulation and voicing.

Articulation

Place of articulation is the place in the vocal tract where the disruption of airflow occurs. As shown in Figure 8.1, a bilabial consonant such as /b/ disrupts the airflow at the lips, whereas /h/ disrupts the column of air at the rear of the vocal tract, at the glottis. *Manner of articulation* is how the airflow coming up from the lungs is disrupted. If the column of air is completely stopped and then released, it is called a *stop consonant*, such as the consonant sounds in *bat* and *tub*. A *fricative consonant*, such as the /f/ in *fine*, involves only a partial blockage of airflow. Finally, **voicing** refers to whether the vocal cords begin to vibrate immediately with the obstruction of airflow (for example, the /b/ in *bat*) or whether the vibration is delayed until after the release of air (the /p/ in *pat*).

Vowels, by contrast, involve no airflow disruption. Instead, they differ on two dimensions: placement in the mouth (front, centre or back) and tongue position in the mouth (high, middle or low).

Phonemes

Let's develop a few more conscious intuitions about phonemes. Stop for a moment and put your hand in front of your mouth. Say the word *pot* and then *spot*. Did you notice a difference between the two /p/ sounds? Most speakers produce a puff

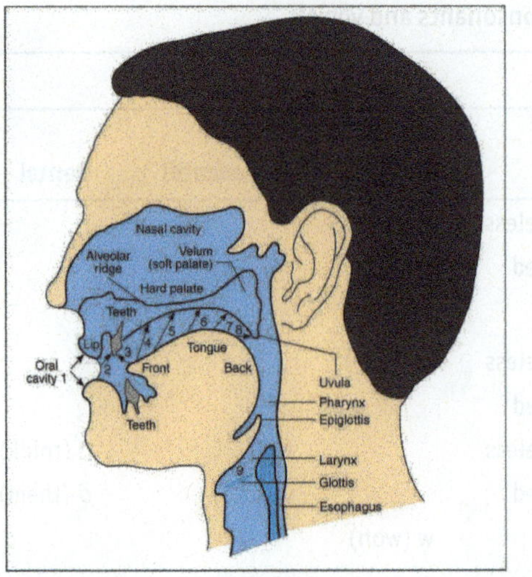

Figure 8.1

Source: Based on Fromkin and Rodman (1974)[34].

of air with the /p/ sound as they say *pot*; we puff very little (if at all) for the /p/ in *spot* if it is spoken normally. Given this, you would have to agree that these two /p/ sounds are different at a purely physical level. And yet you hear them as the same sound in these two words. Figure 8.2 shows hypothetical spectrograph patterns for two families of syllables, the /b/ family on the top and the /d/ family on the bottom. Note how remarkably different 'the same' phoneme can be.

For psycholinguistics, the two /p/ sounds, despite their physical differences, are both instances of the same phoneme, the same basic sound group. That is, the fact that these two sounds are treated as if they were the same in English means that they represent one phoneme. So, let's redefine the term *phoneme* as the category or group of language sounds that are treated as the same, despite physical differences

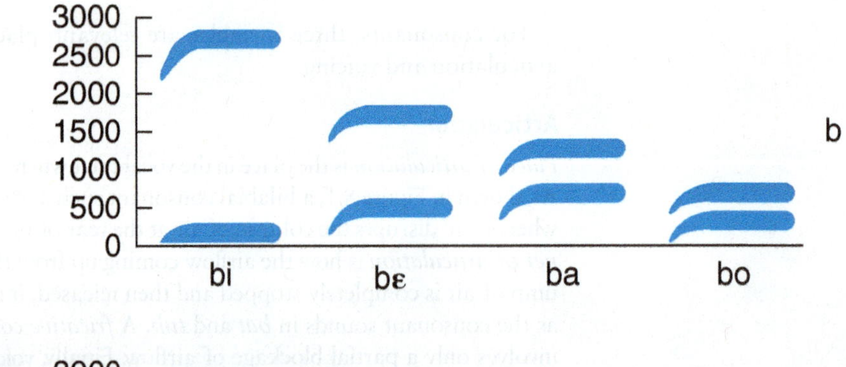

Figure 8.2 Spectrographs of different phonemes

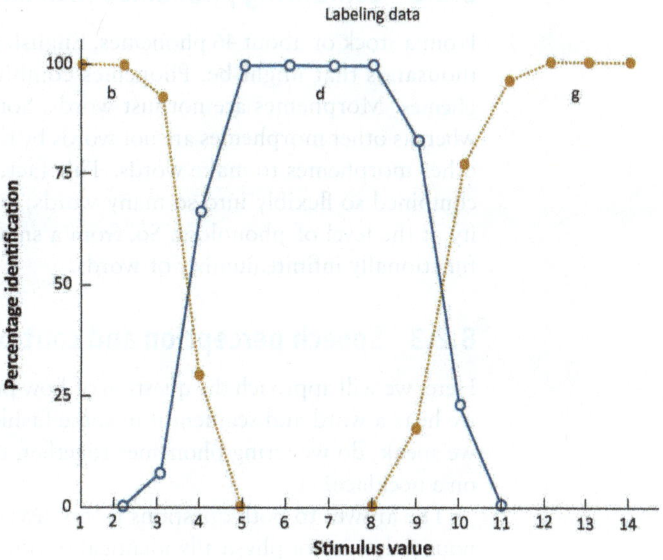

Figure 8.3 Illustration of phoneme boundaries

One person's labelling data for synthesised consonants ranging from /b/ to /g/. Note that small changes in the stimulus value (e.g. from values 3 to 4) can result in a complete change in labelling, whereas larger changes (e.g. from values 4 to 8) that do not cross the phoneme boundary do not lead to a change in labelling.

among the sounds. In other words, the English word *spot* does not change its meaning when pronounced with the /p/ sound in *pot*. A classic illustration of phoneme boundaries is shown in Figure 8.3, from a study by Liberman, Harris, Hoffman, and Griffith (1957).[35]

When the presented sound crossed a boundary, such as between stimulus values 3 and 5 and between 9 and 10 in Figure 8.3, identifications of the sound switched rapidly from /b/ to /d/ and then from /d/ to /g/. Variations within the boundaries did not lead to different identifications; despite the variations, all the sounds from values 5 to 8 were identified as /d/.

There are two critical ideas here. First, all of the sounds falling within a set of boundaries are perceived as the same despite physical differences among them. This is called categorical perception. Because English speakers discern no real difference between the hard /k/ sounds in *cool* and *keep*, they are perceived categorically, as belonging to the same category, the /k/ phoneme. Second, different phonemes are the sounds that are perceived as different by speakers of the language. The physical differences between /s/ and /z/ are important in English. Changing from one to the other gives you different words, such as *ice* and *eyes*. Thus, the /s/ and /z/ sounds in English are different phonemes.

An interesting side effect of such phonemic differences is that you can be insensitive to differences of other languages if your own language does not make that distinction. Spanish does not use the /s/ versus /z/ contrast, so native speakers of Spanish have difficulty distinguishing or pronouncing *ice* and *eyes* in English. Conversely, the hard /k/ sounds at the beginning of *cool* and *keep* are interchangeable in English; they are the same phoneme. But, this difference is phonemic in Arabic. The Arabic words for *heart* and *dog* differ only in their initial sounds, exactly the two different hard /k/ sounds in *cool* and *keep*.

8.2.2 Combining phonemes into morphemes

From a stock of about 46 phonemes, English generates all its words, however many thousands that might be. Phonemes combine into meaningful units called **morphemes**. Morphemes are not just words. Some morphemes are individual words, whereas other morphemes are not words by themselves, but must be combined with other morphemes to make words. This fact, that a small number of units can be combined so flexibly into so many words, is the linguistic universal of productivity at the level of phonology. So, from a small set of phonemes we can generate a functionally infinite number of words.

8.2.3 Speech perception and context

Here, we will approach the question of how people produce and perceive speech. Do we hear a word and segment it in some fashion into its separate phonemes? When we speak, do we string phonemes together, one after another, like stringing beads on a necklace?

The answer to both questions is 'no'. Even when the 'same' sound is being pronounced, it is not physically identical to other instances of that 'same' sound. The sounds *change* – they change from speaker to speaker and from one time to the next within the same speaker. Most prominently, they change or vary from one word to another, depending on the preceding and following sounds.

Rules of combining phonemes into words

Recall also that the essential ingredient of productivity is rules. We now turn to the rules of combining phonemes into words.

Phoneme combinations

Let's work with a simple example of phoneme combinations. There are three phonemes in the word *bat*: the voiced stop consonant /b/, the short vowel sound /ae/, and the final voiceless /t/. Substitute the voiceless /p/ for /b/, and you get *pat*. Now rearrange the phonemes in these words, and you will discover that some of the arrangements do not yield English words, such as **abt*, **tba* and **atp*. Why? What makes **abt* and **atp* illegal strings in English? Although it is tempting to say that syllables like **abt* cannot be pronounced, a moment's reflection suggests this is false. After all, many such 'unpronounceable' strings in English are pronounced in other languages. For example, the initial *pn-* in the French word for *pneumonia* is pronounced, whereas English makes the *p* silent. Instead, the rule is more specific. English usually does not use a 'voiced–voiceless' sequence of two consonants within the same syllable. It only seldom uses any two-consonant sequence when both are in the same 'manner of articulation' category. Of course, if the two consonants are in different syllables, then the rule does not apply.

Phonemic competence and rules

Why does the phoneme composition rule of the previous section seem to be an unusual explanation? It is because our knowledge of English phonology and pronunciation is not particularly verbalisable. You can look at the phoneme combinations in the previous card and think of words that combine consonants, and then come up with tentative pronunciation rules. But this is different from knowing the rules in an easily accessed and expressible fashion. And yet you are a true expert at deciding which phoneme sequences can and cannot be used in English.

Your implicit knowledge of how sounds are combined tells you that **abt* is illegal because it violates a rule of English pronunciation.

This extensive knowledge of the rules of permissible English sound combinations is your phonemic competence. These rules tell you what is permissible: *bat* is, but **abt* is not. No one ever explicitly taught you these rules; you abstracted them from your experience as you acquired language. This competence tells you that a string of letters like 'pnart' is legal only when the *p* is silent but that 'snart' is a legal string – not a word, of course, but a legal combination of sounds. Speakers of the language have this phonemic competence as part of their knowledge of language – an implicit, largely unverbalisable part to be sure, but a part nonetheless.

This variability in sounds is the *problem of invariance*. This term is somewhat peculiar because the problem in speech perception is that sounds *are not* invariant; they change all the time. You saw an illustration of this in Figure 8.2 where the initial /b/ and /d/ sounds looked very different in the spectrographic patterns depending on the vowel that followed. A second illustration of the problem of invariance is in Figure 8.4, which shows the influence of each of the three phonemes in the word *bag*. To pronounce *bag*, do you simply articulate the /b/, then /ae/, then /g/? No! As the figure shows, the /ae/ sound influences both /b /and /g/, the /g/ phoneme (dotted lines) exerts an influence well back into the /b/ sound, and so on.

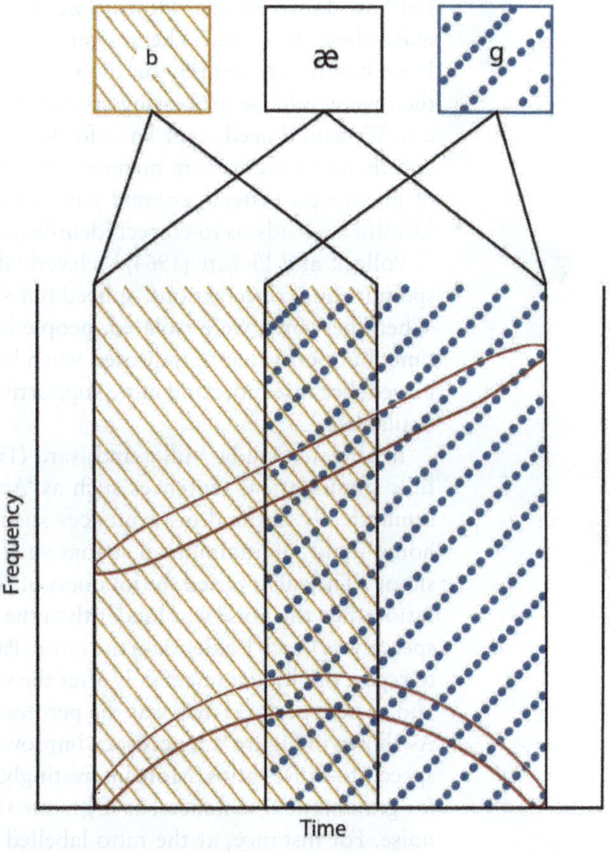

Figure 8.4 Coarticulation and the problem of invariance

Coarticulation is illustrated for the three phonemes in the word *bag*; solid diagonals indicate the influence of the /b/ phoneme, and dotted diagonals the influence of /g/.

The term for this is **coarticulation**: more than one sound is articulated at the same time. As you type the word *the* on a keyboard, your right index finger starts moving towards *h* before your left index finger has struck the *t*. In like fashion, your vocal tract begins to move towards the /ae/ before you have articulated /b/ and towards /g/ before even finishing the /b/. This is another illustration of the problem of invariance where each phoneme changes the articulation of each other phoneme and does so depending on what the other phonemes are. The problem of invariance is made clearer by considering what we do when we whisper. Whispering changes some of the vocal characteristics of the phonemes. For example, voiced phonemes become voiceless. Yet, we typically have little trouble understanding what is being whispered to us.

In short, the sounds of language, the phonemes, vary widely as we speak them. Yet we tolerate a fair degree of variability for the sounds within a phoneme category, both when listening and decoding from sound to meaning and also when speaking, converting meaning into spoken sound. This categorical perception of phonemes in spoken language is a decision-making process that requires some cognitive control to take into account a variety of factors to make this categorisation decision. Studies using functional MRI (fMRI) scans have found that the left inferior frontal sulcus (B.A. 44) is critically involved in this process (Myers, Blumstein, Walsh & Eliassen, 2009[36]), supporting the idea that some mental control is need to make these decisions.

8.2.4 The effect of context

But how do we do this? How do we tolerate the degree of variability – how do we make these decisions? The answer is context and conceptually driven processing. If we had to rely entirely on the spoken signal to figure out what was being said, then we would be processing speech in an entirely data-driven, bottom-up fashion. We would need some basis for figuring out what every sound in the word was and then retrieve it from memory. This is almost impossible, given the variability of phonemes. Instead, *context* – in this case the words, phrases and ideas already identified – leads us to correct identification of new incoming sounds.

Pollack and Pickett (1964)[37] cleverly demonstrated this. They recorded several spontaneous conversations, spliced out single words, then played them to people. When the words were isolated, people identified them correctly only 47% of the time. But performance improved when longer and longer segments of speech were played, because more and more supportive syntactic and semantic context was then available.

In a related study, Miller and Isard (1963)[38] presented three kinds of sentences: fully grammatical sentences such as 'Accidents kill motorists on the highways', semantically anomalous sentences such as 'Accidents carry honey between the house', and ungrammatical strings such as 'Around accidents country honey the shoot'. They also varied the loudness of the background noise, from the difficult −5 ratio, when the noise was louder than the speech, to the easy ratio of +15, when the speech was much louder than the noise. Participants shadowed (i.e., they were asked to repeat out loud immediately after the word was presented) the strings they heard, and correct performance was the percentage of their shadowing that was accurate. As shown in Figure 8.5, accuracy improved going from the difficult to easy levels of speech-to-noise ratios. More interestingly, the improvement was especially dramatic for grammatical sentences, as if grammaticality helped counteract the background noise. For instance, at the ratio labelled 0 in the figure, 63 per cent of the grammatical sentences were shadowed accurately, compared with only 3 per cent of the

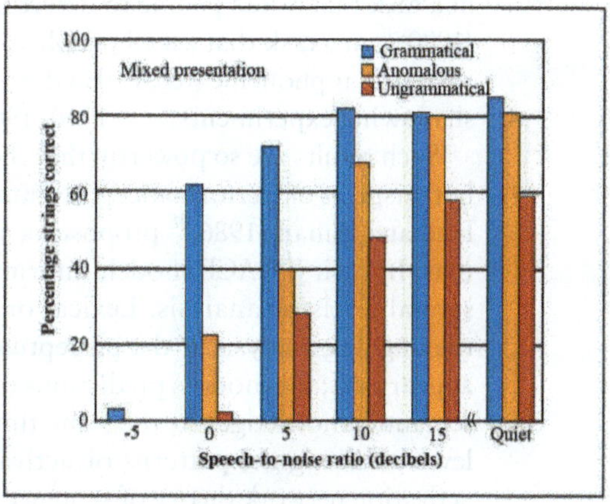

Figure 8.5 Percentage of strings shadowed correctly
Source: From Miller and Isard (1963)[39].

ungrammatical strings. Indeed, even at the easiest ratio of +15, less than 60 per cent of the ungrammatical strings could be repeated correctly.

People use their linguistic knowledge even to the point of hearing things that are not there. In a study by Richard Warren (1970),[40] people heard sentences in which part of the sentence was removed from the recording and replaced with a cough. For example, in the sentence 'The state governors met with their respective legislatures convening in the capital city', the 's' sound was replaced with a cough. The vast majority of people did not report that any speech sounds were missing and could not report the location of the cough when asked to do so on a printed version of the sentence later on. DeWitt and Samuel (1990)[41] reported a similar finding with music, with people reporting hearing musical notes or tones that were obscured by noise. So, when people are listening, they are actively using their knowledge to interpret what they hear.

8.2.5 Top-down and bottom-up processes

More recent evidence is largely consistent with these early findings. That is, there is a combination of data-driven and conceptually driven processing in speech recognition, a position called the *integrative* or *interactive approach* (Rapp & Goldrick, 2000[42]). At a general level, a variety of conceptually distinct language processes, from the perception of the sounds up through integration of word meanings, operate simultaneously, each having the possibility of influencing the ongoing activity of other processes. While features of the speech signal are analysed perceptually, a listener's other linguistic knowledge is also called into play at the same time. These higher levels of knowledge and analysis operate in parallel with the phonemic analysis and help to identify the sounds and words (Dahan, 2010; Dell & Newman, 1980; Pitt & Samuel, 1995; Samuel, 2001[43]). Moreover, to overcome the relative dearth of invariant information in the speech signal, it also appears that language perception relies heavily on characteristic knowledge of the speaker, such as whether the person speaks with a lisp (Kraljik, Samuel & Brennan, 2008[44]).

As a concrete example, imagine a sentence that begins 'The grocery bag was . . . '. You are processing the *bag* segment of this speech signal. Having already processed the previous word to at least some level of semantic interpretation, you have developed a useful context for the sentence. To be simple about it, the grocery topic limits the number of possibilities that can be mentioned later in the sentence. Similar evidence of the role of context was reported by Marslen-Wilson and Welsh (1978)[45] in a task that asked people to detect mispronunciations, and by Dell and Newman (1980)[46] in a task that asked people to monitor spoken speech for the occurrence of a particular phoneme (recall also the demonstrations of context effects in Treisman's shadowing experiments, e.g. 1960, 1964[47]).

Such results are so powerful that theories of speech recognition must account for both aspects of performance, the data-driven and the conceptually driven. McClelland and Elman (1986)[48] proposed a specific connectionist model that does exactly that. In their TRACE model, information is continually being passed among the several levels of analysis. Lexical or semantic knowledge, if activated, can alter the ongoing analysis at the perceptual level by 'telling' it what words are likely to appear next; the model's predictions of what words are likely to appear are based on semantic knowledge. At the same time, phonemic information is passed to higher levels, altering the patterns of activation there (see Dell, 1986[49], for a spreading activation network theory of sentence production, and Tyler, Voice & Moss, 2000[50], for a useful review).

8.2.6 Embodiment in speech perception

The perception of speech is critical for language. Intuitively this may seem like an odd place for aspects of embodied cognition to show up. However, speech perception is actually where one of the first embodied theories of cognition came from (although it was not labelled as such at the time). This is the **motor theory of speech perception** (see, e.g. Liberman, Cooper, Shankweiler & Studdert-Kennedy, 1967; Liberman & Mattingly, 1985[51]).

According to the motor theory of speech perception, people perceive language, at least in part, by comparing the sounds that they are hearing with how they themselves would move their own vocal apparatus to make those sounds. That is, we create embodied representations of how those sounds might be spoken to help us perceive speech. There are several lines of evidence for this idea (for an excellent review, see Galantucci, Fowler & Turvey, 2006[52]). According to Galantucci et al., people find it much easier to understand synthesised speech if it takes issues of coarticulation into account, rather than simply presenting a string of phonemes. Also, the parts of the cortex that are more active during speech perception overlap substantially with those involved in speech production. This is similar to the idea of mirror neurons that fire when primates observe the actions of others. Finally, people find it easier to comprehend speech if they can see the person talking, which gives them more information about how the sounds are being made. It is also true that people can better understand song lyrics if they can see a person singing (Jesse & Massaro, 2010[53]). This theory does not explain all aspects of speech perception, such as how people who could never speak can understand spoken language. However, it does illustrate how the structure of our bodies, and how we use parts of our bodies in the environment (in this case moving the air around with our vocal apparatus) influence cognition.

Related to the idea that motor programs are involved in speech perception, some evidence also exists that people activate mental motor programs just by thinking about words to themselves. Take the example of tongue twisters, which involve difficult movements of the vocal apparatus when they are spoken aloud. What is interesting is that people are likely to show evidence of articulation difficulty, such as reading times, when they are simply asked to read tongue twisters. That is, even when the people were only 'saying' (inner speech) the twisters silently to themselves in their minds, the language processing system takes into account and simulates the muscle movements that would be involved if the person were actually speaking, and these simulated movements produce the normal tongue twister difficulty (Corley, Brocklehurst & Moat, 2011; but see Oppenheim & Dell, 2010[54]).

8.2.7 The puzzle of apparent segments in speech

As if the preceding sections were not enough to convince you of the need for conceptually driven processing, consider one final feature of the stream of spoken speech. Despite coarticulation, categorical perception and the problem of invariance, we naively believe that words are somehow separate from each other in the spoken signal – that there is a physical pause or gap between spoken words, just as there is a blank space between printed words.

This is not true. Our intuition is entirely wrong. Analysis of the speech signal shows that there is almost no consistent relationship between pauses and the ends of words. If anything, the pauses we produce while speaking are longer within words than between words. They bear no particular relationship to the ends of words. There must be other kinds of information that human cognition uses to decode spoken language.

How can our intuitions about our own language, that words are articulated as separate units, be so wrong? (Note that our intuitions about foreign languages – they sound like a continuous stream of speech – are more accurate.) How do we segment the speech stream and come to know what the words and phrases are? Part of the answer is our knowledge of words in the language and the fact that some phoneme combinations simply cannot or do not form words (Norris, McQueen, Cutler & Butterfield, 1997[55]). Another part of the answer to these questions is syntax, the second level of language analysis and the topic we address next.

8.3 Syntax

LEARNING OBJECTIVE
Summarise the rules governing syntax

At the second level of analysis we have *syntax*, the arrangement of words as elements in a sentence to show their relationship to one another. We have already studied how sounds combine to form meaningful words. At this level of analysis, we are interested in how the words are sequenced to form meaningful strings, the study of syntax.

Three elements of syntax

The three elements of syntax help us order words to create acceptable, well-formed sentences.

Word order

Unlike the school grammar idea, the psycholinguistic study of syntax is *descriptive*; that is, it takes as its goal a description of the rules of how words are arranged to form sentences. Let's take a simple example, one that taps into your syntactic competence.

Which is better, sentence 1 or 2?

(1) Beth asked the man about his headaches.

(2) About the Beth headaches man asked his.

Your 'school grammar' taught you that every sentence must have a subject and a verb. According to that rule, sentence 2 is just as much a sentence as 1. Your syntactic competence, on the other hand, tells you that sentence 2 is ill-formed, unacceptable. You can even specify some of the rules that are being violated; for example, definite articles such as *the* do not usually precede a person's name, and two nouns usually do not follow one another in the same phrase or clause.

The point here is that the meaning of a sentence is far more than the meanings of the words. The 'far more' here is the arrangement or sequencing of the words. We are speaking of syntactic word order rules for English (Gershkoff-Stowe & Goldin-Medow, 2002, argue that word order is more than just syntax, and that it reflects a more general property of human thought[56]), which, again, is an SVO language. More than some languages (e.g. Latin), English relies heavily on word order. Consider 'red fire engine' versus 'fire engine red' (or even 'red engine fire'). There is even some evidence that the syntactic structure of one's language spills over into other behaviours, including seemingly unrelated behaviours such as the structure of visual artworks (Segel & Boroditsky, 2011[57]). Despite the fact that *red* and *fire engine* can be nouns, our word-order knowledge tells us that the first word in these phrases is to be treated as an adjective, modifying the following noun. Thus by varying word order alone, 'red fire engine' is a fire engine of the usual colour, and 'fire engine red' is a particular shade of red.

Phrase order

There is more to syntax than just word order, however. We also rely on the ordering of larger units such as phrases or clauses to convey meaning. Consider the following sentences:

(3) Bill told the men to deliver the piano on Monday.

(4) Bill told the men on Monday to deliver the piano.

In these examples, the positioning of the phrase 'on Monday' helps us determine the intended meaning, whether the piano was to be delivered on Monday or whether Bill had told the men something on Monday. Thus, the sequence of words, phrases and clauses contains clues to meaning, clues that speakers use to express meaning and clues that listeners use to decipher meaning.

Number agreement

Yet another part of syntax involves the adjustments we make depending on other words in the sentence. In particular, a part of every sentence is a subject and a verb. It is required, furthermore, that the subject and verb agree in number – if the subject of the sentence is singular, you must use a singular verb, as in 'The car has a flat tyre' (obviously, pronouns have to be coordinated in terms of number, too). As Bock (1995[58], p. 56) noted, agreement helps listeners know what the topic of a sentence is going to be about. For example, consider 'The mother of the girls who was . . . ' versus 'The mother of the girls who were . . . '. You are pretty sure that the first sentence is going to be about the mother, and the second one about the girls just because of the number agreement of *was* versus *were*. So number agreement, like word and phrase order, is a clue to meaning, part of the spoken and written language that we rely on when we comprehend. (See Bock & Miller, 1991, and Hartsuiker, Anton-Mendez & van Zee, 2001, for experimental work on number agreement errors; e.g. when you make the verb agree in number with the nearest noun rather than the subject noun, as in 'The difficulty with all of these issues are that . . . '.[59])

If you have a connotation associated with the word *syntax*, it probably is not the psycholinguistic sense of *grammar* but the 'school grammar' sense instead. In school, if you said, 'He ain't my friend no more', your teacher might have responded, 'Watch your grammar.' To an extent, this kind of school grammar is irrelevant to the psycholinguistic study of syntax. Your teacher was being *prescriptive* by teaching you what is proper or prestigious according to a set of cultural values. In another way, though, school grammar does relate to the psycholinguistic study of language; language is for expressing ideas, and anything that clarifies this expression, even arbitrary rules about *ain't* and double negatives, improves communication. (And finally, your teacher was sensitive to another level of language: People DO judge others on the quality of their speech.)

In general, we need to understand what these syntactic clues are and how they are used. We need to explore the syntactic rules that influence comprehension. We begin by looking at the underlying syntactic structure of sentences, taking a piece-by-piece approach to Chomsky's important work.

8.3.1 Chomsky's transformational grammar

At a general level, Chomsky intended to 'describe the universal aspects of syntactic knowledge' (Whitney, 1998[60]), that is, to capture the syntactic structures of language. He noted that language has a hierarchical phrase structure: The words do not simply occur one after the other. Instead, they come in groupings, such as 'on Monday', 'the men', and 'deliver the piano'. Furthermore, these groupings can be altered, either by moving them from one spot to another in the structure or by modifying them to express different meanings (e.g. by changing the statement into a question). These two ideas – words come in phrase structure groupings, and the groupings can be modified or transformed – correspond to the two major syntactic rule systems in Chomsky's theory.

Example illustrating the phrase structure grammar

Let's start with the phrase structure grammar that generates the overall structure of sentences.

1. S → NP + VP
2. NP → D + N
3. VP → V + NP
4. N → superhero, criminal,... etc.
5. V → caught,... etc.
6. D → the, a,... etc.

Rewrite rules of the grammar

To illustrate this point, consider the following sentence:

The superhero caught a criminal.

In a phrase structure grammar, the entire sentence is symbolised by an *S*. In this grammar, the sentence *S* can be broken down into two major components, a noun phrase (*NP*) and a verb phrase (*VP*). Thus the first line of the grammar illustrated in the above figure shows *S* → *NP* + *VP*, to be read, 'The sentence can be rewritten as a noun phrase plus a verb phrase.' In the second rule, the *NP* can be rewritten as a determiner (*D*), an article such as *the* or *a*, plus a noun (*N*): *NP* → *D* + *N*. In other words, a noun phrase can be rewritten as a determiner and a noun. In rule 3 we see the structure of a verb phrase; a *VP* is rewritten as a verb (*V*) plus an *NP*: VP → *V* + *NP*.

S → NP + VP	(by rule 1)
S → D + N + VP	(by rule 2)
S → D + N + V + NP	(by rule 3)
S → D + N + V + D + N	(by rule 2)
S → the + N + V + D + N	(by rule 6)
S → the + superhero + V + D + N	(by rule 4)
S → the + superhero + caught + D + N	(by rule 5)
S → the + superhero + caught + a + N	(by rule 6)
S → the + superhero + caught + a + criminal	(by rule 4)

Sentence generation by the rules

As the above figure shows, six *rewrite rules* are necessary for generating the sentence.

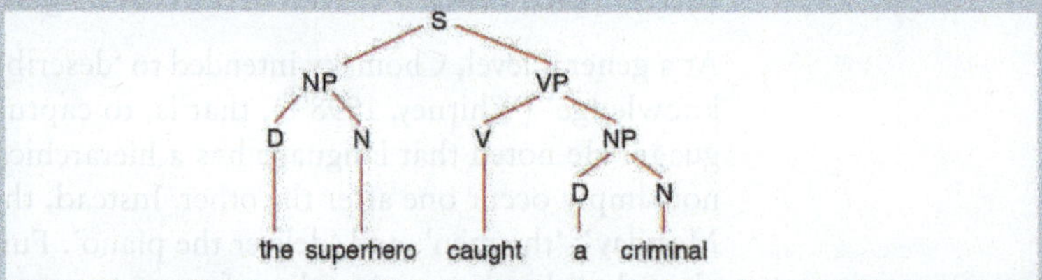

Key: D = Determiner, N = noun, NP = noun phrase, S = Sentence, V = verb, VP = verb phrase

A tree diagram or hierarchical representation

A different but equivalent depiction of the grammar is shown in above figure, in which a tree diagram shows the most general components at the top and the specific words at the bottom. An advantage of the tree diagram is that it reveals the hierarchical structure of the sentence as well as the internal structure and interrelations.

{(The superhero) (caught (a criminal))}

A 'bracket equivalent' diagram of the sentence

Finally, a bracket equivalent is shown in the above figure.

Source: From Lachman, Lachman & Butterfield (1979).[61]

Phrase structure grammar

An important point in Chomsky's system is that the **phrase structure** grammar accounts for the constituents of the sentence, the word groupings and phrases that make up the whole utterance, and the relationships among those constituents.

The inadequacy of phrase structure grammar alone

Chomsky's theory relied heavily on a phrase structure approach because it captures an important aspect of language – its productivity. This kind of grammar is generative; by means of such phrase structure rules, an entire family of sentences can be generated. Furthermore, the phrase structure grammar is joined with two other components, the *lexical entries* (the words of a sentence) and the *lexical insertion rules* (the rules for putting the words into their slots). These components generated the first representation of the sentence, the **deep structure** representation. In Chomsky's view, the deep structure is an abstract syntactic representation of the sentence being constructed with only bare-bones lexical entries (words).

The deep structure is critical for two reasons:

1. It is the representation passed to the transformational fix-it rules to yield the surface structure of the sentence.
2. The deep structure is also submitted to a semantic component that 'computes' meaning. This takes the deep structure and produces a semantic representation that reflects the underlying meaning.

Because of the separate treatment of the semantic component, a sentence's true meaning might not be reflected accurately in the **surface structure**. A surface structure might be **ambiguous**, or have more than one meaning. For instance, consider two classic examples of ambiguous sentences:

(5) Visiting relatives can be a nuisance.

(6) The shooting of the hunters was terrible.

A moment's reflection reveals the ambiguities. These alternative meanings are revealed when we **parse** the sentences, when we divide the sentences into phrases and groupings, much the way the phrase structure grammar does. The two meanings of sentence 5 – the two deep structures – correspond to two different phrase structures. For sentence 5, the ambiguity boils down to the grammatical function of *visiting*, whether it is used as an adjective or a verb. These two grammatical functions translate into two different phrase structures (*verb* + *noun* versus *adjective* + *noun*).

Applying the transformational rules

By applying different transformations, we can form an active declarative sentence, a passive voice sentence, a question, a negative, a future or past tense, and so on. With still other transformations, phrases can exchange places, and words can be inserted and deleted.

Consider the following sentences:

(7a) Pierre bought a fine French wine.

(7b) A fine French wine was bought by Pierre.

In Chomsky's view, sentences 7a and 7b differ only in their surface structures. One deep structure (the core meaning) is transformed in two different fashions. Likewise, for a simple deep structure idea such as {(boy kisses girl)}, the transformational grammar could generate any of the following, depending on which particular grammatical transformations were selected:

(8a) The boy kissed the girl.

(8b) The girl was kissed by the boy.

(8c) Was the girl kissed by the boy?

More elaborate rules are also applied by this transformational component, including rules that allow us to combine ideas, such as the idea that {(boy kisses girl)} and the idea that {(girl is pretty)}:

(9a) The boy kissed the pretty girl.

(9b) The boy kissed the girl who was pretty.

(9c) The girl whom the boy kissed was pretty.

(9d) Will the girl who is pretty be kissed by the boy?

Thus one surface structure for the {(girl) (is) (pretty)} idea is merely 'the pretty girl'; an equivalent structure, in terms of meaning, is 'the girl who is pretty'. On the other hand, sentences 9c and 9d are the most difficult to comprehend, largely because of the passive voice and the embedded relative 'who' clauses.

Sentence 6, however, has only one phrase structure; there is only one way to parse it: {[the shooting of the hunters] [was terrible]}. Thus sentence 6 is ambiguous at the level of surface structure. Because phrase structure rules can generate such ambiguous sentences, Chomsky felt this illustrated a limitation of the pure phrase structure approach: there must be something missing in the grammar. If it were complete, it would not generate ambiguous sentences.

A second difficulty Chomsky pointed out involves examples such as the following:

(7a) Pierre bought a fine French wine.

(7b) A fine French wine was bought by Pierre.

According to phrase structure rules, there is almost no structural similarity between these two sentences. Yet they mean nearly the same thing. The phrase structure approach does not capture people's intuitions in which active and passive paraphrases are identical at the level of meaning.

Transformational rules

Chomsky's solution to such problems was to postulate a second component to the grammar, a set of **transformational rules** that handle the many specific surface forms that can express an underlying idea. These transformational rules convert the deep structure into a surface structure, a sentence ready to be spoken.

8.3.2 Limitations of transformational grammar

A great deal of early psycholinguistic research was devoted to structural aspects of language. For example, many studies focused on testing the *derivational complexity hypothesis*. This hypothesis suggests that the difficulty of comprehending a sentence is directly related to the number of grammatical transformations applied. So, if a deep structure has two transformations applied to it, it is more difficult to comprehend than if only one transformation is applied. Some results tended to support this theory (e.g. Palermo, 1978[62]). However, on the whole, psychology became dissatisfied with this approach. Work by Fodor and Garrett (1966)[63] was especially instrumental in dimming the enthusiasm. They noted that much of the support for the derivational complexity hypothesis failed to control potentially important factors. For instance, a derivationally more complex sentence generally has more words in it than a simpler one (contrast sentences 8a and 8c).

Moreover, there was a *metatheoretical point of view*. To oversimplify a bit, the major components were said to be the syntactic rules for generating first a deep then a surface structure. Meaning was literally off to the side. This depicts the difficulty psychology had with linguistic theory: it seemed that meaning was secondary to syntax. It is almost as if the theory, as it was applied to language use, suggested that we first make up our minds about which phrase constituents we are going to use and only then decide what we are going to talk about. To psychologists concerned with how we use language to express meaning, this theory seemed wrong.

This oversimplification made it seem as if Chomsky encouraged linguists to avoid meaning. It was not that extreme, as Chomsky repeatedly emphasised the importance of both syntax and semantics. He pointed out that even a perfectly grammatical, syntactically acceptable sentence may be semantically anomalous. His most famous example is 'Colourless green ideas sleep furiously.' The sentence is grammatically acceptable – consider a sentence with completely parallel syntax, such as 'Tired young children sleep soundly.' But Chomsky's sentence has no meaning in any ordinary sense.

Still, psychologists felt Chomsky's work never dealt with meaning satisfactorily. Furthermore, trying to apply his theory to the actual use of language – turning his competence-based theory into a performance theory of language production and comprehension – only made it more apparent that a different approach was needed.

8.3.3 The cognitive role of syntax

From a psychological perspective, what is the purpose of syntax? Why follow syntactic rules? Essentially, we use syntax to determine or find meaning. If an infinite number of sentences are possible, then the one sentence being said to us right now could be about *anything*. Syntax helps listeners extract meaning and helps speakers convey it.

Bock's (1982)[64] article on a cognitive psychology of syntax discusses several important issues that psycholinguistics must explain. She notes that the syntactic burden falls more heavily on the speaker than the listener. When you have to produce a sentence rather than comprehend it, you must create a surface structure, a string of words and phrases to communicate your idea as well as possible. Thus, syntax is a feature of language related to the speaker's mental effort.

Automatic processing

Two points Bock raises should illustrate some issues in the psycholinguistic study of syntax. First, consider the issues of automatic and conscious processes as they apply to language production. *Automatic processes* are the product of a high degree of practice or overlearning. Bock noted that several aspects of syntactic structure are consistent with the idea of automaticity. For instance, children rely heavily on regular word orders even if the native language they are learning has irregular word order. By relying repeatedly on the same syntactic frames, children can generate and use them more automatically. Similarly, adults tend to use only a few syntactic structures with regularity, suggesting that those structures can be called into service rapidly and automatically.

Interestingly, the syntax you use can be influenced by a previous sentence, quite literally *syntactic priming* (Bock, 1986; West & Stanovich, 1986[65]). Bock's later work (Bock & Griffin, 2000)[66] found evidence that a particular syntactic construction can prime later ones up to lag 10 (i.e. with 10 intervening sentences), and there is some evidence that it can last up to a week later (Kaschak, Kutta & Schatschneider, 2011[67]). Syntactic priming has even been found in written language and American Sign Language (Branigan, Pickering & Cleland, 1999; Branigan, Pickering, Stewart & McLean, 2000; Hall, Ferreira & Mayberry, 2015[68]). Interestingly, syntactic priming is not affected by anterograde amnesia, which affects declarative memory (Ferreira, Bock, Wilson & Cohen, 2008[69]).

Planning and production of speech

Language does not emerge from thought fully formed and ready to go. There are a series of steps that go into the transformation from idea to spoken utterance. Ferreira and Swets (2002)[70] demonstrated this tendency for language order to be influenced by memory retrieval in a clever experiment by asking people to state the answer to easy and hard addition problems, in sentence frames like 'The answer is __'. They found that people delayed nearly a half a second more before they started talking when the problem was hard (e.g. 23 + 68) than when it was easy (e.g. 21 + 22). Clearly speech production is sensitive to the ease of memory retrieval.

Our planning and execution of speech are also sensitive to grammatical complexity and presumably to the possibility that a listener (or a speaker) might lose track of information if too much time passes. As an example, Stallings, MacDonald, and O'Seaghdha (1998)[71] showed a particular kind of syntactic adjustment used for complex noun phrases, 'heavy NPs' in their words ('heavy' because they are long). Specifically, we tend to shift heavy NPs to the end of a sentence and insert other material in between the subject and NP, but not when the noun phrase is short. Consider the simple sentence 'The boy found the textbook in his car.' The noun phrase ('the textbook') is short, so it does not need to be shifted. But, if there is more to say about the textbook, you might say, 'The boy found in his car the textbook that had been lost for so long', shifting the textbook phrase to the end and putting 'in his car' in the middle. But you probably would not shift the short noun phrase to the end, as in 'The boy found in his car the textbook.' Moving 'the textbook' to the end is not needed here because the listener's working memory is not being overtaxed. But the heavy NP 'the textbook that had been lost for so long' is sufficiently long that working memory might lose essential information – the connections between the boy, the car and finding the book – if the phrase separated those ideas by too many intervening ideas. More generally, as syntactic complexity increases, this increasingly taxes working memory (e.g. Fedorenko, Gibson & Rohde, 2006[72]).

Table 8.3 Fromkin's (1971)[73] model for the planning and production of speech

Stage	Process
1	Identify meaning; generate the meaning to be expressed.
2	Select syntactic structure; construct a syntactic outline of the sentence, specifying word slots.
3	Generate intonation contour; assign stress values to different words slots.
4	Insert content words; retrieve appropriate nouns, verbs, adjectives, and so on from the lexicon and insert into word slots.
5	Add function words and affixes; fill out the syntax with function words (articles, prepositions, etc.), prefixes, suffixes.
6	Specify phonetic segments; express the sentence in terms of phonetic segments according to (pronunciation) rules.

These effects tell us something interesting about the cognitive mechanisms that create sentences. Earlier theories of sentence planning, such as Fromkin's (1971; see above table[74]), described planning as sequential: first you identify the meaning to be conveyed, then you select the syntactic frame, and so on. However, more recent research shows how interactive and flexible the planning process is (Ferreira, 1996;

Griffin & Bock, 2000[75]). Difficulties in one component, e.g. word retrieval, can prompt a return to an earlier planning component, say to rework the syntax (Ferreira & Firato, 2002[76]), or it can prompt you to delay the sentence. By selecting an alternative syntax, the speaker buys more time for retrieving the intended word (see also Kempen & Hoehkamp, 1987[77]). Such a highly interactive system runs counter to strictly hierarchical or sequential approaches, such as Chomsky's.

In general, we begin our utterances when the first part of the sentence has been planned, but before the syntax and semantics of the final portion have been worked out (see Bachoud-Levi, Dupoux, Cohen & Mehler, 1998, and Griffin, 2003, for comparable effects[78]). The time it takes to begin speaking (e.g. Bock, Irwin, Davidson & Levelt, 2003[79]) and hesitations in our spoken speech are clues to the nature of planning and memory retrieval, as are the effects of momentary changes in priming, lexical access and working memory load (Bock & Miller, 1991; Lindsley, 1975[80]). In fact, several reports detail how the false starts, hesitations and restarts in speaking often reflect both the complexity of the intended sentence and a genuine online planning process that unfolds as the sentence is developed (Clark & Wasow, 1998; Ferreira, 1996; Ferreira & Dell, 2000; see Bock, 1996, for a review of methods of studying language production[81]).

More recent work has taken Chomsky to task even further. In a connectionist model of language processing, Chang, Dell, and Bock (2006)[82] have challenged Chomsky's idea that language has a strong genetic component, as compared to the strong learning stance taken by the behaviourists. Chomsky suggested that although you may need to learn your own language, all humans have a strong genetic bias to learn some language. He theorised that aspects of transformational grammar were somehow part of that genetic process. By contrast, the Chang et al. (2006) model assumes that language processing has a strongly learned component, similar to other memory processes. Part of how a language is learned in their model is by comparing its predictions for what will be said to what is actually said, and then adjusting the connection weights based on any discrepancy (see also Griffiths, Steyvers & Tenebaum, 2007, for a predictive model of word meaning in the context of sentences[83]).

Another intteresting aspect of the Chang et al. (2006)[84] connectionist model is the idea that language may operate in a parallel fashion to vision. In other words, there are two routes in visual processing: one for processing what something is and one for processing where it is (Ungerleider & Haxby, 1994[85]), with the what system taking the ventral visual pathway towards the temporal lobe, and the where system taking the dorsal pathway towards the parietal lobe. An idea in the Chang et al. model is that there is a network for processing the meaning aspect of language and a separate system for the sequencing of the words. Together, these two systems converge to predict the type of word that will come next, allowing the system to learn and adjust to new input, such as new words or new ways of using words (e.g. I googled you the other day and was surprised by how many hits there were).

Planning

In Bock's second point, she reviewed evidence of an important interaction between syntax and meaning. In general, we tailor the syntax of our sentences to the accessibility of the lexical or semantic information being conveyed, known as the **given-new strategy** (Clark & Clark, 1977[86]). Phrases that contain more accessible information, or given information, tend to occur earlier in sentences. This is information that is either well known or has been recently discussed in a discourse (and so is more available). By comparison, less accessible, newer concepts tend to come later, possibly to provide extra time for retrieval (but see Clifton & Frazier, 2004, for an alternative account[87]).

8.3.4 Prosody

Syntax is an important aspect of understanding the structure of an utterance, in terms of the role that different words play and how they relate to an underlying meaning. However, word order is not the only way to convey information about meaning and intent. Another major clue, particularly for spoken language, is **prosody**, which is the change in pitch (either higher or lower) across the phonemes and morphemes of an utterance to convey different meanings.

Importance of prosody

Language is typically not delivered in a monotone, but involves a moving up and down in pitch, somewhat like a melody. In fact, prosody in language may use some of the same cognitive processes as music (Perrachione, Fedorenko, Vinke, Gibson & Dilley, 2013[88]).

Conveys meaning in language

Prosody conveys meaning in language, although a different kind of meaning than the semantic meanings of the words themselves. For example, the sentences 'Those are my shoes' and 'Those are my shoes?' contain exactly the same words but are spoken with different prosodies. One conveys a statement and the other a question, often with a rise in pitch at the end of the sentence. If you listen to yourself say these two sentences, you will notice a difference in their pitch pattern, even though they contain the same words and syntactic structure. These differences in pitch information are prosody and convey different information about the intent of the speaker.

Serves to help speakers place emphasis on concept

Prosody also serves to help speakers place emphasis on one concept or another and helps guide the attention and comprehension of listeners (e.g. Kurumada, Brown, Bibyk, Pontillo & Tanenhaus, 2014[89]). For example, consider the sentence 'It looks like a zebra'. People are likely to interpret that sentence as referring to something that is actually a zebra, but may be hard to see. In comparison, in the sentence 'It LOOKS like a zebra', in which the word looks is prosodically emphasised, people are likely to interpret the sentence as referring to something that looks something like a zebra, but actually is not (e.g. it could be an okapi).

8.4 Lexical factors

LEARNING OBJECTIVE
Analyse lexical factors to understand the meaning of language

We now turn to lexical and semantic factors, which relate to the level of meaning in language. In particular, we refer to retrieval from the *mental lexicon*, the mental dictionary of words and their meanings. After rapid perceptual and pattern recognition processes, the encoded word provides access to the word's entry in the lexicon and to the semantic representation of the concept. The evidence you have read about throughout this course, such as results from the Stroop and the lexical decision tasks, attests to the close relationship between a word and its meaning and the seemingly automatic accessing of one from the other. Recall in the Stroop task that seeing the word *red* printed in green ink triggers an interference process with naming the ink colour, clear evidence that *red* is processed to the level of meaning (MacLeod, 1992[90]). Likewise, the lexical decision task does not require that you access the word's meaning but only that you identify a letter string as a genuine word. Nonetheless, identifying *doctor* as a word primes your decision to *nurse*.

8.4.1 Morphemes

A *morpheme* is the smallest unit of language that has meaning. To return to an earlier example, the word *cars* is composed of two morphemes: *car* refers to a concept and a physical object, and *-s* is a meaningful suffix, denoting 'more than one of'. Likewise, the word *unhappiness* is composed of three morphemes: *happy* as the base concept, the prefix *un* meaning 'not', and the suffix *-ness* meaning 'state or quality of being'. In general, morphemes that can stand on their own and serve as words are called *free morphemes*, such as *happy*, *car* and *legal*, whereas morphemes that need to be linked to a free morpheme are called *bound morphemes*, such as *un-*, *-ness* and *-s*. Although the concept of a morpheme is important, there is some debate as to whether the meaning of more common words such as *unhappiness* may be stored directly in memory or 'computed' from the three morphemes (see Carroll, 1986[91]; Whitney, 1998[92]).

8.4.2 Lexical representation

Think about the word *chase* as an example of how free morphemes might be represented in the mental lexicon. The representation of *chase* must specify its meaning – indicate that it means 'to run after or pursue, in hopes of catching'. Like other semantic concepts, *chase* can be represented in reference to related information, like *run, pursue*, the idea of *speed*. Given this, along with what you know about events in the real world from schemas and scripts, you can easily understand a sentence like:

(11) The policeman chased the burglar through the park.

From a more psycholinguistic perspective, however, you know more about *chase* than just its basic meaning. For one thing, you know it is a verb specifying a kind of physical action. Related to that, you have a clear idea of how *chase* can be used in sentences, the kinds of things that can do the chasing, and the kinds of things that can be chased (e.g. McKoon & Macfarland, 2002[93]). Imagine, then, that your lexical representation of *chase* also includes this knowledge; *chase* requires some animate thing to do the chasing, some other kind of thing to be chased, and a location where the chasing takes place.

Embodied aspect to lexical memory

Lexical knowledge can include information that can capture embodied characteristics of cognition, with certain parts of the brain becoming more active for certain types of words. When listening to verbs, various parts of the motor cortex may be activated, depending on the type of action being described (e.g. Andres, Finocchiaro, Buiatti & Piazza, 2015[94]). For example, the hand part of the motor cortex may be more active for a verb such as *write*, but the foot part of the motor cortex is more active for a verb such as *kick*.

As an example of this sort of embodied aspect to lexical memory, Willems, Hagoort, and Casasanto (2010)[95] had people perform a lexical decision task while in a fMRI scanner. The results showed that when people responded to action verbs, the appropriate motor areas of the cortex became more active. Moreover, for right-handers, this was more on the left hemisphere motor areas (which control the right side of the body), whereas for the left-handers, the opposite was true. So, when a word strongly implied using a specific part of the body, the lexical information associated with that word involved information about how to move that part of the body.

In another study, Willems, Labruna, D'Espisito, Ivry, and Casasanto (2011)[96] again had people do a lexical decision task, but this time in conjunction with transcranialmagnetic stimulation treatment in which the hand portion of the brain was stimulated. They found that when people were given verbs that involved using the hand, such as *throw* or *write*, they responded faster under stimulation, compared to being given verbs that did not involve using the hand, such as *earn* or *wander*.

Thus, activating the part of the brain that controls that part of the body made it easier to identify and respond to words that also strongly involve that part of the body. This suggests that our mental lexicon contains motor information about how we do things.

8.4.3 Polysemy

Whereas our understanding of words like *chase* is very clear, it is not too long before we run into cases of **polysemy**, the fact that many words in a language may have multiple meanings. The task of the language processing system is to figure out which meaning is the intended one. Though a word may be polysemous, not all meanings are equal. Generally, there is one primary meaning that people typically would think of first when they hear the word or that would likely be listed first in a dictionary. This is the *dominant* meaning of a word. Other meanings then would be the *subordinate* meanings. So, take a simple word like *run*. The dominant meaning has something to do with using our legs to move fast. However, there are many subordinate meanings, too, such as having a run in our stockings, a film having a run in the cinema, having our nose run, to cut and run (retreat), to run our engine, to watch paint colours run, and so on. The way we distinguish which specific meaning to use from the mental lexicon would depend on the context a word is in.

Polysemy and priming

Let's consider an example of how context can resolve polysemy to determine the intended meaning. As one example, the word *count* is ambiguous by itself. Putting the word in a sentence may not help: 'We had trouble keeping track of the count.' You still cannot tell the intended meaning. What is missing is context, some conceptual framework to guide the interpretation of the polysemous word. With an adequate context, you can determine which sense of the word *count* is intended in these two sentences:

My dog wasn't included in the final count.

The vampire was disguised as a handsome count.

These sentences, taken from Simpson's (1981, 1984)[97] work on polysemy, point out the importance of context: context can help determine the intended meaning. With neutral contexts such as the 'We had trouble' sentence, word meanings are activated as a function of their dominance. The number sense of *count* is dominant, so that meaning is more activated. But a context that biases the interpretation in any way results in a stronger activation for the biased meaning. With *vampire*, you activated the meaning of *count* related to nobility and Count Dracula (see also Balota & Paul, 1996; Klein & Murphy, 2002; Piercey & Joordens, 2000; but cf. Binder, 2003; Binder & Rayner, 1998[98]).

The resolution of lexical ambiguity with polysemous words is important for successful comprehension. If you do not get the intended meaning of a word, then you will not get the intended message. It appears that ambiguity resolution works, in part, in a two-stage process. When people encounter an ambiguous word, they activate all the meanings, at least to some degree. Then in the second stage, they deactivate the inappropriate ones based on the information from the rest of the discourse context. However, not everyone does this equally well. Work by Gernsbacher and Faust (1991)[99] shows that good readers suppress inappropriate meanings faster. By comparison, poor readers maintain multiple meanings for a much longer period, which may contribute to their problems.

Context and ERPs

Let's consider another example of the effects of context, an offshoot of the Kounios and Holcomb work with event-related potentials (ERPs)[100]. In one study, Holcomb, Kounios, Anderson, and West (1999; see also Laszlo & Federmeier, 2009; Lee & Federmeier, 2009; Sereno, Brewer & O'Donnell, 2003[101]) recorded ERPs in a simple sentence comprehension task. People saw sentences one word at a time and were asked to respond after seeing the last word, with 'yes' if the sentence made sense and 'no' if it did not. The experimental sentences varied along two dimensions, whether the last word was concrete or abstract and whether it was congruent with the sentence meaning or anomalous (i.e. made no sense). As an example, 'Armed robbery implies that the thief used a weapon' was a concrete–congruent sentence. Substituting *rose* for *weapon* made the sentence concrete but anomalous. Likewise, 'Lisa argued that this had not been the case in one single instance' was an abstract–congruent sentence, and substituting *fun* for *instance* made it abstract–anomalous.

Figure 8.6 shows some of the ERP patterns obtained. In the left panel, you see the 'normal' ERP patterns for the congruent, sensible sentences.

The three profiles, from top to bottom, came from the three midline electrode sites shown in the schematic drawing (frontal, central and parietal). In the right panel are the ERP patterns when the sentences ended in an anomalous word. Notice first in the left panel that the solid and dotted functions, for concrete and abstract sentences, tracked each other very closely: whatever neural mechanisms operated during comprehension, they generated similar ERP patterns. But now make a left-to-right comparison of the patterns, seeing the differences in the right panel when the sentences ended in a nonsensical, anomalous word (*rose* in the armed robbery sentence, for example). Here, there were marked changes in the ERP profiles. For example, at the central location, there was a steady downward trend (in the positive direction, in terms of electrical potentials) for sensible sentences but a dramatic reversal of direction for anomalous words.

In short, the neural mechanisms involved in comprehension generated dramatically different patterns when an anomalous word was encountered. The mismatch between the context, the already-processed meaning of the sentence, and the final word yielded not only an overt response (the response indicating 'no, that sentence makes no sense'), but also a neural response signifying the brain-related activity that detected the anomalous ending of the sentence. (Don't get confused about directions here. The functions underneath the gridline are electrically positive, so deflection upward in these graphs is a deflection toward the negative, a deflection going in a negative direction; this is what the *N* in **N400** signifies, a 'negative going' pattern.)

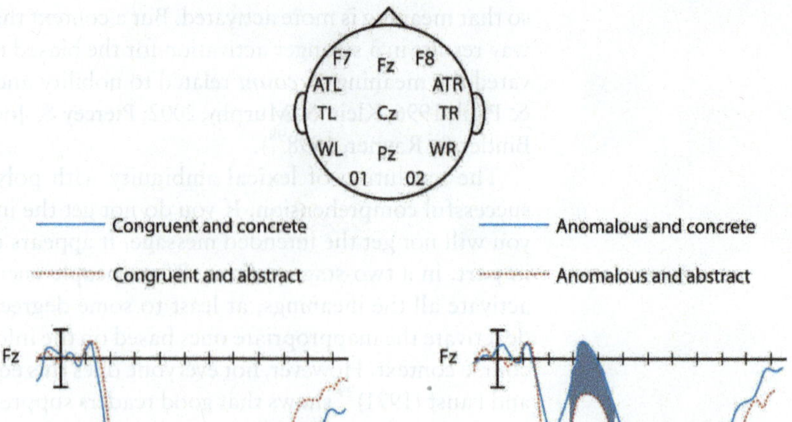

Figure 8.6 ERPs in simple comprehension task

ERP profiles at three midline sites [frontal (Fz), central (Cz), parietal (Pz)] for sentences ending with a congruent (left panel) or contextually anomalous (right panel) word.

Even at the level of neural functioning, there is a rapid response to nonsensical ideas that follow sensible context, a kind of 'something's wrong here' response that the brain makes some 400 ms after the nonsensical event.

8.5 Semantics

LEARNING OBJECTIVE
Compare semantics and syntax to understanding intended meaning

In this section, we consider issues of how the language processing system knows what role a word or concept is playing in a sentence. This approach is called **case grammar**. The ideas originally came from Fillmore (1968)[102]. The basic idea is that the semantic analysis of sentences involves figuring out what semantic role is being played by each word or concept in the sentence and computing sentence meaning based on those semantic roles. Two sample sentences illustrate this:

(12) The key will open the door.

(13) The janitor will open the door with the key.

Fillmore pointed out that syntactic aspects of sentences – which words serve as the subject, direct object, and so on – are irrelevant to sentence meaning. For example, in sentences 12 and 13 the word *key* plays different grammatical roles: subject of the sentence in 12 but object of the preposition in 13. For Fillmore, focusing on this difference misses a critical point for language. Regardless of its different grammatical roles, the key is doing exactly the same thing in both cases, playing the same semantic role of *instrument*. A purely syntactic analysis misses this, but a semantic analysis captures it perfectly.

8.5.1 Case grammar

Fillmore's theory of case grammar proposes that sentence processing involves a semantic parsing that focuses on the semantic roles played by the content words in the sentences. These semantic roles are called *semantic cases*, or simply case role. Thus, *door* is the recipient or patient of the action of *open* in sentences 12 and 13; *janitor* is the agent of *open*; *key* is the instrument; and so on. Stated simply, each content word plays a semantic role in the meaning of the sentence. That role is the word's semantic case.

Analysing case grammar

The significant – indeed, critical – point about such a semantic parsing is that it relies on people's existing semantic and lexical knowledge, their knowledge of what kinds of things will open, who can perform the opening, and so on. Reconsider the *chase* sentence 11, 'The policeman chased the burglar through the park', and three variations, thinking of the content words in terms of their semantic roles:

(14) The mouse chased the cat through the house.

(15) His insecurities chased him even in his sleep.

(16) The book chased the flower.

Your lexical and semantic knowledge of *chase* is that some animate being does the chasing, the agent case. Some other thing is the recipient of the chasing, the patient, but that thing need not be animate, just capable of moving rapidly (e.g. you can chase a piece of paper being blown by the wind). On this analysis, it is clear that sentence 11 conforms to the normal situation stored in memory, so it is easy to comprehend.

However, sentence 14 mismatches the typical state of affairs between mice and cats. Nonetheless, either of these creatures can serve as the required animate agent of the relation *chase*, so sentence 14 is sensible. Because of other semantic knowledge, you know that sentence 15 violates the literal meaning of *chase* but could still have a non-literal, metaphorical meaning. But your semantic case analysis provides the reason sentence 16 is unacceptable. A book is inanimate, so it mismatches the required animate agent role for chase; *book* cannot play the role of agent for *chase*. Likewise, *flower* seems to violate the movable restriction on the patient case for *chase*.

Work by Bresnan (1978; Bresnan & Kaplan, 1982[104]) and Jackendoff (1992)[105] has amplified and extended work on case grammars. For example, Jackendoff's theory of a cognitive grammar (1992; see above figure[106]) aims to build a *conceptual structure*, an understanding of the sentence. We use language and language rules to get from the spoken or written sentence to a meaningful mental structure or understanding.

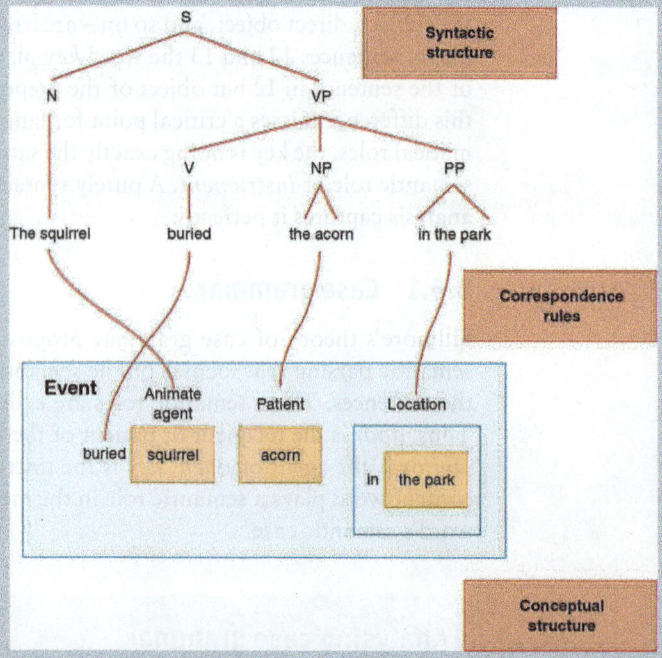

Figure 8.7 Jackendoff's (1992)[103] conceptual semantics approach

Each lexical entry includes the meaning of the word and, for a verb, a list of the arguments or semantic cases that go along with it. Thus, the lexical entry *chase* would state that chase requires an animate agent, some recipient or patient, and so on.

Likewise, for *give*, the case arguments would state that an animate agent and recipient are needed for the *give* relation, and some object is the thing being given (for an excellent summary of these positions, see Whitney, 1998). Accordingly, when we perceive words, we look up the concepts in the lexicon. This look-up process accesses not only the word's meaning but also its syntactic and semantic case roles and any other restrictions. Each word in the sentence is processed as it is encountered, with content words being assigned to their semantic roles. If all goes well, the sentence conveys an exact, specified meaning that is captured accurately by the analysis of the cognitive grammar.

8.5.2 Interaction of syntax and semantics

Note that semantic factors do not stand alone in language, just as syntactic factors are not independent of semantics. Syntax is more than just word and phrase order rules; it is a clue to understanding sentences. For example, O'Seaghdha's (1997)[107] evidence shows separable effects of syntactic assignment and semantic integration of word meanings, with syntactic processes occurring before semantic integration. His results, based on response times (RTs), are largely consistent with those in other studies (e.g. Peterson, Burgess, Dell & Eberhard, 2001,[108] on how we process idioms), including ERP studies of syntactic and semantic processing (Ainsworth-Darnell, Shulman & Boland, 1998; Friederici, Hahne & Mecklinger, 1996; Osterhout, Allen, McLaughlin & Inoue, 2002[109]). And syntax in speech production is sensitive to a word's accessibility. Words that can be easily retrieved right now tend to appear earlier in a sentence.

Semantic focus

Likewise, semantic factors refer to more than just word and phrase meanings because different syntactic devices can be clues to meaning. To anticipate just a bit, note how syntactic differences in the following sentences influence the semantic interpretation:

(17a) I'm going downtown with my sister at four o'clock.

(17b) It's at four o'clock that I'm going downtown with my sister.

(17c) It's my sister I'm going downtown with at four o'clock.

Sentences 17b and 17c differ subtly from 17a in the focus of the utterance. The focus of each sentence is different, so each means something slightly different. Imagine how inappropriate sentence 17c would be, for instance, as a response to the question 'Did you say you're going downtown with your sister at three o'clock?' Our judgements about appropriateness make an important point: our theories of language performance must be as sophisticated as our own knowledge of language. We are sensitive to the focus or highlighted aspects of sentences and subtleties of the ordering of clauses, so a theory of language must reflect this in a psychologically relevant way.

Semantics can overpower syntax

Semantic features can do more than alter the syntax of sentences. Occasionally semantics can overpower syntax.

Conceptually driven processing of language

Let's focus on a classic study by Fillenbaum (1974).[110] Fillenbaum presented several kinds of sentences and asked people to write paraphrases that preserved the original meaning. Ordinary *threat* sentences such as 'Don't print that or I'll sue you' were then reordered into *perverse* threats, such as 'Don't print that or I won't sue you.' Regular *conjunctive* sentences such as 'John got off the bus and went into the store' were then changed into *disordered* sentences, such as:

John went into the store and got off the bus

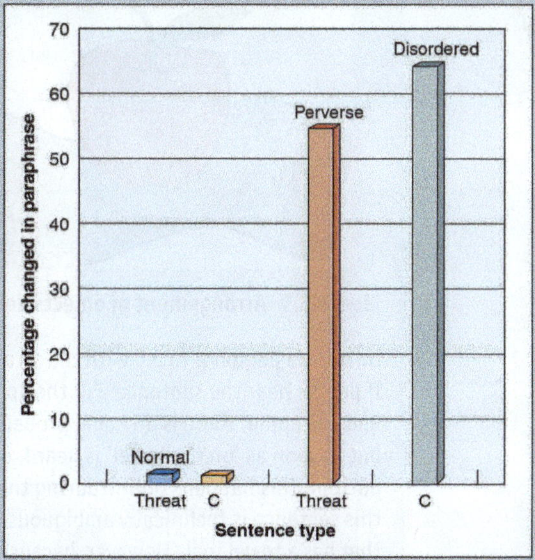

Figure 8.8 Fillenbaum's (1974)[111] results

Two kinds of normal sentences were shown, threats and conjunctives (labelled C) such as 'John got off the bus and went into the store.' Threats were then altered to be perverse, and conjunctives were disordered (e.g. threat C, 'John dressed and had a bath').

When Fillenbaum scored paraphrases of reordered sentences, he found remarkably high percentages of changes, as shown in Figure 8.8. More than 50 per cent of people 'normalised' the perverse threatening sentences, making them conform to the more typical state of affairs, and more than 60 per cent normalised the disordered conjunctive sentences.

Fillenbaum then asked people to reread their paraphrases to see whether there was even a 'shred of difference' from the originals. More than half the time, people saw no discrepancies. Apparently, their general knowledge was influential enough that it overpowered the syntactic and lexical aspects of the sentences. (Try these examples: 'Nice we're having weather, isn't it?' and 'Ignorance is no excuse for the law.')

Sometimes the information that guides language processing can also come from the context and the environment as well as what a person already knows. Essentially, according to such *interactionist* views of language processing, people are simultaneously, in parallel, deriving alternative interpretations of language they hear or read.

Over time (a very short period), cognition uses the constraints derived from the preceding context and the environment to help determine which interpretation is preferred and which not to use (Altmann, 1998; Altmann & Steedman, 1988[112]). This is most clearly illustrated in studies of eye movements as people listen to spoken instructions.

Figure 8.9 **Arrangement of objects depicting two apples and a box**

Imagine a person is faced with the arrangement of objects as shown in Figure 8.9. If people hear the sentence 'Put the apple on the towel in the box', after hearing 'put the apple' there is an equal probability that a person will look at either apple, but as soon as 'on the towel' is heard, eye fixations converge on the apple at the bottom. This happens online during the comprehension of the sentence even though this sentence is technically ambiguous because the sentence could refer to a box that has a towel in it. However, because there is no such box in the environment, people do not experience any ambiguity and are unlikely to notice it. In short, the psycholinguistic approach to lexical and semantic factors in language relies on conceptually driven processing.

8.5.3 Evidence for the semantic grammar approaches

A major prediction of *semantic grammar theory* can be stated in two parts. First, comprehenders begin to analyse a sentence immediately, as soon as the words are encountered. Second, this analysis assigns each word to a particular semantic case role, with each assignment contributing its part to overall sentence comprehension.

Predictions of semantic grammar theory

Language is processed online, as it happens. This can sometimes lead people to initially make mistakes in parsing language, which cognition then needs to correct. As an example, read sentence 18:

(18) After the musician had bowed the piano was quickly taken off the stage.

Your analysis of this sentence proceeds easily and without disruption; it is a fairly straightforward sentence. Now read sentence 19:

(19) After the musician had played the piano was quickly taken off the stage.

What is different about sentence 19? The verb *played* suggests that the *piano* is the semantic recipient of *play*. When you read *played*, your semantic role assignment for piano was *recipient*. But then you read was quickly and realised you had made a mistake in interpretation. Sentences such as 19 are called **garden path sentences**. The early part of the sentence sets you up so that the later phrases in the sentence do not make sense given the way you assigned case roles in the first part. Figuratively speaking, the sentence leads you down the garden path. When you realise your mistake, you have to retrace your steps back up the path to reassign earlier words to different cases. Additional examples (from Singer, 1990)[113] of this effect are shown in sentences 20 and 21:

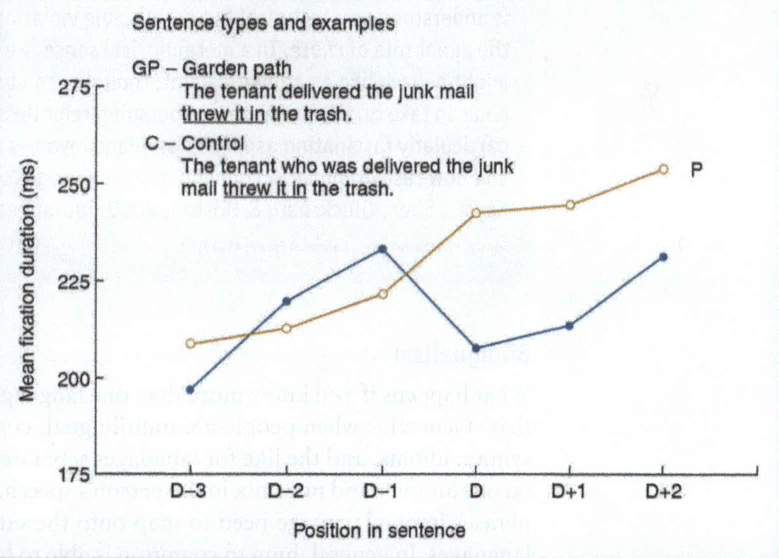

Figure 8.10 A depiction of the effect of garden path sentences on reading time
The curves show eye fixations on phrases before and after D, the point in the sentence where the ambiguity is noticed and disambiguated. The top curve shows the data from garden path sentences; eye fixation time grows noticeably longer for these curves at D, when the ambiguity is noticed (the D phrase is underlined in the sample sentences). The bottom curve shows data from the control sentences and no increase in reading time at point D.

(20) The groundsman chased the girl waving a stick in her hand.

(21) The old train the young.

Many research reports have studied how people comprehend garden path sentences as a way of evaluating case grammar theory (Frazier & Rayner, 1982[114]; Mitchrell & Holmes, 1985[115]; but see McKoon & Ratcliff, 2007, for an account based on semantic plausibility[116]). For the most part, the results have been supportive. For example, when people read such sentences, their eyes tend to fixate longer on the later phrases, signalling their error in comprehension (e.g. on 'was quickly taken off' in sentence 19).

As shown Figure 8.10, people spent 40 ms to 50 ms longer when they encountered their error (at point D in the figure; D stands for the disambiguating part of the sentence that reveals the earlier misinterpretation). This is a recovery time effect; it takes additional time to recover from the initial role assignment when that turns out to be incorrect (see Christianson, Hollingworth, Halliwell & Ferreira, 2001; and Ferreira, Henderson, Anes, Weeks & McFarlane, 1996, for comparable results with spoken language[117]). Interestingly, the more committed you are, the more you dig in, to an initial interpretation, the harder it becomes to change your interpretation (Tabor & Hutchins, 2004[118]).

In addition, there is important work, also using the eye fixations and fMRI, on how we parse – figure out – the syntax of a sentence and the degree to which parsing can be overridden or at least affected by semantic context and other factors (see Clifton et al., 2003; Mason, Just, Keller & Carpenter, 2003; Rayner & Clifton, 2002; Tanenhaus & Trueswell, 1995[119]).

A final point is that the case restrictions sometimes can be violated intentionally, although there are still constraints on that violation. For instance, consider sentence 15 again: 'His insecurities chased him even in his sleep.' Such a sentence is understood as a technical but permissible violation of the animate restriction for the agent role of *chase*. In a metaphorical sense, we can 'compute' how insecurities might behave like an animate agent; thoughts can behave as if they were animate and can take on the properties of pursuing relentlessly, catching, and so on. A particularly fascinating aspect of language involves such figurative uses of words and how case grammar accommodates such usage (see Glucksberg & Keysar, 1990; Keysar, Shen, Glucksberg & Horton, 2000; Tourangeau & Rips, 1991[120]).

Source: Data from Rayner, Carlson & Frazier (1983)[121].

Bilingualism

What happens if you know more than one language? How does cognition deal with that? Generally, when people are multilingual, cognition needs to keep the words, syntax, idioms, and the like for languages separate enough that they do not intrude on one another and intermix in the person's speech. At the same time, the words and phrases in one language need to map onto the same underlying ideas in the other languages. In general, human cognition is able to handle this task quite well. People who learn a second language early in life even show semantic priming effects across languages (Perea, Duñabeitia & Carreiras, 2008[122]), suggesting that the different words tap into the same underlying semantic knowledge base.

The need to keep two or more languages separate, but still be fluent in both, requires increased attention and cognitive control to mentally keep the languages from interfering with one another. This need for bilinguals to engage in such cognitive control spills over to other cognitive abilities that require cognitive control, such as showing larger negative priming effects, because bilinguals are more effective at suppressing salient but incorrect information (Treccani, Argyri, Sorace & Della Salla, 2009[123]). Thus, there seems to be an overall cognitive benefit to knowing more than one language. Bilinguals show superior performance on intelligence tests compared to otherwise similar monolinguals (Lambert, 1990[124]), including doing better on tests of their primary language (Bialystok, 1988; van Hell & Dijkstra, 2002[125]). Certain evidence exists that bilingualism can modestly protect people from cognitive declines associated with ageing (Bialystok, Craik & Luk, 2012[126]). Yet, the evidence for a cognitive and intellectual advantage for bilingualism is mixed, with some studies finding no evidence of a difference (Ljungberg, Hansson, Andrés, Josefsson & Nilsson, 2013; Paap & Greenberg, 2013; Ratiu & Azuma, 2015[127]). Regardless, it is clear that bilingualism does not impair cognition and intelligence as was once thought.

Although bilingualism can have advantages for cognition, learning a second language as an adult is more difficult and requires more cognitive attention and effort. For example, some evidence suggests that when people are immersed in a second language they learned as an adult, their first language is actively suppressed (Linek, Kroll & Sunderman, 2010[128]), which is not observed with people who grew up speaking more than one language.

8.6 Brain and language

LEARNING OBJECTIVE
Describe how the brain comprehends language

One of the most fruitful areas of research on the brain–cognition relation is work on language processing. So, in this section, we discuss aspects of language processing that have strong neural components. These include a consideration of people with intact brains, as well as the disruption of language processing in people who have suffered some sort of brain damage.

8.6.1 Language in the intact brain

With the advent of contemporary brain imaging methods, we have begun to learn an extraordinary amount about how the brain processes language from neurologically intact people. An intriguing observation from early studies on lateralisation (see Chapter 2) was that there is a left hemisphere dominance for language, which is facilitated by the use of the right hand in writing (using the right hand activates the left hemisphere motor centre). From an evolutionary perspective it makes sense to have language processing centres and motor centres for writing situated in the same hemisphere (left hemisphere) to allow faster and more efficient synapse communication.

A different line of research, using brain imaging investigates people's sensitivity to the syntactic structure of sentences. Osterhout and Holcomb (1992)[129] presented sentences to people and recorded the changes in their brain wave patterns (ERPs) as they comprehended. In particular, they examined ERP patterns for sentences that violated syntactic or semantic expectations, comparing these with the patterns obtained with control sentences. When sentences ended in a semantically anomalous fashion ('The woman buttered her bread with socks'), a significant N400 ERP pattern was observed, much as reported in Kounios and Holcomb's (1992)[130] study of semantic relatedness (see Figure 8.11). But when the sentence ended in a syntactically anomalous fashion ('The woman persuaded to catch up'), a strong **P600** pattern occurred (a positive electrical potential) 600 ms after the anomalous word *to* was seen (Figure 8.11).

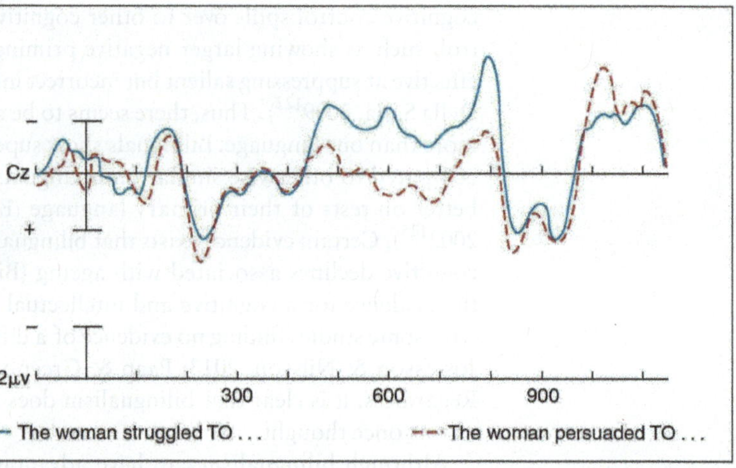

Figure 8.11 ERP Patterns for sentences violating syntactic or semantic expectations
Mean ERPs to syntactically acceptable sentences (solid curve) and syntactically anomalous sentences (dotted curve). The P600 component, illustrated as a downward dip in the dotted curve, shows the effect of detecting the syntactic anomaly. In this figure, positive changes go in the downward direction.

This confirms the important and seemingly separate role of syntactic processing during language comprehension.

Research techniques for language processing

A wealth of evidence illustrates the importance to cognitive science of such imaging and neuropsychological techniques and strongly suggests that the research on language processing will increasingly feature techniques such as imaging and ERP methods.

Learning language

As people acquire the meaning of words of a language, they are using both *experiential* (information derived from people's senses) and distributional (statistical occurrences of a word) knowledge to figure out word meanings and roles in the language (Andrews, Vigliocco & Vinson, 2009[131]). That is, people are using knowledge of what they actually experience during individual events, as well as how often and in which context various words and word combinations may occur. McCandliss, Posner and Givon (1997)[132] captured this basic idea as they taught people a new, miniature artificial language and recorded ERPs during learning. Early in training, words in the new language showed ERP patterns typical of nonsense material. People were focusing entirely on the experiential information of hearing the novel 'words'. However, after 5 weeks of training, the ERP patterns looked like those obtained with English words in which people had enough experience to develop a distributional understanding of the novel language to begin using that type of knowledge as well. Furthermore, left hemisphere frontal areas reacted to semantic aspects of the language (distributional), whereas posterior areas were sensitive to the visual characteristics of the words, the orthography (experiential). Thus, changes occur in the neural, and presumably cognitive, mechanisms involved as one learns a new language.

Syntactic processing

Rosler, Pechmann, Streb, Roder, and Hennighausen (1998)[133] did an ambitious study on syntactic processing using ERPs. Sentences were presented word by word. To assess people's comprehension, they had to answer a question about each sentence 5 seconds after they saw the last word. The sentences were all grammatical, but some of them differed from normal canonical word order. The ERP patterns demonstrated a variety of effects. For example, the patterns differed appreciably when the sentences violated canonical word order and when elements in the sentence cued people that an unusual word order would follow. These results were particularly compelling in that the ERP patterns tapped into purely mental processes that are not revealed by outward, behavioural measures such as response time or accuracy.

Right hemisphere language

Although the left hemisphere is typically credited with being the primary source of language processing, the right hemisphere also does important work. A general characterisation of the role of the right hemisphere in the cortex is that it serves to process information in a more holistic way rather than in the more analytic manner characteristic of the left hemisphere. In other words, the right hemisphere is more adept at processing information in a coarse-grained fashion, whereas the left hemisphere is more adept at processing information at a fine-grained level (Beeman, 1998[134]). One role of the right hemisphere is in making more distant, remote semantic connections between words. For example, the connection between *tiger* and *stripes* is relatively direct and close, but the connection between *tiger* and *beach* is more remote and it requires some creativity to see it. A study by Coulson, Federmeier, Van Petten, and Kutas (2005)[135] illustrates this differential operation of the left and right hemispheres. Researchers recorded ERP patterns in the left and right hemispheres as people read sentences.

Coulson et al. found that the left hemisphere is more involved in integrating lexical information with sentence-level information such as whether a word is sensible in the context of a given sentence (e.g. responses to the word tyre after reading the sentence 'They were truly stuck, since she didn't have a spare'), but not so much in the lexical relations of the words to each other. By comparison, the right hemisphere is more involved in this sort of word-to-word associative processing (e.g. the fact that *spare* and *tyre* are more associated than *spare* and *pencil*).

Individual differences

Although some consistencies exist in how language is processed neurologically, differences also exist between people. Reichle, Carpenter, and Just (2000)[136] used fMRI to look at brain activity while people verified sentence–picture stimuli (e.g. 'The star is above the plus' followed by a picture that did or did not match the sentence). When people used a verbal strategy to make their decisions, brain regions associated with language processing (especially Broca's area) were active. When people used a visual imagery strategy, regions in the parietal lobe were active – the same regions that are active when visual–spatial reasoning tasks are given. Interestingly, the language area activity was lower when high-verbal people were tested and likewise for visual areas in people with high visual–spatial abilities, as if high verbal or spatial ability reduced the amount of brain work needed to do the task. Thus, differences in how people are using language can show up in neurological measures.

8.6.2 Aphasia

A large literature exists on brain-related disorders of language, based on people who, through the misfortune of illness or brain injury, have lost the ability to use language. Formal studies of such disorders date back to the mid-1800s, although records dating back to 3500 BC mention language loss caused by brain injury (see McCarthy & Warrington, 1990[137]).

The disruption of language caused by a brain-related disorder is called **aphasia**. Aphasia is always the product of some physical injury to the brain sustained either in an accident or a blow to the head or in diseases and medical syndromes such as stroke. A major goal in neurology is to understand the aphasic syndromes more completely so that people who suffer from aphasia may be helped more effectively. From the standpoint of cognitive neuroscience, the language disruptions of aphasic patients can also help us to understand language and its neurological basis.

Although there are many kinds of aphasias, with great variety in their effects and severity, three basic forms are the most common:

1. Broca's aphasia
2. Wernicke's aphasia
3. Conduction aphasia.

Broca's aphasia

As described by Kertesz (1982)[138], **Broca's aphasia** is characterised by severe difficulties in producing speech. It is also called *expressive* or *production aphasia*. Patients with Broca's aphasia show speech that is hesitant, effortful and phonemically distorted. Aside from stock phrases such as 'I don't know', such patients generally respond to questions with one-word answers. If words are strung together, there are few if any grammatical markers, such as bound morphemes like *-ing*, *-ed* and *-ly*. In less severe cases, the aphasia may be limited to more complex aspects of language production, such as the production of verb inflections (Faroqi-Shah & Thompson, 2007[139]). Interestingly, such patients typically show less (or even no) impairment of comprehension for both spoken and written language.

Table 8.4 provides a list and short explanation of these disruptions and some others you have already encountered.

Table 8.4 Brain-related disruptions of language and cognition

Disorder	Disruption of
Language-related	
Broca's aphasia	Speech production, syntactic features
Wernicke's aphasia	Comprehension, semantic features
Conduction aphasia	Repetition of words and sentences
Anomia (anomic aphasia)	Word finding, either lexical or semantic
Pure word deafness	Perceptual or semantic processing of auditory word comprehension
Alexia	Reading, recognition of printed letters or words
Agraphia	Writing
Other symbolic-related	
Acalculia	Mathematical abilities, retrieval or rule-based procedures
Perception, movement-related	
Agnosia	Visual object recognition
Prosopagnosia	(Visual) face recognition
Apraxia	Voluntary action or skilled motor movement

French neurosurgeon Pierre Broca first described this syndrome in the 1860s and also identified the damaged area responsible for the disorder. The site of the brain damage, an area toward the rear of the left frontal lobe, is therefore called **Broca's area**. Broca's area lies adjacent to a major motor control centre in the brain and is shown in Figure 8.12.

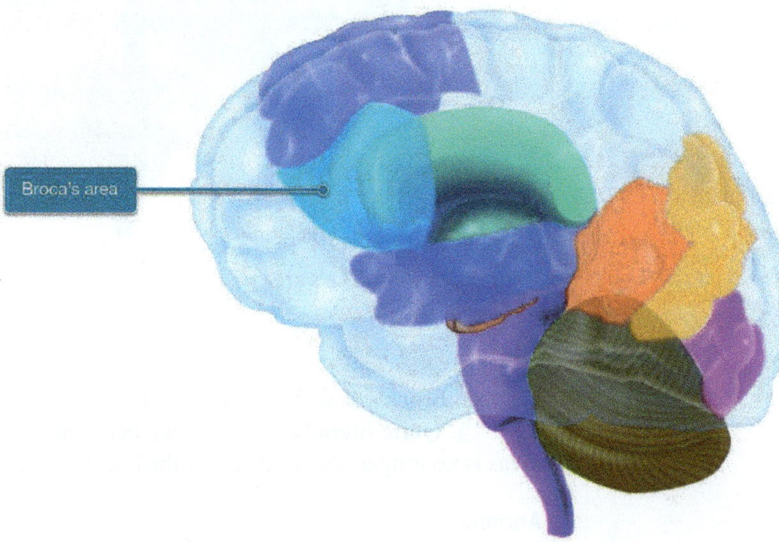

Figure 8.12 Broca's area
Source: miha de/Shutterstock

Wernicke's aphasia

Loosely speaking, the impairments in **Wernicke's aphasia** are the opposite of those in Broca's aphasia. In patients affected by Wernicke's aphasia, comprehension is impaired, as are repetition, naming, reading and writing, but the syntactic aspects of speech are preserved. It is sometimes called *receptive* or *comprehension aphasia*. In this syndrome, there may be unrecognisable content words; recognisable but often inappropriate semantic substitutions; or *neologisms*, invented nonsense words. What is striking is that victims of this disorder are sometimes unaware of the aphasia.

German investigator Carl Wernicke identified this disorder, and the left-hemisphere region that is damaged, in 1874. This region is thus known as Wernicke's area. The area, toward the rear of the left temporal lobe, is adjacent to the auditory cortex, in the left temporal lobe, and is shown in Figure 8.13.

This is a very different area with very different abilities than Broca's area in the frontal lobe. Note also that this disorder demonstrates a *double dissociation*, a basic distinction at the level of brain organisation between syntax and semantics (see Breedin & Saffran, 1999, for a case study showing loss of semantic knowledge but preserved syntactic performance[140]).

Conduction aphasia

Much less common than Broca's and Wernicke's aphasias, **conduction aphasia** is a narrower disruption of language ability. Both Broca's and Wernicke's areas seem to be intact in conduction aphasia, and people with conduction aphasia can understand and produce speech quite well. Their language impairment is an inability to repeat what they have just heard. In intuitive terms, the intact comprehension and production systems seem to have lost their normal connection or linkage. Indeed, the site of the brain lesion in conduction aphasia appears to be the primary pathway

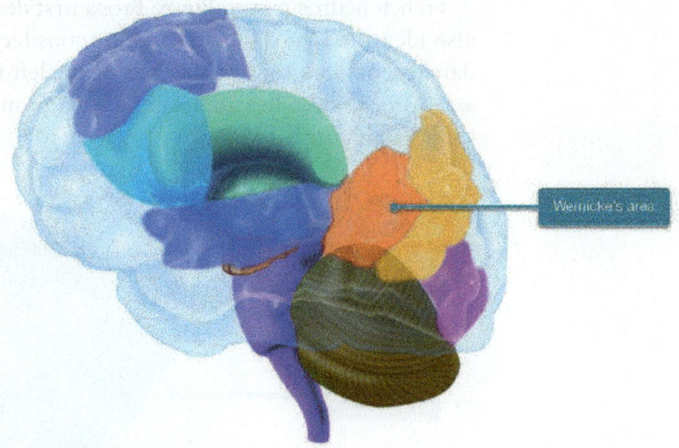

Figure 8.13 **Wernicke's area**
Source: miha de/Shutterstock

between Broca's and Wernicke's areas, called the *arcuate fasciculus* (Geschwind, 1970[141]). Quite literally, the pathway between the comprehension and production areas is no longer able to conduct the linguistic message.

Anomia

Another type of aphasia deserves brief mention here because it relates to the separation of the semantic and lexical systems discussed earlier. **Anomia, or anomic aphasia**, is a disruption of word finding, an impairment in the normal ability to retrieve a semantic concept and say its name. In anomia, some aspect of the normally automatic semantic or lexical components of retrieval has been damaged. Although moderate word-finding difficulty can result from damage almost anywhere in the left hemisphere, full-fledged anomia seems to involve damage especially in the left temporal lobe (Coughlan & Warrington, 1978; see McCarthy & Warrington, 1990, for details[142]). There is a superficial similarity between anomia and the tip-of-the-tongue (TOT) effect. Several researchers (Geschwind, 1967; Goodglass, Kaplan, Weintraub & Ackerman, 1976[143]) have found no evidence among anomic patients of the partial knowledge that characterises a TOT effect. Evidence also indicates that anomia can involve retrieval blockage only for the lexical component of retrieval, leaving semantic retrieval of the concept intact (Kay & Ellis, 1987[144]). This finding, along with that of other cases (e.g. Ashcraft, 1993[145]), suggests preserved semantic retrieval but a blockage in finding the lexical representation that corresponds to the already retrieved semantic concept.

Other aphasias

As Table 8.4 shows, a variety of highly specific aphasias are also possible. Although most of these are quite rare, they nonetheless give evidence of the separability of several aspects of language performance. For instance, in **alexia** (or *dyslexia*), there is a disruption of reading without any necessary disruption of spoken language or aural comprehension. In **agraphia**, conversely, the patient is unable to write. Amazingly, a few reports describe patients with alexia but without agraphia – in other words, patients who can write but cannot read what they have just written (Benson & Geschwind, 1969[146]). In **pure word deafness**, a patient cannot comprehend spoken language, although he or she is still able to read and produce written and spoken language.

There is documentation for even more specific forms of aphasia than these – for instance, difficulties in retrieval of verbs in written but not spoken language (Berndt

& Haendiges, 2000[147]) and difficulties in naming just visual stimuli, without either generalised visual *agnosia* or generalised anomia (Sitton, Mozer & Farah, 2000[148]). Thus, cognitive processing is quite complex, made of many different parts that need to work together seamlessly for language to work. When any one of those parts is faulty, unique language processing deficits can occur.

Right hemisphere damage

Despite the fact that most of the aphasias discussed here involve processing in the left hemisphere of the cortex, there is also evidence of the right hemisphere's contribution to language comprehension and production. (See Beeman & Chiarello, 1998, for a useful overview of the complementary right- and left-hemisphere language processes.[149]) Work by Beeman (1993, 1998[150]) suggests that a problem of people with right hemisphere damage is an inability to activate an appropriately diverse set of information from long-term memory from which inferences can be derived. In one study, after reading a text, people were given a lexical decision task. Some of the words in the task were related to inferences that needed to be drawn for comprehension. For example, read the following short text: 'Then he went into the bathroom and discovered that he had left the bathtub water running. He had forgotten about it while watching the news. The mess took him a long time to mop up.' After reading, you are then given a lexical decision probe word like *overflow*. Because the tub overflowing was not mentioned in the text, responding to *overflow* faster than baseline would be evidence of having inferred it based on reading the text itself. The results showed that normal controls responded 49 ms faster relative to neutral control words – they drew the appropriate inference. However, the patients with right hemisphere damage responded 148 ms more *slowly* to these words.

Review Table 8.5 for a listing of the typical impairments in both aphasias, including speech samples.

Table 8.5 Classic impairments in Broca's and Wernicke's aphasias

	Broca's aphasia	Wernicke's aphasia
Quality of speech	Severely impaired; marked by extreme effort to generate speech, hesitant utterances, short (one-word) responses	Little if any impairment; fluent speech productions, clear articulation, no hesitations
Nature of speech	Agrammatical; marked by loss of syntactic markers and inflections and use of simple noun and verb categories	Neologistic; marked by invented words (neologisms) or semantically inappropriate substitutions; long strings of neologistic jargon
Comprehension	Unimpaired compared with speech production. Word-finding difficulty caused by production difficulties	Severely impaired; marked by lack of awareness that speech is incomprehensible; comprehension impaired also in non-verbal tasks (e.g. pointing)
Speech samples	Experimenter asks the patient's address. 'Oh dear. Um. Aah. O! O dear. Very-there-were-ave. avedevershер avenyer.' (Correct address was Devonshire.)	Experimenter asks about the patient's work before hospitalisation. 'I wanna tell you this happened when happened when he rent. His-his kell come down here and is – he got ren something. It happened. In these ropliers were with him for hi-is friend – like was. And he roden all of these arranjen from the pedis on from iss pescid.'

8.6.3 Generalising from cases of brain damage

Although it is a mistake to believe that our eventual understanding of language will be reducible to a catalogue of biological and neurological processes (e.g. Mehler, Morton & Jusczyk, 1984[151]), knowledge of the neurological aspects of language is useful for something beyond the rehabilitation and treatment of aphasia. What do studies of such abnormal brain processes tell us about normal cerebral functioning and language?

Well, for one, the different patterns of behavioural impairments in Broca's and Wernicke's aphasias, stemming from different physical structures in the brain, imply that these two physical structures are responsible for different aspects of linguistic skill. Furthermore, these selective impairments reinforce the notion that syntax and semantics are two separable but interactive aspects of normal language (e.g. O'Seaghdha, 1997; Osterhout & Holcomb, 1992[152]). The double dissociations show that different independent modules govern comprehension and speech production. Other dissociations indicate yet more independent modules of processing, such as separate ones corresponding to reading and writing.

An intriguing inference from such studies is that the specialised regions signal an innate, biological basis for language – that the human nervous system is specifically adapted to learn and use language, as opposed to simply being able to do so. Several theorists have gone so far as to discuss possible evolutionary mechanisms responsible for lateralisation, hemispheric specialisation, the dissociation of syntax and semantics revealed by Broca's and Wernicke's aphasias, and even cognition in general (Corballis, 1989; Geary, 1992[153]; Lewontin, 1990[154]). These are fascinating lines of reasoning on the nature of language and cognition as represented in the brain.

Summary: Language

8.1 Linguistic universals and functions

- Language is our shared symbolic system of communication, unlike naturally occurring animal communication. True language involves a set of characteristics, linguistic universals, that emphasise the arbitrary connections between symbols and referents, the meaningfulness of the symbols, and our reliance on rules for generating and comprehending language.
- Three traditional levels of analysis – phonology, syntax and semantics – are joined by two others in psycholinguistics, the levels of conceptual knowledge and beliefs. Linguists focus on an idealised language competence as they study language, but psycholinguists are also concerned with language performance. Therefore, the final two levels of analysis take on greater importance as we investigate language users and their behaviour.
- To some degree, we can use people's linguistic intuitions, their linguistic competence, to discover what is known about language; language performance, on the other hand, is also affected by memory lapses and the like.
- The Sapir–Whorf linguistic relativity hypothesis claims that language controls or determines thought, making it impossible to think of an idea if there was no word for it in the language. The weak version of this hypothesis is generally accepted now; language exerts an influence on thought by making it more difficult to think of an idea without having a word to name or express it.

8.2 Phonology

- Phonology is the study of the sounds of language. Spoken words consist of phonemes, the smallest units of sound that speakers of a language can distinguish. Surprisingly, a range of physically different sounds are classified as the same phoneme; we tolerate some variation in the sounds we categorise as 'the same', called categorical perception.
- Categorical perception is particularly important in the study of speech recognition because the phonemes in a word exhibit coarticulation, overlapping effects among successive phonemes, such that an initial sound is influenced by the sounds that follow and the later sounds are influenced by what came before.
- Speech recognition relies heavily on conceptually driven processes. This includes our knowledge of the sentence and word context, our estimates of how we would produce the sounds ourselves, and our knowledge of what the words in our language are.
- As an illustration of embodied cognition, according to the motor theory of speech perception, part of the way that people go about understanding spoken language may be

to try to mentally simulate those sounds as if they were actually being spoken.

8.3 Syntax

- Syntax involves the ordering of words and phrases in sentence structure and features such as active versus passive voice. Chomsky's theory of language is a heavily syntactic scheme with two sets of syntactic rules. Phrase structure rules are used to generate a deep structure representation of a sentence, and then transformational rules convert the deep structure into the surface structure, the string of words that makes up the sentence.
- There are a variety of syntactic clues to the meaning of a sentence, so an understanding of syntax is necessary to psycholinguists. On the other hand, psycholinguistics has developed its own theories of language, at least in part because of linguists' relative neglect of semantic and performance characteristics.
- Studies of how we plan and execute sentences reveal a highly interactive set of processes, rather than a strictly sequential sequence. We pause, delay and rearrange sentences as a function of planning and memory-related factors like accessibility and working memory load.
- Meaning in language is conveyed not only by word order (syntax) but also by the changes in pitch (prosody) of the phonemes that are spoken. Prosody can convey information such as the nature of an utterance (e.g. the difference between a statement and a question) or which concepts the speaker wishes to emphasise.

8.4 Lexical factors

- Semantic factors in language can sometimes override syntactic and phonological effects. The study of semantics breaks words down into morphemes, the smallest meaningful units in language; *cars* contains the free morpheme *car* and the bound morpheme *-s* signifying a 'plural.'
- Speech errors that people make can be used to help reveal the processes by which language is produced. These speech errors follow regularities that are likely to be produced by otherwise consistent and stable cognitive processes.

8.5 Semantics

- As the study of language comprehension has matured, the dominant approach to semantics claims that we perform a semantic parsing of sentences, assigning words to their appropriate semantic case roles as we hear or read.
- Garden path sentences, in which later phrases indicate an error in interpretation, have provided rich information about how syntax and semantics are processed online during comprehension and how we recover from comprehension errors.
- The cognitive processing consequences of bilingualism are not well understood. Whereas some evidence suggests that knowing more than one language can improve intelligence, other studies find no evidence of a benefit. Regardless, there appears to be no evidence that multilingualism compromises cognition and intelligence as was once thought.

8.6 Brain and language

- Extensive evidence from studies with brain-damaged people and more modern work using imaging and ERP methods reveal several functional and anatomical dissociations in language ability.
- ERP methods show how language development unfolds over time, the time course of syntactic processing, the lateralised processing of language between the left and right hemispheres, and the strategies people use to comprehend language.
- The syntactic and articulatory aspects of language seem centred in Broca's area, in the left frontal lobe, whereas comprehension aspects are focused more on Wernicke's area, in the posterior left hemisphere junction of the temporal and parietal lobes.
- The study of these and other deficits, such as anomia and right hemispheric damage, converges with evidence from imaging and ERP studies to illustrate how various aspects of language performance act as separable, distinct components within the overall broad ability to produce and comprehend language.

Chapter 8 quiz

Question 1

Why did Chomsky reject the idea of behaviourism in language development?

A. Chomsky proposed that language development is seen as an innate process whereby the brain structures allow us a natural capacity to learn.

B. Chomsky proposed that language development was seen as an adaptation, through natural selection processes.

This is incorrect. Pinker supported the views of Chomsky, however, he saw language as unique to humans whereby it has been developed through evolution.

C. Chomsky proposed that we use several brain regions to process language and that we acquire language due to the interaction our brains have with our own environment.

D. Chomsky proposed that language is a purposeful activity and that we use it to express our own thoughts and ideas, and to make things happen.

Question 2

Which one of these is not a part of Hockett's Linguistic Universals?

A. Visual-spatial channel
B. Vocal-auditory channel
C. Interchangeability
D. Displacement

Question 3

Why might it be difficult for someone to learn English as a second language?

A. English has half the number of phonemes as other languages.
B. All phonemes are phonetically different within English making it difficult to learn.
C. Sounds in English may not sound the same in another language.
D. Language production in English does not need a context to understand the word.

Question 4

A child is presented with the following sentence 'Calo asked teacher the about homework his'. Someone who learns language would need to consider the first element of syntax to be able to understand this sentence. Which element of syntax would this be?

A. Number agreement.
B. Word order
C. Phrase order.
D. Transformational grammar.

Question 5

What does 'polysemy' refer to?

A. A word that has multiple meanings.
B. A word that sounds the same as another word but it is spelled differently.
C. A word that elicits the N400 brain region in ERP research.
D. A word that is spelled the same as another word, but they sound different.

9 Language in context

Learning objectives

9.1 Understand the conceptual level of language analysis

9.2 Evaluate the measures that can be used to improve reading comprehension

9.3 Assess how current research integrates multiple factors in understanding reading comprehension

9.4 Evaluate the cognitive effects of conversational interactions

Key figures

Professor Rebecca Treiman

Rebecca Treiman is a Professor in Child and Developmental Psychology and works on understanding spelling and reading in young people and adults. She is the head of Reading and Language Lab at Washington University in St Louis

Introduction

Comprehension – this is a concept that requires a bit of explanation. What does the word *comprehension* mean? Basically, the expanded meaning includes both the fundamental language processes and the additional processes we use when comprehending realistic extended tracts of language, say a passage in a book or a connected, coherent conversation, or even a perceived event. How do we comprehend and understand? What do we *do* when we read, understand and remember connected sentences? These are issues that will be discussed in this chapter.

9.1 Conceptual and rule knowledge

LEARNING OBJECTIVE
Understand the conceptual level of language analysis

Earlier in the course, we discussed the first three levels of language analysis: the phonological, syntactic, and lexical and semantic levels. Now we will dig into Miller's (1977)[1] fourth and fifth levels: the conceptual and belief levels.

Conceptual and belief levels of language analysis

The conceptual and belief levels aim more squarely at issues of comprehension and understanding. Here's the sentence Miller uses to illustrate these conceptual and belief levels:

(1) Mary and John saw the mountains while they were flying to California.

If this sentence were spoken aloud, your comprehension would begin with phonological processes, translating the stream of sounds into words. Your syntactic knowledge would parse the sentence into phrases and assist the semantic level of analysis as you determined the case role for each important word: *Mary* and *John* are the agents of see, the word *mountains* is assigned the *patient* or *recipient* role, *they* is the agent of *fly* in the second main clause, and so on.

So far so good. But this sentence is more challenging than that. It is ambiguous, having more than one meaning. There is the obvious meaning, that Mary and John looked out the plane window and saw mountains during a flight to California. But there is also the possibility that *they* refers to the mountains. *They* merely denotes something plural, after all, so syntactically, the *they* could refer to the mountains.

Those of you who noticed this ambiguity probably rejected it immediately for the obvious reason: Mountains do not fly. We are getting close to Miller's point. Knowing that mountains do not fly is part of your semantic, **conceptual knowledge**. Look in as many dictionaries as you'd like, and you will not find 'mountains don't fly' in any of them. Accordingly, your comprehension of sentence 1 must also have included a conceptual level of analysis, in which you compared your interpretation with semantic knowledge.

Miller also argues that **beliefs** are important for comprehension. I could tell you 'No, I'm not saying that Mary and John were flying to California. I'm saying that it was the mountains that were flying.' Although you can understand that I might think mountains can fly, you would not change your mind about the issue; your belief in your own knowledge and your feeling that I am lying or playing a trick (or am just plain crazy) are important for comprehension. A purely linguistic analysis of language misses this critical aspect of comprehension as well. Consider, for example, the prominence and importance of your beliefs to comprehension and memory when you hear advertisements or speeches in political campaigns.

Rules are yet another part of the knowledge that must be accounted for. Beyond case rules, additional rules are operating when we deal with more complex passages of text or with connected conversation. Some rules have the flavour of strategies. For example, we tend to interpret sentence 2 as focusing on Tina, largely because she is mentioned first in the sentence (Gernsbacher, 1990[2]):

(2) Tina gathered the kindling as Lisa set up the tent.

Several lines of evidence speak to this idea, that we provide a focus to our sentences by using mechanisms such as first mention and certain kinds of reference (e.g. 'There was this guy who . . . ' instead of 'A guy . . . ').

Other rules deal with reference – building bridges between words referring to the same thing. Still more rules parade under the name **pragmatics** and refer to a variety of extralinguistic factors. And finally, high-level rules – rules that specify how people in a conversation structure their remarks and understand the remarks of others – operate in conversational interactions. As always, just because you cannot state the rule or were never explicitly taught it does not mean that it is not there. It simply means that the rules are part of your implicit, tacit knowledge.

9.1.1 Comprehension research

Much of the traditional evidence about comprehension relies on people's linguistic intuitions, their (leisurely) judgements about the acceptability of sentences, or simple measures of recall and accuracy. The Sachs (1967)[3] study is a classic example of early comprehension research with a straightforward conclusion: 'My experiment demonstrated that the specific wording of an utterance is forgotten within seconds after it is heard. In contrast, the meaning of that utterance can be retained for a very long period.' Recall that as people were reading a passage, they were interrupted and tested on a target sentence, either 0, 80 or 160 syllables after the end of the target. Their recognition of the sentence was very accurate at the immediate interval. But beyond that, they were accurate only at rejecting the choice that changed the sentence meaning. So, people could not accurately discriminate between a verbatim target sentence and a paraphrase: if the choice preserved the original meaning, then people mistakenly 'recognised' it. Clearly, these results showed that memory for meaningful passages does not retain verbatim sentences for long but does retain meaning quite well.

9.1.2 Levels of comprehension

Comprehension is a complex process involving several different levels.

One of the ways of characterising these different levels is the van Dijk and Kintsch's (1983)[4] levels of representation theory. As a reminder, at one level is the **surface form**. This is our verbatim mental representation of the exact words and syntax used. At an intermediate level is the **propositional textbase**, which captures the basic idea units present in a text. Finally, there is the level of the **situation model** (Johnson-Laird, 1983; Van Dijk & Kintsch, 1983; Zwaan & Radvansky, 1998[5]), which is a mental representation that serves as a simulation of a real or possible world as described by a text.

These three levels of comprehension involve different cognitive processes that operate at different time scales, with surface-level cognitive processing occurring more rapidly than situation model-level processing (Huang & Gordon, 2011[6]).

As you read, try to understand how comprehension may depend on these different levels. For example, research on order of mention in establishing discourse reference will depend on the surface form. Work on bridging inferences requires processing at the textbase level. Finally, work showing how people monitor various aspects of experience involves the situation model.

9.1.3 Metacomprehension

Reading a passage and having some understanding of what you have read does not mean you have actually learned something or will remember it later. Yet, we need to use our **metacomprehension** abilities (e.g. Dunlosky & Lipko, 2007[7]) to monitor how well we understand and will remember information later.

Measures of metacomprehension

Metacomprehension is important because it can influence how much we may study information later, and to which information we devote our time. To this end, we need ways to measure or evaluate metacomprehension.

A popular measure of metacomprehension is judgements of learning (JOLs) (Arbuckle & Cuddy, 1969[8]). These are estimates people are asked to make of how well they feel they have learned some material they just read. Research on JOLs typically compares people's estimates of how well they have learned information with how they actually do. Unfortunately, in many cases, the relationship between JOLs and actual performance is quite low – in other words, people are typically not very good at estimating whether they have learned something or not. Consequently, when you plan your studying, say for an upcoming exam, you may not spend the time you need on certain material because you think you know it better than you actually do. Your test performance would improve if you could better monitor what you have and have not learned.

In addition to difficulty in judging whether something has been learned, people also have metacomprehension problems in choosing how to plan or distribute their study time. For example, memory is worse when people study using massed practice (cramming) rather than distributed practice. In other words, many people are unaware that massed practice is a poor learning strategy. Another metacomprehension error that people make is spending their study time focusing on very difficult material. This strategy is inefficient (Nelson & Leonesio, 1988[9]). This *labour-in-vain effect* occurs when people spend large amounts of time trying to learn information that is too far beyond their current level of knowledge, but end up with little to no new learning.

A better strategy is to spend time learning information that falls within the *region of proximal learning* (e.g. Metcalfe, 2002[10]). This is information just beyond a person's current level of understanding. So, what we have here is a bit like the 'Goldilocks and the Three Bears' story. It is a waste of time to study information that one knows well (this material is too easy). Also, as the labour-in-vain effect shows, trying to study information that is far too difficult will not help either (this material is too hard). However, learning that occurs in the region of proximal learning is just beyond what a person currently knows, so that person can draw on existing knowledge as a scaffolding to integrate new information (this material is just right).

Some classroom settings are set up to take advantage of the region of proximal learning, provided people have the prerequisites and keep up with the material in the class. Here are three more ways to improve metacomprehension (cf. Dunlosky & Lipko, 2007; Griffin, Wiley, & Thiede, 2008[11]):

1. Before making a judgement about whether you have learned something, wait a few minutes (Dunlosky & Nelson, 1991[12]). Often, when you are making judgements of learning, you are assessing whether you can retrieve the information from long-term memory into working memory. But when you make these judgements right after reading material, there is still a lot of information in working memory, so you are overconfident in your ability to remember the information.
2. It is helpful to reread the material. Reading the material before a lecture and then going to the lecture can give you the same benefit and will save you time and effort later.
3. You can boost the accuracy of your JOLs by generating summaries or even lists of key words. This strategy helps you reveal to yourself what you do and do not yet know.

9.1.4 Online comprehension tasks

As work on comprehension developed, researchers needed a task that measures comprehension as it happens, or an **online comprehension task**. Such tasks involve the same approach discussed throughout this course: find a dynamic, time- or action-based task that yields measurements of the underlying mental processes as they occur. Contrast performance in a variety of conditions, pitting factors against each other to see how they affect comprehension speed or difficulty. Then draw conclusions about the underlying mental processes, based on the performance measures.

Methods to assess online comprehension

There are three other ways commonly used to assess online comprehension.

Sample stimuli and test words

Sample stimuli and test words for online comprehension tasks.

Task	Sentence	Yes	Related	Unrelated	No
Was this word in the sentence?	Ken really liked the boxer.	Ken			Bill
Naming	Ken really liked the boxer.		Dog/fight	Plate	
Lexical decision	Ken really liked the boxer.		Dog/fight	Plate	

In one commonly used method to assess what a person is thinking about while reading, a text appears on a computer screen and is immediately followed by a probe word. Sometimes a person must make a 'yes' or 'no' response, indicating whether the word was in the just-read sentence. Sometimes the person must simply name the word or perform a lexical decision task on it. Look at the table above to see some sample stimuli and test words for these tasks. Response times (RTs) would lead us to inferences about the nature and operation of cognition. For example, if the ambiguous word *boxer* activates both the dog and the fighter meanings, then we might expect RT to *dog* and *fight* to be about the same, but faster than to the neutral word *plate*. But if *boxer* is interpreted only in one of its two senses, then *dog* would be faster than *fight* (or the other way around, depending on which meaning is dominant).

Think-aloud verbal protocol

Another way to assess online comprehension is the think-aloud verbal protocol method (e.g. Magliano, Trabasso & Graesser, 1999[13]). In this method, people are asked to verbalise their thoughts as they read a passage of text. The *verbal protocols* can then be analysed later to assess the conscious thoughts people were having as they read, including the following:

- How do they link up a current portion of text with events that occurred earlier?
- Were they making predictions about what would happen next?
- Did they notice an inconsistency in the text?

The data generated from think-aloud verbal protocols can provide insights into the aspects of a text that might be fruitful candidates for further research. For example, this information can be used to focus investigation of which aspects of a text will yield interesting reading time data, or which kind of information to test using a probe task.

Neural imaging

Finally, as in many other areas of cognitive psychology, there has been an increase in the use of neural imaging to aid our investigations. For online comprehension, these measures often require the temporal resolution necessary to capture understanding across relatively brief periods of time, as would be done with event-related potential (ERP) and functional magnetic resonance imaging (fMRI) recordings. Using these methods, we can reveal aspects of comprehension that otherwise might be difficult to uncover.

9.1.5 Comprehension as mental structure building

The mental representations that are created during comprehension are built up and updated in systematic ways. Specially, theoretical approaches, such as Gernsbacher's (1990)[14] **structure-building** framework, assume that language is essentially a set of instructions for how to build one's understanding of the world. This concerns how to start, how to update it and how to remove irrelevant information.

A convenient way to organise thinking about comprehension is to use Gernsbacher's (1990)[15] structure-building framework.

Gernsbacher's structure-building framework

The basic theme is that comprehension is a process of building mental structures. Laying a foundation, mapping information onto the structure, and shifting to new structures are the three principal components, whereas enhancement and suppression modify the relative salience of the structures created.

Laying a foundation

As we read, we begin to build a mental structure that captures the meaning of a sentence. A foundation is initiated as the sentence begins and typically is built around the first mentioned character or idea. This is equivalent to saying that sentence 3 is about Dave and studying:

(3) Dave was studying hard for his statistics midterm.

Mapping information

As more elements appear, they are added to the structure, by the process called **mapping**. Mapping here simply means that additional word and concept meanings are added to the 'DAVE' structure by specifying Dave's activities. For instance, the prepositional phrase 'for his statistics midterm' is processed. Because the concept 'MIDTERM' is coherent in the context of studying, these concepts are added to the structure. Also, any inferences you draw are added to the structure. For instance, when your knowledge about 'MIDTERM' was activated, you drew the inference that Dave probably was enrolled in a statistics course.

Shifting to a new structure

We continue trying to map incoming words to the current structure on the assumption that those words belong to the structure under construction right now.

But at some point, a different idea is encountered that signals a change in focus or topic shift. As an example, consider this continuation of the Dave story:

> (4) Because the professor had a reputation for giving difficult exams, the students knew they'd have to be well prepared.

When you read 'Because the professor', a coherence process detects the change in topic or focus. One clue is the word *because* or other connectives (e.g. *later, although, meanwhile)*. Another clue involves the introduction of a new character and the inferences you need to draw to determine who the professor is. You inferred that Dave must be enrolled in a statistics class, and because midterms are exams given in college classes that are taught by professors, the professor must be the one who teaches that statistics class. At such moments, you close off or finish the 'DAVE' structure and begin a new one about the professor. Although the 'DAVE' structure still retains its prominence in memory, you are now working on a new current structure, mapping the incoming ideas (e.g. reputation, difficult exams) onto it. And at the end of that phrase, you will have constructed two related structures, one for each meaning (the phrase beginning 'the students' will trigger yet another structure to be built, yielding three substructures).

Enhancement and suppression

Finally, let's consider the two control mechanisms by adding one more sentence to the Dave story:

(3) Dave was studying hard for his statistics midterm.

(4) Because the professor had a reputation for giving difficult exams, the students knew they'd have to be well prepared.

(5) Dave wanted an A on that test.

As noted earlier, reading sentence 4 results in a new substructure and a change in focus. Still, the new substructure is related to the first one where the two ideas in sentence 4 map onto those from sentence 3. *Exams* refers to the same concept as *midterm*, and *professor* maps onto the *statistics course* implied by sentence 3. Such mappings reflect the activation of related concepts, especially those mapped into the foundation of the first structure (Millis & Just, 1994[16]). This activation combines with the activation from *midterm* and *statistics course* because of their semantic relatedness. This is the process of **enhancement**, that the many related concepts are now boosted or enhanced in their level of activation. This enhancement process is the priming process in semantic memory. It is the degree of enhancement and activation among concepts that predicts which ones will be remembered better or responded to more rapidly. The more frequently the same set of concepts is enhanced across a sentence, the more coherent the passage is.

However, the enhancement of some concepts implies that others lose activation. So although sentence 4 enhances the activation of concepts related to 'PROFESSOR', 'EXAMS' and so on, there is also **suppression** of concepts that are now out of the main discourse focus. In other words, activated concepts that become unrelated to the focus decrease in activation by the process of suppression.

Figure 9.1 is an illustration of these competing tendencies. As the professor clause is processed (ideas 3 and 4), the activation level for 'DAVE' is suppressed because it is no longer the main discourse focus. Then, as the story unfolds further (ideas 5 to 7), the concept 'DAVE' regains its enhancement, and the 'PROFESSOR' dwindles. Thus, the original Dave structure from sentence 3 receives renewed enhancement when

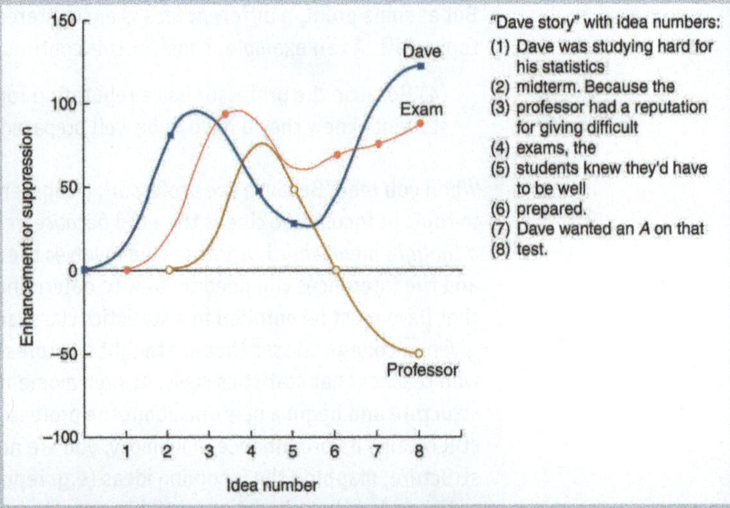

Figure 9.1 Hypothetical concept activation curves

As the concepts 'DAVE' AND 'EXAMS' from the Dave story in the text are introduced or mentioned again, their activation becomes enhanced. When the focus shifts, old concepts ('PROFESSOR') become suppressed; their activation dwindles.

you read in sentence 5 that Dave wanted an A on the test (O'Brien, Albrecht, Hakala, & Rizzella, 1995; for an extension of enhancement and suppression to the topic of metaphor comprehension, see Gernsbacher, Keysar, Robertson, & Werner, 2001[17], and Kintsch, 2000[18]). And, in the meantime, less important concepts are suppressed (e.g. McNamara & McDaniel, 2004[19]).

Neural basis of language comprehension

Determining who is the actor and who is the undergoer

In an fMRI study by Bornkessel et al. (2005),[20] they use the 'argument hierarchy construction' to understand the brain regions involved in syntax and semantics during a language comprehension task. Three factors were manipulated: the order of arguments, the type of verb used, and morphological ambiguity. Monolingual German participants read 104 case ambiguous and 52 case unambiguous sentences and were asked to judge if the new second sentence correctly described the content of the previous experimental sentence using a 'yes' or 'no' response on a switchboard (e.g. '. . . that Peter helps teachers' or '. . . that teachers help Peter'). As the cognitive demand and complexity increased there was greater activation in a network of brain regions including the inferior frontal, posterior superior temporal, premotor and parietal areas. Different components of this network were activated depending on the structure examined. For example, the left posterior superior temporal sulcus was activated when morphological and syntactic information was processed, while activation in the left inferior frontal gyrus was linked to the linear organisation of sentences. Based on these findings, it was suggested that, in German, the posterior superior temporal and inferior frontal regions play distinct roles in extracting information about who is the actor from morphosyntactic structures.

9.2 Reading

LEARNING OBJECTIVE
Evaluate the measures that can be used to improve reading comprehension

For years, the standard way to study reading was for people to read a passage of text and then take a memory test, such as a multiple-choice or recall test. Such tasks certainly have face validity; they test memory for the text, because much of our reading is for learning and remembering what we read.

But this approach is lacking because it does not gather online measures of comprehension; it measures only what people remember *after* reading (not that this is not important, there's just more to what's going on). Earlier, you saw a graph of the hypothetical activation levels for concepts in a set of sentences (Figure 9.1). We would like to know directly how concepts vary in their activation levels across a passage because that tells us a great deal about online reading comprehension.

9.2.1 Gaze duration

In reading research that assesses gaze duration, the equipment used is an **eye tracker**, a camera- and computer-based apparatus that records eye movements and the exact words that are fixated on during reading. More details about eye-tracking methodology are provided in Chapter 14 (Research methods in human cognition) Research on reading comprehension using eye-tracking methodology typically requires people to view a passage of text on the screen, while the eye tracker records the eye movements and durations as people read the words in the passage. In this task, the researcher knows which word is being processed on a moment-by-moment basis and how long the eyes dwell on each word. So, **gaze duration** is a prime measure of what is happening when people read (see Kambe, Duffy, Clifton, & Rayner, 2003, and Rayner, 1998, for thorough discussions of alternatives[21]). Time-based eye movement data provide a window on the process of comprehension and reading.

Table 9.1 Summary of Gernsbacher's structure-building framework

Interactive	
Process/control mechanisn	**Explanation/ function**
Laying a foundation	
Mapping information	
Shifting	
Enhancement	
Suppression	
	Increase the activation of coherent, related information.
Start over	

Making eye movements

The eyes move in rapid sweeps – *saccades* – and then stop to focus on a word – *fixations*. Fixations in reading (English) last about 200–250 ms, and the average saccade size is from 7 to 9 letter spaces. However, as Figure 9.2 shows, there is considerable variability in these measures (Rayner, 1998[22]).

Immediacy and eye-mind assumptions

Two assumptions that have guided much of the work using eye movements are the immediacy and the eye-mind assumptions (Just & Carpenter, 1980, 1987, 1992[23]). The **immediacy assumption** states that readers try to interpret each content word of

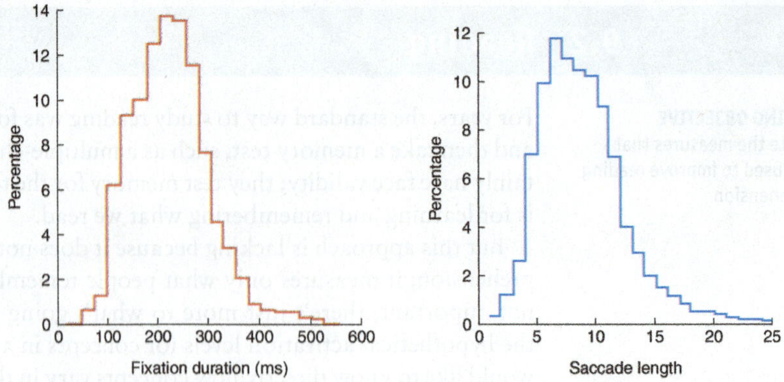

Figure 9.2 Variability in saccade and fixation measures

Frequency distributions for fixation durations (top) in ms, and saccade length (bottom) in number of character spaces.

Source: From Rayner (1998)[24].

a text as they encounter that word. In other words, we do not wait until we take in a group of words, say in a phrase, before we start to process them. Instead, we begin interpreting and comprehending as soon as we encounter a word. The **eye–mind assumption** is the idea that the pattern of eye movements directly reflects the complexity of the underlying cognitive processes.

Although these assumptions are robust, they do have some limitations. For example, eye gazes often take in more than one word, depending on the length of the words, the size of the text fonts and the span of the perceptual beam. So, there is not always a direct one-to-one relationship between an eye fixation and the words being processed. Moreover, eye gazes do not reflect only the processing of the current word; they can also reflect some spillover processing of previous words and some anticipatory processing of upcoming words (Kliegl, Nuthmann & Engbert, 2006[25]). Despite these limits, eye tracking is still a powerful tool for the reading researcher.

Also, outside of reading, you can use eye gaze to understand other aspects of comprehension, such as the understanding of spoken language. For example, if you listen to and follow a set of directions, say to pick up and move an object from one place to another, your eye movements track the spoken instructions very closely. As you hear, 'Put the apple in the box', your eyes fixate immediately on those objects in the visual scene (Spivey, Tanenhaus, Eberhard & Sedivy, 2002; see Crosby, Monin & Richardson, 2008, for an application of eye tracking to social cognition[26]).

9.2.2 Basic online reading effects

An example of online reading research examined regressive eye movements (using eye-tracking methodology – see Chapter 14) back to a portion of text that had been read earlier. Just (1976)[27] was interested in such eye movements when the referents in the sentence could not be immediately determined: if an initial assignment of a character to a case role was wrong, then what happened? Was there a regressive eye movement back to the correct referent? People read sentences such as 6 and 7, and eye movements were monitored:

(6) The tenant complained to his landlord about the leaky roof. The next day, he went to the attic to get his luggage.

(7) The tenant complained to his landlord about the leaky roof. The next day, he went to the attic to repair the damage.

In sentence 6, when *luggage* was encountered, eye movements bounced up immediately to the word *tenant*. In sentence 7, eye movements bounced up to *landlord*.

These eye movements provided evidence of the underlying mental processes of finding antecedents and determining case roles.

Another study provides a demonstration of the detail afforded by eye trackers. Look at Figure 9.3, taken from Just and Carpenter (1987; see also Just & Carpenter, 1980[28]). You see two sentences taken from a larger passage. Above the words are two numbers. The top number indicates the order in which people fixated on the elements in the sentence; 1 to 9 in the first sentence and 1 to 21 in the second. The number below is the gaze duration (in milliseconds). So, for example, the initial word in sentence 1, *Flywheels*, was fixated on for 1,566 ms. The next word, *are*, was fixated on only 267 ms. The fourth word, *of*, was not fixated on at all by this person, so neither a gaze number nor time is presented there. This is not unusual because many function words, such as *of* and *the*, are so automatically processed that they may not need to be fixated on (Angele & Rayner, 2013[29]). In fact, you can raerragne the lteters in wodrs, usch as is odne in this snetence, and people have little trouble extracting the meaning. There is some disruption to reading, but not as much as if different letters were substituted for correct letters (Rayner, White, Johnson, & Liversedge, 2006[30]). So, reading does not *require* a strict adherence to the printed form.

In the Just and Carpenter study, these passages were technical writing, in which a new concept, such as a flywheel, is introduced, defined and explained. The average reading rate was about 225 words per minute, slower than for simpler material, such as newspaper stories or novels. At a general level, every content word was fixated. According to Just and Carpenter, this is the norm for all kinds of text. In fact, about 85 per cent of the content words are fixated. Short function words, however, like *the* or *of* often tend not to be fixated; Rayner and Duffy (1988)[31] estimate that function words are fixated only about 35 per cent of the time. Readers also tend to skip some content words if the passage is very simple for them (say, a children's story given to an adult), if they are skimming or speed reading, or if a word is very predictable (Rayner & Well, 1996[32]).

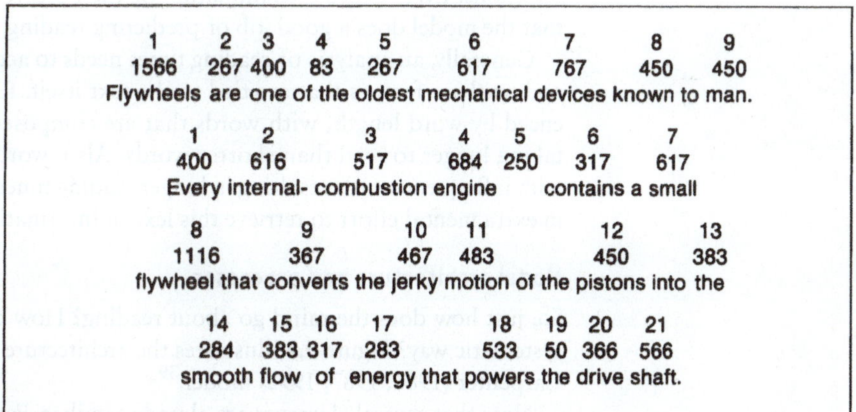
1 2 3 4 5 6 7 8 9
1566 267 400 83 267 617 767 450 450
Flywheels are one of the oldest mechanical devices known to man.
1 2 3 4 5 6 7
400 616 517 684 250 317 617
Every internal- combustion engine contains a small
8 9 10 11 12 13
1116 367 467 483 450 383
flywheel that converts the jerky motion of the pistons into the
14 15 16 17 18 19 20 21
284 383 317 283 533 50 366 566
smooth flow of energy that powers the drive shaft.

Figure 9.3 Eye fixations of a college student reading a scientific passage

Gazes within each sentence are sequentially numbered above the fixated words with the durations (in milliseconds) indicated below the sequence numbers.

Source: Just and Carpenter (1980)[33].

saccade

Saccades Fast, automatic movement of both eyes.

fixation

Clusters of eye gaze when the eyes are 'locked' on an object. Eye movements between fixations are saccades.

As noted already, gaze durations are quite variable, and the duration of a **saccade** is about 100 ms, followed by a **fixation** of 200–250 ms. These estimates come from situations in which the viewer is merely gazing out on a scene. In reading studies, however, people do not move their eyes as far, averaging 2 degrees of angle versus 5 degrees in scene perception. Hence, saccades during reading are shorter – they take about 30 ms. Although word fixations may be brief, readers often make repeated fixations on the *same* word. In some studies, successive fixation times are summed together. Alternatively, investigators report both the first-pass fixations and total fixation duration.

9.2.3 Benefits of online reading

A strength of online reading measures is that they provide evidence at two levels of comprehension. First, there are word-level processes operating at the surface form level. These are crucial to an understanding of reading. For instance, several studies attest to the early use of syntactic features of a sentence when we comprehend not just major syntactic characteristics such as phrase boundaries but even characteristics such as subject–verb agreement (Pearlmutter, Garnsey, & Bock, 1999[34]) and pronoun gender (McDonald & MacWhinney, 1995[35]). Reichle, Pollatsek, Fisher, and Rayner (1998)[36] provide an account of such word-level processes with their E-Z Reader models of eye movement control in reading (for a review of several computational models of reading, see Norris, 2013[37]).

Second, reading time measures can also be used to examine larger, macroscopic processes, such as at the textbase and situation model levels. We will hold off a discussion of situation model processing until the next section. At the textbase level, Table 9.2 presents Just and Carpenter's (1980)[38] analysis of the 'flywheel' passage. To the left of each line is a category label; each sector was categorised as to its role in the overall paragraph structure. To the right are two columns of numbers – observed gaze durations for a group of people and estimated durations – based on the 'READER' model's predictions. For example, the 1,921 ms observed for sector 1 is the sum of the separate gaze durations (measured using eye-tracking methodology – see Chapter 14) for that sector (averaged across people). Different kinds of sectors take different amounts of time. For instance, definition sectors have more difficult words and are longer than other sector types, so they show longer gaze durations. Even a casual examination of the observed and predicted scores shows that the model does a good job of predicting reading times.

Generally, an analysis of reading times needs to account for several surface form and textbase factors that are tied to the text itself. Reading time is strongly influenced by word length, with words that are composed of more letters or syllables taking longer to read than shorter words. Also, word frequency plays a vital role, with infrequent words resulting in longer reading times as the reader needs to engage in extra mental effort to retrieve this lexical information from memory.

Model architecture and processes

So, just how does the mind go about reading? How can we address this issue in a systematic way? Figure 9.4 illustrates the architecture and processes of the Just and Carpenter (1980, 1987, 1992) model.[39]

Note that several elements are already familiar. For instance:

- Working memory is where different types of knowledge – visual, lexical, syntactic, semantic, and so on – are combined. Not surprisingly, working memory capacity is important in reading comprehension (e.g. Kaakinen, Hyona, & Keenan, 2003[41]).
- Long-term memory contains a wide variety of knowledge used during reading. Each of these types of knowledge can match the current contents of working memory and update or alter those contents. In simple terms, what you know combines with what you have already read and understood. Together, these elements permit comprehension of what you are reading.

Table 9.2 Sector-by-sector analysis of 'flywheel' passage

		Gaze duration (ms)	
Category	**Sector**	**Observed**	**Estimated**
Topic	Flywheels are one of the oldest mechanical devices	1,921	1,999
Topic	known to man.	478	680
Expansion	Every internal-combustion engine contains a small flywheel	2,316	2,398
Expansion	that converts the jerky motion of the pistons into the smooth flow of energy	2,477	2,807
Expansion	that powers the drive shaft.	1,056	1,264
Cause	The greater the mass of a flywheel and the faster it spins,	2,143	2,304
Consequence	the more energy can be stored in it.	1,270	1,536
Subtopic	But its maximum spinning speed is limited by the strength of the material	2,400	2,553
Subtopic	it is made from.	615	780
Expansion	If it spins too fast for its mass,	1,414	1,502
Expansion	any flywheel will fly apart.	1,200	1,304
Definition	One type of flywheel consists of round sandwiches of fiberglass and rubber	2,746	3,064
Expansion	providing the maximum possible storage of energy	1,799	1,870
Expansion	when the wheel is confined in a small space	1,522	1,448
Detail	as in an automobile.	769	718
Definition	Another type, the 'superflywheel,' consists of a series of rimless spokes.	2,938	2,830
Expansion	This flywheel stores the maximum energy	1,416	1,596
Detail	when space is unlimited.	1,289	1,252

Source: Just and Carpenter (1980)[40].

- Finally, wrap-up is an integrative process that occurs at the end of a sentence or clause. During wrap-up, readers tie up any loose ends. For instance, any remaining inconsistencies or uncertainties about reference are resolved here.

9.2.4 Factors that affect reading

An in-depth description of all the variables that influence reading is not possible here – there are simply too many. Here is just a brief list of some of the important factors:

- The effects of word frequency, syntactic structure and context (Altmann, Garnham, & Dennis, 1992; Inhoff, 1984; Juhasz & Rayner, 2003; Schilling, Rayner, & Chumbley, 1998[42]).
- The effects of sentence context on word identification (Paul, Kellas, Martin, & Clark, 1992; Schustack, Ehrlich, & Rayner, 1987; Simpson, Casteel, Peterson, & Burgess, 1989[43]), including ERP work showing how rapidly we resolve anaphoric references (Van Berkum, Brown, & Hagoort, 1999[44]).
- The effects of ambiguity (Frazier & Rayner, 1990; Rayner & Frazier, 1989[45]) and figurative language (Frisson & Pickering, 1999[46]).

- The effects of topic, plausibility and thematic structure on reading (O'Brien & Myers, 1987; Pickering & Traxler, 1998; Rayner, Warren, Juhasz, & Liversedge, 2004; Speer & Clifton, 1998; Taraban & McClelland, 1988[47]), especially the relatedness of successive paragraphs and the presence of an informative introductory paragraph (Lorch, Lorch, & Matthews, 1985[48]) or title (Wiley & Rayner, 2000[49]).
- The effects of scripted knowledge on word recognition and comprehension (Sharkey & Mitchell, 1985[50]).
- The effects of discourse structure on the understanding of reference (Malt, 1985; Murphy, 1985[51]) and the resolution of ambiguity (Vu, Kellas, Metcalf, & Herman, 2000[52]).

Variables that affect reading times

Serial position is also an important factor. The further along a person is in a passage, the more of a foundation there is from which to build mental structures, thereby making comprehension easier and faster.

Variables that increase reading times

- Surface form effects: sweep of the eyes to start a new line, sentence wrap-up, number of syllables, low frequency or new word, unusual spelling patterns.
- Textbase effects: integration of information (after clauses, sentences, sectors, etc.), topic word, new argument, other error recovery, reference and inference processes, difficulty of passage/topic.

Variables that decrease reading times

- Surface form effects: familiar word, higher word frequency, repetition of infrequent word.
- Textbase effects: appropriate title, supportive context, semantic-based expectation (if confirmed).

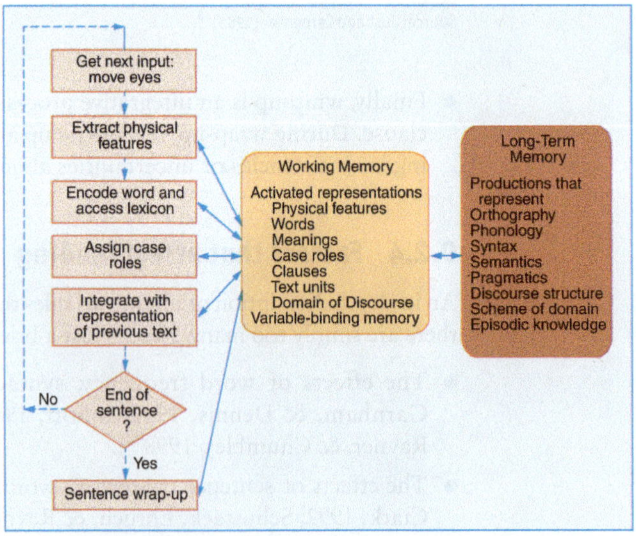

Figure 9.4 The Just and Carpenter (1980)[55] model

This model shows the major structures and processes that operate during reading. Solid lines represent the pathways of information flow; the dashed line shows the typical sequence of processing.

In addition, even phonology plays an important role in reading comprehension, such as research showing that phonological information is activated as rapidly as semantic knowledge in silent reading (Lee, Rayner, & Pollatsek, 1999; Rayner, Pollatsek, & Binder, 1998[53]), especially for readers of lower skill levels who rely more on print-to-sound-to-meaning processes than a direct print-to-meaning route (Jared, Levy, & Rayner, 1999[54]).

9.3 Reference, situation models and events

LEARNING OBJECTIVE
Assess how current research integrates multiple factors in understanding reading comprehension

Although the cognitive mechanisms and processes involved in comprehension at the surface form and textbase levels are critically important, they are not the goal of comprehension. A person who has successfully comprehended something that has been read has derived not only an adequate representation of the text itself. This person also understands the circumstances being described – the reference of the text.

9.3.1 Reference

Reference involves finding the connections between elements in a passage of text, finding the words that refer to other concepts in the sentence. In sentence 3 from earlier, 'Dave was studying hard for his statistics midterm', the word *his* refers to *Dave*. In this situation *Dave* is the **antecedent** of *his* because *Dave* comes before the pronoun. And the act of using a pronoun or possessive later is **anaphoric reference**. So, **reference** is the linguistic process of alluding to a concept by using another name. Commonly we use pronouns or synonyms to refer to the antecedent, although there are other types of reference. For example, using a person's name would be a form of *identity reference* in that it refers to a previous instance of using that person's name.

Reference is as common in language as any other feature we can identify. Part of reference is that it reduces redundancy and repetition. Contrast a normal passage such as 8a with 8b to see how boring and repetitive language would be without synonyms, pronouns, and so on.

(8a) Mike went to the pool to swim some laps. After his workout, he went to his psychology class. The professor asked him to summarise the chapter that he'd assigned the class to read.

(8b) Mike went to the pool to swim some laps. After Mike swam some laps, Mike went to Mike's psychology class. The professor of Mike's psychology class asked Mike to summarise the chapter that Mike's psychology professor had assigned Mike's psychology class to read.

This repetition of identity reference can be detrimental to comprehension. Research has shown a **repeated name penalty**, an increase in reading times when a direct reference is used again (e.g. the person's name) compared to when a pronoun is used (e.g. Almor, 1999; Gordon & Chan, 1995; Gordon & Scearce, 1995[56]). That said, when we produce language, if there is more than one character being discussed or present in a situation, people are less likely to use indirect references, such as pronouns, and are more likely to use a direct reference, such as a person's name (Arnold & Griffin, 2007[57]). This may be more acceptable under these circumstances because there may be some ambiguity as to who is being referenced.

Simple reference

Simple reference picks out an entity or entities in a more or less direct manner. That said, there are several ways of doing this. In natural discourse, different kinds of reference can occur.

Consider three simple forms of reference:

(9) I saw a convertible yesterday. The convertible was red.

(10) I saw a convertible yesterday. The car was red.

(11) I saw a convertible yesterday. It was red.

In sentence 9 the reference is so direct that it requires no inference on the part of the listener; this is *identity reference*, using the definite article *the* to refer back to a previously introduced concept, *a convertible*. *Synonym reference* requires that you consider whether the second word is an adequate synonym for the first, as in sentence 10; can *a convertible* also be referred to as *the car*? *Pronoun reference* requires similar reference and inference steps. In sentence 11, pronoun reference can refer only to the *convertible*, because it is the only concept in the earlier phrase that can be equated with *it*. That is, in English, the word *it* must refer to an ungendered concept, just as *he* must refer to a male, and so on. In some languages, the nouns have gender and pronouns must agree with the gender of the noun; translated literally from French we get, 'Here is the Eiffel Tower. She is beautiful.'

Influence of antecedents

Another important idea is that there is some evidence that the order in which antecedents are encountered influences the likelihood that they will be linked to later reference. Two effects of this type are the **advantage of first mention** and the **advantage of clause recency**. In the advantage of first mention, characters and ideas that were mentioned first have a special significance. For example, in a study by Gernsbacher and Hargreaves (1988),[58] after reading a sentence such as 'Tina gathered the kindling as Lisa set up the tent', people responded faster to *Tina* than to *Lisa* (see Figure 9.5). This effect involves remembering which entity was first and which was second. Patients with bilateral hippocampal damage do not show this effect (Kurczek, Brown-Schmidt & Duff, 2013[59]), suggesting we need some kind of memory for information order to perform normally. Clark's (1977)[60] useful list is shown in Table 9.3.

Table 9.3 Types of reference and implication

Direct reference
Identity. Michelle bought a computer. The computer was on sale.
Synonym. Michelle bought a computer. The machine was on sale.
Pronoun. Michelle bought a computer. It was on sale for 20% off.
Set membership. I talked to two people today. Michelle said she had just bought a computer.
Epithet. Michelle bought a computer. The stupid thing doesn't work.
Indirect reference by association
Necessary parts. Eric bought a used car. The tires were badly worn.
Probable parts. Eric bought a used car. The radio doesn't work.
Inducible parts. Eric bought a used car. The salesperson gave him a good price.
Indirect reference by characterisation
Necessary roles. I taught my class yesterday. The time I started was 1:30.
Optional roles. I taught my class yesterday. The chalk tray was empty.
Other
Reasons. Rick asked a question in class. He hoped to impress the professor.
Causes. Rick answered a question in class. The professor had called on him.
Consequences. Rick asked a question in class. The professor was impressed.
Concurrences. Rick asked a question in class. Vicki tried to impress the professor too.

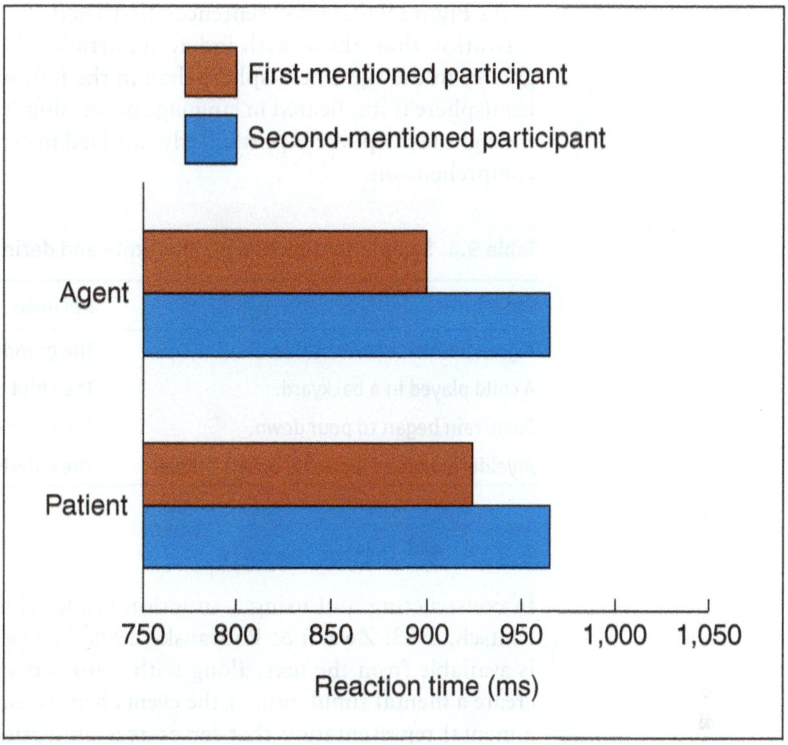

Figure 9.5 Advantage of first mention

Mean response time to names that had appeared in the studied sentences when the name was the first- or second-mentioned participant and when the name played the agent or patient case role in the sentence.

Source: Data from Gernsbacher and Hargreaves (1988)[61].

Advantage of clause recency

Conversely, there is also a time, at the end of the sentence, when the most recent character has an advantage – this is the *advantage of clause recency*. Again, for the sentence 'Tina gathered the kindling as Lisa set up the tent,' if you are probed immediately after it, Lisa has a slight advantage due to recency, but this advantage is shortlived, showing an advantage at about 50–60 ms, but disappearing by 150 ms (Gernsbacher, Hargreaves, & Beeman, 1989[62]).

Other work has found that even the articles used can influence reference. Definite articles (such as *the*) convey given information and make sentences seem more coherent and sensible as compared to when indefinite articles (e.g. *a*, *an* and *some*) are used (Robertson et al., 2000; see Table 10.4 for sample sentences[63]), and sentences with definite articles are remembered better later (Haviland & Clark, 1974[64]).

For Gernsbacher (1997),[65] *the* is a cue for discourse coherence, enabling us to map information more efficiently and accurately. In one study (Robertson et al., 2000[66]), people read sentences, followed by a recognition test (to make sure people tried to comprehend the sentences). Overall, sentences using *the* showed greater evidence of coherence than those with the indefinite *a, an* and *some*. More importantly, this study tested people using fMRI and measured their levels of activity of different brain regions.

As Figure 9.6 shows, sentences that used the definite article showed greater activation than those with indefinite articles. Moreover, these activations were greater in the right hemisphere than in the left, whereas more commonly the left hemisphere is implicated in language processing (e.g. Polk & Farah, 2002[67]). Thus, the right hemisphere is particularly involved in establishing coherence in language comprehension.

Table 9.4 Sample sentences with indefinite and definite articles

Indefinite	Definite
A grandmother sat at a table.	The grandmother sat at the table.
A child played in a backyard.	The child played in the backyard.
Some rain began to pour down.	The rain began to pour down.
An elderly woman led some others outside.	The elderly woman led the others outside.

9.3.2 Situation models

In constructing and using a situation model (Johnson-Laird, 1983; van Dijk & Kintsch, 1983; Zwaan & Radvansky, 1998[69]), a person combines information that is available from the text, along with prior semantic and episodic knowledge, to create a mental simulation of the events being described. This is a *situation model*, a mental representation that serves as a simulation of a real or possible world as described by a text. The important idea is that comprehension is a search after meaning (Graesser, Singer, & Trabasso, 1994[70]). Although comprehension does use some passive activation of semantic and episodic memories (e.g. McKoon & Ratcliff, 1992[71]), we also actively build situation models that elaborate on the causal

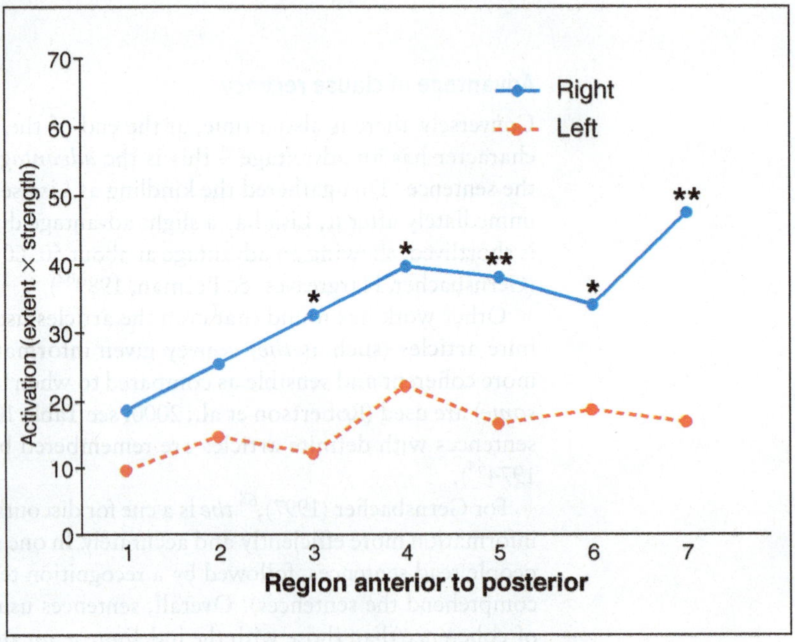

Figure 9.6 Greater activation in sentences using definite article

Activation levels for sentences presented with definite versus indefinite articles (levels in the figure are difference scores, showing how much greater the activations for definite than indefinite article sentences were), for seven left and right hemisphere locations in the brain.
Source: From Robertson et al. (2000)[68].

structure of the event a person is trying to understand. To improve our understanding of situation models, we examine three processes in the use of situation models in comprehension:

1. The use of inferences to elaborate on the information provided by a text.
2. The influence of language structure, namely grammatical aspects, on situation model construction.
3. The updating of the situation model as shifts in the described events are encountered.

Inference making

Instead of specifying everything for comprehension, we rely on people to know the meanings of words and about syntactic devices that structure our discourse, and to share our general conceptual knowledge of the world (e.g. to know that swimming laps can be a workout, or that professors assign chapters for their students to read; you remember a similar point from an earlier module, in discussing scripts). In fact, if you do specify everything exactly, you are breaking an important conversational rule and people will be unhappy with you.

Inferences and language comprehension

People spontaneously draw inferences as they comprehend language. These inferences are then incorporated into the situation models of what is being heard or read. Therefore, people frequently misremember information as having been heard or read, when in fact it was not. For this research box we provide an example of an activity to demonstrate the effect. Please note that all research-related activities should adhere to ethical guidelines on human participation. Please do not carry out any demonstration tasks with human participants unless you have ethical approval from your institution.

In the following example, along with each sentence is an inference that people are likely to make (in parentheses). Read these sentences with your class peers and perform a distractor task, such as solving maths problems for 3–5 minutes. When the distractor period is over, try to recall the sentences. You should find that people will likely report the inferences they made while they were comprehending. That is, they will 'recall' more information than you actually read to them. These inferences – false memories, in a sense – are now part of their memory.

1. The housewife spoke to the manager about the increased meat prices. (complained)
2. The paratrooper leaped out of the door. (jump out of a plane/helicopter)
3. The cute girl told her mother she wanted a drink. (asked)
4. The weatherman told the people of the approaching tornado. (warned)
5. The karate champion hit the cement block. (broke)
6. The absent-minded professor didn't have his car keys. (lost or forgot)
7. The safecracker put the match to the fuse. (lit)
8. The hungry python caught the mouse. (ate)
9. The man dropped the delicate glass pitcher. (broke)
10. The clumsy chemist had acid on his coat. (spilled)
11. The barnacle clung to the sides. (ship)
12. Dennis sat in Santa's chair and asked for an elephant. (lap)

Grammatical aspect

When thinking about language comprehension in terms of the situation model, one can view language as being a set of instructions for how comprehenders should build a situation model in their minds (Zwaan, 1999[72]). So, the things we say suggest to other people the elements of an event that should be part of their understanding. If you were to say to your friends, 'I left my backpack at the library', this would tell them how to build their situation model and what to put in it. First, there is a spatial-temporal framework for that event. The place is the library, and the thing that was done (leaving the backpack there) was done in the past. Second, there is an entity involved, your backpack, and it is (or was) located within the library. Moreover, you may infer some unspoken information, such as that your backpack may be at a table on the second floor (where you usually study), and perhaps even create a mental image derived from this situation model of those circumstances.

Given that language is a set of instructions for understanding described circumstances, then saying different things, even when the differences seem subtle, can result in different understandings. A good illustration of this is *grammatical aspect*, which is how an event action or state is conveyed by a verb, as in the idea of a verb tense. For example, the sentence 'Sam walked to the store' conveys the perfective aspect, and the sentence 'Sam was walking to the store' contains an imperfective aspect. For the perfective aspect, the action *walk* is described as having already been completed. For the imperfective aspect, the action is described as ongoing. As such, you would create two different situation models for these two sentences. For the first, you might create a situation model in which Sam is in the store; for the second, you would create a situation model in which Sam is walking along a pavement on his way to the store (e.g. Anderson, Matlock, & Spivey, 2013; Magliano & Schleich, 2000[73]).

A study by Madden and Zwaan (2003)[74] offers an example of the influence of verb aspect on cognition. Researchers presented people with a series of pictures to view. Some of these pictures involved an ongoing action (e.g. a person driving), whereas other pictures involved a completed action (e.g. a driver having arrived home). After viewing each picture, people were given sentences to read and verify whether those sentences were satisfactory descriptions of the preceding picture. The sentences contained either a perfective verb ('The man made the fire') or an imperfective verb ('The man was making the fire'). People were asked to make these responses as quickly as possible. Madden and Zwaan found that for the completed event pictures, people responded faster to the perfective than to the imperfective sentences. Thus, the state of the described event as it is mentally represented is captured, to some degree, in the language used in a description.

Situation model creation

A simple type of inference making is what Clark (1977)[75] termed a **bridging inference**, which is a process of constructing a connection between concepts. A bridging inference binds two units of language together. For example, determining that a reference like the epithet *the stupid thing* refers to the same entity as a *computer* is a bridging inference. It builds a connection between these two forms of reference, indicating that they refer to the same discourse entity. In bridging inference, the language producer uses reference to indicate the intended kinds of implications. For their part, comprehenders interpret the statement in the same fashion, computing the references and drawing the inferences needed. When the implication and inferences are intended, they are called **authorised inferences**. Alternatively, unintended implications and inferences are called **unauthorised inferences**, as when I say, 'Your hair looks pretty today,' and you respond, 'So you think it was ugly yesterday?' (see also McKoon & Ratcliff, 1986[76]).

The examples in Table 9.3 make it clear that the bridges we need to build for comprehension vary in their complexity, from simple and direct to difficult and remote. Even on intuitive grounds, consider how the following sentences differ in the ease of comprehension:

(12) Marge went into her office. It was very dirty.

(13) Marge went into her office. The floor was very dirty.

(14) Marge went into her office. The African violet had bloomed.

Whereas sentence 12 is a simple case of pronoun reference, sentence 13 refers back to *office* with the word *floor*. Because an office necessarily has a floor, it is clear the implication in sentence 13 is that it was Marge's office floor that was dirty. One property you retrieve from semantic memory is that an office has a floor. Thus, if you comprehend that the office floor was dirty, you draw this inference. But it is an even longer chain of inference to draw the inference in sentence 14 that Marge has an African violet in her office; a floor is necessary, but an African violet is not. Overall, the integration of this semantic knowledge with the information in the text is part and parcel of creating a situation model.

It seems likely that the structure of concepts in semantic memory activation influences the ease with which information is inferred during situation model construction (e.g. Cook & Myers, 2004[77]). So, more predictable pieces of information are processed faster (McKoon & Ratcliff, 1989; O'Brien, Plewes, & Albrecht, 1990[78]). Marge's office necessarily has a floor as well as a desk, a chair, some shelves, and so on. It is conceivable that it has some plants, but that fact is optional enough that sentence 14 would take more time to comprehend.

Influences of embodied cognition

Further evidence that people are drawing on their semantic knowledge, and that this knowledge has an embodied character, was found in a study by Zwaan, Stanfield, and Yaxley (2002).[79] People read short descriptions of situations and then were shown pictures of objects. The task was to indicate whether the pictured object had been in the description they read. The critical manipulation was whether the picture either matched or mismatched the perceptual characteristics of the object in the description. For example, the critical sentence could be either 'The ranger saw the eagle in the sky' or 'The ranger saw the eagle in its nest' followed by a picture of either an eagle with its wings outstretched or perched (see Figure 9.7).

Zwaan et al. found that people responded faster when the picture matched the described state. Even though they saw an eagle in both pictures, the eagle with its wings outstretched 'matched' the 'eagle in the sky' description better, so people responded faster (see also Connell & Lynott, 2009, for the activation of perceptual

Figure 9.7a Example of a picture of an eagle in flight
Source: Nature's Charm/Shutterstock

Figure 9.7b Example of a picture of an eagle on a perch
Source: Marr Ragen/Shutterstock

colour information[80]). Thus, people seemed to be activating perceptual qualities of objects during comprehension. Conversely, if people view a picture before reading, their response times are faster if the sentence matches the picture than if it doesn't (Wassenburg & Zwaan, 2010[81]).

Speech act

People are also aware of the intended consequence of someone saying something, called a **speech act** (Searle, 1969).[82] For example, if you ask your roommate to turn down the stereo, the *speech* itself is the set of words you say, but the *speech act* is your intention, getting your roommate to let you study for an upcoming exam. Not only do people spontaneously derive the implied speech acts of what other people say, but they may misremember what was said in terms of the speech act itself. For instance, in a study by Holtgraves (2008b),[83] people read a series of short vignettes, some of which conveyed speech acts. For example, suppose people read the following story: 'Gloria showed up at the office wearing a new coat. When her co-worker Heather saw it, she said to her, 'Gloria, I like your new coat.' The last sentence conveys the speech act of complimenting Gloria. What Holtgraves found was that people were more likely to mistakenly remember that they had read 'I'd like to compliment you on your new coat', an utterance that describes the speech act. However, a different group of people read a different version of the story, in which the last two sentences were: 'When her co-worker Heather saw it, she said *to* her friend Stacy: "I like her new coat."' In this condition, people were less likely to misremember having read, 'I'd like to compliment her on her new coat.' Because there was no actual complimenting speech act to Gloria, people did not store this information in memory and so did not make the memory error. Interestingly, the processing of speech act information appears to be dominated more by right hemisphere than left hemisphere (Holtgraves, 2012[84]), consistent with the idea that understanding a speech act involves going beyond the actual words themselves.

Realise that we do not automatically and spontaneously draw all possible inferences while we read. Although some inferences are directly and typically drawn, such as simple and straightforward references, others are more complex and may not be drawn, and possibly should not be drawn. If you did draw such inferences, your cognitive resources would be quickly overwhelmed (Singer, Graesser & Trabasso, 1994[85]). For example, when you read a sentence like 12 or 13, you are likely not to draw an inference that Marge decided to clean her office, although if the next sentence in the story said that, you would certainly understand it. Most of the inferences you make are *backward* inferences. You are trying to understand what has already been described and how it all goes together. You make *forward* inferences – trying to predict what will happen next – under much rarer circumstances (Millis & Graesser, 1994[86]).

Moreover, the creation of the situation model needs to account for constraints of embodiment. For example, in a study by de Vega, Robertson, Glenberg, Kaschak, and Rinck (2004; see also Radvansky, Zwaan, Federico, & Franklin, 1998[87]), people were asked to read a series of passages. Embedded in those passages were critical sentences that described two actions that a character was doing, either at the same time or in sequence. If a person was described as doing two things that require the same parts of the body, such as 'While chopping wood with his large axe, he painted the fence white', reading times were slower, as if readers were trying to figure out how this could be done. However, reading times were faster when either different parts of the body were being used, such as 'While whistling a lively folk melody, he painted the fence white', or the actions were done in sequence, such as 'After chopping wood with his large axe, he painted the fence white.' Thus, we account for the limits of our human bodies, and the way actions happen in time, to help us comprehend what we are reading.

Individual differences

Interestingly, reference and inference processes depend significantly on individual characteristics of the reader, particularly on the reader's skill. For instance, Long and De Ley (2000)[88] found that less skilled readers resolve ambiguous pronouns just as well as more skilled readers, but they do so only when they are integrating meanings together. The more skilled readers resolve the pronouns earlier, probably when they first encounter a pronoun.

Several studies have also examined inferences as a function of working memory capacity (e.g. Fletcher & Bloom, 1988[89]). One study by Singer, Andrusiak, Reisdorf, and Black (1992)[90] explored individual differences in bridging as a function of working memory capacity and vocabulary knowledge. The gist of this work is that the greater your working memory capacity and vocabulary size, the greater the likelihood that information necessary for an inference will still be in working memory and can be used (see also Long, Oppy, & Seely, 1997; Miyake, Just, & Carpenter, 1994[91]).

Evidence for individual differences in comprehension has also revealed itself in neurological measures. In a study by Virtue, van den Broek and Linderholm (2006),[92] people read sentences that had causal constraints that were either weak (e.g. As he arrived at the bus stop, he saw his bus was already five blocks away) or strong (As he arrived at the bus stop, he saw his bus was just pulling away). During reading, the researchers presented lexical decision probes that corresponded to likely inference that the readers might make (e.g. *run*, in this case). Of significance, this presentation was done to the left and right hemispheres by presenting the words on either the right or left half (respectively) of the computer screen. The data showed that the right hemisphere was more involved in generating remote associations (associated concepts that are not closely semantically related to the concepts in the sentences). Moreover, people with high working memory capacity activated fewer remote associations than low-span people. Essentially, people with a high working memory span were more focused in the amount of knowledge they activated during comprehension.

Updating

Situations that we experience or read about often are in a state of flux. Things are always changing, and the events may differ from one moment to the next. Thus, the cognitive processes involved in comprehension must be able to shift the current understanding to adapt to these ongoing changes.

Situation model updating

Several updating processes alter a person's situation model in the face of information about how the situation has changed.

To provide a framework for understanding how these changes can occur, we will use Zwaan's Event Indexing Model (Zwaan, Langston, & Graesser, 1995; Zwaan, Magliano, & Graesser, 1995; Zwaan & Radvansky, 1998[93]). Per this theory, people actively monitor multiple event dimensions during reading to assess whether there has been a meaningful change along any of them.

In the original version of the theory, five dimensions were proposed: space, time, entity, intentionality (goals) and causality. When there is a disruption along any one of these dimensions, people update their situation models, and this updating process takes time. For example, a break along the space and time dimensions could happen if a story protagonist were to move to a new location or there were a jump

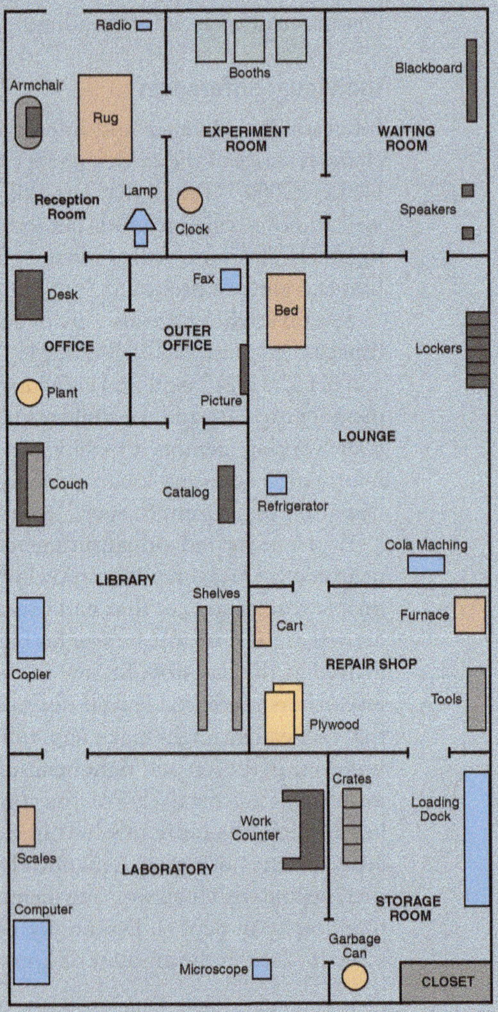

Figure 9.8 Updating along the spatial dimension

This figure shows an example of the map of the research centre memorised by people in Morrow, Greenspan and Bower's (1987)[96] study of spatial updating during language comprehension.

in time (e.g. a week later . . .). Similarly, if a new character were introduced into a story, the person would need to update the entity dimension, and so on. The mental processing involved in updating any of these dimensions appears to operate largely independently of the others (Curiel & Radvansky, 2014[94]), suggesting that people are actively tracking multiple dimensions of experience when trying to understand stories. Further research has shown that people monitor more than just these five dimensions. For example, people also track emotional information (e.g. Komeda & Kusumi, 2006[95]).

They then read narratives about people moving about in that space. During reading, people were occasionally probed with pairs of object names, such as sink–furnace. The task was to indicate whether the two objects were in the same room (in this case, they were not). The critical factor was the distance on the map between the story protagonist's current location and the location of the probe objects. Moreover, these memory probes came after motion sentences in which the person moved from one room to another, such as 'He walked from the laboratory into the wash room.' Based on this, four conditions were defined. The Location Room was the room that the person just moved to – the wash room in this case. The Path Room was the (unmentioned) room that the person walked through to get to the Location Room – the storage area here. The Source Room was the room

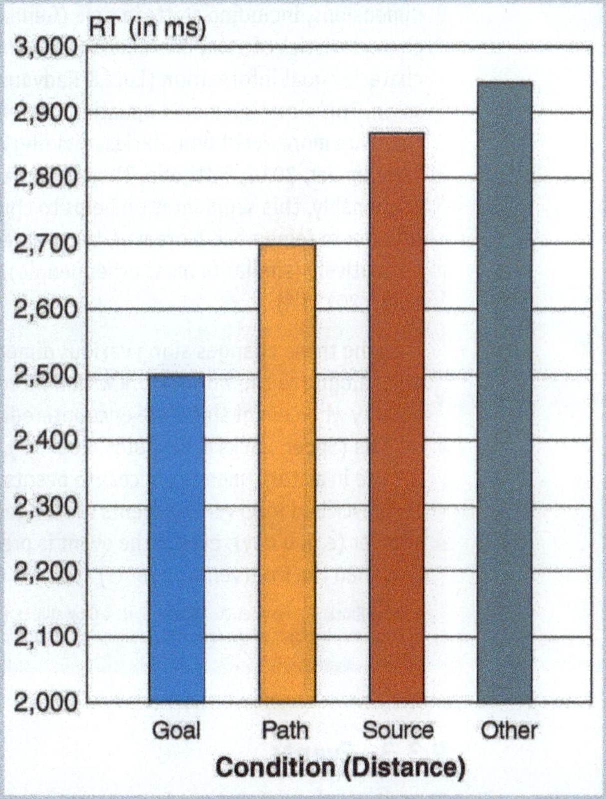

Figure 9.9 **Response time data from a study by Morrow et al. (1987)[97]**

While reading a passage about a character moving about the building in Figure 9.9, people were interrupted with memory probes. In this case, the probes were two objects, and the person's task was to indicate whether they were in the same room or not. In this task, the Goal Room was the room the story character had just moved to (the goal of the movement), the Path Room was a room along the character's path of travel, the Source Room was the room that the movement started from (the source from which the movement began), and the Other Room was just some other room in the building.

the person was in just before moving to the Location Room – the laboratory here. Finally, there was an 'Other Room' condition, which corresponded to probes from any other room in the building.

The results from one study, shown in Figure 9.10, reveal that response times to the memory probes increased with an increase in distance between the protagonist and the objects. It is as if people were mentally scanning their situation models from the protagonist's current place in the building to another room.

The farther away that other room was, the longer it took people to scan. People actively updated their situation models as there were changes in spatial locations in the texts. When people read that a story character has moved from one room to another, they update their situation model so that the spatial framework that is at the focus of comprehension is now different. What is particularly compelling here are the response times to the probes from the Path Room condition. This room was not even mentioned, yet people were scanning their situation model in a way that activated that information. If people were simply activating knowledge of the rooms that the protagonist was in, then this would not have occurred. But because people were mentally simulating the environment as they read, this activation of an intermediate location emerged.

Similar influences of situation model updating can be seen for other situation model dimensions, including shifts in time (Gennari, 2004; Zwaan, 1996[98]), monitoring characteristics of story characters (Albrecht & O'Brien, 1995[99]), processing character goal information (Lutz & Radvansky, 1997; Suh & Trabasso, 1993[100]), and so on. This situation model updating is useful because memory is improved when there are more event boundaries, dividing action into multiple events (Huff, Meitz & Papenmeier, 2014; Pettijohn, Thompson, Tamplin, Krawietz & Radvansky, 2016[101]). Presumably, this segmentation helps to chunk and organise the information, making it easier to remember. Moreover, the greater the degree to which people segment normatively (similar to most other people), the better their memory will be (Sargent et al., 2013[102]).

Tracking these changes along various dimensions is part of an attempt to create an analogue to the world. Work with fMRI recordings has shown increases in brain activity when event shifts are encountered and people need to update their situation models (Speer, Zacks & Reynolds, 2007[103]). For example, looking at the dimension of time in a story, memory access to events earlier in a sequence is more difficult if the described intervening events are longer in duration (e.g. a year) than if they are shorter (e.g. a day), even if the event is presented as a flashback (i.e. later in the text than the intervening events) (Claus & Kelter, 2006[104]).

Source: Rinck, M., Hahnel, A., Bower, G. H. & Glowalla, U. (1997). The metrics of spatial situation models. *Journal of Experimental Psychology: Learning, Memory, and Cognition, 23*(3), 622–63.

9.3.3 Events

Our discussion of comprehension up to now has largely focused on language comprehension of either written or spoken language, and this reflects the thrust of research in this area. However, this is not the only type of comprehension that people can engage in. We also comprehend events that we see or are involved in.

Comprehending events

To better understand event comprehension, it helps to look at comics, video and interactive experiences.

Comics

Narrative events often are thought of as being written or spoken descriptions, and certainly some narratives are. However, we encounter many different narratives in the world. A common type of narrative is comics. This includes comic strips, comic books and graphic novels. What is interesting is that these narratives convey information visually as well as verbally in a printed format. How do they compare to written and spoken narratives in terms of the cognitive processes that are used to understand and remember them? The answer is: quite well.

Like written narratives, comics have a structure to convey information, much like the syntactic structure of language. In fact, Cohn (2013b)[105] has developed a theory of visual narrative grammar he applies to comics and other visual narratives. For example, agents typically appear prior to patients in comics (Cohn & Paczynski, 2013[106]). Thus, there is a mental syntax, or set of rules, that we follow to comprehend visual narratives of this type. That said, unique aspects of processing and comprehension occur when a person reads comics, particularly on a rather complicated page in a graphic novel (Cohn 2013a[107]).

Video

Another example of extending research on comprehension beyond strictly linguistic materials are studies using video. Works by Cutting (2014; Cutting, Brunick & Candan, 2012; Cutting & Iricinschi, 2015[108]), Magliano (Magliano, Miller & Zwaan, 2001; Magliano, Taylor & Kim, 2005[109]), and Zacks (Zacks et al., 2001; Zacks, Speer, Swallow, Braver & Reynolds, 2007[110]) show that people are actively comprehending events viewed on video or film. For example, in a study by Magliano et al. (2001),[111] rather than having people read texts, people watched narrative films such as *Star Trek II: The Wrath of Khan*. As people watched these movies, they indicated when they thought the situation depicted by the film had changed. The results showed that people made these indications at the same points as situation model theory suggested they would update their understanding.

For example, people indicated there was a change in the film if a new character entered a scene, if there was a change in spatial location, if something unexpected happened, and so on – exactly as the theory predicts. Thus, comprehension is a general process and not unique to language.

9.4 Conversation and gesture

LEARNING OBJECTIVE
Evaluate the cognitive effects of conversational interactions

We turn now to the comprehension of conversation and gesture. We focus on *conversation* – normal, everyday language interactions, such as an ordinary talk among friends. The issues we consider, however, apply to all kinds of linguistic interactions: how professors lecture and students comprehend, how people converse on the telephone, how an interviewer and a job applicant talk, how we reason and argue with one another (Rips, 1998)[112], and so on. Furthermore, we look at how we expand on what we say by moving our hands about, making gestures, and by examining the cognitive role of these gestures.

During a conversation, speakers develop a rhythm as each person takes successive turns speaking. Non-verbal interaction can occur during a turn, such as when a listener nods to indicate attention or agreement.

9.4.1 The structure of conversations

Conversations have a form and structure. They are not chaotic and random and we expect them to go certain ways. These expectations are part of the cognitive process of how conversations emerge. This section covers the structure of conversations and the implications for our understanding of cognition.

Characteristics of conversations

Let's examine two characteristics of conversations, the issues of turn taking and social roles, to get started and introduce some of the more cognitive effects of interest.

Taking turns

Conversations are structured by a variety of cognitive and social variables and rules governing 'the what and how' of our contributions. To begin, we take turns. Typically, there is little overlap between participants' utterances. Generally, two people speak simultaneously only at the change of turns, when one speaker is finishing and the other is beginning. In fact, interchanges in conversation often come in an *adjacency pair*, a pair of turns that sets the stage for another part of the conversation. For instance, if Ann wants to ask Betty a question, there can be an adjacency pair of utterances in which Ann sets the stage for the actual question:

Ann: Oh, there's one thing I wanted to ask you.

Betty: mhm

Ann: In the village, they've got some of those . . . rings. . . . Would you like one?

(From Svartvik & Quirk, 1980, cited in Clark, 1994.[113])

The neutral 'mhm' is both an indication of attention and a signal that Ann can go ahead and ask the question (Duncan, 1972[114]).

The rules we follow for turn taking are straightforward (Sacks, Schegloff & Jefferson, 1974[115]). First, the current speaker oversees selection of the next speaker. This is often done by directing a comment or question toward another participant ('What do you think about that, Fred?'). Second, if the first rule is not used, then anyone can become the current speaker. Third, if no one else takes the turn, the current speaker may continue but is not obliged to.

Speakers use a variety of signals to indicate whether they are finished with their turn. For example, a long pause at the end of a sentence is a turn-yielding signal, as are a comment directed at another participant, a drop in the pitch or loudness of the utterance, and establishing direct eye contact with another person; the latter is often merely a non-verbal way of selecting the next speaker. If the current speaker is not relinquishing the conversational turn, however, these signals are withheld. Other 'failure to yield' signals include trailing off in mid-sentence without completing the grammatical clause or the thought, withholding such endings as 'you know', or even looking away from other participants during a pause (Cook, 1977[116]).

In addition to overt signals of when turn taking may occur, there are neurological underpinnings. Margaret and Thomas Wilson (Wilson & Wilson, 2005[117]) have suggested that conversational turn taking may be tied to neurological oscillators that help us keep track of time. The idea is that these oscillators become synchronised to one another based on the rate at which people are producing syllables. These oscillators give us a neurologically based intuition about the pace of the conversation and when it would be appropriate to step in and take our turn.

Social roles and wettings

The social roles of conversational partners, along with conversational settings, influence the contributions made by participants (Kemper & Thissen, 1981[118]). Formal settings among strangers or mere acquaintances lead to more structured, rule-governed conversations than informal settings among friends (Blom & Gumperz, 1972[119]). Conversations with a 'superior' – for instance, your boss or a police officer – are more formal and rule-governed than those with peers (e.g. Brown & Ford, 1961; Edwards & Potter, 1993, and Holtgraves, 1994, discuss the social and interpersonal aspects of such situations[120]). More generally, when involved in conversation, people consider a wide variety of information, not only the words and syntax, but also the context and who is speaking (Van Berkum, 2008[121]).

9.4.2 Cognitive conversational characteristics

Conversations are structured by cognitive factors. We focus on three: the conversational rules we follow, the issue of topic maintenance, and the online theories of conversational partners.

Conversational rules

Grice (1975; see also Norman & Rumelhart, 1975[122]) suggested four **conversational rules** or maxims, rules that govern our conversational interactions with others, all derived from the **cooperative principle**. This is the idea that each participant in a conversation implicitly assumes all speakers are following the rules and that each contribution to the conversation is a sincere, appropriate contribution. In a sense, we enter a pact with our conversational partner, pledging to abide by certain rules and adopting certain conventions to make our conversations manageable and understandable (Brennan & Clark, 1996; Wilkes-Gibbs & Clark, 1992[123]). This includes issues of syntax, where we choose syntactic structures that mention important discourse focus information early in our sentences (Ferreira & Dell, 2000[124]) and use syntactic structures that are less ambiguous (Haywood, Pickering, & Branigan, 2005[125]). We use intonation and prosody that help disambiguate an otherwise ambiguous syntactic form (Clifton, Carlson, & Frazier, 2006)[126]; we decide on word choice, as in situations when two conversational partners settle on a mutually acceptable term for referring to some object (Metzing & Brennan, 2003[127]; Shintel & Keysar, 2007[128]); and we often use gestures to amplify or disambiguate our speech (Goldin-Meadow, 1997; Kelly, Barr, Church & Lynch, 1999; Özyürek, 2002[129]).

As Table 9.5 shows, the four maxims specify in detail how to follow the cooperative principle. (Two further rules have been added to the list for purposes that will become clear in a moment.)

A simple example or two should help you understand the point of these maxims. When a speaker violates or seems to violate a maxim, the listener assumes there is a reason and may not detect a violation (Engelhardt, Bailey & Ferreira, 2006[132]). That is, the listener assumes that the speaker is following the overarching cooperative

Table 9.5 Grice's (1975)[130] conversational maxims, with two additional rules

The cooperative principle	
Relevance	Your utterances should be relevant to the discourse (e.g. stay on topic; don't make statements about things that others are not interested in).
Quantity	As needed, provide as much information as is necessary (e.g. don't give too much information; don't go beyond or give short shrift to what you know; don't give too much information).
Quality	Have what you say be truthful (e.g. don't give misleading information; don't lie; don't exaggerate).
Manner and tone	Aim for clarity (e.g. avoid saying things that are unnecessarily ambiguous or obscure); keep it brief, but polite, and don't interrupt someone else.
Two additional rules	
Relations with conversational partner	Infer and respond to partner's knowledge and beliefs (e.g. tailor contributions to partner's level; correct misunderstandings).
Rule violations	Signal or mark intentional violations of rules (e.g. use linguistic or pragmatic markers [stress, gestures]; use blatant violations; signal the reason for the violation).

From Grice (1975)[131]; see also Norman and Rumelhart (1975).

principle and so must have intended the remark as something else, maybe sarcasm or maybe a non-literal meaning (Kumon-Nakamura, Glucksberg & Brown, 1995[133]). As an example, imagine studying in the library when your friend asks:

(15) Can I borrow a pencil?

This is a straightforward speech act, a simple request you could respond to directly. But if you had just lent a pencil to your friend, and he said,

(16) Can I borrow a pencil with lead in it?

the question means something different. Assuming your friend was being cooperative, you now must figure out why he broke the quantity maxim about overspecifying: All pencils have lead in them, and mentioning the lead is a violation of a rule. You infer that it was a deliberate violation, where the friend's authorised implication can be expressed as 'The pencil you lent me doesn't have any lead in it, so would you please lend me one I *can* use?' In general, people are adept at decoding speech acts and knowing what others are trying to achieve by what they say (Holtgraves, 2008a[134]).

Topic maintenance

We also follow the conversational rules in terms of **topic maintenance**, making our contributions relevant to the topic and sticking to it. Topic maintenance depends on two processes: comprehension of the speaker's remark and expansion, contributing something new to the topic.

Schank (1977; see also Litman & Allen, 1987[135]) provides an analysis of topic maintenance and topic shift, including a consideration of what is and is not a permissible response, called simply a *move*, after one speaker's turn is over. The basic idea is that the listener comprehends the speaker's comment and stores it in memory. As in reading, the listener must infer what the speaker's main point was or what the discourse focus was. If the speaker, Ben, says,

(17) I bought a new car in Baltimore yesterday,

then Ed, his conversational partner, needs to infer Ben's main point and expand on that in his reply. Thus, sentence 18 is legal because it apparently responds to the speaker's authorised implication, whereas sentence 19* is probably not a legal move (denoted by the * sign):

(18) Ed: Really? I thought you said you couldn't afford a car.

(19*) Ed: I bought a new shirt yesterday.

Sentence 18 intersects with two main elements for sentence 17, 'BUY' and 'CAR', so it is probably an acceptable expansion. Sentence 19* intersects with 'BUY', but the other common concept seems to be the time case role 'YESTERDAY', an insufficient basis for most expansions. Thus, in general a participant's responsibility is to infer the speaker's focus and expand on it in an appropriate way. That is the relevance maxim: sticking to the topic means you must infer it correctly. Ed seems to have failed to draw the correct inference.

On the other hand, maybe Ed did comprehend Ben's statement correctly. If so, then he deliberately violated the relevance maxim in sentence 19*. But it is such a blatant violation that it suggests some other motive. Ed may be expressing disinterest in what Ben did or may be saying indirectly that he thinks Ben is bragging. And if Ed suspects Ben is telling a lie, then he makes his remark even more blatant, as in sentence 20:

(20) Yeah, and I had lunch with the Queen of England.

Online theories during conversation

A final point involves the theories we develop of our conversational partners, something called **theory of mind (ToM)**. This refers to the ability that we have to recognise that another person's knowledge and beliefs are different than our own (Fodor, 1992; Frith & Frith, 2005[136]) In developmental psychology ToM has been studied extensively in typically developing children and those with neurodiversity characteristics (e.g. Asperger's syndrome, autism). In the context of conversations, it plays a crucial role in how people communicate and their socio-cognitive development more widely. For example, children's theory of mind becomes more complex as they grow older and therefore their conversational ability more elaborate and flexible (De Rosnay & Hughes, 2006[137]). In its simplest form also called **direct theory**, this is the mental model of what the conversational partner knows and is interested in, what the partner is like. We tailor our speech so we are not being too complex or too simplistic, and so we are not talking about something of no interest to the listener. Some clear examples of this involve adult–child speech, where a child's smaller vocabulary and knowledge prompt adults to modify and simplify their utterances in several ways (DePaulo & Bonvillian, 1978; Snow, 1972; Snow & Ferguson, 1977[138]). But sensitivity to the partner's knowledge and interests is present to some degree in all conversations – although not perfectly, of course. We do not talk in our college classes the way we would to a group of 5-year-olds, nor do we launch into conversations with security guards or cashiers about our research. Horton and Gerrig (2002)[139] call this 'audience design', which is awareness of the need to design our speech to the characteristics of our audience (e.g. Lockridge & Brennan, 2002[140]). Alternatively, if we do not know much about another, we may assume that person knows what we know (Nickerson, 2001[141]) and then revise our direct theory as we observe how well that person follows our remarks (Clark & Krych, 2004[142]).

Audience design has implications beyond conversations. When we tell stories, we modify what we tell our listeners based on who they are and our social relationship to them. This retelling is not the same as recall. We modify the information we report to fit the social situation. In retelling stories, we often exaggerate some parts, minimise others, add information that was not there originally, and leave some bits out, all to suit our audience and the broader message we are trying to convey (Marsh, 2007[143]).

There is another layer of theories during a conversation, an interpersonal level related to 'face management', or public image (Holtgraves, 1994, 1998[144]). Let's call this the **second-order theory of mind**. This second-order theory is an evaluation of the other participant's direct theory: what you think the other participant believes about you.

Examples of direct and second-order theories

Let's develop examples of these two theories to illustrate their importance. Imagine that you are registering for classes next semester and say to your friend Frank that you have decided to take Psychology of Personality. What would your reaction be if Frank responded to you with these statements?

(21) Why would you want to take that? It's just a bunch of experiments with rats, isn't it?

(22) Yeah, I'm taking Wilson's class next term too. John told me he's going to assign some books he thinks I'll really like.

(23) Maybe you shouldn't take Wilson's class next term. Don't you have to be pretty smart to do all that reading?

In sentence 21, you assume your friend made the remark sincerely; that it was intended to mean what it says. Because you know that research on laboratory animals had little to do with the field of personality, you conclude that your friend knows a lot less about personality theory than you do. This becomes part of your direct theory.

For sentence 22, you probably interpret Frank's remark as boastful, intended to show that he's on a first-name basis with the professor. Frank has authorised that inference by using a more familiar term of address. You update both your direct theory of Frank and your second-order theory. You update your direct and second-order theories after sentence 23, too, but the nature of the updates is different: you have been insulted by the implication in Frank's response, something like 'He thinks I'm not smart enough to take the class.'

Here are some examples of direct and second-order theories:

Setting: For all three conversations, Chris's first sentence and direct theory are the same.

Chris: 'I think I'll take Personality with Dr Wilson next term.'

Chris's direct theory: Frank is interested in the courses I'm taking.

Conversation 1

Frank replies: 'Personality? Ah, that's just a bunch of experiments with rats, isn't it?'

Chris's updated direct theory: Frank doesn't know much about personality research.

Conversation 2

Frank replies: 'Yeah, I am too. John told me he's going to assign some books he thinks I'll really like.'

Chris's updated direct theory: Frank knows the professor on a first-name basis, and he's bragging about it by calling him John.

Chris's second-order theory: Frank thinks I'll be impressed that he calls the professor John.

Conversation 3

Frank replies: 'Hmm, maybe you shouldn't take that class. Don't you have to be pretty smart to do all that reading?'

Chris's updated direct theory: Frank is a jerk; he just insulted me?

Chris's second-order theory: Frank thinks I'm not smart.

9.4.3 Empirical effects in conversation

Let's conclude with some evidence about the conversational effects we have been discussing. One of the most commonly investigated aspects of conversation involves **indirect requests**, such as when we ask someone to do something ('Close the window'; 'Tell me what time it is') by an indirect and presumably politer statement ('It's drafty in here'; 'Excuse me, but do you have the correct time?').

Indirect requests

Clark (1979)[145] reported an impressive investigation of indirect requests. The study involved telephone calls to some 950 merchants in the San Francisco area in which the caller asked a question that the merchant normally would be expected to deal with on the phone (e.g. 'What time do you close?'; 'Do you take credit cards?'; 'How much does something cost?'). The caller would write down a verbatim record of the call immediately after hanging up. A typical conversational interaction was as follows:

(24) Merchant: 'Hello, Scoma's Restaurant.'
Caller: 'Hello. Do you accept any credit cards?'
Merchant: 'Yes we do; we even accept Carte Blanche.'

Of course, the caller's question here was indirect: 'Yes' isn't an acceptable answer to 'Do you accept any credit cards?' because the authorised implication of the question was 'What credit cards do you take?' Merchants almost always responded to the authorised implication rather than merely to the literal question. Furthermore, they tailored their answers to be informative while not saying more than necessary (obeying the second rule, on quantity), as in 'We accept *only* Visa and MasterCard,' or 'We accept *all* major credit cards.' Such responses are both informative and brief.

Such research has been extended to include not only indirect requests but also a variety of indirect statements and replies to questions. For instance, Holtgraves (1994)[146] examined comprehension speed for indirect requests as a function of whether the speaker was of higher status than the listener (e.g. boss and employee) or whether they were of equal status (two employees). People read a short scenario (e.g. getting a conference room ready for a board of directors' meeting), which concluded with one of two kinds of indirect statements. *Conventional statements* were normal indirect requests, such as 'Could you go fill the water glasses?' *Negative state remarks* were more indirect, merely stating a negative situation and only indirectly implying that the listener should do something (e.g. 'The water glasses seem to be empty'). People showed no effects of status when comprehending regular indirect requests; it did not matter whether it was a peer or the boss who said, 'Could you go fill the water glasses?' But comprehension time increased significantly with negative state remarks made by peers. In other words, when the boss says, 'The water glasses seem to be empty', we comprehend the conventional indirect request easily. But when a peer says it, we need additional time to comprehend.

Indirect replies

Holtgraves (1998)[147] also focused on indirect replies, especially the idea of making a 'face-saving' reply. His participants read a description of a situation, such as:

(25) Nick and Paul are taking the same history class. Students in this class must give a 20-minute presentation to the class on some topic.

They then read a sentence that gave positive (26) or negative (27) information about Nick's presentation or a sentence that was neutral (28):

(26) Nick gave his presentation and it was excellent. He decides to ask Paul what he thought of it: 'What did you think of my presentation?'

(27) Nick gave his presentation and it was truly terrible. He decides to ask Paul what he thought of it: 'What did you think of my presentation?'

(28) Nick gave his presentation and then decided to ask Paul what he thought of it: 'What did you think of my presentation?'

If you were Paul and faced the prospect of telling Nick that his presentation was awful, wouldn't you look for some face-saving response? This is exactly how people responded when they comprehended Paul's responses. In the excuse condition, Paul says:

(29) It's hard to give a good presentation,

in effect giving Nick a face-saving excuse for his poor performance. Another possible move is to change the topic, to avoid embarrassing Nick, as in:

(30) I hope I win the lottery tonight.

Holtgraves (1998)[148] collected several measures of comprehension, including overall comprehension time for the critical sentences 29 and 30. The comprehension times, shown in Figure 9.10 (from Experiment 2), were very clear. When people had heard positive information – the talk was excellent – it took them a long time to comprehend either the excuse (29) or topic change (30) response. But having heard negative information – the talk was terrible – was nearly the same as having heard nothing about the talk. People comprehended the excuse or topic change responses more rapidly, and there was no major difference between no information and negative information. People interpreted the violations of the relevance maxim as attempts to save face and avoid embarrassment.

Another role of indirect speech is to provide some element of *plausible deniability* (Lee & Pinker, 2010[150]). Indirect requests and responses may carry implicit

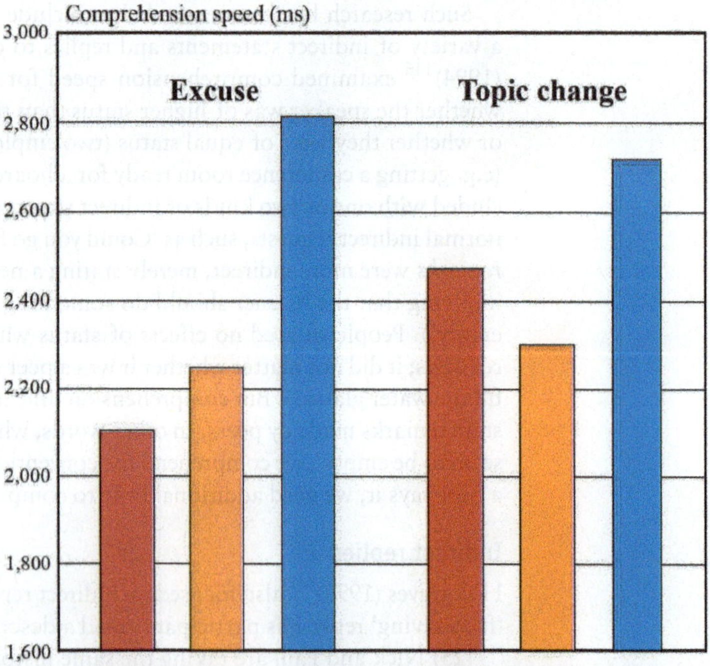

Figure 9.10 Comprehension times from Holtgraves' (1998)[149] study

Participants read settings in which negative information, positive information or neutral information was offered about a character, followed by a conversational move in which the speaker made an excuse for the character or changed the topic. In both cases, it took longer to comprehend the remark when positive information about the character had just been encountered.

meta-linguistic awareness
Being able to reflect on language and evaluate it.

balanced bilingual
People who are equally fluent in both languages, not only verbally but also in written and expressive language.

Cognitive correlates of bilingualism

Early research on bilingualism suggested that bilinguals may have cognitive advantages in areas of language acquisition, including symbolic representation and **meta-linguistic awareness**. Over the decades, there have been conflicting findings mainly due to lack of consideration of other factors such as socio-economic status, whether a person is a **balanced bilingual** or not, with many studies showing advantages for bilinguals on cognitive tasks, while other studies reporting negative, null, or mixed effects of bilingualism. In a meta-analysis of 63 studies (over 6,000 participants), Adesope et al. (2010)[151] found that bilinguals outperform monolinguals on a range of linguistic tasks including abstract and symbolic representation and metalinguistic awareness and some non-linguistic tasks such as attentional control and problem-solving. There were several factors that influenced the effect size of the findings (which ranged from small to large) and these included the age of second language acquisition, the combination of spoken languages (e.g. English-Chinese did not show a superiority effect over monolinguals) and the methodology used to assess these effects.

requests and replies that would be inappropriate otherwise. For example, suppose you are told by a not-so-savoury relative of the defendant of a trial, 'I hear you're the foreman of the jury. It's an important responsibility. You have a wife and kids. We know you'll do the right thing.' Thus, while the literal wording is acceptable, the implied meaning (a threat) is not. Thus, indirect speech can play many roles in human communication.

9.4.4 Metaphors and idioms

There is a great deal of language that is not intended to be interpreted literally. Instead, the expectation is for people to read the intended meaning *into* what is being said. So, although language can be viewed as instructions for how to create one's situation models, these instructions are not always straightforward and direct.

Two additional forms of language

Thus, we consider two additional forms of language in which what is said literally is not what is intended. These are metaphors and idioms.

Metaphors

A metaphor is an expression using language to compare one thing with another, or to use language and knowledge of one domain to understand another domain. So, when you hear people say, 'My job is my jail', they typically do not literally mean that they are physically imprisoned by their employment. Instead, they are comparing their job to the confining, restrictive and inescapable experience of being in jail. This is using language to make an analogy, much as we have seen analogical thinking with other types of cognition. It is just much more explicit and out in the open here. The standard way of thinking about how people cognitively process metaphors is that they first derive the literal meaning of a sentence. This process results in an interpretation that is literally false and may not even make sense. After drawing the literal interpretation, people compare it with what is going on in the current context in which the utterance was produced. At this point, it becomes clear that a literal interpretation does not work. So, people then try to derive a non-literal meaning, using the metaphor as the basis of an analogy.

Psycholinguistically, although there may be cases in which people go through the steps outlined above, it appears that most of the time people are deriving the literal meaning of the metaphor while simultaneously, in parallel, obtaining the metaphoric meaning (Glucksberg, Gildea & Bookin, 1982)[152]. This may involve an actual retrieval of a metaphor from memory if it has been encountered before or deriving a metaphorical interpretation by making a mental analogy using the literal circumstances of the language and comparing it with the set of circumstances confronting the person. This derivation of metaphoric analogy appears to be largely automatic and cannot be disrupted to lead people to interpret a metaphor only literally (Glucksberg, 2003)[153].

Idioms

Idioms are like metaphors in that the literal language in the expression does not convey the intended meaning of the phrase. Instead of having the listener make an analogy by using a metaphor, we ask the listener to remember a stored meaning for an idiom, a frozen or unchangeable expression that conveys a particular idea. Sometimes it is easy to figure out the meaning of the idiom based on the words in the expression. Youcan locate 'the foot of the bed' easily even if you don't already know that idiom. But you probably don't know how we got the expression 'icked the bucket' or why it means that someone has died. How it came to mean this is not entirely clear, although several theories exist. Moreover, this idiom needs to be expressed in this way. That is why idioms are described as frozen expressions. Although some idioms seem open to a degree of syntactic flexibility, others seem to function as if they were a single word, showing little or no flexibility (Gibbs & Nayak, 1989)[154]. Most English speakers would find it odd, or simply wrong, to hear 'Jim, well, he punted the pail last night.' People learning a new language often have trouble with idioms because they catch on that what is being said is not what is literally conveyed, but has some other, culturally specific meaning.

Although idioms need to be presented in their entirety to convey the idea, they are not initially processed as wholes. Instead, people first try to process them word by word, as with normal language. Only at some later point, say midway through the phrase, does the person process the idiom in entirety to capture its meaning (Gibbs, Navak & Cutting, 1989)[155]. This is surely when normal comprehension processes are starting to run into difficulties, noticing that a literal interpretation is not possible, and possibly when the idiom is retrieved from memory. Of course, people who are learning a new language have trouble with idioms because they often are not familiar with the idioms and so attempt a literal interpretation of the words. So, say what you mean, and mean what you say, except when you don't.

9.4.5 Gesture

When we speak, we not only move our lips, tongues, throats, and so on, we may also move our arms and hands. This movement, or gesture, is done to facilitate communication to listeners and excludes sign language and non-communicative mannerisms, such as touching one's hands to one's face (McNeill, 1992[156]). That is, if they are not *beat gestures* that mark out important words or ideas, the movements we make with our hands communicate information to augment the words we are using. For example, when describing your route to school, you may make gestures to convey information about turns you made, obstructions you encountered, speeders you saw, and so on.

Importance of gesture

Although gesturing may serve a more holistic function when we use it while speaking, when we gesture in the absence of spoken speech, it takes on more linguistic characteristics, much as sign languages do (Goldin-Meadow, 2006[157]).

Involves strong social component

Work by researchers such as Bavelas, Gerwing, Sutton, and Prevost (2008)[158] shows that people even gesture when they are talking on the telephone and know the person on the other end cannot see them (although they gesture less than when they can see the other person). So, there must be something cognitively important about gesturing that facilitates language production in conversation. It should also be noted that people gesture less when they are speaking into a tape recorder (Bavelas et al., 2008[159]) or when listeners do not appear to be attending to what they are saying (Jacobs & Garnham, 2007[160]). So, this impact of gesture has a strong social component and is not purely psycholinguistic.

Helps learning and thinking

Gesture may even help learning and thinking (for a review, see Goldin-Meadow & Beilock, 2010[161]). The gestures a person produces when trying to solve a problem may reveal knowledge that the person has that is in a nascent, implicit stage. In some way, the gestures reveal to the person that knowledge is present, but not fully developed in a way that can be used consciously. So, if people are encouraged to gesture, they may make the knowledge conscious faster and facilitate learning. In a study by Broaders, Cook, Mitchell and Goldin-Meadow (2007),[162] third- and fourth-grade children were told to gesture while they solved maths problems. They then showed an increased ability to develop new strategies and solve those types of maths problems and were faster than children who did not gesture. Moreover, when two people are solving problems together, such as assembling a piece of furniture, they can solve the problems more effectively when one person is communicating using gestures, and the person assembling can use the gestures (Lozano & Tversky, 2006[163]).

Communicates embodied information

Essentially, gestures communicate embodied information that cognition can process more directly than linguistic information alone (Hostetter & Alibali, 2008; Kelly, Ozyurek & Maris, 2010[164]). Assessments of later memory show that people who gesture while learning have better situation models of the circumstances described by a text, but poorer memory for the surface form (verbatim) text (Cutica & Bucciarelli, 2013[165]). That is, people who gesture have a better memory of what the text is about, but a poorer memory of specifically what the text was.

Summary: Comprehension

9.1 Conceptual and rule knowledge

- A variety of online tasks have been devised to investigate comprehension, such as tasks involving reading times, the use of probes, think-aloud verbal protocols and neuroimaging evidence.
- Successful comprehension is best achieved when people can self-monitor what they are and are not learning through judgements of learning. However, these judgements are often poor estimates of how much has actually been learned. Estimates can be improved by delaying these judgements, rereading and providing summaries of the material.
- Comprehension involves processing at many levels, including the surface form, propositional textbase and

situation model levels. Evidence of processing at each of these levels can be derived across many different aspects of understanding.

9.2 Reading

- Tremendous progress has been made in understanding the mental processes of reading, largely by using the online measures of comprehension, such as reading times and gaze durations.
- Modern models of reading make predictions about comprehension based on a variety of factors; for instance, word frequency and recency in the passage influence surface form and textbase processing, respectively.
- Online measures of language comprehension provide a unique window into human cognition. Using these sorts of measures, we can gain moment-to-moment insights not only into the effectiveness of processing, but also into the very contents of people's minds.

9.3 Reference, situation models and events

- Reference in language involves the idea of bridging together different elements of a statement. The source of knowledge that permits speakers to include reference in their messages and listeners to infer the basis for those bridges is not just our knowledge of syntax and word meanings, but the entirety of semantic memory and much top-down processing.
- Situation models are created by combining information from the language itself as well as inferences people draw based on their prior semantic and episodic knowledge. This inference making is at the heart of comprehension and reflects information in the language, such as grammatical aspect.
- The capacity and operation of working memory are known to be important factors in understanding individual differences in reading comprehension.
- Situation model updating occurs when people detect a meaningful change in an event dimension. This updating process is cognitively effortful, resulting in increases in reading times and brain activity.
- Comprehension occurs not only for language that people hear or read, but also for other aspects of experience, including narrative comics, videos and interactive experiences.

9.4 Conversation and gesture

- Conversations follow a largely implicit set of conversational rules. Some of these involve turn taking and social status and conventions, but many more govern the nature or topic of participants' contributions. Topic shifts involve selecting some part of a person's utterance to form the basis for a new contribution but then adding some new information. Schank's work is a particularly important analysis of this process of topic shifting.
- Participants in a conversation develop theories of mind of the other speakers, such as direct theory, as well as theories of what the other speakers think of them, called second-order theories. When we converse, we tailor our contributions to these theories and follow a set of conversational rules, the unspoken contract between conversational partners. When a rule is violated intentionally, usually to make some other point (e.g. sarcasm), we mark our violation so that its apparent illegality as a conversational move is noticed and understood.
- Empirical work on conversational interaction often tests general notions about direct theories, the politeness rule or indirect requests. Although we sometimes attempt to manipulate another person's direct theory of us, research also shows that the initially planned utterance is usually from a very egocentric perspective, whereas later adjustments may take the other person's perspective into account. The standard, everyday use of non-literal language is also seen with the use of metaphors and idioms.
- Gestures made during conversation are a way that simulated spatial and action information can be communicated. Making gestures is part of the social act of conversation, although it may sometimes occur when our partner cannot see us, as when we are talking on the telephone. Gesture can serve as a working memory aid and help people solve problems.

Chapter 9 quiz

Question 1

In van Dijk and Kintsch's (1983) levels of representation theory, the intermediate level is known as:

A. Surface form
B. Situation model
C. Metacomprehension
D. Propositional textbase

Question 2

Which types of methods are most commonly used to help us understand language components that would ordinarily be difficult to uncover?

A. EKG and fMRI.
B. ERP and fMRI.
C. CT Scan and ERP.
D. CT Scan and PET scan.

Question 3

Why is serial position important in affecting reading times?

A. As a person reasons more words, the more opportunity there is to build word structures meaning that it is easier to read.
B. If a person reads slowly, there is less opportunity to build word structures, delaying reading times.
C. If there are more syllables in a sentence, this makes reading time decrease
D. If a person is reading a sentence with no relevant context, this will decrease reading times.

Question 4

Which one of the following is an example of a direct reference within language processing?

A. Jennifer bought a PlayStation. The PlayStation was on sale.
B. Eric bought a used car. The tyres were badly worn.
C. Michael bought a used car. The radio doesn't work.
D. Jawad bought a used car. The salesperson gave him a good price.

Question 5

Which perspective within psychology has been extensively researched in the context of conversations?

A. Structure binding framework
B. The immediacy assumption
C. Theory of Mind
D. Anaphoric reference

10 Reasoning and decision-making

Learning objectives

10.1 Understand the limitations of human reasoning

10.2 Describe the characteristics of decision-making

10.3 Explain how heuristics and biases can influence decision-making

10.4 Assess how decisions are influenced by the context in which they are made

10.5 Explain how heuristics can aid decision-making

10.6 Discuss the range of approaches to human decision-making

10.7 Evaluate the limitations of human reasoning and decision-making

Key figures

Professor Philip Johnson-Laird

Professor Philip Johnson-Laird pioneered research on higher cognitive functions, such as reasoning. His career spans a wealth of publications, including books on reasoning and thinking processes, and his theory on mental models has spawned a large body of research in cognitive psychology.

Professor Ruth Byrne

Professor Ruth Byrne is Professor of Cognitive Science at the Department of Psychology at Trinity College Dublin, University of Dublin. Her research has explored reasoning and thinking in both healthy and clinical populations. She has authored a large number of research publications and books, including a book about thinking, reasoning and decision-making in autism.

Professor Gerd Gigerenzer

Professor Gerd Gigerenzer has advanced our understanding of heuristics and decision-making, providing alternative views about the adaptive nature and function of heuristics-based thinking and decision-making.

Introduction

In June 2016 the British people cast their vote on whether they wanted their country to remain in the European Union or leave – an important milestone in Britain's modern history. The vast majority of opinion poll companies at the time predicted that the vote for Remain would prevail, although others indicated that Leave was a more probable outcome from the Brexit referendum. Eventually, 51.9 per cent of the voters supported Leave, whereas 48.1% voted for Remain. Three years later, following high-level political discussion stalemate and negotiations between the UK and the European Union, opinion polls started to show that public opinion was slowly moving towards favouring Remain and rejecting the idea of Brexit. A 2023 poll with a representative sample of the general public showed that 53 per cent of respondents now thought that Brexit was wrong and only 34 per cent still backed the Leave vote. Empirical research also showed that various factors, such as judgements about the costs and benefits of Brexit, emotional responses to Brexit and the EU, as well as towards immigration issues, influenced the ways people decided to vote in the Brexit referendum (Clarke, Goodwin & Whiteley, 2017[1]).

From the perspective of cognitive psychology, the Brexit referendum and the shifting public opinion over the years raise some important questions: how do we reason and make decisions, when we have little or incomplete information available to inform our decision-making? How do our judgements and reasoning patterns influence the decisions we make, the less and the more important, in a given situation? How can the social context influence our decisions? These are some of the key questions we discuss in this chapter.

In Chapter 4 we hinted at the notion of dual mental processes, one being automatic, fast and intuitive, and the other being reflective, slow and deliberate. A clear picture of cognitive psychology would be incomplete without a consideration of the slower, more deliberate kinds of thinking. One of the main tenets in reasoning and decision-making research is that we are often biased and overly influenced by our general knowledge of the world, including our past behaviour and experiences, our expectations about future outcomes, as well as the behaviour or expectations of other people (especially those who are important to us, for given decisions). These influences affect how we reason and how we make judgements and decisions every day. Another important tenet is that, instead of being objective, we search for evidence that confirms our decisions, beliefs and hypotheses far more than is logical. In general, we tend to be much less sceptical than we ought to.

How do we make choices, even about life-changing matters, when we have limited information available?

Source: Alexeyfedoren/123RF

Before we delve into decision-making processes, we discuss two classic forms of human reasoning (syllogistic and conditional), and how our reasoning ability is prone to errors, especially under conditions of uncertainty and when relevant information from our memory is partial or inaccessible. Overall, this research provides convincing examples of the uncertainty of human reasoning and the often surprising inaccuracies in our knowledge.

10.1 Formal logic and reasoning

LEARNING OBJECTIVE
Understand the limitations of human reasoning.

At some point during their education – often in a course on philosophy and logic – many students are exposed to the classic forms of reasoning. The two forms we discuss here are *syllogistic* and *conditional reasoning*. However, there are others, such as relational reasoning (e.g. Krawczyk, 2012[2]). Overall, we are not particularly good at solving reasoning problems when they are presented in an abstract form. We do better when the problems are presented in terms of concrete, real-world concepts, but still fall short. That said, in some situations, our world knowledge almost prevents us from seeing the 'pure' (i.e. logical) answers (e.g. Markovits & Potvin, 2001[3]), and errors are made. In short, we are not particularly logical thinkers.

When reading about these formal logic problems, do not let yourself be lulled into the feeling that these are just abstract, academic thought problems. The basic form of these problems has the potential to underlie much of our everyday thinking and decision-making. For example, Sudoku puzzles are built on logical, not mathematical, reasoning even though they involve numbers (Khemlani & Johnson-Laird, 2012[4]). When a doctor is trying to determine a diagnosis, a police inspector is trying to solve a crime, or you are trying to figure out which postgraduate course to enrol in, the principles of formal logic are at work.

Reasoning is important for clinical diagnosis
Source: Serhi Bobyk/Shutterstock

10.1.1 Categorical syllogisms

A categorical syllogism is a three-statement logical form, with the first two parts stating the premises taken to be true, and the third stating a conclusion based on those premises.

Biases of syllogistic reasoning

Let's look at some of the biases that have been observed when people try to reason syllogistically. First, there is the *figural effect*, which is a bias to arrive at conclusions in which the terms are in the same order as in the figure of the premises. For example, if people learn that '*Some A are B*' and '*All B are C*' there is a bias to conclude that '*Some A are C*'. Whereas if people learn that '*Some B are A*' and '*All C are B*', there is a bias to conclude that '*Some C are A*'. So, people are drawing conclusions not based on a rational assessment of the information, but simply based on the *order* in which the terms appear in the sentences (Johnson-Laird et al., 2015[5]).

Another difficulty or confusion people have is the belief bias, which is illustrated by the following example:

(1d) *All cava poos are animals.*

All animals are wild.

Therefore, all cava poos are wild.

Source: James Thoms/Alamy Stock Photo

The difficulty is that, although it seems to fly in the face of your world knowledge, the conclusion is logically true. If we assume that the premises are true, then the conclusion must be. However, the truth of the premises is another issue. Because the conclusion follows from the premises, the syllogism is valid. Of course, it is easy to think of counterexamples in the real world of cava poos; hardly any cava poos are wild, after all. In this case, the premise 'All animals are wild' violates your world knowledge. However, that is beside the point here. The point is to assess the validity of the conclusion given the premises.

The belief bias is the human tendency to ignore the logical form of an argument and focus instead on prior knowledge (Evans, Barston & Pollard, 1983; see also Copeland, Gunawan & Bies-Hernandez, 2011, for a similar source credibility effect[6]). For the rules of syllogistic reasoning, the *truth* of the premises is *separate* from the *validity* of the logical argument. What matters is that the conclusion does or does not validly follow from the premises. So, applying syllogistic reasoning to real-world problems is at least a two-step process. First, determine whether the

Syllogism premises and conclusions

The goal of syllogistic reasoning is to understand how the premises combine to yield logically true conclusions (if any). Syllogisms may be presented in an abstract form using just letters, such as

(1a) *All A are B.*
All B are C.
Therefore, all A are C.

'All A are B' states that the set *A* is a subset of the group *B*, that *A* is included in the set *B*. The third statement is the conclusion. By applying the rules of categorical reasoning, we can determine that the conclusion *'All A are C'* is true, so the conclusion follows logically from the premises.

Inserting words into the syllogism can help you understand this. For instance,

(1b) *All poodles are dogs.*
All dogs are animals.
Therefore, all poodles are animals.

As abstract and obtuse as syllogisms may first seem, they reflect the types of reasoning in which people engage all the time. For example, say you are shopping and thinking to yourself, 'Well, I know that ammonia is good for cleaning windows, and that all window cleaners are found in aisle 12 at the store', then you can conclude that the ammonia can be found in aisle 12. Written syllogistically, this would be:

(1c) *All ammonias are good for cleaning windows.*
All window cleaners are found in aisle 12.
Therefore, all ammonias are found in aisle 12.

Looking a little deeper, syllogism premises and conclusions can convey four different *moods*, namely *all*, *some*, *no* and *some not*, as in 'All A are B', 'Some A are B', 'No A are B' and 'Some A are not B'. Moreover, classic syllogisms can also be expressed in one of four different figures, which refer to the location of the B term in the syllogism.

These four figures are shown here:

Figure 1	Figure 2	Figure 3	Figure 4
A—B	B—A	A—B	B—A
B—C	C—B	C—B	B—C

Putting together the four moods with the four figures, you can see that there are 64 possible syllogisms, of which 27 have a valid conclusion. For the rest, no valid conclusion can be drawn. This is important because, as we will see, the errors that people make, and the theories about how people reason, draw on these moods and figures.

syllogism itself is valid; second, if the syllogism is valid, determine the truth of the premises.

Now consider another example:

(2a*) *All A are B.*
Some B are C.
Therefore, some A are C.

Note that in formal logic, *some* means 'at least one and possibly all', although some people have trouble understanding this (Schmidt & Thompson, 2008[7]). That said, not all the errors we make when solving syllogisms arise from relying on *Gricean maxims* of what is 'likely meant' by the premises (Newstead, 1995[8]). Try inserting words into this example to see whether the conclusion is correct.

(2b*) *All polar bears are animals.*
Some animals are white.
Therefore, some polar bears are white.

Despite the idea that word substitutions can lead to an empirically correct statement, this syllogism is invalid because the two premises do not invariably lead to

Euler circle illustrations for three categorical syllogisms

Euler circles can help you determine whether a syllogism is valid.

The 'All–All' form

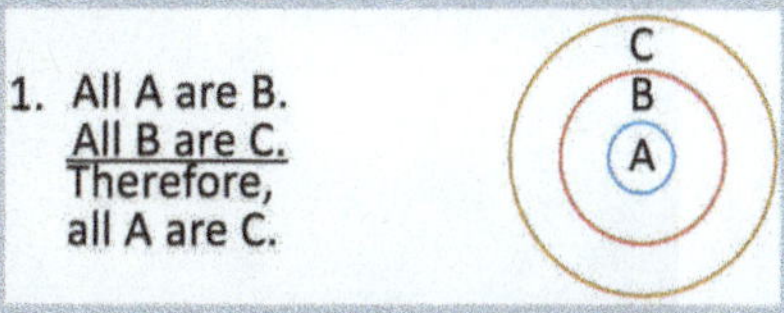
1. All A are B.
All B are C.
Therefore,
all A are C.
C
B
A

In the above illustration, the 'All–All' form shows that it is necessarily true that '*All A* are *C*'. The circles, which represent the class of things *A*, *B* and *C*, are nested such that *A* is a subset of *B* and *B* is a subset of *C*. There is no other way to represent the premises except by concentric circles (when *A* and *B* are identical, their boundaries overlap completely, and the diagram shows one circle labelled both *A* and *B*).

The 'All–Some' form

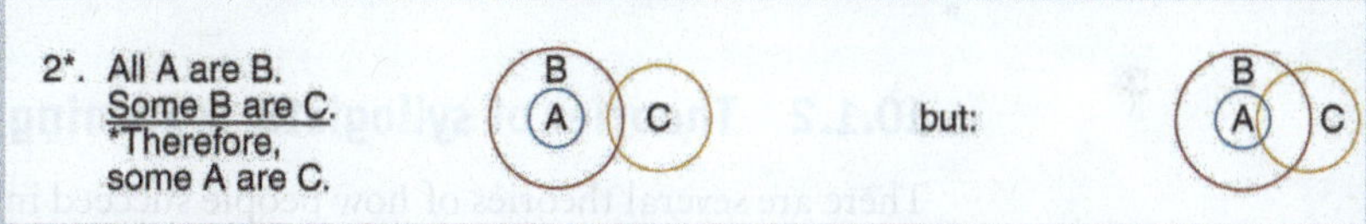
2*. All A are B.
Some B are C.
*Therefore,
some A are C.
B
A
C
but:
B
A
C

If a diagram can be constructed that shows the conclusion doesn't hold for all cases, then the conclusion is false. The first diagram in 2* shows why 2* is incorrect; it is not necessarily true that some A are C. The second diagram in 2* shows that an arrangement can be found that seems to support the argument.

In the above figure, the *invalidness* of syllogism 2 is illustrated by the first diagram. Here, a portion of *B* that does not contain *A* is exactly the portion that overlaps with C. So it isn't necessarily true that some *A* are *C*. The second diagram for this problem, however, illustrates the 'Some polar bears are white' conclusion, one that is true of the real world even though the syllogism is invalid.

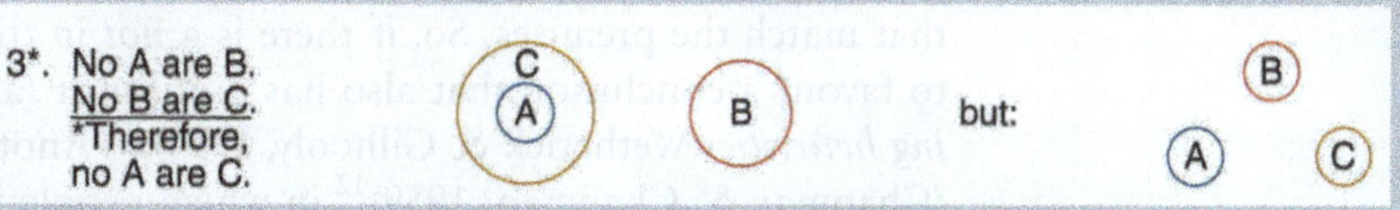
3*. No A are B.
No B are C.
*Therefore,
no A are C.
C
A
B
but:
B
A
C

The 'No–No' form. The first diagram in 3* shows why 3* is incorrect; it is not necessarily true that no A are C. The second diagram in 3* shows an arrangement that does seem to support the conclusion.

The above syllogism is like the previous one; it is invalid, but the second diagram seems to show that it is true.

a valid conclusion. The entire form of the syllogism is invalid (the reason for the asterisk). The incorrectness of the conclusion in (2a) stems from the qualifier *some*. Although the conclusion may be true in the world here, when you use another concrete example, you can see that this conclusion is not necessarily the case for *all* examples, as shown by the following:

(2c*) *All polar bears are animals.*
Some animals are brown.
Therefore, some polar bears are brown.

Source: Gecko1968/Shutterstock

10.1.2 Theories of syllogistic reasoning

There are several theories of how people succeed in mentally solving syllogisms (or fail to do so). Following Khemlani and Johnson-Laird (2012),[9] we discuss three basic types of theories: heuristics, rules and mental models.

Heuristics are general rules of thumb that often provide the correct answer but are not guaranteed to do so. As we will discuss later in the chapter, heuristics can also have a pervasive effect on judgement and decision-making, in various domains, from the social (e.g. how we judge other people) to the professional (e.g. how professionals make important decisions). The idea behind **heuristic theories** of reasoning is that people solve logical problems not by using well-reasoned steps, using instead shortcuts based on the nature of the problem itself. For example, according to the **atmosphere heuristic** (Woodworth & Sells, 1935[10]), people favour conclusions that match the premises. So, if there is a *not* in the premises, then there is a bias to favour a conclusion that also has a *not* in it (a similar heuristic is the *matching heuristic*; Wetherick & Gilhooly, 1995).[11] Another heuristic is **illicit conversion** (Chapman & Chapman, 1959)[12] in which people inappropriately swap the terms of a statement, such as mentally converting 'All *B* are *A*' to 'All *A* are *B*.' Chater and Oaksford (1999)[13] have proposed a **probability heuristics theory** in which people make decisions using a collection of heuristics that use the information value of the premises. Regardless of the particular theory, all capture the finding that we sometimes do not put forth the kind of thinking needed to assess the validity of logical conclusions, but instead rely on superficial characteristics of the problem itself. This is the opposite of what most of us think of as rational thinking, but that is what many of us do.

Another group of theories are **mental rules theory** of human reasoning that assume that people are using rules, similar to the types of mental rules people use when processing language, to assess the validity of various logical conclusions (e.g. Rips, 1994[15]). These mental rules are thought to capture an otherwise sound mental logic. As long as these rules are allowed to run and function without interruption, people can come to valid logical conclusions. Problems arise because either people are distracted and apply their mental resources elsewhere, or, more often in experiments

Mental model theories

mental model theory (Johnson-Laird, 1983, 2013[14]) has suggested that people create mental simulations of the circumstances derived from the premises. People then develop various conclusions from those mental models. Many problems are difficult because they require people to construct multiple mental models to derive the appropriate conclusion or to determine that no valid conclusion can be drawn because there are mutually incompatible circumstances that apply to those premises.

To illustrate this process, assume that people are given the syllogism 2a. For the first premise, *All A are B*, people might create a mental model with tokens standing for *A* that are all associated with *Bs*, such as:

A	B
A	B

Then, when people get the next premise, *Some B are C*, they might augment their mental model so that it has some *Bs* associated with *C*, and some not, such as:

A	B	C
A	B	
		C

At this point, a person might conclude *Some A are C*. However, this conclusion is invalid because there is another state of affairs not captured by the first mental model. Specifically, it might be the case that there are some *B* that are not *A*, and some *B* that are *C*. Taking this into consideration allows a person to create a mental model more like:

A	B	
A	B	
	B	C
		C

When one is armed with both of these mental models, it is clear there is no statement that can be made that is true of both of them, so no valid conclusion can be drawn.

Therefore, mental model theory assumes that people are trying to capture the world as it might be described by the premises, and mistakes are made when people fail to think of other circumstances that could also be true, but which might be inconsistent with the original way of thinking about the premises. Thus, it is so important to try to think of counterexamples and see whether there is any negative evidence, rather than just go on positive evidence. Overall, these different types of theories capture different aspects of human reasoning. The trick is identifying how a given person is reasoning in any specific set of circumstances.

on mental logic, problems stem from difficulties in converting the language of the problem into the appropriate mental logic form.

A final group of theories assumes that when people are given the premises of a syllogism, they create mental representations, called **mental model**, of the states of affairs they describe. An externally written analogue to this process would be the Euler circles described earlier. However, Euler circles cannot capture what people are doing mentally because, often, this method must be explicitly taught. It is not something we do naturally or spontaneously.

Can we reason with our 'inner eye'?

Think about a relational reasoning problem that goes like:

Tom is taller than Betty.
Anisha is shorter than Betty.
Who is tallest?

How would you approach this problem? Would you abstractly represent the relations given in the premises to reach a logical conclusion? Or would you use your 'inner eye' to mentally visualise Tom, Betty and Anisha and then determine who is tallest? Proponents of visual mental imagery theory suggest that mental imagery assists reasoning and problem-solving because it helps people mentally represent the premises of the problem in visuo-spatial terms. Others, however, have criticised this view and even demonstrated that visual mental imagery can hinder logical reasoning (Knauff, 2013[16]), and effect that is not observed in people who were born blind (Knauff & May, 2006[17]). Hamburger et al. (2018)[18] used transcranial magnetic stimulation (TMS) to explore the role of visual imagery on reasoning outcomes. The researchers were particularly interested in the neural basis of reasoning, and if reasoning accuracy is affected when the brain areas responsible for visual processing (i.e. visual cortex) are stimulated. If visual imagery indeed hinders reasoning, then disrupting the processing of the visual cortex (V1) via TMS should lead to lower error rates in reasoning problems. Participants were asked to complete a series of relational reasoning tasks by making inferences through different premises, as shown below.

Premise 1: *The dog is cleaner than the cat.*
Premise 2: *The dog is dirtier than the ape.*
Inference: *The cat is dirtier than the ape?*

TMS was applied following the premises and before presenting the conclusion as shown in Figure 10.1.

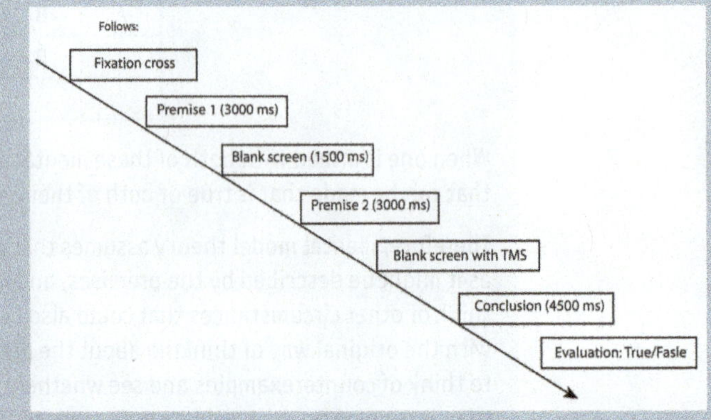

Figure 10.1
If visual imagery hinders reasoning, then disrupting visual imagery with TMS should improve performance in reasoning tasks.

In support of their hypothesis, Hamburger et al. (2018)[19] found that disrupting the function of the V1 via TMS, made fewer errors in solving the reasoning problems: 'People reasoned better without visual mental images' (p. 2283).

10.1.3 Conditional reasoning

Conditional reasoning: *if P then Q*

Conditional reasoning is a second kind of logical reasoning. Conditional reasoning problems contain two parts, a *conditional clause*, a statement that expresses a relationship (*if P then Q*), followed by some *evidence* pertaining to the conditional clause (*p*, for example). *Conditional reasoning* involves a logical determination of whether the evidence supports, refutes or is irrelevant to the stated *if–then* relationship.

The *conditional* in these problems is the *if–then* statement. The *if* and *then* clauses are respectively known as the **antecedent** and the **consequent**. The *if* states a possible cause, and the *then* states an effect of that possible cause. So, *if P then Q* means 'if *P* is true, then *Q* is true'; for example, if it rains (*P*) then the streets will be wet (*Q*). So far so good.

After the *if–then*, you get a second statement, some evidence about the truth or falsity of one part of the *if–then* relationship. The aim is to take this evidence and decide what follows logically from it. In other words, is the conditional *if–then* statement true or false given this evidence, or is the evidence irrelevant to it?

The evidence that follows an *if P, then Q* conditional results in one of the four possible outcomes. For 'If it rains, the streets will be wet,' the four possibilities are:

P: That is, *P is true, it's raining.*

not P: That is, *P is not true, it's not raining.*

Q: That is, *Q is true, the streets are wet.*

not Q: That is, *Q is not true, the streets are not wet.*

Putting this together yields four possibilities:

Conditional	If *P*, then *Q*.	If *P*, then *Q*.	If *P*, then *Q*.	If *P*, then *Q*.
Evidence	*P*.	Not *P*.	*Q*.	Not *Q*.
Conclusion	Therefore, *Q*.	(No conclusion)	(No conclusion)	Not *P*.

For a conditional *if–then* statement, if the antecedent P is true, then its consequence (the *consequent*) Q must be true. As an expanded example consider the following, with all four possibilities worked out in Table 10.1:

If I am a first year Psychology undergraduate student, then I must take Introduction to Research Methods and Statistics.

Valid arguments

As Table 10.1 shows, only two possibilities lead to a valid conclusion. In the first one, when given evidence that *P* is true, 'I am a first year Psychology undergraduate student', then the consequent *Q* must be true, 'I do have to take Introduction to Research Methods and Statistics'. This is *affirming the antecedent* (saying that the antecedent is true). The classic name for this is *modus ponens*. Likewise, if the evidence is that *Q* is not true, 'I do not have to take Introduction to Research Methods

Table 10.1 Conditional reasoning

From	Name	Example
If *P*, then *Q*.	*Modus ponens*:	If I am a first year Psychology undergraduate student, I must take Introduction to Research Methods and Statistics.
Evidence: *P*.	affirming the	Evidence: I am a first year Psychology undergraduate student.
Therefore, *Q*.	Antecedent	Therefore, I must take Introduction to Research Methods and Statistics.
	(valid inference)	
If *P*, then *Q*.	Denying the	If I am a first year Psychology undergraduate student, I must take Introduction to Research Methods and Statistics.
Evidence: not *P*.	Antecedent	Evidence: I am not a first year Psychology undergraduate student.
*Therefore, not *Q*.	(invalid inference)	*Therefore, I do not have to take Introduction to Research Methods and Statistics.
If *P*, then *Q*.	Affirming the	If I am a first year Psychology undergraduate student, I have to take Introduction to Research Methods and Statistics.
Evidence: *Q*.	Consequent	Evidence: I have to take Intro to Logic.
*Therefore, *P*.	(invalid inference)	*Therefore, I am a first year Psychology undergraduate student.
If *P*, then *Q*.	*Modus tollens*:	If I am a first year Psychology undergraduate student, I have to take Introduction to Research Methods and Statistics.
Evidence: not *Q*.	denying the consequent	Evidence: I do not have to take Introduction to Research Methods and Statistics.
Therefore, not *P*.	(valid inference)	Therefore, I am not a first year Psychology undergraduate student.

and Statistics', it must therefore be that *P* is not true, 'I am not a first year Psychology undergraduate student.' This is *denying the consequent* (saying that the consequent is not true) and is called *modus tollens*.

In comparison, the other two arguments do not lead to a valid conclusion. By denying the antecedent you cannot conclude that the consequent is false; similarly, by affirming the consequent you cannot conclude that the antecedent is true. Continuing with the example, 'If I am a first year Psychology undergraduate student, then I must take Introduction to Research Methods and Statistics'. Denying the antecedent with 'I am not a first year Psychology undergraduate student' does not lead to the conclusion that 'I do not have to take Introduction to Research Methods and Statistics'. It could be that two groups must take Introduction to Research Methods and Statistics (e.g. all first year Psychology undergraduate students and all Sociology and Business students). So, not being a first year Psychology undergraduate student doesn't necessarily mean you don't have to take Introduction to Research Methods and Statistics. Likewise, affirming the consequent with 'I must take Introduction to Research Methods and Statistics' does not permit the conclusion that 'I'm a first year Psychology undergraduate student'; you might as well be a Business student.

Evidence on conditional reasoning

Generally, we are good at inferring the validity of a consequent given evidence that the antecedent is true (affirming the antecedent, *modus ponens*). Early research by Rips and Marcus (1977)[20] found that 100 per cent of participants drew this conclusion. The other valid inference, denying the consequent (*modus tollens*), is more difficult. Only 57 per cent in Rips and Marcus's study drew this conclusion. More recent studies have explored the neurocognitive basis of conditional reasoning. Following a meta-analysis of 32 reasoning experiments using neuroimaging techniques (fMRI), Wertheim and Ragni (2020)[21] found that a widely distributed network was activated in conditional reasoning tasks, mainly involving the inferior parietal lobule (IPL), the dorsomedial prefrontal cortex (DMPFC), the rostrolateral prefrontal cortex (RLPFC), the supplementary motor area (SMA), and regions of the cerebellum. The researchers suggested that this pattern of activation indicates the role of working memory and information integration during conditional reasoning and inferences.

Now, let's consider the conditional reasoning errors we make, which can be classified into three broad categories: those involving the form of the problem, those involving search for evidence, and memory-related phenomena.

Sample stories and tests from Rader and Sloutsky (2002)

About 60 per cent of people (incorrectly) 'recognised' the affirm-the-consequent conclusion as having been in the story, not that different from the 61 per cent who (correctly) recognised the *modus ponens* conclusion. So, both kinds of conclusions are routinely drawn as we read, even though one of them is incorrect (see Bonnefon & Hilton, 2004, for how the desirability of the consequent influences our predictions about the truth of the antecedent[22]).

Participants read four sentences. Sentence 1 was the same for both groups, but Sentence 2 differed between version A and version B. Sentence 3 was the same for both groups, but Sentence 4 differed for the two groups, containing either an inference or no inference. In other words, one-quarter of the participants saw version A and an inference, one-quarter saw Version A and no inference, one-quarter saw version B and an inference, and one-quarter saw version B and no inference.

1. Frank woke up on his couch after taking a long nap and realised that he didn't know what time it was.
2. (Version A) He thought that if it was cold outside, then it was night.

   OR

   (Version B) He thought that if it was night, then it was cold outside.
3. Still feeling sleepy, Frank arose to open a window.
4. (Inference condition) He discovered that it was cold outside.

   OR

   (No Inference condition) He wondered whether it was cold outside.

   Inference test: yes/no – Was this information in the story?

The time of day was night.		
	Percentage saying 'yes' to the inference test	
	Inference	No inference
Modus ponens (version A)	61%	24%
If it's cold, then it's night.		
Yes/no – The time of day was night.		
Affirm – the consequent (version B)	59%	21%
If it's night, then it's cold.		
If it's night, then it's cold.		
Yes/no – The time of day was night.		

Source: results from Rader and Sloutsky (2002)[23].

Form errors

People sometimes draw invalid conclusions by using one of the two invalid forms, either denying the antecedent or affirming the consequent. Rader and Sloutsky (2002)[24] found that we commonly draw such invalid conclusions when comprehending discourse. They gave people short scenarios that contained an *if–then* conditional, then tested recognition memory for either words or ideas in the stories (i.e. was this information in the story?).

Another form error is subtler. People tend to reverse the elements in the *if* and *then* and then go on to evaluate the evidence against the now-reversed conditional. This is called *illicit conversion*. So, given *If P, then Q* and evidence Q, people tend to switch the conditional to **If Q, then P*. They then decide that the evidence *Q* implies that *P* is true. This is wrong because the order of *P* and *Q* is meaningful. The *if* often specifies some cause, and the *then* specifies an effect. So, we cannot draw correct conclusions if we reverse the roles of the cause (*P*) for some effect (*Q*). For example, the statement 'If the clock says 9:15, then you are late for work' cannot be validly converted to 'If you are late for work, then the clock says 9:15'.

Search errors

A second kind of error involves a search for evidence. We often rely on a first impression or on the first example – the first mental model – that comes to mind (Evans, Handley, Harper & Johnson-Laird, 1999[25]). Unfortunately, this is often a *search for positive evidence*, also called the confirmation bias, where we often (inappropriately) seek *only* information that confirms a conclusion we have already drawn or a belief we already have. Knowing an outcome and the conditions that might lead to it causes people to overestimate the likelihood of that outcome. Because we find it easier to draw backward causal inferences, we mistakenly think that the prior events are more likely to lead to the actual outcome (Koriat, Fiedler & Bjork, 2006[26]), which is an analogue to the hindsight bias discussed later in this chapter.

As a demonstration, consider a classic study (reported in Wason & Johnson-Laird, 1972[27]) shown at the top of Figure 10.2, the Wason card problem. Four cards are visible and each card has a letter on one side and a number on the other. The task

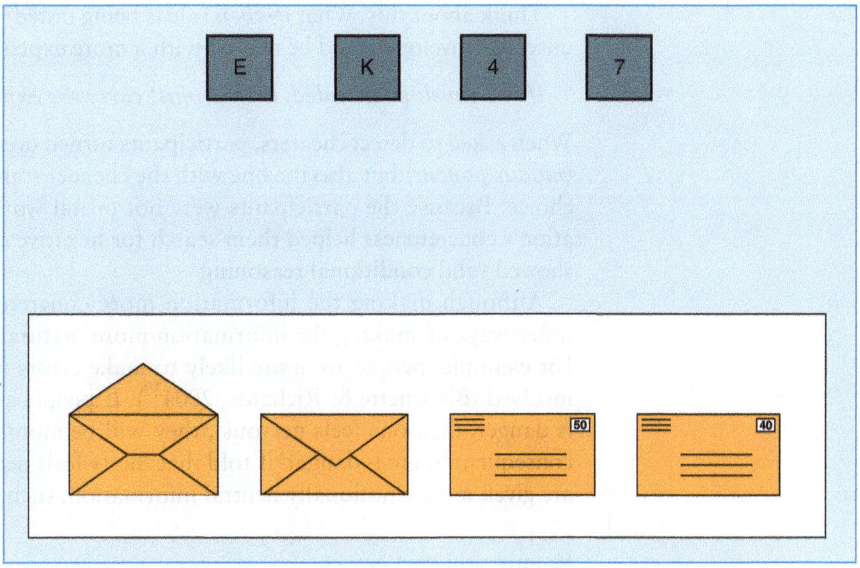

Figure 10.2 The Wason card problem
At the top are the four cards in the Wason card problem. Which card or cards would you turn over to obtain conclusive evidence about the following rule: a card with a vowel on it will have an even number on the other side? At the bottom of the illustration are four envelopes. Which envelopes would you turn over to detect postal cheaters, under the rule that an unsealed envelope can be stamped with the less expensive stamp?

is to pick the card or cards you would turn over to gather conclusive evidence on the following rule:

If a card has a vowel on one side, then it has an even number on the other side.

Think about this statement and decide how you would test the rule before you continue reading.

Of the people tested, 33 per cent turned over only the *E* card, a correct choice conforming to *modus ponens* (affirming the antecedent). However, a thorough test of the rule's validity requires that another card be turned over (the rule might be rephrased 'Only if a card has a vowel on one side will it have an even number on the other side'). Only 4 per cent of the people turned over the correct combination to check on this, the *E* card (*modus ponens*) and the 7 card (*modus tollens*). Turning over the 7, which might yield negative evidence (*not Q*), was rarely considered. Instead, people preferred turning over the *E* and the 4 card: 46 per cent of the people did this. Note two points. First, turning over the 4 is invalidly affirming the consequent (the rule doesn't say anything about the other side of a consonant; it could be odd or an even). Second, turning over the *E* is a search for positive evidence, the (tentative) 'yes' conclusion that *P* is true. Our bias is either to stop after turning over the *E* (positive evidence) or to continue searching for more positive evidence (turning over the 4). The poor performance on this task has spawned a great deal of work to understand the cognitive processes that give rise to it, including creating computational models (Klauer, Stahl & Erdfelder, 2007[28]) that suggest that people are not considering the cards one by one. Rather, they are looking at the configurations of the cards.

In a different situation, however, Johnson-Laird, Legrenzi and Legrenzi (1972)[29] found that 21 of 24 participants made both of the correct choices when the situation was made more concrete. In that study, people were trying to find cheaters on the postal regulations, where unsealed envelopes could be mailed with a less expensive stamp than sealed envelopes.

Think about this. What *if–then* rule is being tested? Because either a sealed or an unsealed envelope could be mailed with a more expensive stamp, the rule must be:

If the envelope is sealed, then it must carry the expensive stamp.

When asked to detect cheaters, participants turned over not only the sealed envelope (*modus ponens*) but also the one with the cheaper stamp, that is, the *modus tollens* choice. Because the participants were not postal workers, it is clear that the situation's concreteness helped them search for negative evidence; in the process, they showed valid conditional reasoning.

Although making the information more concrete can improve performance, other ways of making the information more 'naturalistic' can impede reasoning. For example, people are more likely to make errors if more intense emotions are involved (Blanchette & Richards, 2004[30]). If people are given the premise 'If there is danger, then one feels nervous', they will be more likely to invalidly affirm the consequent 'there is danger' if told that 'Betty feels nervous' as compared to if they are given more emotionally neutral information, such as the stamp problem.

Memory-related errors

The third category of errors in conditional reasoning involves memory limitations. As noted earlier, Johnson-Laird has suggested that we reason by constructing *mental models*, mental representations of meanings of the terms in reasoning problems (Johnson-Laird & Byrne, 2002; Johnson-Laird, Byrne & Schaeken, 1992[31]). It is difficult to flesh out a set of meanings in abstract conditional reasoning problems of the 'If *P* then *Q*' variety, but it is far easier in concrete, meaningful problems such as:

If it was foggy, then the match was cancelled.
It was foggy.
Therefore, the match was cancelled.

Furthermore, if additional terms appear in the problem, additional mental models must be derived, two additional ones in the following case:

If it was foggy, then the match was cancelled.
The match was not cancelled.
Therefore, it was not foggy.

The use of mental models during reasoning is supported by evidence showing that the time to respond to a conditional logical problem is a function of the amount of effort needed to mentally construct and manipulate relevant mental models (Vergauwe, Gauffroy, Morsanyi, Dagry & Barrouilet, 2013[32]). When additional models are needed, the load on working memory mounts and can interfere with reasoning. In a series of experiments, De Neys et al. (2005)[33] showed that performance in conditional reasoning tasks was impaired when the load on working memory resources was higher. The same is true when the phrasing of the problem places a greater load on comprehension (Thompson & Byrne, 2002[34]). And finally, Evans et al. (1999)[35] point out that if a conclusion matches the first mental model derived from the problem, it is particularly easy to accept the conclusion, leading to fallacies or errors in reasoning.

10.1.4 Hypothesis testing

Part of the importance of conditional reasoning is its connection to scientific hypothesis testing. Consider a typical experimental hypothesis:

If theory A *is true, then data resembling* X *should be found in the experiment.*

If data resembling X are indeed found, there is a strong tendency to conclude that theory A must be true: it affirms the consequent. What's wrong with this? Well, affirming the consequent is an error in logical reasoning and could lead to mistakenly concluding that, based on this evidence, the antecedent is true. This error is seductive. Of course, it might be true that theory *A* is correct. But it is also possible that theory *A* is incorrect and that some other (correct) theory would also predict data X. People are strongly biased to draw conclusions based on what they know to be true about premises and not by what could be false about them (Espino, Santamaria & Byrne, 2009[36]).

Because of the invalidity of affirming the consequent, and because we want to test hypotheses, our experiments test a *different* hypothesis than 'Theory *A* is correct'. As you learned (or will learn) in statistics, we test the null hypothesis in hopes that the evidence will be inconsistent with the predicted null outcome. Note the form of such a test:

If the null hypothesis is true (if P*), then there will be no effect of the variable (then* Q).

If we obtain evidence that there *is* an effect of the variable, then we have evidence that the consequent is not true. We can then conclude, via denying the consequent (*modus tollens*), that the antecedent is not true. In other words, we reject the null hypothesis, deciding that it is false. Although people can make a variety of errors in such situations, especially when the *if–then* relationship is more complex (Cummins, Lubart, Alksnis & Rist, 1991[37]), the typical mistake is to search for positive, confirming evidence (Klayman & Ha, 1989[38]). In similar vein, another strategy is simply to make a judgement as to the relevance or strength of the arguments and base a decision on that (e.g. Medin, Coley, Storms & Hayes, 2003; Rips, 2001[39]).

Overall, people have a bias to adopt what they view as the best explanation of a cause–effect relationship based on the plausibility of the various elements of the situation. When this requires that they place greater emphasis on confirmatory evidence, they do so, but when it involves placing greater emphasis on disconfirming evidence, they also do that (Koslowski, Marasia, Vermeylen & Hendrix, 2013[40]). Moreover, the degree of emphasis people place on positive and negative evidence also reflects the source of that information, whether it comes randomly or from a helpful communicator (Voorspoels, Navarro, Perfors, Ransom & Storms, 2015[41]).

Thus, we have the capacity to use positive and negative evidence appropriately. However, the emphasis that we place on this information varies greatly depending on how plausible we think the evidence is and the circumstances surrounding how we learn about it. For example, suppose you are trying to learn what triggers rashes on some patients' skin. If you hear evidence about the weather, you may reject this as implausible, unless you later learn that a bacterium is responsive to exposure to sunlight. Similarly, you are less likely to place emphasis on weather information if you think you are getting random facts about the situation compared to if you are getting the information from a cooperative and knowledgeable source.

10.2 Decision-making

LEARNING OBJECTIVE
Describe the characteristics of decision-making

In the opening of the chapter we described how Britons decided about the future of their country – an important decision with life-changing implications for the generations to come. Other important decisions we may make in our lifetimes include, which course of study to select, whether to get married, whether to start a family, get a pet, buy a car, and so on. We also regularly make other, more trivial decisions, such as whether to order pepperoni pizza or margarita on a given night, or whether to buy one brand of cosmetic products or another. But how do we choose among several alternatives? Which factors influence the ways we make decisions? Is decision-making only a matter of cognitive processes, such as retrieving information from our memory?

In a sense, a lot of the cognitive psychology studies we have reviewed in different chapters of this book relate to decision-making processes, although the decisions were often fairly simple – for example, deciding 'yes' or 'no' in semantic or lexical decision tasks. At its base, *decision-making* can be viewed as a search for evidence where the ultimate decision depends on some criterion or rule for evaluating the evidence, or the likelihood of either desirable or undesirable outcomes. How we make decisions as a function of such evidence and how we evaluate the evidence itself reflect the intersection of reasoning, decision-making, and problem-solving, which we discuss in more detail in the next chapter.

For example, research on reasoning and decision-making, as well as problem-solving, investigates processes that are slow and deliberate, also known as system 2 processes. Another similarity relates to *familiarity:* the domains of reasoning and problem-solving we investigate are not well known or understood by people, or they involve material that is not highly familiar. *Uncertainty* is another common feature of decision-making and problem-solving. There is often no certain answer to the problems or at least no good way of deciding whether a particular solution is the correct approach. Despite this, taking a principled and careful look at how people make decisions can provide enormous benefits and guide further research. Lastly, people make decisions and base their reasoning on a variety of strategies, some good and some not. This is also a characteristic of much problem solving. Because of these similarities, investigations of such strategies are often impossible to categorise clearly as reasoning on the one hand or problem-solving on the other.

10.3 Heuristics and biases in decision-making

LEARNING OBJECTIVE
Explain how heuristics and biases can influence decision-making

10.3.1 Heuristics in decision-making

Reasoning processes often underlie decision-making. For example, you may reason whether an electric car is suitable for your current needs and, based on this judgement, decide whether to invest money in buying an electric car or a more conventional, petrol one.

In this sense, reasoning and judgement are linked to decision-making, with decision-making often being informed by, or following from, judgement and reasoning processes. To better understand this relationship, let's consider some reasoning and judgement tasks first.

Answer these questions:

1. If you toss a fair coin, what is the probability of getting heads?
2. If you toss a fair coin, what is the probability of getting heads two times in a row?

Most people know that the probability of tossing heads is .50 – that's a simple 50/50 situation. But you may be hazy about the correct algorithm to apply to the second problem, two heads in a row. It is fairly simple and requires two basic probability statements.

First, the chance of getting heads once is 50/50, or stated as a probability, .50.

Heuristics in decision-making can help you to make decisions, such as whether to invest in an electric vehicle or stay with petrol.

Source: Matej Kastelic/Shutterstock

Second, the probability of any particular sequence of independent events (like coin tosses) is the basic probability of the event (.50) multiplied by itself once for each event in the sequence, that is .50 × .50 for two heads in a row.

More formally, the formula is $p(e)^2$, the basic probability of the event $p(e)$ raised to the nth power, where n is the number of events in the sequence. So the answer here is .25 for two heads, $.50^2$, or simply .50 × .50.

We tend to be poor at probability questions. In a survey of an undergraduate class that used those questions, 89 per cent got question 1 correct, but only 42 per cent got the second question right; 37 per cent said that the correct answer was .50, the same as in question 1 (Ashcraft, 1989)[42]. It is clear these people did not know the formula.

Likewise, we generally don't know the algorithm for answering this kind of question:

If each of 10 people at a business meeting shakes hands (once) with each other person, then how many handshakes are exchanged?

If you do not know how to compute this, then make an estimate, and then introspect for a moment on how you came up with that estimate. If you guessed, then what guided your guess? The algorithm for this problem is also fairly easy. If N is the number of people, then the number of handshakes is:

$$N\times\left(\frac{N-1}{2}\right)=10\times\frac{9}{2}=45 \text{ handshakes}$$

Notice that the algorithm provides a systematic and orderly procedure that is guaranteed to yield the correct result (assuming you do the math correctly). Algorithmic methods, in all these settings, follow the *normative model*, the method provided by mathematics and probability. Heuristics, in contrast, seem very human. They are not necessarily systematic or orderly and rely heavily on educated guessing. This is referred to as the *descriptive model*, just a description of how the question was answered. A large part of the research on reasoning and decision-making looks at

Daniel Kahneman is a Nobel Prize laureate (Economic Sciences) and pioneered the field of heuristics and decision-making.
Source: Roger Parkes/Alamy Stock Photo

how different the normative and descriptive models are, meaning how people diverge from the normative method. Under certain circumstances, or for particular kinds of questions, heuristics seem prone to distortions, inaccuracies and omissions – the descriptive model diverges from what's normative or 'right'.

By far, some of the most influential work done on decision-making and heuristics has been the classic work of Tversky and Kahneman (1973, 1974, 1980; Kahneman, Slovic & Tversky, 1982; Kahneman & Tversky, 1972, 1973; Shafir & Tversky, 1992[43]). Tversky and Kahneman's work on heuristics and fallacies has had an impact in such diverse areas as law, medicine and business. Most prominently, it has affected the field of economics, and Daniel Kahneman received the Nobel Prize in Economics in 2002 (see Kahneman, 2003a, for a first-person account of that work, and 2003b for his personal history of the collaboration with the late Amos Tversky; see Kahneman & Tversky, 2000, for a compendium of chapters on this approach[44]). In a way, it isn't surprising that this topic is of interest to many different fields – after all, think of how many situations and settings involve decision-making. The Nobel Prize signifies more than just the relevance of the topic; it indicates noteworthy achievement in tackling and explaining a large and important set of ideas: how humans reason and make decisions.

Kahneman and Tversky's research focused extensively on a set of heuristics and biases that appears to characterise everyday decision-making about uncertain events. In some of the situations they studied, these researchers found an algorithm can be applied to arrive at a correct answer. Many of these situations involve probabilistic reasoning. Knowledge of the algorithms doesn't necessarily mean that people understand them, can use them spontaneously, or can recognise when to apply them. Indeed, Kahneman and Tversky found that a sample of graduate students in psychology, all of whom had been exposed to statistical algorithms, did well on simple problems but still relied on a heuristic when given more complex situations. That said, some studies have shown good transfer and improved reasoning after relevant training (Agnoli, 1991; Agnoli & Krantz, 1989; Fong & Nisbett,

1991; Lehman, Lempert & Nisbett, 1988[45]). Other situations that have been studied involve estimates of likelihood when precise probabilities cannot be assigned or have not been supplied, although elements of statistical and probabilistic reasoning are still appropriate, and in situations contrasting verbal and numerical descriptions (for instance, 'rain is likely' versus 'there's a 70% chance of rain'; Windschitl & Weber, 1999[46]). Finally, some settings involve very uncertain or even impossible situations; for example, asking people to predict the outcome of a hypothetical event, such as the outcome of World War II if Germany had developed the atomic bomb before the United States.

10.3.2 The representativeness heuristic

In this section, we begin by covering a series of classic heuristics that have been studied in research on decision-making, including the representativeness, availability, simulation and undoing heuristics. We also cover a more contemporary approaches that provide alternative ways of thinking about the role of heuristics in human decision-making.

The laws of large and small numbers

The representativeness heuristic embodies a bias of insensitivity to sample size. When people reason, they fail to account for the size of the sample or group on which the event is based. They seem to believe that both small and large samples should be equally similar to the population from which they were drawn. In other words, people believe in the law of small numbers. Now the *law of large numbers* – that a large sample is more representative of its population – is true. But people erroneously believe that there is also a law of small numbers (see also Bar-Hillel, 1980[47]), but that is not the case.

Examples of classic heuristics in decision-making

As you read, try to develop your own examples of situations that are like the given ones. You will be surprised at how often we use heuristics in everyday judgements and decision-making. Refer to the Appendix at the end of the chapter for more details on these situations.

If you toss a coin six times in a row, which of the following outcomes is more likely: HHHTTT or HHTHTT? Most of us would say – and quite rapidly at that – that the second alternative is more likely. But if you stop and think about it, you realise that each of these is *exactly* as likely as the other, because each is one of the possible ways six coin tosses can occur (the total number of outcomes is 2^6, i.e. 64 sequences of heads and tails). We tend to think of the alternating pattern HHTHTT as a representative of a whole class in which most outcomes have alternations between heads and tails. The thinking here, illogical but understandable, is that a random process ought to look random (e.g. Burns & Corpus, 2004[48]), and if it doesn't, there might be some underlying systematic process (Hahn & Warren, 2009[49]). The sequence of three heads then three tails looks non-random and so seems less likely. Likewise, because the likelihood for six tosses is three heads and three tails (in the long run), almost any sequence with three of each will appear more representative than sequences with more of one outcome than the other. (See Pollatsek, Konold, Well & Lima, 1984, and Nickerson, 2002, for evidence on people's beliefs about random sampling processes, and the ability to produce and perceive randomness.[50])

The mistake we make indicates how we reason in similar situations (the math for this problem is explained in the Appendix at the end of the module).

For Kahneman and Tversky (1972),[51] we judge the likelihood of uncertain events based on the event's representativeness. The **representativeness heuristic** is a judgement rule in which an estimate of probability is determined by one of two features: how similar the event is to the population of events it came from or whether the event seems similar to the process that produced it. In other words, we judge whether event *A* belongs to class *B* based on the extent to which *A* is representative of *B*, the degree to which it resembles *B*, or the degree to which it resembles the kind of process that *B* is known to be.

Now, let's consider a situation more like Kahneman and Tversky's first criterion, where the event is similar in essential characteristics to the population of events from which it is drawn. Their example of such a situation is as follows:

> In a certain town, there are two hospitals. In one, about 45 babies are born each day; in the other only about 15. As you know, about 50 per cent of all babies are boys, although on any day, of course, this percentage may be higher or lower. Across one year, the hospitals recorded the number of days on which 60 per cent or more of the babies were male. Which hospital do you think had more such days? *(Decide on your answer before reading further.)*
>
> (After Kahneman & Tversky, 1972, p. 443.)

Overall, most people believed that both hospitals would have about the same number of days on which 60 per cent or more (or less) of the babies would be male ('extreme days'). Let's explain this. People know there will be variations around the expected percentage of 50 per cent boy/50 per cent girl and that 60 per cent is somewhat extreme. Because 'somewhat extreme' events occasionally happen, a small and a large hospital both having 60 per cent or more male babies is viewed as representative of a larger population. Note the implicit and incorrect assumption here that 'extreme' means the same thing for the two hospitals. In fact, the correct answer is that the smaller hospital will have more extreme days. This is because extreme or unlikely outcomes are more likely with small sample sizes. So, with fewer events, the likelihood is greater for variations from the expected proportion. Thus, it is more likely that the small hospital will have more extreme days than the large hospital because 60 per cent proportion is being computed on an average of 15 births instead of 45. Another way of saying this is that, given the 50-50 odds, 60 per cent is not as extreme an occurrence out of 15 opportunities as it is out of 45; 60 per cent or more male babies is not as extreme for the small hospital as it is for the large one.

Biases in the representativeness heuristic

Let's take a look at several of the biases that stem from the representativeness heuristic.

Ignoring base rates (ignoring prior odds) (based on Johnson & Finke, 1985)[52]

Questions

(a) Why are more fighter pilots first-born than second-born sons?

(b) Why do more apartment building fires occur on the first three floors than the second three floors?

(c) In football, do more interceptions occur on first down or second down?

The bias: In all three questions, people tend to ignore base rates. To answer the questions correctly, we should consider:

(a) How many first-born versus second-born sons are there? (b) How many apartment buildings even have more than three floors? (c) How many times do teams make first downs compared to second downs?

Base rates and stereotypes

Question

Frank is a meek and quiet person whose only hobby is playing chess. He was near the top of his college class and majored in philosophy. Is Frank a librarian or a businessman?

The bias

The personality description seems to match a librarian stereotype, whether the stereotype is true or not. Second, we fail to consider base rates, that is, the relative frequencies of the two professions. In other words, there are far more businessmen than librarians, a base rate that tends to be ignored because of the stereotype 'match'.

Gambler's fallacy

Question

You've watched a (fair) coin toss come up heads five times in a row. If you bet $10 on the next toss, would you choose heads or tails?

The bias

The gambler's fallacy is that the next toss is more likely to be tails, because 'it's time for tails to show up'. Of course, the five previous tosses have no bearing at all on the sixth toss, assuming a fair coin. The bias is related to the law of small numbers, in particular, that we expect randomness even on the 'local' or short-run outcomes. Thus getting tails after five heads seems more representative of the random process that produces the outcomes, so we mistakenly prefer to bet $10 on tails.

Stereotypes

stereotyping
A generalised belief of judgement about other people based on their social group membership.

Another bias resulting from the representativeness heuristic affects our judgements of other people, which often serves as the cognitive foundations of stereotyping. Kahneman and Tversky (1973)[53] reported evidence on estimations based on personality descriptions. They had people estimate the probability that a described person was a member of one or another professional group. They found that estimations were influenced by the similarity of a description to a widely held stereotype (i.e. what people belonging into this professional group should like or behave like). Consider first the premises of the judgement task posed by the researchers: 100 people are in a room, 70 of them lawyers, 30 of them engineers. Given this situation, answer the following question:

1. A person named Bill was randomly selected from this roomful of 100 people. What is the likelihood that Bill is a lawyer?

Simple probability tells us that the chances of selecting a lawyer are .70. Consider two other situations. There are still the same 70 lawyers and 30 engineers. But now you are given a description of two randomly selected people and are asked 'What is the likelihood that this person is an engineer?':

2. 'Dick is a 30-year-old man. He is married with no children. A man of high ability and high motivation, he promises to be quite successful in his field. He is well liked by his colleagues.'
3. 'Jack is a 45-year-old man. He is married and has four children. He is generally conservative, careful and ambitious. He shows no interest in political and social issues and spends most of his free time on his many hobbies, which include home carpentry, sailing and mathematical puzzles.'

base-rate information
Information about the frequency or prevalence of an event or a phenomenon in a given context (e.g. the mean prevalence of smokers in a country, the percentage of males and female employees in a company).

Here, people did *not* judge the probabilities to be the same as the prior odds – in other words, they discarded **base-rate information**. Instead, they assumed that the personality descriptions contained relevant information and adjusted their estimates accordingly. Particularly, people responded that the probability was close to .50, that is, about a 50-50 chance that Dick and Jack were engineers. Description 3 resembles the stereotype for engineers. People de-emphasised the prior odds (.30) and based their judgements on the description that seemed representative of engineers, allowing an influence of the stereotype. Description 2 was written to be uninformative about Dick's profession. And yet, people still changed their estimates. People tend to view any evidence as a basis for changing and, they hope, improving their prediction (Fischhoff & Bar-Hillel, 1984; Griffin & Tversky, 1992[54]).

According to normative model, one way to deal with these problems is to assess the usefulness or relevance of the additional information to decide how much weight to give it; its usefulness or 'diagnosticity'. Then the estimate is adjusted by the appropriate weighting. This is *Bayes' theorem*, which states that estimates should be based on two kinds of information, the base rate of the event and the 'likelihood ratio', which is an assessment of the usefulness of the new information.

10.3.3 The availability heuristic

What proportion of business leaders are women? What proportion of undergraduate students have a TikTok account? How much safer it is to travel by train or a car? Questions such as these ask you to estimate the frequency or probability of real-world events, even though you are unlikely to have more than a few shreds of relevant information stored in memory. Short of doing the fact-finding necessary to know the real answers, how do we make such estimates? The simplest way of making estimates is to try to recall relevant information from memory. Event frequency is coded in memory (Brown & Siegler, 1992; Hasher & Zacks, 1984[55]), perhaps automatically. If the retrieval of examples is easy, we infer that the event must be frequent or common. If the retrieval is difficult, then we estimate that it must not be frequent. Interestingly, frequency estimates affect your eventual judgements about the information: if it is repeated often enough, even false statements become 'truer' (Brown & Nix, 1996[56]). The easiness with which events or information are recalled from memory is related to the second heuristic that Tversky and Kahneman (1973)[57] discussed, the availability heuristic – 'ease of retrieval' is what the term *availability* means here. In Chapter 8, *availability* meant whether some information was stored in memory, and *accessibility* referred to whether the information could be retrieved. Clearly, *availability* in the Kahneman and Tversky sense is referring to *accessibility* in the memory retrieval sense. Kahneman (2003a)[58] has acknowledged that his original choice of terms was confusing in this sense and now refers to this as the *accessibility heuristic*. We continue to use the original term *availability*, however, to be consistent with the 50-year history of its usage in the decision-making literature.

Aspects of the availability heuristic

In short, when people make estimates of likelihood, their estimates are influenced by the ease with which they can remember relevant examples. Because our judgements are based on what we can remember easily, any factor that leads to storing information in memory can influence our judgements. If reasonably accurate and undistorted information is in memory, then the availability heuristic does a good job. But if it contains information that is inaccurate, incomplete or influenced by factors other than objective frequency, then it biases and distorts our reasoning. As a simple example, if your friend's Volvo needs repeated trips to the mechanic, you may view Volvos as unreliable. The availability heuristic has biased your judgement.

Another aspect of the availability heuristic is that people may judge events as more frequent or important because they are more familiar and can be accessed readily in memory. As such, this unwarranted influence of how much you have experienced something is called a **familiarity bias**. Familiarity influences decision-making beyond just how frequent something is. For example, food additives with unfamiliar, low-frequency, hard-to-pronounce names are rated as being riskier than additives whose names are more familiarly structured and easier to pronounce (Song & Schwarz, 2009[59]).

Other examples of the availability heuristic in everyday life are not difficult to imagine. Consider people's feelings about travelling by air. Statistically, one is far safer travelling by commercial airliner than by private car (by one estimate, some 25 times safer, based on normalised passenger-miles travelled). People who have no particularly relevant information in memory, and whose only information comes from casual attention to news media and the like, judge that one is much safer when travelling by car (e.g. Hertwig, Barron, Weber & Erev, 2004)[60]. This bias can be attributed to the factor of salience or vividness. News accounts of an airline accident are far more vivid and given far more attention than those of car accidents. Also, even though aeroplane crashes are rare, the number of victims involved is often dramatic enough that the event makes a stronger impression. In a revealing demonstration, Gigerenzer (2004)[61] examined US highway traffic fatality data for the 3 months following the 9/11 terrorist attacks on the hypothesis that people's 'dread fear' of flying after the attack might have led them to drive instead. As predicted, traffic fatalities were higher in those months than they had been for those months during the previous 5 years. (Ironically and sadly, approximately 350 additional lives were lost because of increased driving during that 3-month period.)

10.3.4 The simulation heuristic

A variation on the availability heuristic is the simulation heuristic. Here people predict future events or are asked to imagine a different outcome of an event or action. The simulation involves a mental construction or imagining of outcomes, a forecasting of how some event will turn out or how it might have turned out under other circumstances. The ease with which these outcomes can be imagined is critical. If a sequence of events can easily be imagined, then the events are viewed as likely. Alternatively, if it is difficult to construct a plausible scenario, the hypothetical outcome is viewed as unlikely. An example of this was given earlier, when you were asked to imagine possible outcomes if Germany had developed the atomic bomb before the United States – an outcome vividly depicted in the TV series *The*

Man in the High Castle, based on Philip Dick's novel. Given the role of the atomic bomb in ending World War II, people would give that far more weight than if they were asked about the development of some other device, say a long-range bomber or submarine.

Here is another alternative by Kahneman and Tversky (1982, p. 203) in which imagining a different outcome may be difficult:[62]

> Mr Crane and Mr Tees were scheduled to leave the airport on different flights, at the same time. They travelled from town in the same limousine, were caught in a traffic jam, and arrived at the airport 30 minutes after the scheduled departure time of their flights. Mr Crane is told that his flight left on time. Mr Tees is told that his flight was delayed and just left 5 minutes ago. Who is more upset, Mr Crane or Mr Tees?

As you would expect, almost everyone decides that Mr Tees is more upset; in Kahneman and Tversky's study, 96 per cent of the people made this judgement. However, from an objective standpoint, Mr Crane and Mr Tees are in identical positions: both missed their planes, and because of the traffic jam, both expected to do so. The reason Mr Tees is viewed as being more upset is that it was more 'possible', in some sense, for him to have caught his flight. So, it is easier to imagine an outcome in which the limousine arrives a few minutes earlier than it is to imagine one in which it arrives a half-hour earlier. As such, we believe the traveller who 'nearly caught his flight' will be more upset.

10.3.5 Elimination by aspects

Another important heuristic that people use when coming to a decision is **elimination by aspects** (Tversky, 1972[63]). Often the choices that you are presented with have multiple features or aspects. For example, when deciding which used car to buy, you may look at several features of several cars, such as price, petrol consumption, sound system, and so on. Similarly, if you are trying to decide which apartment to rent, you may also consider different features of several alternatives, such as rent, location, size, and so on.

A rational way to approach these sorts of situations would be to look at all the relevant features for all the options, tally up their various values, and then select the option that provides the best fit for the person across all available features. This approach is problematic because there often is a large number of options and features to consider. Imagine looking over and test-driving every single used car in town, or imagine visiting every single apartment that is available for rent in town prior to making a decision. What a lot of time and work!

To short-circuit this process, people can use the elimination by aspects heuristic. Basically, people go through the various features, one at a time, starting with those of most importance to them. Each option that does not meet some criterion for a given feature is then dropped from consideration. This process is repeated feature by feature until the number of options is drawn down to a manageable size or leads to a single option.

As an example of how this would work, if you were looking for a used car, you might first eliminate any cars that are too expensive or too cheap. Then, you might rule out any that are the wrong style (e.g. you don't want a convertible), followed by eliminating any with poor fuel consumption, and so on. This would continue until you had the number of used cars down to a small amount. Similarly, you might narrow down the set of apartments to look at by first eliminating those that have rents that are too high or too low. Then you would rule out those in parts of town where you do not want to live, those that are too high in a building, those that are part of complexes, and so on, until you get the number of apartments to consider down to a small group. Although this process will often get you to the ideal or close to ideal choice given what is available, it is not guaranteed to. It is possible that the best choice for you, given all the features to consider, might be eliminated early on.

10.3.6 The undoing heuristic

A more complete example of the simulation heuristic is illustrated in the simulation stories below, including the **undoing** of an outcome by changing what led up to it. This is called **counterfactual reasoning**, when a line of reasoning deliberately contradicts the facts in a 'what if' kind of way (e.g. '*What would have happened if Germany had developed the bomb first*'). This is the process of judging that some event 'nearly happened', 'could have occurred', 'might have happened if only' and so on (Roese & Epstude, 2017[64]).

Typical bias

Other factors have also been implicated in counterfactual reasoning. Byrne and McEleney (2000)[65] suggested that people focus on actions, not failures to act, when they undo events. They had people read a story about Joe and Paul. Joe got an offer to trade his stock in Company *B* for stock in Company *A*, which he did, although he ultimately lost money on Company *A*. Paul got a comparable offer to trade his stock in Company *A*, but he decided not to take the offer. Staying with Company *A*, he ultimately lost the same amount of money as Joe did, even though they both started with the same amount. Despite the equal loss by both, 87.5 per cent of the people claimed that Joe, the one who acted, would feel worse about his decision ('If only I hadn't traded my stock'). In a companion article, McCloy and Byrne (2000)[66] developed a scenario in which a character is late for an appointment, because of either controllable or uncontrollable factors – the character stopped to buy a hamburger or was delayed because a tree had fallen in the street. People undid the controllable factors far more frequently, although they distinguished among delaying factors in terms of interpersonal and social norms of how acceptable or polite the factor was; stopping to visit his parents on the way to the appointment was viewed less negatively than stopping for the hamburger.

Stories for the simulation heuristic

Read the stories now and decide how you would complete the 'if only' phrase before continuing.

1 Mr Jones was 47 years old, the father of three, and a successful banking executive. His wife had been ill at home for several months.

2a. On the day of the accident, Mr Jones left his office at the regular time. He sometimes left early to take care of home chores at his wife's request, but this was not necessary on that day. Mr Jones did not drive home by his regular route. The day was exceptionally clear, and Mr Jones told his friends at the office that he would drive along the shore to enjoy the view.

3 The accident occurred at a major intersection. The light turned amber as Mr Jones approached. Witnesses noted that he braked hard to stop at the crossing, although he could easily have gone through. His family recognised this as a common occurrence in Mr Jones' driving. As he began to cross after the light changed, a light truck charged into the intersection at top speed and rammed Mr Jones' car from the left. Mr Jones was killed instantly.

4a. It was later ascertained that the truck was driven by a teenage boy who was under the influence of drugs.

5 As commonly happens in such situations, the Jones family and their friends often thought and often said, 'If only' during the days that followed the accident. How did they continue this thought? Please write one or more likely completions.

Route version

2b. On the day of the accident, Mr Jones left the office earlier than usual to attend to some household chores at his wife's request. He drove home along his regular route. Mr Jones occasionally chose to drive along the shore, to enjoy the view on exceptionally clear days, but that day was just average.

Time version (substitute 2b for 2a)

4b. It was later ascertained that the truck was driven by a teenage boy named Tom Searler.
Tom's father had just found him at home under the influence of drugs. This was a common occurrence, as Tom used drugs heavily. There had been a quarrel, during which Tom grabbed the keys that were lying on the living room table and drove off blindly. He was severely injured in the accident.

'Boy' focus version (substitute 4b for 4a)

Percentage of people responding to the ''If only' stem in the five different response categories

'If only' completion focuses on:	Route version	Time version
Route	51%	13%
Time	3%	26%
Crossing	22%	31%
Boy	20%	29%
Other	4%	1%
	n = 65	*n* = 62

Result

The results are shown in the above table. Many people – 51 per cent – chose to change the unusual event in the Mr Jones story, that he took a different route home than normal; in the participants' responses, they basically said, 'If only he had taken his normal route home, the accident could have been avoided.' This is a **downhill change**, when we alter an unusual story element, substituting a more typical or normal element in its place. In general, such changes focus on an unusual event and thus 'normalise' the story. Other kinds of changes, for instance, inserting an unusual event (after he left work, Mr Jones had a flat tyre), were rarely supplied. We are biased towards downhill changes for at least three reasons:

1. Downhill changes are more easily imagined – the ease of retrieval factor again (Koehler & Macchi, 2004[67]).
2. Downhill changes seem more plausible; it is more plausible that Mr Jones left on time than it is that he left early and then had a flat tyre.
3. We have a tendency to judge unusual events as being the cause of unanticipated outcomes (Roese, 1997; see Reyna, 2004, and Trabasso & Bartolone, 2003, for gist- or comprehension-based explanations of reasoning[68]).

A puzzle in the Mr Jones story was that people seldom focused on the actual cause of the accident, the teenage boy. Kahneman and Tversky speculate that this was caused by a focus rule: we tend to maintain properties of the main object or focus of the story unless a different focus is provided. In support of this, when people read a version of the story that focused more on the boy, they were more likely to undo the boy's actions. However, undoing Mr Jones's behaviour – having him take his normal route home – essentially claims that Mr Jones was responsible for the accident because it was his behaviour that people altered. It is a case of 'blaming the victim'. This is especially common when there was an unusual event in the story that could have been altered via a downhill change. For instance, Goldinger, Kleider, Azuma, and Beike (2003) offer clear examples:[69]

> Paul normally leaves work at 5:30 and drives directly home. One day, while following this routine, Paul is broadsided by a driver who violated a stop sign and receives serious injuries. (p. 81)

When people consider how much compensation Paul should receive for his injuries and how much punishment is appropriate for the other driver, they examine Paul's behaviour closely. In this scenario, however, Paul tends not to be blamed for the accident. But if the scenario is changed to:

> Paul, feeling restless at work, leaves early to see a movie . . . . Paul is broadsided by a driver . . .

and so forth, then we tend to view him as less deserving of compensation. In addition, in an echo of the social norms result of McCloy and Byrne (2000),[70] Goldinger et al. (2003)[71] pointed out the following: *if* Paul receives an emergency call to return home and then is broadsided, 'the accident now appears exceptionally tragic, and compensation awarded to him increases' (p. 81). This reversal, observed by Miller and McFarland (1986),[72] depends on how free Paul was to choose what to do and how socially acceptable his choices were.

Related to this idea, people are more willing to say that an action was intentional when they are told it will be done in the future than if it has already happened in the past (Burns, Caruso & Bartels, 2012[73]). Essentially, people are perceived as having more control over future actions than past actions, even if the actions are identical. Also, regarding the future, people are more realistic when they are planning how they would do things differently in the future as compared to how they would change things in the past, even when what would need to be changed remains largely the same (Ferrante, Girotto, Stragà & Walsh, 2013[74]).

Hindsight bias

The simulation heuristic provides a nice explanation of **hindsight bias** (Fischhoff & Bar-Hillel, 1984[75]), the after-the-fact judgement that some event was very predictable, even though it wasn't. This is the 'I knew it all along' effect. In thinking about the now-finished event, the scenario under which that event could have happened is easy to imagine – after all, it just happened (Kneer & Skoczeń, 2023; Sanna & Schwartz, 2006[76]). The connection between the initial situation and the outcome is very available after the fact. This availability makes other possible connections seem less plausible than they otherwise would (e.g. Hell, Gigerenzer, Gauggel, Mall & Muller, 1988; Hoch & Loewenstein, 1989[77]). Hindsight bias can even distort our perceptual memories. Gray, Beilock and Carr (2007)[78] report that batters misremember how well they thought they would hit a baseball when they have been hitting well as compared to when they have struggled. That is, the current success with batting causes the player to experience hindsight bias by misremembering the batting as being better than it actually was.

Interestingly, hindsight bias even influences memory for events. People routinely 'remember' their original position to be more consistent with their final decision than it really was (Erdfelder & Buchner, 1998; Holyoak & Simon, 1999[79]) and even reconstruct story elements so that they are more consistent with the final outcome. As a demonstration of this, Carli (1999)[80] had two groups of people read a story (about Pam and Peter in Experiment 1, and Barbara and Jack in Experiment 2) with no ending (control group), a happy ending (Jack proposes marriage to Barbara), or a tragic ending (Jack rapes Barbara). The stories had information that was consistent with both scenarios; for example, 'Barbara met many men at parties' and 'Pam wanted a family very much.' After finishing the stories, people were questioned about the story events. The groups that heard either one of the stories with an ending agreed far more than the control group that they would have predicted that ending all along – but of course the endings were completely different. As a follow-up, in a later memory test, both groups mistakenly 'remembered' information that had not been in the story but was consistent with the ending they had read. (See Harley, Carlsen & Loftus, 2004, and Mather, Shafir & Johnson, 2000, for an extension of hindsight bias to identifying pictures, important for eyewitness testimony situations, and remembering blind dates, respectively.[81])

In many ways, the simulation heuristic, which leads to hindsight bias, may come closer to what people mean by such terms as *everyday thinking* and *contemplating* than anything else we have covered so far. For example:

If I stop for a cup of coffee, I might miss my bus.

If I hadn't been so busy yesterday, I would have remembered to go to the ATM for money.

In looking back, I guess I could have predicted that waiting until senior year to take statistics was a bad idea.

Such thinking often is the reason we decide to do one thing versus another; we think through the possible outcomes of different actions, then base our decision on the most favourable one. Thus, *mental simulation*, taking certain input conditions then forecasting possible outcomes, is an important way to understand cognitive processes related to planning. A general warning here is important, based on studies of how people generate possible outcomes of future events (Hoch, 1984, 1985[82]). If you begin planning by thinking only of the desirable outcomes, you may blind yourself to possible undesirable outcomes. By starting with positives, you become more confident that your plan will have a good outcome. Overly optimistic predictions at the outset bias our ability to imagine negative outcomes and inflate our view of the likelihood of a positive outcome. ('Hey, what could go wrong if I wait until next week to start my term paper?'; see Petrusic & Baranski, 2003, on how confidence affects decision-making.[83])

When future positive outcomes are perceived as more probable than negative ones

Optimism bias (also known as *unrealistic optimism*) describes the tendency to anticipate more positive personal outcomes in the future as compared to a more objective standard, or to the comparative likelihood of other people experiencing a future positive outcome (Shepperd, Waters, Weinstein & Klein, 2015[84]). For example, a smoker may perceive that their personal prospect of suffering from lung cancer is much lower than the average likelihood of lung cancer among other smokers, and married couples in Western countries may perceive that their marriage will be long-lasting and happy despite the high numbers of divorces documented in most Western countries (Sharot, 2011[85]). Sharot and colleagues (Sharot & Garrett, 2016; Sharot, Korn & Dolan, 2011[86]) explored the cognitive and neural aspects of the optimistic bias. In a series of studies, they found that people are more likely to update their current beliefs about future outcomes when these outcomes are positive and presented as more probable. Accordingly, when people are presented with information indicating a high probability of more negative future outcomes, they are not as effective in updating their current beliefs. Sharot et al. (2011)[87] demonstrated that this tendency is explained by hemispheric asymmetry in the activation of the inferior frontal gyrus (IFG) – a brain area associated with the regulation of response biases in decision-making tasks. Specifically, updating current beliefs after being presented with more positive future prospects was associated with higher activation of the left IFG, but failure to update current beliefs on the face of more negative future outcomes was associated with reduced activation of the right IFG. This research reveals how neural function can facilitate cognitive biases in judgement and decision-making.

10.4 Framing and risky decisions

LEARNING OBJECTIVE
Assess how decisions are influenced by the context in which they are made

Not only are decisions made under conditions of uncertainty, they may also be made with an element of risk. The consequences of these decisions can come with costs and benefits. A *cost* is a negative outcome for a decision-maker, whereas a *benefit* is a positive outcome. For example, imagine that people are asked to decide whether to take some cold medicine to relieve their symptoms. If they take it, the benefit is that they will alleviate many of their symptoms, but the cost is that they will feel sleepy. Alternatively, if they do not take the medicine, then they will have their cold symptoms in full force, but they will be more alert. So, this is a decision with costs and benefits. We make decisions by weighing the outcome of various choices and the degree to which we personally value different outcomes.

If people were rational decision makers, they would pick the option that leads to the best outcome for themselves, right? Sure. However, as you have already seen, people are not rational when making decisions that involve some element of risk. What influences the choice a person makes is a function of the *framing* of a situation, in terms of what information the person places an emphasis on.

Another way that people are influenced by assessing the amount of risk is by being attracted to 100 per cent values. So if a treatment is described as being 100 per cent effective on 70 per cent of pathogens, it is preferred by people over one that is 70 per cent effective on 100 per cent of pathogens, even though the two have equivalent effectiveness (Li & Chapman, 2009[88]). Thus, this illustrates again how people use heuristics to make decisions about probability and risk, even when they can calculate those values quickly and easily.

10.4.1 Risk aversion and seeking

The example of framing illustrates an important aspect of how people make decisions that involve uncertainty and risk, as well as the gains and losses that can follow from a decision. The choices that people make depending on the framing of a problem, and the emphasis on either whether something is gained (people will live) or something is lost (people will die), can be captured by a couple of principles.

The first is that people tend to show *risk aversion for gains*. So when they are thinking about what they will gain, they have a bias to avoid making risky decisions. Thus, people prefer Programme A for problem version 1 (see the box below, 'Two versions of a problem illustrating the framing of a choice') because the framing of the problem is for a gain (how many people will be saved). Version 1 presents the number of people saved as a sure thing.

The second principle is that people also tend to show risk seeking for losses, so when they are thinking about what they will lose, they have a bias to make risky decisions to avoid a loss. Thus, people prefer Programme B for problem version 2 because the framing of the problem is for a loss (how many people will die). Version 2 presents the number of people who will die as a probability or a risk. Because risk aversion for gains and risk seeking for losses is a heuristic, it doesn't take much cognitive effort, and this basic pattern is observed even when people are placed under a working memory load, such as having to do a secondary task in addition to the primary one of making a decision (Whitney, Rinehart & Hinson, 2008)[89].

Two versions of a problem illustrating the framing of a choice

Version 1

Imagine that the country is preparing for the outbreak of an unusual disease expected to kill 6,000 people. Two alternative programmes to combat the disease have been proposed, and only one of them can be implemented. Assume that the exact scientific estimate of the consequences of the programs are as follows:

If Programme A is adopted, 2,000 people will be saved.
If Programme B is adopted, there is 1/3 probability that 6,000 people will be saved, and 2/3 probability that no people will be saved.
Which of the two programmes would you favour?

Version 2

Imagine that the country is preparing for the outbreak of an unusual disease expected to kill 6,000 people. Two alternative programmes to combat the disease have been proposed, and only one of them can be implemented. Assume that the exact scientific estimate of the consequences of the programs are as follows:

If Programme A is adopted, 4,000 people will die.
If Programme B is adopted, there is 1/3 probability that nobody will die, and 2/3 probability that 6,000 people will die.
Which of the two programmes would you favour?

Tversky and Kahneman found that when they gave people version 1 of the problem, the majority (72 per cent) in their study preferred Programme A. However, when they gave people version 2 of the problem, the majority (78 per cent) preferred Programme B. These results are so interesting because the outcomes of Programmes A and B in two versions are identical. For example, if Programme A is used, then 2,000 people will be saved and 4,000 will die.

So, why is there a difference in people's preferences and decisions in the two cases? Well, this difference is caused by how the issues are being framed. For version 1, the emphasis is on how many people live, whereas in version 2, the emphasis is on how many people die. Thus, the two versions present identical information within different framing contexts, which lead to different outcomes. This is an important issue to consider when you encounter the results of polling or opinion data. Sometimes, the positions people seem to take on issues of social importance can vary widely depending on the phrasing of questionnaire items. Therefore, questioners sometimes seem to ask the same question several different ways. They may be trying to mitigate any framing effects to get at what people generally think. They are not trying to be annoying; they are trying to be careful.

10.4.2 Outcome magnitude

If people were cold and rational, they would be able to see the true value of their various choices. But even the value that we place on something is coloured by the context or framing in which the thing is embedded. Consider these two problems:

1. Imagine that you are about to purchase a jacket for £80. While you are looking at yourself in the mirror on the salesroom floor, another customer tells you that the same jacket is on sale for £60 at another store 20 minutes away. Would you make the trip to the other store?
2. Imagine that you are about to purchase a jacket for £500. While you are looking at yourself in the mirror on the salesroom floor, another customer tells you that the same jacket is on sale for £480 at another store 20 minutes away. Would you make the trip to the other store?

Overall, people are more likely to say that they would make the trip across town for the £80 jacket, but not for the £500 jacket, even though it is to save the exact same £20! Tversky and Kahneman (1981)[90] also appealed to the idea of framing to explain this thinking. Essentially, the initial cost of the jacket sets up the framework to make a decision. It is an anchor from which people will adjust their assessments. People are more likely to make the trip across town for the £80 jacket because the £20 is a much larger proportion of the cost, compared to the £500 jacket. It is also noteworthy that people are more willing to spend extra effort on the jacket that is less valuable compared to the more valuable jacket.

Another example of framing and decision-making has to do with whether options are added to or subtracted from something. For example, Biswas (2009)[91] reported that people were more willing to pay for options on a car if the initial state of the car was with all the options present, and they had to select the ones that they did not want, compared to if none of the options were present and they needed to indicate which ones they wanted to add. Thus, the initial state of the car, in terms of the options, set the framework from which they based the other decisions.

Overall, though these framing effects are persistent across many people and situations, steps can be taken to lessen their impact. For example, if the impact of a decision is presented graphically, such as by presenting circles to represent people, and colouring the ones corresponding to people who will die, then framing effects can be attenuated or eliminated (Garcia-Retamero & Dhami, 2013). That said, it is important to keep in mind that images can be distorted to bias decisions as well.

10.5 Adaptive thinking and 'fast and frugal' heuristics

LEARNING OBJECTIVE
Explain how heuristics can aid decision-making

Although the research by Tversky and Kahneman and its legacy, especially in the field of behavioural economics, sees heuristics as maladaptive and as the cause of suboptimal and ill-informed decision-making, other researchers praise the 'Homo

Heuristicus' and argue that once we learn how heuristics work, we can use them to improve decision-making (Gigerenzer, Reb & Luan, 2022[92]). In other words, these researchers describe heuristics-based decision-making as *adaptive thinking*. The biggest proponent of this adaptive thinking view is Gerd Gigerenzer (e.g. 1996; but see Dougherty, Franco-Watkins & Thomas, 2008, and Hilbig, 2010, for arguments against some fast and frugal heuristics[93]), Director of the Harding Center for Risk Literacy at the University of Potsdam, and formerly Director of the Center for Adaptive Behaviour and Cognition (ABC) at the Max Planck Institute for Human Development. For Gigerenzer, it is a mistake to assume that the correct answer to any decision-making problem must be the normative answer supplied by classic probability theory. Instead, it is important to assess how well people's heuristics actually do in guiding behaviour. People use heuristics not just because of memory limitations, incomplete algorithm knowledge, and so forth, but because they work; because they are adaptive in the sense of leading to successful behaviour. This is "adaptive" in an evolutionary sense, an explicit argument in Gigerenzer's approach. If a biological factor is adaptive, in evolutionary terms, it leads to greater success of the individual, hence greater spread of the feature into succeeding generations. In similar fashion, an adaptive heuristic will be successful, so it will be used more widely. Gigerenzer also advocates devoting more attention to the reasoning processes that are used, as opposed to the heavy focus on how answers deviate from the normative model. People use heuristics because they are tractable and robust (Gigerenzer, 2008)[94]. *Tractable* means that people can mentally track everything they need to use the heuristic, in contrast to algorithms that require people to track (often) much larger amounts of information. Calling heuristics *robust* means that they provide reasonable answers under a wide range of circumstances.

Gigerenzer (2008)[95] provides some compelling illustrations of the value of heuristics. He notes that the 1990 Nobel Prize winner in economics, Harry Markowitz, derived an algorithm for maximising the allocation of funds into various financial assets. However, even Markowitz did not use this principle for his own retirement savings, but instead used a simple 1/*N* heuristic (where *N* is the number of possible funds that the investments could go into). Based on this heuristic, your retirement funds are equally distributed across the number of available funds. Pretty simple, huh? This is the heuristic used by about half of the people in the real world. Although many financial wizards sneer at this strategy, when it was pitted against 12 different optimal return strategies, the 1/*N* heuristic beat them all. Bravo, heuristics!

Another example Gigerenzer (2008)[96] gives deals with organ donation. He notes that the rate of organ donation is only 28 per cent in the United States but is 99 per cent in France. His explanation has to do with a go-with-the-default heuristic. In the United States, the default is that people do not donate their organs when they die. They need to try to be organ donors (so that other people may live), for example, by filling out a form or having the option checked on a driver's licence. In the absence of this, the default is 'not an organ donor'. On the other hand, the default in France is to be an organ donor.

Part of the appeal of Gigerenzer's approach is that it is 'positive' – it doesn't emphasise errors or deviations from some norm, but searches for the usefulness of heuristics. A second appealing feature is that the 'fast and frugal' heuristics are simple. Third, the approach seems more open to input from general cognitive principles. For example, it acknowledges and considers memory limitations, incomplete knowledge, time limitations, and so on, as well as the general cognitive processes you have been studying. There is also the idea that human cognition is tuned to environmental statistics, and there is often a need to make assumptions based on small numbers of experiences (Hahn & Warren, 2009[97]). So, if six coin flips in a row end in heads, although this is a completely acceptable outcome from a random process, it may also reflect some causal systematicity in the environment, such as a weighted coin.

Adaptive thinking heuristics

Here are some of the adaptive fast and frugal heuristics that people may use in their decision-making. Each is used in different circumstances to help people arrive at a satisfactory decision in a short period of time.

Satisfificing heuristic

An important adaptive thinking heuristic that has been on the scene for some time is Simon's (1979)[98] satisficing heuristic (for a review see Artinger et al., 2022[99]). This principle is that we make a decision by taking the first solution that satisfies some criterion we may have – it is the 'good enough' heuristic. For example, if you are looking for a place to eat dinner while travelling, rather than checking out every single eatery in town, you simply pick the first one that satisfies your criterion, such as 'cheap fast-food place'. People can use this heuristic to make reasonably optimal decisions.

The recognition heuristic

One of Gigerenzer's simplest heuristics is the recognition heuristic, whereby you base a decision on whether you recognise the thing to be judged. For instance, if I ask you, 'Which city is larger, Kansas City or Junction City?' you might choose Kansas City because you have never heard of the alternative. Of course, it is illogical in some sense to base a decision on ignorance or missing knowledge – 'I've never heard of it, so it must be smaller' doesn't come across as a sound, iron-clad basis for deciding. On the other hand, to the extent that we notice things and store information in memory about them, the fact that we don't know about something could be informative. This is Gigerenzer's point – when having heard of something correlates with the decision criterion, here the population of the city, then not having heard of something is useful.

Another heuristic is the 'take the best' heuristic (Gigerenzer & Goldstein, 1996[100]), whereby you decide between alternatives based on the first useful information you find. Essentially, you search the alternatives for some characteristic – Gigerenzer calls this a 'cue' – that is relevant and that you might know about, and check to see if one option or the other has a positive value for that cue. So if I ask which is larger, Kansas City or Wichita, simple recognition will not work, because you have probably heard of both. You then try to retrieve a cue to help you decide, a cue that correlates with city populations – maybe something like 'Does the city have a major-league sports team?' If you know that Kansas City does but Wichita doesn't, you decide Kansas City is larger based on that bit of positive evidence. Of course, if I asked you about two cities that both had sports teams, then you would have to search for yet another cue and continue doing so until one of the cities differed – or until you finally ran out of cues and merely guessed. Figure 10.3 shows a flow diagram of the 'take the best' heuristic. The heuristic is simple – you search for some evidence favouring one over the other alternative, and you make your decision as soon as you find such evidence.

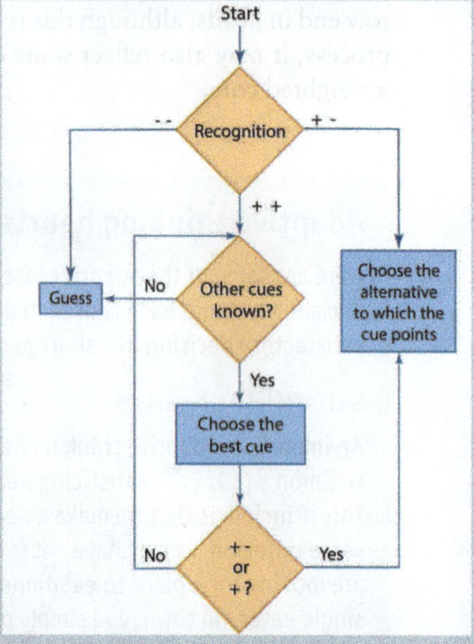

Figure 10.3 Flowchart of the 'take the best' heuristic

Processing starts with the comparison of two objects, A and B. If neither is recognised (− −), guess. If one is recognised and one is not (+ −), choose the recognised alternative. If both are recognised (+ +), search for a cue that might discriminate. If one cue is positive and the other is negative or null (+ −) and (+ ?), choose the alternative with the positive cue. If both are positive or both are negative, return to search for another cue.

This approach has generated several heuristics that differ in some ways from those outlined by Kahneman and Tversky and others.

10.5.1 Some fast and frugal heuristics in detail

Here we will look at some of these fast and frugal heuristics in more detail to give you a better understanding of how they would work. These are included in the following section – namely *satisficing*, *recognition* and *take the best* – to better illustrate this approach. It should also be noted that although these heuristics can be useful, people sometimes become overly dependent on one, such as persisting with one when it has become clear that it is no longer optimal and that a switch in strategies would be better (Bröder & Schiffer, 2006[101]).

The adaptive thinking approach does have its detractors. For instance, Newell and Shanks (2004)[102] have questioned whether the simple recognition heuristic is as powerful as Goldstein and Gigerenzer (2002)[103] claimed, showing how recognition is discounted when other cues of higher validity are presented. Similarly, Newell and Shanks (2003)[104] showed that the 'take the best' heuristic is used especially when the mental 'cost' (e.g. in terms of time) of information is high. However, when the cost is low (for instance, when it takes very little time to retrieve additional information), people seem to rely on that heuristic less frequently.

10.5.2 The ongoing debate

Gigerenzer and Goldstein (1996; also Gigerenzer, 1996, and Goldstein & Gigerenzer, 2002[105]) present considerable modelling and survey data to show how these heuristics do a good job of making decisions. Other supportive work has begun to appear as well (for instance, Burns, 2004, in a challenge to the well-known Gilovich, Vallone & Tversky, 1985, paper on the 'hot hand' in basketball[106]).

Linda and the conjunction fallacy

Let's get into time-travel and go back to 1983, when Tversky and Kahneman published their research about the conjunction fallacy, the mistaken belief that a compound outcome of two characteristics is more likely than either one of the characteristics by itself. Participants in the study were presented with different probability reasoning 'problems', one of which was the so-called 'Linda Problem' – before you read it, consider that at the time, societal norms around equality, diversity and inclusion were very different from what they are today.

> Linda is 31 years old, single, outspoken and very bright. She majored in philosophy. As a student, she was deeply concerned with issues of discrimination and social justice, and also participated in anti-nuclear demonstrations.

After reading this description about Linda, participants had to select one of the following probable outcomes about her:

> *Linda is a teacher in elementary school.*
> *Linda works in a bookstore and takes yoga classes.*
> *Linda is active in the feminist movement.*
> *Linda is a social worker in a psychiatric hospital.*
> *Linda is a member of the League of Women Voters.*
> *Linda is a bank teller.*
> *Linda is an insurance salesperson.*
> *Linda is a bank teller and is active in the feminist movement.*

Tversky and Kahneman (1983)[107] found that people endorse the compound alternative 'bank teller and active in the feminist movement' as more likely than either 'bank teller' or 'active in the feminist movement'. Such a judgement, from a purely probabilistic standpoint, is odd. It particularly illustrates the conjunction fallacy. According to strict probability theory, this should be impossible – making up some numbers to illustrate, if the chances are .20 that Linda is a bank teller, and .30 that she is active in the feminist movement, then the conjunction of those two characteristics should never be larger than .20. In fact, in stripped-down form, the probability ought to be .06; that is,

> In a room with 100 people, 20 are bank tellers, 80 are something else. Furthermore, 30 of the people are feminists, and 70 are not. What is the probability of randomly selecting someone who is both a bank teller and a feminist?

Heuristic	Biases and fallacies
	Insensitivity to sample size
	Belief in the law of small numbers
	Stereotype bias, belief bias, confirmation bias
	Familarity bias, salience or vividness bias
	Overly optimistic predictions inflate our confidence and prevent thinking of possible negative outcomes
	Bias to undo unusual event, bias to focus on action, bias to focus on controllable events, hindsight bias
	Blaming the victim
	Familiarity bias
	Reliance on ignorance or lack of knowledge

And yet people routinely say that the compound 'bank teller and feminist' is more likely (has a higher rank) than the simpler probability that Linda is a bank teller. Considerable significance has attached to this fallacy in Kahneman and Tversky's work – it comes close to epitomising the errors, the departures from the normative model, found in human reasoning. A cautionary note, however, is that the Linda Problem may reflect the gendered roles and stereotypes of the time this research was carried out.

10.6 Other explanations

LEARNING OBJECTIVE
Discuss the range of approaches to human decision-making

Considering humans from a broader perspective than simply 'lousy probabilists' reveals other reasonable interpretations of these rankings and judgements in decision-making. Indeed, Moldoveanu and Langer (2002)[108] supply a whole variety of explanations to justify this choice. For instance, many people treat the statement 'Linda is a bank teller and is active in the feminist movement' not as a conjunction of two characteristics but as a *conditional probability* – in other words, as if it said, '*Given* that Linda is a bank teller, what is the likelihood that she is active in the feminist movement?' The probabilities for a conditional probability are very different than those for a conjunction.

A second reason for these interpretations involves the idea of how consistent personality tends to be – in other words, whatever job a person may end up in, there still should be some consistency in the person's personality, something like 'once an activist, always an activist'. Such a comment also exemplifies an idea on conversational rules and the cooperative principle. If you were supposed to judge the Linda Problem solely on the probabilities of being a bank teller and being active in the feminist movement, then there is no communicative need to supply all of the background information on the character. In other words, 'Why did you tell me about Linda's activism as a college student if I was supposed to ignore it?' The fact that personality information was supplied tells people that it is important and should be factored into the answer. So people rather naturally take that into account and come up with a plausible scenario – 'OK, maybe she ended up as a bank teller, but she can still be a social activist by being active in the feminist movement.' In short, they may be developing a personality-based model of Linda.

More recently, there have been some attempts to address how reasoning and decision-making account for people's understanding of the causal structures of situations when making decisions (e.g. Garcia-Retamero, Wallin & Dieckmann, 2007; Hagmayer & Sloman, 2009; Kim, Yopchick & de Kwaadsteniet, 2008; Kynski & Tenebaum, 2007[109]), as opposed to the more statistical approach of Kahneman and Tversky, and Gigerenzer's adaptive thinking approach. For example (see Kynski & Tenebaum, 2007[110]), using standard base rate information, there is the finding that people who use sunscreen are more likely to develop skin cancer than those who do not, which runs counter to most people's intuitive judgements. This is because most people who don't use sunscreen are not out in the sun to begin with. It is important to note that using sunscreen does not cause skin cancer; rather, sunbathing does. When the problem is recast with this additional information, we see that among those people who sunbathe (a smaller group), those who use sunscreen are less likely to get skin cancer. Thus, when people are provided with the appropriate causal structure, their decisions processes can be very good.

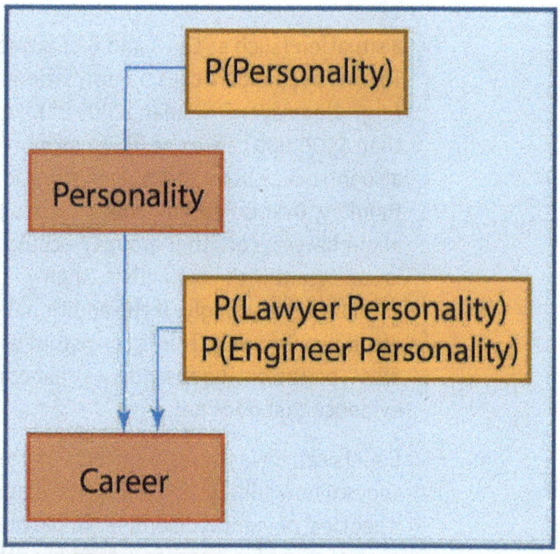

Figure 10.4 Causal model used by Kynski and Tenebaum (2007)[111]
An illustration of the concept that people use ideas of causal information and relations to help them make decisions. In this case, people are assuming there is a causal relationship between a person's personality and his or her career choice.

This approach can also explain why people make some of the errors that they do. For example, Figure 10.4 illustrates the causal model that many people have of personality and career choice, and how it relates to the lawyer–engineer problem. When statistical principles are applied to this model, then the results are similar to the estimates provided by people. That is, people account for the causes that produce various outcomes, as well as their combined influence, then make their decisions.

Example illustrating the Bayesian theorem

OK, if that doesn't clear things up, let's go over a more concrete example. Let's set *A* to be the probability of being audited by the IRS, and let's set *B* to be the probability of making more than $500,000 a year. What we want to know is what the probability is that a person will be audited if he or she makes over $500,000 a year, $P(A|B)$. For the sake of argument, let's say that the probability of being audited by the IRS, $P(A)$, is 5% or .05 (we are making up numbers here). Moreover, let's assume that the probability of making over $500,000 a year, $P(B)$, is 1% or .01. Finally, let's suppose that the probability of a person who makes more than $500,000 a year being audited by the IRS, $P(B|A)$, is 10% or .10. So, armed with this information, we can derive the following: Probability of an IRS audit if you make over $ 500, 000 = (.10 * .05)/.01, which is .05 or 5%.

Thus, you can see that, based on this fictional example, a person who makes more than $500,000 a year would be more likely to be audited than someone who does not.

Bayes' theorem is important because some cognitive psychologists have tried to make the argument that human reasoning may do something very close to this (Chater, Tenenbaum & Yuille, 2006[112]), even if it is done at a largely implicit and unconscious level (e.g. Hoffrage, Krauss, Martignon & Gigerenzer, 2015[113]). However, there is some disagreement about this (Bowers & Davis, 2012[114]).

In addition to Bayes' theorem in and of itself, cognitive theories of decision-making often assume that people are making causal inferences about how two elements of a situation (such as the A and B of Bayes' theorem) play into their decision-making. That is, people take causal information into account when making their decisions (e.g. Hagmayer & Sloman, 2009[115]). For example, in what way does making more than $500,000 cause the IRS to be more likely to audit someone? Our beliefs can also introduce bias, certainly in situations that rely on Bayes' theorem (e.g. Evans, Handley, Over & Perham, 2002[116]). It's easy to see how a belief – say in a stereotype about lawyers (or other groups) – could influence your judgements. If the evidence is consistent with your beliefs, then you give weight to the evidence and adjust your estimates accordingly. If the additional evidence is inconsistent with your belief, you could just ignore it. This is consistent with what you read earlier about *confirmation bias* – pay attention to evidence that confirms your belief, but pay little attention to evidence that does not.

Beliefs can have pervasive effects on reasoning. For example, Kim and Ahn (2002)[117] showed how clinical psychologists' diagnoses are affected by their individual 'theories' of mental disorders. In particular, if a clinician views symptom *X* as a central cause for disorder *Y*, then hypothetical patients who present symptom *X* are diagnosed as having disorder *Y*. Yet another clinician may view symptom *X* as being peripheral and so diagnoses a different disorder. In both cases, the belief guides the diagnosis rather than the 'supposedly' theoretical symptom lists in the *DSM (Diagnostic and Statistical Manual of Mental Disorders)* that clinicians are taught to use. Moshinsky and Bar-Hillel (2002)[118] have documented how beliefs can even affect judgements of when different world events happened. Israeli students are often surprised to learn that an event in US history happened at about the same time as another event in Europe. The source of the surprise is that the United States is representative of the 'New' World, so events that happened there are interpreted as being more recent than events in Europe, the 'Old' World. In a sense, these students view the United States as more representative of the recent or 'not long ago' category. Hence, they judge events in the New World as being more recent than those in the Old World.

10.6.1 Bayesian theories

Another way of approaching human decision-making is to take as a starting point that human decision-making is Bayesian, that is, it follows Bayes' theorem. This probability theorem accounts for the various base rates of different circumstances that contribute to a given situation. Bayes' theorem can be stated formulaically as:

$$P(A \mid B) = (P(B \mid A) * P(A)) / P(B)$$

In English, we would read this as 'The probability of *A* given *B* is equal to the probability of *B* given *A* times the probability of *A*, all divided by the probability of *B*.' Is this clear?

10.6.2 Quantum theory

More recently, cognitive psychologists have been using theoretical principles from quantum mechanics in the field of particle physics to address issues of human decision-making. OK . . . wait . . . what? How does this even begin to make sense? It is true that the cognitive processes involved in human minds and brains are very different from quarks and such, and these cognitive scientists acknowledge the difference. Yet these researchers are attracted to and find useful the methods

Principles of quantum theory

Several principles of quantum theory apply to accounts of human decision-making. These include the complementarity, uncertainty and superimposition principles.

Complementarity

Complementarity is the idea that different questions about the world are incompatible in that they provide different perspectives (which we saw in our discussion of framing effects), and these incompatible questions provide different viewpoints for understanding the world. Moreover, a grasp of these different perspectives is necessary for a complete understanding of what a person could be thinking.

Uncertainty

Uncertainty is the idea introduced when thinking about the circumstances of making a decision if there are two different perspectives. Should you find out something about one perspective, this influences what you can find out about the other. For example, the order in which things are done is important, and the outcome of one part of a decision affects the values of other parts. Suppose you are trying to decide where to have lunch with a friend. To make this decision, you need to consider both where you would prefer to go and where your friend would prefer to go. Although you may have some idea of where you want to go, getting information about where your friend wants to go introduces some uncertainty into your preferences that would not have been there before, and vice versa. Moreover, the order in which the two of you voice your preferences can influence the final decision.

Superimposition

Superimposition is the idea that, rather than a decision-maker being in one of two states concerning an issue, a person may be in an indefinite state between various options until a measurement is taken, such as asking the person to pick a choice.

for analysing probabilistic states and outcomes, which quantum mechanics and human decision-making have in common. Thus, although the underlying processes may be different, the ways of thinking about and analysing them may be analogous, and so warrant the use of an approach from a very different field. This application of quantum theory has also been extended to studies of episodic memory (Brainerd, Wang, Reyna & Nakamura, 2015[119]) and semantic memory (Aerts, Sozzo & Veloz, 2015[120]).

If you remember from our discussion of sensory and perceptual processes, a heavy emphasis is placed on signal detection theory. Signal detection theory arose from work in the field of communications: signals travelling over wires or radio waves. Psychologists found a way to take that approach and reapply it to the study of cognition. A similar sort of thing is going on here. Essentially, quantum theories of decision-making take the probabilistic mathematical framework of quantum theory and apply it to cases of human decision-making (Bruza, Wang & Busemeyer, 2015; Busemeyer & Wang, 2015; Moreira & Wichert, 2016[121]).

Bruza et al. (2015)[122] provide an account of the Linda Problem from a quantum theory view. Specifically, a quantum cognition approach focuses on the complementarity of Linda's being a bank teller and a feminist. First, a person decides the probability that Linda is a bank teller (let's say the person decides that Linda *is* a teller). Then, from this point, a person decides whether Linda is a feminist. This is no longer an independent decision based on what was decided beforehand, even though traditional probability theory treats these decisions as independent. Because they are not independent in the mind of the decision-maker, it is possible for the subjective probability to increase from that point given that Linda's characteristics match that of a feminist. The quantum theory approach can even account for phenomena such as preference reversals, which are a challenge for more traditional accounts of human decision-making (Yukalov & Somette, 2015[123]).

10.7 Limitations in reasoning

LEARNING OBJECTIVE
Evaluate the limitations of human reasoning and decision-making

A central fact in studies of decision-making and reasoning bears repeating. We use heuristics because of limitations in our cognition. We have bounded rationality and can only process so much information at a time. Moreover, there are limitations in our knowledge, both of relevant facts and of relevant algorithms. You had to estimate the number of handshakes in an earlier question because you didn't know the pertinent algorithm. In addition, there are limitations in the reasoner, sometimes as ordinary as unwillingness to make the effort needed but sometimes in working memory.

10.7.1 Limited domain knowledge

Everyday examples of how limited knowledge affects decision-making and reasoning are abundant. Kempton (1986)[124], for example, looked at reasoning based on analogies, particularly how we develop analogies based on known events and situations to reason about unknown or poorly understood domains. Focusing on a mechanical device, Kempton studied people's understanding of home heating, particularly their understanding of a furnace thermostat. The results indicated that some people's (incorrect) mental model is that a thermostat works like a water faucet: turn it up higher to get a faster flow of heat. Likewise, many people's behaviour suggests that they believe that the call button on a lift works like a doorbell: if the lift doesn't arrive reasonably soon, press the button again (see Shafir & Tversky, 1992)[125].

It should be obvious that our mental models – our cognitive representations that we use during reasoning and decision-making – can vary from true and complete knowledge (*expertise*, in other words) all the way down to no knowledge or information at all, *ignorance*. (People's awareness that they do not know something is quite interesting itself; see Gentner & Collins, 1981, and Glucksberg & McCloskey, 1981.[126]) The most interesting situation to study is when knowledge is incomplete or inaccurate. Indeed, the fact that we are so concerned with how people estimate under uncertainty, and with their errors in reasoning, implies that complete and certain knowledge usually is not available to people.

Naive physics

Some of the most intriguing research has been in naive physics, which is people's conceptions of the physical world, in particular, their understanding of the principles of motion. A compelling aspect of the effects observed in naive physics studies is that the motion of physical objects is not a rarefied, unusual kind of knowledge. We have countless opportunities in our everyday experience to witness the behaviour of objects in motion and to derive an understanding of the principles of motion. Anyone who has ever thrown a ball has had such opportunities. And yet the mental model we derive from that experience is flawed and is different from the perceptual–motor model that governs throwing a ball (Krist, Fieberg & Wilkening, 1993; Schwartz & Black, 1999[127]).

A second compelling aspect to the research concerns the nature of the mental model itself. People's erroneous understanding of bodies in motion is amazingly similar to the so-called *impetus theory*, which states that setting an object in motion puts some impetus or 'movement force' into the object, with the impetus slowly dissipating across time (e.g. Catrambone, Jones, Jonides & Seifert, 1995; see Cooke & Breedin, 1994, and Hubbard, 1996, on the notion of impetus[128]). The punchline here is that the impetus theory was the accepted explanation of motion during the 13th and 14th centuries, a view abandoned by physics when Newton's laws of gravity and motion were advanced some 300 years ago. The correct mental model, basically, is that a body in motion continues in a straight line unless some other force, such as gravity, is applied. If some other force is applied, then that force combines with the continuing straight-line movement. Thus, when the ball leaves the tube, or when the string breaks, the ball moves in a straight line. No 'curved force' continues to act on it because no such thing as 'impetus' has been given to it. Likewise, the horizontal movement of the ball dropped from the aeroplane continues until the ball hits the ground. This movement is augmented by a downward movement caused by gravity; the ball accelerates vertically as it continues its previous horizontal motion. (If you demonstrated a naive belief in impetus, you might take some consolation in the fact that, across recorded history, people have believed in impetus theory longer than they have in Newton's laws.)

Experience and knowledge

Only a little research addresses the nature of our experience and the kind of information we derive from it. In some of the naive physics problems, especially the plane problem, an optical illusion seems partly responsible for the 'straight down belief' (McCloskey, 1983; Rohrer, 2003[129]). Beyond that, some inattentiveness on the reasoner's part, or perhaps difficulty in profiting from real-world feedback, may also account for part of the inaccuracy. To be sure, some of the difficulties we experience involve the difficulty of the problems themselves (Proffitt, Kaiser & Whelan, 1990[130]). For example, Medin and Edelson (1988)[131] found that, depending on the structure and complexity of the problem, people may use base rate information appropriately, may use it inappropriately, or may ignore it entirely. We do know that instruction and training influence reasoning. Taking a physics class improves your knowledge of the rules of motion, but it doesn't completely eliminate the misbeliefs (Donley & Ashcraft, 1992; Kozhevnikov & Hegarty, 2001[132]). Likewise, instruction in statistics, probability and hypothesis testing improves your ability to reason accurately in those domains (Agnoli & Krantz, 1989; Fong & Nisbett, 1991; Lehman et al., 1988[133]). In short, acquiring a fuller knowledge of the domain is an important part of making more accurate decisions.

Types of reasoning

The types of reasoning errors we have been encountering here don't just apply to physics and statistics problems, but even extend to everyday objects. Lawson (2006[135]) gave people partial drawings of bicycles, such as those shown in Figure 10.5, and asked people to select the drawings from a set of options that correctly depict the proper position of the bicycle frame, pedals and chain. She found that, although bicycles are very familiar objects to most people, an average of 39 per cent made errors in their selections. Performance was better for expert cyclists, although even

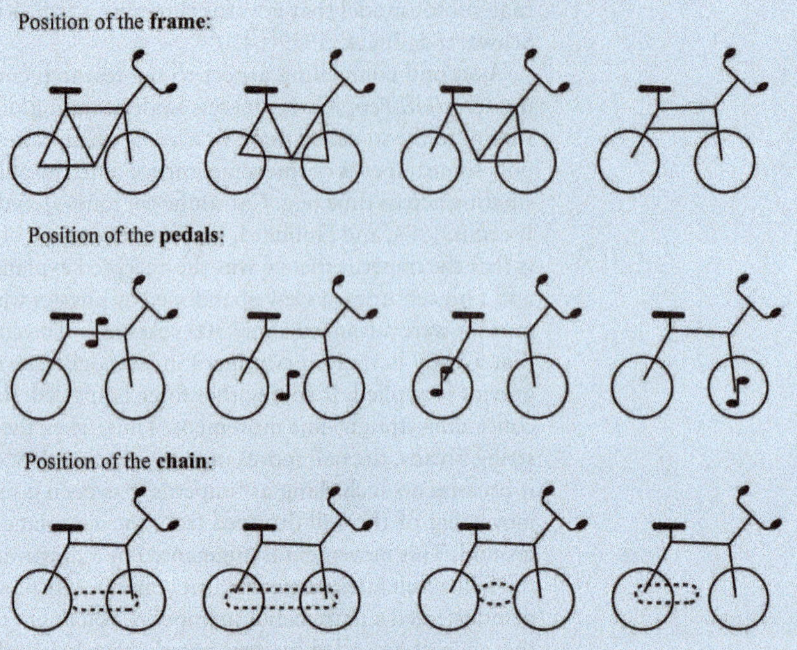

Figure 10.5 Response choices for a study by Lawson (2006)[134]

In this study by Lawson, people needed to select the correct location of a bicycle's frame, pedals and chain.

they made 15 per cent errors. So, from this we can learn that even when people have a great deal of familiarity with a common object, they may still have trouble reasoning about the structure of that object and how it works.

10.7.2 Limitations in processing resources

Several studies attest to the role of processing resources in adequate decision-making and problem-solving. Cherniak (1984; see also Tversky & Kahneman, 1983[136]) has studied the *prototypicality heuristic*, a strategy in which we generate examples to reason out an answer rather than follow the correct, logical procedures of deductive reasoning. The heuristic is useful in the sense that it reduces people's errors when they are working under time constraints. However, it depends on a limitation, possibly of working memory, possibly of time in which to perform the task, or possibly in the willingness to do the slow, effortful work of following the algorithmic procedure.

Examples illustrating limitations in processing resources

As an example, what is the answer to the following?

$$8 \times 7 \times 6 \times 5 \times 4 \times 3 \times 2 \times 1$$

Be honest – even though you know how to multiply, you didn't really multiply out all those values, did you? You probably estimated; you used a heuristic. One way we know this is by comparing your estimate to a different problem,

$$1 \times 2 \times 3 \times 4 \times 5 \times 6 \times 7 \times 8$$

In Tversky and Kahneman's (1983)[137] data, people's estimates for the first problem averaged 2,250, but for the second problem, estimates averaged 512. Heuristic processing was clearly involved, because the estimates depended on whether the arithmetically identical sequences began with a large or small number and because both estimates were wildly inaccurate: The correct answer to 8! is 40,320.

Although it is tempting to say the limitation was in your lack of willingness to solve the problem (a nice way of saying laziness), a full solution would take considerable working memory resources if you attempted to solve the problem mentally. Even if you were only approximating in your calculations, a sequence of problem-solving and keeping-track steps would tax working memory. Surely such a limited system as working memory would interfere with this type of processing.

An intriguing study demonstrates an entirely compatible result – that limited working memory resources compromise the ability to reason and make decisions in difficult situations. The study, by Waltz et al. (1999),[138] tested normal people, six patients with brain damage in the prefrontal cortex, and five with damage to the anterior temporal lobes. People were given two reasoning tasks. In the first, people read transitive inference problems such as 'Beth is taller than Tina. Tina is taller than Amy', where anywhere from two to four propositions were included (the 'Beth' example has two propositions), and had to arrange the sentences in order of height (so the 'Beth' sentence would be at the top). Problems presented in this order were at Level 1 complexity, and performance was contrasted with Level 2 complexity, in which the propositions were in scrambled order (e.g. for a four-proposition problem, 'Beth is taller than Tina. Sally is taller than Laura. Tina is taller than Joyce. Laura is taller than Beth'). The critical difference between conditions was that Level 1 problems, because they were in correct transitive order, never required more than one relation ('Beth is taller than Tina') to be held in mind at one time, whereas the Level 2 sentences required simultaneous consideration and integration of two relations (and that happened two to four times per trial, depending on how many propositions were shown).

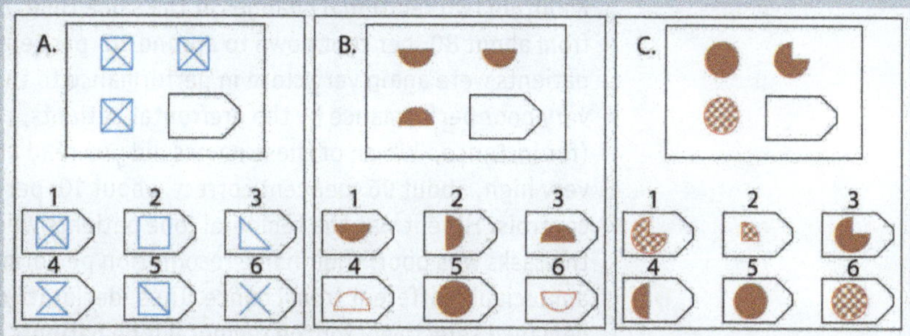

Figure 10.6 Examples of reasoning problems taken from Raven's standard progressive matrices test

The task asks people to pick the one correct choice out of six possibilities at the bottom to complete the pattern at the top. A. Level 0 problem (non-relational). B. Level 1 problem (one-relation problem). C. Level 2 problem (two-relation problem). Correct answers are choice 1 in A, choice 3 in B, and choice 1 in C.

In the other half of the experiment, people solved a set of matrix problems from Raven's Standard Progressive Matrices test (Raven, 1941[139]). Figure 10.6 illustrates three such problems. The problem in Figure A was a 'non-relational' Level 0 problem; the correct answer merely involved finding the matching pattern among the six choices at the bottom. Figure B shows a one-relation, Level 1 problem (technically, the relation is 'reflect the pattern across the x-axis,' that is, flip the pattern from bottom to top; the answer is choice 3), and Figure C, a two-relation, Level 2 problem (the two relations are from solid to checked, and remove the upper-right quarter; the correct answer is choice 1).

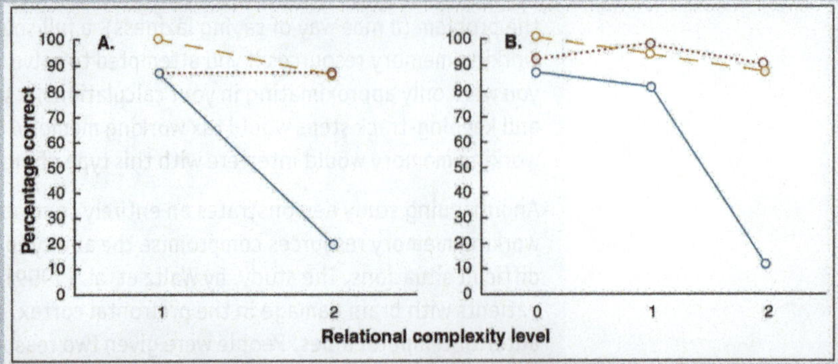

Figure 10.7 Percentage correct in the transitive inference (A) and Raven's matrices (B) problems, as a function of relational complexity (Levels 0, 1 and 2), for normals (dashed lines), temporal lobe patients (dotted lines) and prefrontal lobe patients (solid lines)

Waltz et al.'s (1999)[140] results for transitive inference are shown in the left half of Figure 10.7. Normal control participants (dashed line) showed only a modest decline in the Level 2 condition and did only slightly better than the patients with temporal lobe damage (the dotted line). However, patients with frontal lobe damage (solid line) dropped from more than 80 per cent correct on Level 1 problems to around 20 per cent correct on Level 2 problems – seemingly unable to consider and integrate multiple propositions at the same time. In the right half of the figure, it is clear that the frontal lobe patients were unable to maintain two relational changes at the same time, their performance dropping from about 80 per cent down to around 10 per cent at Level 2. The temporal lobe patients were again very close in performance to the normal controls. Despite this very poor performance by the prefrontal patients, their accuracy on memory tests (for instance, 'Which of these names did you read about, Amy or Susan?') was very high, about 96 per cent correct, about 10 per cent higher than the normal controls. Here it was the temporal lobe patients whose declarative memory for the tasks was poor; their name recognition performance was at 56 per cent, not appreciably different from chance. Thus, declarative memory for the tasks was damaged selectively for the temporal lobe patients, but reasoning performance was selectively damaged for the prefrontal patients (see also Wagar & Thagard, 2004[141]). The results are compatible with the idea that the frontal lobes are important for reasoning and, in particular, for maintaining relational information in working memory while the reasoning task is being done. The Waltz et al. results extend that research: 'In other words, relational integration may be the "work" done by working memory. We thus view our results as being consistent with the idea that the DLPFC, which was severely damaged in all our prefrontal patients, is critical for working memory' (1999, p. 123; see also Hinson, Jameson & Whitney, 2003[142]).

10.7.3 Artificial intelligence and biases in reasoning and decision-making

In this chapter we presented a detailed overview of the biases in human reasoning and decision-making and the errors that result from these biases. Proponents of Gigerenzer's position to biases in decision-making suggest that we learn how to use our flaws to our benefit. Proponents of machine learning, however, have long strived to develop expert systems that would be able to reason and make decisions, free from the biases and errors that are characteristic to humans. Artificial intelligence (AI) uses sophisticated algorithms to do so, and over the last few years its use has expanded considerably. From completing the sentences in our texts and emails, to supporting editing (or even developing) professional documents, AI will gradually become integrated into a large number of daily professional and leisure activities. Nevertheless, even AI is not free from flaws. The very limitations of its creators' cognitive abilities seem to haunt AI developments. **algorithmic bias** describes the tendency of automated, data-driven decision-making systems to produce biased outcomes that breach the rules and norms of social justice; equality, diversity and inclusivity; and even moral conduct. It occurs when algorithms unjustifiably favour one social group over another, creating inequalities (e.g. excluding certain social groups from benefits, or favouring a dominant ethnic group over minorities). As a result, algorithmic biases can lead to biased decision-making and have negative unintended consequences on the society and the effective function of organisations and institutions (Kordzadeh & Ghasemaghei, 2022[143]). Essentially, algorithmic bias perpetuates in information systems the limitations, errors and biases of human decision-making. According to IBM (2023),[144] some real life consequences of algorithmic bias include:

- Racial profiling and targeting disadvantaged ethnic minority groups in crime analysis and informing policing.
- Ageism and sexism biases in how professions and career opportunities are represented and promoted online.
- Underrepresenting women and ethnic group minorities in healthcare systems (e.g. computer-aided diagnosis), leading to inaccurate diagnoses and treatment plans.

Research has identified different sources of bias that lead to algorithmic bias errors across the stages of developing AI systems, from data mining to model

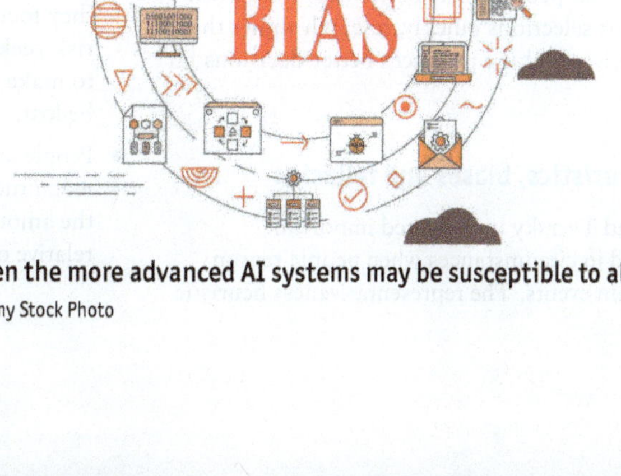

Figure 10.8 Even the more advanced AI systems may be susceptible to algorithmic bias

Source: teguh jati/Alamy Stock Photo

development and learning, deployment and assessment. Baker and Hawn (2022)[145] reviewed different types of biases that may contribute to algorithmic bias, including historical bias (e.g. predicting outcomes for the general population by using unrepresentative data from smaller elite groups) and representational bias (failing to predict outcomes for groups that may underrepresented in the existing data, such as ethnic minorities), to name just a few. Understanding the sources of bias along the design and deployment phases of AI and using relevant mitigation strategies can lead to more effective and bias-free AI systems. There is still a long way to go, however. For the moment, neither human nor artificial intelligence is entirely bias-free.

Summary: Reasoning and decision-making

10.1 Formal logic and reasoning

- Human reasoning is not especially logical, as shown in formal syllogistic and conditional reasoning problems. In syllogism tasks, conditional reasoning *if–then* problems and hypothesis testing, people often fail to search for negative evidence. Instead, they frequently look for positive evidence, called confirmation bias, and are often influenced, both positively and negatively, by semantic knowledge. When a more sceptical attitude is adopted, and when the reasoning involves more concrete concepts, reasoning accuracy improves.
- There are a number of theories of how humans deal with logical reasoning. Heuristic theories assume that people largely use shortcuts based on the information in the premises. Probability heuristic theory assumes that people reason using subjective estimates of the likelihood of various premises. Mental rules theories assume that people use sets of implicit rules to reason, similar to the rules people may use when comprehending and producing language. Finally, mental model theory assumes that people create mental simulations of the circumstances described by the premises to reason.

10.2 Decisions

- People make decisions in many different ways. Sometimes they use algorithms, following a normative model, that are guaranteed to produce the correct answer. However, because they do not have enough time or the issues are not well defined, they may use heuristics or rules of thumb, following a descriptive model.
- Although there has been some suggestion in the popular press that people make better decisions when they make their selections quickly, research shows that slow, deliberative thinking produces better decisions far more often.

10.3 Classic heuristics, biases and fallacies

- Kahneman and Tversky investigated important heuristics used in circumstances when people reason about uncertain events. The representativeness heuristic guides people to judge outcomes as likely if they seem representative of the type of event being evaluated; for instance, a random-looking sequence of coin tosses is judged more representative of coin toss outcomes almost regardless of the true probabilities involved. Included among the reasoning effects predicted by this heuristic are various stereotyping results.
- In the availability heuristic, people judge the likelihood of events by how easily they can remember examples or instances. These judgements therefore can be biased by any other factor that affects memory, such as salience or vividness.
- In the simulation heuristic, people forecast or predict how some outcome could have been different. These forecasts are influenced by how easily the alternative outcomes can be imagined. Interestingly, when people complete 'if only' statements, the changes they include tend to normalise the original situation by removing an unusual event and substituting a more common one. Such normalisations can be affected by the *focus* of the situation.
- With elimination by aspects, people can take a very large set of options and quickly reduce it down to a manageable size. People drop out options that do not meet various criteria that are set as important by a person.

10.4 Framing and risky decisions

- The choices that people make can be influenced by whether the choices are framed as either costs or benefits. In general, people are risk-aversive for gains, meaning that they will go with the safer choice when they focus on what can be gained. However, they are risk-seeking for losses in that they will be more likely to make a riskier choice if they focus on what will be lost.
- People are also somewhat blind at times to the absolute magnitude of a change, such as the cost of an item, or the amount saved. Instead, people focus more on the relative or proportional size of an outcome.

10.5 Adaptive thinking and 'fast and frugal' heuristics

- Work on 'fast and frugal' heuristics reveals that simple one-reason decision-making heuristics often do a very good job. These heuristics come from the adaptive thinking approach to decision-making, the source of a current debate about the basis for reasoning under uncertainty.

10.6 Other explanations

- Recent work on causal reasoning suggests that people use more normative reasoning than has otherwise been suspected if it is assumed that people have a reasonably correct causal model of the situation they are trying to make decisions about. Some of these newer approaches to human decision-making use ideas derived from the Bayes' theorem, whereas others use ideas developed from the field of quantum physics.

10.7 Limitations in reasoning

- In everyday reasoning, we rely on mental models of the device or event to make our judgements. These mental models are sometimes quite inaccurate. In the best-known research, people's mental models of physical motion lead them to incorrect predictions (e.g. the trajectory of a ball dropped from an airplane).
- Ongoing research is focused on the kinds of limitations that lead to incorrect reasoning and decision-making, including limited domain knowledge and limitations in working memory processes.

Chapter 10 quiz

Question 1

Which of the following describes a problem in which you need to know an algorithm to find the solution?

A. Of the 32 students in the class, about half were female. When students formed two study groups, what is the likelihood there were eight of each gender in each group?
B. Five groups of five students introduced themselves to their own group members, and then to all other group members. How many introductions did each person give?
C. The hockey team numbered 22 members altogether, including staff and players. What percentage of the staff is male if four members are female?
D. Sixteen girls and four boys tried out for the cheerleading team. If 50% of each gender made the team, how many total of each gender will be at the first practice?

Question 2

Syllogism refers to a logical argument that applies deductive reasoning to arrive at a conclusion. What is a bias of this type of reasoning?

A. The number of sentences
B. The order of the sentences
C. The number of sentences
D. Euler circles

Question 3

Why are heuristics important within reasoning and decision making?

A. They can help us to make decisions quicker as they provide us with mental shortcuts to solve a problem.
B. They prolong the decision-making process meaning that decisions will be more accurate.
C. They allow us to see all of the information that is correct.
D. They enable effortful thinking, leading to improved consideration of likely outcomes.

Question 4

You are at a local bookshop and a book is £10.99. A friend informs you that the bookstore 20 miles away across town has the book discounted to £5.20. You decide to drive to the other bookstore to purchase the cheaper book. What process, within decision-making, has influenced this?

A. Adaptive thinking
B. Imitate the majority
C. Conjunction fallacy
D. Outcome magnitude

Question 5

Which factor needs to be consistent to allow us to make interpretations, reasonings, and finally, make appropriate decisions?

A. Memory
B. Mathematics ability
C. Knowledge
D. Personality

Appendix: Algorithms for coin tosses and hospital births

Coin tosses

To begin with the obvious, the probability of a head on one coin toss is .50. Flipping a coin twice and keeping track of the possible sequences yields a .25 probability for each of the four possibilities HH, HT, TH, TT. In general, when the simple event has a probability of .50, the number of possibilities for a sequence of n events is 2 raised to the nth power. Thus, the number of distinct sequences for six coin tosses is 2^6, a total of 64 possibilities.

Two of the 64 possibilities are pure sequences, HHHHHH and TTTTTT. Two more are double sequences, HHHTTT and TTTHHH. All the remaining 60 possibilities involve either or both of the following characteristics: more of one outcome (e.g. heads) than the other and at least one alternation between the two outcomes at a position *other than* halfway through the sequence. Thus, the probability of a pure sequence is 2/64, as is the probability of a double sequence. Getting any one of the other 60 possibilities has a likelihood of 1/64. However, getting a 'random-like' outcome – that is, any outcome other than straight or double – has a probability of 60/64.

Hospital births

Many statistics texts contain tables of the binomial distribution, the best way to understand the hospital births example. Because most of these tables go up only to a sample size of 20, we will use a revised hospital example, comparing hospitals with three versus nine births per day (note that the 1:3 ratio is the same as the original example, 15:45). The probabilities for the original example are more extreme than these, but they will be in the same direction.

Just as with coin tosses, we are dealing with an event whose basic probability is .50, the likelihood that a newborn infant is male (ignoring the fact that male births are actually slightly more common than 50 per cent). What is the probability that, in three births, all three will be boys? According to the binomial tables (Table 10.A), this probability is .1250. This is the probability that on any randomly selected day, the three-birth hospital will have all boys, $p = .1250$. Across the 365 days in a year, we expect an average of 45.625 such days ($365 \times .1250$).

The temptation now is to consider the likelihood of exactly three boys in the nine-birth hospital. But this is not the relevant comparison. The relevant comparison to the all boys probability in the three-birth hospital would be all boys in the nine-birth hospital. This puts the comparison on the same footing as the original problem, 60 per cent as the 'extreme' cutoff.

The probability of exactly nine boys out of nine births is .0020, two chances in a thousand. For a whole year, we expect only 0.73 such days ($365 \times .0020$). Now it should be clearer. The criterion of 'extreme', all boys, is much more likely in the smaller sample than in the larger one, $p = .1250$ versus .0020. Multiplied out, the prediction is 45 days for the small hospital and .70 days for the large one.

By extension, and using the appropriate binomial values, the 15-birth hospital should have about 111 days per year with 60 per cent or more boys, contrasted with 42 such days per year for the 45-birth hospital.

Table 10.A Binomial probabilities for exact number of relevant outcomes, where the simple probability of the outcome is .50

$n = 3$		$n = 9$		$n = 15$		$n = 45$	
Exact number of relevant outcomes:							
0	0.125	0	0.002	9	0.1527	27	0.0488
1	0.375	1	0.0176	10	0.0916	28	0.0314
2	0.375	2	0.0703	11	0.0417	29	0.0184
3	0.125	3	0.1641	12	0.0139	30	0.0098
		4	0.2461	13	0.0032	31	0.0047
		5	0.2461	14	0.0005	32	0.0021
		6	0.1641	15	0	33	0.0008
		7	0.0703	etc.		34	0.0003
		8	0.0176			35	0.0001
		9	0.002			36	0

11 Problem-solving

Learning objectives

11.1 Describe problem-solving as a topic of inquiry for cognitive psychology

11.2 Analyse the approach adopted by cognitive psychologists to study the problem-solving process

11.3 Describe the work done by Gestalt psychologists to study problem-solving

11.4 Compare insight and analogy as methods of problem-solving

11.5 Outline the means–end analysis method of problem-solving

11.6 Summarise a set of problem-solving techniques

Key figures

Professor John Kounios
Professor Kounios's research has advanced our understanding of insight and creativity. His studies have focused on the cognitive and neural aspects of creativity and insight.

Professor Mark Beeman
Professor Beeman's research focuses on the neural and neurophysiological underpinnings of problem-solving and creative thinking. Together with Professor Kounios he co-authored the best seller The Eureka Factor: Aha Moments, Creative Insight, and the Brain.

11.1 Understanding problem-solving

LEARNING OBJECTIVE
Describe problem-solving as a topic of inquiry for cognitive psychology

11.1.1 Problem-solving in action

A favourite example of 'problem-solving in action' is the following true story:

> When I (M.H.A.) was a graduate student, I attended a departmental colloquium at which a candidate for a faculty position was to present his research. As he started his talk, he realised that his first slide was projected too low on the screen. A flurry of activity around the projector ensued, one professor asking out loud, '*Does anyone have a book or something?*' Someone volunteered a book, the professor tried it, but it was too thick; the slide image was now too high. '*No, this one's too big. Anyone got a thinner one?*' he continued. After several more seconds of hurried searching for something thinner, another professor finally exclaimed, '*Well, for Pete's sake, I don't*

believe this!' He marched over to the projector, grabbed the book, opened it halfway, and put it under the projector. He looked around the lecture hall and shook his head, saying, *'I can't believe it. A roomful of PhDs, and no one knows how to open a book!'*

This chapter examines **problem-solving**, which involves resolving a challenge or a problem following a goal-directed sequence of steps, mental operations to work through a problem, as well as action planning, monitoring and control. As with decision-making and reasoning, in problem-solving a person is confronted with a difficult, time-consuming task: a problem is presented, the solution to this problem may not be immediately obvious, and the person is often uncertain about what to do next. We are interested in all aspects of the person's activities, from initial understanding of the problem to the steps that lead to a final solution, and, in some cases, how a person decides that a problem has been solved.

Our interest in the psychological study of problem-solving needs no further justification than this: we confront countless problems in our daily lives, some of which are trivial (e.g. which route to choose to cycle to university to avoid potholes), but some are important for us to figure out and solve (e.g. how to resolve conflict between friends or with a significant other). We attempt to solve problems by mentally analysing the situation, devising a plan of action and then carrying out that plan. Therefore, the mental processing involved in problem-solving is, by definition, part of cognitive psychology.

11.1.2 Types of problems

A useful way to understand everyday problems, and to be able to determine their resolution, is to consider their *specificity* and *expertise*. In terms of *specificity*, problems can be either well-defined or poorly defined. **Well-defined problems** are characterised by a low level of uncertainty about the possible solutions that may be applied by the problem-solver. Chess, poker and board or card games represent well-defined problems with clearly defined sets of moves, strategies and solutions. On the other hand, poorly or **ill-defined problems** present little insight into their possible solutions. Let's assume that you set a goal of becoming more sociable. Which strategy would you choose from a multitude of options? How would you evaluate the outcomes of the strategy you chose? By the number of new social contacts you

What seems to be a difficult problem to one person can seem easy to another
Source: Godong/Alamy Stock Photo

made within a given timeframe? By the number of invitations you received to attend social events? Given that your sociability may be affected by other factors outside your control (e.g. availability and accessibility of social networks and contacts in a given setting) it can be difficult to predict the success of your problem-solving strategy. For this reason, cognitive psychologists study well-defined problems to determine one's problem-solving capacity, because well-defined problems offer a reliable way for measuring successes and errors in the problem-solving strategies and plans people use.

In addition to specificity, problem-solving may be determined by one's expertise in a given domain. To illustrate, a novice Candy Crush player may spend hours (or days!) to progress through a *'hard'* level, but a more experienced player who has already mastered several *'nightmarishly hard'* levels can pass a *'hard'* level in a matter of minutes. What seems to be a difficult problem to one person, might be an easy task to resolve for another.

This indicates that problem-solving can be, to some extent, determined by individual differences. Accordingly, problems can be distinguished between **knowledge-rich** and **knowledge-lean** ones, the former requiring a certain level of expertise to be resolved, while the latter do not require any specific expertise. Chronicle et al. (2004)[1] assessed problem-solving among undergraduate students at the University of Lancaster in a series of experiments, using the six-coin problem: a task requiring participants to rearrange six coins that were originally arranged in a straight line, in such a way that each coin touches exactly two others. The six-coin problem did not require any level of expertise, and a specific number of clearly defined rules were to be followed (e.g. coins should not be lifted). Therefore, the six-coin problem was well-defined and knowledge-lean. Table 11.1 summarises the different types of problem based on their level of specificity and knowledge richness.

Table 11.1 Different types of problem based on their level of specificity and knowledge richness

	Knowledge richness	
Specificity	**Knowledge-lean**	**Knowledge-rich**
Ill-defined	Get physically fit	Become more successful at work, win a Nobel Prize
Well-defined	Rubik's cube, six-coin problem	Solve a difficult math equation, win *University Challenge*

We will begin our coverage of problem-solving by first defining the basic terms and ideas that pervade much of the research in this area. After that, we will look at the classic problem-solving research of the Gestalt psychologists. The Gestalt movement co-existed with behaviourism early in the 20th century but never achieved the central status that behaviourism did. In retrospect, however, it was an important influence on cognitive psychology. We will discuss various approaches to understanding different types of problem-solving: by analogy and by insight, with reference to creative problem-solving (CPS). We will finally consider an overview of different techniques for improving problem-solving capacity.

11.2 Basics of problem-solving

LEARNING OBJECTIVE
Analyse the approach adopted by cognitive psychologists to study the problem-solving process

Cognitive psychology has adopted a **reductionistic approach** to the study of problem-solving. That is, cognitive psychologists work to break the process of problem-solving down into its various components and processes, and then understand how they work together. For instance, Newell and Simon's analysis of a cryptarithmetic

reductionistic approach Understanding complex behaviours or processes by breaking them down into their various components.

problem (1972)[2] is a microscopic analysis and interpretation of every statement made by one person as he solved a problem, all 2,186 words and 20 or so minutes of problem-solving activity. In Newell and Simon's description, 'A person is confronted with a *problem* when he wants something and does not know immediately what series of actions he can perform to get it' (p. 72). The 'something' can be renamed for more general use as a **goal**, the desired end point of the problem-solving activity.

11.2.1 Characteristics of problem-solving

Problem-solving consists of goal-directed activity, moving from some initial configuration or state through a series of intermediate steps until finally the overall goal has been reached: an adequate or correct solution. The difficulty is determining which intermediate states are on a correct pathway ('Will step *A* get me to step *B* or not?') and in devising operations or moves that achieve those intermediate states ('How do I get to step *B* from here?').

An intuitive illustration of such a nested solution structure is presented in Figure 11.1, a possible solution route to the locked-car problem. Note that during the solution, the first two plans led to barriers or blocks, thus requiring that another plan be devised. The problem solver finally decided on another plan, breaking a window to get into the locked car. This decision is followed by a sequence of related acts: the search for some heavy object that will break a window, the decision as to which window to break, and so forth. Each of these decisions is a subgoal nested within the larger subgoal of breaking into the car, itself a subgoal in the overall solution structure.

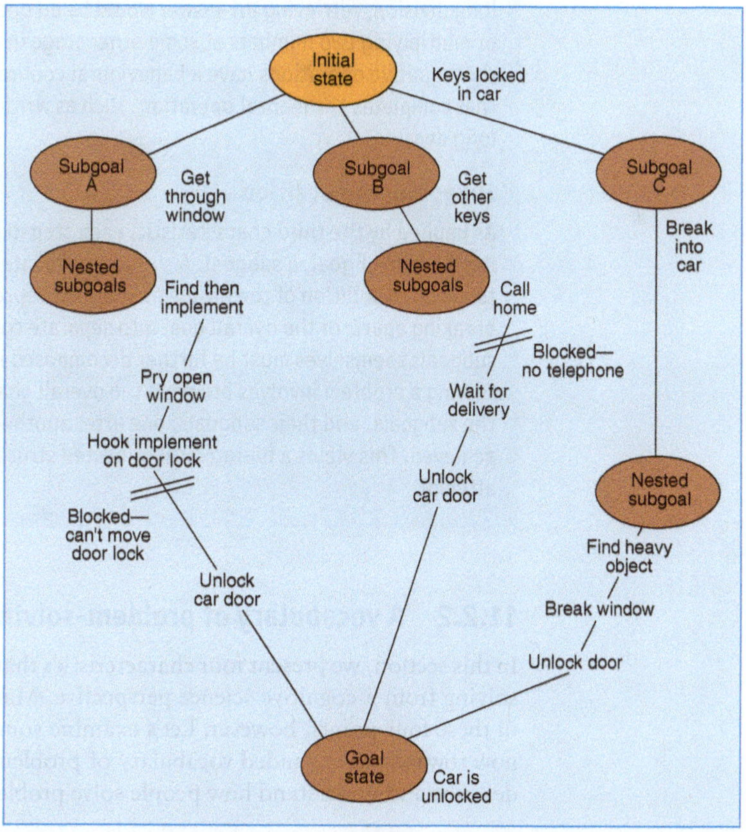

Figure 11.1 A representation of part of the problem space for getting into a locked car
Note the barriers encountered under plans A and B.

Four aspects of problem-solving

Let's start by listing several characteristics that define what is and is not a genuine instance of problem-solving.

Goal-directedness

The overall activity we are examining is directed towards achieving some goal or purpose. As such, we exclude daydreaming, for instance; it is mental, but it is not goal-directed. Alternatively, if you have locked your keys in your car, both physical and mental activities are involved. The goal-directed nature of those activities, your repeated attempts to get into the locked car, makes this an instance of true problem-solving.

Sequence of operations

An activity must have a *sequence of operations* or steps to be problem-solving. A simple retrieval from memory, say, remembering that 2×3 is 6, is not problem-solving because it is not a slow, discernible sequence of separate operations. Alternatively, doing a long division problem or solving the locked-car problem definitely involves a sequence of mental operations, so these are instances of problem-solving.

Cognitive operations

Solving the problem involves the application of various cognitive operations. Different operators can be applied to different problems, where each operator is a distinct cognitive act, a permissible step or move in the problem space. For long division, retrieving an answer would be an operator, as would be subtracting or multiplying two numbers at some other stage in problem solution. Often, the cognitive operations have a behavioural counterpart, some physical act that completes the mental operation, such as writing down a number during long division.

Subgoal decomposition

As implied by the third characteristic, each step in the sequence of operations is itself a kind of goal, a subgoal. A **subgoal** is an intermediate goal along the route to eventual solution of the problem. Subgoals represent the decomposition, or breaking apart, of the overall goal into separate components. In many instances, subgoals themselves must be further decomposed into smaller subgoals. Thus, solving a problem involves breaking the overall goal into subgoals, then pursuing the subgoals, and their subgoals, one after another until the final solution is achieved. This yields a hierarchical or nested structure to the problem-solving attempt.

11.2.2 A vocabulary of problem-solving

In this section, we present four characteristics that define what qualifies as problem-solving from a cognitive science perspective. Many important ideas are embedded in these four points, however. Let's examine some of these points further, looking now towards an expanded vocabulary of problem-solving, a set of terms we use to describe and understand how people solve problems.

The problem space

The term **problem space** is critical. Anderson (1985)[3] defines it as the various states or conditions that are possible. More concretely, the problem space includes the initial, intermediate and goal states of the problem. It also contains the problem-solver's knowledge at each of these steps, both knowledge that is currently being applied and knowledge that could be retrieved from memory and applied. Any available external devices, objects or resources can also be included in the description of the problem space. Thus, a difficult arithmetic problem that must be completed mentally has a different problem space from the same problem as completed with pencil and paper.

To illustrate, VanLehn (1989)[4] describes one man's error in the 'three men and a rowboat' problem. The man focused only on the arithmetic of the problem and said essentially, '400 pounds of people, 200 pounds per trip, it'll take two trips of the boat'. When he was reminded that the boat couldn't row itself back to the original side, he adopted a different problem space:

> 'Three men want to cross a river. They find a boat, but it is a very small boat. It will only hold 200 pounds. The men are named Large, Medium, and Small. Large weighs 200 pounds, Medium weighs 110 pounds, and Small weighs 80 pounds. How can they all get across? They might have to make several trips in the boat.'
>
> (VanLehn, 1989, p. 532)[5]

In some problem contexts, we can speak of problem-solving as a search of the problem space or, metaphorically, a search of the solution tree, in which each branch and twig represents a possible pathway from the initial state of the problem. For problems that are 'wide open', that is, those with many possibilities that must be checked, there may be no alternative but to start searching the problem space, node by node, until some barrier or block is reached. As often as not, however, there is information in the problem that permits us to restrict the search space to a manageable size. Metaphorically, this information permits us to *prune* the search tree.

A general depiction of this is in Figure 11.2.

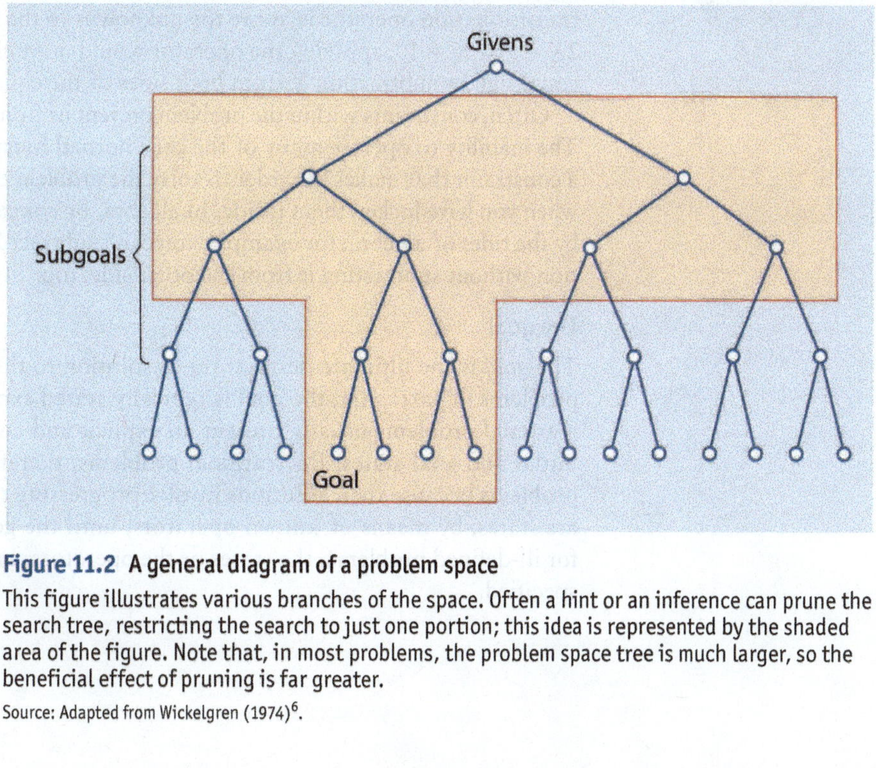

Figure 11.2 A general diagram of a problem space

This figure illustrates various branches of the space. Often a hint or an inference can prune the search tree, restricting the search to just one portion; this idea is represented by the shaded area of the figure. Note that, in most problems, the problem space tree is much larger, so the beneficial effect of pruning is far greater.

Source: Adapted from Wickelgren (1974)[6].

The initial state of the problem is the top node, and the goal state is some terminal node at the bottom. For 'wide open' problems, each branch may need to be searched until a dead end is encountered. For other problems, information may be inferred that permits a restriction in the branches that are searched (the shaded area of the figure). Clearly, if the search space can be limited by pruning off the dead-end branches, then problem-solving efficiency increases.

The look ahead

When problem-solving, people are more likely to be successful if they have a plan about how they are going to solve the problem. In problem-solving research, this is called lookahead (VanLehn, 1989[7]) and refers to how many steps in the future a problem-solver is considering.

lookahead
When a problem-solver considers the different steps in the future involved in solving the problem.

An example of failing to plan would be, when cooking, not warming up the oven while doing other preliminary steps, but turning on the oven at the point when something needs to be placed in it. Another example of lookahead failure would be cutting off a tree branch while sitting on it. Research has shown that people often do not adequately look ahead to the consequences of their actions when problem-solving. Because they are more likely to be impulsive in making a move that initially seems to bring them closest to the solution, their efforts often do not result in success (Ormerod, MacGregor, Chronicle, Dewald & Chu, 2013[8]).

The operators

operator are the set of legal moves or actions that can be done during problem solution. The term *legal* means permissible in the rules of the problem. For example, an illegal operator in the three men in a rowboat problem is having the men swim across the river or loading the boat with too heavy a load. An illegal operator in getting the keys out of your locked car is teleporting into the car. You just can't do that.

For transformation problems (Greeno, 1978[9]), applying an operator transforms the problem into a new or revised state from which further work can be done. In general, a legal operator moves you from one node to the next along some connecting pathway in the search space. For instance, in solving algebraic equations, one transformation operator is 'move the unknowns to the left'. Thus, for the equation $2X + 7 = X + 10$, applying the operator would move the single X to the left of the equal sign by subtracting X from both sides of the equation.

Often, constraints within the problem prevent us from applying certain operators. The inability to operate many of the car's normal functions when on the outside is a constraint that makes it harder to solve the problem of getting your keys out of it when you have locked them inside. In algebra, by contrast, constraints are imposed by the rules of algebra; for example, you can't subtract X from one side of the equation without subtracting it from the other side, too.

The goal

The *goal* is the ultimate destination or solution to the problem. For recreational problems in particular, the goal is typically stated explicitly in the problem. Recreational problems usually present an explicit and complete specification of the initial and goal states. Recreational problems, therefore, represent well-defined problems because their solutions involve progressing through the legal intermediate states, by means of known operators, until the goal is reached. By contrast, for ill-defined problems, the states or the operators, or both, may be only vaguely specified.

Sample recreational problems

This section provides several recreational problems for you to work through to give you a better feel for and understanding of the problem-solving process. These include the Buddhist monk, drinking glasses and six pennies problems.

Buddhist monk

The Buddhist monk problem in the interactive exercise below is distressingly vague in its specification of the goal (as are the problems 'write a term paper that will earn you an **A**,' 'write a computer program that does *X* in as economical and elegant a fashion as possible', and so on). Here is the problem.

One morning, exactly at sunrise, a Buddhist monk began to climb a tall mountain. The narrow path, no more than a foot or two wide, spiralled around the mountain to a glittering temple at the summit. The monk ascended the path at varying rates of speed, stopping many times along the way to rest and to eat the dried fruit he carried with him. He reached the temple shortly before sunset.

After several days of fasting and meditation, he began his journey back along the same path, starting at sunrise and again walking at variable speeds with many pauses along the way. His average descending speed was, of course, greater than his average climbing speed. Show that there is a spot along the path that the monk will occupy on both trips at precisely the same time of day.

Hint: Although the problem seems to ask for a quantitative solution, think of a way of representing the problem using visual imagery.

Drinking glasses

Six drinking glasses are lined up in a row. The first three are full of water, the last three are empty. By handling and moving only one glass, change the arrangement so that no full glass is next to another full one, and no empty glass is next to another empty one.

Hint: How else can you handle a glass of water besides moving it to another location?

Six pennies

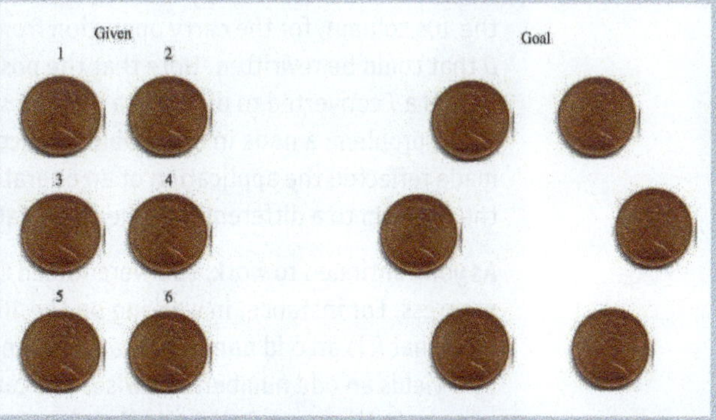

Show how to move only two pennies in the left diagram to yield the pattern on the right.

Hint: From a different perspective, some of the pennies might already be in position.

DONALD + GERALD example of recreational problem

Let's consider one recreational problem to pin down some of these terms and ideas.

The DONALD + GERALD problem

This is a *cryptarithmetic* problem in which letters of the alphabet have been substituted for the digits in an addition problem.

Your task is to reverse the substitutions – to figure out which digits go with which letters to yield a correct addition problem. The restriction is that the digits and letters must be in one-to-one correspondence (only one digit per letter and vice versa). Plan on spending 15 minutes or so, about the amount of time it takes people on their first attempt. Make notes on paper as you work so you can go back later to retrace and analyse your attempt to solve the problem. (Incidentally, this is the cryptarithmetic problem Newell and Simon's single person worked on.)

$$\frac{DONALD + GERALD}{ROBERT} \qquad (\textit{Hint: } D + 5)$$

Now that you have worked on the problem and have found (or come close to) the solution, we can use the insights you developed to fill in our definitions of terms. To begin with, the initial state consists of the statement of the problem, including the rules, restrictions and the hint you are given. These, along with your own knowledge of arithmetic (and pencil and paper), are your problem-solving tools.

Each conceivable assignment of letters to digits makes up the problem space, and each substitution operator you might apply constitutes a branch or pathway on the search tree (a substitution operator here is an operator that substitutes a digit for a letter). The hint $D = 5$ prunes the search tree by a tremendous amount. Without the hint, you can only start working on the problem by trying arbitrary assignments, then working until an error shows up, then returning to an earlier node and reassigning the letters to different digits. (Even without the hint, however, there is only one solution.)

You no doubt started working on the problem by replacing the *D*s in the 1s column with 5s, then immediately replacing the *T* with a 0. You also probably wrote a 1 above the 10s column, for the carry operation from $5 + 5$. A quick scan revealed one more *D* that could be rewritten. Note that the position you were in at this point, with three *D*s and a *T* converted to digits, is a distinct step in the solution, an intermediate state in the problem, a node in the problem space. Furthermore, each substitution you made reflected the application of an operator, a cognitive operation that transforms the problem to a different intermediate state.

As you continued to work, you were forced to infer information as a way of making progress. For instance, in working on the 10s column, $L + L +$ the carried 1, you can infer that *R* is an odd number because any number added to itself and augmented by 1 yields an odd number. Likewise, you can infer from the $D + G$ column that *R* must be in the range 5 to 9 and that $5 + G$ cannot produce a carried 1. Putting these together, *R* must be a large odd number, and *G* must be 4 or less. Each of these separate inferences, each mental conclusion you draw, is also an instance of a cognitive operation, a simple mental process or operator that composes a step in the problem-solving sequence. Each of these, furthermore, accomplishes progress towards the immediate subgoal, find out about *L*.

Greeno (1978)[10] calls this process a **constructive search**. Rather than blindly assigning digits and trying them out, people usually draw inferences from the other columns and use those to limit the possible values the letters can take. This approach is typical in arrangement problems, the third of Greeno's categories, in which some combination of the given components must be found that satisfies the constraints in the problem. In other kinds of arrangement problems, say anagrams, a constructive search heuristic would be to look for spelling patterns and form candidate words from those familiar units. The opposite approach, sometimes known as **generate and test**, merely uses some scheme to generate all possible arrangements, then tests those one by one to determine whether the problem solution has been found.

A related aspect of problem-solving here (it can be postponed, but your solution will be more organised if it is done now) is quite general and almost constitutes good advice rather than an essential feature of performance. Some mechanism or system for keeping track of the information you know about the letters is needed, if only to prevent you from forgetting inferences you have already drawn. Indeed, such an external memory aid can go a long way towards making your problem-solving more efficient. In some instances, it may even help you generalise from one problem variant to another (as in the Tower of Hanoi problem – see Section 11.5.2).

Is there a single problem-solving modality in the human brain?

Cognitive scientists have examined whether different problem-solving tasks, from verbal and mathematical to visuo-spatial problems, can be explained by activation of either domain-general or context-specific neural networks in the human brain. Bartley et al. (2018)[11] meta-analysed data from 280 problem-solving experiments using neuroimaging and identified a 'common network' supporting problem-solving across contexts. Specifically, areas of the fronto-parietal network were activated across mathematical, verbal and visuo-spatial problem-solving. Bartley et al. (2018)[12] also identified domain-specific brain areas for each type of problem-solving.

11.3 Gestalt psychology and problem-solving

LEARNING OBJECTIVE
Describe the work done by Gestalt psychologists to study problem-solving

More than 2,000 years ago, the Ancient Greek mathematician and polymath Archimedes was tasked by a local ruler to solve a problem: detect whether the ruler's allegedly golden crown was tainted with silver. As the anecdote has it, Archimedes came up with a solution to the problem by accidentally observing water displacement after submerging himself in water while taking a bath – the so-called '*Eureka*' moment. Fast-forward to the early 18th century when a British polymath and mathematician allegedly experienced a similar moment of scientific revelation (so-called '*Aha*' moment) about the earth's gravity after an apple fell on his head while he was sitting under an apple tree. Although their accuracy has been challenged by contemporary historians and scientists, both anecdotes signify that humans (and, possibly other animals too!) can experience problem-solving as a sudden, inspirational moment. This idea was firstly examined in the mid-1900s by Gestalt psychologists.

As we noted previously, **Gestalt** is a German word that translates poorly into English; the one-word translations 'whole', 'shape' and 'field' fail to capture what the term means. Roughly speaking, a Gestalt is a whole pattern, a form or a

Archimedes examining the method of measuring an object's volume (e.g. how much an object weighs) by displacement (i.e. before and after the object has been submerged into water).

Source: World History Archive/Alamy Stock Archive

configuration. It is a cohesive grouping, a perspective from which the entire field can be seen. A variety of translations have been used, but none ever caught on, which prompted Boring (1950)[13] to remark that Gestalt psychology 'suffered from its name'. So, we use the German term *Gestalt*, rather than an inadequate translation. In perception, the Gestalt principles show that humans tend to perceive and deal with integrated, cohesive wholes.

11.3.1 Early Gestalt research

The connection between Gestalt psychologists and problem-solving is best explained by an anecdote (see Boring, 1950, pp. 595–597[14]). In 1913, Wolfgang Köhler, a German psychologist, went to the Spanish island of Tenerife to study 'the psychology of anthropoid apes' (p. 596). Trapped there by the outbreak of World War I, Köhler experimented with visual discrimination among several animal species. In the course of this research, he began to apply Gestalt principles to animal perception. His ultimate conclusion was that animals do not perceive individual elements in a stimulus, but that they perceive relations among stimuli. Furthermore, 'Köhler also observed that the perception of relations is a mark of intelligence, and he called the sudden perception of useful or proper relations *insight*' (p. 596).

Still stranded on the island, Köhler continued to examine 'insight learning'. He presented problems to chimpanzees and searched for evidence of genuine problem-solving in their behaviour. By far, the most famous of his subjects was a chimpanzee named Sultan (Köhler, 1927[15]). In a simple demonstration, Sultan was able to use a long pole to reach through the bars of his cage and get a bunch of bananas. Köhler made the situation more difficult by giving Sultan two shorter poles, neither of which was long enough to reach the bananas. After failing to get the bananas, and sulking in his cage for a while, Sultan (as the story goes) suddenly went over to the poles and put one inside the end of the other, thus creating one pole that was long enough to reach the bananas.

For Sultan, this is problem-solving because the behaviour was goal-directed (to get the bananas), there was a sequence of operations (the pole parts needed to be put together before reaching for the bananas), and there were cognitive operations (Sultan needed to think about the situation). For Sultan, the problem space involved the confines of his cage, the distance of the bananas from him, and the pole segments. The operations were all of the things he could do inside his cage, and with the poles.

Köhler found this to be an apt demonstration of *insight*, a sudden solution to a problem by means of an insightful discovery. In another situation, Sultan discovered how to stand on a box to reach a banana that was otherwise too high to reach. In yet another, he discovered how to get a banana that was just out of reach through the cage bars: he walked *away* from the banana, out a distant door and around the cage. All these problem solutions seemed to illustrate Sultan's perception of relations and the importance of insight in problem-solving.

Grande builds a three-box structure to reach the bananas, while Sultan watches from the ground. Insight, sometimes referred to as an 'Aha' experience, was the term Köhler used for the sudden perception of useful relations among objects during problem-solving.

Source: Köhler, W. (1927). The mentality of apes. London: Routledge & Kegan Paul

11.3.2 Difficulties in problem-solving

Other Gestalt psychologists, most notably Duncker and Luchins, pursued research with humans. Two major contributions of this work are essentially the two sides of the problem-solving coin. One involved a set of negative effects related to rigidity or difficulty in problem-solving; the other, insight and creativity during problem-solving.

Functional fixedness

Two articles on functional fixedness, one by Maier (1931)[16] and one by Duncker (1945),[17] identify and define this difficulty. **Functional fixedness** is a tendency to use objects and concepts in the problem environment in only their customary and usual ways.

It is probably not surprising that problem-solvers experience functional fixedness. After all, we comprehend the problem situation by means of our world knowledge, along with whatever procedural knowledge we have that might be relevant. When you find 'PLIERS' in semantic memory, the most accessible properties involve the normal use for pliers. Far down on your list would be characteristics related to their weight or aspects of their shape that would enable you to tie a string to them. Likewise, 'BOX' probably is stored in semantic memory in terms of 'container' meanings—that a box can hold things, that you put things into a box—and not in terms of 'platform or support' meanings (see Greenspan, 1986[18], for evidence on retrieval of central and peripheral properties). Simply from the standpoint of routine retrieval from memory, then, we can understand why people experience functional fixedness.

Problems demonstrating functional fixedness

Sometimes problems seem like they should be easy, but then they turn out to be difficult. This difficulty may stem from thinking about things in a typical or standard way. The following functional fixedness problems illustrate two ways in which research on problem-solving has addressed this issue.

The two-string problem

Maier (1931),[19] for instance, had people work on the two-string problem. Two strings are suspended from the ceiling, and the goal is to tie them together. The problem is that the strings are too far apart for a person to hold one, reach the other, then tie them together. Also available are several other objects, including a chair, some paper and a pair of pliers. Even standing on the chair does not get the person close enough to the two strings.

In Maier's results, only 39 per cent of the people came up with the correct solution during a 10-minute period. The solution (if you haven't tried solving the problem, do so now) involves using an object in the room in a novel way. A correct solution is to tie the pliers to one string, swing it like a pendulum, then catch it while holding the other string.

Thus, the functional fixedness in this situation was failing to think of using the pliers in any other way than their customary function; people were fixed on the normal use for pliers and failed to appreciate how they could be used as a weight for a pendulum.

The candle problem

A similar demonstration is shown in the above figure, the candle problem from Duncker (1945).[20] The task is to find a way to mount the candle on a wall using just the objects illustrated. Can you solve the problem? If you haven't come up with a solution after a minute or two, here's a hint: Can you think of another use for a box besides using it as a container? In other words, the idea of functional fixedness is that we generally think only of the customary uses for objects, whereas successful problem-solving may involve finding novel uses. By thinking of the box as a platform or means of support, you can then solve the problem (empty the box, thumbtack it to the door or wall, then mount the candle in it).

The candle problem used by Duncker

Using only the pictured objects, figure out how to mount the candle to the wall.

Problem-solving in situations that elicit functional fixedness typically involves people failing to notice some minor or obscure feature, but one which is critical to solving the problem. This would be the weight of the pliers in the two-string problem and the rigidness of the box bottom (when inverted) in the candle problem. Problem-solving (and creativity) can involve noticing and applying such obscure features. McCaffrey (2012)[21] asked people to assess a series of objects to address the questions 'Can this be decomposed further?' and 'Does this description imply a use?' Afterward, people had to solve a series of problems that typically elicit functional fixedness. People who had just practised assessing various parts of objects were more likely to notice an otherwise obscure feature of objects and solve the problems.

Negative set

A related difficulty in problem-solving is **negative set** (or simply *set effects*). This is a bias or tendency to solve problems in a particular way, using a single specific approach, even when a different approach might be more productive. The term *set* is a rough translation of the original German term **Einstellung**, which means something like 'approach' or 'orientation.'

The water jug problem

A classic demonstration of set effects comes from the water jug problem, studied by Luchins (1942).[22] In this problem, you are given three jugs, each of a different capacity, and are to measure out a quantity of water using just the three jugs.

Problem	Capacity of jug *A*	Capacity of jug *B*	Capacity of jug *C*	Desired quantity
1	5 cups	40 cups	18 cups	28 cups
2	21 cups	127 cups	3 cups	100 cups

Measuring quantities with water jugs

As a simple illustration, consider the first problem in above table. You need to measure out 28 cups of water and can use containers that hold 5, 40 and 18 cups (jugs *A*, *B* and *C*). The solution is to fill *A* twice, then fill *C* once, each time pouring the contents into a destination jug. This approach is an addition solution because you add the quantities together.

For the second problem, a subtraction solution is appropriate: Fill *B* (127), subtract jug *C* from it twice (−3, −3), then subtract jug A (−21), yielding 100.

Problem	Capacity of jug *A*	Capacity of jug *B*	Capacity of jug *C*	Desired quantity
1	21	127	3	100
2	14	163	25	99
3	18	43	10	5
4	9	42	6	21
5	20	59	4	31
6	23	49	3	20
7	15	39	3	18
8	28	76	3	25
9	18	48	4	22
10	14	36	8	6

Luchins' water jug problem

Luchins' (1942)[23] demonstration of negative set involved sequencing the problems so that people developed a particular set or approach for measuring out the quantities. The problems in the above table illustrate such a sequence. Go ahead and work the problems now before you read any further.

Outcome

If you were like most people, your experience on Problems 1 through 7 led you to develop a particular approach or set: specifically, $B - 2C - A$: Fill jug *B*, subtract *C* from it twice, then subtract *A* from it to yield the necessary amount (subtracting *A* can be done before subtracting $2C$, of course).

People with such a set or Einstellung generally failed to notice the simpler solution possible for Problems 6 to 10, simply $A - C$. That is, about 80 per cent of the people who saw all 10 problems used the lengthy $B - 2C - A$ method for these problems. Compare this with the control participants, who saw only Problems 6 to 10: only 1 per cent of people used the longer method. Clearly, the control people had not developed a set for using the lengthy method, so they were better able to find the simpler solution.

Consider Problem 8 now. Only 5 per cent of Luchins's control people failed to solve Problem 8. This was remarkable because 64 per cent of the 'negative set' people, those who saw all 10 problems, failed to solve it correctly. These people had such a bias to use the method they had already developed that they were unable to generate a method that would solve Problem 8 ($B - 2C - A$ does not work on this problem).

Greeno's (1978)[24] description here is useful: by repeatedly solving the first seven problems with the same formula, people learned an integrated algorithm. This algorithm was strong enough to bias their later solution attempts and prevent them from seeing the simple solution, $28 - 3 = 25$.

Consistent with this idea – that if people develop a routine way of solving problems then they are more likely to experience Einstellung – there is evidence that experts at a task are more prone to this than others (Bilalić, McLeod & Gobet, 2008[25]). They are more likely to have developed a set of routines for solving certain kinds of problems.

Also, people are less likely to show mental set effects if they are given actual water jugs as compared to just paper-and-pencil problems (Vallee-Tourangeau, Euden & Hearn, 2011[26]). Moreover, if the initial problems given are more varied, then people can more easily generalise their solution to a variety of other problems (e.g. Chen & Mo, 2004[27]).

As the slide projector problem in the introduction suggests, functional fixedness and negative set are common occurrences. The occurrence of mental set in problem-solving can result from people focusing their attention on information that is consistent with an initial solution attempt for a problem, at the expense of other possible solutions. For example, eye tracking data reveal that people look more often at information consistent with their first solution attempt than at other information (Bilalić, McLeod & Gobet, 2010[28]). Possibly because we eventually find an adequate solution to our everyday problems despite the negative set or without overcoming our functional fixedness (e.g. eventually locating a thinner book), we are less aware of these difficulties in our problem-solving behaviour. The classic demonstrations, however, illustrate dramatically how rigid such behaviour can be and how barriers to successful problem-solving can arise.

11.4 Insight and analogy

LEARNING OBJECTIVE
Compare insight and analogy as methods of problem-solving

On a more positive side of problem-solving are the topics of insight and problem-solving by analogy. These are ways that people arrive at a solution, either more consciously or more unconsciously, by using the knowledge that they have of the problem domain.

11.4.1 Insight

insight is a deep, useful understanding of the nature of something, especially a difficult problem. As the Archimedes and Newton anecdotes indicate, insight occurs suddenly – the '*Aha!*' reaction – possibly because a novel approach to the problem is taken or a novel interpretation is made (Sternberg, 1996[29]), or even just because you have overcome an impasse (for research on the various sources of difficulty in insight problems, see Chronicle et al., 2004, and Kershaw & Ohlsson, 2004[30]).

Sometimes, the necessary insight for solving a problem comes from an analogy: an already-solved problem is similar to a current one, so the old solution can be adapted to the new situation.

Insight problems

Puzzle over the insight problems provided here for a moment to see whether you have a sudden 'Aha!' experience when you realise how to solve the problems.

Chain links

A woman has four pieces of chain. Each piece is made up of three links. She wants to join the pieces into a single closed ring of chain. To open a link costs 2 cents and to close a link costs 3 cents. She has only 15 cents. How does she do it?

Hint: You don't have to open a link on each piece of chain.

Four trees

A landscape gardener is given instructions to plant four special trees so that each one is exactly the same distance from each of the others. How would you arrange the trees?

Hint: We don't always plant trees on flat lawns.

Prisoner's escape

A prisoner was attempting to escape from a tower. He found in his cell a rope that was half long enough to permit him to reach the ground safely. He divided the rope in half and tied the two parts together and escaped. How could he have done this?

Hint: Is there only one way to divide a rope in half?

Bronze coin

A stranger approached a museum curator and offered him an ancient bronze coin. The coin had an authentic appearance and was marked with the date 544 BC. The curator had happily made acquisitions from suspicious sources before, but this time he promptly called the police and had the stranger arrested. Why?

Hint: Imagine that you lived in 544 BC. What did it say on your coins?

Nine dots

Connect the nine dots with four connected straight lines without lifting your pencil from the page as you draw.

Hint: How long a line does the problem permit you to draw?

Bowling pins

The 10 bowling pins shown above are pointing toward the top of the page. Move any three of them to make the arrangement point down toward the bottom of the page.

Hint: Pins 1, 2, 3, and 5 form a diamond at the top of the drawing. Consider where the diamond might be for the arrangement that points down.

Source: Adapted from Metcalfe (1986)[31] and Metcalfe & Wiebe (1987)[32].

Metcalfe and Wiebe (1987; also Metcalfe, 1986[33]) studied how people solved such problems and compared that with how they solved algebra and other routine problems. They found two interesting results. First, people were rather accurate in predicting whether they would be successful in solving routine problems but not in predicting success with insight problems. Second, solutions to the insight problems seemed to come suddenly, almost without warning. This result is shown in Figure 11.3.

As they worked through the problems, people were interrupted and asked to indicate how 'warm' they were, that is, how close they felt they were to finding the solution. For routine algebra problems, 'warmth' ratings grew steadily as people worked through the problems, reflecting their feeling of getting closer and closer to the solution. But there was little or no such increase for the insight problems even 15 seconds before the solution was found.

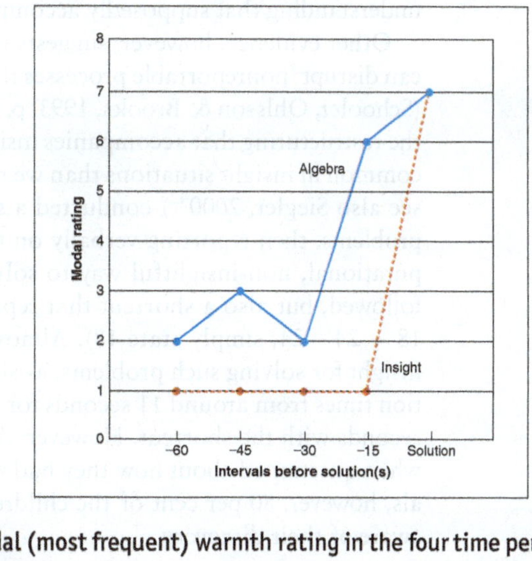

Figure 11.3 Modal (most frequent) warmth rating in the four time periods leading up to a problem solution

Source: Data from Metcalfe & Wiebe (1987).

Letting go and improving insight

Although these results support the idea that insight arrives suddenly, insight problems can be thought of in simpler terms, say overcoming functional fixedness or negative set (as in prisoner's escape and nine-dot), taking a different perspective (bronze coin), and the like (Smith, 1995[34]).

A neuroimaging study by Kounios et al. (2006)[35] provides some support for this idea. Using both electroencephalographic and functional magnetic resonance imaging (fMRI) recordings, they found increased cortical activity centred on the frontal lobes (particularly the anterior cingulate cortex, B.A.s 24, 32 and 33) when people produced insight solutions as compared to normal problem-solving. Kounios et al.'s theory is that this part of the frontal lobe suppresses the irrelevant information (an attentional process) that tends to dominate a person's thinking up to that point. Suppression of these dominant thoughts allows more weakly activated ideas, such as those remote associations drawn by the right hemisphere, to come to the fore, possibly providing the solution to a problem. In other words, part of a person's thought processes is working on the problem along with the steps that are being worked on at the forefront of consciousness (which are going nowhere). When these dead-end thoughts are moved aside, alternative solutions can then present themselves.

This release from irrelevant modes of thinking, seen in the neuroimaging data, can be extended to a process called incubation. With **incubation**, when people have difficulty solving a problem, they may stop working on it for a while. Then at some point, the solution or key to a solution may present itself to them (Sio & Ormerod, 2009[36]). Although this can work at times, it appears that incubation is most useful when people have originally been provided with misleading information, by either others or themselves, that steers them away from the correct solution. During incubation, the representations for these misleading ideas lose strength, so that later the more successful alternatives can then present themselves (Vul & Pashler, 2007[37]). Interestingly, insight is more likely to occur when people are mind wandering while doing relatively undemanding tasks, as compared to doing a demanding task, simply resting, or not taking a break at all (Baird et al., 2011[38]).

In some circumstances, insight may mean that we have drawn a critical inference that leads to a solution; for example, there is more than one way to divide a rope in half (Wickelgren, 1974[39]). Weisberg (1995)[40] reports that some people solve insight problems like those mentioned earlier without any of the sudden restructuring or understanding that supposedly accompanies insight.

Other evidence, however, suggests that verbalisation can interfere with insight, can disrupt 'nonreportable processes that are critical to achieving insight solutions' (Schooler, Ohlsson & Brooks, 1993, p. 166[41]). Furthermore, being unable to report the restructuring that accompanies insight, or the actual insight itself, may be more common in insight situations than we realise. For instance, Siegler and Stern (1998; see also Siegler, 2000[42]) conducted a study of second-graders solving arithmetic problems, then reporting verbally on their solutions. There was the regular computational, non-insightful way to solve the problems, which the second-graders followed, but also a shortcut that represented an insight (e.g. for a problem like 18 + 24 − 24, simply state 18). Almost 90 per cent of the sample discovered the insight for solving such problems, as shown by the dramatic decrease in their solution times from around 11 seconds for the computational method to a mean of 2.7 seconds with the shortcut. However, the children were unaware of their discovery when questioned about how they had solved the problems. Within another five trials, however, 80 per cent of the children's verbal reports indicated that they were aware of their discovery.

11.4.2 Analogy

analogy
A relationship between similar situations, problems or concepts.

In general, an **analogy** is a relationship between two similar situations, problems, or concepts. Understanding an analogy means putting the two situations into some kind of alignment so that the similarities and differences are made apparent (Gentner & Markman[43] 1997).

Take a simple example, the analogy 'MERCHANT : SELL :: CUSTOMER : ______'. Here, you must figure out the structure for the first pair of terms and then project or map that structure onto the second part of the analogy. Because 'SELL' is the critical activity of 'MERCHANT', the critical activity relationship is then mapped onto 'CUSTOMER', and retrieval from memory yields 'BUY'.

Researchers argue that analogies provide excellent, widely applicable methods for solving problems. That is, if you are confronted with a difficult problem, a useful heuristic is to find a similar or related situation and build an analogy from it to the current problem. Such reasoning and problem-solving may help us understand a variety of situations, such as how students should be taught in school, how people adopt professional role models, and how we empathise with others (Holyoak & Thagard, 1997; Kolodner, 1997[44]). Furthermore, it has long been held that important scientific ideas, breakthroughs and explanations often depend on finding analogies – for instance, that neurotransmitters fit into the receptor sites of a neuron much the way a key fits into a lock (see Gentner & Markman, 1997, for a description of reasoning by analogy in Kepler's discovery of the laws of planetary motion[45]).

Curiously, analogical problem-solving is better when people receive the information by hearing about it rather than reading it (Markman, Taylor & Gentner, 2007[46]), perhaps reflecting the more natural use of spoken over written language. Also, analogies are more effective if the analogy source is more abstract, rather than concrete (Day, Motz & Goldstone, 2015[47]), perhaps making it easier to apply to a new situation. Finally, people are more likely to use analogies to solve problems if they have had an opportunity to sleep (Monaghan et al., 2015[48]). As with insight, perhaps this period of rest allows for remote associations to be made, leading people to use analogies.

Analogy problems

Although insight involves the seemingly sudden awareness of the solution to a problem, often as a result of unconscious processes, analogy involves the use of the solutions to prior problems that have a similar, underlying, abstract structure. This use of analogy can be conscious and explicit or, like insight, unconscious and implicit. Although people can certainly use analogies, especially when they are explicitly pointed out, they do not use them as often as would benefit them.

Different analogy problems

Gick and Holyoak (1980)[49] had people read the parade problem, a somewhat different army fortress story, or no story at all. They then asked them to read and solve a second problem, the classic Duncker (1945)[50] radiation problem.

The parade problem

A dictator ruled over a small country. He ruled from a strong fortress. The fortress was located in the centre of the country, surrounded by numerous towns, villages and farms. Like spokes on a wheel, many roads radiated out from the fortress. As part of a celebration of his glorious grab of power, this dictator demanded from one of his generals a large, over-the-top parade of his military might. The general's troops were assembled for the march in the morning, the day of the anniversary, at the end of one of these roads that led up to the dictator's fortress. At that time, the general was given a report by one of his captains that brought him up short. As commanded by the dictator, this parade needed to be far more spectacular and impressive than any other parade that had ever been seen in the land (or else). The dictator demanded that everyone in every region of the country see and hear his army at the same time. Given this demand, it seemed nearly impossible for the general to have the whole country see the parade as requested.

Solution to the parade problem

The general, however, knew just what to do. He divided his army up into small groups and dispatched each group to the head of a different road. When all was ready he gave the signal, and each group marched down a different road. Each group continued down its road to the fortress, so that the army finally arrived together at the fortress at the same time. In this way, the general was able to have the parade seen and heard through the entire country at once, and thus please the dictator.

The radiation problem

Suppose you are a doctor faced with a patient who has a malignant tumour in his stomach. It is impossible to operate on the patient, but unless the tumuor is destroyed the patient will die. There is a kind of ray that can be used to destroy the tumour. If the rays reach the tumour all at once at a sufficiently high intensity, the tumour will be destroyed. Unfortunately, at this intensity the healthy tissue that the rays pass through on the way to the tumour will also be destroyed. At lower intensities the rays are harmless to healthy tissue, but they will not affect the tumour either. What type of procedure might be used to destroy the tumour with the rays without destroying the healthy tissue?

Solution to the radiation problem

The ray can be divided into several low-intensity rays, no one of which will destroy the healthy tissue. If these several rays are positioned at different locations around the body and focused on the tumour, their effect will combine, thus being strong enough to destroy the tumour.

The attack dispersion problem

A small country was controlled by a dictator. The dictator ruled the country from a strong fortress. The fortress was situated in the middle of the country, surrounded by farms and villages. Many roads radiated outward from the fortress like spokes on a wheel. A general arose who raised a large army and vowed to capture the fortress and free the country of the dictator. The general knew that if his entire army could attack the fortress at once, it could be captured. The general's troops were gathered at the head of one of the roads leading to the fortress, ready to attack. However, a spy brought the general a disturbing report. The ruthless dictator had planted mines on each of the roads. The mines were set so that small bodies of men could pass over them safely because the dictator needed to be able to move troops and workers to and from the fortress. However, any large force would detonate the mines. Not only would this blow up the road and render it impassable, but the dictator would then destroy many villages in retaliation. It therefore seemed impossible to mount a full-scale direct attack on the fortress.

Solution to the attack dispersion problem

The general, however, knew just what to do. He divided his army up into small groups and dispatched each group to the head of a different road. When all was ready he gave the signal, and each group marched down a different road. Each group continued down its road to the fortress, so that the army finally arrived together at the fortress at the same time. In this way, the general was able to capture the fortress and thus overthrow the dictator.

The radiation problem is interesting for a variety of reasons, including the fact that it is rather ill-defined and thus comparable to many problems in the real world. Duncker's participants produced two general approaches that led to dead ends: trying to avoid contact between the ray and nearby tissue, and trying to change the sensitivity of surrounding tissue to the effects of the ray. But the third approach, reducing the intensity of the rays, was more productive, especially if an analogy from some other, better understood situation was available.

Gick and Holyoak's (1980) results

Gick and Holyoak (1980)[51] used the radiation problem to study analogy. In fact, we have just simulated one of their experiments here by having you read the parade story first and then the radiation problem. In case you didn't notice, there are strong similarities between the problems, suggesting that the parade story can be used to develop an analogy for the radiation problem.

Gick and Holyoak found that 49 per cent of people who first solved the parade problem realised it could be used as an analogy for the radiation problem. A different initial story, in which armies are attacking a fortress, provided a stronger hint about the radiation problem. Fully 76 per cent of these participants used the attack analogy in solving the radiation problem. In contrast, only 8 per cent of the control group, which merely attempted to solve the radiation problem, came up with the dispersion solution (i.e. multiple pathways).

When Gick and Holyoak provided a strong hint, telling people that the attack solution might be helpful as they worked on the radiation problem, 92 per cent of them used the analogy, and most found it 'very helpful'. By contrast, only 20 per cent of the people in the no-hint group produced the dispersion solution, even though they too had read the attack dispersion story. In short, only 20 per cent spontaneously noticed and used the analogous relationship between the problems.

Summary of Gick and Holyoak's (1980) results

The simulation below summarises Gick and Holyoak's results concerning the influence of analogical processing on problem-solving.

Group	Order of stories	Percentage of people who used the analogy on the radiation problem
Group *A*	Parade, radiation	49%
Group *B*	Attack dispersion, radiation	76%
Group *C*	No story, radiation	8%

Study 1 (Experiment II originally; after Gick & Holyoak, Table 10)

People in groups *A* and *B* are given a general hint that their solution to one of the earlier stories may be useful in solving the radiation problem.

Group	Order of stories	Percentage of people who used the analogy on the radiation problem
Group *A* (hint)	Attack dispersion, radiation	49%
Group *B* (no hint)	Attack dispersion, radiation	20%

Study 2 (Experiment IV originally)

People in group *A* are given the general hint (as above). People in group *B* are given no hint whatsoever.

Multiconstraint theory

Holyoak and Thagard (1997)[52] proposed a theory of analogical reasoning and problem-solving, based on such results. The theory, called the **multiconstraint theory**, predicts how people use analogies in problem-solving and what factors govern the analogies people construct.

A final point is that most of the work on analogy, like many studies of problem-solving, has focused on the conscious, explicit use of analogies. However, some evidence suggests that people may use analogies in a more unconscious, implicit manner as well. In a study by Day and Gentner (2007)[53], people read given pairs of texts. When the events described by the second text were analogous to those described in the first (in terms of their relational structure), people read the second text faster. That is, people were able to use their unconscious knowledge of the event structure from the first text to help them understand the second text. When asked, people showed no awareness of this relationship between the two texts. So, in some sense, by having people read the first text, the relational structure of the event was primed, and this made the processing of the second text easier.

11.4.3 Neurocognition in analogy and insight

Some exciting work has been reported on the cognitive neuropsychology of analogical reasoning and insight. Wharton et al. (2000)[54] identified brain regions associated with the mapping process in analogical reasoning. In their study, people saw a source picture of geometric shapes, followed by a target picture. They had to judge whether the target picture was an *analogue pattern* – whether it had the same system of relations as the source picture. In the control condition, they judged whether the target was literally the same as the source. See Figure 11.4 for sample stimuli. In the top stimulus, the correct target preserves both the *spatial relations* in the source (a shape in all four quadrants) and the *object relations* (the patterned figures on the main diagonal are the same shape, and the shapes on the minor diagonal are different). Response times to analogy trials were in the 1,400 to 1,500 ms range and approximately 900 to 1,000 ms in the literal condition; accuracy was at or above 90 per cent in both kinds of trials.

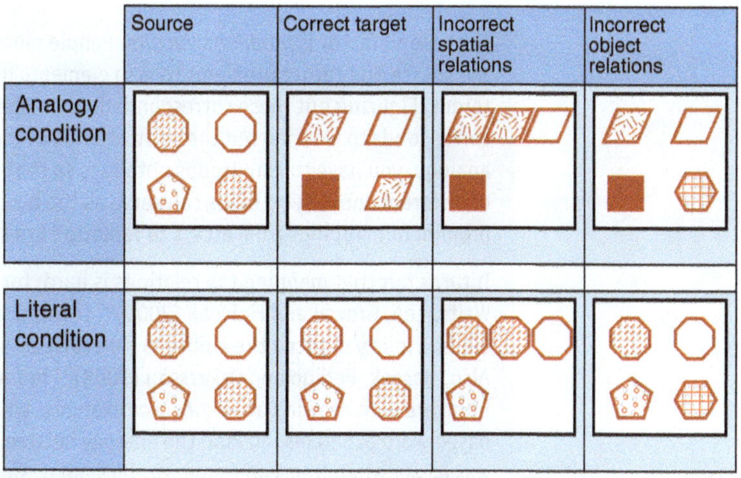

Figure 11.4 A depiction of analogy condition and literal condition trials

The first column shows the source stimuli, the second shows the correct choice, and the third and fourth show incorrect choices for the stated reasons.

Source: Wharton et al. (2000)[55], Figure 2, p. 179.

Factors governing analogy

In particular, the multiconstraint theory states that people are constrained by three factors when they try to use or develop analogies.

Problem similarity

The first factor is *problem similarity*. There must be some degree of similarity between the already-understood situation, the source domain, and the current problem, the target domain. In the parade story, for example, the fortress and troops are similar to the tumour and rays. Similarity between source and target has been shown to be important. Chen, Mo and Honomichl (2004),[56] for example, found that similarities from well-known folktales to new problems were especially important for finding problem solutions, even if participants did not report remembering the folktale. Alternatively, Novick (1988)[57] found that novices focus especially on similarities, even when they are only superficial, which can interfere with performance.

Problem structure

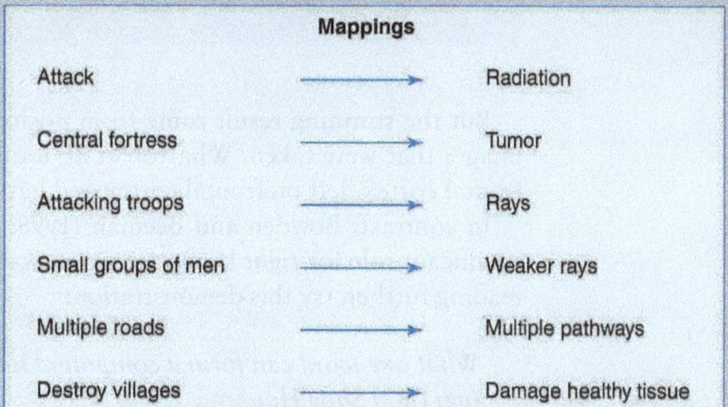
Mappings
Attack
Radiation
Central fortress
Tumor
Attacking troops
Rays
Small groups of men
Weaker rays
Multiple roads
Multiple pathways
Destroy villages
Damage healthy tissue

Prominent mappings between the attack and radiation problems

The second factor is *problem structure*. People must find a parallel structure between the source and target problems to map elements from the source to elements in the target. Figuring out these correspondences or mappings is important because it corresponds to working out the relationships of the analogy. In the attack–radiation analogy, you have to map troops onto rays so that the important relationship of different converging roads can serve as the basis for the solution. The most prominent mappings from attack to radiation are shown in **above figure.**

It turns out that mapping the relations is hard; for instance, in a dual-task setting, Waltz, Lau, Grewal and Holyoak (2000)[58] found that holding on to a working memory load seriously reduced the ability to find correct mappings between two problems. Also, Bassok, Pedigo and Oskarsson (2008)[59] found that semantic memory can interfere with drawing analogies. For example, when people are doing addition-based word problems, making the analogy between the word problem and addition was easier when items are semantically similar, such as apples and oranges (which are separate things that can easily be added together), but not when there is an inconsistency, such as records and songs. In this case, the knowledge that songs are parts of records implies a hierarchical, part–whole relationship, so it is more difficult to make the additional analogy where is it easier to think of things that can be treated on more equal footing.

Purpose of the analogy

The third factor that constrains people is the *purpose of the analogy*. The person's goals and the goal stated in the problem are important. This is deeper than merely the general purpose of trying to solve the problem. Notice that the goals in the attack and radiation stories match, whereas the goals do not match for parade and radiation (parade involves sending troops *out* from the central fortress, for display purposes, but radiation involves sending rays *in* towards the tumour). This mismatch may have been responsible for the low use of parade as a source for the analogy to radiation. Likewise, Spellman and Holyoak (1996)[60] report a study in which college students drew analogies between two soap opera plots. When different purposes or goals were given in the instructions, the students developed different analogies; that is, their problem-solving by analogy was sensitive to purposes and overall goals. Kurtz and Loewenstein (2007)[61] reported that people are more likely to draw on analogies from previous problems when they are comparing two other problems than if they are working on a single problem alone. This suggests that the processing goals and tactics, in this case direct comparison, in use can influence whether people actually use analogies or not.

But the stunning result came from positron emission tomography (PET) scan images that were taken. Wharton et al. found significant activation in the medial frontal cortex, left prefrontal cortex and left inferior parietal cortex.

In contrast, Bowden and Beeman (1998; Beeman & Bowden, 2000[62]) found a significant role for *right* hemisphere processing in solving insight problems. Before reading further, try this demonstration:

What one word can form a compound word or phrase with each of the following: Palm Shoe House?

What one word can form a compound word or phrase with each of the following: Pie Luck Belly?

People were given such word triples – called 'compound remote associates' – and had to think of a fourth word that combines with each of the three initial words to yield a familiar word pair. On many trials, people fail to find an immediate solution and end up spending considerable time working on the problem. They also report that when they finally solve the problem, the solution came to them as an insight – an 'Aha!' type of solution.

In the Bowden and Beeman (1998)[63] study, people saw the problems and then after 15 seconds were asked to name a new word that appeared on the screen (if they solved the problem before the 15-second period was up, they were given the word immediately). When the target word was unrelated to the three words seen before (e.g. *planet*), there was the typical effect – that targets presented to the right visual field, hence the left hemisphere of the brain, were named faster than those presented to the left visual field–right hemisphere. But when the target was the word that solved the insight problem (*tree* in the first problem, *pot* in the second one), there was a significant priming effect. As shown in Figure 11.5, time to pronounce was shorter for the solution words than for the unrelated words. And the priming effect – the drop-off from 'unrelated' to 'solution' – was greater for targets presented to the right hemisphere than to the left (in other words, presented to the left visual field so going first to the right hemisphere).

Putting it differently, semantic priming in the right hemisphere was more prominent than in the left hemisphere for these problems: people were faster to name *pot* when it was presented to the right hemisphere, presumably because it had been primed by the initial three words. As the authors noted, these results fit nicely with other results concerning the role in language comprehension that the right hemisphere plays, especially the part having to do with drawing inferences (Bowden, Beeman & Gernsbacher, 1995[65]). Further evidence from neuroimaging studies (both EEG and fMRI) highlighted the role of the right hemisphere in insight problem-solving. Compared to analytic processing (analogy) problem-solving, when people solved identical problems by insight, fMRI indicated higher activation of the right anterior superior temporal gyrus, and EEG indicated higher electrical activity over the right temporal lobe (Jung-Beeman et al., 2004[66]).

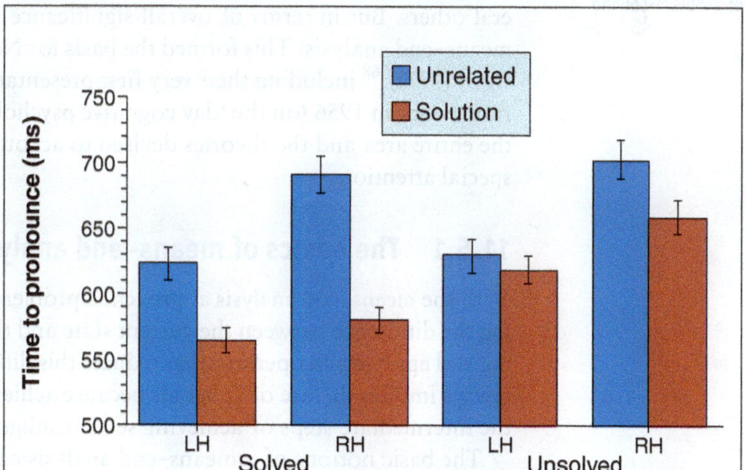

Figure 11.5 **Time to pronounce the target word for solved and unsolved trials**

Bars labelled LH refer to target words presented to the right visual field, left hemisphere of the brain; RH means left visual field, right hemisphere. The figure shows priming effects for solution words, especially in the right hemisphere.

Source: Adapted from Bowden and Beeman (1998)[64].

Synchronised brains and creativity in teamwork

Contemporary research is seeking to understand creativity in real-life settings, with some studies focusing on teamwork and creative problem-solving. Mayseless, Hawthorne and Reiss (2019)[67] assessed the neural profile of teamwork creative problem-solving using *hyperscanning* – a novel neuroimaging approach that allows researchers to record and analyse brain activity from multiple interacting individuals. Mayseless et al. (2019) had participants working in pairs to design a either a new product that would nudge people to vote (creative design task, experimental condition), or building a 3D model plane together (control task). While working on these tasks participants were wearing caps that allowed researchers to measure cortical blood flow, a functional neuroimaging method also known as *functional near-infrared spectroscopy* (fNIRS). This measure indicated inter-brain (or neural) synchrony, that is the degree of correlation of the neural activity between the interacting partners in the dyadic creative task.

The researchers found that there was increased inter-brain synchronisation in areas related to cognitive control and mentalising (e.g. inferior frontal gyrus, superior temporal gyrus), and the mirror neuron system, and that this pattern of neural activity was positively correlated with cooperative behaviour among the interacting partners and creative problem-solving (i.e. designing innovative voting systems). Such studies provide insights into creative problem-solving, but also into the neural activity that facilitates this process, especially within interacting partners.

11.5 Means–end analysis

LEARNING OBJECTIVE
Outline the means–end analysis method of problem-solving

Several problem-solving heuristics have been discovered and investigated. You have already read about analogy, and the final section of the module illustrates several others. But in terms of overall significance, no other heuristic comes close to means–end analysis. This formed the basis for Newell and Simon's groundbreaking work (1972),[68] including their very first presentation of the information-processing framework in 1956 (on the 'day cognitive psychology was born'). Because it shaped the entire area and the theories devised to account for problem-solving, it deserves special attention.

11.5.1 The basics of means–end analysis

With the **means–end analysis** approach, a problem is solved by repeatedly determining the difference between the current state and the goal or subgoal state, then finding and applying an operator that reduces this difference. Means–end analysis nearly always implies the use of subgoals because achieving the goal state usually involves the intermediate steps of achieving several subgoals along the way.

The basic notions of a means–end analysis can be summarised in a sequence of five steps:

1. Set up a goal or subgoal.
2. Look for a difference between the current state and the goal or subgoal state.
3. Look for an operator that will reduce or eliminate this difference. One such operator is the setting of a new subgoal.

4. Apply the operator.
5. Apply steps 2 to 4 repeatedly until all subgoals and the final goal are achieved.

At an intuitive level, means–end analysis and subgoals are familiar and represent 'normal' problem-solving. If you have to write a term paper for class, you break the overall goal down into a series of subgoals: Select a topic, find relevant material, read and understand the material, and so on. Each of these may contain its own subgoals.

11.5.2 The Tower of Hanoi

One of the most thoroughly investigated problems is the Tower of Hanoi problem. This problem shows clearly the strengths and limitations of the means–end approach.

The three-disc version

Work on the Tower of Hanoi problem carefully, using the three-disc version in the simulation below. Try to keep track of your solution so you will understand how it demonstrates the usefulness of a means–end analysis. So that you will become familiar with the problem and be able to reflect on your solution, do it several times again after you have solved it. See whether you can become skilled at solving the three-disc problem by remembering your solution and being able to generate it repeatedly. (By the way, an excellent heuristic for this problem is to solve it physically; draw the pegs on a piece of paper and move three coins of different sizes around to find the solution.)

The seven-step solution for the Tower of Hanoi Problem

Having done the three-disc version of the problem, consider your solution in terms of subgoals and means–end analysis.

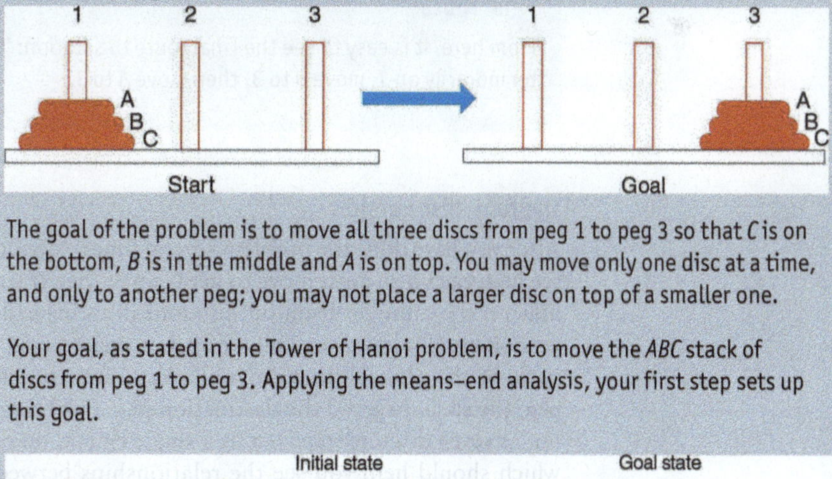

The goal of the problem is to move all three discs from peg 1 to peg 3 so that *C* is on the bottom, *B* is in the middle and *A* is on top. You may move only one disc at a time, and only to another peg; you may not place a larger disc on top of a smaller one.

Your goal, as stated in the Tower of Hanoi problem, is to move the *ABC* stack of discs from peg 1 to peg 3. Applying the means–end analysis, your first step sets up this goal.

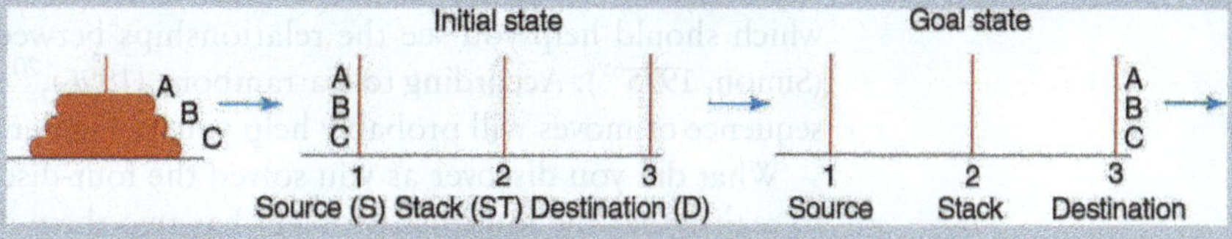

Difficulty

The second step reveals a difficulty: there is a difference between your current state and the goal, simply the difference between the starting and ending configurations. You look for a method or operator that reduces this difference and then apply it. As you no doubt learned from your solution, your first major subgoal is 'Clear off disc *C*'. This entails getting *B* off *C*, which entails another subgoal, getting *A* off *B*.

Note: The pegs have been renamed as 'source', 'stack' and 'destination.'

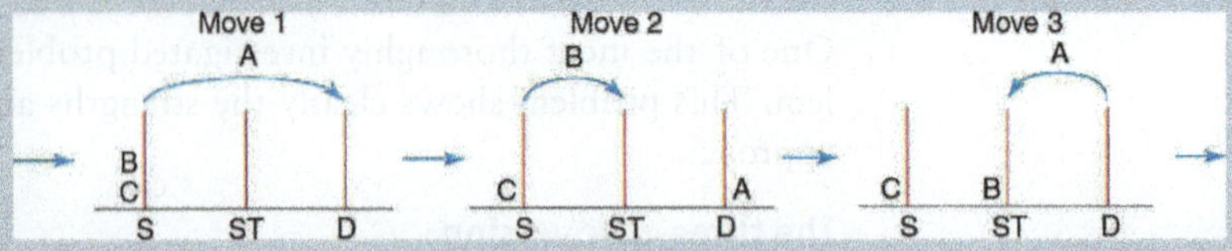

Operator

The next step involves a simple operator that satisfies the most immediate subgoal, 'Move *A* to 3', which permits satisfying the next subgoal, 'getting *B* off *C*'. So the next operator is 'move *B* to 2'; it can't go on top of *A* (rule violation), and it can't stay on top of *C* because that prevents achieving a subgoal. Now peg 3 can be cleared by moving *A* to 2. This allows the major subgoal to be accomplished, putting *C* on 3.

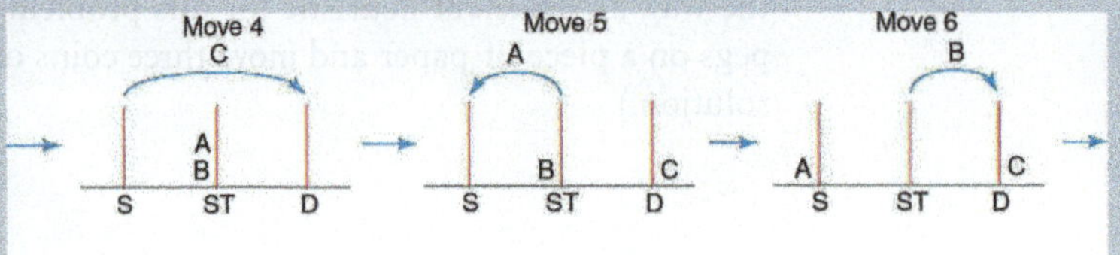

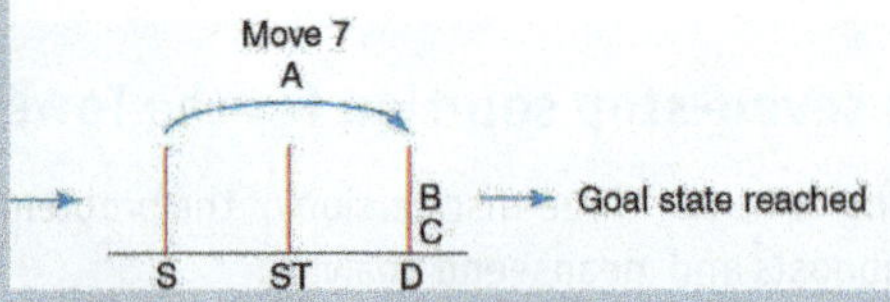

Final route

From here, it is easy to see the final route to solution: 'Unpack' *A* from 2, putting it temporarily on 1, move *B* to 3, then move *A* to 3.

The four-disc version

After you have done the problem several times, solving it becomes easy. You come to see how each disc must move to get *C* on 3, then *B*, and finally *A*. Spend some time now on the same problem but use four discs instead of three. Don't work on this version blindly, however. Think of it as a variation on the three-disc problem, where parts of the new solution are 'old'. As a hint, try renaming the pegs as the source peg, the stack peg and the destination peg. Furthermore, think of the seven moves not as seven discrete steps but as a single chunk, 'moving a pyramid of three discs', which should help you see the relationships between the problems more clearly (Simon, 1975[69]). According to Catrambone (1996),[70] almost any label attached to a sequence of moves will probably help you remember the sequence better.

What did you discover as you solved the four-disc problem? Most people come to realise that the four-disc problem has two three-disc problems embedded in it, separated by the bridging move of *D* to 3. That is, to free *D* so it can move to peg 3 you must first move the top three discs out of the way, moving a 'pyramid of three discs', getting *D* to peg 3 on the eighth move. Then the *ABC* pyramid has to move again to get them on top of *D* – another seven moves. Moving the discs entails the same order of moves as in the simpler problem, although the pegs take on different functions: for the four-disc problem, peg 2 serves as the destination for the first half of the solution, then as the source for the last half.

The four-disc Tower of Hanoi Problem, with solution

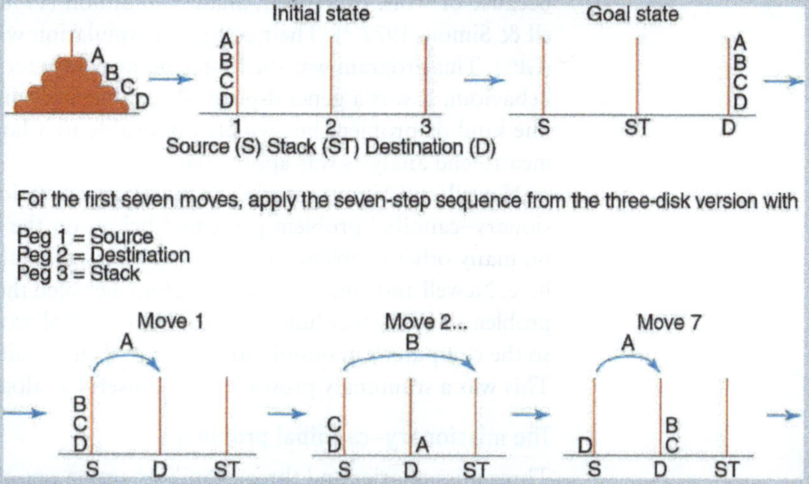

Initial state

The variation from the three-disc version is that the pegs must switch roles.

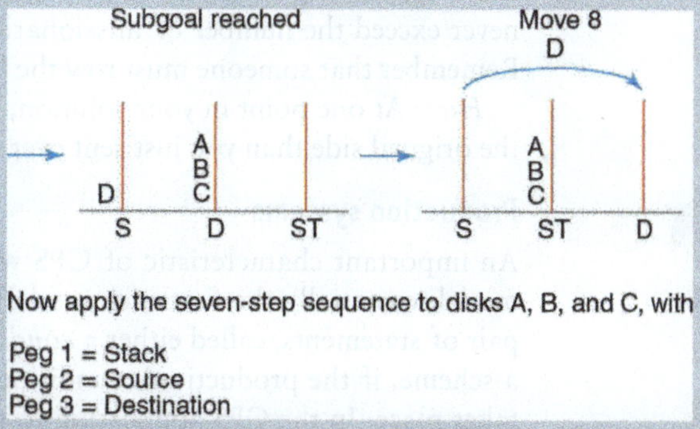

Subgoal

In the beginning, the subgoal is to move a pyramid of three discs so that D can move to peg 3. After that, the subgoal again is to move a three-disc pyramid.

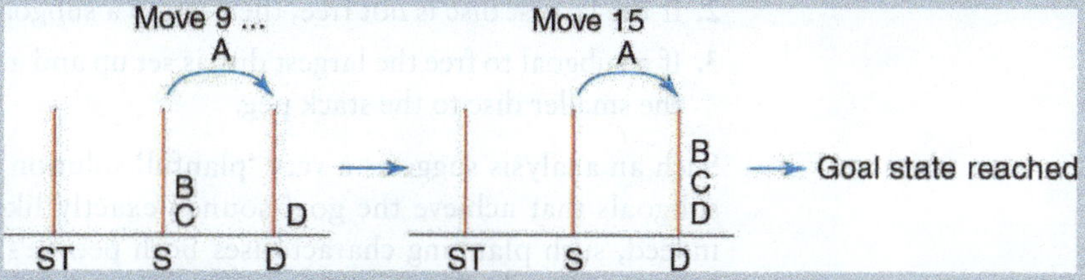

Goal state

In both the first and second halves, the pegs must switch roles for the problem to be solved.

Because the three-disc solution is embedded in the four-disc problem – and, likewise, the four-disc solution is embedded in the five-disc problem – this is known as a recursive problem, where *recursive* simply means that simpler components are embedded in the more difficult versions.

11.5.3 General problem-solver

Means–end analysis was an early focus of research on problem-solving, largely because of work by Newell, Shaw, and Simon (1958; Ernst & Newell, 1969; Newell & Simon, 1972[71]). Their computer simulation was called **general problem-solver (GPS)**. This program was the first genuine computer simulation of problem-solving behaviour. It was a general-purpose, problem-solving program, not limited to just one kind of problem but widely applicable to a large class of problems in which means–end analysis was appropriate.

Newell and Simon ran their simulation on various logical proofs, on the missionary–cannibal problem presented below, on the Tower of Hanoi problem, and on many other problems to demonstrate its generality. Notice the critical analogy here. Newell and Simon drew an analogy between the way computer programs solve problems and the way humans do: human mental processes are of a symbolic nature, so the computer's manipulation of symbols is a fruitful analogy to those processes. This was a stunningly provocative and useful analogy for the science of cognition.

The missionary–cannibal problem

Three missionaries and three cannibals are on one side of a river and need to cross to the other side. The only means of crossing is a boat, and the boat can hold only two people at a time. Devise a set of moves that will transport all six people across the river, bearing in mind the following constraint: the number of cannibals can never exceed the number of missionaries in any location, for the obvious reason. Remember that someone must row the boat back across each time.

Hint: At one point in your solution, you will have to send more people back to the original side than you just sent over to the destination.

Production systems

An important characteristic of GPS was its formulation as a **production system** model, essentially the first such model proposed in psychology. A **production** is a pair of statements, called either a *condition–action* pair or an *if–then* pair. In such a scheme, if the production's conditions are satisfied, the action part of the pair takes place. In the GPS application to the Tower of Hanoi, three sample productions might be

1. If the destination peg is clear and the largest disc is free, then move the largest disc to the destination peg.
2. If the largest disc is not free, then set up a subgoal to free it.
3. If a subgoal to free the largest disc is set up and a smaller disk is on it, then move the smaller disc to the stack peg.

Such an analysis suggests a very 'planful' solution by GPS: setting up a goal and subgoals that achieve the goal sounds exactly like what we call planning. And indeed, such planning characterises both people's and GPS's solutions to problems, not just the Tower of Hanoi but all kinds of transformation problems. GPS had what amounted to a planning mechanism, a mechanism that abstracted the essential features of situations and goals, then devised a plan that would produce a problem-solving sequence of moves. Provided with such a mechanism and the particular representational system necessary to encode the problem and the legal operators, GPS yielded an output that resembles the solution pathways taken by human problem-solvers.

Limitations of general problem-solver

Later investigators working with the general principles of GPS found some cases when the model did not do a good job of characterising human problem-solving. Consider the missionary–cannibal problem presented earlier; the solution pathway is presented in Figure 11.6.

The problem is difficult, most people find, at step 6, where the only legal move is to return one missionary and one cannibal back to the original side of the river. Having just brought two missionaries over, this return trip seems to be moving away from the overall goal. That is, returning one missionary and one cannibal seems to be incorrect because it appears to increase the distance to the goal: it is the only return trip that moves two characters back to the original side. Despite this being the only available move (other than returning the same two missionaries who just came over), people have difficulty in selecting this move (Thomas, 1974[72]).

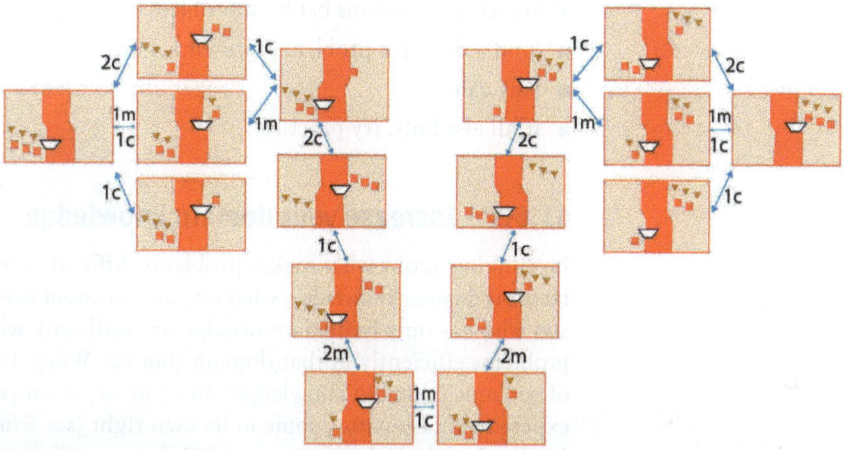

Figure 11.6 An illustration of the steps needed to solve the missionary–cannibal problem

The left half of each box is the 'start side' of the river, and the right half is the 'destination side'. The numbers and letters next to the arrows represent who is travelling on the boat.

Source: Based on Glass & Holyoak (1986)[73].

GPS did not have this difficulty because sending one missionary and one cannibal back was consistent with its immediate subgoal. On the other hand, at step 10, GPS is trying to fulfil its subgoal of getting the last cannibal to the destination side and seemingly can't let go of this subgoal. People, however, realise that this subgoal should be abandoned: anyone can row back over to bring the last cannibal across and, in the process, finish the problem (Greeno, 1974[74]). GPS was simply too rigid in its application of the means–end heuristic, however: it tried to bring the last cannibal across and then send the boat back again.

Beyond general problem-solver

Newell and Simon's GPS model, and models based on it, often provided a good description of human problem-solving performance (Atwood & Polson, 1976[75]) and offered a set of predictions against which new experimental results could be compared (Greeno, 1974[76]). Despite some limitations (Hayes & Simon, 1974[77]), the model demonstrated the importance of means–end analysis for an understanding of human problem-solving.

11.6 Improving your problem-solving

LEARNING OBJECTIVE
Summarise a set of problem-solving techniques

Sprinkled throughout the module have been hints and suggestions about how to improve your problem-solving. Some of these are based on empirical research and some on intuitions that people have had about problem-solving.

Let's close the module by pulling these hints and suggestions together and offering a few new ones. Here is a list of these suggestions for improving problem-solving:

- Increase your domain knowledge.
- Automate some components of the problem-solving solution.
- Follow a systematic plan.
- Draw inferences.
- Develop subgoals.
- Work backwards.
- Search for contradictions.
- Search for relations between problems.
- Find a different problem representation.
- Stay calm.
- If all else fails, try practice.

11.6.1 Increase your domain knowledge

In thinking about what makes problems difficult, Simon suggests that the likeliest factor is **domain knowledge**, what one knows about the topic. Not surprisingly, a person who has only limited knowledge or familiarity with a topic is less able to solve problems efficiently in that domain (but see Wiley, 1998[78], on some disadvantages of too much domain knowledge). In contrast, extensive domain knowledge leads to expertise, a fascinating topic in its own right (see Ericsson & Charness, 1994, and Medin, Lynch, Coley & Atran, 1997, for example[79]).

Much of the research supporting this comes from Simon's work with chess (Chase & Simon, 1973; Gobet & Simon, 1996; see also Reeves & Weisberg, 1993[80]). In several studies of chess masters, an important but not surprising result was obtained: chess masters need only a glimpse of the arrangement of chess pieces to remember the arrangement, far beyond what novices or players of moderate skill can do. This advantage holds, however, only when the pieces are in legal locations (i.e. sensible within the context of a real game of chess). When the locations of the pieces are random, then there is no advantage for the skilled players. This advantage of expertise in remembering legal board positions is attributed to experts' more skilled perceptual encoding of the board, literally more efficient eye movements and fixations while looking at the board (Reingold, Charness, Pomplun & Stampe, 2001[81]).

11.6.2 Automate some components of the problem-solving solution

A second connection also exists between the question 'What makes problems difficult?' and the topics you have already studied. Kotovsky, Hayes and Simon (1985)[82] tested adults on various forms of the Tower of Hanoi problem and on problem *isomorphs*, problems with the same form but different details. Their results showed that a heavy working memory load was a serious impediment to successful problem-solving: if a person had to hold three or four nested subgoals in working memory all at once, performance deteriorated.

Thus, a solution to this memory load problem was to *automate* the rules that govern moves, just as you were supposed to master and automate the seven-step sequence in the Tower of Hanoi. This frees working memory to be used for higher-level subgoals (Carlson, Khoo, Yaure & Schneider, 1990[83]). This is the same reasoning you encountered early in the course, where automatic processing uses few if any of the limited conscious resources of working memory.

11.6.3 Follow a systematic plan

Especially in long multistep problems, it is important to follow a *systematic plan* (Bransford & Stein, 1993; Polya, 1957[84]). Although this seems straightforward, people do not always generate plans when solving problems, although doing so can dramatically improve performance (Delany, Ericsson & Knowles, 2004[85]). A plan helps you keep track of what you have done or tried and keeps you focused on the overall goal or subgoals you are working on. For example, on DONALD + GERALD, you need to devise a way to keep track of which digits you have used, which letters remain and what you know about them. If nothing else, developing and following a plan helps you avoid redoing what you have already done. Keep in mind that people often make errors when planning how long a task will take, but can plan their time better if they break the task down into the problem subgoals, estimate the time needed for each of those and then add those times together (Forsyth & Burt, 2008[86]).

11.6.4 Draw inferences and develop subgoals

Wickelgren's (1974)[87] advice is to *draw inferences* from the givens, the terms and the expressions in a problem before working on the problem itself.

It can also help you abandon a misleading representation of the problem and find one more suitable to solving the problem (Simon, 1995[88]).

Beware of *unwarranted inferences*, the kinds of restrictions we place on ourselves that may lead to dead ends. For instance, for the nine-dot problem, an unwarranted inference is that you must stay within the boundaries of the nine dots.

Wickelgren also recommends a *subgoal heuristic* for problem-solving, that is, breaking a large problem into separate subgoals. This is the heart of the means–end approach. There is a different slant to the subgoal approach, however, that bears mention. Sometimes in our real-world problem-solving, there is only a vaguely specified goal and, as often as not, even more vaguely specified subgoals. How do you know when you have achieved a subgoal, say when the subgoal is 'find enough articles on a particular topic to write a term paper that will earn an A'?

Simon's (1979)[89] *satisficing* heuristic is important here; satisficing is a heuristic in which we find a solution to a goal or subgoal that is satisfactory although not necessarily the best possible one. For some problems, the term paper problem included, an initial satisfactory solution to subgoals may give you additional insight for further refinement of your solution. For instance, as you begin to write your rough draft, you realise there are gaps in your information. What seemed originally to be a satisfactory solution to the subgoal of finding references turns out to be insufficient, so you can recycle back to that subgoal to improve your solution. You might only discover this deficiency by going ahead and working on that next subgoal, the rough draft. If you do this appropriately, it can often save you from wasting time on blind alleys, as in the two trains and 15 pennies problems.

Two trains and 15 pennies problems

Two trains

Two train stations are 50 miles apart. At 2pm one Saturday afternoon, the trains start toward each other, one from each station. Just as the trains pull out of the stations, a bird springs into the air in front of the first train and flies ahead to the front of the second train. When the bird reaches the second train it turns back and flies toward the first train. The bird continues to do this until the trains meet. If both trains travel at the rate of 25 miles per hour and the bird flies at 100 miles per hour, how many miles will the bird fly before the trains meet?

Hint: Don't think about how far the bird is flying; think of how far the trains will travel and how long that will take.

Fifteen pennies

Fifteen pennies are placed on a table in front of two players. Players must remove at least one but not more than five pennies on their turns. The players alternate turns of removing pennies until the last penny is removed. The player who removes the last penny from the table is the winner. Is there a method of play that will guarantee victory?

Hint: What do you want to force your opponent to do to leave you with the winning move? What will the table look like when your opponent makes that move?

11.6.5 Work backward and search for contradictions

Another heuristic is *working backwards*, in which a well-specified goal may permit a tracing of the solution pathway in reverse order, thus working back to the givens. The 15 pennies problem is an illustration, a problem that is best solved by working backwards. Many maths and algebra proofs can also be worked backwards or in a combination of forward and backward methods.

In problems that ask 'Is it possible to?' or 'Is there a way that?' you should *search for contradictions* in the givens or goal state. Wickelgren uses the following illustration: Is there an integer x that satisfies the equation $x^2 + 1 = 0$? A simple algebraic operation, subtracting 1 from both sides, yields $x^2 = -1$, which contradicts the known property that any squared number is positive. This heuristic can also be helpful in multiple-choice exams. That is, maybe some of the alternatives contradict some idea or fact in the question, or some fact you learned in the course. Either will enable you to rule out those choices immediately.

11.6.6 Search for relations among problems

In *searching for relations* among problems, you actively consider how the current problem may resemble one you have already solved or know about. The four-disc (and more) Tower of Hanoi problems are examples of this, as are situations in which you search for an analogy (Bassok & Holyoak, 1989[90]; Ross, 1987[91]). Don't become impatient. Bowden (1985)[92] discovered that people often found and used information from related problems, but only if sufficient time was allowed for them to do so.

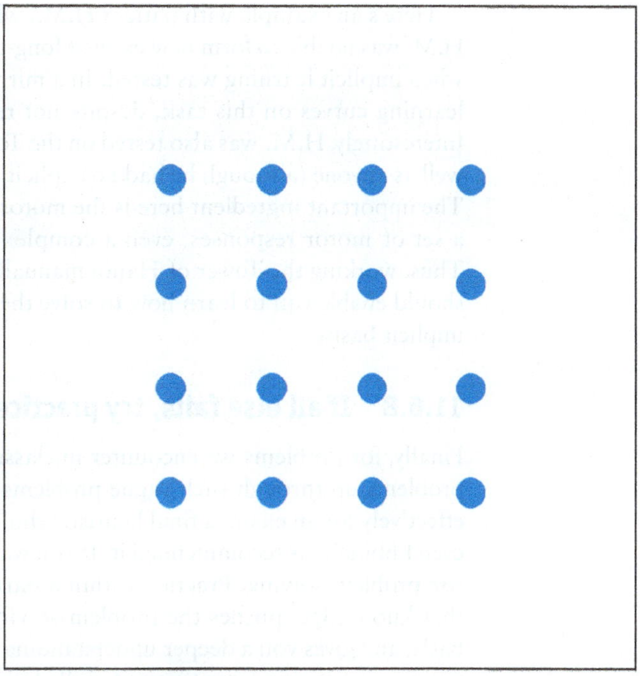

Figure 11.7 The 16 dots problem

Take a paper and draw 16 dots as shown in the figure. Now, without lifting your pencil, join all 16 dots with six straight lines.

11.6.7 Find a different problem representation

Another heuristic involves the more general issue of *problem representation*, or how you choose to represent and think about the problem you are working on. Often, when you get stuck on a problem, it is useful to go back to the beginning and *reformulate* or *reconceptualise* it. For instance, as you discovered in the Buddhist monk problem, a quantitative representation of the situation is unproductive. Return to the beginning and try to consider other ways to think about the situation, such as a visual imagery approach, especially a mental film that includes action. In the Buddhist monk problem, superimposing two such mental films permits you to see him walking up and down at the same time, thus yielding the solution. Likewise, animated diagrams, with arrows moving in towards a point of convergence, helped participants solve the radiation problem in Pedone, Hummel and Holyoak's (2001)[93] study, as compared to either static diagrams or a series of diagrams showing intermediate points in problem solution (see also Reed & Hoffman, 2004[94]).

For other kinds of problems, try a numerical representation, including working the problem out with some examples, or a physical representation, using objects, scratch paper, and so forth. Simon (1995)[95] makes a compelling point that one representation of a problem may highlight a particular feature while masking or obscuring a different, possibly important, feature. According to Ahlum-Heath and DiVesta (1986),[96] verbalising your thinking also helps in the initial stages of problem-solving.

Earlier, it was suggested that you can master the Tower of Hanoi problem more easily if you use three coins of different sizes. This is more than just good advice.

Here's an example with patient H.M., who had profound anterograde amnesia. H.M. was unable to form new explicit long-term memories but performed normally when implicit learning was tested. In a mirror tracing study, H.M. showed normal learning curves on this task, despite not remembering the task from day to day. Interestingly, H.M. was also tested on the Tower of Hanoi task, and he learned it as well as anyone (although he had no explicit memory of ever having done it before). The important ingredient here is the motor aspect of the tower problem: learning a set of motor responses, even a complex sequence, relies on implicit memory. Thus, working the Tower of Hanoi manually by moving real discs or coins around should enable you to learn how to solve the problem from both an explicit and an implicit basis.

11.6.8 If all else fails, try practice

Finally, for problems we encounter in classroom settings, from algebra or physics problems up through such vague problems as writing a term paper and studying effectively for an exam, a final heuristic should help. It is well known in psychology; even Ebbinghaus recommended it. If you want to be good at problem-solving, *practise* problem-solving. Practice within a particular knowledge domain strengthens that knowledge, pushes the problem-solving components closer to an automatic basis, and gives you a deeper understanding of the domain. Although it isn't flashy, practice is a major component of skilled problem-solving and of gaining expertise in any area (Ericsson & Charness, 1994[97]).

Gaining expertise and improving problem-solving comes with more practice.
Source: Roman Tiraspolsky / Alamy Stock Photo

In Ericsson and Charness's (1994)[98] review, people routinely believe that stunning talent and amazing accomplishments result from inherited, genetic, or 'interior' explanations, when the explanation usually is dedicated, regular, long-term practice. This relationship between practice and performance level is seen in an analysis of practice and expertise data by Ericsson, Krampe and Tesch-Römer (1993)[99] shown in Figure 11.8.

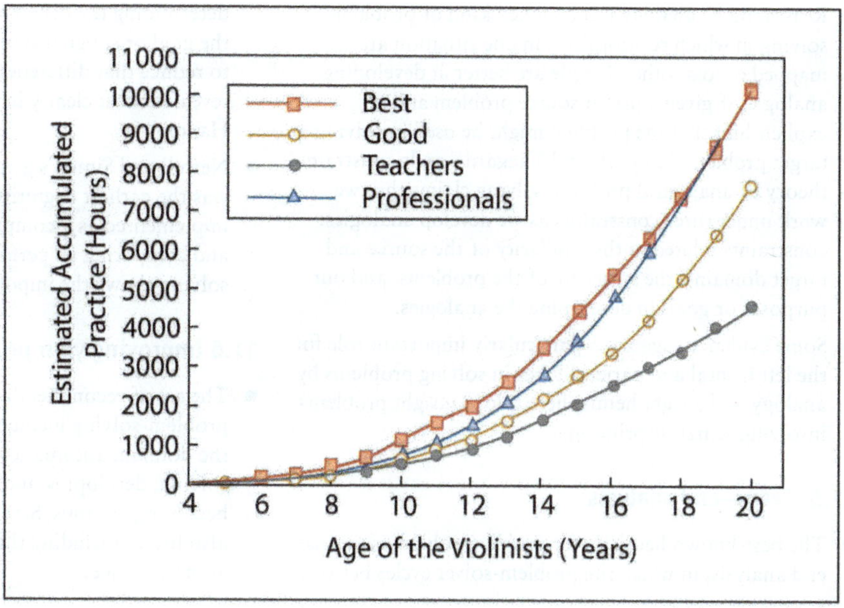

Figure 11.8 **Illustration of the relationship between amount of practice over the course of years and the level of expertise**

Source: Ericsson, Krampe and Tesch-Romer (1993)[100].

As can be seen, the people who had higher levels of expertise also were the ones who engaged in more practice. So, practice is important to becoming an expert. However, it is unclear whether there is also some innate characteristic such as motivation, interest or talent that could also be driving those people to practise more. Regardless, if you want to become highly skilled at something, your elementary school clarinet teacher was right – you really do need to practise.

Summary: Problem-solving

11.1 Studying problem-solving

- Unlike many topics of study in cognitive psychology, problem-solving is slower and more deliberative. As such, verbal protocols of what people are thinking, which are normally actively avoided, can be used to gain insight into the steps that people take to arrive at a solution.
- Cognitive psychology often studies how people solve 'recreational' problems, simple brain teasers, as a way of understanding problem-solving. A common kind of data collected is the verbal protocol, a transcription of the person's verbalisations as the problem is being solved.

11.2 Basics of problem-solving

- We are solving a problem when our behaviour is goal-directed and involves a sequence of cognitive steps or stages. The sequence involves separate cognitive operations, where each goal or subgoal can be decomposed into separate, smaller subgoals. The overall problem, including our knowledge, is called the problem space, within which we use lookahead, apply operators, draw inferences and conduct a constructive search for moves that bring us closer to the goal.

11.3 Gestalt psychology and problem-solving

- The early Gestalt psychologists studied problem-solving and discovered two major barriers to successful performance: functional fixedness and negative set. Köhler also studied chimpanzees and found evidence for insight during problem-solving.

11.4 Insight and analogy

- Insight is a deep understanding of a situation or problem, often thought to occur suddenly and without warning. Although there is some debate as to the nature of insight, insights may be discovered and used unconsciously and only later be available to consciousness.

- Reasoning by analogy is a complex kind of problem-solving in which relationships in one situation are mapped onto another. People are better at developing analogies if given a useful source problem and an explicit hint that the problem might be used in solving a target problem. Holyoak and Thagard's multiconstraint theory of analogical problem-solving claims that we work under three constraints as we develop analogies: constraints related to the similarity of the source and target domains, the structure of the problems, and our purposes or goals in developing the analogies.
- Some evidence suggests a particularly important role for the left frontal and parietal lobes in solving problems by analogy and a right hemisphere role in insight problems involving semantic priming.

11.5 Means–end analysis

- The best-known heuristic for problem-solving is means–end analysis, in which the problem-solver cycles between determining the difference between the current state and the goal or subgoal state and applying legal operators to reduce that difference. The importance of subgoals is revealed most clearly in problems such as the Tower of Hanoi.
- Newell and Simon's general problem-solver (GPS) was the earliest cognitive theory of problem-solving, implemented as a computer simulation. Studying GPS and comparing its performance with human problem-solving shows the importance of means–end analysis.

11.6 Improving your problem-solving

- The set of recommendations for improving your problem-solving includes increasing your knowledge of the domain, automaticity of components in problem-solving, developing and following a plan, and not becoming anxious. Several special-purpose heuristics are also listed, including the mundane yet important advice about practice.

Chapter 11 quiz

Question 1

Within the definition of problem solving, how many aspects does this have?

A. 3
B. 4
C. 2
D. 1

Question 2

In early work by Köhler (1927), it was suggested that chimpanzees can present problem solving behaviour. Why was this the case?

A. The chimpanzees demonstrated that they could use long and short sticks to see which of them could reach a banana.
B. The chimpanzees demonstrated that they could work as a team to reach the bananas that were outside of the cage.
C. The chimpanzees did not understand how to use the sticks to reach the bananas and they had to be shown.
D. The chimpanzees were seen to already have knowledge of what the sticks were for, therefore they knew how to use these to collect the bananas.

Question 3

You are provided with 2 paired words of 'BIKE' and 'RIDE'. You are then presented with another word, 'CAR', and you have to decide upon a word to be paired this. What can you use to help with this problem?

A. Means-end analysis
B. An analogy
C. An insight
D. A sub-goal.

Question 4

The general problem solver (GPS) was a computer simulation designed to simulate problem solving behaviour. What was a key issue with this?

A. The GPS gave multiple solutions when only one was needed.
B. The GPS gave the most complicated solution.
C. The GPS did not always provide a clear solution even though it was designed to.
D. The GPS was able to include more sub-goals but this was not always the quickest solution.

Question 5

Problem solving can be improved by:

A. Drawing inferences and developing sub-goals.
B. Having no systematic plan
C. Always following steps forward
D. Select they key parts of a problem

12 Social cognition

Learning objectives

12.1 Identify the brain regions involved in social cognition

12.2 Discuss the cognitive and neural processes involved in self-referential processing

12.3 Describe the cognitive and neural aspects of mentalising and empathy

12.4 Summarise neurocognitive models of social rejection and loneliness

Key figures

Professor Simone Shamay-Tsoory

Professor Simone Shamay-Tsoory, Professor of Neuroscience, University of Haifa, Israel. Her work has elucidated the neural pathways underlying important social cognitive processes, such as theory of mind and empathy.

Professor Naomi Eisenberger

Professor Naomi Eisenberger, Professor of Social Psychology and Director at the Social & Affective Neuroscience Lab, UCLA. Her research has uncovered the neural aspects of social rejection.

Introduction

> Social species, by definition, form organizations that extend beyond the individual. These structures evolved hand in hand with behavioural, neural, hormonal, cellular, and genetic mechanisms to support them because the consequent social behaviours helped these organisms survive, reproduce, and care for offspring sufficiently long that they too reproduced, thereby ensuring their genetic legacy. (Cacioppo et al., 2011, p. 17[1])

In the award-winning Korean drama series *Squid Game*, different people experiencing financial hardship participate in a secret contest involving a series of children's games. Winners will be awarded a large financial prize, while losers will die. Of interest to this chapter is that, in situations like those depicted in the *Squid Game*, contestants' survival relies on their profound ability to mentally represent and infer other people's intentions and mental states, as well as the ability to *track changes* in others' mental states and intentions, as today's friend can become tomorrow's foe. Absent the mortal threat embedded in *Squid Game*, everyday social

Understanding others is important for survival in *Squid Game*.
Source: FlixPix/Alamy Stock Photo

interactions also draw on our ability to represent and make attributions about other people's (unobservable) mental states and, accordingly, regulate and adapt our behaviour (e.g. from how you respond to an aggressive driver who shouts at you while cycling, to how you respond to a work colleague asking you out for dinner).

social neuroscience
Interdisciplinary field of research that investigates the roles of neural and biological processes in social behaviour.

The opening quote by the late John Cacioppo, the founder of **social neuroscience**, invites us to think about cognition as instrumental to something bigger than just information-processing capacity: survival. In fact, this is what the cognitive processes we discussed so far *are for*. Nevertheless, for social species, such as humans, survival means more than the capacity to reason and engage in complex problem-solving. It also requires the ability to think intelligently about the self *and* others: how other people in our social surrounding think and feel, and whether their intentions towards us are malevolent or friendly. The cognitive processes that underlie this ability and enable adaptive social function are the subject area of **social cognition**.

social cognition
Social cognition is concerned with how we perceive, store, process and use information from our social environment, and how we plan and execute actions based on such information.

Integral to this process is the activation of mental representations about the self (our self-concept, identity, goals and intentions) and others. Once thought to be a subfield of social psychology, the emergence of interdisciplinary research in the field of human cognition over the last 25 years (e.g. cognitive neuroscience: the study of biological and neural processes underlying cognition; affective neuroscience: the study of biological and neural processes underlying emotion; social cognitive neuroscience: the study of social behaviour using cognitive neuroscience methods) has made it apparent that social cognition is much broader and entails the study of automatic and deliberate processes that tap onto different cognitive systems, from perception, memory and attention to learning and decision-making. This chapter presents empirical evidence and theories that are relevant to two main functions or areas of social cognition: (a) how we understand ourselves and (b) how we understand others. The former is relevant to how we come to understand ourselves through self-evaluation and reflection upon our core traits and attributes (i.e. self-referential processing). The latter is relevant to social cognitive processes that allow us to engage in adaptive social functioning. These involve mentalising (understanding other people's *mental* states) and empathy (understanding other people's *affective* states). Before delving into these topics, though, we start off our discussion by presenting the neural basis of social cognition. Understanding ourselves and other people in our social milieus is important for adaptively navigating and interacting with our social world (Lou, Changeux & Rosenstad, 2017[2]). Before delving into these two important topics, however, we start off our discussion by identifying the neural basis of social cognition.

12.1 The neural basis of social cognition

LEARNING OBJECTIVE
Identify the brain regions involved in social cognition

Mapping the brain regions implicated in social cognitive processes can be challenging, mainly because different brain regions are implicated in different social cognitive processes, a single process can involve a distributed neural network (i.e. this process activates different neural areas and is not localized in a particular cortical area), and because our understanding of the neural basis of social cognition is still evolving. Table 12.1 presents the brain regions involved in different social cognitive processes that are relevant to understanding the self (meta-cognition and self-evaluation) and understanding others (perceiving social rewards, engaging in imitative behaviour and displaying empathy).

Table 12.1 presents different brain regions associated with social cognitive processes that reflect self-understanding and other-understanding. The former is

Table 12.1

	Understanding the self		Understanding others		
	Metacognition	*Self-processing*	*Social rewards*	*Imitative behaviour*	*Empathy*
Medial prefrontal cortex (mPFC)	X			X	X
Dorsomedial prefrontal cortex (dmPFC)		X	X		X
Ventromedial prefrontal cortex (vmPFC)		X	X		
Anterior prefrontal cortex (aPFC)	X	X			
Dorsolateral prefrontal cortex (dlPFC)				X	
Orbitofrontal cortex (OFC)					X
Medial orbitofrontal cortex (mOFC)			X		X
Temporo-parietal junction (TPJ)				X	X
Superior Temporal Sulcus (STS)					X
Posterior superior temporal sulcus (pSTS)				X	
Inferior frontal gyrus (IFG)				X	X
Anterior cingulate cortex (ACC)		X		X	X
Doral mid-anterior cingulate cortex (dmACC)					X
Inferior parietal lobe (IPL)					X
Ventral striatum (VS)	X		X		X
Anterior insula (AI)		X			X
Posterior insula (PI)					X
Amygdala					X

concerned with such processes as metacognition and self-referential processing (excluding self-referential processing of verbal and motor behaviour). The latter is concerned with processes that serve adaptive social functioning and involve a large network of neural regions and distributed processes (e.g. social reward learning, imitative behaviour, empathy). It is noteworthy that this is not an exhaustive mapping of brain regions implicated in the social cognitive processes described, but rather it aims to provide a snapshot of related evidence from cognitive, affective and social neuroscience. The key takeaway is that different brain regions, from subcortical structures to regions of the prefrontal and medial cortex, are implicated in different social cognitive processes. While some brain regions are selectively activated in specific processes (e.g. processing of self-referential stimuli), others may serve more than one process (e.g. overlapping activations for the processing of both self-referential and social stimuli unrelated to the self). For example, Herold et al. (2016)[3] demonstrated that the orbitofrontal cortex (OFC) and the precuneus were activated in the processing of both self-referential and social stimuli.

In the sections that follow you will see that social cognition is associated with activation in the prefrontal and medial cortices, as well as subcortical structures (e.g. the amygdala). It would be helpful to remember these areas as parts of the broader neural systems they belong to, namely:

- **Default mode network (DMN):** includes areas in the frontal, temporal and parietal cortex, and shows higher activation tasks involving daydreaming, mind wandering, semantic and episodic memory (Figure 12.1). The DMN is also activated when we engage in introspection and self-referential thinking.
- **Sensorimotor network (SMN):** includes the supplementary motor areas and the precentral and postcentral gyrus (Figure 12.2). The SMN is primarily activated

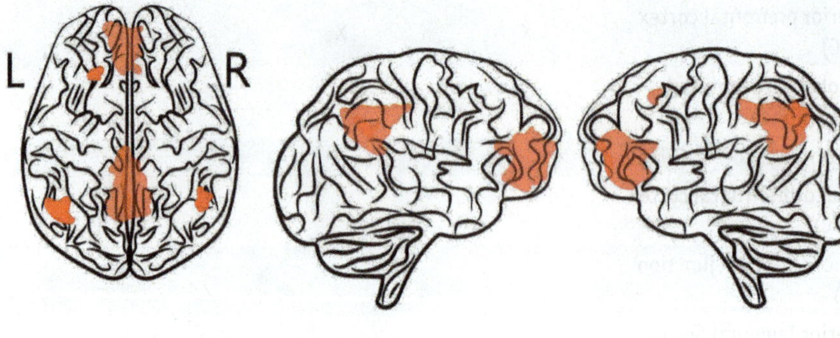

Figure 12.1 The default mode network
Source: Tabish Pathan/Alamy Stock Photo

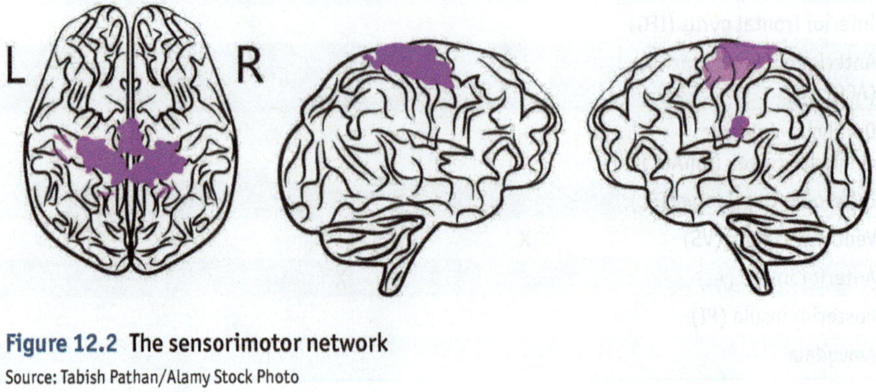

Figure 12.2 The sensorimotor network
Source: Tabish Pathan/Alamy Stock Photo

in tasks involving sensorimotor tasks (e.g. movement coordination), but it is also involved in self-referential processing (e.g. agency of action).

- **The salience network (SN):** includes the anterior insula (AI) and the dorsal anterior cingulate cortex (dACC). It also involves activation in subcortical structures, such as the ventral striatum, the amygdala and midbrain regions. The SN is involved in self-awareness, social behaviour and the processing of socially threatening stimuli (e.g. social exclusion).

12.2 Understanding the self

LEARNING OBJECTIVE
Discuss the cognitive and neural processes involved in self-referential processing

The concept of the 'self' has been the subject of philosophical and scientific inquiry for centuries (if not for millennia). Of relevance to social cognition and related research on the cognitive and neural underpinnings of the 'self' is the understanding that the self is represented at different levels of experience, with each level implicating more than one brain regions and neural networks. As Table 12.2 illustrates, the self can be represented and studied in relation to bodily movements and motor action, mental experiences and higher-order processes – and other levels of explanation or domains can be added to the list. As Northoff et al. (2006)[4] pointed out, a common thread running through the different levels of explanation and domains of the self, is self-referential processing: our ability to process stimuli as related to ourselves and unrelated to others. Let's begin our inquiry into self-referential processing by looking into how brain injury can impair our ability to refer to our experiences and traits. It is noteworthy that the study of brain injury helps us better understand typical function in participants without brain injury. In turn, this allows cognitive scientists to develop experiments and investigations into typical function which can then help in understanding atypical function in the absence of brain injury.

Table 12.2 Localisation of the self in the human brain cortex (adapted by Northoff et al., 2006).

Cortical areas	Processes	Domain of the 'self'
Sensory cortex	Sensorimotor processing	Bodily or motor
Medial cortex	Self-referential processing	Experiential or mental
Lateral cortex	Higher-order processing	Autobiographical, emotional, verbal, etc.

12.2.1 The remembered self

self-understanding
Knowledge about our own characteristics, traits, motivations, strengths and weaknesses.

self-reflection
Our ability to introspect, interpret and evaluate our thoughts, emotions and actions.

How do you know if you are a kind or empathetic person? How do you know that you are not an evil person guided only by self-interest? Being able to answer these questions would indicate a good level of **self-understanding**. Furthermore, performing self-evaluations can be based either on introspection and **self-reflection**, or on external information, such as other people's evaluations (Crone & Fuligni, 2020[5]). According to Klein and Loftus (2015)[6] at a cognitive level, self-evaluation capacity requires the activation of mental representations about the self and accessing self-referential memories (e.g. how often you have expressed empathy for another person's misfortune?). Episodic and semantic memory are implicated in respectively retrieving specific or more abstract and generalised self-referential information, regardless of whether this originates in self-reflection or external sources, such as other people's evaluations of ourselves. Nevertheless, as the following case studies indicate, the exact role played by semantic and episodic memory in self-evaluations remains unclear.

For example, let's examine patient W.J., a young female undergraduate student who sustained retrograde amnesia after sustaining traumatic brain injury (Klein, Loftus & Kihlstrom, 1996[7]). Although W.J. could not recall any personal event in the 6 months preceding her injury (indicating deficits in episodic memory), she retained intact memory for more general facts, such as the classes she had attended at college during this period (which involves semantic memory). The researchers investigated whether W.J. could recall self-referential knowledge during and after her amnesia was resolved. To do so, W.J. completed a series of tasks, including a self-rating personality test. The personality self-ratings took place shortly after W.J. sustained her injury and four weeks later, when her amnesia was resolved. The researchers found that the personality ratings completed between the two periods were very similar, suggesting intact capacity to make reliable self-evaluations. They concluded that semantic and episodic memory differentially affected self-referential knowledge. Self-referential knowledge was accessible because semantic memory was intact, despite profound episodic memory impairment.

Klein, Rozendal, and Cosmides (2002)[8] further explored the role of semantic and episodic memory in self-referential knowledge. They examined patient D.B, an older adult suffering from impairments in both episodic and semantic memory following anoxia (i.e. reduced oxygen supply to the brain) induced by heart attack while he was playing basketball. D.B. had severe retrograde and anterograde amnesia, being unable to recall specific events from his life prior to and after his heart attack. Semantic knowledge was also significantly impaired, with D.B. being able to recall only partial general facts about his life (e.g. he could tell the place but not the date of his birth). The researchers gave D.B. a personality traits questionnaire to complete to indicate which traits he thought he possessed. The questionnaire was completed twice, a week apart between the baseline and follow-up measure. The researchers also gave a similar questionnaire to his daughter to complete, indicating which personality traits she thought her father possessed. The responses of D.B. and his daughter were then compared to a match control pair, that is, two neurologically healthy older adults and their children who were approximately the same age as D.B.'s daughter. Interestingly, the consistency shown in the self-reported personality traits of D.B. (as indicated by Pearson's ($r = 0.69$) was similar to that reported by the healthy older adults ($r = 74$). That means that both D.B. and the neurologically healthy participants were fairly consistent in how they described their personality traits across the two measures. Accordingly, D.B.'s self-reported personality trait ratings were significantly associated with those given by his daughter ($r = .64$), and the same pattern emerged for the healthy adults and their children's ratings ($r = .62$).

The researchers then reversed the process. They asked D.B. (and the matched healthy older adults) to give ratings of their children's personality. Accordingly, they asked the children of D.B. and the healthy adults to complete a self-rating of personality traits. The results were striking. The correlation between D.B.'s rating and his daughters' rating of her personality was low and non-significant ($r = .23$, $p > 0.05$). On the other hand, the correlation between the ratings given by the healthy older adults and their children's ratings was high and statistically significant ($r = .61$, $p < 0.05$). The researchers also revealed that while he was completing his own personality rating D.B. faced no emotional or cognitive turmoil, although this was not the case when he completed the personality rating for his daughter.

Klein et al. (2002)[9] concluded that the debilitating brain damage sustained by D.B. and the resulting impairments in episodic and semantic memory did not affect his ability to recall self-referential knowledge, although it affected his ability to recall knowledge about well-known others (i.e., his daughter). Further analysis of neuropsychological evidence led to the conclusion that trait self-knowledge involves different neural systems, which, although they may be functionally independent, still interact (Klein & Lax, 2010[10]). For example, the processing of self-referential personality traits has been related to activation in several brain areas, such as the dorsomedial PFC, the premotor cortex, the caudate and the cerebellum (Fossati et al., 2004[11]). This can explain why self-knowledge about one's own personality is preserved when brain injury and cognitive impairment severely affect other areas of self-knowledge, such as episodic memory and semantic memory.

Are your personal memories . . . yours?

In the 2005 film *The Island*, Lincoln 6 Echo (played by Scottish actor Ewan McGregor) is tormented by dreams and invading memories of a past life he could not fully remember. Living in a sterilised futuristic colony with little recollection of his past life, he dreams of luxury boats and fast cars – experiences he felt he had had but did not somehow own. Can there be such a dissociation between lived past experiences and a feeling of ownership? Research in patients with amnesia tells us that it is possible. Klein and Nicholls (2012)[12] reported the case of R.B., an electronic design engineer in his early 40s who sustained severe brain injury following an accident while he was riding his bike. While hospitalised, R.B. displayed a wide range of cognitive deficits, such as mild aphasia (language impairment induced by brain injury), amnesia and difficulty maintaining attention. After returning home, his episodic memory seemed intact. He had no difficulty recalling particular incidents from his life and other self-referential knowledge. Interestingly, though, he felt that he did not *own* these memories, as if they belonged another person.

For example, R.B. could recall himself as a small child at the beach with his family, but in recalling these memories he felt as if he was looking at the pictures of somebody else's summer holidays. The researchers argued that knowing ourselves does not only rely on the ability to recall self-referential, episodic, memories. It also requires a sense of personal agency to emerge so that the recalled memories feel 'owned'. This implies that the psychological construct of the *self* and, more specifically, the ways we come to understand ourselves encompass more cognitive processes and systems than just episodic memory. As the following discussion indicates, this is the case from studies using neuroimaging methods with healthy populations.

12.2.2 The multidimensional self

While cognitive neuropsychology research is important for revealing the cognitive processes and systems that play a role in understanding the self, a look into the neural basis of the self can also help us better understand if the self reflects a single, unitary construct, or if it has different dimensions (multidimensional self), and if this distinction is reflected at structural (e.g. are different brain areas involved?) or functional levels (e.g. how different brain areas are connected to and interact with each other?). Of particular interest is the function of the medial frontal cortex (MFC), a brain region involved in a wide range of psychological processes, from cognitive control to self-evaluations. A meta-analysis of approximately 10,000 fMRI studies indicated that the anterior part of the MFC was associated with episodic memory and that the dorsomedial prefrontal cortex (dmPFC) – a sub-region of the MFC – was strongly associated with self-referential thought (de la Vega et al., 2016[13]). This indicates the central role of the MFC in social cognition, and especially in the development and understanding of the *self*.

Further research uncovered the brain regions implicated in the processing of the individual and the social self. For clarity, by *individual self* we mean the sense of self originating in introspection and reflection, and by *social self* we refer to the sense of self stemming from social interactions and/or the processing of external evaluations of ourselves (e.g. what others think of us). Van de Groep et al. (2021)[14] had participants complete an internal and an external self-evaluation task. In the internal self-evaluation task, participants were asked to indicate whether positive and negative traits shown on a screen were characteristic of them. In the external self-evaluation task, participants received positive, negative or neutral feedback by supposedly unknown same-age peers. For the internal self-evaluation task, the results showed that participants displayed a self-serving bias, reporting that they possessed more positive (than negative) traits. Accordingly, the medial PFC was significantly more active for the positive than the negative traits. Furthermore, for the external self-evaluation task, the anterior medial PFC region was activated when both internal (i.e. positive trait self-ratings) and external (i.e. positive feedback from others) positive self-evaluations were presented. These findings suggest that while some PFC regions are particularly responsive to self-evaluations resulting from introspection (e.g. dmPFC), other regions (e.g. aPFC) are activated by self-evaluations stemming from both individually and socially inferred aspects of the self.

In considering the multidimensionality of the self and its neural correlates, another important distinction is between intrinsic and extrinsic self-processing. Whereas **intrinsic self-processing** is concerned with personally relevant information processing (e.g. perceiving my ideas and thoughts as mine), **extrinsic self-processing** is concerned with the sense of personal agency over one's thoughts, actions and their consequences (e.g. I intentionally chose my actions, which result in certain consequences).

intrinsic self-processing
The perception of personally relevant information, and is conceptually closer to *identity*.

extrinsic self-processing
The perception of agency over one's thoughts, actions and their consequences.

Which network is associated with intrinsic self-professing and extrinsic self-processing?

Intrinsic self-processing has been associated with activation of brain regions of the default mode network (DMN; Figure 12.1), and extrinsic self-processing has been associated with the activation of the sensorimotor network (SMN; Figure 12.2) (Ebisch & Aleman, 2016[15]).

DiPlinio, Perrucci, Aleman and Ebisch (2020)[16] investigated the differential effects of intrinsic and extrinsic self-processing on brain activation by manipulating perceived identity and agency. While placed in an fMRI scanner, participants listened to either words they had personally selected (high agency) or pseudo-words that they did not select (low agency), thereby reflecting extrinsic self-processing. Furthermore, the words were presented either in the voice of the participants (high identity) or in the voice of an unknown other (low identity), thereby reflecting extrinsic self-processing. In the high-identity condition, there was greater activation of the ventromedial PFC (vmPFC), whereas the high-agency condition corresponded to activation of areas associated with action planning and performance (e.g. supplementary motor cortex, premotor cortex, intraparietal sulcus and the cerebellum). These findings further support the distinction between intrinsic and extrinsic self-processing modes and identify the corresponding anatomical and functional features of each mode. Of importance to our discussion is the fact that the 'self' seems to employ different brain regions depending on the task at hand, with introspection processes (e.g. self-monitoring) involving the DMN (e.g. vmPFC) regions and action-oriented processes involving regions of the SMN.

12.3 Understanding others

LEARNING OBJECTIVE
Describe the cognitive and neural aspects of mentalising and empathy

In this section, we discuss two core social cognitive processes that are central to our social understanding ability: mentalising and empathising. Mentalising (also known as **Theory of Mind/ToM**) represents the ability to cognitively infer and attribute mental states to other people, such as beliefs, desires, emotions and intentions, in order to be able to understand and predict their behaviour (Frith & Frith, 2010[17]). Empathising (or, simply, empathy), on the other hand, reflects our ability to understand and respond to other people's emotional states. Although sharing common features with mentalising ability, we shall see that empathy represents a distinct social cognitive process.

12.3.1 Mentalising: understanding how other people think and feel

Gallagher and Frith (2003)[18] posited three key prerequisites for mentalising:

- Recognise that other people have goals and that their behaviour is guided by such goals.
- Recognise that other people's perspective and worldviews can be different from our own.
- Understanding other people's behaviour requires that we consider both their perspective and ours – suggesting a distinction between self-other inferences.

A large body of research has shown that mentalising has two dissociable components: cognitive and affective. The former is involved in processing and inferring other people's beliefs and intentions, whereas the latter is relevant to recognising others' emotions (e.g. the emotional tone of sarcasm; Sebastian et al., 2012).[19] According to the neurobiological model of ToM proposed by Abu-Akel and Shamay-Tsoory (2011),[20] attention selection neural systems (e.g. temporoparietal junction or TPJ) overlap anatomically and functionally with neural systems involved in ToM processes. Mental representations about the self and others are hypothesised to originate in the TPJ and then relayed to more specialised brain areas for processing (i.e. cognitive or affective ToM), through the dorsal attentional system (precuneus and posterior cingulate cortex) for self-relevant processing, or the ventral attentional system (superior temporal sulcus) for other-relevant processing. Table 12.3 summarises the brain areas involved in cognitive (cognitive execution loop) and affective ToM (affective execution loop).

Table 12.3 Neural networks activated during cognitive and affective ToM processing, based on the neurobiological model by Abu-Akel and Shamay-Tsoory (2011)[21]

Cognitive ToM	Affective ToM
Dorsal striatum	Amygdala
Dorsal temporal pole (dTP)	Ventral striatum
Dorsal anterior cingulate cortex (dACC)	Ventral temporal pole (vTP)
Dorsal medial prefrontal cortex (dmPFC)	Ventral anterior cingulate cortex (vACC)
Dorsal lateral prefrontal cortex (dlPFC)	Orbitofrontal cortex (OFC)
	Ventral medial prefrontal cortex (vmPFC)
	Inferolateral frontal cortex (IFLC)

Executive functions (e.g. working memory updating, attention shifting and inhibitory control) are relevant to mentalising processes because both mentalising and executive functions develop rapidly during the first five years of life, and deficits in one co-occurs with deficits in the other, as indicated by research in neurodevelopmental disorders. Wade et al. (2018)[22] performed a meta-analysis of studies presenting neuroimaging and behavioural results about executive functions and mentalising and their temporal association (e.g. whether executive functions precede or follow the development of mentalising abilities, or if both mentalising and executive function rely on overlapping and commonly shared networks in the brain). One of the key findings was that ToM processing and executive functions share a common neurodevelopmental trajectory, with both being rapidly developed during the first five years of life. Furthermore, although executive functions and ToM processing involve distinct brain areas, certain tasks that involve both ToM processing and executive function activity (e.g. cognitive control) co-activate common brain regions (e.g. the vmPFC and the IPL). A longitudinal study of children aged 6–11 years showed further that working memory and attention shifting, but not inhibitory control, predicted ToM processing 12 months later (Austin et al., 2014[23]).

Yeh et al. (2015)[24] provided further evidence about the association between executive functions and ToM processing by investigating patients with PFC damage. The researchers compared the performance of patients with vmPFC and dlPFC damage with a group of healthy participants in a series of tests, including the Mini Mental State Examination and tasks related to ToM processing, executive functions and empathy. The results showed that the patients with PFC damage demonstrated deficits in affective ToM, compared with the healthy participants. Accordingly, deficits in affective ToM were significantly associated with impairments in executive functions in patients with PFC damage – similar findings have been reported in the past, with at least 25 per cent of the deficits in mentalising among patients with dementia being attributed to executive function impairments (Rankin et al., 2005[25]).

Research examining the association between mentalising and executive functions in healthy adults has focused on inhibitory control and working memory. Inhibiting dominant responses, such as foregoing immediate personal gratification for the sake of higher-order, group-level rewards (i.e. rewards serving other people and not only the self), is a key aspect of self-regulation and facilitates adaptive social behaviour. As such, aspects of mentalising ability (e.g. understanding others' desires and goals and regulating individual behaviour accordingly) should be related to inhibitory control. Research has partly supported this association, however. Launay et al. (2015)[26] had participants complete a mentalising measure and four tasks of inhibitory control. To assess mentalising, participants were asked to read short stories involving the actions and interactions of fictional characters and provide answers related to understanding the fictional characters' mental states. The inhibitory control tasks included a go/no go task measuring response inhibition; a typical Stroop task assessing cognitive inhibition, where participants had to respond to the name of the words displayed in different ink colours; an automatic inhibition task involving motor movement in response to visual cues displayed on a screen (e.g. moving either the first or the second finger depending on the numbers presented on the screen); and a temporal discounting task assessing the tendency to choose smaller short-term over larger long-term rewards. Mentalising was unrelated to performance in three out of the four inhibitory control tasks. Only temporal discounting was significantly associated with scores in the mentalising task. The researchers suggested that inhibitory control, as reflected by the temporal discounting task, may partly share common underlying cognitive processes with mentalising: the ability to understand and respond adaptively to other people's mental states may occasionally require the ability to forego short-term immediate gratification to fulfil longer-term collective goals.

Other research further examined the specific role of working memory in mentalising. The Social Brain Hypothesis (Dunbar, 2009[27]) suggests that primates have developed large heads compared to the size of their bodies because of the evolutionary demand to process large amounts of complex social information. Stemming from this hypothesis, Meyer and Lieberman (2012)[28] posited that, whenever we find ourselves in the company of others, a special kind of working memory (dubbed '*social working memory*') helps us keep track and remember the vast amount of incoming social information, such as friends' perspectives over an issue (e.g. how worried we should be about climate change), their traits (e.g. being argumentative and opinionated) and their political orientation. Meyer et al. (2012)[29] developed a social working memory task to identify the brain areas activated when people engage in social information processing. While placed in an fMRI scanner, participants were presented with the names of their close friends (encoding phase) on a screen and asked to read them. This task varied the number of friends presented: two, three or four friends' names. This was followed by the presentation of a given trait word (e.g. funny). During the delay phase, participants were asked to mentally rank the friends who were presented on the screen based on the trait just shown (e.g. from the most to the least funny). Lastly, in the retrieval phase, participants were presented with a probe statement about the friends' ranking on the trait (e.g. is Rebecca the second funniest?), and indicated whether the statement was true or false. The researchers found that during the social working memory task, the brain areas involved in **domain-general**, working memory (e.g. dlPFC, supplementary motor area) were activated. Nevertheless, the brain areas involved in mentalising (e.g. dorsomedial PFC, posterior cingulate cortex, and TPJ) were also activated, suggesting that social working memory involves distinct neural pathways that are not activated by non-social working memory tasks. Taken together, the extant research seems to support an association between mentalising ability and executive functions, especially working memory and certain types of inhibitory control, that are relevant to regulating social behaviour and interactions.

12.3.2 Empathising: understanding (and *feeling*) how other people feel

Empathy is a psychological construct that can be described both as a trait and as an ability that can be learned and improved through practice. In a general sense, empathy reflects our responses to the observed experiences of others (e.g. a person experiencing pain while getting a vaccine). Although the dimensionality of empathy (e.g. whether empathy is a unitary or a multidimensional construct) has been debated, there is converging evidence supporting the distinction between cognitive and affective (or emotional) empathy. Cognitive empathy involves perspective-taking, reflecting upon another person's experience and understanding how they feel. Cognitive Theory of Mind (ToM) or mentalising, therefore, appears to share common characteristics with cognitive empathy, and some studies use the terms 'cognitive empathy', 'mentalising' and 'cognitive Theory of Mind' interchangeably. Affective empathy, on the other hand, is more intuitive and less reflective and involves sharing emotional experiences, as in emotional contagion (e.g. the quick spread of emotions between persons or groups of people). A defining feature of affective empathy is the display of 'appropriate' emotional responses to other people's emotional states or experiences. In this sense, 'appropriateness' reflects the congruence between our emotional states and the emotional states of people we observe. When our emotional responses are incongruent or 'inappropriate', such as displaying joy at someone else's sorrow, this is not a manifestation of affective empathy, but may rather indicate callousness or psychopathic traits (Rueda et al., 2015[30]).

Studies using self-reported empathy measures across different cultures have shown that cognitive and affective empathy represent psychometrically distinct factors, and that females tend to score higher on both empathy facets than males, even more so in affective empathy (Anastacio et al., 2016; D'Ambrosio et al., 2009; Geng et al., 2012; Trentini et al., 2022[31]). Lower scores in cognitive and affective empathy are also associated with various forms of aggressive behaviour, including cyberbullying in adolescents (Del Rey et al., 2016[32]). Also, alcohol dependence and associated interpersonal problems are related to affective empathy deficits, but not to cognitive empathy (Maurage et al., 2011[33]). Accordingly, higher affective empathy is inversely associated with substance use problems, including alcohol and other drugs (Winters et al., 2021[34]). Further supporting the dissociation between cognitive and affective empathy, Rueda et al. (2015)[35] showed that people with autism spectrum disorders (ASD) display significantly lower cognitive empathy scores than individuals without ASD, but no such differences were evident in affective empathy.

Cognitive processes, such as executive functions, are implicated in empathy. Research with healthy young adolescents has shown that working memory, inhibition and shifting were positively associated with self-reported cognitive empathy, but only working memory was associated with self-reported affective empathy (Mairon et al., 2023[36]). A meta-analysis of 18 studies further showed that all three executive function subcomponents (working memory, inhibition and shifting) were significantly associated with cognitive empathy. However, affective empathy was significantly associated only with inhibition (Yan et al., 2020[37]). Taken together, these findings further support the idea that domain-general cognitive processes, such as working memory, attentional shifting and inhibitory control contribute to our ability to take perspective and understand other people's internal states. The empirical evidence is less conclusive about the contribution of executive functions to our ability to share other people's emotional experiences and respond to them using a matching emotional response.

Cognitive and affective empathy also have distinct neuroanatomical characteristics. Early lesion studies have showed that individuals with localised damage in the ventromedial cortex (e.g. orbitofrontal and/or ventromedial frontal lobe) displayed selective deficits in cognitive empathy and mentalising, while affective empathy was unaffected. On the other hand, individuals with inferior frontal gyrus damage (e.g. Broadman areas 44) displayed selective deficits in affective empathy (Shamay-Tsoory et al., 2009[38]). Research with patients who had suffered stroke also showed that affective empathy deficits were associated with infarction of the temporal lobe and anterior insula (Leigh et al., 2013[39]). A meta-analysis of fMRI studies on empathy (Fan et al., 2011[40]) showed that empathy is supported by a core empathy network that is activated by different types of stimuli and tasks, from those eliciting pain to those assessing empathic processing of pain inflicted on others. It is noteworthy that this meta-analysis conceptualised cognitive and affective empathy based on the level of awareness of the empathic task used in the analysed studies. For example, affective-perceptual empathy reflected empathic processing elicited automatically by experimental tasks where participants were not consciously aware of the experimental goal or purpose. By contrast, cognitive-evaluative empathy reflected empathic processing elicited by tasks where participants consciously attended to or evaluated the emotions of others. The core empathy network primarily includes the dorsal anterior cingulate cortex and the anterior middle cingulate cortex, the supplementary motor areas and the anterior insula, bilaterally. Tasks involving affective empathy tend to activate the right anterior insula, whereas tasks involving cognitive empathy tend to activate the anterior middle cingulate cortex.

Decety (2015)[41] provided a more comprehensive overview of the brain areas implicated in empathic processing, which included the following:

- Ventromedial prefrontal cortex (vmPFC)
- Anterior cingulate cortex (ACC)
- Somatosensory areas
- Insula
- Striatum
- Amygdala
- Hypothalamus
- Ventral tegmental area
- Brainstem.

12.3.3 Integrating social understanding networks and processes

Given the importance of mentalising and empathising in social understanding, efforts have been made, at theoretical and empirical levels, to provide an integrative approach to these two core social cognitive processes. At a theoretical level, Shamay-Tsoory et al. (2010)[42] provided an integrative model presenting sequential pathways linking cognitive with affective mentalising, taking into account the roles of cognitive and affective empathy (Figure 12.3). According to this model, cognitive mentalising is a prerequisite for affective mentalising. This means that being able to infer one's emotions (affective ToM) requires that we first have an understanding of that person's beliefs, desires and motivations (cognitive ToM). Affective ToM, in turn, requires that we have the ability to understand and share the emotional processes of others – that is, cognitive and affective empathy, respectively.

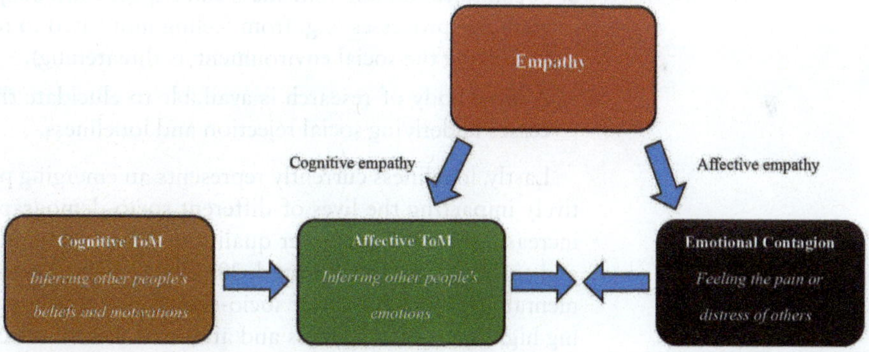

Figure 12.3 Integrative model of ToM and empathy
Source: Adapted by Shamay-Tsoory et al. (2010).[43]

This model has received partial support in empirical studies assessing the neural correlates of cognitive and affective ToM, and associations with self-reported cognitive and affective empathy. For example, Sebastian et al., (2012)[44] demonstrated that cognitive and affective ToM activated both distinct and shared brain areas, suggesting that at least some aspect of affective ToM requires cognitive ToM processing. Furthermore, Arioli et al. (2021)[45] investigated the distinct and overlapping brain areas activated by tasks requiring mentalising and empathy, by performing a meta-analysis of more than 100 studies using neuroimaging techniques (Arioli et al., 2021[46]). The results further corroborated the key tenets of the integrative model by Shamay-Tsoory et al. (2010)[47] showing that cognitive and affective ToM activated both shared and distinct brain areas, with affective cognitive ToM and empathic processing providing input to affective ToM. However, Arioli et al. (2021)[48] further argued that affective ToM is facilitated by a distributed neural network of interconnected brain areas and processes, not accounted for by Shamay-Tsoory et al. (2010).[49] A more comprehensive review and discussion of the extant neuroimaging research on ToM and empathy processes, as well as an alternative explanation about the hierarchical structuring of these processes and the corresponding neural networks that support them, is presented by Schurz et al. (2021).[50] One of the key conclusions to remember is that mentalising and empathising cannot be characterised as purely cognitive, neural or social. Nor can they be seen as processes that recruit distinct brain areas with singular inputs and outputs. Rather, mentalising and empathising represent complex social cognitive processes that implicate different cognitive functions and are facilitated by distributed neural networks with different levels of functional interconnectivity.

12.4 Responding to adverse social signals

LEARNING OBJECTIVE
Summarise neurocognitive models of social rejection and loneliness

The last part of this chapter is devoted to understanding how we respond to social signals, particularly those that indicate (potential) social threat or disruption in our social relationships. If social cognition serves the purpose of adaptive social functioning, then attendant social cognitive processes should be in place to motivate corrective action when the ties with our social world are threatened. We therefore focus our discussion on social rejection and loneliness for the following reasons:

1. Both social rejection and loneliness represent adverse social experiences with acute and long-term consequences on psychological well-being and social behaviour.
2. Social rejection and loneliness can trigger both adaptive and maladaptive social cognitive processes (e.g. from feeling motivated to restore social relationship, to perceiving the social environment as threatening).
3. A large body of research is available to elucidate the cognitive and neural processes underlying social rejection and loneliness.

Lastly, loneliness currently represents an emerging public health challenge negatively impacting the lives of different socio-demographic groups, by significantly increasing the risk for lower quality of life, mental and physical health problems, and early death (Holt-Lunstad, 2017[51]). Females, younger people, and people with mental disorders and lower socio-economic status are at higher risk for experiencing high levels of loneliness and attendant health effects (McQuaid et al., 2021[52]). Thus, better understanding the neuro-cognitive mechanisms underlying loneliness can accordingly inform the development of relevant interventions aimed at reducing its negative effects.

12.4.1 The neurocognitive basis of social rejection

Social rejection can be experienced in different ways: breaking up, unfriending (online and offline), being excluded from a social group at work, and so forth. For someone to feel socially excluded, they need to perceive that other people or social groups they were (or aspired to be) associated with have intentionally excluded them. Interestingly, social exclusion is not experienced only when we are shunned from valued social groups or relationships. Research tells us that even under conditions of minimal affiliation, social exclusion can be experienced at a profound level. In their seminal study, Eisenberger, Lieberman and Williams (2003)[53] had participants placed in an fMRI scanner while playing a virtual ball-tossing game (Cyberball) ostensibly with two other participants located in different rooms. In reality, this was a cover story – the actual participants of the study were playing Cyberball against a computer. The researchers designed three different conditions. In the inclusion condition, participants played Cyberball with the ostensive two 'other' players and the ball was tossed between all three of them for the entire duration of the trial. In the implicit exclusion condition, participants were told that they could not participate in the Cyberball game because of a technical glitch, but they were allowed to observe it. In the explicit exclusion condition, though, participants experienced exclusion shortly after the Cyberball game began. After receiving just seven throws at the beginning of the trial, the participants observed the two supposed other players tossing the ball between them but not passing it back to the participant. In other words, the participants felt excluded for the most part of the Cyberball game and were not given a logical explanation (e.g. technical glitch) for it. The feeling of social rejection could only be attributed to the intentions of the supposed other players and not to external forces (e.g. technical glitch). Self-reported measures indicated that participants in the explicit social exclusion condition indeed felt excluded and ignored by the 'other players'. In line with the researchers' hypothesis, explicit social rejection activated brain regions involved in the emotional component of physical pain (i.e. pain-induced emotional distress), namely the anterior cingulate cortex (ACC) and the right ventral PFC (rvPFC) (see Figure 12.4).

The researchers further argued that this evidence is indicative of 'social pain' – the pain experienced when we are being shunned – which, at least partly, shared

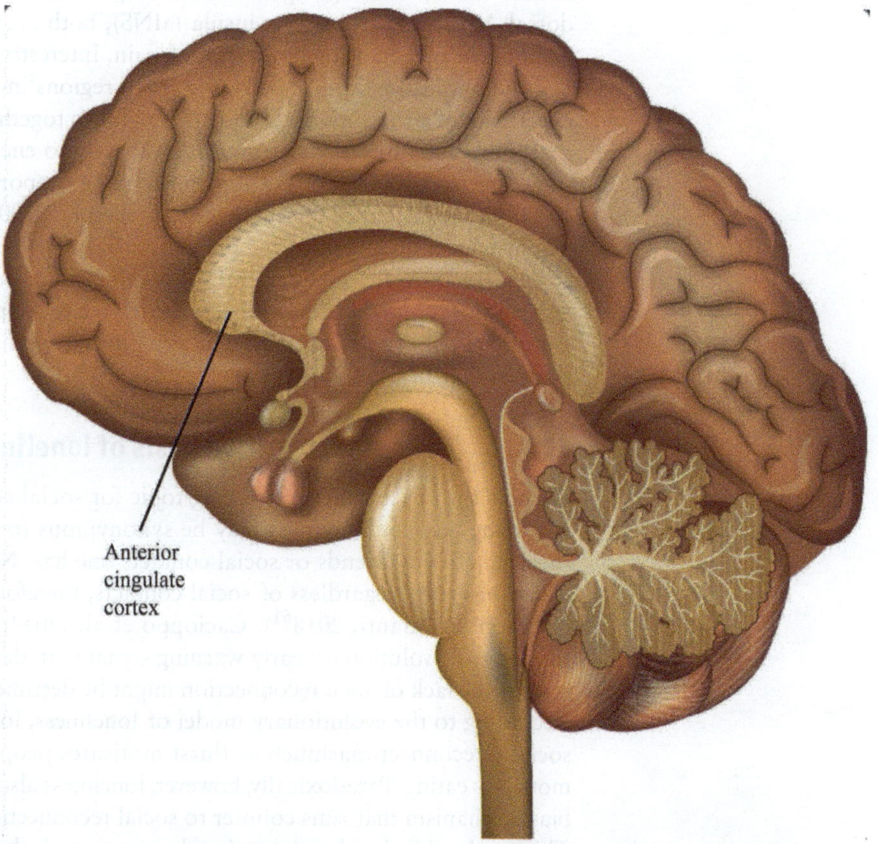

Figure 12.4 The anterior cingulate cortex (ACC) and the right ventral PFC (rvPFC).

Source: miha de/Shutterstock

common neuroanatomical features with physical pain (e.g. being stabbed; Eisenberger, 2015).[54]

Further research examined whether social pain activated brain regions associated with the sensory component of physical pain (e.g. as in being stubbed with a knife), such as the secondary somatosensory cortex (S2) and the dorsal posterior insula (dpINS). Kross et al. (2011)[55] argued that the early studies on social pain (e.g. Eisenberger et al., 2003[56]) relied on computerised tasks or used conditions that may not accurately resemble how people experience social pain in their daily social interactions. They further posited that experiencing an unwanted break-up in a romantic relationship could actually elicit activation in the somatosensory pain matrix in the brain, which is responsive to acutely noxious (painful) stimuli. To examine this hypothesis, Kross et al. (2011)[57] recruited participants who had recently experienced a painful and unwanted break-up in their romantic relationship. While placed in an fMRI scanner, participants completed two tasks, namely a social rejection and a physical pain task. In the social rejection task, participants were shown either pictures of their friends or pictures of their former partner. In the former, participants were asked to reflect upon the most recent positive experience they had with that person, but in the latter, they were asked to recall the painful break-up experience they had. In the physical pain task, participants were exposed to either noxious hot thermal stimulation (i.e. placing their forearm in hot water) or non-noxious thermal stimulation (i.e. placing their forearm in lukewarm water). At the end of each task, participants also completed self-reports indicating how distressing the task was (i.e. exposure to ex-partners vs. friends and to hot vs. lukewarm water). The results replicated early research on social pain, by indicating that the critical trials in both tasks (i.e. being exposed to the ex-partner and the hot water) activated the dorsal ACC and the anterior insula (aINS), both areas involved in the emotional component (i.e. emotional distress) of pain. Interestingly, both the social rejection and the physical pain task activated brain regions involved in the somatosensory experience of pain, such as S2 and dpINS. Taken together, these two studies indicate that the human brain has evolved the capacity to encode social pain in much the same way as it does for physical pain – lending support to the argument that being socially rejected 'hurts' (Eisenberger & Lieberman, 2004[58]). As expected, such novel and insightful findings elicited great enthusiasm, as well as critique (e.g. Cacioppo et al., 2013[59]). As Eisenberger (2015)[60] highlighted in summarising the research findings on this topic, there is compelling evidence to support the notion that physical and social pain share common neural pathways, perhaps as the result of adaptive changes that enable social survival.

12.4.2 The neurocognitive basis of loneliness

We now turn to another interesting topic for social cognition inquiry: loneliness. For many people, loneliness may be synonymous to social isolation, that is, the sheer number of friends or social contacts one has. Nevertheless, loneliness refers to *feeling* lonely regardless of social contacts, therefore reflecting 'perceived social isolation' (Ypsilanti, 2018[61]). Cacioppo et al. (2014)[62] suggested that loneliness serves as an evolutionary early warning signal that alerts us to reconnect with others, for the lack of such reconnection might be detrimental to our (social) survival. According to the evolutionary model of loneliness, loneliness motivates people to socially reconnect inasmuch as thirst motivates people to drink water or hunger motivates eating. Paradoxically, however, loneliness also spurs an implicit, automatic bias mechanism that runs counter to social reconnection (Spithoven et al., 2017[63]). This mechanism involves attentional processes, so that people with higher loneliness levels become hypervigilant to social threats and are more likely to perceive their social environments as more threatening (e.g. by orienting attention to socially threatening vs. positive signals). Such biases may lead to further social withdrawal, thereby perpetuating the loneliness experience (Qualter et al., 2013[64]).

Research has supported the function of attentional biases in loneliness. Bangee et al. (2014)[65] had university students watch footage depicting other young people having either positive (e.g. smiling, conversing in a friendly manner) or negative (e.g. being ignored by a group of same-age peers) social interactions. Using eye-tracking technology (see Chapter 14) the researchers captured participants' eye fixation patterns while they were watching the footage. The results showed that participants with significantly higher loneliness scores fixated more on the negative social interaction footage, as compared to the participants who did not experience high levels of loneliness (i.e. non-lonely). Accordingly, first fixation analysis showed that while non-lonely participants' visual attention was fixated to the positive social interactions, lonely participants were fixated to the socially threatening stimuli. Bangee and Qualter (2018)[66] further examined whether this attentional bias is generalised across negative social stimuli. Using eye-tracking they recorded the eye movements of participants while they were watching either facial expressions of positive or negative emotions (i.e. generalised stimuli) or visual scenes that depicted physical threat (e.g. violence), social threat (e.g. rejection), positive social interactions (e.g. social relationships) or neutral stimuli (e.g, a green field). The results showed that lonely participants displayed attentional bias to angry faces and socially threatening scenes depicting social rejection and did not differ from non-lonely participants in any other respect.

Studies using electrophysiological measures (EEG/ERP) have further established attentional biases in lonely individuals. Cacioppo et al. (2016)[67] asked participants to wear a head cap with electrodes that measured cortical electric activity while they viewed a series of images that varied in their depiction of social (vs. non-social) and threatening (vs. non-threatening) stimuli (see Figure 12.5 for the trial structure). Participants were also asked to press a button to provide valence ratings of the images

Figure 12.5 Trial structure
Source: Adapted by Cacioppo et al. (2016)[68].

as pleasant or unpleasant. The results showed that, compared to non-lonely participants, socially threatening stimuli (e.g. image of someone being ignored by others) evoked event-related brain microstates in lonely participants that involved the activation of brain regions associated with attention, self-presentation and threat, such as the PFC, dlPFC, associative visual cortex, the inferior and superior temporal gyrus, the insula and the amygdala.

In a variation of the aforementioned study, Cacioppo, Balogh and Cacioppo (2015)[69] used a social Stroop task where participants were asked to indicate the colour of the ink of each string of letters that depicted a socially threatening word (e.g. ALONE), a socially non-threatening word (e.g. LIKED), a positive emotional word (e.g. HAPPY) and a negative emotional word (e.g. SAD). While performing the Stroop task, participants were wearing electrodes that recorded their electrical brain activity. The results of the study showed that the brains of lonely participants differentiated negative social words (e.g. EXCLUDED) more quickly (approx. 200 ms) than the brains of non-lonely participants. Furthermore, compared to negative non-social words (e.g. MISERY), the negative social words evoked activation of brain regions involved in the orientation and executive control of visual attention. The findings by Cacioppo et al. (2015)[70] further supported the hypothesis that loneliness is more associated with hypervigilance (i.e. early allocation of attention) to negative social words than to negative non-social words.

Can we eliminate loneliness by tackling automatic cognitive processes?

People who experience higher levels of loneliness not only display higher attention allocation to potential social threats, but they also expect more negative outcomes from social relationships and perceive greater negative intentions in others (Qualter et al., 2013[71]). Such biases can be automatically activated and perpetuate the cycle of loneliness because they contribute to greater social withdrawal in people experiencing them.

Riddleston et al. (2023)[72] developed a cognitive bias modification task to help people with loneliness alter the ways they process and interpret social information. Participants were first asked to perform a short mental imagery training and were subsequently assigned to a control or an interventions group. Participants in the intervention group listened to 50 audio clips (in 5 blocks of 10 clips) describing different ambiguous social situations (e.g. 'Your friends are posting often on Instagram. Their pictures tend to make you feel . . . '). In each block, eight of the ambiguous situations were resolved in benign or positive way (e.g. by describing a positive social outcome), one remained unresolved, and one was resolved in a negative ways (e.g. by describing a negative social outcome). The audio clips were followed by prompt questions like 'Does looking at other people's posts on Instagram make you feel good?' and the binary response option yes/no. Participants responding 'yes' received 'correct' feedback and those responding 'no' received 'incorrect' feedback. Participants in the control group followed exactly the same procedure, but in all cases the social situations described by the audio clips remained ambiguous and were not resolved either positively or negatively. In other words, positive resolutions were reinforced in the social situations presented only in the intervention group. Participants in both groups completed a self-reported measure of perceived social threats in social situations and a loneliness questionnaire prior to (baseline) and after the intervention (post-training).

Pre- and post-training differences were observed, with participants in the intervention group reporting lower scores in social threat interpretations (i.e. reduced cognitive bias to social threats) and loneliness, and higher scores in positive interpretations of social situations.

Being the first study to examine cognitive bias modification in the context of loneliness, this research provides the foundation for future intervention development. How would you design such an intervention to tackle automatic/implicit attentional biases to social threats such as those described in this chapter? What methods would you use to assess the effectiveness of your intervention (behavioural or neuroimaging, or both)?

12.4.3 Reflections and considerations

The research discussed in this chapter lends support, at least partly, to the social brain hypothesis (Dunbar, 2009[73]). Our brains have evolved not only to serve self-referential processing, but also to support us in inferring other people's emotional and thinking states and, accordingly, to regulate our behaviour for adaptive social outcomes: bonding, friendships, long-term romantic relationships, and so on. In fact, social relationships appear to be highly demanding at a cognitive level, both for nurturing and for protecting them against threats. Our capacity to detect threats in our social relationships is profound. As the findings from social rejection and loneliness research indicate, being rejected is not just emotionally distressing, it is 'painful' – figuratively and, at least to a certain extent, literally. Our brains have developed to use the same neural networks that register physical pain to detect and respond to social pain that emerges from unwanted disruptions in our 'social body'. People who experience loneliness (i.e. unfulfilled social relationships) seem to be particularly attuned to social rejection – their attention is automatically drawn to cues in the environment that may signify risk of social rejection.

As a concluding remark, it is apparent that social cognition represents a complex process involving biological, neural, cognitive and social components. Cognitive psychology and its sister disciplines (e.g. cognitive neuroscience) play a key role in furthering our understanding of how and why our brains respond in the way they do to social stimuli, the self being an integral part of it.

Summary: Social cognition

12.1 The neural basis of social cognition

- Different brain regions are implicated in different social cognitive processes.
- A single process can involve a distributed neural network, and our understanding of the neural basis of social cognition is still evolving.
- Social cognitive processes focusing on self-referential thought and action involve regions of the medial frontal cortex (e.g. mPFC, dfPFC, vmPFC, aPFC), as well as the anterior cingulate cortex, the anterior insula and the ventral striatum.
- Social cognitive processes focusing on understanding other people's mental states involve a wide range of brain regions, including the medial frontal and the temporoparietal cortex, the inferior frontal gyrus, the anterior and posterior insula, and the amygdala.

12.2 Understanding the self

- Evidence from cognitive neuropsychology lends insights into the memory systems that support self-referential knowledge, such as self-evaluations (e.g. whether we possess favourable or unfavourable personality traits).

- Semantic and episodic memory differentially affect self-referential knowledge.
- In some patients with brain injury (e.g. W.J.), self-referential knowledge can be accessible when semantic memory is intact and episodic memory is impaired.
- Even when both episodic and semantic memory are intact, patients with brain injury may feel a lack of ownership of their personal memories.
- Self-referential processing can be distinguished between intrinsic and extrinsic. Intrinsic processing reflects identity, and extrinsic processing reflects agency.
- Neuroimaging studies indicate that intrinsic processing activates areas of the default mode network (DMN), while extrinsic processing is more action-oriented and activates areas in the sensorimotor network (SMN).

12.3 Understanding others

- Mentalising is a complex social cognitive process that allows us to infer other people's emotional, cognitive and motivational states.
- Mentalising ability and executive functions share common developmental trajectory and, partly, common brain regions.
- Research on the social working memory indicates that mentalising requires short-term memory processing of social information that is distinct from the domain-general, canonical working memory.
- Cognitive empathy reflects perspective-taking and understanding other people's affective states.
- Affective empathy reflects emotional sharing and the ability to display matching emotions in response to other people's emotional states.
- Cognitive and affective empathy involve distinct neuroanatomical pathways.
- Empathy is widely distributed in the human brain, involving both cortical and subcortical structures.
- Mentalising and empathy involve shared and distinct neural areas and can support each other in the process of social understanding.

12.4 Responding to adverse social signals

- Social rejection elicits brain activation that is similar to sensing physical pain.
- Common neural pathways for social and physical pain suggest that the brain has evolved to identify and regulate adverse social experiences.
- Loneliness is associated with automatic allocation of attention to social threats that signify potential social exclusion.
- Loneliness activates brain regions in the PFC and in subcortical structures, such as the amygdala.

Chapter 12 quiz

Question 1

Social cognition can be defined as:

A. The ability to perceive, categorise and interpret social behaviour of other people.
B. The ability to look at concepts with more than just information processing capacity.
C. The activation of mental representations about the self.
D. The knowledge of the differences between a person's mental and affective states.

Question 2

Which part of the brain is associated with the processing of both self-referential and social stimuli?

A. Default Mode Network (DMN)
B. The Salience Network (SN)
C. Temporo-parietal junction (TPJ)
D. Orbitofrontal cortex (OFC)

Question 3

Patient W.J. was found to have a case of retrograde amnesia after sustaining a traumatic brain injury. What type of details remained intact before and after the amnesia?

A. Episodic knowledge
B. Social knowledge.
C. Self-referential knowledge.
D. Implicit knowledge

Question 4

The Social Brain Hypothesis (Dunbar, 2009) utilised which aspect of memory?

A. Working memory
B. Semantic memory
C. Long-term memory
D. Episodic memory

Question 5

Why is empathy strongly linked to social cognition?

A. Empathy is needed to enable social cognition.
B. Empathy is often elicited when we see pain from others in social situations, in terms of both people receiving pain and those inflicting pain.
C. Empathy can be characterised as being social, cognitive or neural, therefore it only links to one process at a time.
D. Empathy is seen more in positive social situations whereby someone is more empathetic towards someone who is being more positive.

13 Cognition and emotion

Learning objectives

13.1 Describe how emotions are processed in the brain

13.2 Discuss the effect of emotions on perception and attention

13.3 Explain how emotions influence memory

13.4 Summarise the impact of emotions on language processing

13.5 Analyse the influence of emotion on decision-making

13.6 Discuss emotion regulation and coping

Key figures

Professor James Gross

Professor James Gross is Professor of Psychology at Stanford University, USA. His award-winning research has improved our understanding of emotions and strategies people use to regulate their emotional experiences.

Professor Ralph Adolphs

Professor Ralph Adolphs is Professor of Psychology, Neuroscience and Biology at Caltech, USA. His research has focused on how people recognise, process and respond to emotional cues in social perception and judgement and decision-making contexts.

Introduction

Although the computer metaphor has been a dominant guide for years in cognitive psychology, people are much more than computational machines. One of the big differences between us and our silicon-based creations is in the realm of emotions. We have them, and they don't. Although some might describe much of the research in cognitive psychology as 'cold cognition' because it does not take people's emotions into account, there is more and more research doing just that (see Mather & Sutherland, 2011[1]). What has been found is that emotion can influence cognition in a variety of complex but systematic ways. Understanding the interplay between emotion and cognition has led to the rise of **affectivism** in psychological research (Dukes et al., 2021[2]) where emotions play a key role in understanding cognitive processes such as perception, attention, memory and language (Figure 13.1).

affectivism
The systematic study of the interplay between emotions and cognition, perception and decision-making.

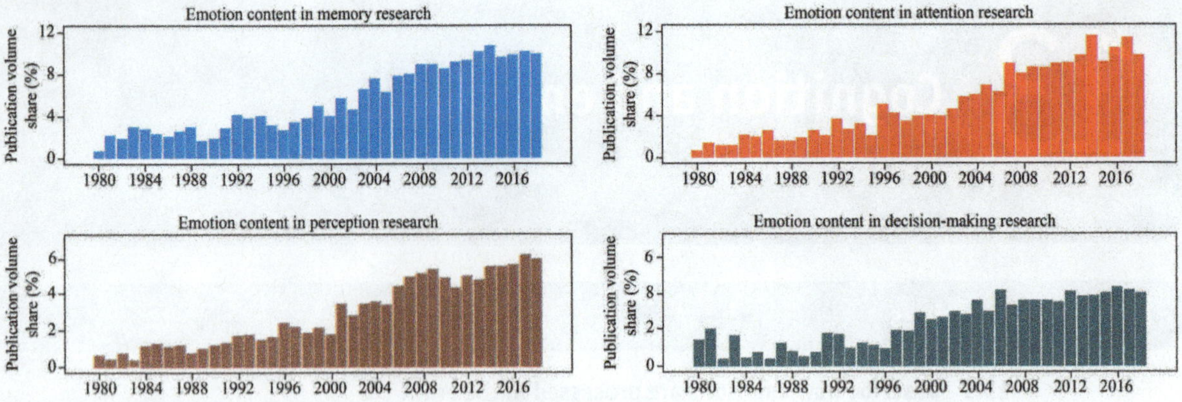

Figure 13.1 The rise of research on affect in different cognitive domains
Source: Adapted from Dukes et al. (2021)[3].

This chapter presents an overview of various ways that emotion affects cognition. In a sense, this module serves to recapitulate many of the topics already covered in the course. We start with a consideration of how emotion can influence seemingly basic perceptual and attention processes. We then look at how emotion influences memory, making it better in some cases and worse in others. After this we consider roles of emotion in language, followed by some coverage of how emotion can influence our ability to make decisions and solve problems.

13.1 What is emotion?

LEARNING OBJECTIVE
Describe how emotions are processed in the brain

Although we all have an intuitive sense about what emotions are, we need to go beyond that here and offer a formal definition to work with. Consider **emotion** to be both the state of mind a person is in at a particular moment, as well as the physiological response a person is experiencing at that time (in terms of heart rate, pupillary dilation, neurotransmitter release, and so on). There are other similar, related terms that mean somewhat different things, such as *affect*, *mood* and *arousal*, but we leave these aside for now. Our purpose here is to look at some basic ideas about how emotion can influence and interact with other aspects of cognition, such as perception, attention and memory.

There are numerous ways of dividing up and classifying different types of emotion. Also, it certainly is the case that we all have a lot of variety and subtlety in the different types of emotions we experience. However, for our purposes, we employ a simple approach of looking at two dimensions of emotion and use these to guide our coverage of how emotion influences cognition.

Many of the cognitive phenomena that we discuss can be understood in terms of where an emotion is along these two dimensions – that is, closer to the positive or negative end of the **valence** dimension, or closer to the low or high end of the **intensity** dimension (arousal). For example, you can see in Figure 13.2 that when arousal is high we experience high-intensity emotions that are both negative (e.g. tense) and positive (excited). Similarly, when we have low arousal, this would be described as being 'calm' or 'bored'.

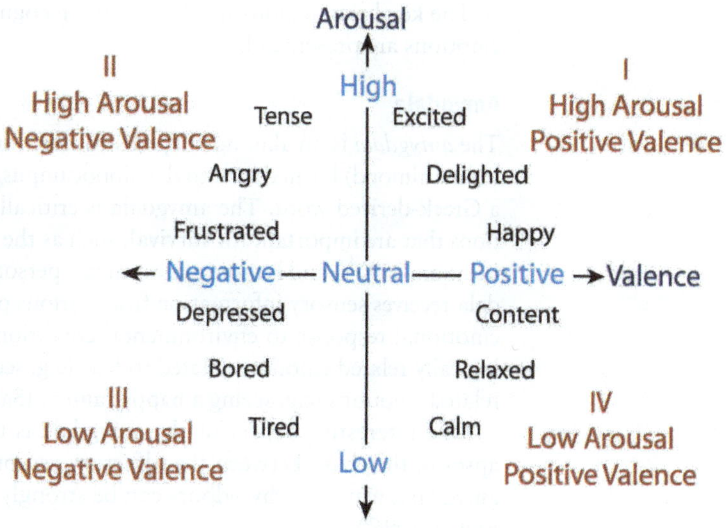

Figure 13.2 Valence and arousal as dimensions of emotional experience

13.1.1 Neurological underpinnings

limbic system
The limbic system is a set of structures in the brain that play a crucial role in emotional processing.

Emotion has both physical and psychological components. It is certainly a visceral experience. We feel it in our bodies: our hearts race, our breathing speeds up. As a mental experience, it has the power to influence brain and cognition in meaningful ways. Emotional experience is associated with several brain structures, collectively referred to as the **limbic system**. The advancement of brain imaging techniques has allowed researchers to identify key sub-cortical regions that support emotion recognition and the localisation of distinct emotional states in the brain. In addition, there is one cortical region that is often considered part of the limbic system and is predominately involved in emotional regulation: the prefrontal cortex (PFC). Input from subcortical regions to the PFC ensures that emotional experiences are constantly monitored and controlled (Figure 13.3).

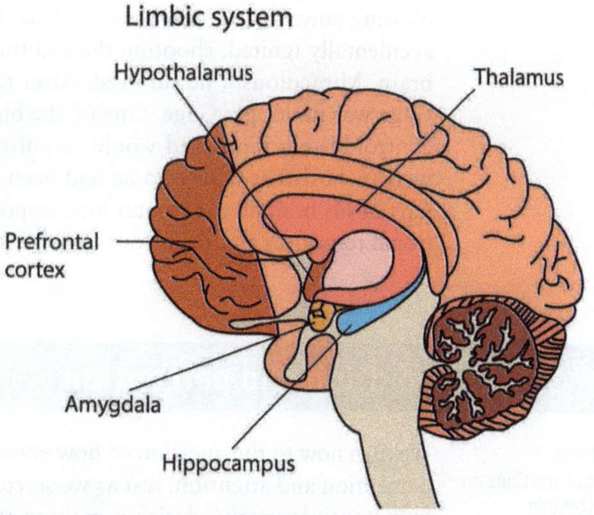

Figure 13.3 The main parts of the limbic system
Source: QBS/Pearson Education Ltd

The key brain regions involved in the recognition, experience and regulation of emotions are presented below.

Amygdala

The *amygdala* is an almond-shaped structure (originating from the Greek work that means almond) located next to the hippocampus, which looks like a 'sea horse', again a Greek-derived word. The amygdala is critically involved in more instinctual emotions that are important for survival, such as the experience of fear (e.g. Davis, 1997; LeDoux, 2000[4]), and is more active when a person is in an emotional state. The amygdala receives sensory information from various parts of the brain, allowing for a fast emotional response to environmental conditions. This is particularly true for biologically related emotions related to fear (e.g. seeing a snake), as opposed to socially related emotions (e.g. seeing a happy family) (Sakaki, Niki & Mather, 2012[5]).

One interesting aspect of the amygdala is that there are very few neural synapses in the chain between the olfactory receptors of your sense of smell and the amygdala, which is why odours can be strongly associated with emotions (Herz & Engen, 1996[6]).

The amygdala in turn sends its signals to the hypothalamus and brain stem, which help regulate the body's arousal state, as well as to areas tied to cognitive processes, such as the prefrontal cortex (attention) and the hippocampus (memory). Thus, emotional responses have the potential to directly influence the context and processing of thoughts in ways that differ from when we are not in an emotionally aroused state.

Prefrontal cortex

In addition to the amygdala, another region of the brain that is important in emotional processing is the **ventromedial prefrontal cortex (vmPFC)**. This part of the brain is involved in the identification and interpretation of emotional stimuli and responses, and the integration of that emotional interpretation with the surrounding context (Roy, Shohamy & Wager, 2012[7]), as well as the regulation and control of those experiences.

The famous case of *Phineas Gage* illustrates the important role of the prefrontal cortex in emotion regulation. Gage suffered a massive trauma that destroyed a large portion of his frontal lobe. This occurred in 1848 in Vermont when Gage was working on a blasting crew for a railroad. He was using a tamping rod to press some blasting power into a hole that had been drilled into a rock formation. The powder accidentally ignited, shooting the rod through Gage's skull, destroying part of his brain. Miraculously, he survived. After the accident, however, people claimed that Gage was no longer Gage. One of the biggest changes was that he was less able to control his emotions and would impulsively act and express himself in ways that were inconsistent with who he had been before the accident. This was because the part of his brain in the frontal lobe responsible for controlling and regulating emotional responses and behaviours was seriously damaged in the accident.

13.2 Emotion and perception

LEARNING OBJECTIVE
Discuss the effect of emotions on perception and attention

We turn now to the question of how emotion can influence cognition. We start with perception and attention, just as we started out this text. Then we move on to topics in memory, language, decision-making and problem-solving.

Let's start by explaining how we process emotional stimuli. The emotional content of an item can influence the accuracy with which it is processed. Looking more deeply, we can see that emotion appears to actually increase the amount of neural activity in perceptual brain areas, such as the *occipital* and *occipital-parietal cortex* areas (Taylor, Liberzon & Koeppe, 2000[8]).

However, it is not the case that there is a uniform boost in the ability to perceive emotional items. Instead, what appears to be happening is that there is an increase in the ability to process broad, global or general characteristics of an emotional stimulus (e.g. a threatening stimulus), but a decline in the ability to perceive details (Bocanegra & Zeelenberg, 2011[9]). That is, when emotions are aroused by a threatening stimulus, the perceptual processes in our cognitive system direct processing efforts to knowing generally what and where that emotional something is, so that we can be safe from it. In doing so, the cognition system is, in some sense, making sure that it does not waste its limited energies and resources on details that are probably trivial, thereby increasing the probability of survival.

In addition to influencing the ease with which a person recognises different objects and entities in the world, emotion can alter the subjective perception of the world when the situation has characteristics that arouse emotions in the person (for a review see Niedenthal & Wood, 2019[10]).

13.2.1 Emotional salience and attention

In addition to the effects of emotion on perceptual processes, there is some evidence that emotion can influence attention. A moment's reflection will reveal that this basic idea is not all that surprising. If something elicits our emotions, we are more likely to pay attention to it, whatever that thing is. What the work on emotion and attention has shown is that this influence can be more extensive and unconscious than a casual moment's reflection might reveal.

An important point is that what is significant to you often has some relation to emotions, either positive or negative. It is vital to keep in mind that emotion and attention use some of the same neural components. These include structures such as the amygdala, portions of the prefrontal lobes, and the anterior cingulate cortex.

Thus, emotion can affect the direction of attention (Vuilleumeir, 2005[11]). For example, people are more likely to direct their attention to emotionally arousing stimuli, such as a seeing a snake in the grass. Attention can also influence how you feel about things. For example, people can develop a negative emotional response towards things that they try to ignore (Fenske & Raymond, 2006[12]).

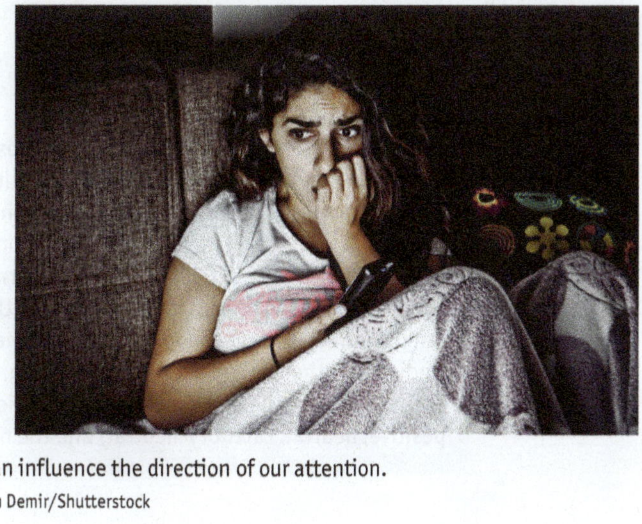

Emotions can influence the direction of our attention.

Source: Gorkem Demir/Shutterstock

13.2.2 Attentional biases for emotional stimuli

As we have seen in Chapter 4, one of the tasks in cognitive psychology that has been used extensively for the study of the operation of attention is *visual search* – trying to find an object in a display of irrelevant distractors. Research using this task has found that emotions, particularly negative emotions such as fear, can influence the visual search processes (e.g. Öhman, Flykt & Esteves, 2001[13]).

In some sense, this seems like a form of attention capture, with attention being preferentially moved to more fearful stimuli. This shifting of attention to emotional items seems to be directed more by the amygdala than by emotional control processes in the frontal lobe (Vuilleumeir & Huang, 2009[14]). Also, it may have more to do with an increase in attentional resources, because even the processing of non-emotional targets in a visual search task is facilitated when emotions are triggered (Becker, 2009[15]).

Emotional processing can even influence attention and cognition in ways that might seem irrelevant at first. An example of this is the **emotional Stroop task** (see Williams, Matthews & MacLeod, 1996, for a review[16]).

In the emotional Stroop task, words are presented in different colours, and people are asked to name those colours. However, rather than having colour words on the list, the critical comparison has to do with words that elicit an emotional response in a person, such as *spider*, as compared to more neutral words, such as *spade*.

What is typically found with this task is that people name the colour of the word more slowly if the word is emotional for them. For example, a person who has a deathly fear of spiders would be slower to say 'green' if the word *spider* were printed in green ink, whereas a person without this fear would not show this effect. The explanation is actually very straightforward.

Even though the person is supposed to focus on the colour of ink, reading the word and accessing its meaning happen automatically. If the word is related to an emotional stressor for the person, it intrudes on the person's cognitive processing and thus takes away resources from the other cognitive processes necessary for focusing on and naming the ink colour. Colour naming is slowed down for the emotional word condition, but not for the neutral word condition. This emotional Stroop task has been used to study numerous psychopathologies, including depression (e.g. Mitterschiffthaler et al., 2008[17]), anxiety (e.g. Dresler, Mériau, Heekeren, & van der Meer, 2009[18]) and post-traumatic stress disorder (PTSD; e.g. Cisler et al., 2011[19]). Also, factors such as whether the Stroop task is performed by native speakers (compared to non-native speakers) influence the magnitude of the inhibitory effects observed.

Emotional Stroop task

The emotional Stroop task has been used in psychological disorders to demonstrate interference effects with substantial variability between groups but also within groups. For example, individuals with PTSD show larger interference effects than individuals with specific phobias ranging from −1 ms (no interference) to 400 ms (e.g. Waters et al., 2005; McKenna & Sharma, 2004[20]). However, even within a group of individuals with specific phobias, the reported interference effects range between −1 ms and 204 ms. Some of the reasons why studies differ in their findings are the method of presentation (e.g. on screen vs. on paper), the design of the study (e.g. block vs. mixed), the emotional valence of the stimuli (negative, positive, neutral, category-neutral) and the number of repetitions of each stimulus (e.g. smaller number of words repeated many times or many words presented less times). There is also some debate around the nature of the interference effects and whether they operate within and/or between trials. For example, if neutral words are mixed with emotional words, there is an interference effect, but if there is a blocked design, this interference effect may disappear.

13.2.3 Emotion and self-control

An important function for attention is to guide thoughts and behaviour. To some degree this is automatic, but it can also take place under conscious control. What role does emotion play in this self-control? One would intuitively think that expressing emotions leads to less self-control; however, there is some evidence that *suppression* of your emotions can lead to attentional control problems.

An example of this is a study by Inzlicht and Gutsell (2007)[21] in which people watched an emotional film. Researchers asked them either to simply watch the movie (the control condition) or to suppress their emotions while viewing the film (the experimental suppression condition). After this, participants did a traditional colour–word Stroop task. The results, as can be seen in Figure 13.4, were that people who had suppressed their emotions took longer (higher response times) to complete the Stroop task.

The explanation was that suppressing one's emotions drained resources from the error-monitoring aspect of attentional control, which is guided by the anterior cingulate cortex. When these attentional resources were drained by the effort to control one's emotions over a long period of time, fewer resources were available for doing other tasks.

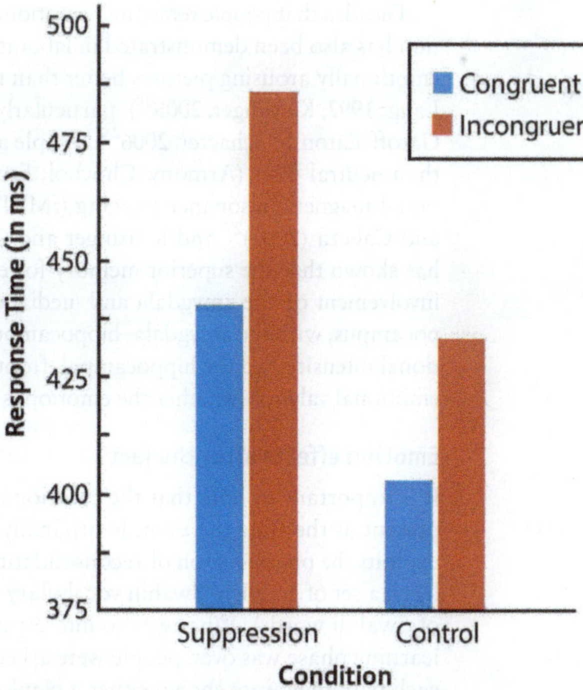

Figure 13.4 Response times to colour–word Stroop task with suppressed and non-suppressed emotions

So, overall, clear evidence indicates that emotional content and responses can influence even basic perceptual and attention processes. This suggests that emotion serves as a primary and fundamental force in guiding and influencing cognition.

13.3 Emotion and memory

LEARNING OBJECTIVE
Explain how emotions influence memory

One of the aspects of cognition where emotion has its largest effects is memory. The influence of emotion on memory is somewhat complicated. In some ways and under certain circumstances, emotion can make memories more durable, but in other ways or under different circumstances, emotion can have the opposite effect on memory.

13.3.1 Recalling emotional experiences

If you think about your own life and the things you remember, one of the things you quickly notice is that your emotional experiences are often the most memorable. This can include negative memories, such as losing a close family member or friend, an emotional break-up or being in a serious car accident, as well as positive memories, such as a new birth in the family, getting a degree or obtaining a scholarship. Thus, emotion can have clear benefits for memory.

Part of what influences memory is the quality of emotion – rather than the *valence* of the emotion (such as whether it is positive or negative) – and how intense the emotional experience is (Talarico, LaBar & Rubin, 2004[22]). In other words, what is important about the influence of emotion on memory is how strong the emotion is, not whether you feel good or bad about the event. That said, some memories may not be more intense but still may be remembered well if the information is viewed as being disgusting as compared to being fearful (Chapman, Johannes, Poppenk, Moscovitch & Anderson, 2013[23]).

The idea that people remember emotional information better than neutral information has also been demonstrated in laboratory work. For example, people remember emotionally arousing pictures better than neutral ones (Bradley, Greenwald, Petry & Lang, 1992; Kensinger, 2009[24]), particularly the details of negative images (Kensinger, Garoff-Eaton & Schacter, 2006[25]). People also remember emotional utterances better than neutral ones (Armony, Chochol, Fecteau & Belin, 2007[26]). Work using functional magnetic resonance imaging (fMRI) scanning, such as that by Dolcos, Labar and Cabeza (2005)[27] and Kensinger and Corkin (2004; see also Kensinger, 2007[28]), has shown that the superior memory for emotional memories appears to reflect the involvement of the amygdala and medial temporal-lobe structures, such as the hippocampus, with the amygdala–hippocampus network being more important for emotional intensity and the hippocampal–frontal lobe network being more important for emotional valence (whether the emotion is positive or negative, happy or sad).

Emotion effects after the fact

It is important to note that the emotional benefit of memory does not need to be present at the time the event is originally experienced. In a fascinating study that exploits the phenomenon of reconsolidation, Finn and Roediger (2011)[29] had people learn a set of English–Swahili vocabulary pairs (essentially the English translations of Swahili words). What was so interesting about this study is that after the initial learning phase was over, people were asked to recall the word pairs. As they recalled each pair, they were shown either a blank screen, a neutral picture or an emotional picture. Then, after another period of time, they were asked to recall the word pairs again. On this final test, memory was better for words that had been followed by the emotional pictures, as can be seen in Figure 13.5.

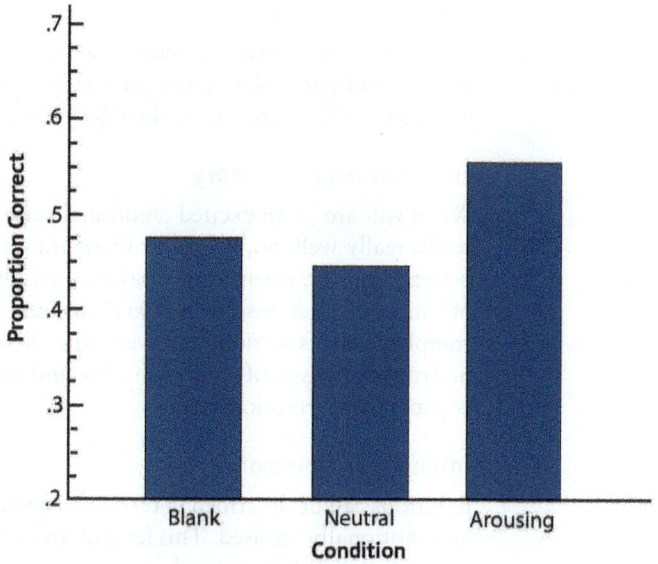

Figure 13.5 Finn and Roediger (2011)[30] study using English-Swahili vocabulary pairs

This figure illustrates the finding that recall memory for English–Swahili word pairs was better when people were looking at an emptional picture than when they were looking at either a blank screen or a neutral picture.

Source: Data from Fin and Roediger (2011)[31]

It seemed that when the word pairs were retrieved the first time, they were in a labile and fluid state, not yet firmly in memory. If an emotional picture was shown with the pairs, the person's emotional response to the picture was incorporated into the memory trace and stored with it during the reconsolidation process. Thus, emotional responses can be incorporated into memory traces after the event has occurred if we experience an emotion when we are remembering the event. Note that Finn and Roediger did not find this benefit when people merely restudied the information, but only when they actively retrieved it, as would be expected in a reconsolidation process.

Why does emotion help memory?

There are probably several reasons why emotional memories are easier to remember. First, emotional events are likely to be things that are important to us. Consequently, people are apt to devote more attention to processing that information relative to something that is more emotionally neutral. Part of this is driven by the recruitment of the amygdala, which is a critical brain structure for processing emotions. There is also some evidence that emotionally charged memories appear to benefit more from the process of memory consolidation offered by sleep compared to emotionally neutral memories (Hu, Stylos-Allan & Walker, 2006[32]).

Emotion may also help memory because emotional information is more distinctive. Much of what we encounter in our day-to-day lives does not elicit much in the way of a strong emotional reaction. Thus, truly emotional information is more likely to be distinctive, resulting in a kind of a **von Restorff effect**, which orients an increase in attention towards the emotion-eliciting stimulus (e.g. Talmi & Garry, 2012; Talmi, Schimmack, Paterson & Moscovitch, 2007[33]).

Emotional context

Emotions influence memory beyond just the fact that emotional memories are recalled more accurately. Another way that your emotions influence memory involves the kind of information that becomes activated or primed in long-term memory.

Flashbulb memories

We often seem to have – or believe we have – extremely accurate and very detailed memories of particular events, especially when the events were surprising and highly emotional. These are often called **flashbulb memories**.

Emotions impair memory

When you are in an excited emotional state, you certainly remember some things really, really well. So, it is clear there are ways in which emotions make memories better. However, when your emotions are running high, there also happen to be a lot of things that get missed, and so there are ways that emotions make memories less complete. In this section, we cover some ideas about how the accuracy of memory is tied to the intensity of your memories, and then discuss some ways in which memory is impaired by emotions.

Intensity and memory

Emotions can be described in terms of how intense they are – that is, how much you are emotionally aroused. This level of arousal is systematically related to how much is remembered. In a general sense, memory follows what is known as the **Yerkes–Dodson law**, which is shown in Figure 13.6.

According to the Yerkes–Dodson law, when you are in a low arousal state, such as when you are tired or bored, your memory for information is not that good. However, as your arousal level goes up, so does your memory performance – but only up to a point. That point is an optimal level of arousal that allows the most learning to occur. Beyond that, things change. At even higher levels of arousal your memory starts to decline. You are too agitated and excited, and the amount of information you can adequately remember goes down. As you can see, at high levels of arousal, emotions can make memory worse.

Although the Yerkes–Dodson law accurately captures what is going on with emotional arousal in a general sense, the actual situation is a bit more complex. At higher levels of arousal, memory does not decline for everything – there are some things for which memory actually continues to improve. This is captured by the **Easterbrook hypothesis**, which states that at higher levels of emotional arousal there is a narrowing of attention onto whatever is eliciting the emotions in a person. *That*

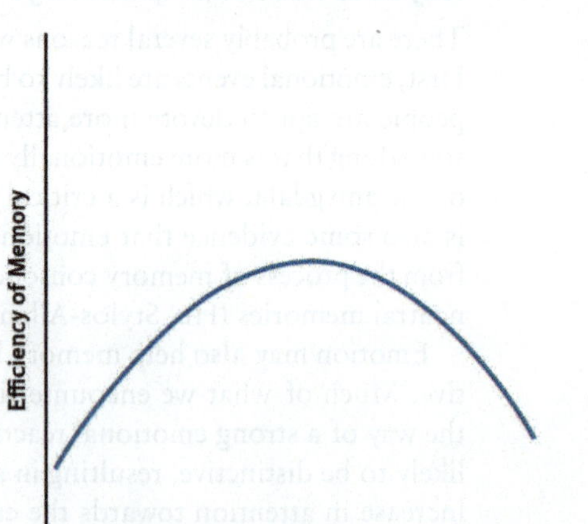

Figure 13.6 Illustration of the Yerkes–Dodson Law
Performance is poor at low levels of intensity, increased as intensity becomes greater, and then decreases for high levels of intensity.

information can be thought of as being at the centre of the event. The more irrelevant details of the event are less emotionally arousing, however, and can be thought of as the peripheral information. According to the Easterbrook hypothesis, when people are in more highly emotion-arousing situations, their attention narrows in on the central information and away from the peripheral information. Consequently, with high levels of emotional intensity, memory for the central details continues to get better, but memory for peripheral details declines. Because there are more peripheral than central details, overall memory is getting worse at high levels of emotional arousal, but things that are most important might be remembered really well.

So, as you can see, emotional information can both make memories better and make memories worse, depending on which aspect of an event is focused on. Memory for central information from an event is heightened by emotion, but memory for peripheral information is harmed by emotion. This differential influence of emotion on memory is exaggerated as time goes on during the consolidation process, particularly the consolidation that occurs during sleep (e.g. Payne & Kensinger, 2010[34]), although this increase in emotional memory is more likely when people expect their memory to be tested (Cunningham, Chambers & Payne, 2014[35]). The neural processing that occurs during rapid eye movement (REM) sleep seems to be particularly important for the consolidation of emotional information in memory because of the increase in cortisol levels during this time, which reinstates the emotional experience.

13.3.2 Recalling autobiographical memories

Autobiographical memories (ABMs) are our recollections of specific personal events and can involve facts (i.e. semantic events such as when you were born) or events (i.e. episodes such as a trip abroad). In this chapter, we will focus on episodic ABMs, which can be influenced by our emotions in two ways. First, our emotions at the time that the event occurred can influence the way it is remembered (e.g. an event can be marked as positive of negative), and secondly, our emotions at the time of retrieval can influence the way information is recalled (i.e. the way we reconstruct an event) (Holland and Kensinger, 2010[36]). There are certain phenomenological characteristics in recalling AMBs, including *vividness*, which refers to the sensory or perceptual detail of the event recalled, and *specificity*, which refers to the level of hierarchy of an event (lifetime periods, general events and event-specific knowledge; Conway & Pleydell-Pearce, 2000[37]). Neuroanatomical studies have shown that the medial temporal lobe and the hippocampal formation support ABMs in the brain, and damage to this area could cause loss of autobiographical events (Moscovitch et al., 2005[38]).

Not all personal events are recalled with the same ease or detail, and we tend to remember emotionally salient events more vividly. These can be positive, for example our school graduation, or negative, such as the loss of a friend. Also, the level of personal involvement and self-relevance are important in memory recollection, and this is why when people focus on how information relates to them, this enhances encoding and the memory becomes more resilient to forgetting.

13.4 Emotion and language

LEARNING OBJECTIVE
Summarise the impact of emotion on language processing

Emotion can also influence language processes in cognition and is a quality that is communicated by language. In this section we cover the transmission and impact of emotional information. This includes emotional tone that is carried by the speech signal, as well as the tracking of emotional information in a situation model and how that influences the cognitive processing of other event dimensions.

13.4.1 Prosody

prosody
The up and down pitch of an utterance that can convey emotional information.

As noted in an earlier chapter, language is not delivered in a monotone, but involves changes in pitch. This change is called *prosody*. **prosody** conveys meaning above and beyond the words used.

The sentences 'Those are my shoes' and 'Those are my shoes?' consist of the same words but are spoken using different prosodies. One conveys a statement and the other a question, often with a rise in pitch at the end of the sentence. One other aspect of meaning that prosody can convey is the speaker's emotional tone. For example, imagine the sentence 'Those are my shoes' spoken in happy, sad and angry tones of voice, and notice the different prosodies used to convey these emotional meanings.

Recall from our earlier discussions of language that, in most people, most language-processing components are lateralised in the left hemisphere. For example, this is the location of Broca's and Wernicke's areas. However, not all language processing is strongly left lateralised, and the processing of emotional prosody is one example of this. Several studies, including research involving fMRI, have found that the brain is right lateralised for the processing of emotional prosody (e.g. Buchanan et al., 2000[39]). This is further supported by evidence from people with damage to the right hemisphere. Patients with right hemisphere damage often have difficulty deriving emotional information from linguistic prosody (e.g. Pell, 1999[40]).

Note that although damage to the right hemisphere can impede the processing of linguistic prosody, damage to other brain structures involved in emotion processing, such as the amygdala, does not (Adolphs & Tranel, 1999[41]). In other words, the processing of emotional prosody by the right hemisphere does not depend on emotional input from the amygdala; rather it involves a determination of emotional content from the pitch information in speech that is heard. Overall, although prosody is a deeply linguistic characteristic, this meaning is processed more by the right hemisphere than the left.

Hearing language that conveys characteristics of emotion does not mean that a listener will be convinced that the person speaking is experiencing that particular emotion. Sometimes, we know that people are only play-acting and that the emotion expressed is not one that the person is actually feeling. What people need to do is take the emotional information conveyed by the prosody, along with other characteristics of the language, to determine whether the emotion is genuinely being experienced or not. Evidence exists that people have some success at doing this and that different regions of the brain are involved in such detection (e.g. Drolet, Schubotz & Fischer, 2012[42]). For example, if vocal expressions of emotion are taken from the radio and played to people, the detection of genuine emotion is above chance, particularly for emotions such as anger and sadness. Moreover, as measured by fMRI recordings, there is greater activity in the medial prefrontal cortex and the temporal-parietal junction. The latter is often found to be important in perspective taking. Thus, people are able to use information in the language signal to determine not only the emotion that is being expressed by the prosody of the spoken words, but also the genuineness of that emotion.

13.4.2 Words and situations

Let's shift our focus at this point away from *how* language is said to *what* is said, and how emotional processing influences language. First, and most obviously, some words are about emotion, such as *happy*, *sad* and *angry*. We have specific words that aim to capture and communicate the experience of emotion, and the fact that such words are common in the language attests to the fact that such ideas are readily transmitted and received. We are quite facile at processing emotional information.

Emotions and events

Thinking further about emotion and how it factors into our understanding of events, emotion can be experienced in two ways. One way is by the person who is feeling the emotion. This is the mental and physiological state of the person depending on the emotion being experienced. The second is by people other than the person experiencing the emotion. For example, some external person might see the emotion-experiencer smiling, scowling, blushing, and so on. These external manifestations of emotion are available to viewers. When comprehending language that involves emotion, how emotion is represented in the situation model may reflect, in part, whose perspective is emphasised by the language: the person experiencing or the one observing the emotion. These different ways of thinking about and processing emotional information may require different kinds of cognitive processes.

In a study by Oosterwijk et al. (2012),[43] people read individual sentences that conveyed emotional information. A given sentence expressed emotion from either an internal perspective of the experiencer, such as 'Hot embarrassment came over her', or the external perspective that some other person would see: 'His nose wrinkled in disgust'. People were simply asked to judge the sensibility of the sentences as quickly and as accurately as possible (some non-sensible sentences were included as well). The researchers found that people showed a processing cost when they had to switch from an internal emotional focus to an external one (or vice versa), from one sentence to the next, but not when the nature of the emotional focus stayed the same across the sentences. This shows that we mentally represent emotional information in our understanding of situations and events depending on the perspective we take and the kind of experience that needs to be captured in the situation model.

Why would it be important to track emotional information in a story, other than the additional experience one might get from reading about a story character's emotional state? It is important to keep in mind that emotions do not exist in a vacuum, but are human reactions to experienced events. More specifically, emotions can be seen as being strongly tied to people's goals. For example, negative emotions come from having one's goals impeded or blocked, such as not being able to buy a bicycle or not having one's parents continue to be healthy. Likewise, positive emotions come from having one's goals achieved, such as getting to go on a hot date or winning the lottery. Goals are important for language comprehension because they help to provide an understanding of the motivations and causes of narrative events as the circumstances unfold.

13.5 Emotion and decision making

LEARNING OBJECTIVE
Analyse the influence of emotion on decision-making

Decision-making is another area of cognition that is affected by emotion. The effectiveness and quality of the decisions a person makes in a heightened emotional state differ from those made in a more relaxed and neutral state. Note that the terms *emotion*, *stress* and *pressure* have different meanings, but for our current purposes, we gloss over some of these.

13.5.1 Stress impairs performance

It is clear to anyone who has ever been under emotional stress, such as being anxious, that decision-making and problem-solving performance can decline under these conditions. Essentially, when people experience anxiety, they tend to crowd their working memory with irrelevant thoughts about whatever it is they are anxious about. For example, people who are maths-anxious (i.e. they avoid doing maths problems, taking maths classes, exploring careers that use a lot of maths) do worse on maths problems because their working memory capacity is consumed by off-topic thoughts that stem from their maths anxiety (Ashcraft & Krause, 2007[44]). These thoughts detract from their limited capacity to devote to the problems, and their performance suffers. This anxiety may also disrupt the mental representations people hold in memory, making them less precise, thereby decreasing performance (Maloney, Ansari & Fugelsang, 2011[45]).

People who are threatened also experience physiological changes due to the stress they are under (e.g. Kassam, Koslov & Mendes, 2009[46]). Specifically, there is a decrease in cardiovascular efficiency, reducing blood flow to the body and brain, causing the body to slow down. This makes thinking, including decision-making, less effective. Part of the reason for this is that when a person is in a threatened state, it may be adaptive (under certain circumstances) to be more immobile and for the body to be prepared for some damage. Although there may be some survival advantages to this in the wild, in the circumstances of our everyday lives, this is a maladaptive response. It hinders our ability to make decisions that would be to our advantage.

Choking under pressure

We have all been there. The pressure is on. We have done well in this situation before. It comes time to step up and do it again. And we choke. When people become anxious because of external pressures, their performance can decline. Choking under pressure is a metaphorical way to express the effect of stress on performance. For example, you may be asked to talk in front of an audience, and you may feel that you can't find the right words because you are stressed. Of course, you are not literally choking, but you may feel your flow is disrupted.

As a process becomes increasingly practised, it becomes more and more automatised, allowing people to do it better and more quickly. Thus, as people gain expertise, they become more fluid in the task as its components become more and more unconscious. For the most part, this is a good thing. However, it can also cause problems, especially when the pressure is on.

In a pressure situation, what skilled people should do is allow their unconscious cognitive processes to play themselves out. However, sometimes these people start to consciously think about what they are doing and how they are doing it. This is the classic case of when a person, such as a basketball player, experiences **choking under pressure**. What can happen under these circumstances is that the athlete's conscious thought processes about what he or she is doing begin to intrude on and compete with more unconscious and automatic cognitive processes.

In a classic demonstration of this, Beilock and Carr (2001)[47] tested two groups of people in a golf-putting situation: expert golf players and golf novices. At first Beilock and Carr had people just putt to gain a sense of how well they did normally. Then they placed these people under pressure, asking them to consciously focus on their putting. What happened, not surprisingly, was that novices did better when they focused on accuracy. Surprisingly, however, golf experts did worse when they focused on their putting. This is because their conscious thoughts disrupted their normally automatic performance on this task. In fact, the experts actually did better if they focused on speed (Beilock, Bertenthal, McCoy & Carr, 2004[48]).

Stereotype threat

The influence of emotion and stress on cognition can come from several different sources. Some sources that people often do not think about are the social and ethnic groups we belong to and identify with, along with the cultural stereotypes of these groups that we carry around with us. When these stereotypes convey a negative view of our abilities in a typical domain, there can be an unconscious mental activation of this knowledge. This activation may happen even when doing something as simple as indicating gender or ethnic group when beginning to fill out a questionnaire. **Stereotype threat** occurs when this unconscious activation of a negative stereotype leads a person to perform worse on a task than he or she would otherwise. This can come about in any way that orients a person to identify with the stereotyped group, such as indicating membership in that group on a form, or being placed in a situation in which the person feels like a minority (e.g. Murphy, Steele & Gross, 2007)[49]. For example, asking people to indicate their gender before taking tests in science or engineering can lead women to do worse on such tests because of a (mistaken) cultural stereotype that women are not as good at these tasks – thereby perpetrating gender biases.

Stereotype threat can influence the types of choices and decisions a person makes (Carr & Steele, 2010[50]). For example, people do worse at solving maths problems resulting from stereotype threat if they belong to a group for which the stereotype claims that they should not be good at maths (e.g. Beilock, Rydell & McConnell, 2007[51]). As with the research on choking under pressure, stereotype threat lowers performance because a person's thoughts, and consequently working memory capacity, are consumed to some degree by counterproductive ideas related to the stereotype. This is the case even if that effort is oriented towards suppressing the stereotype-related thoughts (Schmader, 2010[52]). As a result, less mental capacity is available for doing the task at hand, and performance suffers.

Note that priming can also lead to an increase in performance. For example, a study by Lang and Lang (2010)[53] asked students to do a series of verbal analogy problems. Before actually doing the task, the researchers instructed one group of students to imagine a person who is successful at solving problems and write down several abilities such a person would have, the personality traits of such a person, and how such a person would feel just before starting to solve a problem. Compared to a control group that did not do this, people who spent time imagining what it was like to be such a successful person did better on the analogy test. Thus, it is possible to attenuate test anxiety as well as magnify it.

13.5.2 Stress improves performance

Emotion, stress and pressure can decrease performance, but there are times when they can actually increase performance. There is bad stress and good stress. Performance can improve when the stress being experienced is viewed as challenging and exciting rather than threatening (e.g. Brooks, 2014; Kassam et al., 2009[54]). Under such circumstances, there is an increase in cardiovascular efficiency, and more blood and oxygen are delivered to the body and brain, allowing the body to function at a heightened level. So, people perform better under these conditions.

As an example, consider a study by LePine, LePine and Jackson (2004).[55] The researchers assessed people for their stress level at school, along with other factors such as exhaustion and academic aptitude, and their performance as measured by their grade point average. Importantly, stress was identified as being either negative stress, where school work was viewed as a negative stressor and a threat, or positive stress, where school work was viewed as a positive stressor and a challenge. Their results showed that when there was negative stress, performance was negatively affected – grades were lower. However, when school work was viewed as a challenge, performance was positively affected – students had higher grades.

In addition to physiological changes that can facilitate performance, there are also more cognitive changes where pressure can improve performance. For example, if people acquired a skill in a way that involved more implicit unconscious processes, such as learning to putt under dual-task conditions where attention is divided (Masters, 1992),[56] then most of the cognitive processes involved in the task were unconscious. People trained under more explicit conscious conditions did worse under pressure. However, people who attended divided-attention training and then demonstrated more implicit learning actually improved performance under pressure. To be sure, the implicit learning people had slower learning and worse overall accuracy – but they experienced less disruption, compared to their baseline, when they were put under pressure. Thus, the overall effect of making a task more unconscious is to insulate it from the disruption of being in stressful situations. This is why it is valuable for some professions to constantly drill and practise their skills, in a variety of settings, to the point that they are automatic and unconscious. This way, when people need these skills for real, say in a genuine emergency, they can execute them without succumbing to the detrimental effects of pressure.

13.6 Emotion regulation

LEARNING OBJECTIVE
Discuss emotion regulation and coping

Throughout the chapter we have talked about the interplay of cognition and emotion, highlighting that emotions can serve as facilitators or barriers to various cognitive functions. We will now look at the way emotions can be regulated and discuss prominent theoretical accounts of emotion regulation.

13.6.1 What is emotion regulation?

The term is often used interchangeably with 'mood regulation', 'coping' and 'affect regulation' but in essence it refers to the active goal to change or alter an emotional state. This can involve decreasing (down-regulating) or increasing (up-regulating) the emotional intensity or changing the valence (from unpleasant to pleasant). For example, someone may regulate their anger to feel less angry (see Table 13.1).

Table 13.1 Emotion regulation strategies

	Increase	Decrease
Negative emotion (e.g. anger, fear, sadness, disgust)	Expressing your disgust when exposed to rotten food	Controlling your anger when someone insults you
Positive emotion (e.g. happiness, surprise)	Expressing your happiness when seeing an old friend	Reducing your smile in a funeral

People often refer to emotion regulation to describe their efforts to down-regulate a negative emotion, but it's equally important to discuss up-regulation in positive emotions like joy, love and surprise. Gross (2015)[57] uses a process model of emotion regulation (Gross, 1998[58]) to explain the way people change their emotional states in different phases of an emotional experience (Figure 13.7).

In brief, the model identifies the critical points at which an emotional experience can be altered. For example, if you have the option of two situations (S1 and S2) you might select the one with the least emotional impact (situation selection stage). Alternatively, you can change the situation to reduce the emotional impact (situation modification stage). Attentional deployment refers to the choice between competing situations (e.g. S1 and S2), and cognitive change refers to the attempt to change the meaning of the situation (interpretation or appraisal). Finally, you can choose to avoid situations with high emotional impact altogether (suppression).

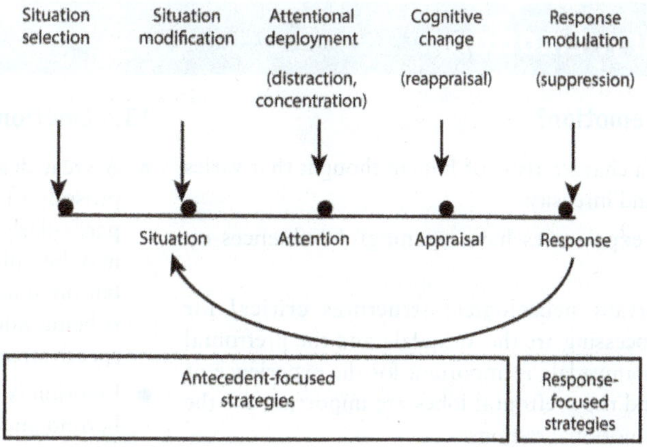

Figure 13.7 Model of emotion regulation

Source: Adapted from Gross (2015)[59]

13.6.2 Emotion regulation and coping

There are numerous interventions to improve emotion regulation, with the most obvious being trying to train people to regulate their emotions and to select adaptive coping styles. Cognitive reappraisal is one if the key adaptive mechanisms of emotion regulation that involves changing the 'meaning' or the 'interpretation' of a situation so that it becomes less emotionally intense. On the other hand, emotional suppression, which involves avoiding or distracting oneself when exposed to an emotionally intense situation, is less adaptive because it doesn't allow individual to 'face' their emotions and change them. Individuals with psychopathology such as social anxiety (SA) and/or major depressive disorder seem to be less able to use cognitive reappraisal than emotional suppression, suggesting a more maladaptive coping mechanism (for a systematic review, see Dryman and Heimberg, 2018[60]).

Emotional fit hypothesis and social exclusion

Batja Mesquita and her team at the Centre for Social and Cultural Psychology at the University of Leuven have systematically studied cultural differences in the way people develop friendships in young people from minority groups. Minority individuals who engage in intercultural contact experience psychological changes that can influence their emotions, personality, self-concept and cognition. The study, one of the first to use a longitudinal approach to acculturation, demonstrates that friendships between minority and majority group members can serve to facilitate emotional change by adopting local norms. It is unclear if this adaptation comes at the expense of emotional alignment with their heritage or immigrant communities. Additionally, the study reveals that emotional alignment with the majority culture can promote social inclusion for minorities. By fostering friendships with majority group members, this emotional fit may grant minorities access to resources and opportunities typically reserved for the majority group. This is an important finding for research on social inclusion because it provides evidence that young people from minorities who integrate and engage with majority groups are more likely to feel included and perhaps less lonely.

Source: Jasini, A., De Leersnyder, J., Ceulemans, E., Gagliolo, M. & Mesquita, B. (2024). Do minorities' friendships with majority culture members and their emotional fit with majority culture influence each other over time? *Social Psychological and Personality Science, 15*(5), 540–549.

Summary: Emotion

13.1 What is emotion?

- Emotion is a characteristic of human thought that varies in valence and intensity.
- Emotional experiences have meaningful influences on cognition.
- Two important neurological structures critical for emotion processing are the amygdala and the prefrontal cortex. The amygdala is important for the experience of emotion, and the prefrontal lobes are important for the control of emotional responses.

13.2 Emotion and perception

- Emotion can influence perceptual processing by channelling cognitive processing resources towards those stimuli in the environment that are more emotional.
- Emotional responses can guide where attention is directed, even to the extent of sending attention to nominally irrelevant information in the environment, which can produce a processing cost, as in the emotional Stroop effect.
- The experience of intense emotions can consume cognitive resources, such as attention, to the point that that there is less available for other tasks and performance shows a deficit.

13.3 Emotion and memory

- There is a general benefit for remembering emotional over neutral information. However, people who have suffered brain damage to areas critical for emotion processing, such as the amygdala, do not exhibit such a benefit.
- The memory benefit for emotional information comes from numerous sources, including its importance, its distinctiveness, the amount of attention and rehearsal it is given, and the superior benefit it seems to derive from neural consolidation processes.
- Emotion can serve as a form of contextual information that can facilitate some memory processes, as with mood-congruent and mood-dependent memories.
- The highly detailed memories we have for very emotional events, and for the contexts in which we learn about such events, are called flashbulb memories. Flashbulb memories, although not perfect and prone to some forgetting, appear to be more durable than normal everyday memories.
- Although emotion can improve some types of memories, it can also impose a cost on later memory. Specifically, at high levels of emotional intensity, attention is captured by central information at the cost of processing of peripheral information. Consequently, memory for more peripheral information is worse under conditions of higher emotional intensity.

13.4 Emotion and language

- A great deal of emotional meaning is conveyed by the prosody of spoken utterances. Brain-damaged people, particularly those with damage to the right hemisphere, may be able to understand the words spoken to them but often lack the ability to process the prosody of what is being said, and so miss out on emotional cues in the speech stream.
- Emotion is infused in a lot of language processing, even beyond auditory cues such as prosody. Some words, even abstract words, carry emotional meaning in their lexical entries, thereby influencing how people process that information.
- People actively track and simulate emotional experiences in the situation models that are created during comprehension. There is evidence that people show processing costs of changes in emotional quality of a character, and in changes in the internal or external experience of an emotion.

13.5 Emotion and decision-making

- When the source of stress is viewed as a threat, there can be declines in performance, causing people to choke under pressure.
- Choking under pressure can occur when the pressure is outcome-based and the focus is on the outcome of the task, or monitoring-based and the focus is on how the task is done. These different types of pressure arise under various kinds of situations.
- Some forms of pressure may be more subtle and unconscious, such as when a person experiences stereotype threat. This occurs when people are reminded of their membership of a group that cultural stereotypes suggest will not do well, and so their performance goes down.
- Stress can also improve performance when it involves activities that are highly over-practised, or when the stress is viewed as a challenge rather than a threat.

13.6 Emotion regulation

- Regulating emotions is part of our human nature and it is an important coping mechanism.
- There are two ways that we regulate emotions; reappraisal and suppression.
- People often use less adaptive coping mechanisms to regulate their emotions because they may temporarily relieve emotional burden, but they are not very effective in the long term.

Chapter 13 quiz

Question 1

Which area/s of the brain are associated with emotional processing?

A. Occipital lobe
B. Limbic system
C. Temporal cortex
D. Lateral ventricles

Question 2

Why is attention important in emotional processing?

A. Attention is preferentially allocated to more fearful stimuli.
B. Attention is preferentially allocated to more positive stimuli.
C. Attention is preferentially allocated to more salient stimuli.
D. Attention is preferably allocated to more colourful stimuli.

Question 3

The Yerkes-Dodson law proposes:

A. We have an optimal level of emotional arousal which allows most learning to occur. After that point, higher levels of arousal mean memory then declines.
B. Higher levels of emotional arousal mean a narrowing of attention onto whatever is eliciting the emotion in a person.
C. The differential influence in emotion on memory is exaggerated as times goes on during the consolidation process, particularly consolidation that occurs during sleep.
D. Emotions at the time of retrieval can influence the way the information is recalled.

Question 4

Why is stress important when considering the relationship between emotions and decision making?

A. Stress acts as a moderator for an emotive situation as it can mean that we make decisions about positive and negative situations in the same way.
B. Stress has a direct link with performance, therefore we should not consider how emotion is linked with this when making decisions.
C. Stress has no survival advantage, and it shows no links to the decision-making process and emotional processing.
D. People who experience stressful situations can also experience physiological changes due to the stress they are under. This can impact how emotions are processed and how decisions are made.

Question 5

Rushi is attending a very important meeting about job losses within the company he works for. Rushi is feeling disgusted, but he is aware that he is in a professional situation, so he does not show his disgust until he is talking to a colleague after the meeting. Rushi is fully engaged with the meeting, takes notes and leaves once the meeting is finished. What is this an example of?

A. Emotional blunting
B. Emotion regulation
C. Emotional coping
D. Emotional suppression

14 Research methods in human cognition

Learning objectives

14.1 Understand the importance of research methods for the study of human cognition

14.2 Describe reaction-time tasks and their applications in cognitive psychology

14.3 Understand visual search tasks and their use in cognitive psychology research

14.4 Describe neural imaging methods and their applications in cognitive psychology

14.5 Understand principles of research integrity and ethics in cognitive psychology

Key figures

Professor Mahzarin Banaji
Professor Banaji made important contributions in understanding and measuring implicit social cognition and co-developed the Implicit Association Test (IAT).

Professor Colin MacLeod
Professor Colin MacLeod co-developed the emotional Stroop task, a modification of the original Stroop task, that has been widely used in research in human cognition.

14.1 The purpose of research methods in cognitive psychology

LEARNING OBJECTIVE
Understand the importance of research methods for the study of human cognition

Whatever career path you choose, from a behavioural science consultant or clinical practitioner to an academic researcher, research methods will always be at the epicentre of your inquiry. Whether it be for the assessment of clinical symptoms, preference for a given product or brand over another, or the assessment of neural activation in response to exposure to certain stimuli, you will need to engage in some sort of psychological measurement: the assessment of observed or unobserved psychological characteristics, mental processes and/or behaviours.

Research methods in the study of human cognition serve the fundamental purpose of understanding how individuals acquire, process, store and use information from the environment and how they are influenced by unconscious forces. This knowledge contributes to developing or refining theories of cognition across different domains, including perception, memory, attention, language and decision-making, as well as interdisciplinary areas, such as machine learning and neuromarketing. In addition, since cognition exhibits considerable variability among individuals and across demographic groups, the selection of robust methods allows the thorough investigation of individual and group differences, through the identification of discerning patterns and factors that contribute to cognitive diversity. This is particularly useful for tailoring interventions, educational strategies and therapeutic approaches to meet the specific needs of diverse populations.

As we will see, research methods in cognitive psychology vary from experimental to neuroimaging techniques. These are broadly categorised into **behavioural** and **neural imaging** methods. Technologies such as functional magnetic resonance imaging (fMRI), electroencephalography (EEG) and transcranial magnetic stimulation (TMS) provide unprecedented insights into the neural mechanisms underlying cognition. This integration of cognitive and neuroscientific methods enhances our understanding of the brain–mind relationship. As our methodologies evolve, the pursuit of understanding human cognition continues to be at the forefront of scientific inquiry, offering profound insights into what drives our thoughts and behaviours. In this chapter we have collated some of the most common methods used in cognitive psychology and provide examples on how these are used in different fields.

14.2 Reaction time-based tasks

LEARNING OBJECTIVE
Describe reaction-time tasks and their applications in cognitive psychology

Reaction time-based tasks play a pivotal role in effectively measuring different aspects of cognition. As suggested by their names, these tasks measure the time it takes for individuals to respond to specific stimuli presented in different modalities (e.g. visual, auditory, haptic, etc). We discuss four widely used reaction time-based tasks: the dot-probe, go/no go, Simon and the Implicit Association Test (IAT) – each shedding light on different aspects of cognitive functioning.

14.2.1 Dot-probe paradigm

The dot-probe task was designed to assess **attentional biases** by presenting pairs of stimuli, typically words or images, on a screen for about 1000–2000 ms. The task measures the speed of attentional orientation to visual stimuli (e.g. a face displaying a particular emotion), by analysing reaction times to probes (a dot)

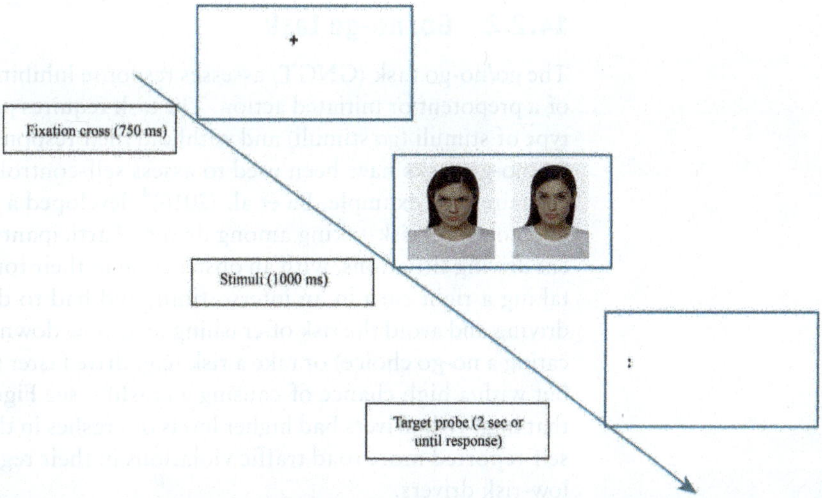

Figure 14.1 Example of a dot-probe paradigm
Participants are asked to respond to the location of the target probe (:) by pressing a button on a keyboard. The stimuli (facial expressions) appearing before the target probe influence response times.
Source: photos from Inagehit Limited Exclusive Contributor/123RF

appearing on the same or the opposite side of the visual stimuli (in our example, a face) presented earlier. The task relies on the assumption that individuals will react more quickly when a probe appears in the location of their attentional focus (congruent probes) rather than a different spatial location (incongruent probes), revealing implicit biases. For example, the visual dot-probe task has been widely used to assess implicit biases in people with social anxiety disorder (SAD). A large body of research has shown that, compared to non-anxious control participants, individuals with SAD display a bias towards threatening facial expressions (for a review see Bantin et al., 2016[1]).

Is trait anxiety always correlated with attentional biases?

Although there is ample research that suggests that affective disorders are associated with attentional biases for threatening stimuli, this may not always be the case. Theoretical accounts suggest that people with trait anxiety have the tendency to attend to aversive stimuli faster than neutral ones, unlike people with low trait anxiety. This attentional bias to threat stimuli has been attributed to increased 'vigilance', suggesting that anxiety is associated with facilitated orienting of attention towards threat-related stimuli. Alternatively, it has been explained using the 'maintenance hypothesis', which suggests that once threat (i.e. negative stimulus) is detected it holds attention longer than neutral stimuli in anxious individuals. A meta-analysis (Kruijt et al., 2019[2]) did not find any evidence for attentional bias towards threat in individuals with clinical anxiety, suggesting that methodological differences may contribute to the discrepancy in findings. A study utilising an emotional dot-probe task to investigate attentional biases in healthy individuals with trait anxiety also did not reveal such findings (Veerapa et al., 2020[3]). The study used an emotional dot-probe task with pairs of negative and neutral scenes, presented for either 1 or 2 seconds and followed by a target placed at the previous location of either negative or neutral stimulus. They did not find evidence of bias in orienting attention in high trait anxiety individuals but found evidence of maintenance using eye-tracking methodology.

14.2.2 Go/no-go task

The go/no-go task (GNGT) assesses response inhibition, which is the suppression of a prepotent or initiated action. The task requires participants to respond to one type of stimuli (*go* stimuli) and withhold their response to another (*no-go* stimuli). Go/no-go tasks have been used to assess self-control and impulsivity in different domains. For example, Ba et al. (2016)[4] developed a go/no-go driving simulation task to assess risk-taking among drivers. Participants were presented with different driving situations, with an obstruction to their route (e.g. an oncoming vehicle taking a right turn in an intersection), and had to decide either to control their driving and avoid the risk of crashing (e.g. slow down or stop the car entirely, indicating a no-go choice) or take a risk (e.g. drive faster to overcome the obstruction, but with a high chance of causing a crash) – see Figure 14.2. The results showed that high-risk drivers had higher levels of crashes in the driving simulator and also self-reported more road traffic violations in their regular driving, as compared to low-risk drivers.

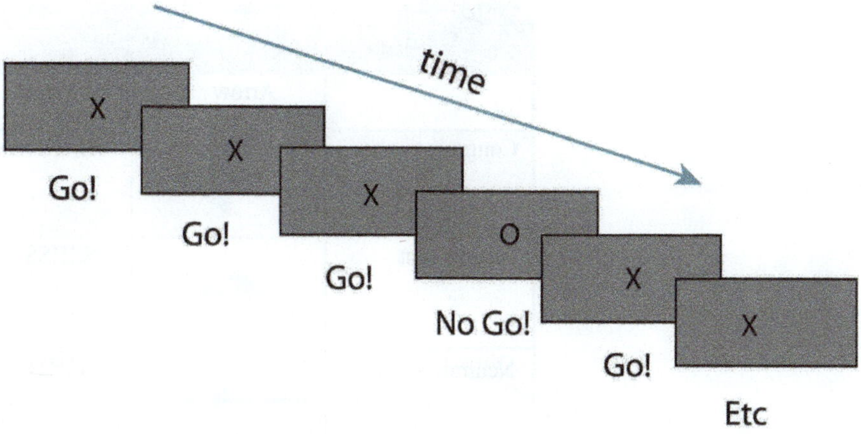

Figure 14.2 An example of a go/no-go task

How many inhibitory control mechanisms do we have?

Different methods are used to assess response inhibition (or inhibitory control) in diverse behavioural domains. The stop signal task (SST) represents an alternative response inhibition measurement method to the GNGT discussed above. Although both the GNGT and the SST require participants not to respond to certain stimuli (i.e. inhibition signals), they differ in the timing of these stimuli. For example, unlike the GNGT, in SSTs the go signal is always presented first. Raud et al. (2020)[5] examined whether the GNGT and the SST are based on the same or on different inhibitory control mechanisms. The researchers presented participants with both the GNGT and the SST and recorded their responses to the go and no-go/stop stimuli, while recording the automatic muscular responses with electromyography (EMG), placed on the participants' arm used to press the relevant keys on a keyboard in response to the presented stimuli, and regional electrical brain activity with an electroencephalogram (EEG) – both the EMG and EEG are discussed in more detail later in the chapter. The study showed that different inhibitory control mechanisms were activated by each task: the GNTG involved response selection at a later stage of inhibition, whereas the SST involved action cancellation at earlier stages of stimuli perception.

14.2.3 Interference tasks (Simon, flanker and Stroop)

The Simon task investigates the interference between stimulus-response compatibility and spatial location. Participants respond to a feature of a stimulus while its location is varied, creating a conflict between the location and the correct response. For example, by examining the interference effect, researchers gain insights into the automaticity of cognitive processes and the interaction between spatial and non-spatial information.

The flanker task involves responding to a stimulus that is 'flanked' by irrelevant stimuli which can affect your response because of their similarity to the target stimuli – see Figure 14.3:

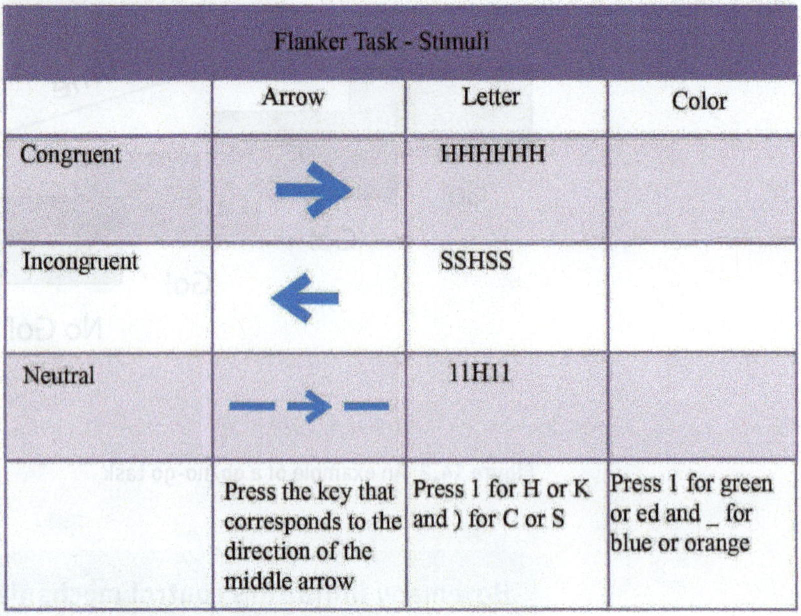

Flanker Task - Stimuli			
	Arrow	Letter	Color
Congruent		HHHHHH	
Incongruent		SSHSS	
Neutral		11H11	
	Press the key that corresponds to the direction of the middle arrow	Press 1 for H or K and) for C or S	Press 1 for green or ed and _ for blue or orange

Figure 14.3a An example of a flanker task

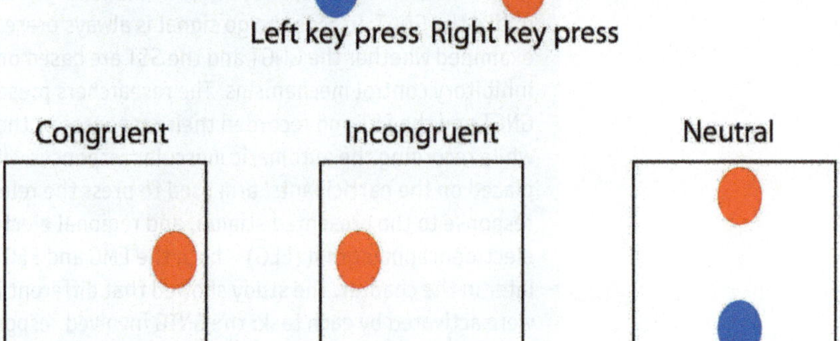

Figure 14.3b

The Stroop colour-naming task (Figure 14.4), introduced by Stroop in 1935, was designed to assess cognitive flexibility and behavioural control by requiring participants to suppress a dominant response. In the congruent condition, the stimulus word matches the stimulus colour (e.g. 'BLUE' written in blue ink), and so participants can rely on well-established reading processes for quick and accurate responses. However, in the incongruent condition, accurate responding requires cognitive control mechanisms to suppress word reading and activate colour-naming processes. The additional time taken to name the ink colour in the incongruent condition, compared to the congruent condition, is termed *Stroop interference*. Despite the popularity of the task and over 1000 articles reporting Stroop interference effects (see review by MacLeod, 1991[6]) the task has been criticised (e.g. Bugg, Jacoby & Toth, 2008[7]).

The Stroop task and emotional Stroop task

One of the most common variations of the traditional Stroop task is the emotional Stoop task, used to assess attentional biases in clinical and sub-clinical populations. Cognitive models assume that attentional biases are not just a by-product of an emotional disorder but play a vital role in the causation and maintenance of the disorder across time. As such, the association between emotions and attentional biases have been widely investigated, using different paradigms. In its simple form, the emotional Stroop task uses disorder-specific words in the incongruent condition to determine interference effects associated with the disorder which have become emotionally salient. For example, in individuals experiencing panic disorder, the heightened sensitivity of bodily sensations is an indicator of collapse or death, leading to heightened anxiety and a reinforced focus on additional bodily sensations. Therefore, words like *death* and *collapse* become more salient.

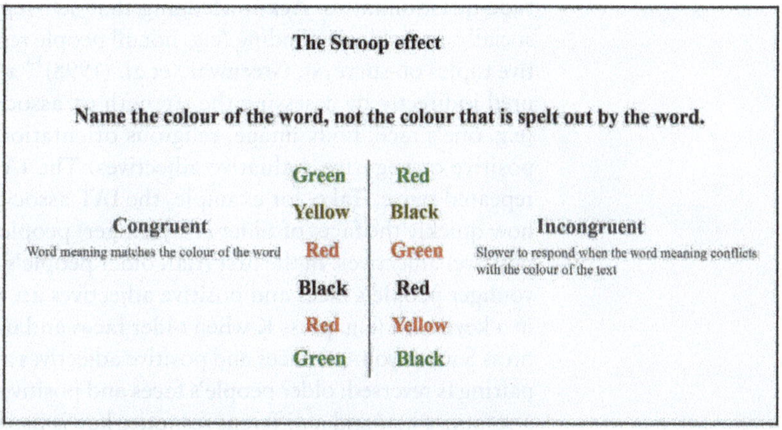

The Stroop effect
Name the colour of the word, not the colour that is spelt out by the word.
Congruent
Word meaning matches the colour of the word
Green
Yellow
Red
Black
Red
Green
Red
Black
Green
Red
Yellow
Black
Incongruent
Slower to respond when the word meaning conflicts with the colour of the text

Figure 14.4 An example of the Stroop task

Did the COVID-19 pandemic change the way we respond to our social world?

During the COVID-19 pandemic the emotional Stroop task was used to evaluate interference effects in individuals with high versus low COVID-related anxiety (Ypsilanti et al., 2021[8]). Examples of words used for each category matched for word length are presented in Table 14.1.

Table 14.1 Words used to assess attentional bias to COVID-related cues

COVID-19 related words	Neutral words	Negative words
Coronavirus	Electricity	Cannibalism
Covid	Ankle	Anger
Virus	Elbow	Panic
Lockdown	Pamphlet	Pressure
Infection	Appliance	Terrified

In this study there was no evidence of a bias in emotional processing of COVID-related words in people with high COVID-19 anxiety, suggesting that automatic affective responses, as measured by the emotional Stroop task, remained unaffected during the pandemic. Although the emotional Stroop task has been used in wide range of disorders (Williams et al., 1996[9]), studies have yielded mixed findings for PTSD (Cisler et al., 2011[10]), depression (Epp et al., 2012[11]), neuroimaging studies (Song et al., 2017[12]) and anxiety (Joyal et al., 2019[13]).

14.2.4 Implicit Association Test (IAT)

The Implicit Association Test (IAT) was designed by Anthony Greenwald and Mahzarin Banaji in the late 1990s as an alternative measure to self-reported attitude questionnaires. Acknowledging that self-reports may be faked or subject to socially desirable responding (e.g. not all people report their true beliefs about sensitive topics on surveys), Greenwald et al. (1998)[14] argued that attitudes can be measured indirectly by assessing the strength of association between a target stimulus (e.g. one's race, body image, religious orientation, sexual orientation) and either positive or negative evaluative adjectives. The IAT involves a sorting task that is repeated twice. Take, for example, the IAT associated with ageism, which assesses how quickly the faces of older (vs. younger) people are associated with negative (vs. positive) adjectives. In the first trial, older people's faces and negative adjectives and younger people's faces and positive adjectives are paired on the same response key in a keyboard (e.g. press K when older faces and negative adjectives are shown, and press S when younger faces and positive adjectives are shown). In the second trial, the pairing is reversed: older people's faces and positive adjectives are paired together on a response key, and a different response key is used pairing negative adjectives with the faces of younger people. If an ageing bias is present, participants will be faster in their responses when the response key pairs older people's faces with negative (vs. positive) adjectives.

The IAT assesses how strongly you have associated target stimuli (e.g. faces of older vs. younger people) with either positive or negative evaluative adjectives.
Source: Maria Ypsilanti

14.2.5 Implicit priming tasks

Implicit priming tasks are also used to examine the influence of different stimuli (e.g. affective, semantic) on reasoning and decision-making. In semantic priming tasks participants are exposed to **prime** that activate mental representations and actions that are semantically related to the prime (Bargh et al., 1996[15]). In an affective priming task, participants are presented with images or words with positive or negative valence and are evaluated on subsequent judgments. The goal is to observe how the emotional context of the primes influences participants' responses to the target, revealing implicit biases that may affect decision-making. In the priming phase, participants are briefly exposed to a series of images or words that evoke positive emotions (e.g. happy faces, positive words) or negative emotions (e.g. sad faces, negative words). They are then presented with a series of target stimuli, such as images depicting elderly people doing a range of activities (e.g. shopping, as shown in the image below) and they are asked to evaluate the target stimuli using a rating scale or by making rapid judgements about their feelings towards the individuals depicted. If participants who were primed with positive stimuli show more positive evaluations of the elderly individuals compared to those primed with negative stimuli, it suggests an implicit positive bias towards older adults. Conversely, if negative primes lead to more negative evaluations, it indicates a potential implicit negative bias. Implicit priming tasks are not without controversy and the mechanisms involved in priming effects have not been fully understood (Bargh 2006[16]).

Implicit priming tasks tasks can be used to identify implicit biases in decision-making.
Source: Fillippos Ypsilanti

14.3 Visual search tasks

LEARNING OBJECTIVE
Understand visual search tasks and their use in cognitive psychology research

14.3.1 Eye-tracking methodology: unveiling cognition through eye gaze

Eye-tracking methodology serves as a powerful tool in cognitive psychology, enabling researchers to investigate visual attention and the cognitive processes underlying it. The primary purpose is to understand how individuals allocate their attention to stimuli in their environment, providing insights into decision-making, problem-solving and information processing. The key mechanisms of eye-tracking involve the precise measurement of eye movements, including fixations and saccades. Fixations represent moments when the eyes are momentarily still, focusing on a specific point, while saccades are rapid eye movements between fixations. These mechanisms help researchers identify patterns of visual attention, gaze duration and the sequence of fixations, contributing to a comprehensive understanding of cognitive processes. There are two main types of eye-trackers: remote eye-trackers which use infrared light to track eye movements from a distance, providing more comfort to the participant; and head-mounted eye-trackers that allow participants to move into the environment, providing a more naturalistic experience. Analysis involves examining heat maps, gaze plots and areas of interest (AOIs) to understand visual attention distribution.

Eye-tracking methodology has contributed greatly to the understanding attentional biases and gaze patterns in individuals with anxiety disorders (Armstrong & Olatunji 2012[17]). The advantage of eye-tracking methodology is that longer exposure times allow researchers to understand orientation of attention (time to fixation), attentional maintenance (first fixation duration) and avoidance patterns that come into play later. For example, it is possible to understand patterns of eye-gaze across time (usually about 7000 ms) to emotionally salient and neutral faces (Figure 14.5) and explore *changes* of attentional allocation.

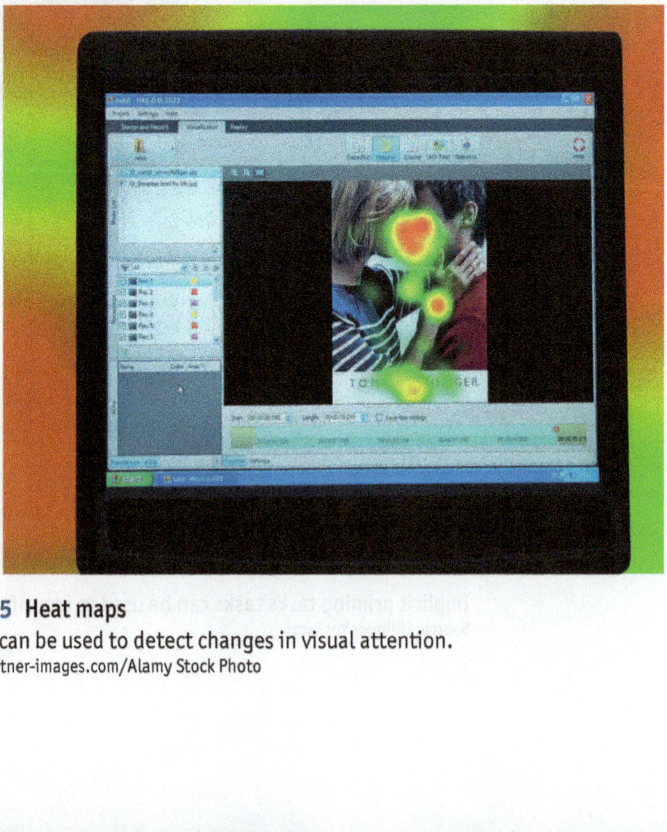

Figure 14.5 Heat maps
Heat maps can be used to detect changes in visual attention.
Source: artpartner-images.com/Alamy Stock Photo

What can eye-tracking reveal about consumer behaviour?

Eye-tracking is extensively employed in consumer behaviour research, particularly in marketing and advertising (Al-Azawai, 2019[18]). It helps assess visual attention to advertisements, product packaging and website layouts, influencing marketing strategies. For example, mobile eye-tracking glasses have been used to record natural eye-gaze patterns in supermarkets to identify marketing strategies in placing products on the shelves. Determinants like location, colour and images can increase eye-gaze duration, making certain products more attractive to consumers.

Mobile eye-tracking can be used to assess consumer responses in real-life dynamic environments.
Source: Ekkasit Keatsirikul/Alamy Stock Photo

14.3.2 Visual selection tasks

In a complex visual environment, individuals scan and process information to locate specific targets that are relevant to their current goal. Visual search tasks offer valuable insights into the mechanisms of visual attention, perception and cognition. Visual search tasks involve actively scanning a visual display to identify and locate a specific target amidst distractors. The targets can vary in complexity, ranging from simple shapes or colours to more intricate objects or symbols. The key characteristics of visual search tasks include the search rate, accuracy and reaction time, all of which contribute to our understanding of cognitive processes. The operationalisation of these tasks relies on feature integration theory, initially developed by Anne Treisman, which suggests that basic features (such as colour, shape and orientation) are processed at a pre-attentive level where visual processing is rapid and parallel across the visual field (things tend to 'pop' out). In the second stage, individuals actively search for specific objects by serial processing which takes more time and effort. There are numerous visual search paradigms in the literature and there are several factors that influence attention during such tasks (Wolfe & Horowitz, 2017[19]).

More recently visual search tasks have been used in conjunction with eye-tracking methodology to gain a better understanding of the strategies used to scan the environment.

14.4 Neurophysiological methods

LEARNING OBJECTIVE
Describe neural imaging methods and their applications in cognitive psychology

14.4.1 Electroencephalography/event-related potentials (EEG/ERPs)

Electroencephalography (EEG) and event-related potentials (ERPs) are invaluable tools in cognitive neuroscience, allowing researchers to understand brain–behaviour relationships. The strength of EEG/ERPs relies on the recording of the temporal aspects of neural processing (i.e. when do action potentials occur in response to external stimuli called events?). EEG is a non-invasive neuroimaging technique that records the electrical activity of the brain through electrodes placed on the scalp. The recorded signals represent electrical potential-generated neurons firing synchronously. EEG provides *high temporal resolution*, capturing rapid changes in brain activity on the order of milliseconds. EEG/ERP is particularly useful in investigating *when* something occurs rather than *where* something occurs and has been widely used in research on attention, memory, language and perception. ERPs are time-locked EEG responses that are elicited by specific events or stimuli. The P100, N100, P200 and P300 components are examples of ERP components, each with specific characteristics and associated cognitive processes. For example, the P100 component typically occurs around 80–120 milliseconds (ms) after the presentation of a visual stimulus mainly in the visual cortex, and represents early processing of visual stimuli, particularly the initial sensory processing of basic visual features such as contours and shapes. On the other hand, the P200 component typically occurs around 150–250 ms after the presentation of a stimulus and is associated with more complex stages of processing and influenced by factors such as attention and stimulus relevance.

ERPs reveal how our brain responds to disgusting stimuli

In a study by Zhang et al. (2015)[20], EEG/ERPs were used to identify differences in responses to stimuli of core and moral disgust. Participants were briefly exposed to a series of images depicting disgusting stimuli (i.e. core disgust such as faeces), most of which inferred physical contamination. They were also shown images of socially disgusting acts (i.e. moral disgust, such as someone hitting someone else) and a set of neutral stimuli (e.g. a chair), and ERPs were recorded to identify differences in the elicited brain activity between the image categories. The results showed differences in temporal responses between core and moral disgust images, suggesting that people process each type of disgust at different timeframes during exposure. Core disgust images elicit neural activities earlier than moral disgust, to protect the person from physical contaminants that may harm them. Moral disgust images elicit activation later in time, possibly because they require more conscious evaluation of a situation.

14.4.2 Heart rate variability (HRV)

Heart rate variability is a non-invasive measure that assesses the variations in the time intervals between successive heartbeats. Unlike traditional heart rate measures that provide a single average value, HRV captures the nuanced fluctuations in the beat-to-beat intervals, reflecting the balance between the sympathetic and parasympathetic branches of the autonomic nervous system. The key components

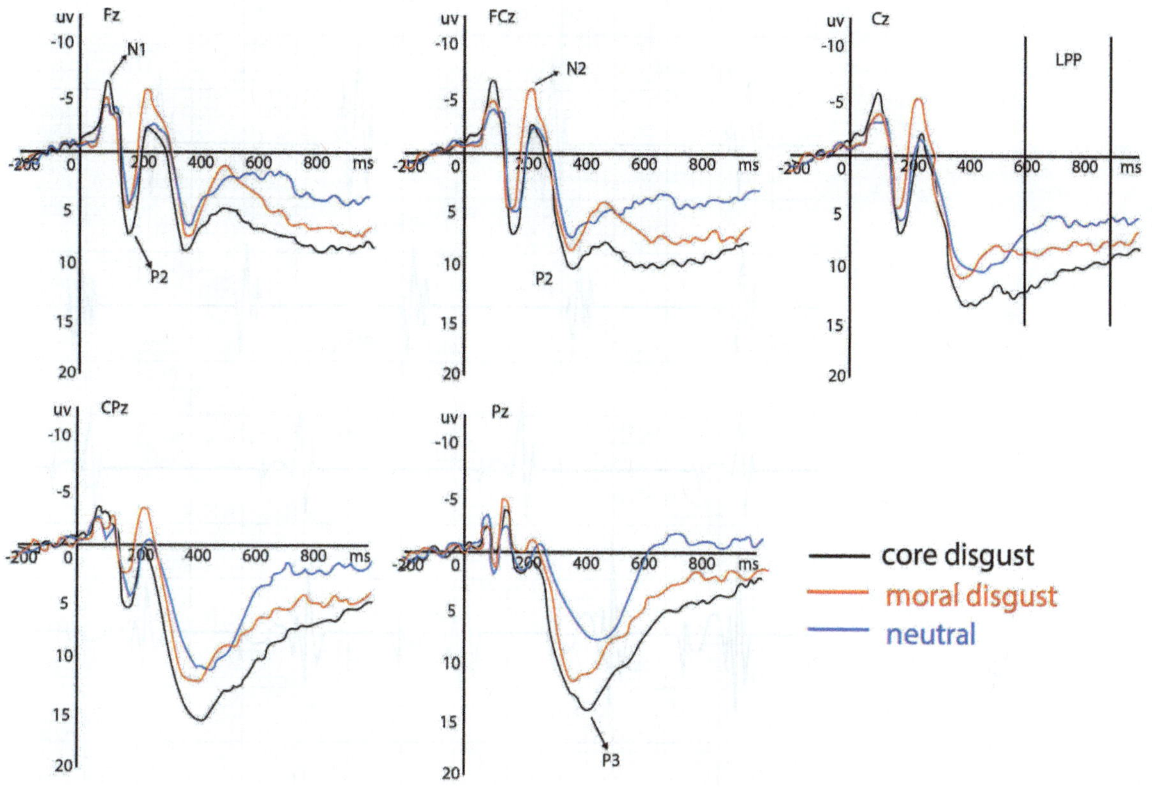

Figure 14.6 Event-related potential (ERP) recordings

ERP recordings indicate larger N100 amplitudes in response to core than to moral disgust stimuli.

of HRV include standard deviation of NN intervals (SDNN), which represent the overall variability of heart rate and the root mean square of successive differences (RMSSD), which reflects short-term variations in heart activity.

HRV methodology has been utilised to investigate the impact of cognitive stressors on autonomic regulation. Increased cognitive load often correlates with decreased HRV, indicating a shift towards sympathetic dominance. In addition, several studies have explored the role of HRV in emotion regulation, demonstrating that higher HRV is associated with better emotional regulation. Finally, HRV measures have been employed in mindfulness research, revealing the positive impact of mindfulness practices on autonomic balance and cognitive functioning.

14.4.3 Galvanic skin response (GSR)

Galvanic skin response (GSR), also known as electrodermal activity (EDA) or skin conductance response (SCR), is a physiological measure that gauges the electrical conductance of the skin in response to emotional or arousing stimuli. This non-invasive and sensitive technique has found applications in various fields, ranging from psychology and neuroscience to market research and lie detection. GSR is based on the premise that the skin is an excellent conductor of electricity due to the presence of sweat glands. When an individual experiences emotional arousal, the autonomic nervous system responds by altering the activity of the sweat glands. This, in turn, affects the electrical conductance of the skin. GSR measures the changes in skin conductance, reflecting the variations in emotional states, arousal

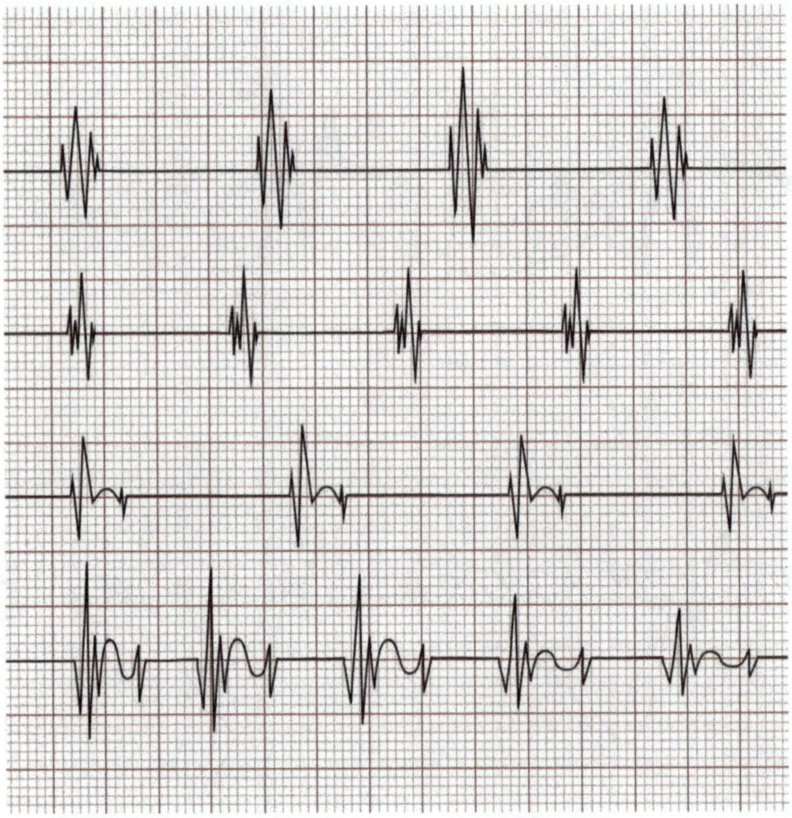

Figure 14.7 HRV measures fluctuations in beat-to-beat intervals
Source: DigitLog/Alamy Stock Photo

levels and sympathetic nervous system activity. GSR is extensively used in psychology to investigate emotional responses. It helps researchers understand how individuals react physiologically to stimuli, providing insights into emotional processing and arousal. Also, GSR has historical significance in lie detection. Changes in skin conductance may occur when individuals attempt deception, making GSR a component in polygraph examinations.

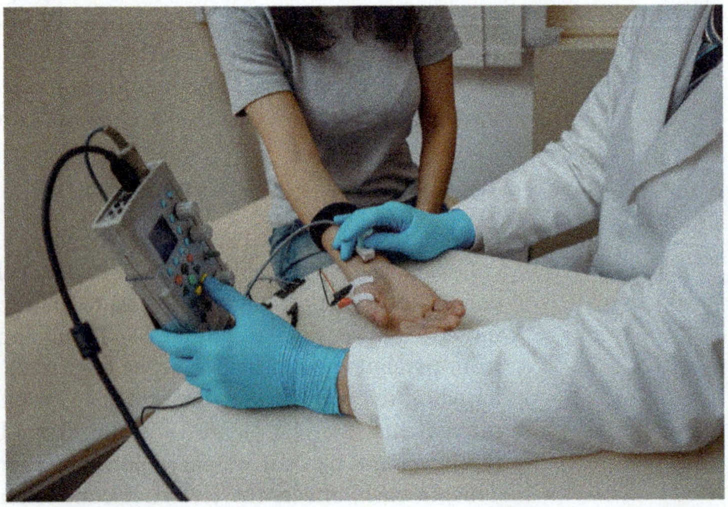

GSR measures the changes in skin conductance, reflecting changes in sympathetic nervous system activity.
Source: romanzaiets/123RF

14.4.4 Electromyography (EMG)

Electromyography (EMG) is a physiological measurement technique that captures and records the electrical activity generated by skeletal muscles. Traditionally associated with studies in kinesiology and motor control, EMG has increasingly found applications in cognitive psychology, shedding light on the intricate relationship between cognitive processes and muscular activity. Facial muscles play a crucial role in expressing emotions. EMG is used to measure the subtle muscle movements associated with facial expressions, providing information about emotional responses and cognitive processing related to affective states. Another application of EMG is to assess muscle tension as an indicator of stress and cognitive load. Increased muscle tension, especially in the facial and neck muscles, can be associated with heightened cognitive demands and emotional stress.

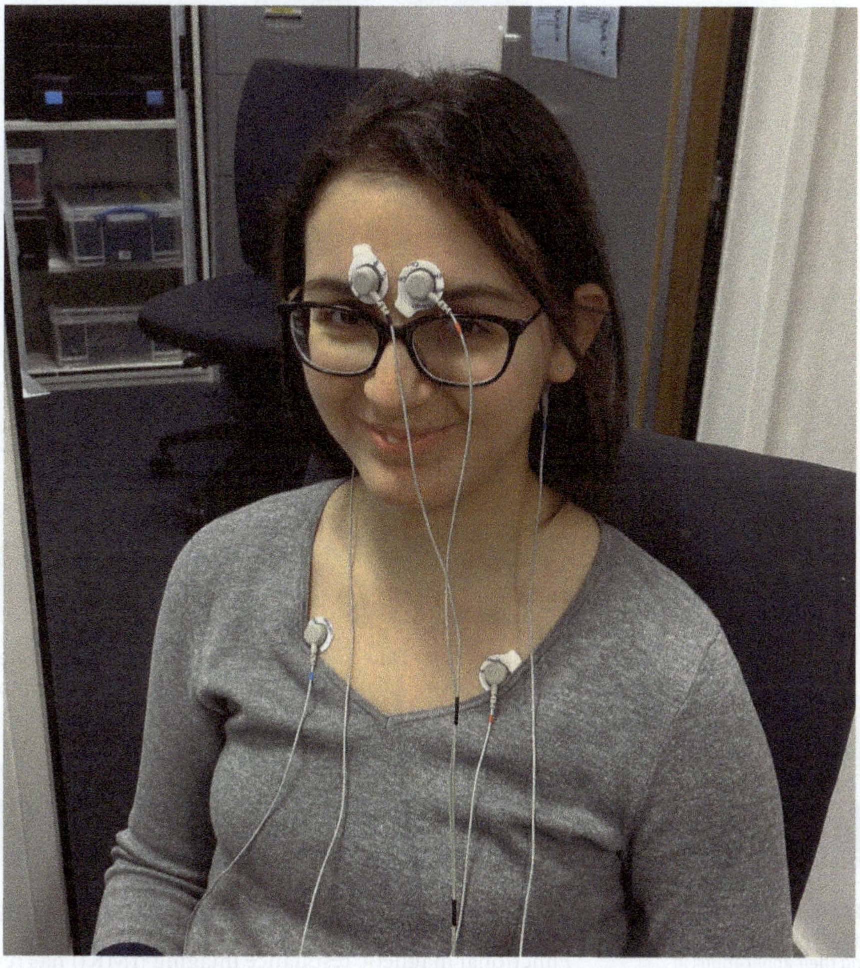

Electromyography in use on a subject.
Source: Chrysi Savvidou

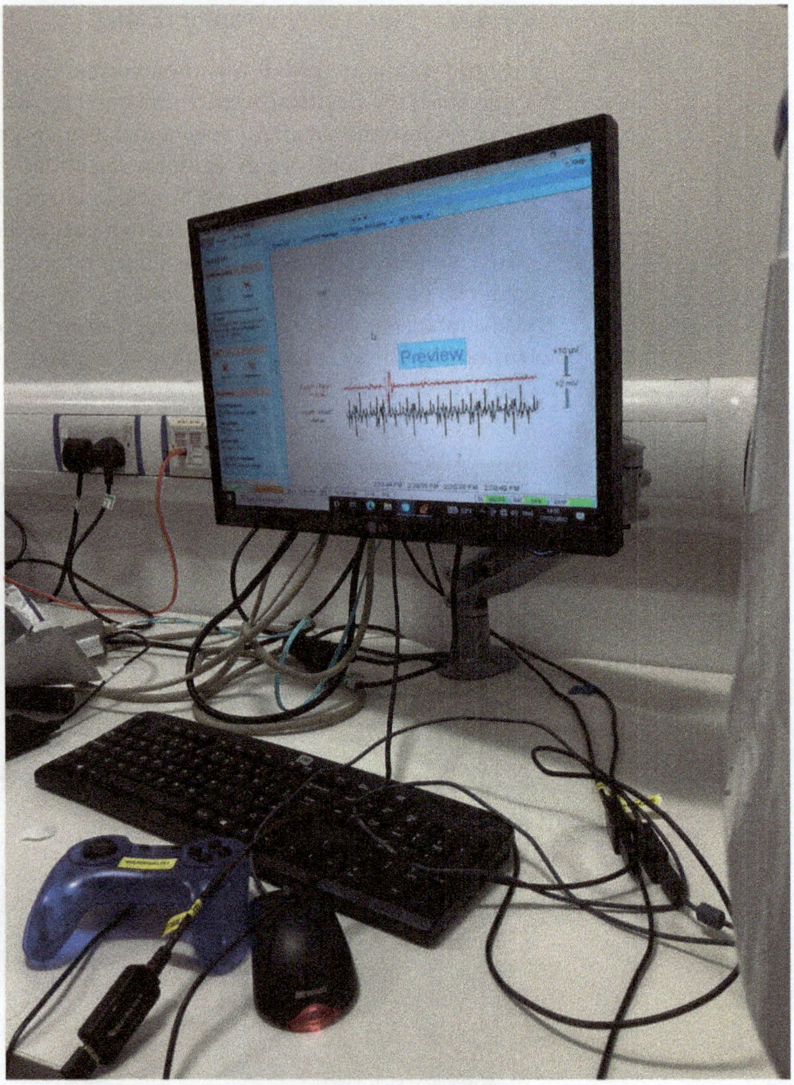

EMG is used to measure the subtle muscle movements associated with facial expressions, providing information about emotional responses and cognitive processing related to affective states.
Source: Chrysi Savvidou

14.5 Neuroimaging and brain stimulation methods

LEARNING OBJECTIVE
Describe neural imaging methods and their applications in cognitive psychology

14.5.1 Functional magnetic resonance imaging (fMRI)

Functional magnetic resonance imaging (fMRI) has revolutionised cognitive neuroscience by allowing researchers to investigate brain activity and functional connectivity non-invasively by measuring changes in blood flow. It is different to a **magnetic resonance imaging (MRI)** in that is records brain activity while someone is performing a particular task.

The approach taken is to use fMRI to show differences in activation between brain regions that are assumed to take part in a specific mental process, although many would argue that there is rarely a single region that is activated when executing a cognitive task. Indeed, *mapping* cognitive process onto the brain has been challenging since there isn't a simple one-to-one method. One reason for this is that cognitive processes probably activate **networks** rather than selective regions in the brain. It could also be that our understanding of mental processes does not depict accurately how the brain works – also known as **mental ontology** (Poldrack, 2008[21]).

Mental ontology
Refers to a set of concepts that define the properties and relationships between mental processes.

In the realm of memory research, fMRI has been instrumental in unravelling the neural substrates and networks involved in encoding, storage and retrieval of memories. In a meta-analysis, Terry et al. (2015)[22] found that different brain regions were activated during an episodic memory task in individuals with Alzheimer's disease and those with mild cognitive impairment.

14.5.2 Transcranial magnetic stimulation (TMS)

Transcranial magnetic stimulation (TMS) is a non-invasive technique that involves the use of a magnetic coil placed on the scalp to generate brief and focused magnetic pulses. These pulses induce electrical currents in the underlying neural tissue, leading to the depolarisation of neurons and subsequent modulation of brain activity. Depending on the parameters of the stimulation, TMS can either enhance or inhibit neuronal excitability. You can directly observe the effects of a single foil in behaviour when a single pulse is delivered to a specific brain region affecting motor movement (involuntary finger movement). Repeated TMS (rTMS) involves the application of repeated pulses over time, allowing for longer-lasting effects on neural activity, causing changes to brain connectivity (similar to training).

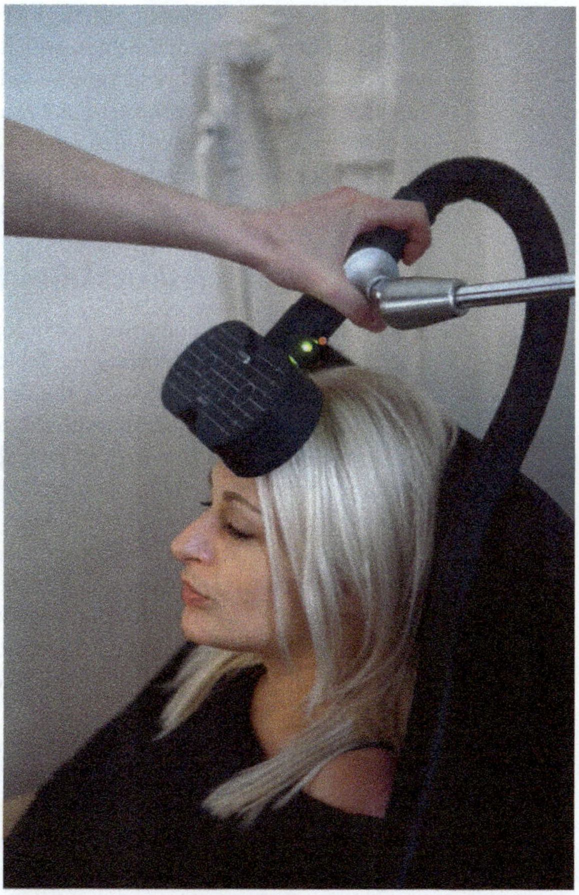

TMS has been used both in basic neuroscience research and in clinical applications for the treatment of different disorders and health conditions.
Source: BSIP SA/Alamy Stock Photo

Clinical applications of TMS

Treatment-resistant depression (TRD). TMS has been FDA-approved for the treatment of major depressive disorder in individuals who have not responded to at least one antidepressant medication. It is typically used when other treatment options, such as medication and psychotherapy, have not been effective.

Obsessive-compulsive disorder (OCD). TMS has shown promise in the treatment of OCD, particularly in cases that are resistant to conventional therapies. Studies have investigated its potential to modulate neural circuits associated with obsessive-compulsive symptoms.

Migraine treatment. TMS has been explored as a potential treatment for migraines. Repetitive TMS (rTMS) has shown some efficacy in reducing the frequency and severity of migraines in certain individuals.

14.6 Research integrity and ethics

LEARNING OBJECTIVE
Understand principles of research integrity and ethics in cognitive psychology

Having discussed the different methods you can use to measure and assess cognitive functions, we now turn to an important aspect of research: *how* to conduct research ethically and with integrity.

The British Psychological Society (BPS) sets out clear principles for research integrity and practice in psychology. All students and practitioners working in psychology, as researchers, academics or practitioners, should acquaint themselves with the BPS Code of Human Research Ethics, which can be found on the BPS website (https://www.bps.org.uk/). In this section we will summarise some of the key principles of research integrity and ethics set out by the BPS, by providing examples of violations of these principles. We will also refer to key ethical issues when developing experiments in cognitive psychology and provide some information on how to complete an ethics form for high- and low-risk studies.

14.6.1 What is research integrity?

Research integrity in cognitive psychology, as defined by the British Psychological Society (BPS), involves meeting the ethical and professional standards that govern the research cycle. This includes the design, conduct, analysis, interpretation and reporting of research. Research should meet standards of quality, integrity and contribution to the development of knowledge and understanding and should not waste resources by using participants unjustly. Importantly, researchers should be honest about their findings, methods and data. This includes accurately reporting results, methodologies and potential limitations of the research. In cognitive psychology, research should use robust methodologies, appropriate and validated measures, and unbiased statistical analyses that report negative and null findings. Any form of research misconduct, including fabrication, falsification and plagiarism, is strictly prohibited. Researchers must adhere to ethical guidelines and report any breaches they observe.

Recently, the requirement by journal articles is that all experimental designs and intervention studies are pre-registered to ensure that research questions and hypotheses are predetermined and remain the same throughout the study. To pre-register a study you can use the Open Science Framework free[23] online platform that supports researchers and their collaborators. Once registered a study can be open to the public to ensure integrity and honesty in the research findings. In addition, now more often than ever, journal articles and funders expect researchers to make their data openly available when they publish articles so that other researchers can utilise them for reviews and meta-analyses, or they can be re-analysed or reused in research.

Findings should be communicated clearly, accurately and without misleading or biased interpretations. Researchers should strive to publish their work in reputable journals and ensure that their findings are accessible to other researchers and the public. This is why funders request that the publications are 'open access', that is, they are immediately available to the public (or have a period of embargo).

Also, research should be done for the general good and all researchers have a social responsibility to respect and address the impact of their research on the public. In doing so, on some occasions researchers may need to disclose the purpose of the study after data collection has taken place (also known as temporary deception). In any case, psychology researchers should always maximise the benefits of their work and minimise the risk to participants and themselves.

14.6.2 The reproducibility crisis in psychology

The reproducibility crisis in psychology refers to growing concerns about the inability of researchers to replicate findings of previous research (Hensel, 2020[24]). This means that when the studies are repeated, often by different researchers using the same methods, the results are not consistently reproduced. This crisis has significant implications for the credibility and reliability of psychological research. Some of the causes of the reproducibility crisis involve a publication bias where researchers and journals prefer to publish positive findings and not **null findings**. By not publishing null findings, other researchers may spend resources exploring a topic, a method or an intervention that has been found ineffective but was not reported or published. Researchers may engage in practices such as p-hacking (manipulating data to achieve statistically significant results) or selective reporting of results or **fishing** data and **salami** publications. Naturally, when a researcher spends time and money on an idea, they hope for their hypotheses to be true. However, null findings are as important as positive findings and need to be reported and published. Other causes of the reproducibility crisis in psychology are small sample sizes (low **power** of studies) and underreported **effect sizes**. Furthermore, insufficient reporting of methods and data makes it difficult for other researchers to replicate studies. Psychological phenomena are often complex and influenced by numerous variables. This complexity can make replication challenging, as slight variations in study design or context can lead to different outcomes. As we discussed in the previous section, promoting open science practices, such as pre-registering studies, sharing data and materials, and publishing in open-access journals, helps to increase transparency and accountability in research.

Personality, stress and the risk for cancer and cardiovascular disease

Cancer and cardiovascular disease represent the top two leading causes of death globally, and understanding their causes is of paramount importance for their prevention and treatment. For over 30 years, related psychological research was influenced by the work of the late prominent psychologist Hans Eysenck and his associate, Ronald Grossarth-Maticek. Starting in the mid-1980s, Eysenck and Grossarth-Maticek published a large number of research articles about the relationship between personality, stress and the risk for cancer and cardiovascular disease. In 1991, Eysenck proclaimed that personality and stress significantly increased the risk for cancer and coronary heart disease (CHD), while dismissing the negative effects of smoking. Some 30 years later, and following controversy led mostly by the British psychiatrist Professor Anthony Pelosi, more than a dozen papers by Eysenck and Grossarth-Maticek were retracted over concerns of data manipulation. David Marks (2019),[25] the chief editor of the *Journal of Health Psychology*, requested further scrutiny of over 60 papers published by Eysenck and Grossarth-Maticek. The inquiry led by King's College London, the employer of Hans Eysenck for more than three decades, concluded that the findings reported by Eysenck and Grossarth-Maticek were implausible and incompatible with contemporary science and understanding of how diseases develop. Cases like this signify the importance of research integrity and reproducibility.

14.6.3 The ethics process

How do we obtain ethical approval for our studies? All organisations that carry out research are expected to have an ethics committee that reviews, scrutinises and approves ethics applications. This is an independent committee, and reviewers remain anonymous throughout the review process. The researcher applied for ethics approval prior to the commencement of any study, whether it involves human participants or not. In essence, no research should be undertaken unless the researcher has ethics approval and can document it. This process ensures that research is carried out using the ethical principles determined by the BPS. Some of the main principles involve:

- **Respect for participants.** Ethical considerations for participants are of paramount importance including respecting the autonomy, privacy, dignity and rights of participants. This includes obtaining informed consent, ensuring confidentiality and minimising harm when designing studies and collecting data.
- **Informed consent and withdrawal.** All potential participants should be thoroughly informed about the purpose of the study, why they have been asked to participate, what will happen during the study, what they are expected to do and where they can address questions should they wish to do so. They should also be informed on how their personal data will be managed, stored and used and any potential risks to participation (physical or psychological risks). Finally, they should be aware that they have the right to withdraw at any point and that they do not need to provide any reason for doing so, and how they can do that. All this information is conveyed to participants through a participant information sheet (PIS) that should be written in an accessible way. All participants should provide consent (written or verbal) after having read the PIS. This involves participants of all ages – including children. If certain participants are unable to provide consent themselves then a carer or guardian can sign a consent form on their behalf, but they should still be made aware of their rights in an accessible way. Importantly, they need to know their right to withdraw and how they can do that without providing a reason. Details about how to obtain informed consent can be found in the BPS Code of Human Research Ethics. If there is any form of deception, participants should be debriefed at the end of the study about the true purpose of the study and should be allowed to withdraw their data if they wish to do so.

Summary: Research methods in human cognition

14.1 Purpose of research methods in cognitive psychology

Research methods in cognitive psychology allow us to systematically investigate processes that underlie human cognition. By employing various experimental and observational techniques, cognitive psychologists can develop theories about how the mind works, test these theories and refine them based on empirical evidence. These methods allow us to isolate specific cognitive functions and to understand the brain's role during cognitive processes. Ultimately, the purpose of these research methods is to gain a deeper and more accurate understanding of cognitive functions, which can inform the development of interventions, educational strategies and technologies to enhance mental health and cognitive performance.

14.2 Reaction time-based tasks

Reaction time-based tasks are fundamental tools in cognitive psychology, used to measure the speed and accuracy of mental processes. These tasks involve recording the time it takes for a participant to respond to a specific stimulus, providing insights into the efficiency and organisation of cognitive functions. These tasks are particularly valuable because they offer a quantifiable and objective means to study cognitive processes that are otherwise difficult to observe directly. Reaction-time experiments have contributed significantly to our understanding of problem-solving and the impact of cognitive load on performance, thereby playing a crucial role in advancing cognitive psychology.

14.3 Visual search tasks

Visual search tasks are a pivotal methodology in cognitive psychology, used to explore how people locate a target among distractors. These tasks involve participants scanning a visual environment to identify a specific target, such as finding a red circle among green circles and red squares. Through these tasks, researchers investigate various aspects of attentional processes, such as selective attention, feature integration and the effects of target–distractor similarity.

14.4 Neurophysiological methods

Neurophysiological methods are essential in cognitive psychology to investigate physiological changes during exposure to sensory stimuli.

14.5 Neuroimaging and brain stimulation methods

Techniques such as magnetic resonance imaging (MRI) and functional MRI (fMRI) allow researchers to measure neural activity with high temporal and spatial resolution. These methods enable the investigation of how different brain regions are involved in processes such as perception, memory, decision-making and language. By correlating specific patterns of brain activity with cognitive tasks, researchers can map the neural circuits that underpin mental functions. Such methods also facilitate the study of brain plasticity, showing how cognitive processes change with learning and experience.

14.6 Research integrity and ethics

Research integrity and ethics are a vital part of the work of every student and academic involved in the field of psychology. The British Psychological Society (BPS) is the monitoring body that sets out the guidelines on how to adhere to ethical principles. The ethics process ensures that all researchers receive appropriate ethics approval before commencing any data collection. This involves all studies, whether they involve humans or not.

Chapter 14 quiz

Question 1

Which type of task was originally designed to measure attentional biases in processing stimuli?

A. Dual tasks
B. Go/No-Go Tasks
C. Dot-probe paradigm
D. Serial recall paradigms

Question 2

You are a researcher looking at how people view the home page on their mobile phones and which apps they view first. Which research method could be used?

A. Eye tracking
B. fMRI
C. EEG
D. HRV

Question 3

What is an advantage of an fMRI (functional magnetic resonance imaging) method?

A. The method is only slightly invasive as it involves a participant being strapped to a bed
B. The method is quick to administer
C. The method has been shown to be an effective treatment for anxiety and depression
D. You can research behaviour in real time through recording brain activity during a particular task.

Question 4

In psychology, a researcher must meet the ethical and professional standards of appropriate governing bodies such as the British Psychological Society. What is this known as?

A. Research integrity
B. Research reproducibility
C. Research respect
D. Research consent

Question 5

A researcher has designed a questionnaire study to be conducted online. They give out the survey through an online link that provides a participant information sheet, the questionnaires and a debrief sheet. Before participants complete the questionnaire, what must they do?

A. Read a thank you message
B. Sign a consent form
C. Inform the researcher that they are taking part
D. Read up on what the questionnaires are

Answers to end of chapter quizzes

Chapter 1 quiz

Question 1

CORRECT ANSWER:

A. Perceptual, memory and comprehension.

- *First your perceptual processes would be involved as you see/hear the question.*
- *Next, memory processes would be involved as look to see if you have heard or seen the words and phrases before. Have you heard some of these terms before or read them? You need to then encode the question key terms and try to locate them in your memory.*
- *The final process is then comprehension. Do you understand the words? Consider whether they make sense.*

Question 2

CORRECT ANSWER:

C. There was a lack of control over other variables now associated with memory.

The work from Ebbinghaus was very early work. As he has no laboratory and relied upon his own resources., he could not control the conditions he recalled memory in unlike the research today.

Question 3

CORRECT ANSWER:

B. In the application of language, it was found that other cognitive processes such as attention, recognition and computation were involved in language development.

This is correct. Researchers have since suggested that language development is underpinned by many cognitive factors. These can include attention and memory processes alongside the ability to compute sentences. Language development was key here as it provided an application.

Question 4

CORRECT ANSWER:

A. Perception

As this is a new task to you, you have been given a recipe. The first thing you would likely do would be to read the recipe (through your visual senses). From there, your perceptual processes would be key as you would need to turn the visual information into a form that you understand.

Question 5

CORRECT ANSWER:

A. Top-down processing involves existing knowledge rather than environmental information.

This is correct as top-down processing is also knows as conceptually driven processing.

Chapter 2 quiz

Question 1

CORRECT ANSWER:

B. *Semantic memory*

This type of memory will help people remember things that involve general information. For example, to ride a bike, you must sit on the bike, place your feet on the pedals and move your feet and legs to push the bike.

Question 2

CORRECT ANSWER:

C. Dendrites gather electrical and chemical information that is received from another neuron.

This is correct. Neurons can have both chemical and electrical communication, therefore dendrites gather this information.

Question 3

CORRECT ANSWER:

A. Hypothalamus

This part of the brain is responsible for stress responses. It is important as the aim of this region is to maintain homeostasis, and it is the link between your endocrine system where hormones are released, and your nervous system.

Question 4

CORRECT ANSWER:

B. The occipital lobe and temporal lobe.

The occipital lobe is primarily responsible for vision, therefore this is likely to be damaged with the vision issues. As there are linguistics issues alongside this, the temporal lobe is also likely to be damaged.

Question 5

CORRECT ANSWER:

D. Basic units match elementary shapes against more complex symbols.

This is the main process of the connectionist model. The model looks for connections between the basic units to the higher level units. An example of this would be word recognition whereby words are broken down into smaller symbols (basic units) before being matched with single letters and, finally, full words.

Chapter 3 quiz

Question 1

CORRECT ANSWER:

A. Detection studies determine what sensory data can be detected, while discrimination studies calculate how much change is required to detect a difference.

Detection studies indicate the properties of physical stimuli that can be sensed and perceived, either within or outside one's conscious awareness. In comparison, studies of discrimination tell us how much a stimulus needs to change to make a noticeable difference to a (human) observer.

Question 2

CORRECT ANSWER:

C. At each step of visual perception (from the eye to the brain), there is some loss of information.

Because there is a great deal of compression of information in the early stages of vision, the message that finally reaches the visual cortex is an already processed and summarised record of the original stimulus

Question 3

CORRECT ANSWER:

B. Synaesthesia

This condition is where the brain has issues processing certain types of stimuli and can mix them up. For example, a person may smell or taste something when they see a certain image. In this case, the brain will have involuntary sensory experiences alongside the normal ones.

Question 4

CORRECT ANSWER:

D. When viewing an image, part of the image is treated as the figure or foreground (the object identified), which is segregated from the visual information upon which it is set (the background).

This is the correct answer. Classic examples illustrating difficulties in determining figure-ground are reversible figures, in which a person shifts back and forth between what is the foregrounded object and what is the background. Examples include the 2024 Olympic Games logo.

Question 5

CORRECT ANSWER:

A. Parallel processing

This is correct. As the individual is completing multiple taks at the same time (driving, listening to music, looking at the sat nav), they have the ability to process multiple pieces of information at the same time. It is very similar when someone carries out a task that they are familiar with such as walking with a friend down the street and having a conversation.

Chapter 4 quiz

Question 1

CORRECT ANSWER:

A. Selective attention

As the student is focussing upon many different aspects (reading, listening, writing, speaking with friends), selective attention would not be appropriate here as the student is completing multiple tasks at the same time.

Question 2

CORRECT ANSWER:

C. Explicit processing is the type of processing that you are consciously aware of whereas implicit processing is the processing you are not consciously aware of.

This is correct. When someone is actively aware of what they are processing, they are using explicit processes and this can, in turn, use explicit memory.

Question 3

CORRECT ANSWER:

B. A person looks in the mirror and only cleans makeup off the left side of their face.

A person who has hemineglect has an attentional disorder where they cannot voluntarily direct attention to half of the perceptual world.

Question 4

CORRECT ANSWER:

A. Personally relevant information can bypass the selective filter.

This is one of the main criticisms with the early selection theory. Personally relevant information may not be detected by the filter. This is evidenced from the work of Moray (1959) who investigated whether people heard their own name in a crowded place.

Question 5

CORRECT ANSWER:

D. Driving an automatic car and when using a manual car, not putting the car into the correct gear.

If a person has been driving an automatic car, their attention will become more automatic for driving that type of car. When a different type of car is used, the driver may make several mistakes, thinking that they are still driving an automatic car.

Chapter 5 quiz

Question 1

CORRECT ANSWER:

C. Remembering the colours of the rainbow

This is correct. In memory, recoding occurs when we chunk individual items together to form a larger, more meaningful pattern. In this case, the individual colors have no meaning but when chunked together, they suggest a rainbow.

Question 2

CORRECT ANSWER:

C. Retroactive interference

This is the correct answer. The newer material of the Maths homework has interference with the memory of the phone number. As both are numbers, they are in the same mode of communication meaning that interference is more likely.

Question 3

CORRECT ANSWER:

A. Encode the probe, Scan and comparison with the memory set, Binary (yes/no) decision, Executive motor response.

This is correct. A participant would view a probe (e.g. a letter), they would then compare this with the memory set they have been given before making the decision. Finally, they would provide their response to the stimulus.

Question 4

CORRECT ANSWER:

B. Investigating whether several unrelated tasks drain the same pool of resources.

Dual task methods are important as they can show interference effects. If one mode of stimuli impacts another, we can suggest that the same pool of resources are used.

Question 5

CORRECT ANSWER:

D. We are unsure of the effectiveness of working memory training as working memory has associations with other cognitive constructs such as intelligence and attention.

This is correct! As working memory can be associated with many other constructs, we can question the effectiveness of the training.

Chapter 6 quiz

Question 1

CORRECT ANSWER:

A. Mnemonic

This is correct. The child has used the first letter of the planets to create a narrative of what elephants can do in their spare time. This will help the child to remember the word with the use of words known as 'peg words'. These are unrelated words to the key information to be remembered, however, they are simple to use.

Question 2

CORRECT ANSWER:

C. There is a positive relationship between the frequency of rehearsal and the rate of recall in the first few words of the serial position curve.

Rundus (1971) suggested that the items at the start of a list need to have greater rehearsal to be remembered more clearly. He demonstrated evidence of this within his research. The greater rehearsal suggested that more words were remembered.

Question 3

CORRECT ANSWER:

C. Rob is studying snakebite prevention tips in preparation for a week-long hike in the desert.

There is an emotional link between what Rob is trying to remember and his destination. For example, he knows if he receives a snake bite, there will be a painful and emotional response. In general, emotive situations can increase the ability to remember situations. Rob may also use mental imagery to imagine the effects of the snake bite.

Question 4

CORRECT ANSWER:

B. The researchers could not remain impartial from the data collection and the results could have been biased with the expectation of remembering items from the past.

Yes, this is correct. In more recent times, when case studies are developed, the researcher will not use themselves or their relatives to remain unbiased. Researchers such as Wagennar recorded his daily events and tried to recall them at a later date and while this research is important in the understanding of autobiographical meory, there can be issues with how the early data was collected.

Question 5

CORRECT ANSWER:

A. The semantic network model viewed memory as a computer program whereby networks are used to connect different nodes, like a computer.

Yes, this is correct. The feature comparison model acts as though the human brain is less like a computer in the way we develop feature lists within memory.

Chapter 7 quiz

Question 1

CORRECT ANSWER:

D. Suggestibility

Rachel has heard information from her friends and although there was no rock music at the festival, she has incorporated this as a memory.

Question 2

CORRECT ANSWER:

A. Research has shown that they can dramatically increase memory performance.

This is correct. While the research is mixed on the type of retrieval cues used, in general the research does suggest that memory can be dramatically improved. This is from both earlier work such as Tulving and Pearlstone (1966) to more recent work.

Question 3

CORRECT ANSWER:

C. Leading questions

This is correct. Although not a strategy widely used, research has shown that solicitors can use leading questions to make eyewitnesses think that something happened when it did not. For example, they could ask the colour of a handbag when the eyewitness did not recall a handbag.

Question 4

CORRECT ANSWER:

D. Anterograde amnesia impacts the formation of new memories whereas retrograde amnesia impacts memories from the past.

As H.M. could remember some childhood experiences, he was not experiencing retrograde amnesia as he could not form new memories.

Question 5

CORRECT ANSWER:

B. The activation of related concepts and meanings.

This is correct. Priming can be seen as awakening the associations within memory, meaning someone may be able to remember more. Priming occurs when an individual's exposure to a certain stimulus influences their response to a subsequent prompt, without any awareness of the connection.

Chapter 8 quiz

Question 1

CORRECT ANSWER:

A. Chomsky proposed that language development is seen as an innate process whereby the brain structures allow us a natural capacity to learn.

This is correct. In psycholinguistics, a key position was suggested from Chomsky who studied language in terms of humans having an innate ability to process and develop language. He suggested that we have an internal system called the Language Acquisition Device within the brain.

Question 2

CORRECT ANSWER:

A. Visual-spatial channel

This is correct. As Hockett's Linguistic Universals focus more upon the auditory side of processing, he did not make specific references to reading information which would require the use of visual and spatial information. This channel does not exist.

Question 3

CORRECT ANSWER:

C. Sounds in English may not sound the same in another language.

This is correct. Each language has their own set of sounds or syllables, and English has a more complex system of blending phonemes.

Question 4

CORRECT ANSWER:

B. Word order

A person must understand that sentences come in certain word orders for them to be understood. In this case the sentence would be 'Calo asked the teacher about his homework'.

Question 5

CORRECT ANSWER:

A. A word that has multiple meanings.

This is correct. For example, the word 'play' could mean a child can play with their friends, and it could also mean a play in a theatre, both of which are different.

Chapter 9 quiz

Question 1

CORRECT ANSWER:

D. Propositional textbase

This is correct. At an intermediate level is the propositional textbase, which captures the basic idea units present in a text.

Question 2

CORRECT ANSWER:

B. ERP and fMRI.

This is correct. Both of these methods look at brain activation in real time and can show how the brain is working during language comprehension.

Question 3

CORRECT ANSWER:

A. As a person reasons more words, the more opportunity there is to build word structures meaning that it is easier to read.

This is correct. More words (and in the correct order) means that a person has the opportunity to build structures with these words and will make understanding easier.

Question 4

CORRECT ANSWER:

A. Jennifer bought a PlayStation. The PlayStation was on sale.

This is correct as the sale of the PlayStation was directly related to Jennifer buying it.

Question 5

CORRECT ANSWER:

C. Theory of Mind

This is correct. The theory plays a crucial role in explaining how people converse and explained their development more widely. It is the mental model that explains how people understand what their conversational partner is interested in and what they are like.

Chapter 10 quiz

Question 1

CORRECT ANSWER:

B. Five groups of five students introduced themselves to their own group members, and then to all other group members. How many introductions did each person give?

This is correct as it refers to a statistics problem which can be solved by a set of rules. You know how many group members there are, so this can be solved with an algorithm.

Question 2

CORRECT ANSWER:

B. The order of the sentences

This is correct. People will perceive information based upon the order of sentences and will form a logical argument based upon the order in which the sentences are displayed. This means that different orders of sentences can have different conclusions.

Question 3

CORRECT ANSWER:

A. They can help us to make decisions quicker as they provide us with mental shortcuts to solve a problem.

This is correct. Heuristics are strategies that can be used to help someone make decisions. For example, comparing statements against each other or using trial and error methods in decision making.

Question 4

CORRECT ANSWER:

D. Outcome magnitude.

This is correct. The outcome is getting a book with over 50% discount meaning that this outcome could be seen as more important than the fuel it would cost to drive across town.

Question 5

CORRECT ANSWER:

D. Personality

This is correct. Personality has been shown to impact reasoning and decision making. For example, if you are a person who takes risks, your decision making may mean you then take more risks than someone who has a less risky personality.

Chapter 11 quiz

Question 1

CORRECT ANSWER:

B. 4

This is correct. Problem solving involves goal directedness, sequence of operations, cognitive operations and sub-goal decomposition.

Question 2

CORRECT ANSWER:

A. The chimpanzees demonstrated that they could use long and short sticks to see which of them could reach a banana.

This is correct. In this early work, one of the key chimpanzees (Sultan) demonstrated that he had goal directed behaviour and that there was a sequence of operations before the problem was solved.

Question 3

CORRECT ANSWER:

B. An analogy

This is correct as you can look at the relationship between the first and second words. You can ride a bike so your analogy would suggest that you can do a similar thing with a car. In this case, you drive the car. This is known as analogical problem-solving.

Question 4

CORRECT ANSWER:

C. The GPS did not always provide a clear solution even though it was designed to.

This is correct. The GPS would often stop and abandon the task when a sub-goal could not be completed instead of locating another way of completing the task.

Question 5

CORRECT ANSWER:

A. Drawing inferences and developing sub-goals.

This is correct and it was advice given from Wicklgren's (1974) approach. This approach can help in finding more than one solution to a problem.

Chapter 12 quiz

Question 1

CORRECT ANSWER:

A. The ability to perceive, categorise and interpret social behaviour of other people.

This is correct. Social cognition is concerned with how we interpret our social environment and other people's interactions, which can then influence our behaviour.

Question 2

CORRECT ANSWER:

D. Orbitofrontal cortex (OFC)

This is correct. Research from Herold et al. (2016) has suggested that this is a key area to process both types of information.

Question 3

CORRECT ANSWER:

C. Self-referential knowledge.

This is correct. Klein et al. (1996) investigated this using a personality test, administered both before and after the amnesia had been resolved. There were no differences in the test results suggesting that personality, self-referential knowledge, was not impacted.

Question 4

CORRECT ANSWER:

A. Working memory

This is correct. This is the area of memory we use on a daily basis, therefore the term 'social working memory' was developed from this.

Question 5

CORRECT ANSWER:

B. Empathy is often elicited when we see pain from others in social situations, in terms of both people receiving pain and those inflicting pain.

This is correct. There are areas within the brain that have been suggested to link empathy and social processing.

Chapter 13 quiz

Question 1

CORRECT ANSWER:

B. Limbic system

This is correct. This brain region that consists of several structures has been shown to be crucial in emotional processing.

Question 2

CORRECT ANSWER:

A. Attention is preferentially allocated to more fearful stimuli.

This is correct. Attention directs us to more fearful stimuli so we can appraise and deal with the stimuli if needed.

Question 3

CORRECT ANSWER:

A. We have an optimal level of emotional arousal which allows most learning to occur. After that point, higher levels of arousal mean memory then declines.

This is correct. The Yerkes-Dodson law suggested links between memory and emotional arousal, suggesting that we have an optimal limit that we reach before emotional arousal becomes a problem.

Question 4

CORRECT ANSWER:

D. People who experience stressful situations can also experience physiological changes due to the stress they are under. This can impact how emotions are processed and how decisions are made.

This is correct. There is a strong link between stress and decision making, in particular in those who have anxiety related disorders.

Question 5

CORRECT ANSWER:

B. Emotion regulation

This is correct. Rushi understands his emotional feeling but he also understands that he is in a professional situation, therefore it is not appropriate to show these feelings in the meeting.

Chapter 14 quiz

Question 1

CORRECT ANSWER:

C. Dot-probe paradigm

This is correct. These tasks measure the speed of attentional orientation of visual stimuli. Research has focussed on people with anxiety and season affective disorders.

Question 2

CORRECT ANSWER:

A. Eye tracking

This is correct. Researchers can see where a person is directing their attention to by looking at eye movements.

Question 3

CORRECT ANSWER:

D. You can research behaviour in real time through recording brain activity during a particular task.

This is correct. MRI looks at brain activity unlike fMRI which looks at the blood flow changes. Other methods such as CT scans are not high in temporal resolution so they cannot be time specific.

Question 4

CORRECT ANSWER:

A. Research integrity

This is correct. A researcher must meet the professional standards during every part of research. This includes the design of the study, recruiting participants, conducting the study and even when writing up a research paper.

Question 5

CORRECT ANSWER:

B. Sign a consent form

This is correct. A participant must consent to taking part, and in an online questionnaire, this can be done with a box to consent.

References

Chapter 1

1. Melton, A. W. (1963). Implications of short-term memory for a general theory of memory. *Journal of Verbal Learning and Verbal Behavior, 2*, 1–21.
2. Neisser, U. (1967). *Cognitive psychology*. New York, NY: Appleton-Century-Crofts.
3. Holleman, G. A., Hooge, I. T., Kemner, C. & Hessels, R. S. (2020). The 'real-world approach' and its problems: A critique of the term ecological validity. *Frontiers in Psychology, 11*, 529490.
4. Mandler, G. (2007). *A History of Modern Experimental Psychology: From James and Wundt to Cognitive Science*. Cambridge, MA: MIT Press.
5. Hothersall, D. (1984). *History of Psychology*. New York, NY: Random House.
6. Descartes, R. (1972). *Treatise on Man* (T. S. Hall, Trans.). Cambridge, MA: Harvard University Press. (Original work published 1637).
7. Schwarz, K. A. & Pfister, R. (2016). Scientific psychology in the 18th century: A historical rediscovery. *Perspectives on Psychological Science, 11(3)*, 399–407.
8. Watson, R. I. (1968). *The great psychologists from Aristotle to Freud*. New York, NY: Lippincott.
9. Benjamin, L. T., Jr., Durkin, M., Link, M., Vestal, M. & Acord, J. (1992). Wundt's American doctoral students. *American Psychologist, 47*, 123–131.
10. Leahey, T. H. (2000). *A History of Psychology: Main Currents in Psychological Thought* (3rd ed.). Englewood Cliffs, NJ: Prentice Hall.
11. Leahey, T. H. (2000). *A History of Psychology: Main Currents in Psychological Thought* (3rd ed.). Englewood Cliffs, NJ: Prentice Hall.
12. Hall, J. F. (1971). *Verbal Learning and Retention*. Philadelphia, PA: Lippincott.
13. Leahey, T. H. (2000). *A History of Psychology: Main Currents in Psychological Thought* (3rd ed.). Englewood Cliffs, NJ: Prentice Hall.
14. Dewsbury, D. A. (2000). Comparative cognition in the 1930s. *Psychonomic Bulletin & Review, 7*, 267–283.
15. Tolman, E. C. (1948). Cognitive maps in rats and men. *Psychological Review, 55*, 189–208.
16. Mandler, J. M. & Mandler, G. (1969). The diaspora of experimental psychology: The Gestaltists and others. In D. Fleming & B. Bailyn (Eds.), *The Intellectual Migration: Europe and America, 1930–1960* (pp. 371–419). Cambridge, MA: Harvard University Press.
17. Baars, B. J. (1986). *The Cognitive Revolution in Psychology*. New York, NY: Guilford Press.
18. Leahey, T. H. (1992). The mythical revolutions of American psychology. *American Psychologist, 47*, 308–318.
19. Lachman, R., Lachman, J. L. & Butterfield, E. C. (1979). *Cognitive Psychology and Information Processing: An Introduction*. Hillsdale, NJ: Erlbaum.
20. Lachman, R., Lachman, J. L. & Butterfield, E. C. (1979). *Cognitive Psychology and Information Processing: An Introduction*. Hillsdale, NJ: Erlbaum.
21. Bruner, J. S., Goodnow, J. J. & Austin, G. A. (1956). *A Study of Thinking*. New York, NY: Wiley
22. Broadbent, D. E. (1958). *Perception and Communication*. London, England: Pergamon.
23. Glaze, J. A. (1928). The association value of nonsense syllables. *The Journal of Genetic Psychology: Research and Theory on Human Development, 35*, 255–269.
24. Bousfield, W. A. (1953). The occurrence of clustering in therecall of randomly arranged associates. *The Journal of General Psychology, 49*, 229–240; Bousfield, W. A. & Sedgewick, C. H. W. (1944). An analysis of sequences of restricted associative responses. *The Journal of General Psychology, 30*, 149–165.
25. Stroop, J. R. (1935). Studies of interference in serial verbal reactions. *Journal of Experimental Psychology, 18*, 643–662.
26. Chomsky, N. (1959). A review of Skinner's *Verbal Behavior. Language, 35*, 26–58.
27. Bousfield, W. A. (1953). The occurrence of clustering in the recall of randomly arranged associates. *The Journal of General Psychology, 49*, 229–240.
28. MacLeod, C. M. (1991). Half a century of research on the Stroop effect: An integrative review. *Psychological Bulletin, 109*, 163–203; Stroop, J. R. (1935). Studies of interference in serial verbal reactions. *Journal of Experimental Psychology, 18*, 643–662.
29. Norman, D. A. (1986). Reflections on cognition and parallel distributed processing. In J. L. McClelland, D. E. Rumelhart, & PDP Research Group (Eds.), *Parallel distributed processing: Explorations in the microstructure of cognition: Vol. 2. Psychological and biological models* (pp. 531–546). Cambridge, MA: MIT Press.
30. Lachman, R., Lachman, J. L. & Butterfield, E. C. (1979). *Cognitive Psychology and Information Processing: An Introduction*. Hillsdale, NJ: Erlbaum.

31. Leahey, T. H. (2003). Herbert A. Smith Nobel Prize in Economic Sciences, 1978. *American Psychologist, 58*, 753–755.

32. Miller, G. A., Galanter, E. & Pribram, K. H. (1960). *Plans and the Structure of Behavior*. New York, NY: Holt.

33. Campbell, J. I. D. & Graham, D. J. (1985). Mental multiplication skill: Structure, process, and acquisition. *Canadian Journal of Psychology, 39*, 338–366.

34. Campbell, J. I. D. & Graham, D. J. (1985). Mental multiplication skill: Structure, process, and acquisition. *Canadian Journal of Psychology, 39*, 338–366.

35. Donders, F. C. (1969). *Over de snelheid van psychische processen* [Speed of mental processes]. *Onderzoekingen gedann in het Psysiologish Laboratorium der Utrechtsche Hoogeschool* (W. G. Koster, Trans.). In W. G. Koster (Ed.), Attention and performance II. *Acta Psychologica, 30*, 412–431. (Original work published 1868)

36. Glanzer, M. & Cunitz, A. R. (1966). Two storage mechanisms in free recall. *Journal of Verbal Learning and Verbal Behavior, 5*, 351–360.

37. Open Science Collaboration (2015). Estimating the reproducibility of psychological science. *Science, 349*(6251). doi: 10.1126/science.aac4716

38. Cohen, J. (1988). *Statistical Analysis for the Behavioral Sciences*. Hillsdale, NJ: Erlbaum.

39. Atkinson, R. C. & Shiffrin, R. M. (1968). Human memory: A proposed system and its control processes. In W. K. Spence & J. T. Spence (Eds.), *The Psychology of Learning and Motivation: Advances in Research and Theory* (Vol. 2, pp. 89–195). New York, NY: Academic Press; Atkinson, R. C. & Shiffrin, R. M. (1971). The control of shortterm memory. *Scientific American, 225*, 82–90.

40. Campbell, J. I. D. & Graham, D. J. (1985). Mental multiplication skill: Structure, process, and acquisition. *Canadian Journal of Psychology, 39*, 338–366.7

41. Meyer, D. E., Schvaneveldt, R. W. & Ruddy, M. G. (1975). Loci of contextual effects on visual word-recognition. In P. M. A. Rabbitt & S. Dornic (Eds.), *Attention and Performance* (Vol. 5, pp. 98–118). London, England: Academic Press.

42. Sternberg, S. (1969). The discovery of processing stages: Extensions of Donders's method. In W. G. Koster (Ed.), Attention and performance II. *Acta Psychologica, 30*, 276–315.

43. Kucera, H., & Francis, W. N. (1967). *Computational analysis of present day American English*. Providence, RI: Brown University Press.

44. Allen, P. A., & Madden, D. J. (1990). Evidence for a parallel input serial analysis (PISA) model of word processing. *Journal of Experimental Psychology: Human Perception and Performance,16*, 48–64.

45. Whaley, C. P. (1978). Word–nonword classification time. *Journal of Verbal Learning and Verbal Behavior, 17*, 143–154.

46. Logan, G. D. (1988). Toward an instance theory of automatization. *Psychological Review, 95*, 492–527.

47. Salthouse, T. A. (1984). Effects of age and skill in typing. *Journal of Experimental Psychology: General, 113*, 345–371.

48. Townsend, J. T. & Wenger, M. J. (2004). The serial–parallel dilemma: A case study in a linkage of theory and method. *Psychonomic Bulletin & Review, 11*, 391–418.

49. Anderson, J. R., Qin, Y., Jung, K.-J. & Carter, C. S. (2007). Information-processing modules and their relative modality specificity. *Cognitive Psychology, 54*, 185–217.

50. Simpson, G. B. (1981). Meaning dominance and semantic context in the processing of lexical ambiguity. *Journal of Verbal Learning and Verbal Behavior, 20*, 120–136.

51. Reed, S. K. (1992). *Cognition: Theory and applications* (3rd ed.). Pacific Grove, CA: Brooks/Cole.

52. Greenberg, S. N., Healy, A. F., Koriat, A. & Kreiner, H. (2004). The GO model: A reconsideration of the role of structural units in guiding and organizing text on line. *Psychonomic Bulletin & Review, 11*, 428–433.

53. Ericsson, K. A., & Simon, H. A. (1980). Verbal reports as data. *Psychological Review, 87*, 215–251.

54. Spellman, B. A. & Busey, T. A. (2010). Emerging trends in psychology and law: An editorial overview. *Psychonomic Bulletin & Review, 17*, 141–142.

Chapter 2

1. Tulving, E. (1989). Remembering and knowing the past. *American Scientist, 77*, 361–367.

2. McCloskey, M. (1992). Cognitive mechanisms in numerical processing: Evidence from acquired dyscalculia. *Cognition, 44*, 107–157; Fagerholm, E. D., Hellyer, P. J., Scott, G., Leech, R. & Sharp, D. J. (2015). Disconnection of network hubs and cognitive impairment after traumatic brain injury. *Brain, 138*(6), 1696–1709.

3. de Haan, E. H., Corballis, P. M., Hillyard, S. A., Marzi, C. A., Seth, A., Lamme, V. A., . . . & Pinto, Y. (2020). Split-brain: What we know now and why this is important for understanding consciousness. *Neuropsychology Review, 30*, 224–233; Martin, R. C. (2000). Contribution from the neuropsychology of language and memory to the development of cognitive theory. *Journal of Memory and Language, 43*, 149–156.

4. Maggio, M. G., De Luca, R., Molonia, F. et al., (2019). Cognitive rehabilitation in patients with traumatic brain injury: A narrative review on the emerging use of virtual reality. *Journal of Clinical Neuroscience, 61*, 1–4.

5. Coffey, B. J., Threlkeld, Z. D., Foulkes, A. S., Bodien, Y. G. & Edlow, B. L. (2021). Reemergence of the language network during recovery from severe traumatic brain

injury: A pilot functional MRI study. *Brain Injury, 35*(12-13), 1552–1562; Smalle, E. H., Rogers, J. & Möttönen, R. (2015). Dissociating contributions of the motor cortex to speech perception and response bias by using transcranial magnetic stimulation. *Cerebral Cortex, 25*(10), 3690–3698.

6. Fernandino, L., Tong, J. Q., Conant, L. L., Humphries, C. J. & Binder, J. R. (2022). Decoding the information structure underlying the neural representation of concepts. *Proceedings of the National Academy of Sciences, 119*(6), e2108091119; Simor, P., Bogdány, T. & Peigneux, P. (2022). Predictive coding, multisensory integration, and attentional control: A multicomponent framework for lucid dreaming. *Proceedings of the National Academy of Sciences, 119*(44), e2123418119
7. Herculano-Houzel, S. (2012). The remarkable, yet not extraordinary, human brain as a scaled-up primate brain and its associated cost. *Proceedings of the National Academy of Sciences, 109*(supplement_1), 10,661–10,668.
8. Decker, A. L. & Duncan, K. (2020). Acetylcholine and the complex interdependence of memory and attention. *Current Opinion in Behavioral Sciences, 32*, 21–28; Roozendaal, B. & Hermans, E. J. (2017). Norepinephrine effects on the encoding and consolidation of emotional memory: improving synergy between animal and human studies. *Current Opinion in Behavioral Sciences, 14*, 115–122.
9. Mishkin, M. & Appenzeller, T. (1987). The anatomy of memory. *Scientific American, 256*, 80–89.
10. Decker, A. L. & Duncan, K. (2020). Acetylcholine and the complex interdependence of memory and attention. *Current Opinion in Behavioral Sciences, 32*, 21–28; Micheau, J. & Marighetto, A. (2011). Acetylcholine and memory: a long, complex and chaotic but still living relationship. *Behavioural Brain Research, 221*(2), 424–429.
11. Lømo, T. (2018). Discovering long-term potentiation (LTP)–recollections and reflections on what came after. *Acta Physiologica, 222*(2), e12921; Nicoll, R. A. (2017). A brief history of long-term potentiation. *Neuron, 93*(2), 281–290.
12. Pittenger, C., & Kandel, E. R. (2003). In search of general mechanisms for long-lasting plasticity: Aplysia and the hippocampus. *Philosophical Transactions of the Royal Society of London, 358*, 757–763.
13. Abraham, W. C. (2006). Memory maintenance: The changingnature of neural mechanisms. *Current Directions in Psychological Science, 15*, 5–8.
14. Hu, P., Stylos-Allan, M. & Walker, M. P. (2006). Sleep facilitates consolidation of emotional declarative memory. *Psychological Science, 17*, 891–898.
15. Rasch, B. & Born, J. (2008). Reactivation and consolidation of memory during sleep. *Current Directions in Psychological Science, 17*, 188–192.
16. Park, D. C. & Huang, C-M. (2010). Culture wires the brain: A cognitive neuroscience perspective. *Perspectives on Psychological Science, 5*, 391–400.
17. Toda, T., Parylak, S.L., Linker, S.B. et al. (2019). The role of adult hippocampal neurogenesis in brain health and disease. *Molecular Psychiatry*, 24, 67–87, https://doi.org/10.1038/s41380-018-0036-2
18. Shors, T. J. (2014). The adult brain makes new neurons, and effortful learning keeps them alive. *Current Directions in Psychological Science, 23*(5), 311–318.
19. Abraham, W. C. (2006). Memory maintenance: The changingnature of neural mechanisms. *Current Directions in Psychological Science, 15*, 5–8.
20. Thompson, R. F. (1986). The neurobiology of learning andmemory. *Science, 233*, 941–947.
21. Le Berre, A. P., Fama, R. & Sullivan, E. V. (2017). Executive functions, memory, and social cognitive deficits and recovery in chronic alcoholism: a critical review to inform future research. *Alcoholism: Clinical and Experimental Research, 41*(8), 1432–1443.
22. Herz, R. S., & Engen, T. (1996). *Odor memory: Review and analysis. Psychonomic Bulletin & Review, 3*, 300–313.
23. Epstein, R., & Kanwisher, N. (1998). A cortical representation of the local visual area. *Nature, 392*, 598–601.
24. Corballis, M. C. (1989). Laterality and human evolution. *Psychological Review, 96*, 492–505.
25. Ferstl, E. C., Neumann, J., Bogler, C., & Von Cramon, D. Y. (2008). The extended language network: a meta-analysis of neuroimaging studies on text comprehension. *Human Brain Mapping, 29*(5), 581-593.
26. Annett, M. (1998). Handedness and cerebral dominance: the right shift theory. *The Journal of Neuropsychiatry and Clinical Neurosciences, 10*(4), 459–469.
27. Papadatou-Pastou, M., Martin, M., Munafò, M.R. et al. (2008). Sex differences in left-handedness: A meta-analysis of 144 studies. *Psychological Bulletin, 134*(5): 677–699.
28. McManus, C. (2019). Half a century of handedness research: Myths, truths; fictions, facts; backwards, but mostly forwards. *Brain and Neuroscience Advances, 3*, 2398212818820513.
29. Jones, G.V. & Martin, M. (2010). Language dominance, handedness, and sex: Recessive X-linkage theory and test. *Cortex, 46*(6), 781–786.
30. Ambday, N. & Bharucha, J. (2009). Culture and the brain. *Current Directions in Psychological Science, 18*, 342–345.
31. Tang, Y., Zhang, W., Chen, K., Feng, S., Ji, Y., Shen, J., et al. (2006). Arithmetic processing in the brain shaped by cultures. *Proceedings of the National Academy of Sciences, 103*, 10,775–10,780.
32. Glenberg, A. (2011). Introduction to the mirror neuron forum. *Perspectives on Psychological Science, 6*, 363–368.

33. Di Pelligrino, G., Fadiga, L., Fogassi, L., Galese, V. & Rizzolatti, G. (1992). Understanding motor events: A neurophysiological study. *Experimental Brain Research, 91*, 176–180.

34. Heyes, C., & Catmur, C. (2022). What happened to mirror neurons? *Perspectives on Psychological Science, 17*(1), 153–168, https://doi.org/10.1177/1745691621990638

35. Mishkin, M., Ungerleider, L. & Macko, K. A. (1983). Object vision and spatial vision: Two cortical pathways. *Trends in Neurosciences, 6*, 414–417; Ungerleider, L. G. & Haxby, J. V. (1994). "What" versus "where" in the human brain. *Current Opinion in Neurobiology, 4*, 157–165.

36. Minsky, M. L. (1986). *The Society of Mind*. New York, NY: Simon & Schuster

37. Decety, J. (2015). The neural pathways, development and functions of empathy. *Current Opinion in Behavioral Sciences, 3*, 1–6.

38. Cacioppo, J. T. & Decety, J. (2011). Social neuroscience: Challenges and opportunities in the study of complex behavior. *Annals of the New York Academy of Sciences, 1224*(1), 162–173.

39. Vaughn, D. A., Savjani, R. R., Cohen, M. S. & Eagleman, D. M. (2018). Empathic neural responses predict group allegiance. *Frontiers in Human Neuroscience, 12*, 302.

40. Vuilleumier, P., Mohr, C., Valenza, N., Wetzel, C. Landis, T. (2003) Hyperfamiliarity for unknown faces after left lateral temporo-occipital venous infarction: a double dissociation with prosopagnosia. *Brain, 126*(4), 889–907, https://doi.org/10.1093/brain/awg086

41. McClelland, J. L., & Rumelhart, D. E. (1981). An interactive activation model of context effects in letter perception: Part 1. An account of basic findings. *Psychological Review, 88*, 375–407.

42. McClelland, J. L., & Rumelhart, D. E. (1981). An interactive activation model of context effects in letter perception: Part 1. An account of basic findings. *Psychological Review, 88*, 375–407.

43. Thomas, M. S. C. & McClelland, J. L. (2008). Connectionist models of cognition. In R. Sun (Ed.), *The Cambridge Handbook of Computational Psychology* (pp. 23–58). Cambridge University Press. https://doi.org/10.1017/CBO9780511816772.005

44. Monaghan, P. & Pollmann, S. (2003). Division of labor between the hemispheres for complex but not simple tasks: An implemented connectionist model. *Journal of Experimental Psychology: General, 132*, 379–399.

45. Guest, O., Caso, A. & Cooper, R. P. (2020). On simulating neural damage in connectionist networks. *Computational Brain & Behavior, 3*, 289–321.

Chapter 3

1. Hildreth, E. C. & Ullman, S. (1989). The computational study of vision. In M. I. Posner (Ed.), *Foundations of Cognitive Science* (pp. 581–630). Cambridge, MA: Bradford.

2. Moyer, R. S. & Bayer, R. H. (1976). Mental comparison and the symbolic distance effect. *Cognitive Psychology, 8*, 228–246.

3. Johnson, D. M. (1939). Confidence and speed in the two-category judgement. *Archives of Psychology, 241*, 1–52.

4. Scarf, D. (2020). Symbolic distance. In J. Vonk, T. K. Shackelford (Eds.), *Encyclopedia of Animal Cognition and Behavior* (pp. 1–3). Springer. https://doi.org/10.1007/978-3-319-47829-6_1515-1

5. Banks, W. P., Clark, H. H. & Lucy, P. (1975). The locus of the semantic congruity effect in comparative judgments. *Journal of Experimental Psychology: Human Perception and Performance, 1*, 35–47.

6. Banks, W. P. (1977). Encoding and processing of symbolic information in comparative judgments. In G. H. Bower (Ed.), *The Psychology of Learning and Motivation* (Vol. 11, pp. 101–159). New York, NY: Academic Press; Banks, W. P., Fujii, M. & Kayra-Stuart, F. (1976). Semantic congruity effects in comparative judgments of magnitude of digits. *Journal of Experimental Psychology: Human Perception and Performance, 2*, 435–447.

7. Banks, W. P. (1977). Encoding and processing of symbolic information in comparative judgments. In G. H. Bower (Ed.), *The Psychology of Learning and Motivation* (Vol. 11, pp. 101–159). New York, NY: Academic Press.

8. Peterson, W. W., Birdsall, T. G. & Fox, W. C. (1954). The theory of signal detectability. *Institute of Radio Engineers Transactions, PGIT-4*, 171–212.

9. Wixted, J. T. (2020). The forgotten history of signal detection theory. *Journal of Experimental Psychology: Learning, Memory, and Cognition, 46*(2), 201–233. https://doi.org/10.1037/xlm0000732

10. Kellen, D., Winiger, S., Dunn, J. C. & Singmann, H. (2021). Testing the foundations of signal detection theory in recognition memory. *Psychological Review, 128(6)*, 1022–1050. https://doi.org/10.1037/rev0000288

11. Lynn, S. K., & Barrett, L. F. (2014). 'Utilizing' signal detection theory. *Psychological Science, 25*(9), 1663–1673.

12. Donald, F. M. & Gould, M. S. (2020). Visual analysis as a predictor of performance in air traffic control trainees. *The International Journal of Aerospace Psychology, 30*(3-4), 236–253.

13. Grill-Spector, K. & Malach, R. (2004). The human visual cortex. *Annual Review of Neuroscience, 27*, 649–677.

14. Oberfeld, D., Hecht, H. & Gamer, M. (2010). Surface lightness influences perceived room height. *The Quarterly Journal of Experimental Psychology, 63*, 1999–2011.

15. Winer, G. A., Cottrell, J. E., Gregg, V., Fournier, J. S. & Bica, L. A. (2002). Fundamentally misunderstanding visual perception: Adults' belief in visual emissions. *American Psychologist, 57*, 417–424.

16. Mack, A. (2003). Inattentional blindness: Looking without seeing. *Current Directions in Psychological Science, 12*, 180–184.

17. Haines, R. F. (1991). A breakdown in simultaneous information processing. In G. Obrecht & L. W. Stark (Eds.), *Presbyopia Research* (pp. 171–175). New York, NY: Plenum Press

18. Mack, A. & Rock, I. (1998). *Inattentional Blindness*. Cambridge, MA: MIT Press.

19. Kennedy, H. & Bullier, J. (1985). A double-labelling investigation of the afferent connectivity to cortical areas V1 and V2 of the macaque monkey *Journal of Neuroscience, 5*, 2815–2830.

20. Kanwisher, N. & Yovel, G. (2006). The fusiform face area: a cortical region specialized for the perception of faces. *Philosophical Transactions of the Royal Society B: Biological Sciences, 361*(1476), 2109–2128.

21. Hochel, M. & Milán, E. G. (2008). Synaesthesia: The existing state of affairs. *Cognitive Neuropsychology, 25*, 93–117; Hupé, J. M. & Dojat, M. (2015). A critical review of the neuroimaging literature on synesthesia. *Frontiers in Human Neuroscience, 9*, 103; Ward, J. (2013). Synesthesia. *Annual Review of Psychology, 64*(1), 49–75.

22. Dixon, M. J., Smilek, D. & Merikle, P. M. (2004). Not all synesthetes are created equal: Projector versus associative synesthetes. *Cognitive, Affective & Behavioral Neuroscience, 4*, 335–343.

23. Grossenbacher, P. G. & Lovelace, C. T. (2001). Mechanisms of synesthesia: Cognitive and physiological constraints. *Trends in Cognitive Sciences, 5*, 36–41.

24. Maurer, D. (1997). Neonatal synaesthesia: Implications for the processing of speech and faces. In S. Baron-Cohen & J. E. Harrison (Eds.), *Synaesthesia: Classic and Contemporary Readings* (pp. 224–242). Malden, MA: Blackwell.

25. REF Ward, J. (2013). Synesthesia. *Annual Review of Psychology, 64*(1), 49–75.

26. Witthoft, N. & Winawer, J. (2013). Learning, memory, and synesthesia. *Psychological Science, 24*(3), 258–265.

27. Rensink, R. A. (2014). Limits to the usability of iconic memory. *Frontiers in Psychology, 5*, 971.

28. Trigg, G. L. & Lerner, R. J. (Eds.). (1981). *Encyclopedia of Physics*. Reading, MA: Addison-Wesley

29. Averbach, E. & Sperling, G. (1961). Short term storage and information in vision. In C. Cherry (Ed.), *Information Theory* (pp. 196–211). London, England: Butterworth; Sperling, G. (1960). The information available in brief visual presentations. *Psychological Monographs, 74*(48).

30. Sperling, G. (1960). The information available in brief visual presentations. *Psychological Monographs, 74*(Whole no. 48).

31. Sperling, G. (1963). A model for visual memory tasks. *Human Factors, 5*, 9–31.

32. Zhang, W. & Luck, S. J. (2009). Sudden death and gradual decay in visual working memory. *Psychological Science, 20*, 423–428.

33. Sperling, G. (1960). The information available in brief visual presentations. *Psychological Monographs, 74*(Whole no. 48).

34. Averbach, E. & Coriell, A. S. (1961). Short-term memory in vision. *Bell System Technical Journal, 40*, 309–328. (Reprinted in *Readings in Cognitive Psychology*, by M. Coltheart, Ed., 1973, Toronto, Canada: Holt, Rinehart & Winston of Canada)

35. Werner, H. (1935). Studies on contour. *American Journal of Psychology, 47*, 40–64; Kahneman, D. (1968). Method, findings and theory in studies of visual masking. *Psychological Bulletin, 70*, 404–426.

36. Neisser, U. (1967). *Cognitive Psychology*. New York, NY: Appleton-Century-Crofts.

37. Rayner, K., Inhoff, A. W., Morrison, P. E., Slowiaczek, M. L. & Bertera, J. H. (1981). Masking of foveal and parafoveal vision during eye fixations in reading. *Journal of Experimental Psychology: Human Perception and Performance, 7*, 167–179.

38. Finke, R. A. & Freyd, J. J. (1985). Transformations of visual memory induced by implied motions of pattern elements. *Journal of Experimental Psychology: Learning, Memory, and Cognition, 11*, 780–794; Irwin, D. E. (1991). Information integration across saccadic eye movements. *Cognitive Psychology, 23*, 420–456; Irwin, D. E. (1992). Memory for position and identity across eye movements. *Journal of Experimental Psychology: Learning, Memory, and Cognition, 18*, 307–317; Loftus, G. R. & Hanna, A. M. (1989). The phenomenology of spatial integration: Data and models. *Cognitive Psychology, 21*, 363–397.

39. Treisman, A. M., Russell, R. & Green, J. (1975). Brief visual storage of shape and movement. In P. M. A. Rabbitt & S. Dornic (Eds.), *Attention and Performance* (Vol. 5, pp. 699–721). New York, NY: Academic Press.

40. Loftus, G. R. & Irwin, D. E. (1998). On the relations among different measures of visible and informational persistence. *Cognitive Psychology, 35*, 135–199.

41. Neisser, U. (1967). *Cognitive Psychology*. New York, NY: Appleton-Century-Crofts.

42. Higgins, E. & Rayner, K. (2015). Transsaccadic processing: Stability, integration, and the potential role of remapping. *Attention, Perception & Psychophysics, 77*(1), 3–27.

43. Irwin, D. E. (1996). Integrating information across saccadic eye movements. *Current Directions in Psychological Science, 5*, 94–100.

44. Irwin, D. E., Yantis, S. & Jonides, J. (1983). Evidence against visual integration across saccadic eye movements. *Perception & Psychophysics, 34*, 49–57.

45. Kahneman, D., Triesman, A. & Gibbs, B. J. (1992). The reviewing of object files: Object-specific integration of information. *Cognitive Psychology, 24*, 175–219.

46. Henderson, J. M. & Anes, M. D. (1994). Roles of object-file review and type priming in visual identification within and across eye fixations. *Journal of Experimental Psychology: Human Perception and Performance, 20*, 826–839.

47. Higgins, E. & Rayner, K. (2015). Transsaccadic processing: Stability, integration, and the potential role of remapping. *Attention, Perception & Psychophysics, 77*(1), 3–27.

48. Wagemans, J., Elder, J. H., Kubovy, M., Palmer, S. E., Peterson, M. A., Singh, M. & von der Heydt, R. (2012). A century of Gestalt psychology in visual perception: I. Perceptual grouping and figure–ground organization. *Psychological Bulletin, 138*(6), 1172–1217; Wagemans, J., Feldman, J., Gepshtein, S., Kimchi, R., Pomerantz, J. R., van der Helm, P. A. & van Leeuwen, C. (2012). A century of Gestalt psychology in visual perception: II. Conceptual and theoretical foundations. *Psychological Bulletin, 138*(6), 1218–1252.

49. Poljac, E., de-Wit, L. & Wagemans, J. (2012). Perceptual wholes can reduce the conscious accessibility of their parts. *Cognition, 123*(2), 308–312.

50. Kubilius, J., Wagemans, J. & Op de Beeck, H. P. (2011). Emergence of perceptual gestalts in the human visual cortex: The case of the configural superiority effect. *Psychological Science, 22*, 1296–1303.

51. Johansson, G. (1973). Visual perception of biological motion and a model for its analysis. *Perception & Psychophysics, 14*(2), 201–211.

52. Neisser, U. (1967). *Cognitive Psychology*. New York, NY: Appleton-Century-Crofts.

53. Selfridge, O. G. (1959). Pandemonium: A paradigm for learning. In D. V. Blake & A. M. Uttley (Eds.), *Proceedings of the symposium on the mechanisation of thought processes*. London, England: H. M. Stationery Office.

54. Pritchard, R. M. (1961). Stabilized images on the retina. *Scientific American, 204*, 72–78.

55. Hubel, D. H. & Wiesel, T. N. (1962). Receptive fields, binocular interaction, and functional architecture in the cat's visual cortex. *Journal of Physiology, 160*, 106–154.

56. Neisser, U., Novick, R. & Lazar, R. (1963). Searching for ten targets simultaneously. *Perceptual and Motor Skills*, 17, 955–961.

57. Larsen, A. & Bundesen, C. (2009). Common mechanisms in apparent motion perception and visual pattern matching. *Scandinavian Journal of Psychology, 50*, 526–534.

58. Wertheimer, M. (1912). *Experimentelle Studien über das Sehen von Bewegung* [Experimental studies on the seeing of motion]. *Zeitschrift für Psychologie und Physiologie der Sinnesorgane, 61*, 161–265.

59. Wertheimer, M. (1912). *Experimentelle Studien über das Sehen von Bewegung* [Experimental studies on the seeing of motion]. *Zeitschrift für Psychologie und Physiologie der Sinnesorgane, 61*, 161–265.

60. Selfridge, O. G. (1959). Pandemonium: A paradigm for learning. In D. V. Blake & A. M. Uttley (Eds.), *Proceedings of the symposium on the mechanisation of thought processes*. London, England: H. M. Stationery Office.

61. Neisser, U. (1964). Visual search. *Scientific American, 210*, 94–102.

62. Busey, T. A. & Parada, E. F. (2010). The nature of expertise in fingerprint examiners. *Psychonomic Bulletin & Review, 17*, 155–160.

63. Morris, A. L. & Harris, C. L. (2002). Sentence context, word recognition, and repetition blindness. *Journal of Experimental Psychology: Learning, Memory, and Cognition, 28*, 962–982.

64. Kanwisher, N. (1987). Repetition blindness: Type recognition without token individuation. *Cognition*, 27, 117–143; Kanwisher, N. (1991). Repetition blindness and illusory conjunctions: Errors in binding visual types with visual tokens. *Journal of Experimental Psychology: Human Perception and Performance, 17*, 404–421.

65. Morris, A. L. & Harris, C. L. (2002). Sentence context, word recognition, and repetition blindness. *Journal of Experimental Psychology: Learning, Memory, and Cognition, 28*, 962–982.

66. Nisbett, R. E. & Masuda, T. (2003). Culture and point of view. *Proceedings of the National Academy of Sciences, 100*, 11163–11170.

67. Gutchess, A. H., Welsh, R. C., Boduroglu, A. & Park, D. C. (2006). Cultural differences in neural function associated with object processing. *Cognitive, Affective & Behavioral Neuroscience*, 6, 102–109.

68. Gartus, A., Klemer, N., & Leder, H. (2015). The effects of visual context and individual differences on perception and evaluation of modern art and graffiti art. *Acta Psychologica, 156*, 64–76.

69. McClelland, J. L. & Rumelhart, D. E. (1981). An interactive activation model of context effects in letter perception: Part 1. An account of basic findings. *Psychological Review, 88*, 375–407; Rumelhart, D. E. & McClelland, J. L. (1986). *Parallel distributed processing: Explorations in the Microstructure of Cognition: Vol. 1. Foundations*. Cambridge, MA: Bradford.

70. Hussain, Z., Sekuler, A. B. & Bennett, P. J. (2011). Superior identification of familiar visual patterns a year after learning. *Psychological Science, 22*, 724–730.

71. Corder, M. (2004). Crippled but not crashed neural networks can help pilots land damaged planes. *Scientific American, 291*, 94–96.

72. McCarley, J. S., Kramer, A. F., Wickens, C. D., Vidoni, E. D. & Boot, W. R. (2004). Visual skills in airport security screening. *Psychological Science, 15*, 302–306; Smith, J. D., Redford, J. S., Washburn, D. A. & Taglialatela, L. A. (2005). Specific-token effects in screening tasks: Possible implications for aviation security. *Journal of Experimental Psychology: Learning, Memory, and Cognition, 31*, 1171–1185.

73. Biederman, I. (1987). Recognition by components: A theory of human image understanding. *Psychological Review, 94*, 115–147; Biederman, I. (1990). Higher-level vision. In E. N. Osherson, S. M. Kosslyn & J. M. Hollerbach (Eds.), *An Invitation to Cognitive Science* (Vol. 2, pp. 41–72). Cambridge, MA: MIT Press.

74. Biederman, I. (1987). Recognition by components: A theory of human image understanding. *Psychological Review, 94*, 115–147.

75. Biederman, I. (1987). Recognition by components: A theory of human image understanding. *Psychological Review, 94*, 115–147.

76. Biederman, I. (1990). Higher-level vision. In E. N. Osherson, S. M. Kosslyn, & J. M. Hollerbach (Eds.), *An invitation to cognitive science* (Vol. 2, pp. 41–72). Cambridge, MA: MIT Press.

77. Biederman, I. & Blickle, T. (1985). *The perception of objects with deleted contours*. Unpublished manuscript, State University of New York, Buffalo.

78. Biederman, I., Glass, A. L. & Stacy, E. W. (1973). Searchingfor objects in real world scenes. *Journal of Experimental Psychology, 97*, 22–27; Palmer, S. E. (1975). The effects of contextual scenes on the identification of objects. *Memory & Cognition, 3*, 519–526.

79. Tanaka, J. T. & Curran, T. (2001). A neural basis for expert object recognition. *Psychological Science, 12*, 43–47.

80. Grill-Spector, K. & Kanwisher, N. (2005). Visual recognition: As soon as you know it is there, you know what it is. *Psychological Science, 16*, 152–160; Bowers, J. S. & Jones, K. W. (2008). Detecting objects is easier than categorizing them. *The Quarterly Journal of Experimental Psychology, 61*, 552–557.

81. Dell'acqua, R. & Job, R. (1998). Is object recognition automatic? *Psychonomic Bulletin & Review, 5*, 496–503.

82. Cave, K. R. & Kosslyn, S. M. (1989). Varieties of size-specific visual selection. *Journal of Experimental Psychology: General, 118*, 148–164.

83. Dell'acqua, R. & Job, R. (1998). Is object recognition automatic? *Psychonomic Bulletin & Review, 5*, 496–503.

84. Sacks, O. (1970). *The Man who Mistook his Wife for a Hat*. New York, NY: Harper & Row.

85. Biederman, I. (2018). Recognition-by-Components: A Theory of Human Image Understanding. In M. Bertamini & M. Kubovy (Eds.), *Human Perception* (pp. 211–246). Oxford: Routledge.

86. Forgus, R. H. & Melamed, L. E. (1976). *Perception: A Cognitive Stage Approach*. New York, NY: McGraw-Hill.

87. Eimas, P. D. (1975). Speech perception in early infancy. In L. B. Cohen & P. Salapatek (Eds.), *Infant Perception: From Sensation to Cognition: Vol. II. Perception of Space, Speech, and Sound* (pp. 193–231). New York, NY: Academic Press.

88. Neisser, U. (1967). *Cognitive Psychology*. New York, NY: Appleton-Century-Crofts.

89. Darwin, C. J., Turvey, M. T. & Crowder, R. G. (1972). An auditory analogue of the Sperling partial report procedure: Evidence for brief auditory storage. *Cognitive Psychology, 3*, 255–267; Moray, N., Bates, A. & Barnett, T. (1965). Experiments on the four-eared man. *The Journal of the Acoustical Society of America, 38*, 196–201.

90. Darwin, C. J., Turvey, M. T., & Crowder, R. G. (1972). An auditory analogue of the Sperling partial report procedure: Evidence for brief auditory storage. *Cognitive Psychology, 3*, 255–267.

91. Darwin, C. J., Turvey, M. T., & Crowder, R. G. (1972). An auditory analogue of the Sperling partial report procedure: Evidence for brief auditory storage. *Cognitive Psychology, 3*, 255–267.

92. Crowder, R. G., & Morton, J. (1969). Precategorical acoustic storage (PAS). *Perception & Psychophysics, 5*, 365–373.

93. Crowder, R. G. (1970). The role of one's own voice in immediate memory. *Cognitive Psychology, 1*, 157–178; Crowder, R. G. (1972). Visual and auditory memory. In J. F. Kavanaugh & I. G. Mattingly (Eds.), *Language by Ear and by Eye: The Relationships between Speech and Reading* (pp. 251–276). Cambridge, MA: MIT Press.

94. Crowder, R. G. (1972). Visual and auditory memory. In J. F. Kavanaugh & I. G. Mattingly (Eds.), *Language by ear and by eye: The relationships between speech and reading* (pp. 251–276). Cambridge, MA: MIT Press.

95. Crowder, R. G. (1972). Visual and auditory memory. In J. F. Kavanaugh & I. G. Mattingly (Eds.), *Language by Ear and by Eye: The Relationships between Speech and Reading* (pp. 251–276). Cambridge, MA: MIT Press.

96. Ayers, T. J., Jonides, J., Reitman, J. S., Egan, J. C. & Howard, D. A. (1979). Differing suffix effects for the same physical suffix. *Journal of Experimental Psychology: Human Learning and Memory, 5*, 315–321.

97. Neath, I., Surprenant, A. M. & Crowder, R. G. (1993). The context-dependent stimulus suffix effect. *Journal of Experimental Psychology: Learning, Memory & Cognition, 19*, 698–703.

Chapter 4

1. Anderson, B. (2011). There is no such thing as attention. *Frontiers in Psychology*, 2(246), 1–8.

2. Evans, J. S. B. (2019). Reflections on reflection: the nature and function of type 2 processes in dual-process theories of reasoning. *Thinking & Reasoning*, 25(4), 383–415.

3. Leonard, C. J., Balestreri, A. & Luck, S. J. (2015). Interactions between space-based and feature-based attention. *Journal of Experimental Psychology: Human Perception and Performance, 41*(1), 11–16.
4. Förster, J. (2012). GLOMOsys: The how and why of global and local processing. *Current Directions in Psychological Science, 21*(1), 15–19; Navon, D. (1977). Forest before trees: The precedence of global features in visual perception. *Cognitive Psychology,* 9(3), 353–383.
5. Buckner, R. L., Andrews-Hanna, J. R. & Schacter, D. L. (2008). The brain's default network. *Annals of the New York Academy of Sciences, 1124*(1), 1–38.
6. Andrews-Hanna, J. R. (2012). The brain's default network and its adaptive role in internal mentation. *The Neuroscientist, 18*(3), 251–270.
7. Hasson, U., Furman, O., Clark, D., Dudai, Y. & Davachi, L. (2008). Enhanced intersubject correlations during movie viewing correlate with successful episodic encoding. *Neuron,57*(3), 452–462; Lerner, Y., Honey, C. J., Silbert, L. J. & Hasson, U. (2011). Topographic mapping of a hierarchy of temporal receptive windows using a narrated story. *The Journal of Neuroscience, 31*(8), 2906–2915; Regev, M., Honey, C. J., Simony, E. & Hasson, U. (2013). Selective and invariant neural responses to spoken and written narratives. *The Journal of Neuroscience, 33*(40), 15978–15988.
8. Mackworth, N. H. (1948). The breakdown of vigilance during prolonged visual search. *The Quarterly Journal of Experimental Psychology, 1*, 6–21.
9. Warm, J. S. (1984). An introduction to vigilance. In J. S. Warm (Ed.), *Sustained Attention in Human Performance* (pp. 1–14). Chichester, England: Wiley.
10. Wiener, E. L. (1984). Vigilance and inspection. In J. S. Warm (Ed.), *Sustained attention in human performance* (pp. 207–246). Chichester, England: Wiley.
11. Langner, R. & Eickhoff, S. B. (2013). Sustaining attention to simple tasks: A meta-analytic review of the neural mechanisms of vigilant attention. *Psychological Bulletin, 139*(4), 870–900.
12. See, J. E., Howe, S. R., Warm, J. S. & Dember, W. N. (1995). Meta-analysis of the sensitivity decrement in vigilance. *Psychological Bulletin, 2*, 230–249.
13. Warm, J. S. (1984). An introduction to vigilance. In J. S. Warm (Ed.), *Sustained Attention in Human Performance* (pp. 1–14). Chichester, England: Wiley.
14. Warm, J. S. & Jerison, H. J. (1984). The psychophysics of vigilance. In J. S. Warm (Ed.), *Sustained Attention in Human Performance* (pp. 15–60). Chichester, England: Wiley.
15. Helton, W. S. & Russell, P. N. (2015). Rest is best: The role of rest and task interruptions on vigilance. *Cognition, 134*, 165–173.
16. MacLean, K. A., Ferrer, E., Aichele, S. R., Bridwell, D. A., Zanesco, A. P., Jacobs, T. L., et al. (2010). Intensive meditation training improves perceptual discrimination and sustained attention. *Psychological Science, 21*, 829–839.
17. Schacter, D. L. (1989). Memory. In M. I. Posner (Ed.), *Foundations of Cognitive Science* (pp. 683–725). Cambridge, MA: MIT Press. Schacter, D. L. (1996). *Searching for Memory*. New York, NY: Basic Books.
18. Masson, M. E. J. (1984). Memory for the surface structure of sentences: Remembering with and without awareness. *Journal of Verbal Learning and Verbal Behavior, 23*, 579–592.
19. Bonebakker, A. E., Bonke, B., Klein, J., Wolters, G., Stijnen, T., Passchier, J., et al. (1996). Information processing during general anesthesia: Evidence for unconscious memory. *Memory & Cognition, 24*, 766–776; Andrade, J. (1995). Learning during anaesthesia: A review. *British Journal of Psychology, 86*, 479–506.
20. Bridgeman, B. (1988). *The biology of behavior and mind*. New York, NY: Wiley.
21. Öhman, A., Flykt, A. & Esteves, F. (2001). Emotion drives attention: Detecting the snake in the grass. *Journal of Experimental Psychology: General, 130*, 466–478.
22. Brosch, T., Sander, D., Pourtois, G. & Scherer, K. R. (2008). Beyond fear: Rapid spatial orienting toward positive emotional stimuli. *Psychological Science, 19*, 362–370.
23. Anderson, B. A. & Yantis, S. (2013). Persistence of value-driven attentional capture. *Journal of Experimental Psychology: Human Perception and Performance, 39*(1), 6–9.
24. Sanocki, T., Islam, M., Doyon, J. K. & Lee, C. (2015). Rapid scene perception with tragic consequences: Observers miss perceiving vulnerable road users, especially in crowded traffic scenes. *Attention, Perception & Psychophysics, 77*(4), 1252–1262.
25. Birmingham, E., Bischof, W. F. & Kingstone, A. (2008). Social attention and real world scenes: The roles of action, competition and social content. *The Quarterly Journal of Experimental Psychology, 61*, 986–998.
26. Kingstone, A., Smilek, D., Ristic, J., Friesen, C. K., & Eastwood, J. D. (2003). Attention, researchers! It is time to take a look at the real world. *Current Directions in Psychological Science, 12*, 176–180.
27. Emery, N. J. (2000). The eyes have it: The neuroethology, function and evolution of social gaze. *Neuroscience and Biobehavioral Review, 24*, 581–604.
28. Estes, Z., Verges, M. & Barsalou, L. W. (2008). Head up, foot down: Object words orient attention to the objects' typical location. *Psychological Science, 19*, 93–97.
29. Abrams, R. A. & Christ, S. E. (2003). Motion onset captures attention. *Psychological Science, 14*, 427–432; Franconeri, S. L. & Simons, D. J. (2003). Moving and looming stimuli capture attention. *Perception & Psychophysics, 65*, 999–1010.

30. Davoli, C. C., Suszko, J. W. & Abrams, R. A. (2007). New objects can capture attention without a unique luminance transient. *Psychonomic Bulletin & Review, 14*, 338–343; Yantis, S. & Jonides, J. (1984). Abrupt visual onsets and selective attention: Evidence from visual search. *Journal of Experimental Psychology: Human Perception and Performance, 10*, 601–621.

31. Lu, S. & Zhou, K. (2005). Stimulus-driven attentional capture by equiluminent color change. *Psychonomic Bulletin & Review, 12*, 567–572.

32. Pratt, J., Radulescu, P. V., Guo, R. M., & Abrams, R. A. (2010). It's alive! Animate motion captures visual attention. *Psychological Science, 21*, 1724–1730.

33. Vachon, F., Hughes, R. W. & Jones, D. M. (2012). Broken expectations: Violation of expectancies, not novelty, captures auditory attention. *Journal of Experimental Psychology: Learning, Memory, and Cognition, 38*, 164–177.

34. Cole, G. G. & Kuhn, G. (2010). Attentional capture by object appearance and disappearance. *The Quarterly Journal of Experimental Psychology, 63*, 147–159.

35. Cowan, N. (1995). *Attention and Memory: An Integrated Framework*. New York, NY: Oxford University Press.

36. Serences, J. T., Shomstein, S., Leber, A. B., Golay, X., Egeth, H. E., & Yantis, S. (2005). Coordination of voluntary and stimulus-driven attentional control in human cortex. *Psychological Science, 16*, 114–122.

37. Yantis, S. (2008). The neural basis of selective attention. *Current Directions in Psychological Science, 17*, 86–90.

38. Posner, M. I., Nissen, M. J. & Ogden, W. C. (1978). Attended and unattended processing modes: The role of set for spatial location. In H. L. Pick & I. J. Saltzman (Eds.), *Modes of Perceiving and Processing Information* (pp. 137–157). Hillsdale, NJ: Erlbaum; Posner, M. I., Snyder, C. R. R. & Davidson, B. J. (1980). Attention and the detection of signals. *Journal of Experimental Psychology: General, 109*, 160–174.

39. Gibson, B. S. & Sztybel, P. (2014). The spatial semantics of symbolic attention control. *Current Directions in Psychological Science, 23*(4), 271–276.

40. Posner, M. I., Snyder, C. R. R. & Davidson, B. J. (1980). Attention and the detection of signals. *Journal of Experimental Psychology: General, 109*, 160–174.

41. Cave, K. R. & Bichot, N. P. (1999). Visuospatial attention: Beyond a spotlight model. *Psychonomic Bulletin & Review, 6*, 204–223.

42. Cave, K. R. & Bichot, N. P. (1999). Visuospatial attention: Beyond a spotlight model. *Psychonomic Bulletin & Review, 6*, 204–223.

43. Treisman, A. & Gelade, G. (1980). A feature integration theory of attention. *Cognitive Psychology, 12*, 97–136.

44. Finlayson, N. J., & Grove, P. M. (2015). Visual search is influenced by 3D spatial layout. *Attention, Perception, & Psychophysics, 77*(7), 2322–2330.

45. Treisman, A. (1982). Perceptual grouping and attention in visual search for features and for objects. *Journal of Experimental Psychology: Human Perception and Performance, 8*, 194–214; Treisman, A. (1988). Features and objects: The Fourteenth Bartlett Memorial Lecture. *The Quarterly Journal of Experimental Psychology, 40A*, 201–237; Treisman, A. (1991). Search, similarity, and integration of features between and within dimensions. *Journal of Experimental Psychology: Human Perception and Performance, 17*, 652–676; Treisman, A. & Gelade, G. (1980). A feature integration theory of attention. *Cognitive Psychology, 12*, 97–136.

46. Klein, R. (2000). Inhibition of return. *Trends in Cognitive Sciences, 4*, 138–147; Posner, M. I. & Cohen, Y. (1984). Components of visual orienting. In H. Bouma & D. G. Bouwhuis (Eds.), *Attention and performance X* (pp. 531–556). Hillsdale, NJ: Erlbaum.

47. Klein, R. (2000). Inhibition of return. *Trends in Cognitive Sciences, 4*, 138–147; Vivas, A. B., Humphreys, G. W. & Fuentes, L. J. (2006). Abnormal inhibition of return: A review and new data on patients with parietal lobe damage. *Cognitive Neuropsychology, 23*, 1049–1064.

48. Dodd, M. D., Van der Stigchel, S. & Hollingworth, A. (2009). Inhibition of return and facilitation of return as a function of visual task. *Psychological Science, 20*, 333–339.

49. Huang, T-R. & Grossberg, S. (2010). Cortical dynamics of contextually cued attentive visual learning and search: Spatial and object evidence accumulation. *Psychological Review, 117*, 1080–1112.

50. Davoli, C. C., Brockmole, J. R. & Goujon, A. (2012). A bias to detail: How hand position modulates visual learning and visual memory. *Memory & Cognition, 40*, 352–359.

51. Biggs, A. T., Brockmole, J. R. & Witt, J. K. (2013). Armed and attentive: Holding a weapon can bias attentional priorities in scene viewing. *Attention, Perception & Psychophysics*, 75(8), 1715–1724.

52. Johnston, J. C., McCann, R. S. & Remington, R. W. (1995). Chronometric evidence for two types of attention. *Psychological Science, 6*, 365–369.

53. Chisholm, J. D. & Kingstone, A. (2015). Action video games and improved attentional control: Disentangling selectionand response-based processes. *Psychonomic Bulletin & Review, 22*(5), 1430–1436.

54. Feng, J., Spence, I. & Pratt, J. (2007). Playing an action video game reduces gender differences in spatial cognition. *Psychological Science, 18*, 850–855.

55. Heimler, B., Pavani, F., Donk, M. & van Zoest, W. (2014). Stimulus-and goal-driven control of eye movements: Action videogame players are faster but not better. *Attention, Perception & Psychophysics, 76*(8), 2398–2412.

56. Chamberlain, R. & Wagemans, J. (2015). Visual arts training is linked to flexible attention to local and

global levels of visual stimuli. *Acta Psychologica, 161*, 185–197.

57. Gobet, F., Johnston, S. J., Ferrufino, G., Johnston, M., Jones, M. B., Molyneux, . . . Weeden, L. (2015). "No level up!" No effects of video game specialization and expertise on cognitive performance. *Frontiers in Psychology, 5*, 1337.

58. Boot, W. R., Blakely, D. P. & Simons, D. J. (2011). Do action video games improve perception and cognition? *Frontiers in Psychology, 2*, 226; Latham, A. J., Patston, L. L. & Tippett, L. J. (2013). The virtual brain: 30 years of video-game play and cognitive abilities. *Frontiers in Psychology, 4*, 629.

59. Harvey, A. J., Kneller, W. & Campbell, A. C. (2013). The effects of alcohol intoxication on attention and memory for visual scenes. *Memory, 21*(8), 969–980.

60. Banich, M. T. (1997). Neuropsychology: *The Neural Bases of Mental Function*. Boston, MA: Houghton Mifflin.

61. Intriligator, J. & Cavanagh, P. (2001). The spatial resolution of visual attention. *Cognitive Psychology, 43*, 171–216.

62. Banich, M. T. (1997). *Neuropsychology: The Neural Bases of Mental Function*. Boston, MA: Houghton Mifflin; Rafal, R. D. (1997). Hemispatial neglect: Cognitive neuropsychological aspects. In T. E. Feinberg & M. J. Farah (Eds.), *Behavioral Neurology and Neuropsychology* (pp. 319–336). New York, NY: McGraw-Hill.

63. Bisiach, E. & Luzzatti, C. (1978). Unilateral neglect of representational space. *Cortex, 14*, 129–133.

64. Duncan, J., Bundesen, C., Olson, A., Humphreys, G., Chavda, S. & Shibuya, H. (1999). Systematic analysis of deficits in visual attention. *Journal of Experimental Psychology: General, 128*, 450–478.

65. Bundesen, C. (1990). A theory of visual attention. *Psychological Review, 97*, 523–547.

66. Monaghan, P., & Shillcock, R. (2004). Hemispheric asymmetries in cognitive modeling: Connectionist modeling of unilateral visual neglect. *Psychological Review, 111*, 283–308.

67. Kunar, M A., Carter, R., Cohen, M. & Horowitz, T. S. (2008). Telephone conversation impairs sustained visual attention via a central bottleneck. *Psychonomic Bulletin & Review, 15*, 1135–1140; Spence, C. & Read, L. (2003). Speech shadowing while driving: On the difficulty of splitting attention between eye and ear. *Psychological Science, 14*, 251–256; Strayer, D. L. & Johnston, W. A. (2001). Driven to distraction: Dual-task studies of stimulated driving and conversing on a cellular phone. *Psychological Science*, 12, 462–466.

68. Lochner, M. J., & Trick, L. M. (2014). Multiple-object tracking while driving: The multiple-vehicle tracking task. *Attention, Perception, & Psychophysics, 76*(8), 2326–2345.

69. Strayer, D. L. & Drews, F. A. (2007). Cell-phone-induced driver distraction. *Current Directions in Psychological Science, 16*, 128–131.

70. Simons, D. J., & Chabris, C. F. (1999). Gorillas in our midst: Sustained inattentional blindness for dynamic events. *Perception, 28*(9), 1059-1074.

71. Simons, D. J., & Chabris, C. F. (1999). Gorillas in our midst: Sustained inattentional blindness for dynamic events. *Perception, 28*(9), 1059-1074.

72. https://theinvisiblegorilla.com/

73. Sanbonmatsu, D. M., Strayer, D. L., Biondi, F., Behrends, A. A. & Moore, S. M. (2016). Cell-phone use diminishes selfawareness of impaired driving. *Psychonomic Bulletin & Review, 23*(2), 617–623.

74. Galván, V. V., Vessal, R. S. & Golley, M. T. (2013). The effects of cell phone conversations on the attention and memory of bystanders. *PloS one, 8*(3), e58579.

75. Cherry, E. C. (1953). Some experiments on the recognition of speech, with one and with two ears. *Journal of the Acoustical Society of America, 25*, 975–979; Cherry, E. C. & Taylor, W. K. (1954). Some further experiments on the recognition of speech with one and two ears. *Journal of the Acoustical Society of America, 26*, 554–559.

76. Johnston, W. A. & Heinz, S. P. (1978). Flexibility and capacity demands of attention. *Journal of Experimental Psychology: General, 107*, 420–435.

77. Egan, P., Carterette, E. C. & Thwing, E. J. (1954). Some factors affecting multichannel listening. *Journal of the Acoustic Society of America, 26*, 774–782; Spieth, W., Curtis, J. F. & Webster, J. C. (1954). Responding to one of two simultaneous messages. *Journal of the Acoustical Society of America, 26*, 391–396; Wood, N. L. & Cowan, N. (1995a). The cocktail party phenomenon revisited: Attention and memory in the classic selective listening procedure of Cherry (1953). *Journal of Experimental Psychology: General, 124*, 243–262.

78. Cavanagh, J. F. & Frank, M. J. (2014). Frontal theta as a mechanism for cognitive control. *Trends in Cognitive Sciences, 18*(8), 414–421.

79. Broadbent, D. E. (1958). *Perception and communication*. London, England: Pergamon

80. Moray, N. (1959). Attention in dichotic listening: Affective cues and the influence of instructions. *The Quarterly Journal of Experimental Psychology, 11*, 56–60.

81. Wood, N. L., & Cowan, N. (1995a). The cocktail party phenomenon revisited: Attention and memory in the classic selective listening procedure of Cherry (1953). *Journal of Experimental Psychology: General, 124*, 243–262.

82. Conway, A. R. A., Cowan, N. & Bunting, M. F. (2001). The cocktail party phenomenon revisited: The importance of working memory capacity. *Psychonomic Bulletin & Review, 8*, 331–335.

83. Lachter, J., Forster, K. I. & Ruthruff, E. (2004). Forty-five years after Broadbent (1958): Still no identification without attention. *Psychological Review, 111*, 880–913.

84. Treisman, A. M. (1960). Contextual cues in selective listening. *The Quarterly Journal of Experimental Psychology, 12*, 242–248.

85. Lewis, J. L. (1970). Semantic processing of unattended messages using dichotic listening. *Journal of Experimental Psychology, 85*, 225–228; Carr, T. H., McCauley, C., Sperber, R. D. & Parmalee, C. M. (1982). Words, pictures, and priming: On semantic activation, conscious identification, and the automaticity of information processing. *Journal of Experimental Psychology: Human Perception and Performance, 8*, 757–777.

86. Koivisto, M. & Revonsuo, A. (2007). How meaning shapes seeing. *Psychological Science, 18*, 845–849; Marsh, R. L., Cook, G. I., Meeks, J. T., Clark-Foos, A. & Hicks, J. L. (2007). Memory for intention-related material presented in a to-be-ignored channel. *Memory & Cognition, 35*, 1197–1204; Most, S. B., Scholl, B. J., Clifford, E. R. & Simons, D. J. (2005). What you see is what you set: Sustained inattentional blindness and the capture of awareness. *Psychological Review, 112*, 217–242.

87. Stothart, C., Mitchum, A. & Yehnert, C. (2015). The attentional cost of receiving a cell phone notification. *Journal of Experimental Psychology: Human Perception and Performance, 41*(4), 893–897.

88. Lavie, N. (2010). Attention, distraction, and cognitive control under load. *Current Directions in Psychological Science, 19*, 143–148.

89. Treisman, A. M. (1960). Contextual cues in selective listening. *The Quarterly Journal of Experimental Psychology, 12*, 242–248; Treisman, A. M. (1964). Monitoring and storage of irrelevant messages in selective attention. *Journal of Verbal Learning and Verbal Behavior, 3*, 449–459.

90. Tipper, S. P. (1985). The negative priming effect: Inhibitory priming with to be ignored objects. *The Quarterly Journal of Experimental Psychology, 37A*, 571–590.

91. Neill, W. T. (1977). Inhibition and facilitation processes in selective attention. *Journal of Experimental Psychology: Human Perception & Performance, 3*, 444–450; Tipper, S. P. (1985). The negative priming effect: Inhibitory priming with to be ignored objects. *The Quarterly Journal of Experimental Psychology, 37A*, 571–590; Frings, C., Schneider, K. K. & Fox, E. (2015). The negative priming paradigm: An update and implications for selective attention. *Psychonomic Bulletin & Review, 22*(6), 1577–1597.

92. Mayr, S. & Buchner, A. (2006). Evidence for episodic retrieval of inadequate prime responses in auditory negative priming. *Journal of Experimental Psychology: Human Perception and Performance, 32*, 932–943; Neill, W. T., Valdes, L. A. & Terry, K. M. (1995). Selective attention and the inhibitory control of cognition. In F. N. Dempster & C. J. Brainerd (Eds.), *New perspectives on interference and Inhibition in cognition* (pp. 207–261). New York, NY: Academic Press.

93. Hasher, L. & Zacks, R. T. (1984). Automatic processing of fundamental information: The case of frequency of occurrence. *American Psychologist, 39*, 1372–1388.

94. Beech, A., Powell, T., McWilliams, J. & Claridge, G. (1989). Evidence of reduced cognitive inhibition in schizophrenics. *British Journal of Psychology, 28*, 109–116.

95. MacQueen, G. M., Tipper, S. P., Young, L. T., Joffe, R. T. & Levitt, A. J. (2000). Impaired distractor inhibition on a selective attention task in unmedicated, depressed subjects. *Psychological Medicine, 30*, 557–564.

96. Kahneman, D. (1973). *Attention and Effort*. Englewood Cliffs, NJ: Prentice Hall.

97. Barnard, P. J., Scott, S., Taylor, J., May, J. & Knightley, W. (2004). Paying attention to meaning. *Psychological Science, 15*, 179–186; Pashler, H. & Johnson, J. C. (1998). Attentional limitations in dual-task performance. In H. Pashler (Ed.), *Attention* (pp. 155–189). Hove, England: Psychology Press.

98. Livesey, E. J., Harris, I. M. & Harris, J. A. (2009). Attentional changes during implicit learning: Signal validity protects target stimulus from the attentional blink. *Journal of Experimental Psychology: Learning, Memory, and Cognition, 35*, 408–422.

99. Kamienkowski, J. E., Navajas, J. & Sigman, M. (2012). Eye movements blink the attentional blink. *Journal of Experimental Psychology: Human Perception and Performance, 38*(3), 555–560.

100. Berman, M. G., Jonides, J. & Kaplan, S. (2008). The cognitive benefits of interacting with nature. *Psychological Science, 19*, 1207–1212; Joye, Y., Pals, R., Steg, L. & Evans, B. L. (2013). New methods for assessing the fascinating nature of nature experiences. *PloS one, 8*(7), e65332.

101. Posner, M. I. & Snyder, C. R. R. (1975). Facilitation and inhibition in the processing of signals. In P. M. A. Rabbitt & S. Dornic (Eds.), *Attention and performance V* (pp. 669–682). New York, NY: Academic Press.

102. Shiffrin, R. M. & Schneider, W. (1977). Controlled and automatic human information processing: II. Perceptual learning, automatic attending, and a general theory. *Psychological Review, 84*, 127–190.

103. Schneider, W. & Shiffrin. R. M. (1977). Controlled and automatic human information processing: I. Detection, search, and attention. *Psychological Review, 84*, 1–66.

104. Logan, G. D. & Etherton, J. L. (1994). What is learned during automatization? The role of attention in constructing an instance. *Journal of Experimental Psychology: Learning, Memory, and Cognition, 20*,

1022–1050; Navon, D. (1984). Resources: A theoretical soup stone? *Psychological Review, 91*, 216–234; Pashler, H. (1994). Dual-task interference in simple tasks: Data and theory. *Psychological Bulletin, 116*, 220–244.

105. Stroop, J. R. (1935). Studies of interference in serial verbal reactions. *Journal of Experimental Psychology, 18*, 643–662.

106. Protopapas, A., Archonti, A. & Skaloumbakas, C. (2007). Reading ability is negatively related to Stroop interference. *Cognitive Psychology, 54*, 251–282.

107. Dunbar, K. & MacLeod, C. M. (1984). A horse race of a different color: Stroop interference patterns with transformed words. *Journal of Experimental Psychology: Human Perception and Performance, 10*, 622–639; MacLeod, C. M. (1991). Half a century of research on the Stroop effect: An integrative review. *Psychological Bulletin, 109*, 163–203.

108. Wingfield, A., Goodglass, H. & Lindfield, K. C. (1997). Separating speed from automaticity in a patient with focal brain atrophy. *Psychological Science, 8*, 247–249.

109. Warren, R. M. & Warren, R. P. (1970). Auditory illusions and confusions. *Scientific American, 223*, 30–36.

110. Hirst, W. & Kalmar, D. (1987). Characterizing attentional resources. *Journal of Experimental Psychology: General, 116*, 68–81.

111. Spelke, E., Hirst, W. & Neisser, U. (1976). Skills of divided attention. *Cognition, 4*, 215–230.

112. Zbrodoff, N. J. & Logan, G. D. (1986). On the autonomy of mental processes: A case study of arithmetic. *Journal of Experimental Psychology: General, 115*, 118–130.

113. Barshi, I. & Healy, A. F. (1993). Checklist procedures and the cost of automaticity. *Memory & Cognition, 21*, 496–505.

114. Barshi, I., & Healy, A. F. (1993). Checklist procedures and the cost of automaticity. *Memory & Cognition, 21*, 496–505.

115. Barshi, I. & Healy, A. F. (1993). Checklist procedures and the cost of automaticity. *Memory & Cognition, 21*, 496–505.

116. Randall, J. G., Oswald, F. L. & Beier, M. E. (2014). Mind-wandering, cognition, and performance: A theory-driven meta-analysis of attention regulation. *Psychological Bulletin, 140*(6), 1411–1431.

117. Smallwood, J., McSpadden, M. & Schooler, J. W. (2007). The lights are on but no one's home: Meta-awareness and the decoupling of attention when the mind wanders. *Psychonomic Bulletin & Review, 14*, 527–533.

118. Barron, E., Riby, L. M., Greer, J., & Smallwood, J. (2011). Absorbed in thought: The effect of mind wandering on the processing of relevant and irrelevant events. *Psychological Science, 22*, 596–601.

119. Franklin, M. S., Broadway, J. M., Mrazek, M. D., Smallwood, J. & Schooler, J. W. (2013). Window to the wandering mind: Pupillometry of spontaneous thought while reading. *The Quarterly Journal of Experimental Psychology, 66*(12), 2289–2294.

120. Reichle, E. D., Reineberg, A. E. & Schooler, J. W. (2010). Toward a model of eye movement control in reading. *Psychological Review, 105*, 125–157; Smilek, D., Carriere, J. S. A. & Cheyne, J. A. (2010). Out of mind, out of sight: Eye blinking as indicator and embodiment of mind wandering. *Psychological Science, 21*, 786–789.

121. Smallwood, J., & Schooler, J. W. (2006). The restless mind. *Psychological Bulletin, 132*, 946–958.

122. Kane, M. J., & McVay, J. C. (2012). What mind wandering reveals about executive-control abilities and failures. *Current Directions in Psychological Science, 21*(5), 348–354.

123. Smallwood, J., Fishman, D. J., & Schooler, J. W. (2007). Counting the cost of an absent mind: Mind wandering as an unrecognized influence on educational performance. *Psychonomic Bulletin & Review, 14*, 230–236.

124. Sayette, M. A., Reichle, E. D. & Schooler, J. W. (2009). Lost in the sauce: The effects of alcohol on mind wandering. *Psychological Science, 20*, 747–752.

125. Sayette, M. A., Schooler, J. W. & Reichle, E. D. (2010). Out for a smoke: The impact of cigarette craving on zoning out during reading. *Psychological Science, 21*, 26–30.

126. Szpunar, K. K., Moulton, S. T., & Schacter, D. L. (2013). Mind wandering and education: From the classroom to online learning. *Frontiers in Psychology, 4*, 495.

Chapter 5

1. Miller, G. A. (1956). The magical number seven, plus or minus two: Some limits on our capacity for processing information. *Psychological Review, 63*, 81–97.

2. Brown, J. A. (1958). Some tests of the decay theory of immediate memory. *The Quarterly Journal of Experimental Psychology, 10*, 12–21.

3. Peterson, L. R., & Peterson, M. J. (1959). Short-term retention of individual items. *Journal of Experimental Psychology, 58*, 193–198.

4. Baddeley, A. D., Thomson, N. & Buchanan, M. (1975). Word length and the structure of short-term memory. *Journal of Verbal Learning and Verbal Behavior, 14*, 575–589.

5. Dosher, B. A. & Ma, J-J. (1998). Output loss or rehearsal loop? Output-time versus pronunciation-time limits in immediate recall for forgetting-matched materials. *Journal of Experimental Psychology: Learning, Memory, and Cognition, 24*, 316–335; Cowan, N.,

Wood, N. L., Wood, P. K., Keller, T. A., Nugent, L. D. & Keller, C. V. (1998). Two separate verbal processing rates contributing to short-term memory span. *Journal of Experimental Psychology: General, 127*, 141–160.

6. Sattler, J. M. (1982). *Assessment of Children's Intellectual and Special Abilities* (2nd ed.). Boston, MA: Allyn & Bacon.
7. Wilson, M. & Emmorey, K. (2006). Comparing sign language and speech reveals a universal limit on short-term memory capacity. *Psychological Science, 17*, 682–683. Wilson, M. & Fox, G. (2007). Working memory for language is not special: Evidence for an articulatory loop for novel stimuli. *Psychonomic Bulletin & Review, 14*, 470–473.
8. Baddeley, A. D. (1992a). Is working memory working? The Fifteenth Bartlett Lecture. *The Quarterly Journal of Experimental Psychology, 44A*, 1–31; Baddeley, A. D. (1992b). Working memory. *Science*, *255*, 556–559; Baddeley, A. D. & Hitch, G. (1974). Working memory. In G. H. Bower (Ed.), *The Psychology of Learning and Motivation* (Vol. 8, pp. 47–89). New York, NY: Academic Press.
9. Cowan, N. (2015). George Miller's magical number of immediate memory in retrospect: Observations on the faltering progression of science. *Psychological Review, 122*(3), 536–541.
10. Cowan, N. (2010). The magical mystery four: How is working memory capacity limited, and why? *Current Directions in Psychological Science, 19*, 51–57.
11. Chase, W. G. & Ericsson, K. A. (1982). Skill and working memory. In G. H. Bower (Ed.), *The Psychology of Learning and Motivation, 16*, 1–58). New York, NY: Academic Press.
12. Endress, A.D. & Szabó, S. (2017). Interference and memory capacity limitations. *Psychological review, 124*(5), 551.
13. Brown, J. A. (1958). Some tests of the decay theory of immediate memory. *The Quarterly Journal of Experimental Psychology, 10*, 12–21.
14. Peterson, L. R., & Peterson, M. J. (1959). Short-term retention of individual items. *Journal of Experimental Psychology, 58*, 193–198.
15. Waugh, N. C. & Norman, D. A. (1965). Primary memory. *Psychological Review, 72*, 89–104.
16. Wright, A. A. & Roediger, H. L. III. (2003). Interference processes in monkey auditory list memory. *Psychonomic Bulletin & Review, 10*, 696–702.
17. Unsworth, N., Heitz, R. P., & Parks, N. A. (2008). The importance of temporal distinctiveness for forgetting over the short-term. *Psychological Science, 19*, 1078–1081.
18. Altmann, E. M. & Gray, W. D. (2002). Forgetting to remember: The functional relationship of decay and interference. *Psychological Science, 13*, 27–33; Altmann, E. M. & Schunn, C. D. (2012). Decay versus interference: A new look at an old interaction. *Psychological Science, 23*(11), 1435–1437; Berman, M. G., Jonides, J. & Lewis, R. L. (2009). In search of decay in verbal short-term memory. *Journal of Experimental Psychology: Learning, Memory, and Cognition, 35*(2), 317–333.
19. Keppel, G. & Underwood, B. J. (1962). Proactive inhibition in short-term retention of single items. *Journal of Verbal Learning and Verbal Behavior, 1*, 153–161.
20. Wickens, D. D. (1972). Characteristics of word encoding. In A. W. Melton & E. Martin (Eds.), *Coding Processes in Human Memory* (pp. 191–215). New York, NY: Winston; Wickens, D. D., Born, D. G. & Allen, C. K. (1963). Proactive inhibition and item similarity in short-term memory. *Journal of Verbal Learning and Verbal Behavior, 2*, 440–445.
21. Wickens, D. D., Born, D. G., & Allen, C. K. (1963). Proactive inhibition and item similarity in short-term memory. *Journal of Verbal Learning and Verbal Behavior, 2*, 440–445.
22. Wickens, D. D. (1972). Characteristics of word encoding. In A. W. Melton & E. Martin (Eds.), *Coding processes in human memory* (pp. 191–215). New York, NY: Winston.
23. Murdock, B. B., Jr. (1962). The serial position effect of free recall. *Journal of Experimental Psychology, 64*, 482–488.
24. Glanzer, M. & Cunitz, A. R. (1966). Two storage mechanisms in free recall. *Journal of Verbal Learning and Verbal Behavior, 5*, 351–360.
25. Glanzer, M. (1972). Storage mechanisms in recall. In G. H. Bower & J. T. Spence (Eds.), *The Psychology of Learning and Motivation* (Vol. 5, pp. 129–193). New York, NY: Academic Press.
26. Glanzer, M. & Cunitz, A. R. (1966). Two storage mechanisms in free recall. *Journal of Verbal Learning and Verbal Behavior, 5*, 351–360.
27. Sternberg, S. (1966). High-speed scanning in human memory. *Science, 153*, 652–654; Sternberg, S. (1969). The discovery of processing stages: Extensions of Donders's method. In W. G. Koster (Ed.), Attention and performance II. *Acta Psychologica, 30*, 276–315; Sternberg, S. (1975). Memory scanning: New findings and current controversies. *The Quarterly Journal of Experimental Psychology, 27*, 1–32.
28. Sternberg, S. (1969). The discovery of processing stages: Extensions of Donders's method. In W. G. Koster (Ed.), Attention and performance II. *Acta Psychologica, 30*, 276–315.
29. Sternberg, S. (1969). The discovery of processing stages: Extensions of Donders's method. In W. G. Koster (Ed.), Attention and performance II. *Acta Psychologica, 30*, 276–315.
30. Sternberg, S. (1969). The discovery of processing stages: Extensions of Donders's method. In W. G. Koster (Ed.), Attention and performance II. *Acta Psychologica,*

30, 276–315; Sternberg, S. (1975). Memory scanning: New findings and current controversies. *The Quarterly Journal of Experimental Psychology, 27*, 1–32.

31. Baddeley, A. D. (1976). *The Psychology of Memory*. New York, NY: Basic Books.

32. McClelland, J. L. (1979). On the time relations of mental processes: An examination of systems of processes in cascade. *Psychological Review, 86*, 287–330.

33. Baddeley, A. D. (1976). *The Psychology of Memory*. New York, NY: Basic Books; Baddeley, A. D. & Hitch, G. (1974). Working memory. In G. H. Bower (Ed.), *The Psychology of Learning and Motivation* (Vol. 8, pp. 47–89). New York, NY: Academic Press; Baddeley, A. D. & Lieberman, K. (1980). Spatial working memory. In R. Nickerson (Ed.), *Attention and Performance VIII*. Hillsdale, NJ: Erlbaum.

34. Baddeley, A. D., & Hitch, G. (1974). Working memory. In G. H. Bower (Ed.), *The psychology of learning and motivation* (Vol. 8, pp. 47–89). New York, NY: Academic Press.

35. Warrington, E. K. & Shallice, T. (1969). The selective impairment of auditory verbal short-term memory. *Brain, 92*, 885–896; Shallice, T. & Warrington, E. K. (1970). Independent functioning of the verbal memory stores: A neuropsychological study. *The Quarterly Journal of Experimental Psychology, 22*, 261–273; Warrington, E. K. & Weiskrantz, L. (1970). The amnesic syndrome: Consolidation or retrieval? *Nature, 228*, 628–630.

36. Baddeley, A. D. & Hitch, G. (1974). Working memory. In G. H. Bower (Ed.), *The Psychology of Learning and Motivation* (Vol. 8, pp. 47–89). New York, NY: Academic Press; Baddeley, A. D. & Wilson, B. (1988). Comprehension and working memory: A single case neuropsychological study. *Journal of Memory and Language, 27*, 479–498; Vallar, G. & Baddeley, A. D. (1984). Fractionation of working memory: Neuropsychological evidence for a phonological short-term store. *Journal of Verbal Learning and Verbal Behavior, 23*, 151–161.

37. McCarthy, R. A. & Warrington, E. K. (1984). A two route model of speech production: Evidence from aphasia. *Brain, 107*, 463–485.

38. Baddeley, A. D. (2000a). The episodic buffer: A new component of working memory? *Trends in Cognitive Sciences, 4*, 417–423; Baddeley, A. D. & Hitch, G. (1974). Working memory. In G. H. Bower (Ed.), *The Psychology of Learning and Motivation* (Vol. 8, pp. 47–89). New York, NY: Academic Press; Salame, P. & Baddeley, A. D. (1982). Disruption of short-term memory by unattended speech: Implications for the structure of working memory. *Journal of Verbal Learning and Verbal Behavior, 21*, 150–164.

39. Baddeley, A. D. (2000a). The episodic buffer: A new component of working memory? *Trends in Cognitive Sciences, 4*, 417–423.

40. Baddeley, A. D., & Hitch, G. (1974). Working memory. In G. H. Bower (Ed.), *The psychology of learning and motivation* (Vol. 8, pp. 47–89). New York, NY: Academic Press.

41. Smith, E. E. & Jonides, J. (1999). Storage and executive processes in the frontal lobes. *Science, 283*, 1657–1661; Smith, E. E. (2000). Neural bases of human working memory. *Current Directions in Psychological Science, 9*, 45–49.

42. Jonides, J., Lacey, S. C. & Nee, D. E. (2005). Processes of working memory in mind and brain. *Current Directions in Psychological Science, 14*, 2–5.

43. Logan, G. D. (2003). Executive control of thought and action: In search of the wild homunculus. *Current Directions in Psychological Science, 12*, 45–48.

44. Kane, M. J. & Engle, R. W. (2002). The role of prefrontal cortex in working-memory capacity, executive attention, and general fluid intelligence: An individual-differences perspective. *Psychonomic Bulletin & Review, 9*, 637–671.

45. Friedman, N. P., Miyake, A., Young, S. E., DeFries, J. C., Corley, R. P. & Hewitt, J. K. (2008). Individual differences in executive functions are almost entirely genetic in origin. *Journal of Experimental Psychology: General, 137*, 201–225.

46. Jonides, J., Smith, E. E., Koeppe, R. A., Awh, E., Minoshima, S. & Mintun, M. A. (1993). Spatial working-memory in humans as revealed by PET. *Nature, 363*, 623–625.

47. Courtney, S. M., Petit, L. Maisog, C. M., Ungerleider, L. G. & Haxby, J. V. (1998). An area specialized for spatial working memory in human frontal cortex. *Science, 279*, 1347–1351.

48. Miyake, A., Friedman, N. P., Emerson, M. J., Witzki, A. H., Howerter, A., & Wager, T. D. (2000). The unity and diversity of executive functions and their contributions to complex "frontal lobe" tasks: A latent variable analysis. *Cognitive Psychology, 41*, 49–100.

49. Jonides, J., Smith, E. E., Koeppe, R. A., Awh, E., Minoshima, S. & Mintun, M. A. (1993). Spatial working-memory in humans as revealed by PET. *Nature, 363*, 623–625

50. Vandierendonck, A. (2016). A working memory system with distributed executive control. *Perspectives on Psychological Science, 11*(1), 74–100.

51. Vergauwe, E., Barrouillet, P. & Camos, V. (2010). Do mental processes share a domain-general resource? *Psychological Science, 21*, 384–390.

52. Vredeveldt, A., Hitch, G. J. & Baddeley, A. D. (2011). Eyeclosure helps memory by reducing cognitive load and enhancing visualization. *Memory & Cognition, 39*, 1253–1263.

53. Schmeichel, B. J. (2007). Attention control, memory updating, and emotion regularity temporarily

reduce the capacity for executive control. *Journal of Experimental Psychology: General, 136*, 241–255.

54. Banich, M. T. (2009). Executive function: The search for an integrated account. *Current Directions in Psychological Science, 18*, 89–94.

55. Gevins, A., Smith, M. E., McEvoy, L., & Yu, D. (1997). High resolution EEG mapping of cortical activation related to working memory: Effects of task difficulty, type of processing, and practice. *Cerebral Cortex, 7*, 374–385.

56. Baddeley, A. D. (2000b). The phonological loop and the irrelevant speech effect: Some comments on Neath (2000). *Psychonomic Bulletin & Review, 7*, 544–549; Jones, D. M., Macken, W. J. & Nicholls, A. P. (2004). The phonological store of working memory: Is it phonological and is it a store? *Journal of Experimental Psychology: Learning, Memory, and Cognition, 30*, 656–674; Mueller, S. T., Seymour, T. L., Kieras, D. E. & Meyer, D. E. (2003). Theoretical implications of articulatory duration, phonological similarity, and phonological complexity in verbal working memory. *Journal of Experimental Psychology: Learning, Memory, and Cognition, 29*, 1353–1380.

57. Colle, H. A. & Welsh, A. (1976). Acoustic masking in primary memory. *Journal of Verbal Learning and Verbal Behavior, 15*, 17–32.

58. Kozlov, M. D., Hughes, R. W. & Jones, D. M. (2012). Gummed-up memory: Chewing gum impairs short-term recall. *The Quarterly Journal of Experimental Psychology, 65*(3), 501–513.

59. Baddeley, A. D. (1966). Short-term memory for word sequences as a function of acoustic, semantic, and formal similarity. *The Quarterly Journal of Experimental Psychology, 18*, 302–309; Conrad, R. & Hull, A. (1964). Information, acoustic confusion, and memory span. *British Journal of Psychology, 55*, 75–84.

60. Li, X., Schweickert, R. & Gandour, J. (2000). The phonological similarity effect in immediate recall: Positions of shared phonemes. *Memory & Cognition, 28*, 1116–1125.

61. Williamson, V. J., Baddeley, A. D. & Hitch, G. J. (2010). Musicians' and nonmusicians' shortterm memory for verbal and musical sequences: Comparing phonological similarity and pitch proximity. *Memory & Cognition, 38*, 163–175.

62. Shand, M. A. (1982). Sign-based short-term coding of American Sign Language signs and printed English words by congenitally deaf signers. *Cognitive Psychology, 14*, 1–12.

63. Broggin, E., Savazzi, S. & Marzi, C. A. (2012). Similar effects of visual perception and imagery on simple reaction time. *The Quarterly Journal of Experimental Psychology, 65*, 151–164; Kosslyn, S. M., Alpert, N. M., Thompson, W. L., Maljkovic, V., Weise, S. B., Chabris, C. F., et al. (1993). Visual mental imagery activates topographically organized visual cortex: PET investigations. *Journal of Cognitive Neuroscience, 5*, 263–287.

64. Brooks, L. R. (1968). Spatial and verbal components of the act of recall. *Canadian Journal of Experimental Psychology, 22*, 349–368.

65. Cooper, L. A. & Shepard, R. N. (1973). Chronometric studies of the rotation of mental images. In W. G. Chase (Ed.), *Visual information processing* (pp. 75–176). New York, NY: Academic Press; Shepard, R. N. & Metzler, J. (1971). Mental rotation of threedimensional objects. *Science, 171*, 701–703.

66. Wraga, M., Swaby, M. & Flynn, C. M. (2008). Passive tactile feedback facilitates mental rotation of handheld objects. *Memory & Cognition, 36*, 271–281.

67. Flusberg, S. J. & Boroditsky, L. (2011). Are things that are hard to physically move also hard to imagine moving? *Psychonomic Bulletin & Review, 18*, 158–164.

68. Just, M. A., Carpenter, P. A., Maguire, M., Diwadkar, V. & McMains, S. (2001). Mental rotation of objects retrieved from memory: A functional MRI study of spatial processing. *Journal of Experimental Psychology: General, 130*, 493–504.

69. Hubbard, T. L., Hutchison, J. L. & Courtney, J. R. (2010). *Boundary extension: Findings and theories.* The Quarterly Journal of Experimental Psychology, 63, 1467,Ä11494.

70. Hubbard, T. L. (1995). Environmental invariants in the representation of motion: Implied dynamics and representational momentum, gravity, friction, and centripetal force. *Psychonomic Bulletin & Review, 2*, 322–338.

71. Hubbard, T. L. (2005). Representational momentum and related displacements in spatial memory: A review of the findings. *Psychonomic Bulletin & Review, 12*, 822–851.

72. Freyd, J. J. & Finke, R. A. (1984). Representational momentum. Bulletin of the Psychonomic Society, 23, 443–446; Hubbard, T. L. (1990). Cognitive representation of linear motion: Possible direction and gravity effects in judged displacement. *Memory & Cognition, 18*, 299–309.

73. Hubbard, T. L. (1990). Cognitive representation of linear motion: Possible direction and gravity effects in judged displacement. *Memory & Cognition, 18*, 299–309.

74. Hubbard, T. L. (1990). Cognitive representation of linear motion: Possible direction and gravity effects in judged displacement. *Memory & Cognition, 18*, 299–309.

75. Hubbard, T. L. (1996). Representational momentum, centripetal force, and curvilinear impetus. *Journal of Experimental Psychology: Learning, Memory, and Cognition, 22*, 1049–1060.

76. Verfaille, K. & Y'dewalle, G. (1991). Representational momentum and event course anticipation in

the perception of implied motions. *Journal of Experimental Psychology: Learning, Memory, and Cognition, 17*, 302–313.

77. Baddeley, A. D. (2000a). The episodic buffer: A new component of working memory? *Trends in Cognitive Sciences, 4*, 417–423.

78. Allen, R. J., Baddeley, A. D. & Hitch, G. J. (2006). Is the binding of visual features in working memory resource-demanding? *Journal of Experimental Psychology: General, 135*, 298–313.

79. Engle, R. W. & Kane, M. J. (2004). Executive attention, working memory capacity, and a two-factor theory of cognitive control. *Psychology of Learning and Motivation, 44*, 145–200.

80. Shipstead, Z., Harrison, T. L. & Engle, R. W. (2015). Working memory capacity and the scope and control of attention. *Attention, Perception & Psychophysics, 77*(6), 1863–1880.

81. Vredeveldt, A., Hitch, G. J. & Baddeley, A. D. (2011). Eyeclosure helps memory by reducing cognitive load and enhancing visualization. *Memory & Cognition, 39*, 1253–1263.

82. Kane, M. J. & Engle, R. W. (2002). The role of prefrontal cortex in working-memory capacity, executive attention, and general fluid intelligence: An individual-differences perspective. *Psychonomic Bulletin & Review, 9*, 637–671.

83. Gernsbacher, M. A. & Faust, M. E. (1991). The mechanism of suppression: A component of general comprehension skill. *Journal of Experimental Psychology: Learning, Memory & Cognition, 17*, 245–262.

84. Johnson-Laird, P. N. (2013). Mental models and cognitive change. *Journal of Cognitive Psychology, 25*(2), 131–138.

85. Baddeley, A. D. & Hitch, G. (1974). Working memory. In G. H. Bower (Ed.), *The Psychology of Learning and Motivation* (Vol. 8, pp. 47–89). New York, NY: Academic Press.

86. Levy, J., Pashler, H. & Boer, E. (2006). Central interference in driving: Is there any stopping the psychological refractory period? *Psychological Science, 17*, 228–235.

87. Logie, R. H., Zucco, G., & Baddeley, A. D. (1990). Interference with visual short-term memory. *Acta Psychologica, 75*, 55–74.

88. Morey, C. C. & Cowan, N. (2004). When visual and verbal memories compete: Evidence of cross-domain limits in working memory. *Psychonomic Bulletin & Review, 11*, 296–301.

89. Logie, R. H., Zucco, G., & Baddeley, A. D. (1990). Interference with visual short-term memory. *Acta Psychologica, 75*, 55–74.

90. Baddeley, A. D. & Lieberman, K. (1980). Spatial working memory. In R. Nickerson (Ed.), *Attention and performance VIII*. Hillsdale, NJ: Erlbaum.

91. Cowan, N. & Morey, C. C. (2007). How can dual-task working memory retention limits be investigated? *Psychological Science, 18*, 686–688.

92. Engle, R. W. (2001). What is working memory capacity? In H. L. Roediger, J. S. Nairne, I. Neath & A. M. Suprenant(Eds.), *The Nature of Remembering: Essays in Honor of Robert G. Crowder* (pp. 297–314). Washington, DC: American Psychological Association Press; Rosen, V. M. & Engle, R. W. (1997). The role of working memory capacity in retrieval. *Journal of Experimental Psychology: General, 126*, 211–227.

93. Engle, R. W. (2002). Working memory capacity as executive attention. *Current Directions in Psychological Science, 11*, 19–23.

94. Turner, M. L. & Engle, R. W. (1989). Is working memory capacity task dependent? *Journal of Memory and Language, 28*, 127–154.

95. Daneman, M. & Carpenter, P. A. (1980). Individual differences in working memory and reading. *Journal of Verbal Learning and Verbal Behavior, 19*, 450–466.

96. Engle, R. W. (2002). Working memory capacity as executive attention. *Current Directions in Psychological Science, 11*, 19–23.

97. Kane, M. J. & Engle, R. W. (2003). Working-memory capacity and the control of attention: The contributions of goal neglect, response competition and task set to Stroop interference. *Journal of Experimental Psychology: General, 132*, 47–70; Daneman, M. & Merikle, P. M. (1996). Working memory and language comprehension: A meta-analysis. *Psychonomic Bulletin & Review, 3*, 422–433; Engle, R. W. (2002). Working memory capacity as executive attention. *Current Directions in Psychological Science, 11*, 19–23; Miyake, A. & Shah, P. (Eds.). (1999). *Models of Working Memory: Mechanisms of Active Maintenance and Executive Control*. New York, NY: Cambridge University Press; Unsworth, N. & Spillers, G. J, (2010). Working memory capacity: Attention control, secondary memory, or both? A direct test of the dual component model. *Journal of Memory and Language, 62*, 392–406.

98. Unsworth, N. & Engle, R. W. (2007). The nature of individual differences in working memory capacity: Active maintenance in primary memory and controlled search from secondary memory. *Psychological Review, 114*, 104–132.

99. Feng, J., Spence, I. & Pratt, J. (2007). Playing an action video game reduces gender differences in spatial cognition. *Psychological Science, 18*, 850–855.

100. Sanchez, C. A. (2012). Enhancing visuospatial performance through video game training to increase learning in visuospatial science domains. *Psychonomic Bulletin & Review, 19*, 58–65.

101. Moreno, S., Bialystok, E., Barac, R., Schellenberg, E. G., Cepeda, N. J. & Chau, T. (2011). Short-term music training enhances verbal intelligence

and executive function. *Psychological Science, 22*, 1425–1433.

102. Kozhevnikov, M., Louchakova, O., Josipovic, Z. & Motes, M. A. (2009). The enhancement of visuospatial processing efficiency through Buddhist Deity Meditation. *Psychological Science, 20*, 645–653.

103. Chase, W. G., & Ericsson, K. A. (1982). Skill and working memory. In G. H. Bower (Ed.), *The psychology of learning and motivation* (Vol. 16, pp. 1–58). New York, NY: Academic Press.

104. Verhaeghen, P., Cerella, J. & Basak, C. (2004). A working memory workout: How to expand the focus of serial attention from one to four items in 10 hours or less. *Journal of Experimental Psychology: Learning, Memory, and Cognition, 30*, 1322–1337.

105. Copeland, D. E. & Radvansky, G. A. (2004b). Working memory span and situation model processing. *American Journal of Psychology, 117*, 191–213.

106. Miyake, A., Friedman, N. P., Emerson, M. J., Witzki, A. H., Howerter, A. & Wager, T. D. (2000). The unity and diversity of executive functions and their contributions to complex "frontal lobe" tasks: A latent variable analysis. *Cognitive Psychology, 41*, 49–100; Miyake, A., Friedman, N. P., Emerson, M. J., Witzki, A. H., Howerter, A. & Wager, T. D. (2000). The unity and diversity of executive functions and their contributions to complex"frontal lobe" tasks: A latent variable analysis. *Cognitive Psychology*, 41, 49–100; Miyake, A. & Friedman, N.P. (2012) The nature and organization of individual differences in executive functions: four general conclusions. *Current Directions in Psychological Science, 21*(1), 8–14.

107. Au, J., Sheehan, E., Tsai, N., Duncan, G. J., Buschkuehl, M. & Jaeggi, S. M. (2015). Improving fluid intelligence with training on working memory: A meta-analysis. *Psychonomic Bulletin & Review, 22*(2), 366–377.

108. Chooi, W. T. & Thompson, L. A. (2012). Working memory training does not improve intelligence in healthy young adults. *Intelligence, 40*(6), 531–542; Harrison, T. L., Shipstead, Z., Hicks, K. L., Hambrick, D. Z., Redick, T. S. & Engle, R. W. (2013). Working memory training may increase working memory capacity but not fluid intelligence. *Psychological Science, 24*(12), 2409–2419; Waris, O., Soveri, A. & Line, M. (2015). Transfer after working memory updating. *PloS one, 10*(9), e0138734.

109. Bogg, T. & Lasecki, L. (2015). Reliable gains? Evidence for substantially underpowered designs in studies of working memory training transfer to fluid intelligence. *Frontiers in Psychology, 5*, 1589.

110. Benz, S., Sellaro, R., Hommel, B. & Colzato, L. S. (2015). Music makes the world go round: The impact of musical training on non-musical cognitive functions—A review. *Frontiers in Psychology, 6*, 2023; Gordon, R. L., Fehd, H. M. & McCandliss, B. D. (2015). Does music training enhance literacy skills? A meta-analysis. *Frontiers in Psychology, 6*.

111. Conway, A. R. A., Cowan, N. & Bunting, M. F. (2001). The cocktail party phenomenon revisited: The importance of working memory capacity. *Psychonomic Bulletin & Review, 8*, 331–335.

112. Kane, M. J. & Engle, R. W. (2003). Working-memory capacity and the control of attention: The contributions of goal neglect, response competition and task set to Stroop interference. *Journal of Experimental Psychology: General, 132*, 47–70.

113. Sanchez, C. A., &Wiley, J. (2006). An examination of the seductive details effect in terms of working memory capacity. *Memory & Cognition, 34*, 344–355.

114. Rosen, V. M. & Engle, R. W. (1997). The role of working memory capacity in retrieval. *Journal of Experimental Psychology: General, 126*, 211–227.

115. Kane, M. J., & Engle, R. W. (2000). Working memory capacity, proactive interference, and divided attention: Limits on long-term memory retrieval. *Journal of Experimental Psychology: Learning, Memory, and Cognition, 26*, 336–358.

116. Bunting, M. F., Conway, A. R. A. & Heitz, R. P. (2004). Individual differences in the fan effect and working memory capacity. *Journal of Memory and Language, 51*, 604–622; Cantor, J. & Engle, R. W. (1993). Working-memory capacity as long-term memory activation: An individual differences approach. *Journal of Experimental Psychology: Learning, Memory, and Cognition, 19*, 1101–1114; Radvansky, G. A. & Copeland, D. E. (2006a). Memory retrieval and interference: Working memory issues. *Journal of Memory and Language, 55*, 33–46.

117. Unsworth, N. (2007). Individual differences in working memory capacity and episodic retrieval: The dynamics of delayed and continuous distractor free recall. *Journal of Experimental Psychology: Learning, Memory, and Cognition, 33*, 1020–1034.

118. Hambrick, D. Z. & Engle, R. W. (2002). Effects on domain knowledge, working memory capacity, and age on cognitive performance: An investigation of the knowledge-is-power hypothesis. *Cognitive Psychology, 44*, 339–387; Hambrick, D. Z. & Meinz, E. J. (2011). Limits on the predictive power of domain-specific experience and knowledge in skilled performance. *Current Directions in Psychological Science, 20*, 275–279.

119. Moore, A. B., Clark, B. A. & Kane, M. J. (2008). Who shalt not kill? Individual differences in working memory capacity, executive control, and moral judgment. *Psychological Science, 19*, 549–557.

120. Copeland, D. E., & Radvansky, G. A. (2004a). Working memory and syllogistic reasoning. *The Quarterly Journal of Experimental Psychology, 57A*, 1437–1457.

121. Markovits, H. & Doyon, C. (2004). Information processing and reasoning with premises that are empirically false: Interference, working memory, and processing speed. *Memory & Cognition, 32*, 592–601.

122. Beilock, S. L., & DeCaro, M. S. (2007). From poor performance to success under stress: Working memory, strategy selection, and mathematical problem solving under pressure. *Journal of Experimental Psychology, 130*, 701–725.

123. Beilock, S. L. & DeCaro, M. S. (2007). From poor performance to success under stress: Working memory, strategy selection, and mathematical problem solving under pressure. *Journal of Experimental Psychology, 130*, 701–725.

124. Cokely, E. T., Kelley, C. M. & Gilchrist, A. L. (2006). Sources of individual differences in working memory capacity: Contributions of strategy to capacity. *Psychonomic Bulletin & Review, 13*, 991–997; Colflesh, G. J. H. & Conway, A. R. A. (2007). Individual differences in working memory capacity and divided attention in dichotic listening. *Psychonomic Bulletin & Review, 14*, 699–703.

125. Tuholski, S. W., Engle, R. W. & Baylis, G. C. (2001). Individual differences in working memory capacity and enumeration. *Memory & Cognition, 29*, 484–492.

126. Fukuda, K., Vogel, E., Mayr, U. & Awh, E. (2010). Quality, not quantity: The relationship between fluid intelligence and working memory capacity. *Psychonomic Bulletin & Review, 17*, 673–679; Kane, M. J., Hambrick, D. Z., Tuholski, S. W., Wilhelm, O., Payne, T. W. & Engle, R. W. (2004). The generality of working memory capacity: A latent-variable approach to verbal and visuospatial memory span and reasoning. *Journal of Experimental Psychology: General, 133*, 189–217; Salthouse, T. A. & Pink, J. E. (2008). Why is working memory related to fluid intelligence? *Psychonomic Bulletin & Review, 15*, 364–371; Shelton, J. T., Elliott, E. M., Matthews R. A., Hill, B. D. & Gouvier, W. D. (2010). The relationships of working memory, secondary memory, and general fluid intelligence: Working memory is special. *Journal of Experimental Psychology: Learning, Memory, and Cognition, 36*, 813–820.

127. McCabe, D. P. (2010). The influence of complex working memory span task administration methods on prediction of higher level cognition and metacognitive control of response times. *Memory & Cognition, 38*, 868–882.

128. Burgess, G. C., Gray, J. R., Conway, A. R. A. & Braver, T. S. (2011). Neural mechanisms of interference control underlie the relationship between fluid intelligence and working memory span. *Journal of Experimental Psychology: General, 140*, 674–692.

Chapter 6

1. Squire, L. R. (1986). Mechanisms of memory. *Science, 232*(4578), 1612–1619; Squire, L. R. (1993). The organization of declarative and nondeclarative memory. In T. Ono, L. R. Squire, M. E. Raichle; D. I. Perrett & M. Fukuda (Eds.), *Brain Mechanisms of Perception and Memory: From Neuron to Behavior* (pp. 219–227). New York, NY: Oxford University Press.

2. Tulving, E. (1972). Episodic and semantic memory. In E. Tulving & W. Donaldson (Eds.), *Organization of Memory* (pp. 381–403). New York, NY: Academic Press; Tulving, E. (1983). *Elements of Episodic Memory*. Oxford, England: Clarendon. Tulving, E. (1993). What is episodic memory? *Current Directions in Psychological Science, 2*, 67–70.

3. Buchner, A. & Wippich, W. (2000). On the reliability of implicit and explicit memory measures. *Cognitive Psychology, 40*, 227–259; Jacoby, L. L. (1991). A process dissociation framework: Separating automatic from intentional uses of memory. *Journal of Memory and Language, 30*, 513–541; Jacoby, L. L., Toth, J. P. & Yonelinas, A. P. (1993). Separating conscious and unconscious influences of memory: Measuring recollection. *Journal of Experimental Psychology: General, 122*, 139–154.

4. Rubin, D. C. (2007). A basic-systems model of episodic memory. *Perspectives on Psychological Science, 1*, 277–311.

5. Thomas, M. H. & Wang, A. Y. (1996). Learning by the keyword mnemonic: Looking for long-term benefits. *Journal of Experimental Psychology: Applied, 2*, 330–342.

6. Roediger, H. L. (1980). The effectiveness of four mnemonics in ordering recall. *Journal of Experimental Psychology: Human Learning and Memory, 6*(5), 558–567

7. Bower, G. H. & Clark, M. C. (1969). Narrative stories as mediators for serial learning. *Psychonomic Science, 14*(4), 181–182.

8. Hu, Y., Ericsson, K. A., Yang, D. & Lu, C. (2009). Superior selfpaced memorization of digits in spite of a normal digit span: The structure of a memorist's skill. *Journal of Experimental Psychology: Learning, Memory, and Cognition, 35*(6), 1426–1442.

9. Chaffin, R. & Imreh, G. (2002). Practicing perfection: Piano performance as expert memory. *Psychological Science, 13*, 342–349.

10. Melton, A. W. (1963). Implications of short-term memory for a general theory of memory. *Journal of Verbal Learning and Verbal Behavior, 2*, 1–21.

11. Wenger, M. J. & Payne, D. G. (1995). On the acquisition of mnemonic skill: Application of skilled memory theory. *Journal of Experimental Psychology: Applied, 1*, 194–215.

12. Gold, P. E., Cahill, L. & Wenk, G. L. (2002). Ginkgo biloba: A cognitive enhancer? *Psychological Science in*

the Public Interest, 3, 2–11. Gold, P. E., Cahill, L. & Wenk, G. L. (2003). The lowdown on ginkgo biloba. *Scientific American, 288*, 86–92; Greenwald, A. G., Spangenberg, E. R., Pratkanis, A. R. & Eskenazi, J. (1991). Double-blind tests of subliminal self-help audiotapes. *Psychological Science, 2*, 119–122; McDaniel, M. A., Maier, S. F. & Einstein, G. O. (2002). "Brainspecific" nutrients: A memory cure? *Psychological Science in the Public Interest, 31*, 12–38.

13. Ebbinghaus, H. (1885/1913). *Memory: A Contribution to Experimental Psychology* (H. A. Ruger & C. E. Bussenius, Trans.). New York, NY: Columbia University, Teacher's College.(Reprinted 1964, New York, NY: Dover).

14. Nelson, T. O. (1978). Savings and forgetting from long-term memory. *Journal of Verbal Learning and Verbal Behavior, 10*, 568–576. Nelson, T. O. (1985). Ebbinghaus's contribution to the measurement of retention: Savings during relearning. *Journal of Experimental Psychology: Learning, Memory, and Cognition, 11*, 472–479; Schacter, D. L. (1987). Implicit memory: History and current status. *Journal of Experimental Psychology: Learning, Memory, and Cognition, 13*, 501–518.

15. MacLeod, C. M. (1988). Forgotten but not gone: Savings for pictures and words in long-term memory. *Journal of Experimental Psychology: Learning, Memory, and Cognition, 14*, 195–212.

16. Slamecka, N. J. (1968). An examination of trace storage in free recall. *Journal of Experimental Psychology, 4*, 504–513.

17. Wixted, J. T. & Ebbesen, E. B. (1991). On the form of forgetting. *Psychological Science, 2*, 409–415.

18. Erdelyi, M. H. (2010). The ups and downs of memory. *American Psychologist, 65*, 623–633.

19. Schneider, V. I., Healy, A. F. & Bourne, L. E., Jr. (2002). What is learned under difficult conditions is hard to forget: Contextual interference effects in foreign vocabulary acquisition, retention, and transfer. *Journal of Memory and Language, 46*, 419–440.

20. Glenberg, A. M. & Lehmann, T. S. (1980). Spacing repetitions over 1 week. *Memory & Cognition, 8*, 528–538.

21. Zechmeister, E. B. & Shaughnessy, J. J. (1980). When you think that you know and when you think that you know but you don't. *Bulletin of the Psychonomic Society, 15*, 41–44.

22. Pashler, H., Rohrer, D., Cepeda, N. J. & Carpenter, S. K. (2007). Enhancing learning and retarding forgetting: Choices and consequences. *Psychonomic Bulletin & Review, 14*, 187–193.

23. Fenn, K. M. & Hambrick, D. Z. (2013). What drives sleepdependent memory consolidation: Greater gain or less loss? *Psychonomic Bulletin & Review, 20*(3), 501–506; Stickgold, R. & Walker, M. P. (2005). Memory consolidation and reconsolidation: What is the role of sleep? *Trends in Neurosciences, 28*(8), 408–415.

24. Leonesio, R. J. & Nelson, T. O. (1990). Do different metamemory judgments tap the same underlying aspects of memory? *Journal of Experimental Psychology: Learning, Memory, and Cognition, 16*, 464–470; Nelson, T. O. (1988). Predictive accuracy of the feeling of knowing across different criterion tasks and across different subject populations and individuals. In M. Gruneberg; P. Morris & R. Sykes (Eds.), *Practical Aspects of Memory: Current Research and Issues* (Vol. 1, pp. 190–196). New York, NY: Wiley

25. Son, L. K. (2004). Spacing one's study: Evidence for a metacognitive control strategy. *Journal of Experimental Psychology: Learning, Memory, and Cognition, 30*, 601–604.

26. Higham, P. A. & Garrard, C. (2005). Not all errors are created equal: Metacognition and changing answers on multiplechoice tests. *Canadian Journal of Experimental Psychology, 59*, 28–34.

27. Koriat, A., Sheffer, L. & Ma'ayan, H. (2002). Comparing objective and subjective learning curves: Judgments of learning exhibit increased underconfidence with practice. *Journal of Experimental Psychology: General, 131*, 147–162.

28. McCabe, J. (2011). Metacognitive awareness of learning strategies in undergraduates. *Memory & Cognition, 39*, 462–476.

29. Hasher, L. & Zacks, R. T. (1984). Automatic processing of fundamental information: The case of frequency of occurrence. *American Psychologist, 39*, 1372–1388.

30. Greene, R. L. (1986). Effects of intentionality and strategy on memory for frequency. *Journal of Experimental Psychology: Learning, Memory, and Cognition, 12*, 489–495; Hanson, C. & Hirst, W. (1988). Frequency encoding of token and type information. *Journal of Experimental Psychology: Learning, Memory, and Cognition, 14*, 289–297; Jonides, J. & Jones, C. M. (1992). Direct coding for frequency of occurrence. *Journal of Experimental Psychology: Learning, Memory, and Cognition, 18*, 368–378.

31. von Restorff, H. (1933). *Über die Wirkung von Bereichsbildungen im Spurenfeld* [On the effect of sphere formations in the trace field]. *Psychologische Forschung, 18*, 299–342.

32. Cooper, E. H. & Pantle, A. J. (1967). The total-time hypothesis in verbal learning. *Psychological Bulletin, 68*, 221–234; Kelley, M. R. & Nairne, J. S. (2001). von Restorff revisited: Isolation, generation, and memory of order. *Journal of Experimental Psychology: Learning, Memory, and Cognition, 27*, 54–66.

33. Axmacher, N., Cohen, M. X., Fell, J., Haupt, S., Dümpelmann, M., Elger, C. E., et al. (2010). Intracranial EEG correlates ofexpectancy and memory formation in the human hippocampus and nucleus accumbens. *Neuron, 65*, 541–549.

34. Kishiyama, M. M., Yonelinas, A. P., & Lazzara, M. M. (2004). The von Restorff effect in amnesia: The

contribution of the hippocampal system to novelty-related memory enhancements. *Journal of Cognitive Neuroscience, 16*, 15–23.

35. Atkinson, R. C., & Shiffrin, R. M. (1968). Human memory: A proposed system and its control processes. In W. K. Spence & J. T. Spence (Eds.), *The psychology of learning and motivation: Advances in research and theory* (Vol. 2, pp. 89–195). New York, NY: Academic Press.

36. Waugh, N. C. & Norman, D. A. (1965). Primary memory. *Psychological Review, 72*, 89–104.

37. Hellyer, S. (1962). Frequency of stimulus presentation and short-term decrement in recall. *Journal of Experimental Psychology, 64*, 650.

38. Hellyer, S. (1962). Frequency of stimulus presentation and short-term decrement in recall. *Journal of Experimental Psychology, 64*, 650.

39. Karpicke, J. D. (2012). Retrieval-based learning active retrieval promotes meaningful learning. *Current Directions in Psychological Science, 21*(3), 157–163.

40. Karpicke, J. D., & Roediger, H. L. (2007). Repeated during retrieval is the key to long term retention. *Journal of Memory and Language, 57*, 151–162.

41. Rundus, D. (1971). Analysis of rehearsal processes in free recall. *Journal of Experimental Psychology, 89*, 63–77; Rundus, D. & Atkinson, R. C. (1970). Rehearsal processes in free recall: A procedure for direct observation. *Journal of Verbal Learning and Verbal Behavior, 9*, 99–105.

42. Sehulster, J. R. (1989). Content and temporal structure of autobiographical knowledge: Remembering twenty-five seasons at the Metropolitan Opera. *Memory & Cognition, 17*, 590–606.

43. Roediger, H. L. III & Crowder, R. G. (1976). A serial position effect in recall of United States presidents. *Bulletin of the Psychonomic Society, 8*(4), 275–278.

44. Davelaar, E. J., Goshen-Gottstein, Y., Ashkenazi, A., Haarmann, H. J. & Usher, M. (2005). The demise of short-term memory revisited: Empirical and computational investigations of recency effects. *Psychological Review, 112*, 3–42.

45. Brown, G. D. A., Neath, I. & Chater, N. (2007). A temporal ratio model of memory. *Psychological Review, 114*, 539–576.

46. Gerbier, E. & Toppino, T. C. (2015). The effect of distributed practice: Neuroscience, cognition, and education. *Trends in Neuroscience and Education, 4*(3), 49–59; Glenberg, A. M. & Lehmann, T. S. (1980). Spacing repetitions over 1 week. *Memory & Cognition, 8*, 528–538.

47. Benjamin, A. S. & Tullis, J. (2010). What makes distributed practice effective? *Cognitive Psychology, 61*(3), 228–247.

48. Glenberg, A. M. (1976). Monotonic and nonmonotonic lag effects in paired-associate and recognition memory paradigms. *Journal of Verbal Learning and Verbal Behavior, 15*(1), 1–16. Glenberg, A. M. (1979). Component-levels theory of the effects of spacing of repetitions on recall and recognition. *Memory & Cognition, 7*(2), 95–112.

49. Metcalfe, J. & Xu, J. (2016). People mind wander more during massed than spaced inductive learning. *Journal of Experimental Psychology: Learning, Memory, and Cognition, 42*, 978–984.

50. Roediger, H. L. & Karpicke, J. D. (2006). The power of resting memory: Basic research and implications for educational practice. *Perspectives on Psychological Science, 1*, 181–210.

51. Roediger, H. L. & Karpicke, J. D. (2006). The power of resting memory: Basic research and implications for educational practice. *Perspectives on Psychological Science, 1*, 181–210; Gates, A. I. (1917). Recitation as a factor in memorizing. *Archives of Psychology, 40*, 104.

52. McDaniel, M. A., Roediger, H. L. & McDermott, K. B. (2007). Generalized test-enhanced learning from the laboratory to the classroom. *Psychonomic Bulletin & Review, 14*, 200–206.

53. Marsh, E. J., Roediger, H. L., Bjork, R. A. & Bjork, E. L. (2007). The memorial consequences of multiple-choice testing. *Psychonomic Bulletin & Review, 14*, 194–199.

54. Carpenter, S. K. & Pashler, H. (2007). Testing beyond words: Using tests to enhance visuospatial map learning. *Psychonomic Bulletin & Review, 14*, 474–478.

55. Butler, A. C. (2010). Repeated testing produces superior transfer of learning relative to repeated studying. *Journal of Experimental Psychology: Learning, Memory, and Cognition, 36*, 1118–1133.

56. Pastötter, B. & Bäuml, K. H. T. (2014). Retrieval practice enhances new learning: The forward effect of testing. *Frontiers in Psychology, 5*, 3389.

57. Craik, F. I. M. & Lockhart, R. S. (1972). Levels of processing: A framework for memory research. *Journal of Verbal Learning and Verbal Behavior, 11*, 671–684.

58. Craik, F. I. M. & Lockhart, R. S. (1972). Levels of processing: A framework for memory research. *Journal of Verbal Learning and Verbal Behavior, 11*, 671–684.

59. Craik, F. I. & Watkins, M. J. (1973). The role of rehearsal in short-term memory. *Journal of Verbal Learning and Verbal Behavior, 12*(6), 599–607.

60. Craik, F. I. & Tulving, E. (1975). Depth of processing and the retention of words in episodic memory. *Journal of Experimental Psychology: General, 104*(3), 268–294.

61. Henkel, L. A. (2004). Erroneous memories arising from repeated attempts to remember. *Journal of Memory and Language, 50*, 26–46.

62. Baddeley, A. D. (1978). The trouble with levels: A reexamination of Craik and Lockhart's framework for memory research. *Psychological Review, 85*, 139–152.

63. Glenberg, A. & Adams, F. (1978). Type I rehearsal and recognition. *Journal of Verbal Learning and Verbal Behavior, 17*, 455–464; Glenberg, A., Smith, S. M. & Green, C. (1977). Type I rehearsal: Maintenance and more. *Journal of Verbal Learning and Verbal Behavior, 11*, 403–416.

64. Baddeley, A. D. (1978). The trouble with levels: A reexamination of Craik and Lockhart's framework for memory research. *Psychological Review, 85*, 139–152.

65. Curran, T. (2000). Brain potentials of recollection and familiarity. *Memory & Cognition, 28*, 923–938; Yonelinas, A. P. (2002). The nature of recollection and familiarity: A review of 30 years of research. *Journal of Memory and Language, 46*, 441–517.

66. Craik, F. I. M., Govoni, R., Naveh-Benjamin, M. & Anderson, N. D. (1996). The effects of divided attention on encoding and retrieval processes in human memory. *Journal of Experimental Psychology: General, 125*, 181–194; Hicks, J. L. & Marsh, R. L. (2000). Toward specifying the attentional demands of recognition memory. *Journal of Experimental Psychology: Learning, Memory, and Cognition, 26*, 1483–1498.

67. Glenberg, A., Smith, S. M. & Green, C. (1977). Type I rehearsal: Maintenance and more. *Journal of Verbal Learning and Verbal Behavior, 11*, 403–416.

68. Bellezza, F. S. (1992). Recall of congruent information in the self-reference task. *Bulletin of the Psychonomic Society, 30*, 275–278; Gillihan, S. J. & Farah, M. J. (2005). Is the self special? A critical review of evidence from experimental psychology and cognitive neuroscience. *Psychological Bulletin, 131*, 76–97; Rogers, T. B., Kuiper, N. A. & Kirker, W. S. (1977). Selfreference and the encoding of personal information. *Journal of Personality and Social Psychology, 35*, 677–688; Symons, C. S. & Johnson, B. T. (1997). The self-reference effect in memory: A meta-analysis. *Psychological Bulletin, 121*, 371–394.

69. Slamecka, N. J. & Graf, P. (1978). The generation effect: Delineation of a phenomenon. *Journal of Experimental Psychology: Human Learning and Memory, 4*, 592–604.

70. Bertsch, S., Pesta, B. J., Wiscott, R. & McDaniel, M. A. (2007). The generation effect: A meta-analytic review. *Memory & Cognition, 35*, 201–210.

71. deWinstanley, P. A. & Bjork, E. L. (2004). Processing strategies and the generation effect: Implications for making a better reader. *Memory & Cognition, 32*, 945–955.

72. Fawcett, J. M. (2013). The production effect benefits performance in between-subject designs: A meta-analysis. *Acta Psychologica, 142*(1), 1–5; MacLeod, C. M., Gopie, N., Hourihan, K. L., Neary, K. R. & Ozubko, J. D. (2010). The production effect: Delineation of a phenomenon. *Journal of Experimental Psychology. Learning, Memory, and Cognition, 36*, 671–685.

73. Forrin, N. D., MacLeod, C. M. & Ozubko, J. D. (2012). Widening the boundaries of the production effect. *Memory & Cognition, 40*(7), 1046–1055.

74. Engelkamp, J. & Dehn, D. M. (2000). Item and order information in subject-performed tasks and experimenter-performed tasks. *Journal of Experimental Psychology: Learning, Memory, and Cognition, 26*, 671–682; Saltz, E. & Donnenwerth-Nolan, S. (1981). Does motoric imagery facilitate memory for sentences? A selective interference test. *Journal of Verbal Learning and Verbal Behavior, 20*, 322–332.

75. Noice, H. & Noice, T. (1999). Long-term retention of theatrical roles. *Memory, 7*, 357–382; Freeman, J. E. & Ellis, J. A. (2003). The representation of delayed intentions: A prospective subject-performed task? *Journal of Experimental Psychology: Learning, Memory, and Cognition, 29*, 976–992; Shelton, A. L. & McNamara, T. P. (2001). Visual memories from nonvisual experiences. *Psychological Science, 12*, 343–347.

76. Koriat, A. & Pearlman-Avnion, S. (2003). Memory organization of action events and its relationship to memory performance. *Journal of Experimental Psychology: General, 132*, 435–454.

77. Bousfield, W. A. & Sedgewick, C. H. W. (1944). An analysis of sequences of restricted associative responses. *The Journal of General Psychology, 30*, 149–165.

78. Bousfield, W. A. (1953). The occurrence of clustering in the recall of randomly arranged associates. *The Journal of General Psychology, 49*, 229–240.

79. Baddeley, A. D. (1978). The trouble with levels: A reexamination of Craik and Lockhart's framework for memory research. *Psychological Review, 85*, 139–152.

80. Bower, G. H., & Clark, M. C. (1969). Narrative stories as mediators for serial learning. *Psychonomic Science, 14*(4), 181–182.

81. Bower, G. H., Clark, M. C., Lesgold, A. M. & Winzenz, D. (1969). Hierarchical retrieval schemes in recall of categorical word lists. *Journal of Verbal Learning and Verbal Behavior, 8*, 323–343.

82. Ericsson, K. A., Delaney, P. F., Weaver, G. & Mahadevan, R. (2004). Uncovering the structure of a memorist's superior "basic" memory capacity. *Cognitive Psychology, 49*, 191–237.

83. Takahashi, M., Shimizu, H., Saito, S. & Tomoyori, H. (2006). One percent ability and ninety-nine percent perspiration: A study of a Japanese memorist. *Journal of Experimental Psychology: Learning, Memory, and Cognition, 32*, 1195–1200.

84. Anderson, J. R. (1985). *Cognitive Psychology and its Implications* (2nd ed.). New York, NY: Freeman.

85. Johnson, N. F. (1970). The role of chunking and organization in the process of recall. In G. H. Bower (Ed.), *The psychology of learning and motivation* (Vol. 4, pp. 172–247). New York, NY: Academic Press.

86. Mandler, G. (1967). Organization and memory. In K. W. Spence & J. T. Spence (Eds.), *The psychology of learning and motivation* (Vol. 1, pp. 327–372). New York, NY: Academic Press.

87. Mandler, G. (1972). Organization and recognition. In E. Tulving & W. Donaldson (Eds.), *Organization of memory* (pp. 139–166). New York, NY: Academic Press.

88. Ashcraft, M. H., Kellas, G., & Needham, S. (1975). Rehearsal and retrieval processes in free recall of categorized lists. *Memory & Cognition, 3*, 506–512.

89. Bower, G. H., & Clark, M. C. (1969). Narrative stories as mediators for serial learning. *Psychonomic Science, 14*(4), 181–182.

90. Tulving, E. (1962). Subjective organization in free recall of "unrelated" words. *Psychological Review, 69*, 344–354.

91. Kahana, M. J. & Wingfield, A. (2000). A functional relation between learning and organization in free recall. *Psychonomic Bulletin & Review, 7*, 516–521.

92. Brady, T. F., Konkle, T., Alvarez, G. A. & Oliva, A. (2008). Visual long-term memory has a massive storage capacity for object details. *Proceedings of the National Academy of Sciences, 105*(38), 14325–14329.

93. Standing, L. (1973). Learning 10000 pictures. *The Quarterly Journal of Experimental Psychology, 25*(2), 207–222.

94. Paivio, A. (1971). *Imagery and Verbal Processes.* New York, NY: Holt.

95. Bower, G. H. (1970). Analysis of a mnemonic device. *American Scientist, 58*, 496–510; Yuille, J. C. & Paivio, A. (1967). Latency of imaginal and verbal mediators as a function of stimulus and response concreteness-imagery. *Journal of Experimental Psychology, 75*, 540–544.

96. Schnorr, J. A., & Atkinson, R. C. (1969). Repetition versus imagery instructions in the short- and long-term retention of paired associates. *Psychonomic Science, 15*, 183–184.

97. Paivio, A. (1971). *Imagery and Verbal Processes.* New York, NY: Holt.

98. Nairne, J. S., Thompson, S. R. & Pandeirada, N. S. (2007). Adaptive memory: Survival processing enhances retention. *Journal of Experimental Psychology: Learning, Memory, and Cognition, 33*, 263–273; Nairne, J. S., Pandeirada, N. S. & Thompson, S. R. (2008). Adaptive memory: The comparative value of survival processing. *Psychological Science, 19*, 176–180; Nairne, J. S., Pandeirada, N. S. & Thompson, S. R. (2008). Adaptive memory: The comparative value of survival processing. *Psychological Science, 19*, 176–180; Weinstein, Y., Bugg, J. M. & Roediger, H. L. (2008). Can the survival recall advantage be explained by basic memory processes? *Memory & Cognition, 36*, 913–919.

99. Nairne, J. S., Pandeirada, N. S. & Thompson, S. R. (2008). Adaptive memory: The comparative value of survival processing. *Psychological Science, 19*, 176–180.

100. Klein, S. B. (2012). A role for self-referential processing in tasks requiring participants to imagine survival on the savannah. *Journal of Experimental Psychology: Learning, Memory, and Cognition, 38*(5), 1234–1242.

101. Wurm, L. H. (2007). Danger and usefulness: An alternative framework for understanding rapid evaluation effects in perception. *Psychonomic Bulletin & Review, 14*, 1218–1225; Wurm, L. H. & Seaman, S. R. (2008). Semantic effects in naming and perceptual identification but not in delayed naming: Implications for models and tasks. *Journal of Experimental Psychology: Learning, Memory, and Cognition, 34*, 381–398.

102. Tulving, E. & Thompson, D. M. (1973). Encoding specificity and retrieval processes in episodic memory. *Psychological Review, 80*, 352–373; Unsworth, N., Spillers, G. J. & Brewer, G. A. (2012). Dynamics of context-dependent recall: An examination of internal and external context change. *Journal of Memory and Language, 66*, 1–16.

103. Godden, D. B. & Baddeley, A. D. (1975). Context-dependent memory in two natural environments: On land and underwater. *British Journal of Psychology, 66*, 325–331.

104. Godden, D. B., & Baddeley, A. D. (1975). Context-dependent memory in two natural environments: On land and underwater. *British Journal of Psychology, 66*, 325–331.

105. Schab, F. R. (1990). Odors and the remembrance of things past. *Journal of Experimental Psychology: Learning, Memory, and Cognition, 16*, 648–655.

106. Goodwin, D. W., Powell, B., Bremeer, D., Hoine, H. & Stern, J. (1969). Alcohol and recall: State-dependent effects in man. *Science, 163*, 2358–2360.

107. Sachs, J. S. (1967). Recognition memory for syntactic and semantic aspects of connected discourse. *Perception & Psychophysics, 2*, 437–442.

108. Sachs, J. S. (1967). Recognition memory for syntactic and semantic aspects of connected discourse. *Perception & Psychophysics, 2*, 437–442.

109. Masson, M. E. J. (1984). Memory for the surface structure of sentences: Remembering with and without awareness. *Journal of Verbal Learning and Verbal Behavior, 23*, 579–592.

110. Kintsch, W. & Bates, E. (1977). Recognition memory for statements from a classroom lecture. *Journal of Experimental Psychology: Human Learning and Memory, 3*, 150–159.

111. Bates, E., Masling, M. & Kintsch, W. (1978). Recognition memory for aspects of dialogue. *Journal of Experimental Psychology: Human Learning and Memory, 4*, 187–197.

112. Schmidt, S. R. (1994). Effects of humor on sentence memory. *Journal of Experimental Psychology: Learning, Memory, and Cognition, 20*, 953–967.

113. Tillman, B. & Dowling, W. J. (2007). Memory decreases for prose but not for poetry. *Memory & Cognition, 35*, 628–639.

114. Johnson-Laird, P. N. (1983). *Mental models: Towards a Cognitive Science of Language, Inference and Consciousness.* Cambridge, MA: Harvard University Press; Zwaan, R. A. & Radvansky, G. A. (1998). Situation models in language comprehension and memory. *Psychological Bulletin, 123*, 162–185.

115. Horner, A. J., Bisby, J. A., Wang, A., Bogus, K. & Burgess, N. (2016). The role of spatial boundaries in shaping long-term event representations. *Cognition, 154*, 151–164; Pettijohn, K. A., Thompson, A. N., Tamplin, A. K., Krawietz, S. A. & Radvansky, G. A. (2016). Event boundaries and memory improvement. *Cognition, 148*, 136–144.

116. van Dijk, T. A. & Kintsch, W. (1983). *Strategies in Discourse Comprehension.* New York, NY: Academic Press.

117. Kintsch, W., Welsch, D., Schmalhofer, F. & Zimny, S. (1987). Sentence memory: A theoretical analysis. *Journal of Memory and Language, 29*, 133–159.

118. Kintsch, W., Welsch, D., Schmalhofer, F. & Zimny, S. (1987). Sentence memory: A theoretical analysis. *Journal of Memory and Language, 29*, 133–159.

119. Kintsch, W., Welsch, D., Schmalhofer, F., & Zimny, S. (1987). Sentence memory: A theoretical analysis. *Journal of Memory and Language, 29*, 133–159.

120. Kintsch, W., Welsch, D., Schmalhofer, F., & Zimny, S. (1987). Sentence memory: A theoretical analysis. *Journal of Memory and Language, 29*, 133–159.

121. Conway, M. A. & Pleydell-Pearce, C. W. (2000). The construction of autobiographical memories in the self-memory system. *Psychological Review, 107*, 261–288.

122. Anderson, D. J. and Conway, M. A. (1993). Investigating the structure of autobiographical memories. *JoumaJ of Experimental Psychology: Learning. Memory, and Cognition, 19*(5), 1178–1196

123. Anderson, R. J., Dewhurst, S. A. Dean, G. (2017). Direct and generative retrieval of autobiographical memories: The roles of visual imagery and executive processes. *Consciousness and Cognition, 49*, March, 163–171

124. Linton, M. (1975). Memory for real-world events. In D. A. Norman & D. E. Rumelhart (Eds.), *Explorations in Cognition* (pp. 376–404). San Francisco, CA: Freeman; Linton, M. (1978). Real world memory after six years: An in vivo study of very long term memory. In M. M. Gruneberg, P. E. Morris & R. N. Sykes (Eds.), *Practical Aspects of Memory* (pp. 69–76). Orlando, FL: Academic Press.

125. Wagenaar, W. A. (1986). My memory: A study of autobiographical memory over six years. *Cognitive Psychology, 18*, 225–252.

126. Sehulster, J. R. (1989). Content and temporal structure of autobiographical knowledge: Remembering twenty-five seasons at the Metropolitan Opera. *Memory & Cognition, 17*, 590–606.

127. Bahrick, H. P., Hall, L. K. & Berger, S. A. (1996). Accuracy and distortion in memory for high school grades. *Psychological Science, 7*, 265–271.

128. Freud, S. (1899/1938). Childhood and concealing memories. In A. A. Brill (Ed.), *The Basic Writings of Sigmund Freud.* New York, NY: Modern Library.

129. Howe, M. L. & Courage, M. L. (1993). On resolving the enigma of infantile amnesia. *Psychological Bulletin, 113*, 305–326.

130. Magno, E. & Allan, K. (2007). Self-reference during explicit memory retrieval: An event-related potential analysis. *Psychological Science, 18*, 672–677.

131. Nelson, K. (1993). The psychological and social origins of autobiographical memory. *Psychological Science, 4*, 7–14.

132. Nadel, L. & Zola-Morgan, S. (1984). Infantile amnesia: A neurobiological perspective. In M. Moscovitch (Ed.), *Infant Memory.* New York, NY: Plenum Press.

133. Nelson, K. (1993). The psychological and social origins of autobiographical memory. *Psychological Science, 4*, 7–14.

134. Rubin, D. C., Rahhal, T. A. & Poon, L. W. (1998). Things learned in early adulthood are remembered best. *Memory & Cognition, 26*, 3–19.

135. Schrauf, R. W. & Rubin, D. C. (1998). Bilingual autobiographical memory in older adult immigrants: A test of cognitive explanations of the reminiscence bump and the linguistic encoding of memories. *Journal of Memory and Language, 39*, 437–457.

136. Enz, K. F., Pillemer, D. B. & Johnson, K. M. (2016). The relocation bump: Memories of middle adulthood are organized around residential moves. *Journal of Experimental Psychology: General, 145*, 935–940.

137. Berntsen, D. & Rubin, D. C. (2004). Cultural life scripts structure recall from autobiographical memory. *Memory & Cognition, 32*, 427–442; Rubin, D. C. & Berntsen, D. (2003). Life scripts help to maintain autobiographical memories of highly positive, but not highly negative, events. *Memory & Cognition, 31*, 1–14.

138. Ross, M. & Wang, Q. (2010). Why we remember and what we remember: Culture and autobiographical memory. *Perspectives on Psychological Science, 5*, 401–409.

139. Glück, J. & Bluck, S. (2007). Looking back across the life span: A life story account of the reminiscence bump. *Memory & Cognition, 35*, 1928–1939.

140. Collins, K. A., Pillemer, D. B., Ivcevic, Z. & Gooze, R. A. (2007). Cultural scripts guide recall of intensely positive life events. *Memory & Cognition, 35*, 651–659.

141. Copeland, D. E., Radvansky, G. A. & Goodwin, K. A. (2009). A novel study: Forgetting curves and the reminiscence bump. *Memory, 17*, 323–336.

142. Crovitz, H. F. & Shiffman, H. (1974). Frequency of episodic memories as a function of their age. *Bulletin of the Psychonomic Society, 4*, 517–518.

143. Berntsen, D. (2010). The unbidden past: Involuntary autobiographical memories as a basic mode of remembering. *Current Directions in Psychological Science, 19*, 138–142.

144. Berntsen, D. & Hall, N. M. (2004). The episodic nature of involuntary autobiographical memories. *Memory & Cognition, 32*, 789–803.

145. Jones, G. V. & Martin, M. (2006). Primacy of memory linkage in choice among valued objects. *Memory & Cognition, 34*, 1587–1597.

146. Talarico, J. M., LaBar, K. S. & Rubin, D. C. (2004). Emotional intensity predicts autobiographical memory experience. *Memory & Cognition, 32*, 1118–1132.

147. Herz, R. S. & Schooler, J. W. (2002). A naturalistic study of autobiographical memories evoked by olfactory and visual cues: Testing the Proustian hypothesis. *American Journal of Psychology, 115*, 21–32; Willander, J. & Larsson, M. (2006). Smell your way back to childhood. *Psychonomic Bulletin & Review, 13*, 240–244; Willander, J. & Larsson, M. (2007). Olfaction and emotion: The case of autobiographical memory. *Memory & Cognition, 35*, 1659–1663.

148. Loftus, E. F. (1971). Memory for intentions: The effect of presence of a cue and interpolated activity. *Psychonomic Science, 23*, 315–316.

149. Trawley, S. L., Law, A. S. & Logie, R. H. (2011). Event-based prospective remembering in a virtual world. *The Quarterly Journal of Experimental Psychology, 64*, 2181–2193.

150. Scullin, M. K. & McDaniel, M. A. (2010). Remembering to execute a goal: Sleep on it! *Psychological Science, 21*, 1028–1035.

151. Atance, C. M. & O'Neill, D. K. (2001). Episodic future thinking. *Trends in Cognitive Sciences, 5*(12), 533–539; Szpunar, K. K. (2010). Episodic future thought: An emerging concept. *Perspectives on Psychological Science, 5*(2), 142–162; Szpunar, K. K. & Radvansky, G. A. (2016). Cognitive approaches to the study of episodic future thinking. *The Quarterly Journal of Experimental Psychology, 69*(2), 209–216.

152. Klein, S. B., Robertson, T. E. & Delton, A. W. (2010). Facing the future: Memory as an evolved system for planning future acts. *Memory & Cognition, 38*, 13–22.

153. D'Argembeau, A., Renaud, O. & Van der Linden, M. (2011). Frequency, characteristics and functions of future-oriented thoughts in daily life. *Applied Cognitive Psychology, 25*(1), 96–103.

154. Schacter, D. L. & Addis, D. R. (2007). The cognitive neuroscience of constructive memory: Remembering the past and imagining the future. *Philosophical Transactions of the Royal Society of London B: Biological Sciences, 362*(1481), 773–786.

155. Szpunar, K. K. & Schacter, D. L. (2013). Get real: Effects of repeated simulation and emotion on the perceived plausibility of future experiences. *Journal of Experimental Psychology: General, 142*(2), 323–327.

156. Addis, D. R. & Schacter, D. L. (2008). Constructive episodic simulation: Temporal distance and detail of past and future events modulate hippocampal engagement. *Hippocampus, 18*(2), 227–237; Spreng, R. N. & Levine, B. (2006). The temporal distribution of past and future autobiographical events across the lifespan. *Memory & Cognition, 34*(8), 1644–1651.

157. Anderson, R. J., Dewhurst, S. A. & Nash, R. A. (2012). Shared cognitive processes underlying past and future thinking: The impact of imagery and concurrent task demands on event specificity. *Journal of Experimental Psychology: Learning, Memory, and Cognition, 38*(2), 356–265; Berntsen, D. & Bohn, A. (2010). Remembering and forecasting: The relation between autobiographical memory and episodic future thinking. *Memory & Cognition, 38*(3), 265–278; Grysman, A., Prabhakar, J., Anglin, S. M. & Hudson, J. A. (2015). Self-enhancement and the life script in future thinking across the lifespan. *Memory, 23*(5), 774–785; Rasmussen, A. S. & Berntsen, D. (2013). The reality of the past versus the ideality of the future: Emotional valence and functional differences between past and future mental time travel. *Memory & Cognition, 41*(2), 187–200.

158. Collins, A. M. & Quillian, M. R. (1969). Retrieval time from semantic memory. *Journal of Verbal Learning and Verbal Behavior, 8*, 240–247; Collins, A. M. & Quillian, M. R. (1970). Does category size affect categorization time? *Journal of Verbal Learning and Verbal Behavior, 9*, 432–438; Collins, A. M. & Quillian, M. R. (1972). How to make a language user. In E. Tulving & W. Donaldson (Eds.), *Organization of Memory* (pp. 309–351). New York, NY: Academic Press; Collins, A. M. & Loftus, E. F. (1975). A spreading-activation theory of semantic processing. *Psychological Review, 82*, 407–428.

159. Rumelhart, D. E., Lindsay, P. H. & Norman, D. A. (1972). A process model for long-term memory. In E. Tulving & W. Donaldson (Eds.), *Organization of Memory* (pp. 197–246). New York, NY: Academic Press.

160. Rips, L. J., Shoben, E. J. & Smith, E. E. (1973). Semantic distance and the verification of semantic relations. *Journal of Verbal Learning and Verbal Behavior, 12*, 1–20; Smith, E. E., Rips, L. J. & Shoben, E. J. (1974). Semantic memory and psychological semantics. In G. H. Bower (Ed.), *The Psychology of Learning and Motivation* (Vol. 8, pp. 1–45). New York, NY: Academic Press.

161. Smith, E. E., Rips, L. J., & Shoben, E. J. (1974). Semantic memory and psychological semantics. In G. H. Bower (Ed.), *The psychology of learning and motivation* (Vol. 8, pp.1–45). New York, NY: Academic Press.

162. Collins, A. M. & Loftus, E. F. (1975). A spreading-activation theory of semantic processing. *Psychological Review, 82*, 407–428.

163. Martindale, C. (1991). *Cognitive Psychology: A Neural-Network Approach*. Pacific Grove, CA: Brooks/Cole.

164. Martindale, C. (1991). *Cognitive psychology: A neural-network approach*. Pacific Grove, CA: Brooks/Cole.

165. Masson, M. E. J. (1995). A distributed memory model of semantic priming. *Journal of Experimental Psychology: Learning, Memory, and Cognition, 21*, 3–23.

166. McRae, K., de Sa, V. R. & Seidenberg, M. S. (1997). On the nature and scope of featural representations of word meaning. *Journal of Experimental Psychology: General, 126*, 99–130.

167. Seidenberg, M. S. (1993). Connectionist models and cognitive theory. *Psychological Science, 4*, 228–235.

168. McClelland, J. L., Rumelhart, D. E. & Hinton, G. E. (1986). The appeal of parallel distributed processing. In D. E. Rumelhart, J. L. McClelland & PDP Research Group (Eds.), *Parallel Distributed Processing* (Vol. 1, pp. 3–44). Cambridge, MA: MIT Press; Rumelhart, D. E. (1989). The architecture of mind: A connectionist approach. In M. I. Posner (Ed.), *Foundations of Cognitive Science* (pp. 133–159). Cambridge, MA: MIT Press.

169. Warrington, E. K. & Shallice, T. (1984). Category specific semantic impairments. *Brain, 107*, 829–854.

170. Farah, M. J. & McClelland, J. L. (1991). A computational model of semantic memory impairment: Modality specificity and emergent category specificity. *Journal of Experimental Psychology: General, 120*, 339–357.

171. Cree, G. S. & McRae, K. (2003). Analyzing the factors underlying the structure and computation of the meaning of *chipmunk, cherry, chisel, cheese*, and *cello* (and many other such concrete nouns). *Journal of Experimental Psychology: General, 132*, 163–201; Rogers, T. T., Lambon Ralph, M. A., Gerrard, P., Bozeat, S., McClelland, J. L., Hodges, J. R., et al. (2004). Structure and deterioration of semantic memory: A neuropsychological and computational investigation. *Psychological Review, 111*, 205–235.

Chapter 7

1. Schacter, D. L. (1996). *Searching for Memory*. New York, NY: Basic Books.

2. Schacter, D. L. (1999). The seven sins of memory: Insights from psychology and cognitive neuroscience. *American Psychologist, 54*, 182–203.

3. Thorndike, E. L. (1914). *The Psychology of Learning*. New York, NY: Teachers College.

4. McGeoch, J. A. (1932). Forgetting and the law of disuse. *Psychological Review, 39*, 352–370.

5. Schacter, D. L. (1999). The seven sins of memory: Insights from psychology and cognitive neuroscience. *American Psychologist, 54*, 182–203.

6. Jenkins, J. G. & Dallenbach, K. M. (1924). Obliviscence during sleep and waking. *American Journal of Psychology, 35*, 605–612.

7. Drosopoulos, S., Schulze, C., Fischer, S. & Born, J. (2007). Sleep's function in the spontaneous recovery and consolidation of memories. *Journal of Experimental Psychology: General, 136*, 169–183.

8. Jenkins, J. G., & Dallenbach, K. M. (1924). Obliviscence during sleep and waking. *American Journal of Psychology, 35*, 605–612.

9. Wixted, J. T. (2005). A theory about why we forget what we once knew. *Current Directions in Psychological Science, 14*, 6–9.

10. Postman, L. & Underwood, B. J. (1973). Critical issues in interference theory. *Memory & Cognition, 1*, 19–40; Underwood, B. J. (1957). Interference and forgetting. *Psychological Review, 64*, 49–60; Underwood, B. J. & Schulz, R. W. (1960). *Meaningfulness and Verbal Learning*. Philadelphia, PA: Lippincott; Klatzky, R. L. (1980). *Human Memory: Structures and Processes* (2nd ed.). San Francisco, CA: Freeman.

11. Anderson, J. R. (1974). Retrieval of propositional information from long-term memory. *Cognitive Psychology, 6*, 451–474.

12. Anderson, J. R. (1974). Retrieval of propositional information from long-term memory. *Cognitive Psychology, 6*, 451–474.

13. Bunting, M. F., Conway, A. R. A. & Heitz, R. P. (2004). Individual differences in the fan effect and working memory capacity. *Journal of Memory and Language, 51*, 604–622; Radvansky, G. A. & Copeland, D. E. (2006a). Memory retrieval and interference: Working memory issues. *Journal of Memory and Language, 55*, 33–46.

14. Radvansky, G. A. & Zacks, R. T. (1991). Mental models and the fan effect. *Journal of Experimental Psychology: Learning, Memory, and Cognition, 17*, 940–953.

15. Radvansky, G. A., Spieler, D. H. & Zacks, R. T. (1993). Mental model organization. *Journal of Experimental Psychology: Learning, Memory, and Cognition, 19*, 95–114.

16. Radvansky, G. A. & Zacks, R. T. (1991). Mental models and the fan effect. *Journal of Experimental Psychology: Learning, Memory, and Cognition, 17*, 940–953.

17. Radvansky, G. A., & Zacks, R. T. (1991). Mental models and the fan effect. *Journal of Experimental Psychology: Learning, Memory, and Cognition, 17*, 940–953.

18. Radvansky, G. A. & Copeland, D. E. (2006b). Walking through doorways causes forgetting. *Memory & Cognition, 34*, 1150– 1156; Pettijohn, K. A. & Radvansky, G. A. (2016a). Walking through doorways causes forgetting: Environmental effects. *Journal of*

Cognitive Psychology, 28(3), 329–340; Pettijohn, K. A. & Radvansky, G. A. (2016b). Walking through doorways causes forgetting: Event structure or updating disruption? *The Quarterly Journal of Experimental Psychology, 69*(11), 2119–2129; Radvansky, G. A., Krawietz, S. A. & Tamplin, A. K. (2011). Walking through doorways causes forgetting: Further explorations. *The Quarterly Journal of Experimental Psychology, 64*, 1632–1645; Radvansky, G. A., Tamplin, A. K. & Krawietz, S. A. (2010). Walking through doorways causes forgetting: Environmental integration. *Psychonomic Bulletin & Review, 17*, 900–904.

19. Anderson, M. C., Bjork, E. L. & Bjork, R. A. (2000). Retrievalinduced forgetting: Evidence for a recall-specific mechanism. *Psychonomic Bulletin & Review, 7*, 522–530; MacLeod, M. D. & Macrae, C. N. (2001). Gone but not forgotten: The transient nature of retrieval-induced forgetting. *Psychological Science, 12*, 148–152.
20. Anderson, M. C. (2003). Rethinking interference theory: Executive control and mechanisms of forgetting. *Journal of Memory and Language, 49*, 415–445.
21. Storm, B. C. (2011). The benefit of forgetting in thinking and remembering. *Current Directions in Psychological Science, 20*, 291–295.
22. Bäuml, K-H. T. & Samenieh, A. (2010). The two faces of memory retrieval. *Psychological Science, 21*, 793–795.
23. Bjork, E. L. & Bjork, R. A. (2003). Intentional forgetting can increase, not decrease, residual influences of to-be-forgotten information. *Journal of Experimental Psychology: Learning, Memory, and Cognition, 29*, 524–531.
24. Tulving, E., & Pearlstone, Z. (1966). Availability versus accessibility of information in memory for words. *Journal of Verbal Learning and Verbal Behavior, 5*, 381–391.
25. Bransford, J. D. & Stein, B. S. (1984). *The Ideal Problem Solver.* New York, NY: Freeman.
26. Brown, R. & McNeill, D. (1966). The "tip-of-the-tongue" phenomenon. *Journal of Verbal Learning and Verbal Behavior, 5*, 325–337; Burke, D. M., MacKay, D. G., Worthley, J. S. & Wade, E. (1991). On the tip of the tongue: What causes word finding failures in young and older adults? *Journal of Memory and Language, 30*, 542–579; Jones, G. V. (1989). Back to Woodworth: Role of interlopers in the tip-of-the-tongue phenomenon. *Memory & Cognition, 17*, 69–76; Koriat, A., Levy-Sadot, R., Edry, E. & de Marcus, S. (2003). What do we know about what we cannot remember? Accessing the semantic attributes of words that cannot be recalled. *Journal of Experimental Psychology: Learning, Memory, and Cognition, 29*, 1095–1105; Meyer, A. S. & Bock, K. (1992). The tip-of-the-tongue phenomenon: Blocking or partial activation? *Memory & Cognition, 20*, 715–726.
27. Tulving, E. & Pearlstone, Z. (1966). Availability versus accessibility of information in memory for words. *Journal of Verbal Learning and Verbal Behavior, 5*, 381–391.
28. Thomson, D. M. & Tulving, E. (1970). Associative encoding and retrieval: Weak and strong cues. *Journal of Experimental Psychology, 86*, 255–262.
29. Marian, V. & Neisser, U. (2000). Language-dependent recall of autobiographical memories. *Journal of Experimental Psychology: General, 129*, 361–368.
30. Schrauf, R. W. & Rubin, D. C. (2000). Internal languages of retrieval: The bilingual encoding of memories for the personal past. *Memory & Cognition, 28*, 616–623.
31. Noice, H. & Noice, T. (1999). Long-term retention of theatrical roles. *Memory, 7*, 357–382.
32. Slamecka, N. J. (1968). An examination of trace storage in free recall. *Journal of Experimental Psychology, 4*, 504–513.
33. Aslan, A., Bäuml, K-H. & Grundgeiger, T. (2007). The role of inhibitory processes in part-list cuing. *Journal of Experimental Psychology: Learning, Memory, and Cognition, 33*, 335–341.
34. Schacter, D. L. (1996). *Searching for Memory.* New York, NY: Basic Books.
35. Roediger, H. L. III & McDermott, K. B. (1995). Creating false memories: Remembering words not presented in lists. *Journal of Experimental Psychology: Learning, Memory, and Cognition, 21*, 803–814.
36. Deese, J. (1959). On the prediction of occurrence of particular verbal intrusions in immediate recall. *Journal of Experimental Psychology, 58*, 17–22.
37. Roediger, H. L. III & McDermott, K. B. (2000). Tricks of memory. *Current Directions in Psychological Science, 9*, 123–127; Gallo, D. A. (2010). False memories and fantastic beliefs: 15 years of the DRM illusion. *Memory & Cognition, 38*, 833–848.
38. Roediger, H. L. III, & McDermott, K. B. (1995). Creating false memories: Remembering words not presented in lists. *Journal of Experimental Psychology: Learning, Memory, and Cognition, 21*, 803–814.
39. Meade, M. L., Watson, J. M., Balota, D. A. & Roediger, H. L. (2007). The roles of spreading activation and retrieval mode in producing false recognition in the DRM paradigm. *Journal of Memory and Language, 56*, 305–320.
40. Roediger, H. L. III & McDermott, K. B. (1995). Creating false memories: Remembering words not presented in lists. *Journal of Experimental Psychology: Learning, Memory, and Cognition, 21*, 803–814.
41. Bransford, J. D. & Franks, J. J. (1971). The abstraction of linguistic ideas. *Cognitive Psychology, 2*, 331–350; Bransford, J. D. & Franks, J. J. (1972). The abstraction of linguistic ideas: A review. *Cognition: International Journal of Cognitive Psychology, 1*(2), 211–249.

42. Bransford, J. D. & Franks, J. J. (1971). The abstraction of linguistic ideas. *Cognitive Psychology, 2*, 331–350.

43. Bransford, J. D., & Franks, J. J. (1971). The abstraction of linguistic ideas. *Cognitive Psychology, 2*, 331–350.

44. Bransford, J. D. & Franks, J. J. (1971). The abstraction of linguistic ideas. *Cognitive Psychology, 2*, 331–350; Bransford, J. D. & Franks, J. J. (1972). The abstraction of linguistic ideas: A review. *Cognition: International Journal of Cognitive Psychology, 1*(2), 211–249.

45. Loftus, E. F. (2003). Make-believe memories. *American Psychologist, 58*, 867–873; Loftus, E. F. (2004). Memories of things unseen. *Current Directions in Psychological Science, 13*, 145–147.

46. Zaragoza, M. S., McCloskey, M. & Jamis, M. (1987). Misleading postevent information and recall of the original event: Further evidence against the memory impairment hypothesis. *Journal of Experimental Psychology: Learning, Memory, and Cognition, 13*, 36–44.

47. R. F. (1989). Influences of misleading postevent information: Misinformation interference and acceptance. *Journal of Experimental Psychology: General, 118*, 72–85.

48. Loftus, E. F., Donders, K., Hoffman, H. G. & Schooler, J. W. (1989). Creating new memories that are quickly accessed and confidently held. *Memory & Cognition, 17*, 607–616.

49. Lewandowsky, S., Stritzka, W. G. K., Oberauer, K. & Morales, M. (2005). Memory for fact, fiction, and misinformation. *Psychological Science, 16*, 190–195.

50. Fazio, L. K., Barber, S. J., Rajaram, S., Ornstein, P. A. & Marsh, E. J. (2013). Creating illusions of knowledge: Learning errors that contradict prior knowledge. *Journal of Experimental Psychology: General, 142*(1), 1–5.

51. Loftus, E. F. & Palmer, J. C. (1974). Reconstruction of automobile destruction: An example of the interaction between language and memory. *Journal of Verbal Learning and Verbal Behavior, 13*, 585–589.

52. Loftus, E. F., & Palmer, J. C. (1974). Reconstruction of automobile destruction: An example of the interaction between language and memory. *Journal of Verbal Learning and Verbal Behavior, 13*, 585–589.

53. Ayers, M. S. & Reder, L. M. (1998). A theoretical review of the misinformation effect: Predictions from an activation-based memory model. *Psychonomic Bulletin & Review, 5*, 1–21; Loftus, E. F. (1991). Made in memory: Distortions in recollection after misleading information. In G. H. Bower (Ed.), *The Psychology of Learning and Motivation* (Vol. 27, pp. 187–215). New York, NY: Academic Press; Loftus, E. F. & Hoffman, H. G. (1989). Misinformation and memory: The creation of new memories. *Journal of Experimental Psychology: General, 118*, 100–104; Roediger, H. L. III. (1996). Memory illusions. *Journal of Memory and Language, 35*, 76–100.

54. Zaragoza, M. S. & Lane, S. M. (1994). Source misattributions and the suggestibility of eyewitness memory. *Journal of Experimental Psychology: Learning, Memory, and Cognition, 20*, 934–945.

55. Lindsay, D. S., Allen, B. P., Chan, J. C. K. & Dahl, L. C. (2004). Eyewitness suggestibility and source similarity: Intrusions of details from one event into memory reports of another event. *Journal of Memory and Language, 50*, 96–111.

56. Jacoby, L. L., Woloshyn, V. & Kelley, C. (1989). Becoming famous without being recognized: Unconscious influences of memory produced by dividing attention. *Journal of Experimental Psychology: General, 118*, 115–125.

57. Kelley, C. M. & Jacoby, L. L. (1996). Memory attributions: Remembering, knowing, and feeling of knowing. In L. M. Reder (Ed.), *Implicit Memory and Metacognition* (pp. 287–308). Hillsdale, NJ: Erlbaum; Busey, T. A., Tunnicliff, J., Loftus, G. R. & Loftus, E. F. (2000). Accounts of the confidence–accuracy relation in recognition memory. *Psychonomic Bulletin & Review, 7*, 26–48.

58. Loftus, E. F. (1991). Made in memory: Distortions in recollection after misleading information. In G. H. Bower (Ed.), *The Psychology of Learning and Motivation* (Vol. 27, pp. 187–215). New York, NY: Academic Press

59. Payne, D. G., Toglia, M. P. & Anastasi, J. S. (1994). Recognition performance level and the magnitude of the misinformation effect in eyewitness memory. *Psychonomic Bulletin & Review, 1*, 376–382.

60. Sara, S. J. (2000). Retrieval and reconsolidation: Toward a neurobiology of remembering. *Learning and Memory, 7*, 73–84.

61. Chan, J. C. K., Thomas, A. K. & Bulevich, J. B. (2009). Recalling a witnessed event increases eyewitness suggestibility. *Psychological Science, 20*, 66–73.

62. Hyman, I. E., Jr., Husband, T. H. & Billings, F. J. (1995). False memories of childhood experiences. *Applied Cognitive Psychology, 9*, 181–197; Loftus, E. F. & Hoffman, H. G. (1989). Misinformation and memory: The creation of new memories. *Journal of Experimental Psychology: General, 118*, 100–104.

63. Wade, K. A., Garry, M., Don Read, J. & Lindsay, D. S. (2002). A picture is worth a thousand lies: Using false photographs to create false childhood memories. *Psychonomic Bulletin & Review, 9*(3), 597–603.

64. Lindsay, D. S., Hagen, L., Read, J. D.,Wade, K. A., & Garry, M. (2004). True photographs and false memories. *Psychological Science, 15*, 149–154.

65. Chrobak, Q. M., & Zaragoza, M. S. (2008). Inventing stories: Forcing witnesses to fabricate entire fictitious events leads to freely reported false memories. *Psychonomic Bulletin & Review, 15*, 1190–1195.

66. Sun, X., Punjabi, P. V., Greenberg, L. T., & Seamon, J. G. (2009). Does feigning amnesia impair subsequent recall? *Memory & Cognition, 37*, 81–89.

67. Wells, G. L., Olson, E. A. & Charman, S. D. (2002). The confidence of eyewitnesses in their identifications from lineups. *Current Directions in Psychological Science, 11*, 151–154.

68. Roediger, H. L. III & McDermott, K. B. (1995). Creating false memories: Remembering words not presented in lists. *Journal of Experimental Psychology: Learning, Memory, and Cognition, 21*, 803–814.

69. Kelley, C. M., & Lindsay, D. S. (1993). Remembering mistaken for knowing: Ease of retrieval as a basis for confidence in answers to general knowledge questions. *Journal of Memory and Language, 32*, 1–24.

70. Bernstein, D. M. & Loftus, E. F. (2009). How to tell if a particular memory is true or false. *Perspectives on Psychological Science, 4*, 370–374.

71. Mitchell, K. J. & Zaragoza, M. S. (2001). Contextual overlap and eyewitness suggestibility. *Memory & Cognition, 29*, 616–626.

72. Frost, P. (2000). The quality of false memory over time: Is memory for misinformation "remembered" or "known"? *Psychonomic Bulletin & Review, 7*, 531–536.

73. Sederberg, P. B., Schulze-Bonhage, A., Madsen, J. R., Bromfield, E. B., Litt, B., Brandt, A., et al. (2007). Gamma oscillations distinguish true from false memories. *Psychological Science, 18*, 927–932.

74. Roediger, H. L. III & McDermott, K. B. (2000). Tricks of memory. *Current Directions in Psychological Science, 9*, 123–127.

75. Anderson, J. R. & Schooler, L. J. (1991). Reflections of the environment in memory. *Psychological Science, 2*, 396–408.

76. Conway, M. A., Cohen, G. & Stanhope, N. (1991). On the very long-term retention of knowledge acquired through formal education: Twelve years of cognitive psychology. *Journal of Experimental Psychology: General, 120*, 395–409.

77. Brown, A. S. (1998). Transient global amnesia. *Psychonomic Bulletin & Review, 5*, 401–427.

78. Brown, A. S. (2002). Consolidation theory and retrograde amnesia in humans. *Psychonomic Bulletin & Review, 9*, 403–425; Wixted, J. T. (2004). On common ground: Jost's (1897) law of forgetting and Ribot's (1881) law of retrograde amnesia. *Psychological Review, 111*, 864–879.

79. Ribot, T. (1882). Diseases of Memory: *An Essay in the Positive Psychology.* New York, NY: Appleton.

80. Tulving, E. (1989). Remembering and knowing the past. *American Scientist, 77*, 361–367.

81. Squire, L. R. (1987). *Memory and brain.* New York, NY: Oxford University Press.

82. Nyberg, L., McIntosh, A. R. & Tulving, E. (1998). Functional brain imaging of episodic and semantic memory with positron emission tomography. *Journal of Molecular Medicine, 76*, 48–53.

83. Habib, R., Nyberg, L. & Tulving, E. (2003). Hemispheric asymmetries of memory: The HERA model revisited. *Trends in Cognitive Sciences, 7*, 241–245; Nyberg, L., Cabeza, R. & Tulving, E. (1996). PET studies of encoding and retrieval: The HERA model. *Psychonomic Bulletin & Review, 3*, 135–148.

84. Buckner, R. L. (1996). Beyond HERA: Contributions of specific prefrontal brain areas to long-term memory retrieval. *Psychonomic Bulletin & Review, 3*, 149–158.

85. Shallice, T., Fletcher, P., & Dolan, R. (1998). The functional imaging of recall. In M. A. Conway, S. E. Gathercole, & C. Cornoldi (Eds.), *Theories of memory* (Vol. II, pp. 247–258). Hove, England: Psychology Press.

86. Ranganath, C. & Pallar, K. A. (1999). Frontal brain activity during episodic and semantic retrieval: Insights from eventrelated potentials. *Journal of Cognitive Neuroscience, 11*, 598–609; Wiggs, C. L., Weisberg, J. & Martin, A. (1999). Neural correlates of semantic and episodic memory retrieval. *Neuropsychologia, 37*, 103–118.

87. Milner, B., Corkin, S. & Teuber, H. L. (1968). Further analysis of the hippocampal amnesic syndrome: 14–year follow up study of H. M. *Neuropsychologia, 6*, 215–234.

88. Banich, M. T. (1997). *Neuropsychology: The Neural Bases of Mental Function.* Boston, MA: Houghton Mifflin.

89. Schacter, D. L. (1996). *Searching for memory.* New York, NY: Basic Books.

90. Squire, L. R. (1993). The organization of declarative and nondeclarative memory. In T. Ono, L. R. Squire, M. E. Raichle; D. I. Perrett & M. Fukuda (Eds.), *Brain Mechanisms of Perception and Memory: From Neuron to Behavior* (pp. 219–227). New York, NY: Oxford University Press; Gupta, P. & Cohen, N. J. (2002). Theoretical and computational analysis of skill learning, repetition priming, and procedural memory. *Psychological Review, 109*, 401–448; Roediger, H. L. III, Marsh, E. J. & Lee, S. C. (2002). Kinds of memory. In D. Medin (Ed.), *Stevens' Handbook of Experimental Psychology*, (3rd ed., Vol. 2, pp. 1–42). New York, NY: Wiley.

91. Blakemore, C. (1977). *Mechanics of the mind.* Cambridge, England: Cambridge University Press.

92. Milner, B., Corkin, S. & Teuber, H. L. (1968). Further analysis of the hippocampal amnesic syndrome: 14–year follow up study of H. M. *Neuropsychologia, 6*, 215–234.

93. Penfield, W. & Milner, B. (1958). Memory deficit produced by bilateral lesions in the hippocampal zone. *Archives of Neurology and Psychiatry, 79*, 475–497.

94. Zola-Morgan, S., Squire, L., & Amalral, D. G. (1986). Human amnesia and the medial temporal region: Enduring memory impairment following a

bilateral lesion limited to field CA1 of the hippocampus. *Journal of Neuroscience, 6*, 2950–2967.

95. Eichenbaum, H. & Fortin, N. (2003). Episodic memory and the hippocampus: It's about time. *Current Directions in Psychological Science, 12*, 53–57.

96. Barnier, A. J. (2002). Posthypnotic amnesia for autobiographical episodes: A laboratory model for functional amnesia? *Psychological Science, 13*, 232–237.

97. Graf, P. & Schacter, D. L. (1987). Selective effects of interference on implicit and explicit memory for new associations. *Journal of Experimental Psychology: Learning, Memory, and Cognition, 13*, 45–53; Kolers, P. A. & Roediger, H. L. III (1984). Procedures of mind. *Journal of Verbal Learning and Verbal Behavior, 23*, 425–449.

98. Morton, J. (1979). Facilitation in word recognition: Experiments causing change in the logogen models. In P. A. Kolers, M. E. Wrolstad & H. Bouma (Eds.), *Processing of Visible Language* (Vol. 1, pp. 259–268). New York, NY: Plenum.

99. Brown, A. S., Neblett, D. R., Jones, T. C. & Mitchell, D. B. (1991). Transfer of processing in repetition priming: Some inappropriate findings. *Journal of Experimental Psychology: Learning, Memory, and Cognition, 17*, 514–525.

100. Masson, M. E. J. (1984). Memory for the surface structure of sentences: Remembering with and without awareness. *Journal of Verbal Learning and Verbal Behavior, 23*, 579–592.

101. Logan, G. D. (1990). Repetition priming and automaticity: Common underlying mechanisms? *Cognitive Psychology, 22*, 1–35.

102. Jacoby, L. L. & Dallas, M. (1981). On the relationship between autobiographical memory and perceptual learning. *Journal of Experimental Psychology: General, 110*, 306–340.

103. Roediger, H. L. III, Stadler, M. L., Weldon, M. S. & Riegler, G. L. (1992). Direct comparison of two implicit memory tests: Word fragment and word stem completion. *Journal of Experimental Psychology: Learning, Memory, and Cognition, 18*, 1251–1269; Thapar, A. & Greene, R. L. (1994). Effects of level of processing on implicit and explicit tasks. *Journal of Experimental Psychology: Learning, Memory, and Cognition, 20*, 671–679.

104. Goshen-Gottstein, Y. & Kempinsky, H. (2001). Probing memory with conceptual cues at multiple retention intervals: A comparison of forgetting rates on implicit and explicit tests. *Psychonomic Bulletin & Review, 8*, 139–146; Lustig, C. & Hasher, L. (2001). Implicit memory is vulnerable to proactive interference. *Psychological Science, 12*, 408–412.

105. Brown, A. S. (2004). The déja vu illusion. *Current Directions in Psychological Science, 13*, 256–259; Cleary, A. M. (2008). Recognition memory, familiarity, and déja vu experiences. *Current Directions in Psychological Science, 17*, 353–357.

106. Baddeley, A. D., Atkinson, A. L., Hitch, G. J. & Allen, R. J. (2021). Detecting accelerated long-term forgetting: A problem and some solutions. *Cortex, 142*, 237–251.

107. Elliott, G., Isaac, C. L. & Muhlert, N. (2014). Measuring forgetting: A critical review of accelerated long-term forgetting studies. *Cortex, 54*, 16–32.

108. Cassel, A. & Kopelman, M. D. (2019). Have we forgotten about forgetting? A critical review of 'accelerated long-term forgetting' in temporal lobe epilepsy. *Cortex, 110*, 141–149.

109. Tort-Merino, A., Laine, M., Valech, N., et al. (2021). Accelerated long-term forgetting over three months in asymptomatic APOE ε-4 carriers. *Annals of Clinical and Translational Neurology, 8*(2), 477–484; Rodini, M., De Simone, M. S., Caltagirone, C. & Carlesimo, G. A. (2022). Accelerated long-term forgetting in neurodegenerative disorders: A systematic review of the literature. *Neuroscience & Biobehavioral Reviews*, 104815.

110. Roediger, H. L. III, & McDermott, K. B. (1995). Creating false memories: Remembering words not presented in lists. *Journal of Experimental Psychology: Learning, Memory, and Cognition, 21*, 803–814.

Chapter 8

1. Corballis, M. C. (2004). The origins of modernity: Was autonomous speech a critical factor? *Psychological Review, 111*, 543–552.

2. Hockett, C. F. (1960a). Logical considerations in the study of animal communication. In W. E. Lanyon & W. N. Tavolga (Eds.), *Animal Sounds and Communication* (pp. 392–430). Washington, DC: American Institute of Biological Sciences; Hockett, C. F. (1960b). The origin of speech. *Scientific American*, 203, 89–96; Hockett, C. F. (1966). The problem of universals in language. In J. H. Greenberg (Ed.), *Universals of Language* (2nd ed., pp. 1–29). Cambridge, MA: MIT Press.

3. Hockett, C. F. (1960a). Logical considerations in the study of animal communication. In W. E. Lanyon & W. N. Tavolga (Eds.), *Animal sounds and communication* (pp. 392–430). Washington, DC: American Institute of Biological Sciences.

4. Pinker, S. (1994). *The Language Instinct: How the Mind Creates Language*. New York, NY: Morrow.

5. Glass, A. L. & Holyoak, K. J. (1986). *Cognition* (2nd ed.). New York, NY: Random House.

6. Kaschak, M. P. & Glenberg, A. M. (2000). Constructing meaning: The role of affordances and grammatical constructions in sentence comprehension. *Journal of Memory and Language, 43*, 508–529.

7. Chomsky, N. (1959). A review of Skinner's *Verbal Behavior. Language, 35*, 26–58.

8. Chomsky, N. (1957). *Syntactic Structures*. The Hague, The Netherlands: Mouton; Chomsky, N. (1965). *Aspects of a Theory of Syntax*. Cambridge MA: Harvard University Press.

9. Dyer, F. C. (2002). When it pays to waggle. *Nature*, 419, 885–886. Sherman, G. & Visscher, P. K. (2002). Honeybee colonies achieve fitness through dancing. *Nature*, 419, 920–922; von Frisch, K. (1967). The Dance *Language and Orientation of Honeybees*. Cambridge, MA: Harvard University Press.

10. Risueno-Segovia, C., Dohmen, D., Gultekin, Y. B., Pomberger, T. & Hage, S. R. (2023). Linguistic law-like compression strategies emerge to maximise coding efficiency in marmoset vocal communication. *Proceedings of the Royal Society B, 290*(2007), 20231503.

11. Marler, P. (1967). Animal communication signals. *Science, 35*, 63–78.

12. Glass, A. L. & Holyoak, K. J. (1986). *Cognition* (2nd ed.). New York, NY: Random House.

13. Glass, A. L. & Holyoak, K. J. (1986). *Cognition* (2nd ed.). New York, NY: Random House.

14. Liszkowski, U., Schäfer, M., Carpenter, M. & Tomasello, M. (2009). Prelinguistic infants, but not chimpanzees, communicate about absent entities. *Psychological Science*, 20, 654–660.

15. Hopkins, W. D., Russell, J. L. & Cantalupo, C. (2007). Neuroanatomical correlates of handedness for tool use in chimpanzees (*Pan troglodytes*): Implication for theories on the evolution of language. *Psychological Science*, 18, 971–977.

16. Bohn, M., Call, J. & Tomasello, M. (2015). Communication about absent entities in great apes and human infants. *Cognition*, 145, 63–72.

17. Miller, G. A. (1973). Psychology and communication. In G. A. Miller (Ed.), *Communication, Language, and Meaning: Psychological Perspectives* (pp. 3–12). New York, NY: Basic Books.

18. Bekoff, M., Allen, C. & Burghardt, G. (Eds.). (2002). *The Cognitive Animal: Empirical and Theoretical Perspectives on Animal Cognition*. Cambridge, MA: MIT Press.

19. D'Anastasio, R., Wroe, S., Tuniz, C., Mancini, L., Cesana, D. T., Dreossi, D., . . . Capasso, L. (2013). Micro-biomechanics of the Kebara 2 hyoid and its implications for speech in Neanderthals. *PloS one, 8(12)*, e82261; Dediu, D. & Levinson, S. C. (2013). On the antiquity of language: The reinterpretation of Neandertal linguistic capacities and its consequences. *Frontiers in Psychology*, 4, 397.

20. Berwick, R. C., Hauser, M. & Tattersall, I. (2013). Neanderthal language? Just-so stories take center stage. *Frontiers in Psychology, 4*, 671.

21. Chomsky, N. (1957). *Syntactic Structures*. The Hague, The Netherlands: Mouton; Chomsky, N. (1965). *Aspects of a Theory of Syntax*. Cambridge, MA: Harvard University Press.

22. Rosch-Heider, E. (1972). Universals in color naming and memory. *Journal of Experimental Psychology, 93*, 10–21.

23. Wright, O., Davies, I. R. & Franklin, A. (2015). Whorfian effects on colour memory are not reliable. *The Quarterly Journal of Experimental Psychology, 68*(4), 745–758.

24. Regier, T., Kay, P. & Khetarpal, N. (2009). Color naming and the shape of color space. *Language, 85(4)*, 884–892.

25. Barner, D., Li, P. & Snedeker, J. (2010). Words as windows to thought: The case of object representation. *Current Directions in Psychological Science, 19*, 195–200.

26. Boroditsky, L. (2001). Does language shape thought? Mandarin and English speakers' conceptions of time. *Cognitive Psychology, 43*, 1–22; Chen, Y. (2007). Chinese and English speakers think about time differently? Failure of replicating Boroditsky (2001). *Cognition*, 104, 427–436; January, D. & Kako, E. (2007). Re-evaluating evidence for linguistic relativity: Reply to Boroditsky (2001). *Cognition*, 104, 417–426.

27. Greenberg, J. H. (1978). Generalizations about numeral systems. In J. H. Greenberg (Ed.), *Universals of Human Language: Vol. 3*. Word structure (pp. 249–295). Stanford, CA: Stanford University Press; Hunt, E. & Agnoli, F. (1991). The Whorfian hypothesis: A cognitive psychology perspective. *Psychological Review, 98*, 377–389; Roberson, D., Davies, I. & Davidoff, J. (2000). Color categories are not universal: Replications and new evidence from a Stone-Age culture. *Journal of Experimental Psychology: General, 129*, 369–398; Malt, B. C., Sloman, S. A. & Gennari, S. P. (2003). Universality and language specificity in object naming. *Journal of Memory and Language, 49*, 20–42.

28. Fausey, C. M. & Boroditsky, L. (2011). Who dunnit? Cross-linguistic differences in eye-witness memory. *Psychonomic Bulletin & Review, 18*, 150–157.

29. Dolscheid, S., Shayan, S., Majid, A. & Casasanto, D. (2013). The thickness of musical pitch: Psychophysical evidence for linguistic relativity. *Psychological Science, 24(5)*, 613–621.

30. Boroditsky, L. (2011). How language shapes thought. *Scientific American, 304*, 63–65.

31. Whorf, B. L. (1956). Science and linguistics. In J. B. Carroll (Ed.), *Language, Thought, and Reality: Selected Writings of Benjamin Lee Whorf* (pp. 207–219). Cambridge, MA: MIT Press.

32. Palermo, D. S. (1978). *Psychology of Language*. Glenview, IL: Scott, Foresman.

33. Glucksberg, S., & Danks, J. H. (1975). *Experimental psycholinguistics: An introduction.* Hillsdale, NJ: Erlbaum.
34. Fromkin, V. A., & Rodman, R. (1974). *An introduction to language.* New York, NY: Holt, Rinehart & Winston.
35. Liberman, A. M., Harris, K. S., Hoffman, H. S. & Griffith, B. C. (1957). The discrimination of speech sounds within and across phoneme boundaries. *Journal of Experimental Psychology, 54,* 358–368.
36. Myers, E. B., Blumstein, S. E., Walsh, E. & Eliassen, J. (2009). Inferior frontal regions underlie the perception of phonetic category invariance. *Psychological Science, 20,* 895–903.
37. Pollack, I. & Pickett, J. M. (1964). Intelligibility of excerpts from fluent speech: Auditory vs. structural context. *Journal of Verbal Learning and Verbal Behavior*, 3, 79–84.
38. Miller, G. A. & Isard, S. (1963). Some perceptual consequences of linguistic rules. *Journal of Verbal Learning and Verbal Behavior,* 2, 217–228.
39. Miller, G. A., & Isard, S. (1963). Some perceptual consequences of linguistic rules. *Journal of Verbal Learning and Verbal Behavior,* 2, 217–228.
40. Warren, R. M. (1970). Perceptual restoration of missing speech sounds. *Science,* 167, 392–393.
41. Dewitt, L. A. & Samuel, A. G. (1990). The role of knowledgebased expectations in music perception: Evidence from musical restoration. *Journal of Experimental Psychology: General, 119,* 123–144.
42. Rapp, B. & Goldrick, M. (2000). Discreteness and interactivity in spoken word production. *Psychological Review, 107,* 460–499.
43. Dahan, D. (2010). The time course of interpretation in speech comprehension. *Current Directions in Psychological Science, 19,* 121–126; Dell, G. S. & Newman, J. E. (1980). Detecting phonemes in fluent speech. *Journal of Verbal Learning and Verbal Behavior, 20,* 611–629; Pitt, M. A. & Samuel, A. G. (1995). Lexical and sublexical feedback in auditory word recognition. *Cognitive Psychology, 29,* 149–188; Samuel, A. G. (2001). Knowing a word affects the fundamental perception of the sounds within it. *Psychological Science, 12,* 348–351.
44. Kraljik, T., Samuel, A. G. & Brennan, S. E. (2008). First impressions and last resorts: How listeners adjust to speaker variability. *Psychological Science*, 19, 332–338.
45. Marslen-Wilson, W. D. & Welsh, A. (1978). Processing interactions and lexical access during word recognition in continuous speech. *Cognitive Psychology, 30,* 509–517.
46. Dell, G. S. & Newman, J. E. (1980). Detecting phonemes in fluent speech. *Journal of Verbal Learning and Verbal Behavior, 20,* 611–629.
47. Treisman, A. M. (1960). Contextual cues in selective listening. *The Quarterly Journal of Experimental Psychology, 12,* 242–248; Treisman, A. M. (1964). Monitoring and storage of irrelevant messages in selective attention. *Journal of Verbal Learning and Verbal Behavior, 3,* 449–459.
48. McClelland, J. L. & Elman, J. L. (1986). The TRACE model of speech perception. *Cognitive Psychology, 18,* 1–86.
49. Dell, G. S. (1986). A spreading-activation theory of retrieval in sentence production. *Psychological Review, 93,* 283–321.
50. Tyler, L. K., Voice, J. K., & Moss, H. E. (2000). The interaction of meaning and sound in spoken word recognition. *Psychonomic Bulletin & Review, 7,* 320–326.
51. Liberman, A. M., Cooper, F. S., Shankweiler, D. P. & Studdert-Kennedy, M. (1967). Perception of the speech code. *Psychological Review, 74,* 431–461; Liberman, A. M. & Mattingly, I. G. (1985). The motor theory of speech perception revised. *Cognition, 21,* 1–36.
52. Galantucci, B., Fowler, C. A. & Turvey, M. T. (2006). The motor theory of speech perception reviewed. *Psychonomic Bulletin & Review, 13,* 361–377.
53. Jesse, A. & Massaro, D. W. (2010). Seeing a singer helps comprehension of the song's lyrics. *Psychonomic Bulletin & Review, 17,* 323–328.
54. Corley, M., Brocklehurst, P. H. & Moat, H. S. (2011). Error biases in inner and overt speech: Evidence from tongue twisters. *Journal of Experimental Psychology: Learning, Memory, and Cognition, 37,* 162–175; Oppenheim, G. M. & Dell, G. S. (2010). Motor movement matters: The flexible abstractness of inner speech. *Memory & Cognition, 38,* 1147–1160.
55. Norris, D., McQueen, J. M., Cutler, A. & Butterfield, S. (1997). The possible-word constraint in the segmentation of continuous speech. *Cognitive Psychology, 34,* 191–243.
56. Gershkoff-Stowe, L. & Goldin-Medow, S. (2002). Is there a natural order for expressing semantic relations? *Cognitive Psychology, 45,* 375–412.
57. Segel, E. & Boroditsky, L. (2011). Grammar in art. *Frontiers in Psychology, 1,* 244.
58. Bock, K. (1995). Producing agreement. *Current Directions in Psychological Science, 4,* 56–61.
59. Bock, K. & Miller, C. A. (1991). Broken agreement. *Cognitive Psychology, 23,* 45–93; Hartsuiker, R. J., Anton-Mendez, I. & van Zee, M. (2001). Object attraction in subject-verb agreement construction. *Journal of Memory and Language, 45,* 546–572.
60. Whitney, P. (1998). *The psychology of language.* Boston, MA: Houghton Mifflin.
61. Lachman, R., Lachman, J. L., & Butterfield, E. C. (1979). *Cognitive psychology and information processing: An introduction.* Hillsdale, NJ: Erlbaum.

62. Palermo, D. S. (1978). *Psychology of Language*. Glenview, IL:Scott, Foresman.

63. Fodor, J. A. & Garrett, M. (1966). Some reflections on competence and performance. In J. Lyons & R. J. Wales (Eds.), *Psycholinguistic Papers* (pp. 135–154). Edinburgh, Scotland: Edinburgh University Press.

64. Bock, J. K. (1982). Toward a cognitive psychology of syntax: Information processing contributions to sentence formulation. *Psychological Review, 89*, 1–47.

65. Bock, J. K. (1986). Meaning, sound, and syntax: Lexical priming in sentence production. *Journal of Experimental Psychology: Learning, Memory, and Cognition, 12*, 575–586; West, R. F. & Stanovich, K. E. (1986). Robust effects of syntactic structure on visual word processing. *Memory & Cognition, 14*, 104–112.

66. Bock, K., & Griffin, Z. M. (2000). The persistence of structural priming: Transient activation or implicit learning? *Journal of Experimental Psychology: General, 129*, 177–192.

67. Kaschak, M. P., Kutta, T. J. & Schatschneider, C. (2011). Longterm cumulative structural priming persists for (at least) one week. *Memory & Cognition, 39*(3), 381–388.

68. Branigan, H. P., Pickering, M. J. & Cleland, A. A. (1999). Syntactic priming in written production: Evidence for rapid decay. *Psychonomic Bulletin & Review, 6*, 635–640; Branigan, H. P., Pickering, M. J., Stewart, A. J. & McLean, J. F. (2000). Syntactic priming in spoken production: Linguistic and temporal interference. *Memory & Cognition, 28*, 1297–1302; Hall, M. L., Ferreira, V. S. & Mayberry, R. I. (2015). Syntactic priming in American Sign Language. *PloS one, 10*(3), e0119611.

69. Ferreira, V. S., Bock, K., Wilson, M. P. & Cohen, N. J. (2008). Memory for syntax despite amnesia. *Psychological Science, 19*(9), 940–946.

70. Ferreira, F. & Swets, B. (2002). How incremental is language production? Evidence from the production of utterances requiring the computation of arithmetic sums. *Journal of Memory and Language, 46*, 57–84.

71. Stallings, L. M., MacDonald, M. C. & O'Seaghdha, P. G. (1998). Phrasal ordering constraints in sentence production: Phrase length and verb disposition in heavy-NP shift. *Journal of Memory and Language, 39*, 392–417.

72. Fedorenko, E., Gibson, E. & Rohde, D. (2006). The nature of working memory capacity in sentence comprehension: Evidence against domain-specific working memory resources. *Journal of Memory and Language, 54*, 541–553.

73. Fromkin, V. A. (1971). The non-anomalous nature of anomalous utterances. *Language, 47*, 27–52.

74. Fromkin, V. A. (1971). The non-anomalous nature of anomalous utterances. *Language, 47*, 27–52.

75. Ferreira, V. S. (1996). Is it better to give than to donate? Syntactic flexibility in language production. *Journal of Memory and Language, 35*, 724–755; Griffin, Z. M. & Bock, K. (2000). What the eyes say about speaking. *Psychological Science, 11*, 274–279.

76. Ferreira, V. S. & Firato, C. E. (2002). Proactive interference effects on sentence production. *Psychonomic Bulletin & Review, 9*, 795–800.

77. Kempen, G. & Hoehkamp, E. (1987). An incremental procedural grammar for sentence formulation. *Cognitive Science, 11*, 201–258.

78. Bachoud-Levi, A. C., Dupoux, E., Cohen, L. & Mehler, J. (1998). Where is the length effect? A cross-linguistic study of speech production. *Journal of Memory and Language, 39*, 331–346; Griffin, Z. M. (2003). A reversed word length effect in coordinating the preparation and articulation of words in speaking. *Psychonomic Bulletin & Review, 10*, 603–609.

79. Bock, K., Irwin, D. E., Davidson, D. J. & Levelt, W. J. M. (2003). Minding the clock. *Journal of Memory and Language, 48*, 653– 685.

80. Bock, K. & Miller, C. A. (1991). Broken agreement. *Cognitive Psychology, 23*, 45–93; Lindsley, J. R. (1975). Producing simple utterances: How far ahead do we plan? *Cognitive Psychology, 7*, 1–19.

81. Clark, H. H. & Wasow, T. (1998). Repeating words in spontaneous speech. *Cognitive Psychology, 37*, 201–242; Ferreira, V. S. (1996). Is it better to give than to donate? Syntactic flexibility in language production. *Journal of Memory and Language, 35*, 724–755; Ferreira, V. S. & Dell, G. S. (2000). Effect of ambiguity and lexical availability on syntactic and lexical production. *Cognitive Psychology, 40*, 296–340; Bock, K. (1996). Language production: Methods and methodologies. *Psychonomic Bulletin & Review, 3*, 395–421.

82. Chang, F., Dell, G. S. & Bock, J. K. (2006). Becoming syntactic. *Psychological Review*, 113, 234–272.

83. Griffiths, T. L., Steyvers, M. & Tenebaum, J. B. (2007). Topics in semantic representation. *Psychological Review, 114*, 211–244.

84. Chang, F., Dell, G. S. & Bock, J. K. (2006). Becoming syntactic. *Psychological Review, 113*, 234–272.

85. Ungerleider, L. G. & Haxby, J. V. (1994). 'What' versus 'where' in the human brain. *Current Opinion in Neurobiology, 4*, 157–165.

86. Clark, H. H. & Clark, E. V. (1977). *Psychology and Language*. New York, NY: Harcourt Brace Jovanovich.

87. Clifton, C. & Frazier, L. (2004). Should given information come before new? Yes and no. *Memory & Cognition, 32*, 886–895.

88. Perrachione, T. K., Fedorenko, E. G., Vinke, L., Gibson, E. & Dilley, L. C. (2013). Evidence for shared cognitive processing of pitch in music and language. *PloS one, 8*(8), e73372.

89. Kurumada, C., Brown, M., Bibyk, S., Pontillo, D. F. & Tanenhaus, M. K. (2014). Is it or isn't it: Listeners make rapid use of prosody to infer speaker meanings. *Cognition, 133*(2), 335–342.

90. MacLeod, C. M. (1992). The Stroop task: The "gold standard" of attentional measures. *Journal of Experimental Psychology: General, 121*, 12–14.

91. Carroll, D. W. (1986). *Psychology of language*. Pacific Grove, CA: Brooks/Cole.

92. Whitney, P. (1998). *The psychology of language*. Boston, MA: Houghton Mifflin.

93. McKoon, G. & Macfarland, T. (2002). Event templates in the lexical representations of verbs. *Cognitive Psychology, 45*, 1–44.

94. Andres, M., Finocchiaro, C., Buiatti, M. & Piazza, M. (2015). Contribution of motor representations to action verb processing. *Cognition*, 134, 174–184.

95. Willems, R. M., Hagoort, P. & Casasanto, D. (2010). Bodyspecific representations of action verbs: Neural evidence from right- and left-handers. *Psychological Science, 21*, 67–74.

96. Willems, R. M. Labruna, L., D'Espisito, M., Ivry, R. & Casasanto, D. (2011). A functional role for the motor system in language understanding: Evidence from theta-burst transcranial magnetic stimulation. *Psychological Science, 22*, 849–854.

97. Simpson, G. B. (1981). Meaning dominance and semantic context in the processing of lexical ambiguity. *Journal of Verbal Learning and Verbal Behavior, 20*, 120–136; Simpson, G. B. (1984). Lexical ambiguity and its role in models of word recognition. *Psychological Bulletin, 96*, 316–340.

98. Balota, D. A. & Paul, S. T. (1996). Summation of activation: Evidence from multiple primes that converge and diverge within semantic memory. *Journal of Experimental Psychology: Learning, Memory, and Cognition, 22*, 827–845; Klein, D. V. & Murphy, G. L. (2002). Paper has been my ruin: Conceptual relations of polysemous senses. *Journal of Memory and Language, 47*, 548–570; Piercey, C. D. & Joordens, S. (2000). Turning an advantage into a disadvantage: Ambiguity effects in lexical decision versus reading tasks. *Memory & Cognition, 28*, 657–666; Binder, K. S. (2003). Sentential and discourse topic effects on lexical ambiguity processing: An eye movement examination. *Memory & Cognition*, 31, 690–702; Binder, K. S. & Rayner, K. (1998). Contextual strength does not modulate the subordinate bias effect: Evidence from eye fixations and self-paced reading. *Psychonomic Bulletin & Review, 5*, 271–276.

99. Gernsbacher, M. A. & Faust, M. E. (1991). The mechanism of suppression: A component of general comprehension skill. *Journal of Experimental Psychology: Learning, Memory & Cognition, 17*, 245–262.

100. Kounios, J. & Holcomb, P. J. (1992). Structure and process in semantic memory: Evidence from event-related brain potentials and reaction times. *Journal of Experimental Psychology: General, 121*, 459–479.

101. Holcomb, P. J., Kounios, J., Anderson, J. E. & West, W. C. (1999). Dual-coding, context availability, and concreteness effects in sentence comprehension: An electrophysiological investigation. *Journal of Experimental Psychology: Learning, Memory, and Cognition, 25*, 721–742; Laszlo, S. & Federmeier, K. D. (2009). A beautiful day in the neighborhood: An event-related potential study of lexical relationships and prediction in context. *Journal of Memory and Language, 61*, 326–338; Lee, C-L. & Federmeier, K. D. (2009). Wave-ering: An ERP study of syntactic and semantic context effects on ambiguity resolution for noun/verb homographs. *Journal of Memory and Language, 61*, 538–555; Sereno, S. C., Brewer, C. C. & O'Donnell, P. J. (2003). Context effects in word recognition: Evidence for early interactive processing. *Psychological Science, 14*, 328–333.

102. Fillmore, C. J. (1968). Toward a modern theory of case. In D. A. Reibel & S. A. Schane (Eds.), *Modern studies in English* (pp. 361–375). Englewood Cliffs, NJ: Prentice Hall.

103. Jackendoff, R. S. (1992). *Languages of the mind: Essays on mental representation*. Cambridge, MA: MIT Press.

104. Bresnan, J. (1978). A realistic transformational grammar. In J. Bresnan, M. Halle & G. Miller (Eds.), *Linguistic theory and Psychological Reality* (pp. 1–59). Cambridge, MA: MIT Press; Bresnan, J. & Kaplan, R. M. (1982). Introduction: Grammars as mental representations of language. In J. Bresnan (Ed.), *The Mental Representation of Grammatical Relations* (pp. xvii–iii). Cambridge, MA: MIT Press.

105. Jackendoff, R. S. (1992). *Languages of the mind: Essays on Mental Representation*. Cambridge, MA: MIT Press.

106. Jackendoff, R. S. (1992). *Languages of the Mind: Essays on Mental Representation*. Cambridge, MA: MIT Press.

107. O'Seaghdha, P. G. (1997). Conjoint and dissociable effects of syntactic and semantic context. *Journal of Experimental Psychology: Learning, Memory, and Cognition, 23*, 807–828.

108. Peterson, R. R., Burgess, C., Dell, G. S. & Eberhard, K. M. (2001). Dissociation between syntactic and semantic processing during idiom comprehension. *Journal of Experimental Psychology: Learning, Memory, and Cognition, 27*(5), 1223.

109. Ainsworth-Darnell, K., Shulman, H. G. & Boland, J. E. (1998). Dissociating brain responses to syntactic and semantic anomalies: Evidence from event-related potentials. *Journal of Memory and Language, 38*, 112–130; Friederici, A. D., Hahne, A. & Mecklinger, A. (1996). Temporal structure of syntactic parsing: Early and late event-related brain potential effects. *Journal of Experimental Psychology: Learning, Memory, and Cognition, 22*, 1219–1248; Osterhout, L., Allen, M. D., McLaughlin, J. & Inoue, K. (2002). Brain potentials elicited by prose-embedded linguistic anomalies. *Memory & Cognition, 30*, 1304–1312.

110. Fillenbaum, S. (1974). Pragmatic normalization: Further results for some conjunctive and disjunctive sentences. *Journal of Experimental Psychology, 102*, 574–578.

111. Fillenbaum, S. (1974). Pragmatic normalization: Further results for some conjunctive and disjunctive sentences. *Journal of Experimental Psychology, 102*, 574–578.

112. Altmann, G. T. M. (1998). Ambiguity in sentence processing. *Trends in Cognitive Sciences, 2*, 146–157; Altmann, G. T. M. & Steedman, M. (1988). Interaction with context during human sentence processing. *Cognition, 30*, 191–238.

113. Singer, M. (1990). *Psychology of language: An introduction to sentence and discourse processes.* Hillsdale, NJ: Erlbaum.

114. Frazier, L., & Rayner, K. (1982). Making and correcting errors during sentence comprehension: Eye movements in the analysis of structurally ambiguous sentences. *Cognitive Psychology, 14*, 178–210.

115. Mitchell, D. C., & Holmes, V. M. (1985). The role of specific information about the verb in parsing sentences with local structural ambiguity. *Journal of Memory and Language, 24*, 542–559.

116. McKoon, G. & Ratcliff, R. (2007). Interactions of meaning and syntax: Implications for models of sentence comprehension. *Journal of Memory and Language, 56*, 270–290.

117. Christianson, K., Hollingworth, A., Halliwell, J. F. & Ferreira, F. (2001). Thematic roles assigned along the garden path linger. *Cognitive Psychology, 42*, 368–407; Ferreira, F., Henderson, J. M., Anes, M. D., Weeks, P. A., Jr. & McFarlane, D. K. (1996). Effects of lexical frequency and syntactic complexity in spoken-language comprehension: Evidence from the auditory moving-window technique. *Journal of Experimental Psychology: Learning, Memory, and Cognition, 22*, 324–335.

118. Tabor, W. & Hutchins, S. (2004). Evidence for self-organized sentence processing: Digging-in effects. *Journal of Experimental Psychology: Learning, Memory, and Cognition*, 30, 431– 450.

119. Clifton, C., Jr., Traxler, M. J., Mohamed, M. T., Williams, R. S., Morris, R. K. & Rayner, K. (2003). The use of thematic role information in parsing: Syntactic processing autonomy revisited. *Journal of Memory and Language, 49*, 317–334; Mason, R. A., Just, M. A., Keller, T. A. & Carpenter, P. A. (2003). Ambiguity in the brain: What brain imaging reveals about the processing of syntactically ambiguous sentences. *Journal of Experimental Psychology: Learning, Memory, and Cognition, 29*, 1319–1338; Rayner, K. & Clifton, C., Jr. (2002). Language comprehension. In D. L. Medin (Ed.), *Steven's handbook of experimental psychology* (Vol. X, pp. 261–316). New York, NY: Wiley; Tanenhaus, M. K. & Trueswell, J. C. (1995). Sentence comprehension. In J. Miller & P. Eiman (Eds.), *Handbook of Perception and Cognition: Speech, Language, and Communication* (2nd ed., Vol. 11, pp. 217–262). San Diego, CA: Academic Press.

120. Glucksberg, S. & Keysar, B. (1990). Understanding metaphorical comparisons: Beyond similarity. *Psychological Review, 97*, 3–18; Keysar, B., Shen, Y., Glucksberg, S. & Horton, W. S. (2000). Conventional language: How metaphorical is it? *Journal of Memory and Language, 43*, 576–593; Tourangeau, R. & Rips, L. (1991). Interpreting and evaluating metaphors. *Journal of Memory and Language, 30*, 452–472.

121. Rayner, K., Carlson, M., & Frazier, L. (1983). The interaction of syntax and semantics during sentence processing: Eye movements in the analysis of semantically biased sentences. *Journal of Verbal Learning and Verbal Behavior, 22*, 358–374.

122. Perea, M., Dunabeitia, J. A. & Carreiras, M. (2008). Masked associative/semantic priming effects across languages with highly proficient bilinguals. *Journal of Memory and Language, 58*, 916–930.

123. Treccani, B., Argyri, E., Sorace, A. & Della Salla, S. (2009). Spatial negative priming in bilingualism. *Psychonomic Bulletin & Review, 16*, 320–327.

124. Lambert, W. E. (1990). Issues in foreign language and second language education. *Proceedings of the Research Symposium on Limited English Proficient Students' Issues* (1st, Washington, DC, September 10–12).

125. Bialystok, E. (1988). Levels of bilingualism and levels of linguistic awareness. *Developmental Psychology, 24*, 560–567; van Hell, J. G. & Dijkstra, T. (2002). Foreign language knowledge can influence native language performance in exclusively native contexts. *Psychonomic Bulletin & Review, 9*, 780–789.

126. Bialystok, E., Craik, F. I. & Luk, G. (2012). Bilingualism: Consequences for mind and brain. *Trends in Cognitive Sciences, 16*(4), 240–250.

127. Ljungberg, J. K., Hansson, P., Andrés, P., Josefsson, M. & Nilsson, L. G. (2013). A longitudinal study of memory advantages in bilinguals. PloS one, 8(9), e73029; Paap, K. R. & Greenberg, Z. I. (2013). There is no coherent evidence for a bilingual advantage in executive processing. *Cognitive Psychology, 66(2)*, 232–258; Ratiu, I. & Azuma, T. (2015). Working memory capacity: Is there a bilingual advantage? Journal of *Cognitive Psychology, 27*(1), 1–11.

128. Linek, J. A., Kroll, J. F. & Sunderman, G. (2010). Losing access to the native language while immersed in a second language: Evidence for the role of inhibition in second-language learning. *Psychological Science, 20*, 1507–1515.

129. Osterhout, L. & Holcomb, P. J. (1992). Event-related brain potentials elicited by syntactic anomaly. *Journal of Memory and Language, 31*, 785–806.

130. Kounios, J. & Holcomb, P. J. (1992). Structure and process in semantic memory: Evidence from event-related brain potentials and reaction times. *Journal of Experimental Psychology: General, 121*, 459–479.

131. Andrews, M., Vigliocco, G. & Vinson, D. (2009). Integrating experiential and distributional data to learn semantic representations. *Psychological Review, 116(3)*, 463–498.

132. McCandliss, B. D., Posner, M. I. & Givon, T. (1997). Brain plasticity in learning visual words. *Cognitive Psychology, 33*, 88–110.

133. Rosler, F., Pechmann, T., Streb, J., Roder, B. & Hennighausen, E. (1998). Parsing of sentences in a language with varying word order: Word-by-word variations of processing demands are revealed by event-related brain potentials. *Journal of Memory and Language, 38*, 150–176.

134. Beeman, M. J. (1998). Coarse semantic coding and discourse comprehension. In M. Beeman & C. Chiarello (Eds.), *Brain right hemisphere language comprehension: Perspectives from cognitive neuroscience* (pp. 255–284). Mahwah, NJ: Erlbaum.

135. Coulson, S., Federmeier, K. D., Van Petten, C. & Kutas, M. (2005). Right hemisphere sensitivity to word- and sentence level context: Evidence from event-related brain potentials. *Journal of Experimental Psychology: Learning, Memory, and Cognition, 31*, 127–147.

136. Reichle, E. D., Carpenter, P. A. & Just, M. A. (2000). The neural bases of strategy and skill in sentence–picture verification. *Cognitive Psychology, 40*, 261–295.

137. McCarthy, R. A. & Warrington, E. K. (1990). *Cognitive Neuropsychology: A Clinical Introduction.* San Diego, CA: Academic Press.

138. Kertesz, A. (1982). Two case studies: Broca's and Wernicke's aphasia. In M. A. Arbib, D. Caplan & J. C. Marshall (Eds.), *Neural Models of Language Processes* (pp. 25–44). New York, NY: Academic Press.

139. Faroqi-Shah, Y. & Thompson, C. K. (2007). Verb inflections in agrammatic aphasia: Encoding of tense features. *Journal of Memory and Language, 56*, 129–151.

140. Breedin, S. D. & Saffran, E. M. (1999). Sentence processing in the face of semantic loss: A case study. *Journal of Experimental Psychology: General, 128*, 547–562.

141. Geschwind, N. (1970). The organisation of language and the brain. *Science, 170*, 940–944.

142. Coughlan, A. K. & Warrington, E. K. (1978). Word comprehension and word retrieval in patients with localised cerebral lesions. *Brain, 101*, 163–185; McCarthy, R. A. & Warrington, E. K. (1990). *Cognitive Neuropsychology: A Clinical Introduction.* San Diego, CA: Academic Press.

143. Geschwind, N. (1967). The varieties of naming errors. *Cortex*, 3, 97–112; Goodglass, H., Kaplan, E., Weintraub, S. & Ackerman, N. (1976). The 'tip-of-the-tongue' phenomenon in aphasia. *Cortex*, 12, 145–153.

144. Kay, J. & Ellis, A. (1987). A cognitive neuropsychological case study of anomia: Implications for psychological models of word retrieval. *Brain, 110*, 613–629.

145. Ashcraft, M. H. (1993). A personal case history of transient anomia. *Brain and Language, 44*, 47–57.

146. Benson, D. J. & Geschwind, N. (1969). The alexias. In P. Vincken & G. W. Bruyn (Eds.), *Handbook of Clinical Neurology* (Vol. 4, pp. 112–140). Amsterdam, The Netherlands: North-Holland.

147. Berndt, R. S. & Haendiges, A. N. (2000). Grammatical class in word and sentence production: Evidence from an aphasic patient. *Journal of Memory and Language, 43*, 249–273.

148. Sitton, M., Mozer, M. C. & Farah, M. J. (2000). Superadditive effects of multiple lesions in a connectionist architecture: Implications for the neuropsychology of optic aphasia. *Psychological Review, 107*, 709–734.

149. Beeman, M. J. & Chiarello, C. (1998). Complementary rightand left-hemisphere language comprehension. *Current Directions in Psychological Science, 7*, 2–8.

150. Beeman, M. J. (1993). Semantic processing in the right hemisphere may contribute to drawing inferences from discourse. *Brain and Language, 44*, 80–120; Beeman, M. J. (1998). Coarse semantic coding and discourse comprehension. In M. Beeman & C. Chiarello (Eds.), *Brain Right Hemisphere Language Comprehension: Perspectives from Cognitive Neuroscience* (pp. 255–284). Mahwah, NJ: Erlbaum.

151. Mehler, J., Morton, J. & Jusczyk, P. W. (1984). On reducing language to biology. *Cognitive Neuropsychology, 1*, 83–116.

152. O'Seaghdha, P. G. (1997). Conjoint and dissociable effects of syntactic and semantic context. *Journal of Experimental Psychology Learning, Memory, and Cognition, 23*, 807–828; Osterhout, L. & Holcomb, P. J. (1992). Event-related brain potentials elicited by syntactic anomaly. *Journal of Memory and Language, 31*, 785–806.

153. Corballis, M. C. (1989). Laterality and human evolution. *Psychological Review, 96*, 492–505; Geary, D. C. (1992). Evolution of human cognition: Potential relationship to the ontogenetic development of behavior and cognition. *Evolution and Cognition, 1*, 93–100.

154. Lewontin, R. C. (1990). The evolution of cognition. In D. N. Osherson & E. E. Smith (Eds.), *Thinking: An invitation to cognitive science* (Vol. 3, pp. 229–246). Cambridge, MA: MIT Press.

Chapter 9

1. Miller, G. A. (1977). Practical and lexical knowledge. In P. N. Johnson-Laird & P. C. Wason (Eds.), *Thinking: Readings in Cognitive Science* (pp. 400–410). Cambridge, England: Cambridge University Press.

2. Gernsbacher, M. A. (1990). *Language Comprehension as Structure Building*. Hillsdale, NJ: Erlbaum.
3. Sachs, J. S. (1967). Recognition memory for syntactic and semantic aspects of connected discourse. *Perception & Psychophysics, 2*, 437–442.
4. van Dijk, T. A. & Kintsch, W. (1983). *Strategies in Discourse Comprehension*. New York, NY: Academic Press.
5. Johnson-Laird, P. N. (1983). *Mental Models: Towards a Cognitive Science of Language, Inference and Consciousness*. Cambridge, MA: Harvard University Press; van Dijk, T. A. & Kintsch, W. (1983). *Strategies in Discourse Comprehension*. New York, NY: Academic Press. Zwaan, R. A. & Radvansky, G. A. (1998). Situation models in language comprehension and memory. *Psychological Bulletin, 123*, 162–185.
6. Huang, Y. T. & Gordon, P. C. (2011). Distinguishing the time course of lexical and discourse processes through context, co-reference, and quantified expressions. *Journal of Experimental Psychology: Learning, Memory, and Cognition, 37*, 966–978.
7. Dunlosky, J. & Lipko, C. (2007). Metacomprehension: A brief history and how to improve its accuracy. *Current Directions in Psychological Science, 16*, 228–232.
8. Arbuckle, T. Y. & Cuddy, L. L. (1969). Discrimination of item strength at time of presentation. *Journal of Experimental Psychology, 81*, 126–131.
9. Nelson, T. O. & Leonesio, R. J. (1988). Allocation of self-paced study time and the "labor-in vain effect." *Journal of Experimental Psychology: Learning, Memory, and Cognition, 14*, 676–686.
10. Metcalfe, J. (2002). Is study time allocated selectively to a region of proximal learning? *Journal of Experimental Psychology: General, 131*, 349–363.
11. Dunlosky, J. & Lipko, C. (2007). Metacomprehension: A brief history and how to improve its accuracy. *Current Directions in Psychological Science, 16*, 228–232; Griffin, T. D., Wiley, J. & Thiede, K. W. (2008). Individual differences, rereading, and self-explanation: Concurrent processing and cue validity as constraints on metacomprehension accuracy. *Memory & Cognition, 36*, 93–103.
12. Dunlosky, J. & Nelson, T. O. (1994). Does the sensitivity of judgments of learning (JOLs) to the effects of various activities depend on when the JOLs occur? *Journal of Memory and Language, 33*, 545–565.
13. Magliano, J. P., Trabasso, T. & Graesser, A. C. (1999). Strategic processing during comprehension. *Journal of Educational Psychology, 91*, 615–629.
14. Gernsbacher, M. A. (1990). *Language Comprehension as Structure Building*. Hillsdale, NJ: Erlbaum.
15. Gernsbacher, M. A. (1990). *Language comprehension as Structure Building*. Hillsdale, NJ: Erlbaum.
16. Millis, K. K. & Just, M. A. (1994). The influence of connectives on sentence comprehension. *Journal of Memory and Language, 33*, 128–147.
17. O'Brien, E. J., Albrecht, J. E., Hakala, C. M. & Rizzella, M. L. (1995). Activation and suppression of antecedents during reinstatement. *Journal of Experimental Psychology: Learning, Memory, and Cognition, 21*, 626–634; Gernsbacher, M. A., Keysar, B., Robertson, R. R. W. & Werner, N. K. (2001). The role of suppression and enhancement in understanding metaphors. *Journal of Memory and Language, 45*, 433–450.
18. Kintsch, W. (2000). Metaphor comprehension: A computational theory. *Psychonomic Bulletin & Review, 7*, 257–266.
19. McNamara, D. S., & McDaniel, M. A. (2004). Suppressing irrelevant information: Knowledge activation or inhibition? *Journal of Experimental Psychology: Learning, Memory, and Cognition, 30*, 465–482.
20. Bornkessel, I., Zysset, S., Friederici, A. D., Von Cramon, D. Y. & Schlesewsky, M. (2005). Who did what to whom? The neural basis of argument hierarchies during language comprehension. *Neuroimage, 26(1)*, 221–233.
21. Kambe, G., Duffy, S. A., Clifton, C., Jr. & Rayner, K. (2003). An eye-movement-contingent probe paradigm. *Psychonomic Bulletin & Review, 10*, 661–666; Rayner, K. (1998). Eye movements in reading and information processing: 20 years of research. *Psychological Bulletin, 124*, 372–422.
22. Rayner, K. (1998). Eye movements in reading and information processing: 20 years of research. *Psychological Bulletin, 124*, 372–422.
23. Just, M. A. & Carpenter, P. A. (1980). A theory of reading: From eye fixations to comprehension. *Psychological Review, 87*, 329–354; Just, M. A. & Carpenter, P. A. (1987). The *Psychology of Reading and Language Comprehension*. Boston, MA: Allyn & Bacon; Just, M. A. & Carpenter, P. A. (1992). A capacity theory of comprehension. *Psychological Review, 99*, 122–149.
24. Rayner, K. (1998). Eye movements in reading and information processing: 20 years of research. *Psychological Bulletin, 124*, 372–422.
25. Kliegl, R., Nuthmann, A. & Engbert, R. (2006). Tracking the mind during reading: The influence of past, present, and future words on fixation durations. *Journal of Experimental Psychology: General, 135*, 12–35.
26. Spivey, M. J., Tanenhaus, M. K., Eberhard, K. M. & Sedivy, J. C. (2002). Eye movements and spoken language comprehension: Effects of visual context on syntactic ambiguity resolution. *Cognitive Psychology, 45*, 447–481; Crosby, J. R., Monin, B. & Richardson, D. (2008). Where do we look during potentially offensive behavior? *Psychological Science, 19*, 226–228.
27. Just, M. A. (1976, May). *Research strategies in prose comprehension*. Paper presented at the meetings of the Midwestern Psychological Association, Chicago, IL.
28. Just, M. A. & Carpenter, P. A. (1987). *The Psychology of Reading and Language Comprehension*. Boston,

MA: Allyn & Bacon; Just, M. A. & Carpenter, P. A. (1980). A theory of reading: From eye fixations to comprehension. *Psychological Review, 87*, 329–354.

29. Angele, B. & Rayner, K. (2013). Processing the in the parafovea: Are articles skipped automatically? *Journal of Experimental Psychology: Learning, Memory, and Cognition, 39(2)*, 649.

30. Rayner, K., White, S. J., Johnson, R. L. & Liversedge, S. P. (2006). Raeding wrods with jubmled letters: There is a cost. *Psychological Science, 17*, 192–193.

31. Rayner, K. & Duffy, S. A. (1988). On-line comprehension processes and eye movements in reading. In M. Daneman, G. E. MacKinnon & T. G. Waller (Eds.), *Reading research: Advances in Theory and Practice* (pp. 13–66). New York, NY: Academic Press.

32. Rayner, K. & Well, A. D. (1996). Effects of contextual constraint on eye movements in reading: A further examination. *Psychonomic Bulletin & Review, 3*, 504–509.

33. Just, M. A., & Carpenter, P. A. (1980). A theory of reading: From eye fixations to comprehension. *Psychological Review, 87*, 329–354.

34. Pearlmutter, N. J., Garnsey, S. M. & Bock, K. (1999). Agreement processes in sentence comprehension. *Journal of Memory and Language, 41*, 427–456.

35. McDonald, J. L. & MacWhinney, B. (1995). The time course of anaphor resolution: Effects of implicit verb causality and gender. *Journal of Memory and Language, 34*, 543–566.

36. Reichle, E. D., Pollatsek, A., Fisher, D. L. & Rayner, K. (1998). Eye movements during mindless reading. *Psychological Science, 21*, 1300–1310.

37. Norris, D. (2013). Models of visual word recognition. *Trends in Cognitive Sciences, 17*(10), 517–524.

38. Just, M. A. & Carpenter, P. A. (1980). A theory of reading: From eye fixations to comprehension. *Psychological Review, 87*, 329–354.

39. Just, M. A. & Carpenter, P. A. (1980). A theory of reading: From eye fixations to comprehension. *Psychological Review, 87*, 329–354; Just, M. A. & Carpenter, P. A. (1987). *The Psychology of Reading and Language Comprehension*. Boston, MA: Allyn & Bacon; Just, M. A. & Carpenter, P. A. (1992). A capacity theory of comprehension. *Psychological Review, 99*, 122–149.

40. Just, M. A., & Carpenter, P. A. (1980). A theory of reading: From eye fixations to comprehension. *Psychological Review, 87*, 329–354.

41. Kaakinen, J. K., Hyona, J. & Keenan, J. M. (2003). How prior knowledge, WMC, and relevance of information affect eye fixations in expository text. *Journal of Experimental Psychology:Learning, Memory, and Cognition, 29*, 447–457.

42. Altmann, G. T. M., Garnham, A. & Dennis, Y. (1992). Avoiding the garden path: Eye movements in context. *Journal of Memory and Language, 31*, 685–712; Inhoff, A. W. (1984). Two stages of word processing during eye fixations in the reading of prose. *Journal of Verbal Learning and Verbal Behavior, 23*, 612–624; Juhasz, B. J. & Rayner, K. (2003). Investigating the effects of a set of intercorrelated variables on eye fixation durations in reading. *Journal of Experimental Psychology: Learning, Memory, and Cognition, 29*, 1312–1318; Schilling, H. E. H., Rayner, K. & Chumbley, J. I. (1998). Comparing naming, lexical decision, and eye fixation times: Word frequency effects and individual differences. *Memory & Cognition, 26*, 1270–1281.

43. Paul, S. T., Kellas, G., Martin, M. & Clark, M. B. (1992). Influence of contextual features on the activation of ambiguous word meanings. *Journal of Experimental Psychology: Learning, Memory, and Cognition, 18*, 703–717; Schustack, M. W., Ehrlich, S. F. & Rayner, K. (1987). Local and global sources of contextual facilitation in reading. *Journal of Memory and Language, 26*, 322–340; Simpson, G. B., Casteel, M. A., Peterson, R. R. & Burgess, C. (1989). Lexical and sentence context effects in word recognition. *Journal of Experimental Psychology: Learning, Memory, and Cognition, 15*, 88–97.

44. Van Berkum, J. J. A., Brown, C. M. & Hagoort, P. (1999). Early referential context effects in sentence processing: Evidence from event-related brain potentials. *Journal of Memory and Language, 41*, 147–182.

45. Frazier, L. & Rayner, K. (1990). Taking on semantic commitments: Processing multiple meanings vs. multiple senses. *Journal of Memory and Language, 29*, 181–200; Rayner, K. & Frazier, L. (1989). Selection mechanisms in reading lexically ambiguous words. *Journal of Experimental Psychology: Learning, Memory, and Cognition*, 15, 779–790.

46. Frisson, S. & Pickering, M. J. (1999). The processing of metonymy: Evidence from eye movements. *Journal of Experimental Psychology: Learning, Memory, and Cognition, 25*, 1366–1383.

47. O'Brien, E. J. & Myers, J. L. (1987). The role of causal connections in the retrieval of text. *Memory & Cognition, 15*, 419–427; Pickering, M. J. & Traxler, M. J. (1998). Plausibility and recovery from garden paths: An eye-tracking study. *Journal of Experimental Psychology: Learning, Memory, and Cognition, 24*, 940–961; Rayner, K., Warren, T., Juhasz, B. J. & Liversedge, S. P. (2004). The effect of plausibility on eye movements in reading. *Journal of Experimental Psychology: Learning, Memory, and Cognition, 30*, 1290–1301; Speer, S. R. & Clifton, C., Jr. (1998). Plausibility and argument structure in sentence comprehension. *Memory & Cognition, 26*, 965–978; Taraban, R. & McClelland, J. L. (1988). Constituent attachment and thematic role assignment in sentence processing: Influences of content-based expectations. *Journal of Memory and Language, 27*, 597–632.

48. Lorch, R. F., Jr., Lorch, E. P. & Matthews, P. D. (1985). On-line processing of the topic structure of a text. *Journal of Memory and Language, 24*, 350–362.

49. Wiley, J. & Rayner, K. (2000). Effects of titles on the processing of text and lexically ambiguous words: Evidence from eye movements. *Memory & Cognition, 28*, 1011–1021.

50. Sharkey, N. E. & Mitchell, D. C. (1985). Word recognition in a functional context: The use of scripts in reading. *Journal of Memory and Language, 24*, 253–270.

51. Malt, B. C. (1985). The role of discourse structure in understanding anaphora. *Journal of Memory and Language, 24*, 271–289; Murphy, G. L. (1985). Processes of understanding anaphora. *Journal of Memory and Language, 24*, 290–303.

52. Vu, H., Kellas, G., Metcalf, K. & Herman, R. (2000). The influence of global discourse on lexical ambiguity resolution. *Memory & Cognition, 28*, 236–252.

53. Lee, H-W., Rayner, K. & Pollatsek, A. (1999). The time course of phonological, semantic, and orthographic coding in reading: Evidence from the fast-priming technique. *Psychonomic Bulletin & Review*, 6, 624–634; Rayner, K., Pollatsek, A. & Binder, K. S. (1998). Phonological codes and eye movements in reading. *Journal of Experimental Psychology: Learning, Memory, and Cognition, 24*, 476–497.

54. Jared, D., Levy, B. A. & Rayner, K. (1999). The role of phonology in the activation of word meanings during readings: Evidence from proofreading and eye movements. *Journal of Experimental Psychology: General, 128*, 219–264.

55. Just, M. A., & Carpenter, P. A. (1980). A theory of reading: From eye fixations to comprehension. *Psychological Review, 87*, 329–354.

56. Almor, A. (1999). Noun-phrase anaphora and focus: The informational load hypothesis. *Psychological Review, 106*, 748–765; Gordon, P. C. & Chan, D. (1995). Pronouns, passives, and discourse coherence. *Journal of Memory and Language, 34*, 216–231; Gordon, P. C. & Scearce, K. A. (1995). Pronominalization and discourse coherence, discourse structure and pronoun interpretation. *Memory & Cognition, 23*, 313–323.

57. Arnold, J. E. & Griffin, Z. M. (2007). The effect of additional characters on choice of referring expression: Everyone counts. *Journal of Memory and Language, 56*, 521–536.

58. Gernsbacher, M. A. & Hargreaves, D. (1988). Accessing sentence participants: The advantage of first mention. *Journal of Memory and Language, 27*, 699–717.

59. Kurczek, J., Brown-Schmidt, S. & Duff, M. (2013). Hippocampalcontributions to language: Evidence of referential processing deficits in amnesia. *Journal of Experimental Psychology: General, 142(4)*, 1346–1354.

60. Clark, H. H. (1977). Bridging. In P. N. Johnson-Laird & P. C. Wason (Eds.), *Thinking: Readings in cognitive science* (pp. 411–420). Cambridge, England: Cambridge University Press.

61. Gernsbacher, M. A., & Hargreaves, D. (1988). Accessing sentence participants: The advantage of first mention. *Journal of Memory and Language, 27*, 699–717.

62. Gernsbacher, M. A., Hargreaves, D. & Beeman, M. (1989). Building and accessing clausal representations: The advantage of first mention versus the advantage of clause recency. *Journal of Memory and Language, 28*, 735–755.

63. Roberson, D., Davies, I. & Davidoff, J. (2000). Color categories are not universal: Replications and new evidence from a Stone-Age culture. *Journal of Experimental Psychology: General, 129*, 369–398.

64. Haviland, S. E. & Clark, H. H. (1974). What's new? Acquiring new information as a process in comprehension. *Journal of Verbal Learning and Verbal Behavior, 13*, 512–521.

65. Gernsbacher, M. A. (1997). Two decades of structure building. *Discourse Processes, 23*, 265–304.

66. Roberson, D., Davies, I. & Davidoff, J. (2000). Color categories are not universal: Replications and new evidence from a Stone-Age culture. *Journal of Experimental Psychology: General, 129*, 369–398.

67. Polk, T. A. & Farah, M. J. (2002). Functional MRI evidence for an abstract, not perceptual, word-form area. *Journal of Experimental Psychology: General, 131*, 65–72.

68. Robertson, D. A., Gernsbacher, M. A., Guidotti, S. J., Robertson, R. R. W., Irwin, W., Mock, B. J., et al. (2000). Functional neuroanatomy of the cognitive process of mapping during discourse comprehension. *Psychological Science, 11*, 255–260.

69. Johnson-Laird, P. N. (1983). *Mental models: Towards a Ccognitive Science of Language, Inference and Consciousness*. Cambridge, MA: Harvard University Press; van Dijk, T. A. & Kintsch, W. (1983). *Strategies in Discourse Comprehension*. New York, NY: Academic Press; Zwaan, R. A. & Radvansky, G. A. (1998). Situation models in language comprehension and memory. *Psychological Bulletin, 123*, 162–185.

70. Graesser, A. C., Singer, M. & Trabasso, T. (1994). Constructing inferences during narrative text comprehension. *Psychological Review, 101*, 371–395.

71. McKoon, G. & Ratcliff, R. (1992). Inference during reading. *Psychological Review, 99*, 440–466.

72. Zwaan, R. A. (1999). Situation models: The mental leap into imagined worlds. *Current Directions in Psychological Science, 8(1)*, 15–18.

73. Anderson, S. E., Matlock, T. & Spivey, M. (2013). Grammatical aspect and temporal distance in motion descriptions. *Frontiers in Psychology, 4*, 337; Magliano, J. P. & Schleich, M. C. (2000). Verb aspect and situation models. *Discourse Processes, 29(2)*, 83–112.

74. Madden, C. J. & Zwaan, R. A. (2003). How does verb aspect constrain event representations? *Memory & Cognition, 31(5)*, 663–672.

75. Clark, H. H. (1977). Bridging. In P. N. Johnson-Laird & P. C. Wason (Eds.), *Thinking: Readings in cognitive science* (pp. 411–420). Cambridge, England: Cambridge University Press.

76. McKoon, G. & Ratcliff, R. (1986). Inferences about predictable events. *Journal of Experimental Psychology: Learning, Memory, and Cognition, 12*, 82–91.

77. Cook, A. E. & Myers, J. L. (2004). Processing discourse roles in scripted narratives: The influences of context and world knowledge. *Journal of Memory and Language, 50*, 268–288.

78. McKoon, G. & Ratcliff, R. (1989). Inferences about contextually defined categories. *Journal of Experimental Psychology: Learning, Memory, and Cognition, 15*, 1134–1146; O'Brien, E. J., Plewes, P. S. & Albrecht, J. E. (1990). Antecedent retrieval processes. *Journal of Experimental Psychology: Learning, Memory, and Cognition, 16*, 241–249.

79. Zwaan, R. A., Stanfield, R. A. & Yaxley, R. H. (2002). Language comprehenders mentally represent the shapes of objects. *Psychological Science, 13*, 168–171.

80. Connell, L. & Lynott, D. (2009). Is a bear white in the woods? Parallel representation of implied object color during language comprehension. *Psychonomic Bulletin & Review, 16*, 573–577.

81. Wassenburg, S. I. & Zwaan, R. A. (2010). Readers routinely represent implied object rotation: The role of visual experience. *The Quarterly Journal of Experimental Psychology, 63*, 1665–1670.

82. Searle, J. R. (1969). *Speech Acts*. Cambridge, England: Cambridge University Press.

83. Holtgraves, T. (2008b). Conversation, speech acts, and memory. *Memory & Cognition, 36*, 361–374.

84. Holtgraves, T. (2012). The role of the right hemisphere in speech act comprehension. *Brain and Language, 121(1)*, 58–64.

85. Singer, M., Graesser, A. C. & Trabasso, T. (1994). Minimal or global inference during reading. *Journal of Memory and Language, 33*, 421–441.

86. Millis, K. K. & Graesser, A. C. (1994). The time-course of constructing knowledge-based inferences for scientific texts. *Journal of Memory and Language, 33*, 583–599.

87. de Vega, M., Robertson, D. A., Glenberg, A. M., Kaschak, M. P. & Rinck, M. (2004). On doing two things at once: Temporal constraints on action in language comprehension. *Memory & Cognition, 32*, 1033–104; Radvansky, G. A., Zwaan, R. A., Federico, T. & Franklin, N. (1998). Retrieval from temporally organized situation models. *Journal of Experimental Psychology: Learning, Memory, and Cognition, 24*, 1224–1237.

88. Long, D. L. & De Ley, L. (2000). Implicit causality and discourse focus: The interaction of text and reader characteristics in pronoun resolution. *Journal of Memory and Language, 42*, 545–570.

89. Fletcher, C. R. & Bloom, C. P. (1988). Causal reasoning in the comprehension of simple narrative texts. *Journal of Memory and Language, 27*, 235–244.

90. Singer, M., Andrusiak, P., Reisdorf, P. & Black, N. L. (1992). Individual differences in bridging inference processes. *Memory & Cognition, 20*, 539–548.

91. Long, D. L., Oppy, B. J. & Seely, M. R. (1997). Individual differences in readers' sentence- and text-level representations. *Journal of Memory and Language, 36*, 129–145; Miyake, A., Just, M. A. & Carpenter, P. A. (1994). Working memory constraints on the resolution of lexical ambiguity: Maintaining multiple interpretations in neutral contexts. *Journal of Memory and Language, 33*, 175–202.

92. Virtue, S., van den Broek, P. & Linderholm, T. (2006). Hemispheric processing of inferences: The effects of textual constraint and working memory capacity. *Memory & Cognition, 34*, 1341–1354.

93. Zwaan, R. A., Langston, M. C. & Graesser, A. C. (1995). The construction of situation models in narrative comprehension: An event-indexing model. *Psychological Science, 6*, 292–297; Zwaan, R. A., Magliano, J. P. & Graesser, A. C. (1995). Dimensions of situation model construction in narrative comprehension. *Journal of Experimental Psychology: Learning, Memory, and Cognition, 21*, 386–397; Zwaan, R. A. & Radvansky, G. A. (1998). Situation models in language comprehension and memory. *Psychological Bulletin, 123*, 162–185.

94. Curiel, J. M. & Radvansky, G. A. (2014). Spatial and character situation model updating. *Journal of Cognitive Psychology, 26(2)*, 205–212.

95. Komeda, H. & Kusumi, T. (2006). The effect of a protagonist's emotional shift on situation model construction. *Memory & Cognition, 34*, 1548–1556.

96. Morrow, D. G., Greenspan, S. L., & Bower, G. H. (1987). Accessibility and situation models in narrative comprehension. *Journal of Memory and Language, 26*, 165–187.

97. Morrow, D. G., Greenspan, S. L., & Bower, G. H. (1987). Accessibility and situation models in narrative comprehension. *Journal of Memory and Language, 26*, 165–187.

98. Morrow, D. G., Greenspan, S. L., & Bower, G. H. (1987). Accessibility and situation models in narrative comprehension. *Journal of Memory and Language, 26*, 165–187.

99. Gennari, S. P. (2004). Temporal references and temporal relations in sentence comprehension. *Journal of Experimental Psychology: Learning, Memory, and Cognition, 30*, 877–890; Zwaan, R. A.

(1996). Processing narrative time shifts. *Journal of Experimental Psychology: Learning, Memory, and Cognition, 22*, 1196–1207.

100. Albrecht, J. E. & O'Brien, E. J. (1995). *Goal Processing and the Maintenance of Global Coherence*. Hillsdale, NJ: Erlbaum.

101. Lutz, M. E & Radvansky, G. A. (1997). The fate of completed goal information. *Journal of Memory and Language, 36*, 293–310; Suh, S. Y. & Trabasso, T. (1993). Inferences during reading: Converging evidence from discourse analysis, talk-aloud protocols, and recognition priming. *Journal of Memory and Language, 32*, 279–300.

102. Huff, M., Meitz, T. G. & Papenmeier, F. (2014). Changes in situation models modulate processes of event perception in audiovisual narratives. *Journal of Experimental Psychology: Learning, Memory, and Cognition, 40(5)*, 1377–1388; Pettijohn, K. A., Thompson, A. N., Tamplin, A. K., Krawietz, S. A. & Radvansky, G. A. (2016). Event boundaries and memory improvement. *Cognition, 148*, 136–144.

103. Sargent, J. Q., Zacks, J. M., Hambrick, D. Z., Zacks, R. T., Kurby, C. A., Bailey, H. R., . . . & Beck, T. M. (2013). Event segmentation ability uniquely predicts event memory. *Cognition, 129(2)*, 241–255.

104. Speer, N. K., Zacks, J. M. & Reynolds, J. R. (2007). Human brain activity time-locked to narrative event boundaries. *Psychological Science, 18*, 449–455.

105. Claus, B. & Kelter, S. (2006). Comprehending narratives containing flashbacks: Evidence for temporally organized representations. *Journal of Experimental Psychology: Learning, Memory, and Cognition, 32*, 1031–1044.

106. Cohn, N. (2013b). Visual narrative structure. *Cognitive Science, 37(3)*, 413–452.

107. Cohn, N. & Paczynski, M. (2013). Prediction, events, and the advantage of Agents: The processing of semantic roles in visual narrative. *Cognitive Psychology, 67(3)*, 73–97.

108. Cohn, N. (2013a). Navigating comics: An empirical and theoretical approach to strategies of reading comic page layouts. *Frontiers in Psychology, 4*, 186.

109. Cutting, J. E. (2014). Event segmentation and seven types of narrative discontinuity in popular movies. *Acta Psychologica, 149*, 69–77; Cutting, J. E., Brunick, K. L. & Candan, A. (2012). Perceiving event dynamics and parsing Hollywood films. *Journal of Experimental Psychology: Human Perception and Performance, 38(6)*, 1476–1490; Cutting, J. & Iricinschi, C. (2015). Representations of space in Hollywood movies: An event-indexing analysis. *Cognitive Science, 39(2)*, 434–456.

110. Magliano, J. P., Miller, J. & Zwaan, R. A. (2001). Indexing space and time in film understanding. *Applied Cognitive Psychology, 15*, 533–545; Magliano, J. P., Taylor, H. A. & Kim, H. J. (2005). When goals collide: Monitoring the goals of multiple characters. *Memory & Cognition, 33*, 1357–1367.

111. Zacks, J. M., Braver, T. S., Sheridan, M. A., Donaldson, D. I., Snyder, A. Z., Ollinger, J. M., et al. (2001). Human brain activity time-locked to perceptual event boundaries. *Nature Neuroscience, 4*, 651–655; Zacks, J. M., Speer, N. K., Swallow, K. M., Braver, T. S. & Reynolds; J. R. (2007). Event perception: A mind/brain perspective. *Psychological Bulletin, 133*, 273–293.

112. Magliano, J. P., Miller, J. & Zwaan, R. A. (2001). Indexing space and time in film understanding. *Applied Cognitive Psychology, 15*, 533–545.

113. Rips, L. J. (1998). Reasoning and conversation. *Psychological Review, 105*, 411–441.

114. Svartvik, J. & Quirk, R. (Eds.). (1980). *A Corpus of English Conversation*. Lund, Sweden: CWK Gleerup; Clark, H. H. (1994). Discourse in production. In M. A. Gernsbacher (Ed.), *Handbook of Psycholinguistics* (pp. 985–1021). San Diego, CA: Academic Press.

115. Duncan, S. (1972). Some signals and rules for taking speaking turns in conversations. *Journal of Personality and Social Psychology, 23*, 283–292.

116. Sacks, H., Schegloff, E. A. & Jefferson, G. (1974). A simplest systematics for the organization of turntaking for conversation. *Language, 50*, 696–735.

117. Cook, M. (1977). Gaze and mutual gaze in social encounters. *American Scientist, 65*, 328–333.

118. Wilson, M. & Wilson, T. P. (2005). An oscillator model of the timing of turn-taking. *Psychonomic Bulletin & Review, 12*, 957–968.

119. Kemper, S. & Thissen, D. (1981). Memory for the dimensions of requests. *Journal of Verbal Learning and Verbal Behavior, 20*, 552–563.

120. Blom, J. P. & Gumperz, J. J. (1972). Social meaning in linguistic structure: Code-switching in Norway. In J. J. Gumperz & D. Hymes (Eds.), *Directions in Sociolinguistics: The Ethnography of Communication* (pp. 407–434). New York, NY: Holt.

121. Brown, R. & Ford, M. (1961). Address in American English. *Journal of Abnormal and Social Psychology, 62*, 375–385; Edwards, D. & Potter, J. (1993). Language and causation: A discursive action model of description and attribution. *Psychological Review, 100*, 23–41; Holtgraves, T. (1994). Communication in context: Effects of speaker status on the comprehension of indirect requests. *Journal of Experimental Psychology: Learning, Memory, and Cognition, 20*, 1205–1218.

122. Van Berkum, J. J. A. (2008). Understanding sentences in context: What brain waves can tell us. *Current Directions in Psychological Science, 17*, 376–380.

123. Norman, D. A. & Rumelhart, D. E. (Eds.). (1975). *Explorations in Cognition*. San Francisco, CA: Freeman.

124. Brennan, S. E. & Clark, H. H. (1996). Conceptual pacts and lexical choice in conversation. *Journal of Experimental Psychology: Learning, Memory, and Cognition, 22*, 1482–1493; Wilkes-Gibbs, D. & Clark, H. H. (1992). Coordinating beliefs in conversation. *Journal of Memory and Language, 31*, 183–194.

125. Ferreira, V. S. & Dell, G. S. (2000). Effect of ambiguity and lexical availability on syntactic and lexical production. *Cognitive Psychology, 40*, 296–340.

126. Haywood, S. L., Pickering, M. J. & Branigan, H. P. (2005). Do speakers avoid ambiguities during dialogue? *Psychological Science, 16*, 362–366.

127. Clifton, C., Jr., Carlson, K., & Frazier, L. (2006). Tracking the what and why of speakers' choices: Prosodic boundaries and the length of constituents. *Psychonomic Bulletin & Review, 13*, 854–861.

128. Metzing, C. & Brennan, S. E. (2003). When conceptual pacts are broken: Partner-specific effects on the comprehension of referring expressions. *Journal of Memory and Language, 49*, 201–213.

129. Shintel, H., & Keysar, B. (2007). You said it before and you'll say it again: Expectations and consistency in communication. *Journal of Experimental Psychology: Learning, Memory, and Cognition, 33*, 357–369.

130. Goldin-Meadow, S. (1997). When gestures and words speak differently. *Psychological Science, 6*, 138–143; Kelly, S. D., Barr, D. J., Church, R. B. & Lynch, K. (1999). Offering a hand to pragmatic understanding: The role of speech and gesture in comprehension and memory. *Journal of Memory and Language, 40*, 577–592; Özyürek, S. (2002). Do speakers design their cospeech gestures for their addressees? The effects of addressee location on representational gestures. *Journal of Memory and Language, 46*, 688–704.

131. Grice, H. P. (1975). Logic and conversation. In P. Cole & J. L. Morgan (Eds.), Syntax and semantics: Vol. 3. Speech acts (pp. 41–58). New York, NY: Seminar Press.

132. Grice, H. P. (1975). Logic and conversation. In P. Cole & J. L. Morgan (Eds.), *Syntax and semantics: Vol. 3.* Speech acts (pp. 41–58). New York, NY: Seminar Press.

133. Engelhardt, P. E., Bailey, K. G. D. & Ferreira, F. (2006). Do speakers and listeners observe the Gricean maxim of quantity? *Journal of Memory and Language, 54*, 554–573.

134. Kumon-Nakamura, S., Glucksberg, S. & Brown, M. (1995). How about another piece of pie? The allusional pretense theory of discourse irony. *Journal of Experimental Psychology: General, 124*, 3–21.

135. Holtgraves, T. (2008a). Automatic intention recognition in conversation processing. *Journal of Memory and Language, 58*, 627–645.

136. Schank, R. C. (1977). Rules and topics in conversation. *Cognitive Science, 1*, 421–441; Litman, D. J. & Allen, J. F. (1987). A plan recognition model for subdialogues in conversation. *Cognitive Science, 11*, 163–200.

137. Fodor, J. A. (1992). A theory of the child's theory of mind. *Cognition, 44(3)*, 283–296. https://doi.org/10.1016/0010-0277(92)90004-2; Frith, C. and Frith, U. (2005) Theory of Mind. *Current Biology, 15*, R644-R645. https://doi.org/10.1016/j.cub.2005.08.041

138. De Rosnay, M. & Hughes, C. (2006). Conversation and theory of mind: Do children talk their way to socio-cognitive understanding?. *British Journal of Developmental Psychology, 24(1)*, 7–37.

139. DePaulo, B. M. & Bonvillian, J. D. (1978). The effect on language development of the special characteristics of speech addressed to children. *Journal of Psycholinguistic Research, 7*, 189–211; Snow, C. (1972). Mother's speech to children learning language. *Child Development, 43*, 549–565; Snow, C. & Ferguson, C. (Eds.). (1977). *Talking to Children: Language Input and Acquisition*. Cambridge, England: Cambridge University Press.

140. Horton, W. S. & Gerrig, R. J. (2002). Speakers' experiences and audience design: Knowing when and knowing how to adjust utterances to addressees. *Journal of Memory and Language, 47*, 589–606.

141. Lockridge, C. B. & Brennan, S. E. (2002). Addressees' needs influence speakers' early syntactic choices. *Psychonomic Bulletin & Review, 9*, 550–557.

142. Nickerson, R. S. (2001). The projective way of knowing: A useful heuristic that sometimes misleads. *Current Directions in Psychological Science, 10*, 168–172.

143. Clark, H. H. & Krych, M. A. (2004). Speaking while monitoring addressees for understanding. *Journal of Memory and Language, 50*, 62–81.

144. Marsh, E. J. (2007). Retelling is not the same as recalling. *Current Directions in Psychological Science, 16*, 16–20.

145. Holtgraves, T. (1994). Communication in context: Effects of speaker status on the comprehension of indirect requests. *Journal of Experimental Psychology: Learning, Memory, and Cognition, 20*, 1205–1218; Holtgraves, T. (1998). Interpreting indirect replies. *Cognitive Psychology, 37*, 1–27.

146. Clark, H. H. (1979). Responding to indirect speech acts. *Cognitive Psychology, 11*, 430–477.

147. Holtgraves, T. (1994). Communication in context: Effects of speaker status on the comprehension of indirect requests. *Journal of Experimental Psychology: Learning, Memory, and Cognition, 20*, 1205–1218.

148. Holtgraves, T. (1998). Interpreting indirect replies. *Cognitive Psychology, 37*, 1–27.

149. Holtgraves, T. (1998). Interpreting indirect replies. *Cognitive Psychology, 37*, 1–27.

150. Holtgraves, T. (1998). Interpreting indirect replies. *Cognitive Psychology, 37*, 1–27.

151. Lee, J. J. & Pinker, S. (2010). Rationales for indirect speech: The theory of the strategic speaker. *Psychological Review, 117*, 785–807.

152. Adesope, O. O., Lavin, T., Thompson, T. & Ungerleider, C. (2010). A systematic review and meta-analysis of the cognitive correlates of bilingualism. *Review of Educational Research, 80(2)*, 207–245.

153. Glucksberg, S., Gildea, P., & Bookin, H. B. (1982). On understanding nonliteral speech: Can people ignore metaphors? *Journal of Verbal Learning and Verbal Behavior, 21*(1), 85–98.

154. Glucksberg, S. (2003). The psycholinguistics of metaphor. *Trends in Cognitive Sciences, 7*(2), 92–96.

155. Gibbs, R. W., & Nayak, N. P. (1989). Psycholinguistic studies on the syntactic behavior of idioms. *Cognitive Psychology, 21*(1), 100–138.

156. Gibbs, R. W., Nayak, N. P., & Cutting, C. (1989). How to kick the bucket and not decompose: Analyzability and idiom processing. *Journal of Memory and Language, 28*(5), 576–593.

157. McNeill, D. (1992). *Hand and Mind: What Gestures Reveal about Thought*. Chicago, IL: University of Chicago Press.

158. Goldin-Meadow, S. (2006). Talking and thinking with our hands. *Current Directions in Psychological Science, 15*, 34 39.

159. Bavelas, J., Gerwing, J., Sutton, C. & Prevost, D. (2008). Gesturing on the telephone: Independent effects of dialogue and visibility. *Journal of Memory and Language, 58*, 495–520.

160. Bavelas, J., Gerwing, J., Sutton, C. & Prevost, D. (2008). Gesturing on the telephone: Independent effects of dialogue and visibility. *Journal of Memory and Language, 58*, 495–520.

161. Jacobs, N. & Garnham, A. (2007). The role of conversational hand gestures in a narrative task. *Journal of Memory and Language, 56(2)*, 291–303.

162. Goldin-Meadow, S. & Beilock, S. L. (2010). Action's influence on thought: The case of gesture. Perspectives on *Psychological Science, 5*, 664–674.

163. Broaders, S. C., Cook, S. W., Mitchell, Z. & Goldin-Meadow, S. (2007). Making children gesture brings out implicit knowledge and leads to learning. *Journal of Experimental Psychology: General, 136*, 539–550.

164. Lozano, S. C. & Tversky, B. (2006). Communicative gestures facilitate problem solving for both communicators and recipients. *Journal of Memory and Language, 55*, 47–63.

165. Hostetter, A. B. & Alibali, M. (2008). Visible embodiment: Gestures as simulated action. *Psychonomic Bulletin & Review, 15*, 495–514; Kelly, S. D., Özyürek, A. & Maris, E. (2010). Two sides of the same coin: Speech and gesture mutually interact to enhance comprehension. *Psychological Science, 21*, 260–267.

166. Cutica, I. & Bucciarelli, M. (2013). Cognitive change in learning from text: Gesturing enhances the construction of the text mental model. *Journal of Cognitive Psychology, 25(2)*, 201–209.

Chapter 10

1. Clarke, H. D., Goodwin, M. & Whiteley, P. (2017). Why Britain voted for Brexit: An individual-level analysis of the 2016 referendum vote. *Parliamentary Affairs, 70(3)*, 439–464.

2. Krawczyk, D. C. (2012). The cognition and neuroscience of relational reasoning. *Brain Research, 1428*, 13–23.

3. Markovits, H. & Potvin, F. (2001). Suppression of valid inferences and knowledge structures: The curious effect of producing alternative antecedents on reasoning with causal conditionals. *Memory & Cognition, 29*, 736–744.

4. Khemlani, S. & Johnson-Laird, P. N. (2012). Theories of the syllogism: A meta-analysis. *Psychological Bulletin, 138(3)*, 427–457.

5. Johnson-Laird, P. N., Khemlani, S. S. & Goodwin, G. P. (2015). Logic, probability, and human reasoning. *Trends in Cognitive Sciences, 19(4)*, 201–214.

6. Evans, J. St. B. T., Barston, J. L. & Pollard, P. (1983). On the conflict between logic and belief in syllogistic reasoning. *Memory & Cognition*, 11, 295–306; Copeland, D. E., Gunawan, K. & Bies-Hernandez, N. J. (2011). Source credibility and syllogistic reasoning. *Memory & Cognition, 39*, 117–127.

7. Schmidt, J. R. & Thompson, V. A. (2008). "At least one" problem with "some" formal reasoning paradigms. *Memory & Cognition, 36*, 217–229.

8. Newstead, S. E. (1995). Gricean implicatures and syllogistic reasoning. *Journal of Memory and Language, 34(5)*, 644–664.

9. Khemlani, S. & Johnson-Laird, P. N. (2012). Theories of the syllogism: A meta-analysis. *Psychological Bulletin, 138(3)*, 427–457.

10. Woodworth, R. S. & Sells, S. B. (1935). An atmosphere effect in formal syllogistic reasoning. *Journal of Experimental Psychology, 18(4)*, 451–460.

11. Wetherick, N. E. & Gilhooly, K. J. (1995). "Atmosphere," matching, and logic in syllogistic reasoning. *Current Psychology, 14(3)*, 169–178.

12. Chapman, L. J., & Chapman, J. P. (1959). Atmosphere effect re-examined. *Journal of Experimental Psychology, 58*(3), 220–226.

13. Chater, N. & Oaksford, M. (1999). The probability heuristics model of syllogistic reasoning. *Cognitive Psychology, 38*, 191–258.

14. Johnson-Laird, P. N. (1983). *Mental Models: Towards a Cognitive Science of Language, Inference and Consciousness*. Cambridge, MA: Harvard University

Press; Johnson-Laird, P. N. (2013). Mental models and cognitive change. *Journal of Cognitive Psychology, 25(2)*, 131–138.

15. Rips, L. J. (1994). The *Psychology of Proof: Deductive Reasoning in Human Thinking*. Cambridge, MA: MIT Press.

16. Knauff, M. (2013). *Space to Reason: A Spatial Theory of Human Thought*. MIT Press.

17. Knauff, M. & May, E. (2006). Mental imagery, reasoning, and blindness. *Quarterly Journal of Experimental Psychology, 59(1)*, 161-177.

18. Hamburger, K., Ragni, M., Karimpur, H., Franzmeier, I., Wedell, F. & Knauff, M. (2018). TMS applied to V1 can facilitate reasoning. *Experimental Brain Research, 236*, 2277–2286.

19. Hamburger, K., Ragni, M., Karimpur, H., Franzmeier, I., Wedell, F. & Knauff, M. (2018). TMS applied to V1 can facilitate reasoning. *Experimental Brain Research, 236*, 2277–2286.

20. Rips, L. J. & Marcus, S. L. (1977). Supposition and the analysis of conditional sentences. In M. A. Just & P. A. Carpenter (Eds.), *Cognitive Processes in Comprehension* (pp. 185–220). Hillsdale, NJ: Erlbaum.

21. Wertheim, J. & Ragni, M. (2020). The neurocognitive correlates of human reasoning: A meta-analysis of conditional and syllogistic inferences. *Journal of Cognitive Neuroscience, 32(6)*, 1061–1078.

22. Bonnefon, J. F. & Hilton, D. J. (2004). Consequential conditionals: Invited and suppressed inferences from valued outcomes. *Journal of Experimental Psychology: Learning, Memory, and Cognition, 30*, 28–37.

23. Rader, A.W., & Sloutsky, V. M. (2002). Processing of logically valid and logically invalid conditional inferences in discourse comprehension. *Journal of Experimental Psychology: Learning, Memory, and Cognition, 28*, 59–68.

24. Rader, A. W. & Sloutsky, V. M. (2002). Processing of logically valid and logically invalid conditional inferences in discourse comprehension. *Journal of Experimental Psychology: Learning, Memory, and Cognition, 28*, 59–68.

25. Evans, J. St. B. T., Handley, S. J., Harper, C. N. J. & Johnson- Laird, P. H. (1999). Reasoning about necessity and possibility: A test of the mental model theory of deduction. *Journal of Experimental Psychology: Learning, Memory, and Cognition, 25*, 1495–1513.

26. Koriat, A., Fiedler, K. & Bjork, R. A. (2006). Inflation of conditional predictions. *Journal of Experimental Psychology: General, 135*, 429–447.

27. Wason, P. C. & Johnson-Laird, P. N. (1972). *Psychology of Reasoning: Structure and Content*. Cambridge, MA: Harvard University Press.

28. Klauer, K. C., Stahl, C. & Erdfelder, E. (2007). The abstract selection task: New data and an almost comprehensive model. *Journal of Experimental Psychology: Learning, Memory, and Cognition, 33*, 680–703.

29. Johnson-Laird, P. N., Legrenzi, P. & Legrenzi, M. S. (1972). Reasoning and a sense of reality. British *Journal of Psychology, 63*, 395–400.

30. Blanchette, I. & Richards, A. (2004). Reasoning about emotional and neutral materials. *Psychological Science, 15*, 745–752.

31. Johnson-Laird, P. N. & Byrne, R. M. J. (2002). Conditionals: A theory of meaning, pragmatics, and inference. *Psychological Review, 109*, 646–678; Johnson-Laird, P. N., Byrne, R. M. J. & Schaeken, W. (1992). Propositional reasoning by model. *Psychological Review, 99*, 418–439.

32. Vergauwe, E., Gauffroy, C., Morsanyi, K., Dagry, I. & Barrouillet, P. (2013). Chronometric evidence for the dual-process mental model theory of conditional. *Journal of Cognitive Psychology, 25(2)*, 174–182.

33. De Neys, W., Schaeken, W. & d'Ydewalle, G. (2005). Working memory and everyday conditional reasoning: Retrieval and inhibition of stored counterexamples. *Thinking & Reasoning, 11(4)*, 349–381.

34. Thompson, V. A. & Byrne, R. M. J. (2002). Reasoning counterfactually: Making inferences about things that didn't happen. *Journal of Experimental Psychology: Learning, Memory, and Cognition, 28*, 1154–1170.

35. Evans, J. St. B. T., Handley, S. J., Harper, C. N. J. & Johnson- Laird, P. H. (1999). Reasoning about necessity and possibility: A test of the mental model theory of deduction. *Journal of Experimental Psychology: Learning, Memory, and Cognition, 25*, 1495–1513.

36. Espino, O., Santamaria, C. & Byrne, R. M. (2009). People think about what is true for conditionals, not what is false: Only true possibilities prime the comprehension of "if." *The Quarterly Journal of Experimental Psychology, 62(6)*, 1072–1078.

37. Cummins, D. D., Lubart, T., Alksnis, O. & Rist, R. (1991). Conditional reasoning and causation. *Memory & Cognition, 19*, 274–282.

38. Klayman, J. & Ha, Y.-W. (1989). Hypothesis testing in rule discovery: Strategy, structure, and content. *Journal of Experimental Psychology: Learning, Memory, and Cognition, 15*, 596–604.

39. Medin, D. L., Coley, J. D., Storms, G. & Hayes, B. K. (2003). A relevance theory of induction. *Psychonomic Bulletin & Review, 10*, 517–532; Rips, L. J. (2001). Two kinds of reasoning. *Psychological Science, 12*, 129–134.

40. Koslowski, B., Marasia, J., Vermeylen, F. & Hendrix, V. (2013). A disconfirming strategy is not necessarily better than a confirming strategy. *American Journal of Psychology, 126(3)*, 335–354.

41. Voorspoels, W., Navarro, D. J., Perfors, A., Ransom, K. & Storms, G. (2015). How do people learn from negative evidence? Non-monotonic generalizations and sampling assumptions in inductive reasoning. *Cognitive Psychology, 81*, 1–25.

42. Ashcraft, M. H. (1989). *Human memory and cognition*. Glenview, IL: Scott Foresman.

43. Tversky, A. & Kahneman, D. (1973). Availability: A heuristic for judging frequency and probability. *Cognitive Psychology, 5*, 207–232; Tversky, A. & Kahneman, D. (1974). Judgment under uncertainty: Heuristics and biases. *Science, 185*, 1124–1131; Tversky, A. & Kahneman, D. (1980). Causal schemas in judgments under uncertainty. In M. Fishbein (Ed.), *Progress in Social Psychology* (Vol. 1, pp. 49–72). Hillsdale, NJ: Erlbaum; Kahneman, D., Slovic, P. & Tversky, A. (Eds.). (1982). *Judgment under Uncertainty: Heuristics and Biases.* Cambridge, England: Cambridge University Press; Kahneman, D. & Tversky, A. (1972). Subjective probability: A judgment of representativeness. *Cognitive Psychology, 3*, 430–454. Kahneman, D. & Tversky, A. (1973). On the psychology of prediction. *Psychological Review, 80*, 237–251; Shafir, E. & Tversky, A. (1992). Thinking through uncertainty: Nonconsequential reasoning and choice. *Cognitive Psychology, 24*, 449–474.

44. Kahneman, D. (2003a). A perspective on judgment and choice: Mapping bounded rationality. *American Psychologist, 58*, 697–720; Kahneman, D. (2003b). Experiences of collaborative research. *American Psychologist, 58*, 723–730; Kahneman, D. & Tversky, A. (Eds.). (2000). *Choice, Values, and Frames.* New York, NY: Cambridge University Press.

45. Agnoli, F. (1991). Development of judgmental heuristics and logical reasoning: Training counteracts the representativeness heuristic. *Cognitive Development, 6*, 195–217; Agnoli, F. & Krantz, D. H. (1989). Suppressing natural heuristics by formal instruction: The case of the conjunction fallacy. *Cognitive Psychology, 21*, 515–550; Fong, G. T. & Nisbett, R. E. (1991). Immediate and delayed transfer of training effects in statistical reasoning. *Journal of Experimental Psychology: General, 120*, 34–45; Lehman, D. R., Lempert, R. O. & Nisbett, R. E. (1988). The effects of graduate training on reasoning. *American Psychologist, 43*, 431–442.

46. Windschitl, P. D. & Weber, E. U. (1999). The interpretation of "likely" depends on the context, but "70%" is 70%—right? The influence of associative processes on perceived certainty. *Journal of Experimental Psychology: Learning, Memory, and Cognition, 25*, 1514–1533.

47. Bar-Hillel, M. (1980). What features make samples seem representative? *Journal of Experimental Psychology: Human Perception and Performance, 6*, 578–589.

48. Burns, B. D. & Corpus, B. (2004). Randomness and inductions from streaks: "Gambler's fallacy" versus "hot hand." *Psychonomic Bulletin & Review, 11*, 179–184.

49. Hahn, U. & Warren, P. A. (2009). Perceptions of randomness: Why three heads are better than four. *Psychological Review, 116*, 454–461.

50. Pollatsek, A., Konold, C. E., Well, A. D. & Lima, S. D. (1984). Beliefs underlying random sampling. *Memory & Cognition, 12*, 395–401.

51. Kahneman, D. & Tversky, A. (1972). Subjective probability: A judgment of representativeness. *Cognitive Psychology, 3*, 430–454.

52. Johnson, J. T., & Finke, R. A. (1985). The base-rate fallacy in the context of sequential categories. *Memory & Cognition, 13*, 63–73.

53. Kahneman, D., & Tversky, A. (1973). On the psychology of prediction. *Psychological Review, 80*, 237–251.

54. Fischhoff, B. & Bar-Hillel, M. (1984). Diagnosticity and the base-rate effect. *Memory & Cognition, 12*, 402–410; Griffin, D. & Tversky, A. (1992). The weighing of evidence and the determinants of confidence. *Cognitive Psychology, 24*, 411–435.

55. Brown, N. R. & Siegler, R. S. (1992). The role of availability in the estimation of national populations. *Memory & Cognition, 20*, 406–412; Hasher, L. & Zacks, R. T. (1984). Automatic processing of fundamental information: The case of frequency of occurrence. *American Psychologist, 39*, 1372–1388.

56. Brown, A. S. & Nix, L. A. (1996). Turning lies into truths: Referential validation of falsehoods. *Journal of Experimental Psychology: Learning, Memory, and Cognition, 22*, 1088–1100.

57. Tversky, A. & Kahneman, D. (1973). Availability: A heuristic for judging frequency and probability. *Cognitive Psychology, 5*, 207–232.

58. Kahneman, D. (2003a). A perspective on judgment and choice: Mapping bounded rationality. *American Psychologist, 58*, 697–720.

59. Song, H. & Schwarz, N. (2009). If it's difficult to pronounce, it must be risky: Fluency, familiarity, and risk perception. *Psychological Science, 20*, 135–138.

60. Hertwig, R., Barron, G., Weber, E. U., & Erev, I. (2004). Decision from experience and the effect of rare events in risky choice. *Psychological Science, 15*, 534–539.

61. Gigerenzer, G. (2004). Dread risk, September 11, and fatal traffic accidents. *Psychological Science, 15*, 286–287.

62. Kahneman, D. & Tversky, A. (1982). The simulation heuristic. In D. Kahneman, P. Slovic & A. Tversky (Eds.), *Judgment under uncertainty: Heuristics and biases* (pp. 201–208). Cambridge, England: Cambridge University Press.

63. Tversky, A. (1972). Elimination by aspects: A theory of choice. *Psychological Review, 79(4)*, 281–299.

64. Roese, N. J. & Epstude, K. (2017). The functional theory of counterfactual thinking: New evidence, new challenges, new insights. In *Advances in Experimental Social Psychology* (Vol. 56, pp. 1-79). Academic Press.

65. Byrne, R. M. J. & McEleney, A. (2000). Counterfactual thinking about actions and failures to act. *Journal of Experimental Psychology: Learning, Memory, and Cognition, 26*, 1318–1331.

66. McCloy, R. & Byrne, R. M. (2000). Counterfactual thinking about controllable events. *Memory & Cognition, 28*, 1071– 1078.

67. Koehler, J. J. & Macchi, L. (2004). Thinking about low-probability events: An exemplar-cuing theory. *Psychological Science, 15*, 540–546.

68. Roese, N. J. (1997). Counterfactual thinking. *Psychological Bulletin, 121*, 133–148; Reyna, V. F. (2004). How people make decisions that involve risk. *Current Directions in Psychological Science, 13*, 60–67; Trabasso, T. & Bartolone, J. (2003). Story understanding and counterfactual reasoning. *Journal of Experimental Psychology: Learning, Memory, and Cognition, 29*, 904–923.

69. Goldinger, S. D., Kleider, H. M., Azuma, T. & Beike, D. R. (2003). "Blaming the victim" under memory load. *Psychological Science, 14*, 81–85.

70. McCloy, R. & Byrne, R. M. (2000). Counterfactual thinking about controllable events. *Memory & Cognition, 28*, 1071–1078.

71. Goldinger, S. D., Kleider, H. M., Azuma, T. & Beike, D. R. (2003). "Blaming the victim" under memory load. *Psychological Science, 14*, 81–85.

72. Miller, D. T. & McFarland, C. (1986). Counterfactual thinking and victim compensation: A test of norm theory. *Personality and Social Psychology Bulletin, 12*, 513–519.

73. Burns, Z. C., Caruso, E. M. & Bartels, D. M. (2012). Predicting premeditation: Future behavior is seen as more intentional than past behavior. *Journal of Experimental Psychology: General, 141(2)*, 227–232.

74. Ferrante, D., Girotto, V., Straga, M. & Walsh, C. (2013). Improving the past and the future: A temporal asymmetry in hypothetical thinking. *Journal of Experimental Psychology: General, 142(1)*, 23–27.

75. Fischhoff, B. & Bar-Hillel, M. (1984). Diagnosticity and the base-rate effect. *Memory & Cognition, 12*, 402–410.

76. Kneer, M. & Skocze⊠, I. (2023). Outcome effects, moral luck and the hindsight bias. *Cognition, 232*, 105258; Sanna, L. J. & Schwartz, N. (2006). Human judgment: The case of the hindsight bias and its debiasing. *Current Directions in Psychological Science, 15*, 172–176.

77. Hell, W., Gigerenzer, G., Gauggel, S., Mall, M. & Muller, M. (1988). Hindsight bias: An interaction of automatic and motivational factors? *Memory & Cognition, 16*, 533–538; Hoch, S. J. & Loewenstein, G. F. (1989). Outcome feedback: Hindsight and information. *Journal of Experimental Psychology: Learning, Memory, and Cognition, 15*, 605–619.

78. Gray, R., Beilock, S. L. & Carr, T. H. (2007). "As soon as the bat met the ball, I knew it was gone": Outcome prediction, hindsight bias, and the representation and control of action in expert and novice baseball players. *Psychonomic Bulletin & Review, 14*, 669–675.

79. Erdfelder, E. & Buchner, A. (1998). Decomposing the hindsight bias: A multinomial processing tree model for separating recollection and reconstruction in hindsight. *Journal of Experimental Psychology: Learning, Memory, and Cognition, 24*, 387–414; Holyoak, K. J. & Simon, D. (1999). Bidirectional reasoning in decision making by constraint satisfaction. *Journal of Experimental Psychology: General, 128*, 3–31.

80. Carli, L. L. (1999). Cognitive reconstruction, hindsight, and reactions to victims and perpetrators. *Personality and Social Psychology Bulletin, 25*, 966–979.

81. Harley, E. M., Carlsen, K. A. & Loftus, G. R. (2004). The "sawit-all-along" effect: Demonstrations of visual hindsight bias. *Journal of Experimental Psychology: Learning, Memory, and Cognition, 30*, 960–968; Mather, M., Shafir, E. & Johnson, M. K. (2000). Misremembrance of options past: Source monitoring and choice. *Psychological Science, 11*, 132–138.

82. Hoch, S. J. (1984). Availability and interference in predictive judgment. *Journal of Experimental Psychology: Learning, Memory, and Cognition, 10*, 649–662; Hoch, S. J. (1985). Counterfactual reasoning and accuracy in predicting personal events. *Journal of Experimental Psychology: Learning, Memory, and Cognition, 11*, 719–731.

83. Petrusic, W. M. & Baranski, J. V. (2003). Judging confidence influences decision processing in comparative judgments. *Psychonomic Bulletin & Review, 10*, 177–183.

84. Shepperd, J. A., Waters, E. A., Weinstein, N. D., & Klein, W. M. (2015). A primer on unrealistic optimism. *Current Directions in Psychological Science, 24(3)*, 232–237.

85. Sharot, T., Korn, C. W., & Dolan, R. J. (2011). How unrealistic optimism is maintained in the face of reality. *Nature Neuroscience, 14(11)*, 1475-1479.

86. Sharot, T., & Garrett, N. (2016). Forming beliefs: Why valence matters. Trends in *Cognitive Sciences, 20(1)*, 25–33; Sharot, T., Korn, C. W., & Dolan, R. J. (2011). How unrealistic optimism is maintained in the face of reality. *Nature Neuroscience*, 14(11), 1475–1479.

87. Sharot, T., Korn, C. W., & Dolan, R. J. (2011). How unrealistic optimism is maintained in the face of reality. Nature *Neuroscience, 14(11)*, 1475–1479.

88. Li, M. & Chapman, G. B. (2009). "100% of anything looks good": The appeal of one hundred percent. *Psychonomic Bulletin & Review, 16(1)*, 156–162.

89. Whitney, P., Rinehart, C. A., & Hinson, J. M. (2008). Framing effects under cognitive load: The role of working memory in risky decisions. *Psychonomic Bulletin & Review, 15*(6), 1179–1184.

90. Tversky, A. & Kahneman, D. (1981). The framing of decisions and the psychology of choice. *Science, 211*, 453–458.

91. Biswas, D. (2009). The effects of option framing on consumer choices: Making decisions in rational versus experiential processing modes. *Journal of Consumer Behaviour, 8(5)*, 284–299.

92. Gigerenzer, G., Reb, J. & Luan, S. (2022). Smart heuristics for individuals, teams, and organizations. *Annual Review of Organizational Psychology and Organizational Behavior, 9*, 171-198.

93. Gigerenzer, G. (1996). On narrow norms and vague heuristics: A reply to Kahneman and Tversky. *Psychological Review, 103*, 592–596; Dougherty, M. R., Franco-Watkins, A. M. & Thomas, R. (2008). Psychological plausibility of the theory of probabilistic mental models and the fast and frugal heuristics. *Psychological Review, 115(1)*, 199–213; Hilbig, B. E. (2010). Reconsidering "evidence" for fast-and-frugal heuristics. *Psychonomic Bulletin & Review, 17(6)*, 923–930.

94. Gigerenzer, G. (2008). Why heuristics work. *Perspectives in Psychological Science*, 3, 20–29.

95. Gigerenzer, G. (2008). Why heuristics work. *Perspectives in Psychological Science, 3*, 20–29.

96. Gigerenzer, G. (2008). Why heuristics work. *Perspectives in Psychological Science, 3*, 20–29.

97. Hahn, U. & Warren, P. A. (2009). Perceptions of randomness: Why three heads are better than four. *Psychological Review, 116*, 454–461.

98. Simon, H. A. (1979). *Models of Thought*. New Haven, CT: Yale University Press.

99. Artinger, F. M., Gigerenzer, G., & Jacobs, P. (2022). Satisficing: Integrating two traditions. *Journal of Economic Literature, 60(2)*, 598-635.

100. Gigerenzer, G. & Goldstein, D. G. (1996). Reasoning the fast and frugal way: Models of bounded rationality. *Psychological Review, 104*, 650–669.

101. Bröder, A. & Schiffer, S. (2006). Adaptive flexibility and maladaptive routines in selecting fast and frugal decision strategies. *Journal of Experimental Psychology: Learning, Memory, and Cognition, 32*, 904–918.

102. Newell, B. R., & Shanks, D. R. (2004). On the role of recognition in decision making. *Journal of Experimental Psychology: Learning, Memory, and Cognition, 30*, 923–935.

103. Goldstein, D. G. & Gigerenzer, G. (2002). Models of ecological rationality: The recognition heuristic. *Psychological Review, 109*, 75–90.

104. Newell, B. R. & Shanks, D. R. (2003). Take the best or look at the rest? Factors influencing "one-reason" decision making. *Journal of Experimental Psychology: Learning, Memory, and Cognition, 29*, 53–65.

105. Gigerenzer, G. & Goldstein, D. G. (1996). Reasoning the fast and frugal way: Models of bounded rationality. *Psychological Review, 104*, 650–669; Gigerenzer, G. (1996). On narrow norms and vague heuristics: A reply to Kahneman and Tversky. *Psychological Review, 103*, 592–596; Goldstein, D. G. & Gigerenzer, G. (2002). Models of ecological rationality: The recognition heuristic. *Psychological Review, 109*, 75–90.

106. Burns, B. D. (2004). Heuristics as beliefs and as behaviors: The adaptiveness of the "hot hand." *Cognitive Psychology, 48*, 295–331; Gilovich, T., Vallone, R. & Tversky, A. (1985). The hot hand in basketball: On the misperception of random sequences. *Cognitive Psychology, 17*, 295–314.

107. Tversky, A. & Kahneman, D. (1983). Extensional versus intuitive reasoning: The conjunction fallacy in probability judgment. *Psychological Review, 90*, 293–315.

108. Moldoveanu, M. & Langer, E. (2002). False memories of the future: A critique of the applications of probabilistic reasoning to the study of cognitive processes. *Psychological Review, 109*, 358–375.

109. Garcia-Retamero, R., Wallin, A. & Dieckmann, A. (2007). Does causal knowledge help us be faster and more frugal in our decisions? *Memory & Cognition, 35*, 1399–1409; Hagmayer, Y. & Sloman, S. A. (2009). Decision makers conceive of their choices as interventions. *Journal of Experimental Psychology: General, 138*, 22–38; Kim, N. S., Yopchick, J. E. & de Kwaadsteniet, L. (2008). Causal diversity effects in information seeking. *Psychonomic Bulletin & Review, 15*, 81–88; Kynski, T. R. & Tenebaum, J. B. (2007). The role of causality in judgment under uncertainty. *Journal of Experimental Psychology: General, 136*, 430–450.

110. Kynski, T. R. & Tenebaum, J. B. (2007). The role of causality in judgment under uncertainty. *Journal of Experimental Psychology: General, 136*, 430–450.

111. Kynski, T. R., & Tenebaum, J. B. (2007). The role of causality in judgment under uncertainty. *Journal of Experimental Psychology: General, 136*, 430–450.

112. Chater, N., Tenenbaum, J. B. & Yuille, A. (2006). Probabilistic models of cognition: Conceptual foundations. *Trends in Cognitive Sciences, 10(7)*, 287–291.

113. Hoffrage, U., Krauss, S., Martignon, L. & Gigerenzer, G. (2015). Natural frequencies improve Bayesian reasoning in simple and complex inference tasks. *Frontiers in Psychology, 6*, 1473.

114. Bowers, J. S. & Davis, C. J. (2012). Bayesian just-so stories in psychology and neuroscience. *Psychological Bulletin, 138(3)*, 389–414.

115. Hagmayer, Y. & Sloman, S. A. (2009). Decision makers conceive of their choices as interventions. *Journal of Experimental Psychology: General, 138*, 22–38.

116. Evans, J. St. B. T., Handley, S. J., Over, D. E. & Perham, N. (2002). Background beliefs in Bayesian inference. *Memory & Cognition, 30*, 179–190.

117. Kim, N. S. & Ahn, W. (2002). Clinical psychologists' theory based representations of mental disorders predict their diagnostic reasoning and memory. *Journal of Experimental Psychology: General, 131*, 451–476.

118. Moshinsky, A. & Bar-Hillel, M. (2002). Where did 1850 happen first—in America or in Europe? A cognitive account for a historical bias. *Psychological Science, 13*, 20–26.

119. Brainerd, C. J., Wang, Z., Reyna, V. F. & Nakamura, K. (2015). Episodic memory does not add up: Verbatim–gist superposition predicts violations of the additive law of probability. *Journal of Memory and Language, 84*, 224–245.

120. Aerts, D., Sozzo, S. & Veloz, T. (2015). Quantum structure of negation and conjunction in human thought. *Frontiers in Psychology, 6*, 1447.

121. Bruza, P. D., Wang, Z. & Busemeyer, J. R. (2015). Quantum cognition: A new theoretical approach to psychology. *Trends in Cognitive Sciences, 19(7)*, 383–393; Busemeyer, J. R. & Wang, Z. (2015). What is quantum cognition, and how is it applied to psychology? *Current Directions in Psychological Science, 24(3)*, 163–169; Moreira, C. & Wichert, A. (2016). Quantum-like Bayesian networks for modeling decision making. *Frontiers in Psychology, 7*, 11.

122. Bruza, P. D., Wang, Z. & Busemeyer, J. R. (2015). Quantum cognition: A new theoretical approach to psychology. *Trends in Cognitive Sciences, 19(7)*, 383–393.

123. Yukalov, V. I. & Sornette, D. (2015). Preference reversal in quantum decision theory. *Frontiers in Psychology, 6*, 1538.

124. Kempton, W. (1986). Two theories of home heat control. *Cognitive Science, 10*, 75–90.

125. Shafir, E. & Tversky, A. (1992). Thinking through uncertainty: Nonconsequential reasoning and choice. *Cognitive Psychology, 24*, 449–474.

126. Gentner, D. & Collins, A. (1981). Studies of inference from lack of knowledge. *Memory & Cognition, 9*, 434–443; Glucksberg, S. & McCloskey, M. (1981). Decisions about ignorance: Knowing that you don't know. *Journal of Experimental Psychology: Human Learning and Memory, 7*, 311–325.

127. Krist, H., Fieberg, E. L. & Wilkening, F. (1993). Intuitive physics in action and judgment: The development of knowledge about projectile motion. *Journal of Experimental Psychology: Learning, Memory, and Cognition, 19*, 952–966; Schwartz, D. L. & Black, T. (1999). Inferences through imagined actions: Knowing by simulated doing. *Journal of Experimental Psychology: Learning, Memory, and Cognition, 25*, 116–136.

128. Catrambone, R., Jones, C. M., Jonides, J. & Seifert, C. (1995). Reasoning about curvilinear motion: Using principles or analogy. *Memory & Cognition, 23*, 368–373; Cooke, N. J. & Breedin, S. D. (1994). Constructing naïve theories of motion on the fly. *Memory & Cognition, 22*, 474–493; Hubbard, T. L. (1996). Representational momentum, centripetal force, and curvilinear impetus. *Journal of Experimental Psychology: Learning, Memory, and Cognition, 22*, 1049–1060.

129. McCloskey, M. (1983). Naive theories of motion. In D. Gentner & A. L. Stevens (Eds.), Mental models (pp. 299–324). Hillsdale, NJ: Erlbaum; Rohrer, D. (2003). The natural appearance of unnatural incline speed. *Memory & Cognition, 31*, 816–826.

130. Proffitt, D. R., Kaiser, M. K. & Whelan, S. M. (1990). Understanding wheel dynamics. *Cognitive Psychology, 22*, 342–373.

131. Medin, D. L. & Edelson, S. M. (1988). Problem structure and the use of base-rate information from experience. *Journal of Experimental Psychology: General, 117*, 68–85.

132. Kozhevnikov, M. & Hegarty, M. (2001). Impetus beliefs as default heuristics: Dissociation between explicit and implicit knowledge about motion. *Psychonomic Bulletin & Review, 8*, 439–453.

133. Agnoli, F. & Krantz, D. H. (1989). Suppressing natural heuristics by formal instruction: The case of the conjunction fallacy. *Cognitive Psychology, 21, 515–550*; Fong, G. T. & Nisbett, R. E. (1991). Immediate and delayed transfer of training effects in statistical reasoning. *Journal Of Experimental Psychology: General, 120*, 34–45; Lehman, D. R., Lempert, R. O. & Nisbett, R. E. (1988). The effects of graduate training on reasoning. *American Psychologist, 43*, 431–442.

134. Lawson, R. (2006). The science of cycology: Failures to understand how everyday objects work. *Memory & Cognition, 34*, 1667–1675.

135. Lawson, R. (2006). The science of cycology: Failures to understand how everyday objects work. *Memory & Cognition, 34*, 1667–1675.

136. Cherniak, C. (1984). Prototypicality and deductive reasoning. *Journal of Verbal Learning and Verbal Behavior, 23*, 625–642; Tversky, A. & Kahneman, D. (1983). Extensional versus intuitive reasoning: The conjunction fallacy in probability judgment. *Psychological Review, 90*, 293–315.

137. Tversky, A. & Kahneman, D. (1983). Extensional versus intuitive reasoning: The conjunction fallacy in probability judgment. *Psychological Review, 90*, 293–315.

138. Waltz, J. A., Knowlton, B. J., Holyoak, K. J., Boone, K. B., Mishkin, F. S., Santos, M. M., et al. (1999). A system for relational reasoning in human prefrontal cortex. *Psychological Science, 10*, 119–125.

139. Raven, J. C. (1941). Standardization of progressive matrices, 1938. *British Journal of Medical Psychology, 19*, 137–150.

140. Waltz, J. A., Knowlton, B. J., Holyoak, K. J., Boone, K. B. Mishkin, F. S., Santos, M. M., et al. (1999). A system for relational reasoning in human prefrontal cortex. *Psychological Science, 10*, 119–125.

141. Wagar, B. M. & Thagard, P. (2004). Spiking Phineas Gage: A neurocomputational theory of cognitive-affective integration in decision making. *Psychological Review, 111*, 67–79.

142. Waltz, J. A., Knowlton, B. J., Holyoak, K. J., Boone, K. B., Mishkin, F. S., Santos, M. M., et al. (1999). A system for relational reasoning in human prefrontal

cortex. *Psychological Science, 10*, 119–125; Hinson, J. M., Jameson, T. L. & Whitney, P. (2003). Impulsive decision making and working memory. *Journal of Experimental Psychology: Learning, Memory, and Cognition, 29*, 298–306.

143. Kordzadeh, N. & Ghasemaghaei, M. (2022). Algorithmic bias: review, synthesis, and future research directions. *European Journal of Information Systems, 31(3)*, 388-409.

144. https://www.ibm.com/blog/shedding-light-on-ai-bias-with-real-world-examples/

145. Baker, R. S. & Hawn, A. (2022). Algorithmic bias in education. International *Journal of Artificial Intelligence in Education*, 1-41.

Chapter 11

1. Chronicle, E. P., MacGregor, J. N. & Ormerod, T. C. (2004). What makes an insight problem? The roles of heuristics, goal conception, and solution recoding in knowledge-lean problems. *Journal of Experimental Psychology: Learning, Memory, and Cognition*, 30(1), 14.

2. Newell, A. & Simon, H. A. (1972). *Human Problem Solving*. Englewood Cliffs, NJ: Prentice Hall.

3. Anderson, J. R. (1985). *Cognitive Psychology and its Implications* (2nd ed.). New York, NY: Freeman.

4. VanLehn, K. (1989). Problem solving and cognitive skill acquisition. In M. I. Posner (Ed.), *Foundations of Cognitive Science* (pp. 527–579). Cambridge, MA: MIT Press.

5. VanLehn, K. (1989). Problem solving and cognitive skill acquisition. In M. I. Posner (Ed.), *Foundations of Cognitive Science* (pp. 527–579). Cambridge, MA: MIT Press.

6. Wickelgren, W. A. (1974). *How to solve problems*. San Francisco, CA: Freeman.

7. VanLehn, K. (1989). Problem solving and cognitive skill acquisition. In M. I. Posner (Ed.), *Foundations of Cognitive Science* (pp. 527–579). Cambridge, MA: MIT Press.

8. Ormerod, T. C., MacGregor, J. N., Chronicle, E. P., Dewald, A. D. & Chu, Y. (2013). Act first, think later: The presence and absence of inferential planning in problem solving. *Memory & Cognition*, 41(7), 1096–1108.

9. Greeno, J. G. (1978). Natures of problem-solving abilities. In W. K. Estes (Ed.), *Handbook of Learning and Cognitive Processes: Vol. 5. Human information processing* (pp. 239–270). Hillsdale, NJ: Erlbaum.

10. Greeno, J. G. (1978). Natures of problem-solving abilities. In W. K. Estes (Ed.), *Handbook Of Learning and Cognitive Processes: Vol. 5. Human Information Processing* (pp. 239–270). Hillsdale, NJ: Erlbaum.

11. Bartley, J. E., Boeving, E. R., Riedel, M. C., Bottenhorn, K. L., Salo, T., Eickhoff, S. B., ... & Laird, A. R. (2018). Meta-analytic evidence for a core problem solving network across multiple representational domains. *Neuroscience & Biobehavioral Reviews*, 92, 318–337.

12. Bartley, J. E., Boeving, E. R., Riedel, M. C., Bottenhorn, K. L., Salo, T., Eickhoff, S. B., ... & Laird, A. R. (2018). Meta-analytic evidence for a core problem solving network across multiple representational domains. *Neuroscience & Biobehavioral Reviews*, 92, 318–337.

13. Boring, E. G. (1950). *A History of Experimental Psychology* (2nd ed.). New York, NY: Appleton-Century-Crofts. Boroditsky, L. (2001). Does language shape thought?

14. Boring, E. G. (1950). *A History of Experimental Psychology* (2nd ed.). New York, NY: Appleton-Century-Crofts. Boroditsky, L. (2001). Does language shape thought?

15. KÖhler, W. (1927). *The Mentality of Apes*. New York, NY: Harcourt, Brace.

16. Maier, N. R. F. (1931). Reasoning in humans: II. The solution of a problem and its appearance in consciousness. *Journal of Comparative Psychology*, 12, 181–194.

17. Duncker, K. (1945). On problem solving. *Psychological Monographs*, 58(Whole no. 270).

18. Greenspan, S. L. (1986). Semantic flexibility and referential specificity of concrete nouns. *Journal of Memory and Language, 25*, 539–557.

19. Maier, N. R. F. (1931). Reasoning in humans: II. The solution of a problem and its appearance in consciousness. *Journal of Comparative Psychology*, 12, 181–194.

20. Duncker, K. (1945). On problem solving. *Psychological Monographs*, 58(Whole no. 270).

21. McCaffrey, T. (2012). Innovation relies on the obscure: A key to overcoming the classic problem of functional fixedness. *Psychological Science*, 23(3), 215-218.

22. Luchins, A. S. (1942). Mechanization in problem solving. *Psychological Monographs*, 54(Whole no. 248).

23. Luchins, A. S. (1942). Mechanization in problem solving. *Psychological Monographs*, 54(Whole no. 248).

24. Greeno, J. G. (1978). Natures of problem-solving abilities. In W. K. Estes (Ed.), *Handbook of Learning and Cognitive Processes: Vol. 5. Human information processing* (pp. 239–270). Hillsdale, NJ: Erlbaum.

25. Bilalić, M., McLeod, P. & Gobet, F. (2008). Inflexibility of experts—Reality or myth? Quantifying the Einstellung effect in chess masters. *Cognitive Psychology*, 56, 73–102.

26. Vallée-Tourangeau, F., Euden, G. & Hearn, V. (2011). Einstellung defused: Interactivity and mental set. *The Quarterly Journal of Expermental Psychology*, 64, 1889–1895.

27. Chen, Z. & Mo, L. (2004). Schema induction in problem solving:A multidimensional analysis. *Journal of Experimental Psychology: Learning, Memory, and Cognition*, 30, 583–600.

28. Bilalić, M., McLeod, P. & Gobet, F. (2008). Inflexibility of Experts—Reality or myth? Quantifying the

Einstellung effect in chess masters. *Cognitive Psychology*, 56, 73–102.

29. Sternberg, R. J. (1996). *Cognitive Psychology*. Fort Worth, TX: Harcourt Brace.

30. Chronicle, E. P., MacGregor, J. N. & Ormerod, T. C. (2004). What makes an insight problem? The roles of heuristics, goal conception, and solution recoding in knowledge-lean problems. *Journal of Experimental Psychology: Learning, Memory, and Cognition*, 30, 14–27; Kershaw, T. C. & Ohlsson, S. (2004). Multiple causes of difficulty In insight: The case of the nine-dot problem. *Journal of Experimental Psychology: Learning, Memory, and Cognition*, 30, 3–13.

31. Metcalfe, J. (1986). Feeling of knowing in memory and problem solving. *Journal of Experimental Psychology: Learning, Memory, and Cognition*, 12, 288–294.

32. Metcalfe, J., & Wiebe, D. (1987). Intuition in insight and noninsight problem solving. *Memory & Cognition*, 15, 238–246.

33. Metcalfe, J. & Wiebe, D. (1987). Intuition in insight and noninsight problem solving. *Memory & Cognition*, 15, 238–246; Metcalfe, J. (1986). Feeling of knowing in memory and problem solving. *Journal of Experimental Psychology: Learning, Memory, and Cognition*, 12, 288–294.

34. Smith, S. M. (1995). Getting into and out of mental ruts: A theory of fixation, incubation, and insight. In R. J. Sternberg & J. E. Davidson (Eds.), *The Nature of Insight* (pp. 229–251). Cambridge, MA: MIT Press.

35. Kounios, J., Frymiare, J. L., Bowden, E. M., Fleck, J. I., Subramaniam, K., Parrish, T. B., et al. (2006). The prepared mind: Neural activity prior to problem presentation predicts subsequent solution by sudden insight. *Psychological Science*, 17, 882–890.

36. Sio, U. N. & Ormerod, T. C. (2009). Does incubation enhance problem solving? A meta-analytic review. *Psychological Bulletin*, 135(1), 94–120.

37. Vul, E. & Pashler, H. (2007). Incubation benefits only after people have been misdirected. *Memory & Cognition*, 35, 701–710.

38. Baird, B., Smallwood, J., Mrazek, M. D., Kam, J. W., Franklin, M. S., & Schooler, J. W. (2012). Inspired by distraction: Mind wandering facilitates creative incubation. *Psychological Science*, 23(10), 1117-1122.

39. Wickelgren, W. A. (1974). *How to Solve Problems*. San Francisco, CA: Freeman.

40. Weisberg, R. (1995). Prolegomena to theories of insight in problem solving: A taxonomy of problems. In R. J. Sternberg & J. E. Davidson (Eds.), *The nature of insight* (pp. 157–196). Cambridge, MA: MIT Press.

41. Schooler, J. W., Ohlsson, S. & Brooks, K. (1993). Thoughts beyond words: When language overshadows insight. *Journal of Experimental Psychology: General*, 122, 166–183.

42. Siegler, R. S. & Stern, E. (1998). A microgenetic analysis of conscious and unconscious strategy discoveries. *Journal of Experimental Psychology: General*, 127, 377–397; Siegler, R. S. (2000). Unconscious insights. *Current Directions in Psychological Science*, 9, 79–83.

43. Gentner, D. & Markman, A. B. (1997). Structure mapping in analogy and similarity. *American Psychologist*, 52, 45–56.

44. Holyoak, K. J. & Thagard, P. (1997). The analogical mind. *American Psychologist*, 52, 35–44; Kolodner, J. L. (1997). Educational implications of analogy: A view from case-based reasoning. *American Psychologist*, 52, 57–66.

45. Gentner, D. & Markman, A. B. (1997). Structure mapping in analogy and similarity. *American Psychologist*, 52, 45–56.

46. Markman, A. B., Taylor, E. & Gentner, D. (2007). Auditory presentation leads to better analogical retrieval than written presentation. *Psychonomic Bulletin & Review*, 14, 1101–1106.

47. Day, S. B., Motz, B. A. & Goldstone, R. L. (2015). The cognitive costs of context: The effects of concreteness and immersiveness in instructional examples. *Frontiers in Psychology*, 6.

48. Monaghan, P., Sio, U. N., Lau, S. W., Woo, H. K., Linkenauger, S. A. & Ormerod, T. C. (2015). Sleep promotes analogical transfer in problem solving. *Cognition*, 143, 25–30.

49. Gick, M. L. & Holyoak, K. J. (1980). Analogical problem solving. *Cognitive Psychology*, 12, 306–355.

50. Duncker, K. (1945). On problem solving. *Psychological Monographs*, 58(Whole no. 270).

51. Gick, M. L. & Holyoak, K. J. (1980). Analogical problem solving. *Cognitive Psychology*, 12, 306–355.

52. Holyoak, K. J. & Thagard, P. (1997). The analogical mind. *American Psychologist*, 52, 35–44.

53. Day, S. B., & Gentner, D. (2007). Nonintentional analogical inference in text comprehension. *Memory & Cognition*, 35, 39–49.

54. Wharton, C. M., Grafman, J., Flitman, S. S., Hansen, E. K., Brauner, J., Marks, A., et al. (2000). Toward neuroanatomical models of analogy: A positron emission tomography study of analogical mapping. *Cognitive Psychology*, 40, 173–197.

55. Wharton, C. M., Grafman, J., Flitman, S. S., Hansen, E. K., Brauner, J., Marks, A., et al. (2000). Toward neuroanatomical models of analogy: A positron emission tomography study of analogical mapping. *Cognitive Psychology*, 40, 173–197.

56. Chen, Z., Mo, L. & Honomichl, R. (2004). Having the memory of an elephant: Long-term retrieval and the use of analogues in problem solving. *Journal of Experimental Psychology: General*, 133, 415–433.

57. Novick, L. R. (1988). Analogical transfer, problem similarity,and expertise. *Journal of Experimental*

Psychology: Learning, Memory, and Cognition, 14, 510–520.

58. Waltz, J. A., Lau, A., Grewal, S. K. & Holyoak, K. J. (2000). The role of working memory in analogical mapping. *Memory & Cognition*, 28, 1205–1212.

59. Bassok, M., Pedigo, S. F. & Oskarsson, A. T. (2008). Priming addition facts with semantic relations. *Journal of Experimental Psychology: Learning, Memory, and Cognition*, 34, 343–352.

60. Spellman, B. A. & Holyoak, K. J. (1996). Pragmatics in analogical mapping. *Cognitive Psychology*, 31, 307–346.

61. Kurtz, K. J. & Loewenstein, J. (2007). Converging on a new role for analogy in problem solving and retrieval: When two problems are better than one. *Memory & Cognition*, 35, 334–341.

62. Bowden, E. M. & Beeman, M. J. (1998). Getting the right idea: Semantic activation in the right hemisphere may help solve insight problems. *Psychological Science*, 9, 435–440; Beeman, M. J. & Bowden, E. M. (2000). The right hemisphere maintains solution-related activation for yet-to-be-solved problems. *Memory & Cognition*, 28, 1231–1241.

63. Bowden, E. M. & Beeman, M. J. (1998). Getting the right idea: Semantic activation in the right hemisphere may help solve insight problems. *Psychological Science*, 9, 435–440

64. Bowden, E. M., & Beeman, M. J. (1998). Getting the right idea: Semantic activation in the right hemisphere may help solve insight problems. *Psychological Science*, 9, 435–440.

65. Bowden, E. M., Beeman, M. & Gernsbacher, M. A. (1995, March). *Two hemispheres are better than one: Drawing coherence inferences during story comprehension.* Paper presented at the annual meeting of the Cognitive Neuroscience Society, San Francisco, CA.

66. Jung-Beeman, M., Bowden, E. M., Haberman, J., Frymiare, J. L., Arambel-Liu, S., Greenblatt, R., ... & Kounios, J. (2004). Neural activity when people solve verbal problems with insight. *PLoS Biology*, 2(4), e97.

67. Mayseless, N., Hawthorne, G. & Reiss, A. L. (2019). Real-life creative problem solving in teams: fNIRS based hyperscanning study. *NeuroImage*, 203, 116161.

68. Newell, A. & Simon, H. A. (1972). *Human Problem Solving*. Englewood Cliffs, NJ: Prentice Hall.

69. Simon, H. A. (1975). The functional equivalence of problem solving skills. *Cognitive Psychology*, 7, 268–288.

70. Catrambone, R. (1996). Generalizing solution procedures learned from examples. *Journal of Experimental Psychology: Learning, Memory, and Cognition*, 22, 1020–1031.

71. Newell, A., Shaw, J. C. & Simon, H. A. (1958). Elements of a theory of human problem solving. *Psychological Review*, 65, 151–166; Ernst, G. W. & Newell, A. (1969). GPS: *A Case Study in Generality and Problem Solving*. New York, NY: Academic Press; Newell, A. & Simon, H. A. (1972). *Human Problem Solving. Englewood Cliffs*, NJ: Prentice Hall.

72. Thomas, J. C., Jr. (1974). An analysis of behavior in the hobbits–orcs problem. *Cognitive Psychology*, 6, 257–269.

73. Glass, A. L., & Holyoak, K. J. (1986). *Cognition* (2nd ed.). New York, NY: Random House.

74. Greeno, J. G. (1974). Hobbits and orcs: Acquisition of a sequential concept. *Cognitive Psychology*, 6, 270–292.

75. Atwood, M. E. & Polson, P. (1976). A process model for water jug problems. *Cognitive Psychology*, 8, 191–216.

76. Greeno, J. G. (1974). Hobbits and orcs: Acquisition of a sequential concept. *Cognitive Psychology*, 6, 270–292.

77. Hayes, J. R. & Simon, H. A. (1974). Understanding written problem instructions. In L. W. Gregg (Ed.), *Knowledge and Cognition* (pp. 167–200). Hillsdale, NJ: Erlbaum

78. Wiley, J. (1998). Expertise as mental set: The effects of domain knowledge in creative problem solving. *Memory & Cognition, 26*, 716–730.

79. Ericsson, K. A. & Charness, N. (1994). Expert performance: Its structure and acquisition. *American Psychologist*, 49, 725–747; Medin, D. L., Lynch, E. B., Coley, J. D. & Atran, S. (1997). Categorization and reasoning among tree experts: Do all roads lead to Rome? *Cognitive Psychology*, 32, 49–96.

80. Chase, W. G. & Simon, H. A. (1973). Perception in chess. *Cognitive Psychology*, 4, 55–81; Gobet, F. & Simon, H. A. (1996). Recall of random and distorted chess positions: Implications for the theory of expertise. *Memory & Cognition*, 24, 493–503; Reeves, L. M. & Weisberg, R. W. (1993). Abstract versus concrete information as the basis for transfer in problem solving: Comment on Fong and Nisbett (1991). *Journal of Experimental Psychology: General*, 122, 125–128.

81. Reingold, E. M., Charness, N., Pomplun, M. & Stampe, D. M. (2001). Visual span in expert chess players: Evidence from eye movements. *Psychological Science*, 12, 48–55.

82. Kotovsky, K., Hayes, J. R. & Simon, H. A. (1985). Why are some problems hard? Evidence from Tower of Hanoi. *Cognitive Psychology*, 17, 248–294.

83. Carlson, R. A., Khoo, B. H., Yaure, R. G. & Schneider, W. (1990). Acquisition of a problem-solving skill: Levels of organization and use of working memory. *Journal of Experimental Psychology: General*, 119, 193–214.

84. Bransford, J. D. & Stein, B. S. (1984). *The Ideal Problem Solver*. New York, NY: Freeman; Polya, G. (1957). *How to Solve It*. Garden City, NY: Doubleday/Anchor.

85. Delany, P. F., Ericsson, K. A. & Knowles, M. E. (2004). Immediate and sustained effects of planning in a problem-solving task. *Journal of Experimental Psychology: Learning, Memory*, and Cognition, 30, 1219–1234.

86. Forsyth, D. K. & Burt, C. D. B. (2008). Allocating time to future tasks: The effect of task segmentation on planning fallacy bias. *Memory & Cognition*, 36, 791–798.

87. Wickelgren, W. A. (1974). *How to Solve Problems*. San Francisco, CA: Freeman.

88. Simon, H. A. (May, 1995). *Thinking in Words, Pictures, Equations, Numbers: How do we do it and what does it matter*? Invited address presented at the meeting of the Midwestern Psychological Association, Chicago, IL.

89. Simon, H. A. (1979). *Models of thought*. New Haven, CT: Yale University Press.

90. Bassok, M. & Holyoak, K. H. (1989). Interdomain transfer between isomorphic topics in algebra and physics. *Journal of Experimental Psychology: Learning, Memory, and Cognition, 15*, 153–166.

91. Ross, B. H. (1987). This is like that: The use of earlier problems and the separation of similarity effects. *Journal of Experimental Psychology: Learning, Memory, and Cognition, 13*, 629–640.

92. Bowden, E. M. (1985). Accessing relevant information during problem solving: Time constraints on search in the problem space. *Memory & Cognition*, 13, 280–286.

93. Pedone, R., Hummel, J. E. & Holyoak, K. J. (2001). The use of diagrams in analogical problem solving. *Memory & Cognition*, 29, 214–221.

94. Reed, S. K. & Hoffman, B. (2004). Use of temporal and spatial information in estimating event completion time. *Memory & Cognition*, 32, 271–282.

95. Simon, H. A. (May, 1995). *Thinking in words, pictures, equations, numbers: How do we do it and what does it matter*? Invited address presented at the meeting of the Midwestern Psychological Association, Chicago, IL.

96. Ahlum-Heath, M. E. & DiVesta, F. J. (1986). The effect of conscious controlled verbalization of a cognitive strategy on transfer in problem solving. *Memory & Cognition*, 14, 281–285.

97. Ericsson, K. A. & Charness, N. (1994). Expert performance: Its structure and acquisition. *American Psychologist*, 49, 725–747.

98. Ericsson, K. A. & Charness, N. (1994). Expert performance: Its structure and acquisition. *American Psychologist*, 49, 725–747.

99. Ericsson, K. A., Krampe, R. T. & Tesch-ROmer, C. (1993). The role of deliberate practice in the acquisition of expert performance. *Psychological Review*, 100, 363–406.

100. Ericsson, K. A., Krampe, R. T., & Tesch-Römer, C. (1993). The role of deliberate practice in the acquisition of expert performance. *Psychological Review, 100*, 363–406.

Chapter 12

1. Cacioppo, J. T., Hawkley, L. C., Norman, G. J., & Berntson, G. G. (2011). Social isolation. *Annals of the New York Academy of Sciences, 1231*(1), 17–22.

2. Lou, H. C., Changeux, J. P. & Rosenstand, A. (2017). Towards a cognitive neuroscience of self-awareness. *Neuroscience & Biobehavioral Reviews*, 83, 765–773.

3. Herold, D., Spengler, S., Sajonz, B., Usnich, T. & Bermpohl, F. (2016). Common and distinct networks for self-referential and social stimulus processing in the human brain. *Brain Structure and Function*, 221, 3475–3485

4. Northoff, G., Heinzel, A., De Greck, M., Bermpohl, F., Dobrowolny, H., & Panksepp, J. (2006). Self-referential processing in our brain—a meta-analysis of imaging studies on the self. *Neuroimage, 31*(1), 440–457.

5. Crone, E. A., & Fuligni, A. J. (2020). Self and others in adolescence. *Annual Review of Psychology, 71*(1), 447–469.

6. Klein, S. B., & Loftus, J. (1993). The mental representation of trait and autobiographical knowledge about the self. In T. K. Srull & R. S. Wyer, Jr. (Eds.), *The Mental Representation of Trait and Autobiographical Knowledge about the Self* (pp. 1–49). Lawrence Erlbaum Associates, Inc.

7. Klein, S. B., Loftus, J., & Kihlstrom, J. F. (1996). Self-knowledge of an amnesic patient: Toward a neuropsychology of personality and social psychology. *Journal of Experimental Psychology: General, 125*(3), 250.

8. Klein, S. B., Rozendal, K. & Cosmides, L. (2002). A social-cognitive neuroscience analysis of the self. *Social Cognition, 20*(2), 105–135.

9. Klein, S. B., Rozendal, K. & Cosmides, L. (2002). A social-cognitive neuroscience analysis of the self. *Social Cognition, 20*(2), 105–135.

10. Klein, S. B., & Lax, M. L. (2010). The unanticipated resilience of trait self-knowledge in the face of neural damage. *Memory*, 18(8), 918-948.

11. Fossati, P., Hevenor, S. J., Lepage, M., Graham, S. J., Grady, C., Keightley, M. L., ... & Mayberg, H. (2004). Distributed self in episodic memory: neural correlates of successful retrieval of self-encoded positive and negative personality traits. *Neuroimage, 22*(4), 1596–1604.

12. Klein, S. B., & Nichols, S. (2012). Memory and the sense of personal identity. *Mind, 121*(483), 677–702.

13. de la Vega, A., Chang, L. J., Banich, M. T., Wager, T. D., & Yarkoni, T. (2016). Large-scale meta-analysis of human medial frontal cortex reveals tripartite

functional organization. *Journal of Neuroscience*, 36(24), 6553–6562.

14. Van de Groep, I. H., Bos, M. G., Jansen, L. M., Achterberg, M., Popma, A., & Crone, E. A. (2021). Overlapping and distinct neural correlates of self-evaluations and self-regulation from the perspective of self and others. *Neuropsychologia*, *161*, 108000.

15. Ebisch, S. J. & Aleman, A. (2016). The fragmented self: imbalance between intrinsic and extrinsic self-networks in psychotic disorders. *The Lancet Psychiatry*, *3*(8), 784–790.

16. Di Plinio, S., Perrucci, M. G., Aleman, A. & Ebisch, S. J. (2020). I am Me: Brain systems integrate and segregate to establish a multidimensional sense of self. *Neuroimage*, *205*, 116284.

17. Frith, U. & Frith, C. (2010). The social brain: allowing humans to boldly go where no other species has been. *Philosophical Transactions of the Royal Society B: Biological Sciences*, *365*(1537), 165–176.

18. Gallagher, H. L., & Frith, C. D. (2003). Functional imaging of 'theory of mind'. *Trends in Cognitive Sciences*, *7*(2), 77–83.

19. Sebastian, C. L., Fontaine, N. M., Bird, G., Blakemore, S. J., De Brito, S. A., McCrory, E. J. & Viding, E. (2012). Neural processing associated with cognitive and affective Theory of Mind in adolescents and adults. *Social Cognitive and Affective Neuroscience*, *7*(1), 53–63.

20. Abu-Akel, A. & Shamay-Tsoory, S. (2011). Neuroanatomical and neurochemical bases of theory of mind. *Neuropsychologia*, *49*(11), 2971–2984.

21. Abu-Akel, A. & Shamay-Tsoory, S. (2011). Neuroanatomical and neurochemical bases of theory of mind. *Neuropsychologia*, *49*(11), 2971–2984

22. Wade, M., Prime, H., Jenkins, J. M., Yeates, K. O., Williams, T. & Lee, K. (2018). On the relation between theory of mind and executive functioning: A developmental cognitive neuroscience perspective. *Psychonomic Bulletin & Review*, *25*, 2119–2140.

23. Austin, G., Groppe, K. & Elsner, B. (2014). The reciprocal relationship between executive function and theory of mind in middle childhood: A 1-year longitudinal perspective. *Frontiers in Psychology*, *5*, 655.

24. Yeh, Z. T., Lo, C. Y., Tsai, M. D., & Tsai, M. C. (2015). Mentalizing ability in patients with prefrontal cortex damage. *Journal of Clinical and Experimental Neuropsychology*, *37*(2), 128-139.

25. Rankin, K. P., Kramer, J. H. & Miller, B. L. (2005). Patterns of cognitive and emotional empathy in frontotemporal lobar degeneration. *Cognitive and Behavioral Neurology*, *18*(1), 28-36.

26. Launay, J., Pearce, E., Wlodarski, R., van Duijn, M., Carney, J. & Dunbar, R. I. (2015). Higher-order mentalising and executive functioning. *Personality and Individual Differences*, *86*, 6–14.

27. Dunbar, R. I. (2009). The social brain hypothesis and its implications for social evolution. *Annals of Human Biology*, *36*(5), 562–572.

28. Meyer, M. L., & Lieberman, M. D. (2012). Social working memory: neurocognitive networks and directions for future research. *Frontiers in Psychology*, *3*, 571.

29. Meyer, M. L., Spunt, R. P., Berkman, E. T., Taylor, S. E. & Lieberman, M. D. (2012). Evidence for social working memory from a parametric functional MRI study. *Proceedings of the National Academy of Sciences*, *109*(6), 1883–1888.

30. Rueda, P., Fernández-Berrocal, P. & Baron-Cohen, S. (2015). Dissociation between cognitive and affective empathy in youth with Asperger Syndrome. *European Journal of Developmental Psychology*, *12*(1), 85–98

31. Anastácio, S., Vagos, P., Nobre-Lima, L., Rijo, D. & Jolliffe, D. (2016). The Portuguese version of the Basic Empathy Scale (BES): Dimensionality and measurement invariance in a community adolescent sample. *European Journal of Developmental Psychology*, 13(5), 614–623; D'Ambrosio, F., Olivier, M., Didon, D. & Besche, C. (2009). The basic empathy scale: A French validation of a measure of empathy in youth. *Personality and Individual Differences*, *46*(2), 160–165; Geng, Y., Xia, D. & Qin, B. (2012). The Basic Empathy Scale: A Chinese validation of a measure of empathy in adolescents. *Child Psychiatry & Human Development*, *43*, 499–510; Trentini, C., Tambelli, R., Maiorani, S. & Lauriola, M. (2022). Gender differences in empathy during adolescence: Does emotional self-awareness matter?. *Psychological Reports*, *125*(2), 913–936.

32. Del Rey, R., Lazuras, L., Casas, J. A., Barkoukis, V., Ortega-Ruiz, R. & Tsorbatzoudis, H. (2016). Does empathy predict (cyber) bullying perpetration, and how do age, gender and nationality affect this relationship?. *Learning and Individual Differences*, *45*, 275–281.

33. Maurage, P., Grynberg, D., Noël, X., et al. (2011). Dissociation between affective and cognitive empathy in alcoholism: a specific deficit for the emotional dimension. *Alcoholism: Clinical and Experimental Research*, *35*(9), 1662-1668.

34. Winters, D. E., Brandon-Friedman, R., Yepes, G. & Hinckley, J. D. (2021). Systematic review and meta-analysis of socio-cognitive and socio-affective processes association with adolescent substance use. *Drug and Alcohol Dependence*, *219*, 108479.

35. Rueda, P., Fernández-Berrocal, P. & Baron-Cohen, S. (2015). Dissociation between cognitive and affective empathy in youth with Asperger Syndrome. *European Journal of Developmental Psychology*, *12*(1), 85–98.

36. Mairon, N., Abramson, L., Knafo-Noam, A., Perry, A. & Nahum, M. (2023). The relationship between empathy and executive functions among young adolescents. *Developmental Psychology*.

37. Yan, Z., Hong, S., Liu, F. & Su, Y. (2020). A meta-analysis of the relationship between empathy and executive function. *PsyCh journal*, 9(1), 34–43.

38. Shamay-Tsoory, S. G., Aharon-Peretz, J. & Perry, D. (2009). Two systems for empathy: a double dissociation between emotional and cognitive empathy in inferior frontal gyrus versus ventromedial prefrontal lesions. *Brain*, 132(3), 617–627.

39. Leigh, R., Oishi, K., Hsu, J., Lindquist, M., Gottesman, R. F., Jarso, S., Crainiceanu, C., Mori, S. & Hillis, A. E. (2013). Acute lesions that impair affective empathy. *Brain*, 136(8), 2539–2549.

40. Fan, Y., Duncan, N. W., De Greck, M., & Northoff, G. (2011). Is there a core neural network in empathy? An fMRI based quantitative meta-analysis. *Neuroscience & Biobehavioral Reviews*, 35(3), 903-911.

41. Decety, J. (2015). The neural pathways, development and functions of empathy. *Current Opinion in Behavioral Sciences*, 3, 1–6.

42. Shamay-Tsoory, S. G., Harari, H., Aharon-Peretz, J., & Levkovitz, Y. (2010). The role of the orbitofrontal cortex in affective theory of mind deficits in criminal offenders with psychopathic tendencies. *Cortex*, 46(5), 668–677.

43. Sebastian, C. L., Fontaine, N. M., Bird, G., Blakemore, S. J., De Brito, S. A., McCrory, E. J., & Viding, E. (2012). Neural processing associated with cognitive and affective Theory of Mind in adolescents and adults. *Social Cognitive and Affective Neuroscience*, 7(1), 53–63.

44. Arioli, M., Cattaneo, Z., Ricciardi, E. & Canessa, N. (2021). Overlapping and specific neural correlates for empathizing, affective mentalizing, and cognitive mentalizing: A coordinate-based meta-analytic study. *Human Brain Mapping*, 42(14), 4777-4804.

45. Arioli, M., Cattaneo, Z., Ricciardi, E. & Canessa, N. (2021). Overlapping and specific neural correlates for empathizing, affective mentalizing, and cognitive mentalizing: A coordinate-based meta-analytic study. *Human Brain Mapping*, 42(14), 4777–4804.

46. Shamay-Tsoory, S. G., Harari, H., Aharon-Peretz, J., & Levkovitz, Y. (2010). The role of the orbitofrontal cortex in affective theory of mind deficits in criminal offenders with psychopathic tendencies. *Cortex*, 46(5), 668–677.

47. Arioli, M., Cattaneo, Z., Ricciardi, E. & Canessa, N. (2021). Overlapping and specific neural correlates for empathizing, affective mentalizing, and cognitive mentalizing: A coordinate-based meta-analytic study. *Human Brain Mapping*, 42(14), 4777–4804.

48. Shamay-Tsoory, S. G., Harari, H., Aharon-Peretz, J., & Levkovitz, Y. (2010). The role of the orbitofrontal cortex in affective theory of mind deficits in criminal offenders with psychopathic tendencies. *Cortex*, 46(5), 668–677.

49. Schurz, M., Radua, J., Tholen, M. G., Maliske, L., Margulies, D. S., Mars, R. B., Sallet, J. & Kanske, P. (2021). Toward a hierarchical model of social cognition: A neuroimaging meta-analysis and integrative review of empathy and theory of mind. *Psychological Bulletin*, 147(3), 293–327. https://doi.org/10.1037/bul0000303

50. Holt-Lunstad, J. (2017). The potential public health relevance of social isolation and loneliness: Prevalence, epidemiology, and risk factors. *Public Policy & Aging Report*, 27(4), 127–130.

51. McQuaid, R. J., Cox, S. M., Ogunlana, A. & Jaworska, N. (2021). The burden of loneliness: Implications of the social determinants of health during COVID-19. *Psychiatry Research*, 296, 113648.

52. Eisenberger, N. I., Lieberman, M. D. & Williams, K. D. (2003). Does rejection hurt? An fMRI study of social exclusion. *Science*, 302(5643), 290–292.

53. Eisenberger, N. I. (2015). Social pain and the brain: Controversies, questions, and where to go from here. *Annual Review of Psychology*, 66, 601–629.

54. Kross, E., Berman, M. G., Mischel, W., Smith, E. E. & Wager, T. D. (2011). Social rejection shares somatosensory representations with physical pain. *Proceedings of the National Academy of Sciences*, 108(15), 6270–275.

55. Eisenberger, N. I., Lieberman, M. D. & Williams, K. D. (2003). Does rejection hurt? An fMRI study of social exclusion. *Science*, 302(5643), 290–292.

56. Kross, E., Berman, M. G., Mischel, W., Smith, E. E. & Wager, T. D. (2011). Social rejection shares somatosensory representations with physical pain. *Proceedings of the National Academy of Sciences*, 108(15), 6270–6275.

57. Eisenberger, N. I. & Lieberman, M. D. (2004). Why rejection hurts: a common neural alarm system for physical and social pain. Trends in *Cognitive Sciences*, 8(7), 294–300.

58. Cacioppo, S., Frum, C., Asp, E., Weiss, R. M., Lewis, J. W. & Cacioppo, J. T. (2013). A quantitative meta-analysis of functional imaging studies of social rejection. *Scientific Reports*, 3(1), 2027.

59. Eisenberger, N. I. (2015). Social pain and the brain: Controversies, questions, and where to go from here. *Annual Review of Psychology*, 66(1), 601–629.

60. Ypsilanti, A. (2018). Lonely but avoidant—the unfortunate juxtaposition of loneliness and self-disgust. *Palgrave Communications*, 4(1), 1–4.

61. Cacioppo, J. T., Cacioppo, S. & Boomsma, D. I. (2014). Evolutionary mechanisms for loneliness. *Cognition & Emotion*, 28(1), 3–21.

62. Spithoven, A. W., Bijttebier, P. & Goossens, L. (2017). It is all in their mind: A review on information processing bias in lonely individuals. *Clinical Psychology Review*, 58, 97–114.

63. Qualter, P., Rotenberg, K., Barrett, L., Henzi, P., Barlow, A., Stylianou, M. & Harris, R. A. (2013). Investigating

hypervigilance for social threat of lonely children. *Journal of Abnormal Child Psychology, 41*, 325–338.

64. Bangee, M., Harris, R. A., Bridges, N., Rotenberg, K. J. & Qualter, P. (2014). Loneliness and attention to social threat in young adults: Findings from an eye tracker study. *Personality and Individual Differences, 63*, 16–23.

65. Bangee, M. & Qualter, P. (2018). Examining the visual processing patterns of lonely adults. *Scandinavian Journal of Psychology, 59*(4), 351–359.

66. Cacioppo, S., Bangee, M., Balogh, S., Cardenas-Iniguez, C., Qualter, P. & Cacioppo, J. T. (2016). Loneliness and implicit attention to social threat: A high-performance electrical neuroimaging study. *Cognitive Neuroscience*, 7(1-4), 138–159.

67. Cacioppo, S., Bangee, M., Balogh, S., Cardenas-Iniguez, C., Qualter, P. & Cacioppo, J. T. (2016). Loneliness and implicit attention to social threat: A high-performance electrical neuroimaging study. *Cognitive Neuroscience*, 7(1-4), 138–159.

68. Cacioppo, S., Balogh, S., & Cacioppo, J. T. (2015). Implicit attention to negative social, in contrast to nonsocial, words in the Stroop task differs between individuals high and low in loneliness: Evidence from event-related brain microstates. *Cortex, 70*, 213–233.

69. Cacioppo, S., Balogh, S., & Cacioppo, J. T. (2015). Implicit attention to negative social, in contrast to nonsocial, words in the Stroop task differs between individuals high and low in loneliness: Evidence from event-related brain microstates. *Cortex, 70*, 213–233.

70. Qualter, P., Rotenberg, K., Barrett, L., Henzi, P., Barlow, A., Stylianou, M. & Harris, R. A. (2013). Investigating hypervigilance for social threat of lonely children. *Journal of Abnormal Child Psychology, 41*, 325–338.

71. Riddleston, L., Bangura, E., Gibson, O., Qualter, P., & Lau, J. Y. (2023). Developing an interpretation bias modification training task for alleviating loneliness in young people. *Behaviour Research and Therapy, 168*, 104380.

72. Dunbar, R. I. (2009). The social brain hypothesis and its implications for social evolution. *Annals of Human Biology*, 36(5), 562–572.

Chapter 13

1. Mather, M. & Sutherland, M. R. (2011). Arousal-based competition in perception and memory. *Perspectives on Psychological Science, 6*, 114–133.

2. Dukes, D., Abrams, K., Adolphs, R. et al. (2021). The rise of affectivism. *Nature Human Behaviour 5*, 816–820. https://doi.org/10.1038/s41562-021-01130-8

3. Dukes, D., Abrams, K., Adolphs, R. et al. (2021). The rise of affectivism. *Nature Human Behaviour 5*, 816–820. https://doi.org/10.1038/s41562-021-01130-8

4. Davis, M. (1997). Neurobiology of fear responses: The role of the amygdala. *Journal of Neuropsychiatry and Clinical Neurosciences, 9*, 382–402; LeDoux, J. E. (2000). Emotion circuits in the brain. *Annual Review of Neuroscience, 23*, 155–184.

5. Sakaki, M., Niki, K. & Mather, M. (2012). Beyond arousal and valence: The importance of the biological versus social relevance of emotional stimuli. *Cognitive Affective & Behavioral Neuroscience, 12*, 115–139.

6. Herz, R. S. & Engen, T. (1996). Odor memory: Review and analysis. *Psychonomic Bulletin & Review, 3*, 300–313.

7. Roy, M., Shohamy, D. & Wager, T. D. (2012). Ventromedial prefrontal-subcortical systems and the generation of affective meaning. *Trends in Cognitive Sciences, 16*, 147–156.

8. Taylor, S. F., Liberzon, I. & Koeppe, R. A. (2000). The effect of graded aversive stimuli on limbic and visual activation. *Neuropsychologia, 38*, 1415–1425.

9. Bocanegra, B. R. & Zeelenberg, R. (2011). Emotion-induced trade-offs in spatiotemporal vision. *Journal of Experimental Psychology: General, 140*, 272–282.

10. Niedenthal, P. M. & Wood, A. (2019). Does emotion influence visual perception? Depends on how you look at it. *Cognition and Emotion, 33*(1), 77-84

11. Vuilleumeir, P. (2005). How brains beware: Neural mechanisms of emotional attention. *Trends in Cognitive Sciences, 9*, 585–594.

12. Fenske, M. J. & Raymond, J. E. (2006). Affective influences of selective attention. *Current Directions in Psychological Science, 15*, 312–316.

13. Öhman, A., Flykt, A. & Esteves, F. (2001). Emotion drives attention: Detecting the snake in the grass. *Journal of Experimental Psychology: General, 130*, 466–478.

14. Vuilleumeir, P. & Huang, Y-M. (2009). Emotional attention: Uncovering the mechanisms of affective biases in perception. *Current Directions in Psychological Science, 18*, 148–152.

15. Becker, M. W. (2009). Panic search: Fear produces efficient visual search for nonthreatening objects. *Psychological Science, 20*, 435–437.

16. Williams, J. M. G., Mathews, A. & MacLeod, C. (1996). The emotional Stroop task and psychopathology. *Psychological Bulletin, 120*, 3–24.

17. Mitterschiffthaler, M. T., Williams, S. C. R., Walsh, N. D., Cleare, A. J., Donaldson, C., Scott, J., et al. (2008). Neural basis of the emotional Stroop interference effect in major depression. *Psychological Medicine, 38*, 24–256.

18. Dresler, T., Mériau, K., Heekeren, H. R. & van der Meer, E. (2009). Emotional Stroop task: Effect of word arousal and subject anxiety on emotional interference. *Psychological Research, 73*, 364–371.

19. Cisler, J. M., Wolitzky-Taylor, K. B., Adams, T. G., Babson, K. A., Badou, C. L. & Willems, J. L. (2011). The emotional Stroop task and posttraumatic stress

disorder: A meta-analysis. *Clinical Psychology Review, 31*, 817–828.

20. McKenna, F. P., & Sharma, D. (2004). Reversing the emotional Stroop effect reveals that it is not what it seems: the role of fast and slow components. *Journal of Experimental Psychology: Learning, Memory, and Cognition, 30*(2), 382.; Waters, A. J., Sayette, M. A., Franken, I. H., & Schwartz, J. E. (2005). Generalizability of carry-over effects in the emotional Stroop task. *Behaviour Research and Therapy, 43*(6), 715–732.
21. Inzlicht, M. & Gutsell, J. N. (2007). Running on empty: Neural signals for self-control failure. *Psychological Science, 18*, 933–937.
22. Talarico, J. M., LaBar, K. S. & Rubin, D. C. (2004). Emotional intensity predicts autobiographical memory experience. *Memory & Cognition, 32*, 1118–1132.
23. Chapman, H. A., Johannes, K., Poppenk, J. L., Moscovitch, M. & Anderson, A. K. (2013). Evidence for the differential salience of disgust and fear in episodic memory. *Journal of Experimental Psychology: General, 142*(4), 1100–1112.
24. Bradley, M. M., Greenwald, M. K., Petry, M. C. & Lang, P. J. (1992). Remembering pictures: Pleasure and arousal in memory. *Journal of Experimental Psychology: Learning, Memory, and Cognition, 18*, 379–390; Kensinger, E. A. (2009). How emotion affects older adults' memories for event details. *Memory, 17*(2), 208–219.
25. Kensinger, E. A., Garoff-Eaton, R. J. & Schacter, D. L. (2006). Memory for specific visual details can be enhanced by negative arousing content. *Journal of Memory and Language, 54*, 99–112.
26. Armony, J. L., Chochol, C., Fecteau, S. & Belin, P. (2007). Laugh (or cry) and you will be remembered. *Psychological Science, 18*, 1027–1029.
27. Dolcos, F., Labar, K. S. & Cabeza, R. (2005). Remembering one year later: Role of the amygdala and the medial temporal lobe memory system in retrieving emotional memories. *Proceedings of the National Academy of Sciences, 102*, 2626–2631.
28. Kensinger, E. A. & Corkin, S. (2004). Two routes to emotional memory: Distinct neural processes for valence and arousal. *Proceedings of the National Academy of Sciences, 101*, 3310–3315.
29. Finn, B. & Roediger, H. L. (2011). Enhancing retention through reconsolidation: Negative emotional arousal following retrieval enhances later recall. *Psychological Science, 22*, 781–786.
30. Finn, B., & Roediger, H. L. (2011). Enhancing retention through reconsolidation: Negative emotional arousal following retrieval enhances later recall. *Psychological Science, 22*, 781–786.
31. Finn, B., & Roediger, H. L. (2011). Enhancing retention through reconsolidation: Negative emotional arousal following retrieval enhances later recall. *Psychological Science, 22*, 781–786.
32. Hu, P., Stylos-Allan, M. & Walker, M. P. (2006). Sleep facilitates consolidation of emotional declarative memory. *Psychological Science, 17*, 891–898.
33. Talmi, D. & Garry, L. M. (2012). Accounting for immediate emotional memory enhancement. *Journal of Memory and Language, 66*, 93–108; Talmi, J. M., Schimmack, U., Paterson, T. & Moscovitch, M. (2007). The role of attention and relatedness in emotionally enhanced memory. *Emotion, 7*, 89–102.
34. Payne, J. D. & Kensinger, E. A. (2010). Sleep's role in the consolidation of emotional episodic memories. *Psychological Science, 19*, 290–295.
35. Cunningham, T. J., Chambers, A. M. & Payne, J. D. (2014). Prospection and emotional memory: How expectation affects emotional memory formation following sleep and wake. *Frontiers in Psychology, 5*, 862.
36. Holland, A. C. & Kensinger, E. A. (2010). Emotion and autobiographical memory. *Physics of Life Reviews, 7*(1), 88–131.
37. M.A. Conway, C.W. Pleydell-Pearce (2000) The construction of autobiographical memories in the self-memory system *Psychological Review, 107*, 261–288.
38. M. Moscovitch, R.S. Rosenbaum, A. Gilboa, D.R. Addis, R. Westmacott, C. Grady, et al. (2005) Functional neuroanatomy of remote episodic, semantic and spatial memory: A unified account based on multiple trace theory. *Journal of Anatomy, 207*, 35–66.
39. Buchanan, T. W., Lutz, K., Mirzazade, S., Specht, K., Shah, N. J., Zilles, K., et al. (2000). Recognition of emotional prosody and verbal components of spoken language: An fMRI study. *Cognitive Brain Research, 9*, 227–238.
40. Pell, M. D. (1999). Fundamental frequency encoding of linguistic and emotional prosody by right hemisphere– damaged speakers. *Brain and Language, 69*, 161–192.
41. Adolphs, R. & Tranel, D. (1999). Intact recognition of emotional prosody following amygdala damage. *Neuropsychologia, 37*, 1285–1292.
42. Drolet, M., Schubotz, R. I. & Fischer, J. (2012). Authenticity affects the recognition of emotions in speech: Behavioral and fMRI evidence. *Cognitive, Affective & Behavioral Neuroscience, 12*, 140–150.
43. Oosterwijk, S., Winkielman, P., Pecher, D., Zeelenberg, R., Rottveel, M. & Fischer, A. H. (2012). Mental states inside out: Switching costs for emotional and nonemotional sentences that differ in internal and external focus. *Memory & Cognition, 40*, 93–100
44. Ashcraft, M. H. & Krause, J. A. (2007). Working memory, math performance, and math anxiety. *Psychonomic Bulletin & Review, 14*, 243–248.
45. Maloney, E. A., Ansari, D. & Fugelsang, J. A. (2011). The effect of mathematics anxiety on the processing of numerical magnitude. *The Quarterly Journal of Experimental Psychology, 64*, 10–16.

46. Kassam, K. S., Koslov, K. & Mendes, W. B. (2009). Decisions under distress: Stress profiles influence anchoring and adjustment. *Psychological Science, 20*, 1394–1399.

47. Beilock, S. L. & Carr, T. H. (2001). On the fragility of skilled performance: What governs choking under pressure? *Journal of Experimental Psychology: Learning, Memory, and Cognition, 33*, 983–998.

48. Beilock, S. L., Bertenthal, B. I., McCoy, A. M. & Carr, T. H. (2004). Haste does not always make waste: Expertise, direction of attention, and speed versus accuracy in performing sensorimotor skills. *Psychonomic Bulletin & Review, 11*, 373–379.

49. Murphy, M. C., Steele, C. M., & Gross, J. J. (2007). Signaling threat: How situational cues affect women in math, science, and engineering settings. *Psychological Science, 18*, 879–885.

50. Carr, P. B. & Steele, C. M. (2010). Stereotype threat affects financial decision making. *Psychological Science, 21*, 1411–1416.

51. Beilock, S. L., Rydell, R. J. & McConnell, A. R. (2007). Stereotype threat and working memory: Mechanisms, alleviation, and spillover. *Journal of Experimental Psychology: General, 136*, 256–276

52. Schmader, T. (2010). Stereotype threat deconstructed. *Current Directions in Psychological Science, 19*, 14–18.

53. Lang, J. W. B. & Lang, J. (2010). Priming competence diminishes the link between cognitive test anxiety and test performance: Implications for the interpretation of test scores. *Psychological Science, 21*, 811–819.

54. Brooks, A. W. (2014). Get excited: Reappraising pre-performance anxiety as excitement. *Journal of Experimental Psychology: General, 143*(3), 1144–1158; Kassam, K. S., Gilbert, D. T., Swencionis, J. K. & Wilson, T. D. (2009). Misconceptions of memory: The Scooter Libby effect. *Psychological Science, 20*, 551–552.

55. LePine, J. A., LePine, M. A. & Jackson, C. L. (2004). Challenge and hindrance stress: Relationships with exhaustion, motivation to learn, and learning performance. *Journal of Applied Psychology, 89*, 883–891.

56. Masters, R. S. W. (1992). Knowledge, knerves and know-how: The role of explicit versus implicit knowledge in the breakdown of a complex motor skill under pressure. *British Journal of Psychology, 83*, 343–358.

57. James J. Gross (2015) Emotion Regulation: Current Status and Future Prospects, Psychological Inquiry, 26:1, 1–26, DOI: 10.1080/1047840X.2014.940781

58. Gross, J. J. (1998). The emerging field of emotion regulation: An integrative review. *Review of general psychology, 2*(3), 271-299.

59. Dryman, M. T. & Heimberg, R. G. (2018). Emotion regulation in social anxiety and depression: A systematic review of expressive suppression and cognitive reappraisal. *Clinical Psychology Review, 65*, 17–42.

Chapter 14

1. Bantin, T., Stevens, S., Gerlach, A. L. & Hermann, C. (2016). What does the facial dot-probe task tell us about attentional processes in social anxiety? A systematic review. *Journal of Behavior Therapy and Experimental Psychiatry, 50*, 40–51.

2. Kruijt, A. W., Parsons, S., & Fox, E. (2019). A meta-analysis of bias at baseline in RCTs of attention bias modification: No evidence for dot-probe bias towards threat in clinical anxiety and PTSD. *Journal of Abnormal Psychology, 128*(6), 563.

3. Veerapa, E., Grandgenevre, P., El Fayoumi, M., Vinnac, B., Haelewyn, O., Szaffarczyk, S., ... & D'Hondt, F. (2020). Attentional bias towards negative stimuli in healthy individuals and the effects of trait anxiety. *Scientific Reports, 10*(1), 11826.

4. Ba, Y., Zhang, W., Salvendy, G., Cheng, A. S. & Ventsislavova, P. (2016). Assessments of risky driving: A Go/No-Go simulator driving task to evaluate risky decision-making and associated behavioral patterns. *Applied Ergonomics, 52*, 265-274

5. Raud, L., Westerhausen, R., Dooley, N. & Huster, R. J. (2020). Differences in unity: The go/no-go and stop signal tasks rely on different mechanisms. *NeuroImage, 210*, 116582.

6. MacLeod, C. M. (1991). Half a century of research on the Stroop effect: An integrative review. *Psychological Bulletin, 109*, 163–203.

7. Bugg, J. M., Jacoby, L. L. & Toth, J. P. (2008). Multiple levels of control in the Stroop task. *Memory & Cognition, 36*(8), 1484–1494.

8. Ypsilanti, A., Mullings, E., Hawkins, O., & Lazuras, L. (2021). Feelings of fear, sadness, and loneliness during the COVID-19 pandemic: Findings from two studies in the UK. *Journal of Affective Disorders, 295*, 1012–1023.

9. Williams, J. M. G., Mathews, A. & MacLeod, C. (1996). The emotional Stroop task and psychopathology. *Psychological Bulletin, 120*(1), 3.

10. Cisler, J. M., Wolitzky-Taylor, K. B., Adams Jr, T. G., Babson, K. A., Badour, C. L. & Willems, J. L. (2011). The emotional Stroop task and posttraumatic stress disorder: a meta-analysis. *Clinical Psychology review, 31*(5), 817–828

11. Epp, A. M., Dobson, K. S., Dozois, D. J. & Frewen, P. A. (2012). A systematic meta-analysis of the Stroop task in depression. *Clinical Psychology Review, 32*(4), 316–328

12. Song, S., Zilverstand, A., Song, H., d'Oleire Uquillas, F., Wang, Y., Xie, C., ... & Zou, Z. (2017). The influence of emotional interference on cognitive control: A meta-analysis of neuroimaging studies using the emotional Stroop task. *Scientific Reports, 7*(1), 2088

13. Joyal, M., Wensing, T., Levasseur-Moreau, J., Leblond, J., T. Sack, A., & Fecteau, S. (2019). Characterizing emotional Stroop interference in posttraumatic stress disorder, major depression and anxiety disorders: A systematic review and meta-analysis. *PloS One, 14*(4), e0214998.

14. Greenwald, A. G., McGhee, D. E., & Schwartz, J. L. (1998). Measuring individual differences in implicit cognition: the implicit association test. *Journal of Personality and Social Psychology, 74*(6), 1464.

15. Bargh, J. A., Chen, M. & Burrows, L. (1996). Automaticity of social behavior: Direct effects of trait construct and stereotype activation on action. *Journal of Personality and Social Psychology, 71*(2), 230.

16. Bargh, J. A. (2006). What have we been priming all these years? On the development, mechanisms, and ecology of nonconscious social behavior. *European Journal of Social Psychology, 36*(2), 147–168.

17. Armstrong, T. & Olatunji, B. O. (2012). Eye tracking of attention in the affective disorders: A meta-analytic review and synthesis. *Clinical Psychology Review,* 32(8), 704–723.

18. Duerrschmid, K., & Danner, L. (2018). Eye tracking in consumer research. In *Methods in Consumer Research, Volume* 2 (pp. 279–318). Woodhead Publishing.

19. Wolfe, J. M. & Horowitz, T. S. (2017). Five factors that guide attention in visual search. *Nature Human Behaviour, 1*(3), 0058.

20. Zhang, X., Guo, Q., Zhang, Y., Lou, L. & Ding, D. (2015). Different timing features in brain processing of core and moral disgust pictures: An event-related potentials study. *PloS one, 10*(5), e0128531.

21. Poldrack, R. A. (2008). The role of fMRI in cognitive neuroscience: where do we stand? *Current Opinion in Neurobiology, 18*(2), 223–227.

22. Terry, D. P., Sabatinelli, D., Puente, A. N., Lazar, N. A. & Miller, L. S. (2015). A meta-analysis of fMRI activation differences during episodic memory in Alzheimer's disease and mild cognitive impairment. *Journal of Neuroimaging, 25*(6), 849–860.

23. https://osf.io/

24. Hensel, W. M. (2020). Double trouble? The communication dimension of the reproducibility crisis in experimental psychology and neuroscience. *European Journal for Philosophy of Science, 10*(3), 44.

25. Marks, D. F. (2019). The Hans Eysenck affair: Time to correct the scientific record. *Journal of Health Psychology, 24*(4), 409–420.

Glossary

A

accessibility: The degree to which information can be retrieved from memory. A memory is said to be accessible if it is retrievable; memories that are not currently retrievable are said to have become inaccessible (contrast with *availability*).

accuracy: How correct a person is in the responses given; often quantified in terms of both number correct and number incorrect. Separately, or paired with response time, it can give an indication of cognitive functioning.

accuracy: (of response) The level of precision and correctness in response to a target stimulus (e.g. recalling the correct word from a previously presented list of words). Accuracy can relate to speed, that is, the faster you respond the less accurate you may be (speed–accuracy trade off).

acetylcholine: A neurotransmitter that may be involved in strengthening neural connections during long-term potentiation.

action potential: The change in electrical charge that occurs when a neuron 'fires'. Neural firing follows the all-or-none principle, resulting in all action potentials being the same.

advantage of clause recency: The speedup of RT to information in the most recently processed clause.

advantage of first mention: The speedup of RT to information mentioned first in the sentence.

affectivism: The systematic study of the interplay between emotions and cognition, perception and decision-making.

agraphia: A disruption in the ability to write, caused by a brain disorder or injury.

alertness: The nervous system must be awake, responsive and able to interact with the environment.

alexia: A disruption in the ability to read or recognise printed letters or words, caused by a brain disorder or injury.

algorithmic bias: The tendency of automated, data-driven decision-making systems to produce biased outcomes that breach the rules and norms of social justice; equality, diversity and inclusivity; and even moral conduct.

all-or-none principle: The idea that either a neuron fires or it does not, with all action potentials being the same.

ambiguous: Having more than one meaning, said both of words (e.g. bank) and sentences (e.g. 'They are eating apples').

amnesia: Memory loss caused by brain damage or injury. Retrograde amnesia is loss of memory for information before the damage; anterograde amnesia is loss of memory for information after the damage.

amygdala: An almond-shaped structure adjacent to one end of the hippocampus, often involved in emotion processing.

analogy: A relationship between similar situations, problems or concepts.

anaphoric reference: The act of using a pronoun or possessive (or synonym) to refer back to a previously mentioned concept.

animal communication: The ways animals use different channels and modalities to communicate with each other.

anomia: A disruption of word finding or retrieval, caused by a brain disorder or injury.

Anomia, or anomic aphasia: A disruption of word finding or retrieval, caused by a brain disorder or injury (Chs. 2, 7, 9).

antecedent: The if clause in standard conditional reasoning (if–then) tasks. In the statement 'If it rains, then the picnic will be cancelled', the antecedent is 'If it rains'.

antecedent: The concept to which a later word refers; for example, he refers to the antecedent Bill in 'Bill said he was tired'.

aphasia: A loss of some or all of previously intact language skills, caused by brain disorder or damage.

apperceptive agnosia: A form of agnosia in which individual features cannot be integrated into a whole percept or pattern; a basic disruption in perceiving patterns.

Arbitrariness: One of Hockett's (1960a; 1960b) linguistic universals; the connections between linguistic units (sounds, words) and the concepts or meanings referred to by those units are entirely arbitrary; for example, it is arbitrary that we refer to a table by the linguistic unit *table*.

articulatory loop: The part of the phonological loop involved in the active refreshing of information in the phonological store.

articulatory suppression effect: The finding that people have poorer memory for a set of words if they are asked to say something while trying to remember the words.

Associative agnosia: A form of agnosia in which the individual can combine perceived features into a whole pattern but cannot associate the pattern with meaning, cannot link the perceived whole with stored knowledge about its identity.

associative interference: Retrieval interference in memory that results in poorer performance (people are less accurate and/or slower) when information is associated with a target memory, but irrelevant. For example, it is harder to remember several facts about a person compared to just one.

atmosphere heuristic: A logical reasoning heuristic in which people prefer to draw conclusions that use the same terms as the premises (e.g. if there is some in the premises, people prefer a conclusion that uses some).

attentional biases: Biases in the way we process or interpret stimuli from the environment.

attentional blink: A brief slowdown in mental processing due to having processed another very recent event.

attention capture: The spontaneous redirection of attention to stimuli in the world based on physical characteristics.

authorised inferences: Intended or correct. An implication of a speaker's statement is said to be authorised if the speaker intended the implication to be drawn; if the listener draws the intended inference, the inference is said to be authorised (contrast with *unauthorised inference*).

autobiographical memory (ABM): Memories of specific, personally experienced real-world information, such as of one's activities upon learning of the *Challenger* space shuttle disaster; the study of those memories.

automatic: Occurring without conscious awareness orintention and consuming little, if any, of the available mental resources.

automaticity: Occurring without conscious awareness or intention and consuming little, if any, of the available mental resources.

availability: (in memory research) Present in the memory system. Information is said to be available if it is currently stored in memory (contrast with *accessibility*).

axon: The branch-like ending of the axon in the neuron, containing neurotransmitters.

B

balanced bilingual: People who are equally fluent in both languages, not only verbally but also in written and expressive language.

base-rate information: Information about the frequency or prevalence of an event or a phenomenon in a given context (e.g. the mean prevalence of smokers in a country, the percentage of males and female employees in a company).

behavioural: A set of quantitative and/or qualitative methods that systematically measure human behaviour.

behaviourism: The movement or school of psychology in which the organism's observable behaviour was the primary topic of interest; and the learning of new stimulus–response associations, whether by classical conditioning or by reinforcement principles, was deemed the most important kind of behaviour to study.

beliefs: The fifth level of analysis of language, according to Miller, in which the listener's attitudes and beliefs about the speaker influence what is comprehended and remembered.

benefit: Also known as facilitation effect, it reflects faster-than-baseline reaction times resulting from the prior presentation of useful information.

beta movement: Illusory movement that occurs when two or more pictures are viewed in rapid succession, as in a film.

bottom-up, data-driven processing: When mental processing of a stimulus is guided largely or exclusively by the features and elements in the pattern itself, this processing is described as being data-driven (contrast with *conceptually driven processing*).

bridging inference: Clark's (1977) term for the mental processes of reference, implication and inference during language comprehension. Metaphorically, a bridge must be drawn from the back to Gary to comprehend the sentence 'Gary pretended he wasn't interested'.

Broca's aphasia: A form of aphasia characterised by severe difficulties in producing spoken speech; that is, the speech is hesitant, effortful and distorted phonemically (contrast with *Wernicke's aphasia*). The aphasia is caused by damage in Broca's area, a region of the cortex next to a major motor control centre.

Broca's area: Site of brain damage associated with Broca's aphasia; located at the rear of the left frontal lobe, a region of the cortex next to a major motor control centre.

Brodmann's areas: These are numbered areas of the cortex that were identified by an analysis of physical differences in different parts of the brain. These numbers are useful in locating general areas of the cortex.

C

case grammar: An approach in psycholinguistics in which the meaning of a sentence is determined by analysing the semantic roles or cases played by different words, such as which word names the overall relationship and which names the agent or patient of the action. Other cases include time, location and manner.

case role: One of the various semantic roles or functions of different words in a sentence (see also *case grammar*).

categorical perception: The perception of similar language sounds as being the same phoneme, despite the minor physical differences among them; for example, the classification of the initial sounds of cool and keep as both being the /k/ (hard c) phoneme, even though these initial sounds differ physically.

central executive: In Baddeley's working memory model, the mechanism responsible for assessing the attentional needs of the different subsystems and furnishing attentional resources to those subsystems. Any executive or monitoring component of the memory system that is responsible for sequencing activities, keeping track of processes already completed, and diverting attention from one activity to another can be called an executive controller.

cerebral hemispheres: The two major structures in the neocortex. In most people, the left cerebral hemisphere is especially responsible for language and other symbolic processing, and the right for non-verbal, perceptual processing.

cerebral lateralisation: The principle that different functions or actions within the brain tend to be localised in one or the other hemisphere. For instance, motor control of the left side of the body is lateralised in the right hemisphere of the brain.

channel capacity: An early analogy for the limited capacity of the human information-processing system.

choking under pressure: Performance below typical performance levels as a result of emotional stress or pressure at the time of the action.

circularity: A major problem for the depth of processing concerns defining levels of rehearsal independently of retention scores (better memory was held as indicative of deeper processing; thus the depth of processing model could not fail to be consistent with the data).

closure: A Gestalt grouping principle that is used by visual perception to close up gaps in a percept to help identify a whole object.

clustering: The grouping together of related items during recall (e.g. recalling the words *apple, pear, banana, orange* together in a cluster, regardless of their order of presentation) (see also *organisation*).

coarticulation: The simultaneous or overlapping articulation of two or more of the phonemes in a word.

cocktail party effect: Selecting one message in a crowded, noisy environment.

cognition: The collection of mental processes and activities used in perceiving, remembering, thinking and understanding, and the act of using those processes.

cognitive neuroscience: (a.k.a. neurocognition/cognitiveneuropsychology) A hybrid term applied to the analysis of those handicaps in human cognitive functioning that result from brain injury and other neurophysiological effects on cognition.

cognitive psychology: Concerned with the scientific study of diverse mental processes and their relationship with behaviour.

cognitive revolution: The movement away from behaviorism to re-embrace the study of memory and mental activity (instinctive drift; language; perception; judgement; decision-making).

cognitive science: A new term designating the study of cognition from the multiple standpoints of psychology, linguistics, computer science and the neurosciences.

colour opponent theory: A human colour vision theory suggesting that there are three opponent visual channels in the nerve cells of the retina that respond to colours in opposite ways.

common fate: A Gestalt grouping principle that is used by visual perception to group together points that are moving together.

Competence: In linguistics, the internalised knowledge of language and its rules that fully fluent speakers of a language possess, uncontaminated by flaws in performance (contrast with *performance*).

conceptual knowledge: The fourth level of analysis of language in Miller's scheme, roughly equivalent to semantic memory.

conduction aphasia: A disruption of language in which the person is unable to repeat what has just been heard.

connectionism or computational cognitive modelling: The terms refer to a recent development in cognitive theory, based on the notions that the several levels of knowledge necessary for performance can be represented as massive, interconnected networks; that performance consists of a high level of parallel processing among the several levels of knowledge; and that the basic building block of these interconnected networks is the simple connection between nodes stored in memory. For instance, perception of spoken speech involves several levels of knowledge, including knowledge of phonology, lexical information, syntax and semantics. Processing at each level continually interacts with and influences processing at the other levels, in parallel. The connections in connectionist modelling are the network pathways both within and among the levels of knowledge.

connectionist: A computer-based technique for modelling complex systems. Knowledge is represented by the strength of the excitatory or inhibitory connections between massively interconnected nodes.

consequent: In conditional reasoning, the consequent is the then statement; in 'If it rains, then the picnic will be cancelled', the consequent is 'then the picnic will be cancelled'.

consolidation: The more permanent establishment of memories in the neural architecture.

constructive search: A problem-solving strategy of drawing problem-solving inferences based on what is known from other aspects of a problem.

context: The surrounding situation and its effect on cognition, including the concepts and ideas activated during comprehension.

contralaterality: The principle that control of one side of the body is localised in the opposite-side cerebral hemisphere. The fact that the left hand, for instance, is largely under the control of the right cerebral hemisphere illustrates the principle of contralaterality.

controlled attention: The deliberate, voluntary allocation of mental effort or concentration.

control processes: The part of the standard or modal (Atkinson and Shiffrin, 1968, 1971) model of memory responsible for the active manipulation of information in short-term memory.

conversational rules: The rules, largely tacit, that govern our participation in and contributions to conversations.

cooperative principle: The most basic conversational postulate, stating that participants cooperate by sharing information in an honest, sincere and appropriate fashion.

corpus callosum: The fibre of neurons that connects the left and right cerebral hemispheres.

cortex: The band of cortex at the front of the parietal lobes responsible for processing sensory information from throughout the body.

cost: It reflects slower-than-baseline reaction times resulting from the prior presentation of misleading cues.

counterfactual reasoning: A line of reasoning that deliberately contradicts the facts in a 'what if' kind of way; in the simulation heuristic, the changing of details or events in a story to alter the (unfortunate or undesirable) outcome.

critical lure: A word that is highly related to all the other words in a study list but that never was presented to the participant (but which the participant is likely to later report as part of the list).

cued recall: A form of recall in which the person is presented with part of the information as a cue to retrieve the rest of the information.

D

decay: Simple loss of information across time, presumably caused by a fading process, especially in sensory memory; also, an older theory of forgetting from long-term memory.

declarative: Long-term memory knowledge that can be retrieved and then reflected on consciously (see also *explicit memory*).

deep structure: In linguistics and psycholinguistics, the deep structure of a sentence is the meaning of the sentence; a deep structure is presumably the most basic and abstract level of representation of a sentence or idea (contrast with *surface structure*).

default mode network, or DMN: A collection of brain structures that tend to be more active when a person is at rest and not thinking about anything in particular. Hence, it is the part of the brain that is more active by default.

dendrites: The branching, input structures of the neuron.

direct theory: A person's appraisal of or informal theory about the other participant in the conversation, including information about that other person's knowledge, sophistication and personal motives.

displacement: One of Hockett's (1960, 1960b) linguistic universals, referring to the fact that language permits us to talk about times other than the immediate present; language thus permits us to displace ourselves in time, by talking about the past, future, and so on.

dissociation: Pattern of abilities and performance, especially among brain-damaged patients, revealing that one cognitive process can be disrupted while another remains intact. In a double dissociation, two patients show opposite patterns of disruption and preserved function, further evidence that the cognitive processes are functionally and anatomically separate.

distributed practice: This occurs when practice sessions are spaced out in time. This is a more effective way of encoding information.

domain-general: Domain-general cognitive abilities, such as memory, influence performance across contexts and tasks.

domain knowledge: A general term referring to one's knowledge of a specific domain or topic, especially in problem-solving.

dominant: Hemisphere of the brain that is more involved in certain cognitive functions. For example, in a right-handed individual the left hemisphere is dominant for language.

dorsal pathway: The neural pathway across the top of the cortex, stemming from visual processing areas in the occipital lobe, primarily responsible for processing information about where things are in the world.

dorsolateral prefrontal cortex (DLPFC): An area in the prefrontal cortex that plays a functional role in supporting executive functions and cognitive flexibility.

double dissociation: A pattern of impairment that indicates whether cognitive processes or functions operate independently of each other.

downhill change: In the simulation heuristic, an unusual or unexpected aspect of a story or situation that is changed to be more normal or customary. If a story character left work early and was involved in a car accident, a likely downhill change would be to normalize the unusual characteristic and substitute a more customary aspect, such as leaving work on time.

dual-task method: A method in which two tasks are performed simultaneously, such that the attentional and processing demands of one or both tasks can be assessed and varied. Dual-task methodology is commonly used in studies of attention and attention-dependent mental processing.

dysexecutive syndrome: A set of characteristic symptoms, usually following brain damage, that affect executive functions and influence emotional processing and behaviour.

dysfluencies: Error, flaw or irregularity in spoken speech.

E

early selection: Selection or filtering based on early phases of perception (e.g. selection based on physical features of the message such as loudness or location; Broadbent's filter theory).

Easterbrook hypothesis: Easterbrook's idea that at high levels of emotional arousal, there is a narrowing of attention that leads to better memory for details at the focus of an event, but poorer memory for details at the periphery of an event.

ecological validity: The hotly debated principle that research must resemble the situations and task demands that are characteristic of the real world rather than rely on artificial laboratory settings and tasks so that results will generalise to the real world; that is, will have ecological validity.

effect sizes: Magnitude of change that is identified in a study.

Einstellung: Another term for negative set.

elimination by aspects: A decision-making heuristic in which a person makes a choice by eliminating options that do not meet certain criteria.

embodiment: The way we think about and process information reflects the fact that we need to interact with the world using our bodies.

emergent properties: The properties that emerge from collections of elements (e.g. neurons) working together to create a new process of property that the individual elements lack (e.g. reasoning).

emotion: Both the state of mind a person is in at a particular moment, as well as the physiological response a person is experiencing at that time.

emotional Stroop task: A variant of the normal Stroop task in which potentially emotionally arousing words are used (in which people must name the colours they are printed in) to assess the degree to which processing is disrupted by this emotionally arousing content.

empiricism: The philosophical position, originally from Aristotle, that advances observation and observation derived data as the basis for all science.

enactment effect: The finding of improved memory for participant-performed tasks relative to those that are not acted out.

Encoding: To input or take into memory, to convert to a usable mental form, to store into memory. We are said to encode auditory information into sensory memory; if that information is transferred to short-term memory, then it is said to have been encoded into STM.

encoding specificity: Tulving's hypothesis that the specific nature of an item's encoding, including the entire context it was encoded in, determines how effectively the item can be retrieved.

Engle and Kane's controlled attention theory of working memory: A theory of working memory that assumes that working memory is essentially knowledge that is currently being thought about, along with the idea that working memory effectiveness is tied to the ability to control what information is and is not attended to by a person.

enhancement: In Gernsbacher's theory, the boosting of concepts' levels of activation during comprehension.

episodic buffer: The portion of working memory whereby information from different modalities and sources are bound together to form new episodic memories.

episodic buffer: A limited capacity system that integrates information from the phonological loop and the visuo-spatial sketch pad and relevant long-term memory into a an 'episode'.

episodic memory: Tulving's term for the portion of long-term memory in which personally experienced information is stored (contrast with *semantic memory*).

erasure: The masking or loss of information caused by subsequent presentation of another stimulus; usually in sensory memory (see also *masking*).

explicit memory: Long-term memory retrieval or performance that entails deliberate recollection or awareness.

explicit processing: Involving conscious processing, conscious awareness that a task is being performed, and usually conscious awareness of the outcome of that performance.

extrinsic self-processing: The perception of agency over one's thoughts, actions and their consequences.

eye–mind assumption: The assumption that the eye normally remains fixated on a word as long as that word is being actively processed during reading.

eye tracker: A device used to record eye movements and fixations.

F

false alarms: An error in a recognition task in which a response of 'yes' is made to a new stimulus; any 'yes' response in recognition when a 'no' response is correct.

false memory: Memory of something that did not happen.

familiarity bias: In reasoning, the bias in the availability heuristic in which personal familiarity influences estimates of frequency, probability, and so on; judging events as more frequent or important just because they are more familiar in memory.

filtering or selecting: Especially in auditory perception, unwanted, unattended messages are filtered or screened out so that only the attended message is encoded into the central processing mechanism (e.g. Broadbent's filter theory).

fishing: Looking for effects that are not pre-defined with specific hypotheses set out at the beginning of the study.

fixation: Clusters of eye gaze when the eyes are 'locked' on an object. Eye movements between fixations are saccades.

flashbulb memories: Memories of specific, emotionally salient events, reported subjectively to be as detailed and accurate as a photograph but now considered possibly to be more similar to normal, highly accurate memories.

flexibility: The characteristic that enables the meaning of a language symbol to be changed and enables new symbols to be added to the language.

fovea: The highly sensitive region of the retina responsible for precise, focused vision, composed largely of cones.

free recall: A recall task in which subjects may recall the list items in any order, regardless of their order of presentation (contrast with *serial recall*).

frontal lobe: Most forward part of cortex, and important for the control of thought and action.

Functional fixedness: In problem-solving, an inability to think of or consider any but the customary uses for objects and tools.

functionalism: The movement in psychology, closely associated with James, in which the functions of various mental and physical capacities were studied (contrast with *structuralism*).

G

GABA: An inhibitory neurotransmitter involved in weakening connections between neurons during learning.

garden path sentences: A sentence in which an early word or phrase tends to be misinterpreted and thus must be reinterpreted after the mistake is noticed; for example, 'After the musician had played the piano was quickly taken off the stage'.

gaze duration: How long the eyes fixate on a specific word during reading, the principal measure of online comprehension during reading.

general problem-solver (GPS): The first serious computer-based model of problem-solving, by Newell, Shaw, and Simon (1958).

generate and test: A problem-solving strategy of deriving possible solutions to a problem or subgoal and testing each one's effectiveness.

generation effect: The finding that information you generate or create yourself is better remembered compared to information you have only heard or read.

Gestalt: A German term adopted into psychological terminology referring to an entire pattern, form or configuration. The term always carries the connotation that decomposing a pattern into its components in some way loses the essential wholeness of the cohesive pattern.

Gestalt: A German term adopted into psychological terminology referring to an entire pattern, form or configuration. The term always carries the connotation that decomposing a pattern into its components in some way loses the essential wholeness of the cohesive pattern.

gesture: The movement of the hands and arms done to facilitate communication to listeners. This excludes sign language and non-communicative mannerisms, such as touching one's hands to one's face.

given-new strategy: The idea that words and phrases that contain more accessible information (more active in memory), or given information, tend to occur earlier in sentences, whereas new information in a sentence tends to come later.

global–local distinction: When processing visual information, such as a scene or objects, one can focus attention on either the larger whole that the parts make up, or the smaller elements that make up the whole. In studies of global versus local processing, a larger letter (such as an H) may be composed of several smaller letters (such as Rs).

glutamate: An excitatory neurotransmitter involved in strengthening connections between neurons during learning.

goal: In problem solving, the end point or solution to the problem; the ending state toward which the problem-solving attempt is directed.

good continuation: A Gestalt grouping principle that is used by visual perception to structure and organise together elements based on edges and trajectories so that a visually simple solution is obtained.

grammar: In linguistics and psycholinguistics, a set of rules for forming the words or sentences in a language; optimally, the complete set of rules that characterise a language, such that the rules generate only acceptable or legal sentences and do not generate any sentences that are unacceptable.

H

habituation: A gradual reduction of the orienting response back to baseline.

hemineglect: A disorder of attention in which half of the perceptual world, often the left, is neglected to some degree or cannot be attended to.

hemispheric specialisation: The principle that each cerebral hemisphere has specialised functions and abilities.

HERA model: Hemispheric Encoding/Retrieval Asymmetry model is an attempt by Tulving to explain patterns of neural activity data collected during episodic encoding and retrieval along with semantic retrieval. In general, the left hemisphere is more involved in episodic encoding and semantic retrieval, whereas the right hemisphere is more involved in episodic retrieval.

Heuristics: An informal 'rule of thumb' method for solving problems, not necessarily guaranteed to solve the problem correctly but usually much faster or more tractable than the correct algorithm.

heuristic theory: A theory of human mental reasoning in which people reach conclusions not based on logical steps, but based on convenient shortcuts and rules of thumb.

hidden units: The level between the input and output units in a simple three-level connectionist model.

hindsight bias: In reasoning, the bias or attitude that some completed event was very likely to have had just thatoutcome; the likelihood of misremembering a past state of mind as being more consistent with a current state of mind (Chs. 11, 15).

hippocampus: An internal brain structure, just internal to the temporal lobes, strongly implicated in the storing of new information into long-term memory.

I

ill-defined problems: A problem in which the initial, intermediate or final goal state is poorly or vaguely defined or a problem in which the legal operators (moves) are not well specified.

illicit conversion: A logical reasoning heuristic in which people inappropriately swap the terms in a premise, such as mentally converting "Some B are A" to "Some A are C.

immediacy assumption: The assumption that readers try to interpret each content word of a text as that word is encountered during reading.

implanted memories: False memories that are placed in a person from some external source.

implicit memory: Long-term memory performance affected by prior experience with no necessary awareness of the influence.

implicit processing: Processing in which there is no necessary involvement of conscious awareness.

incubation: The time when a person stops actively and consciously thinking about a problem, allowing a solution to present itself.

Independent and non-overlapping stages: The assumption in the strict information processing approach that the stages of processing are independent of one another in their functioning, and that they do not overlap in time. In other words, a stage begins its operations only when a previous stage has finished, and those operations are not changed by previous or subsequent stages.

indirect requests: A question or statement that is not intended to be taken literally but instead is a polite way of expressing the intended meaning; for example, 'Do you have the time?' is an indirect way of asking 'What time is it?'

infantile amnesia: The inability to remember early life events and very poor memory for your life at a very young age.

inhibition: An active suppression of mental representations of salient but irrelevant information so that the activation level is reduced, perhaps below the resting baseline level.

inhibition of return: A process in which recently checked locations are mentally marked by attention as places that the search process would not return to.

input units: The level analogous to the receptors in a simple three-level connectionist model.

insight: Said to be an essential step in creativity and problem-solving, though little, if any, research supports this notion empirically.

integration: When memories from different experiences are combined, they are integrated into a common memory trace. After this occurs, it is often difficult for the person to identify individual experiences.

intensity: The relative strength of an emotional experience.

Interference: When certain memories prohibit the recall of other memories.

intrinsic self-processing: The perception of personally relevant information, and is conceptually closer to *identity*.

introspection: The largely abandoned method of investigation in which subjects look inward and describe their mental processes and thoughts; historically, the method of investigation promoted by Titchener.

involuntary memory: Consists of autobiographical memories that come unbidden, often in response to some environmental cue, such as an odour.

J

just noticeable difference, or JND: The smallest amount of physical change in a stimulus a person can detect.

K

knowledge-lean: Knowledge-lean problems require little or no expertise or prior knowledge to be resolved.

knowledge-rich: knowledge-rich problems require prior knowledge or a certain level of expertise to be resolved.

L

Language: A shared symbolic system for communication.

lateral geniculate nucleus: A neural structure in the thalamus that mediates the transmission of information from the retina to the visual cortex.

late selection: Selection or filtering based on the meaning and importance of information (e.g. selection based on semantic relevance).

levels of processing: Craik and Lockhart's (1972) alternative to the standard three-component memory model. Information subjected only to maintenance rehearsal is not being processed more deeply into the meaning-based levels of the memory system and therefore tends not to be recalled or recognised as accurately as information subjected to elaborative rehearsal.

lexical decision task: A simple yes/no task in which subjects are timed as they decide whether the letter string being presented is a word; sometimes called simply the *word/non-word task* (Chapters 1, 7).

limbic system: The limbic system is a set of structures in the brain that play a crucial role in emotional processing.

linguistic relativity hypothesis: The hypothesis, credited to Whorf (1956), that one's language determines – or at least influences strongly – what one can think about.

Linguistics: The discipline that studies language as a formal system.

linguistic universals: Features and characteristics that are universally true of all human languages (see also *displacement, productivity*).

Lobes (of the brain): Frontal, parietal, occipital and temporal.

Long-term memory: The portion of the memory system responsible for holding information for more than a period of seconds or minutes; virtually permanent storage of information.

long-term potentiation (LTP): The temporary (days, weeks or months) strengthening of connections between neurons as a temporary storage of memories prior to consolidation.

lookahead: When a problem-solver considers the different steps in the future involved in solving the problem.

M

magnetic resonance imaging (MRI): Non-invasive imaging technique that produces three-dimensional images of parts/structures of your body.

mapping: In Gernsbacher's theory, drawing the connections between words and their meanings to the overall meaning of the sentence; in general, the process of determining the connections between two sets of elements, including the relations in analogical problem-solving.

masking: An effect, often in perception experiments, in which a mask or pattern is presented very shortly after a stimulus and disrupts or even prevents the perception of the earlier stimulus (see also *erasure*).

means–end analysis: A major heuristic in problem-solving, assessing the distance between the current state and the goal or subgoal state, then applying some operator that reduces that distance.

memory: The mental processes of acquiring and retaining information for later retrieval; the mental storage system that enables these processes.

memory impairment: A specific interpretation of early eyewitness memory results in which a subsequent piece of information replaces a memory formed earlier, thus impairing memory of the original information.

mental model: The mental representation of a situation or physical device; for example, a person's mental model of the physical motion of bodies or a person's mental model of a thermostat.

mental model theory: A theory of human mental reasoning in which people reach conclusions by creating mental simulations of the circumstances described in the premises.

mental ontology: Refers to a set of concepts that define the properties and relationships between mental processes.

mental rules theory: A theory of human mental reasoning in which people reach conclusions using rules or procedures about how to reach conclusions.

Metacognition: Awareness and monitoring of one's own cognitive state or condition; knowledge about one's own cognitive processes and memory system.

metacognitions: Awareness and monitoring of one's own cognitive state or condition; knowledge about one's own cognitive processes and memory system.

metacomprehension: The ability to monitor how well we are understanding and will remember information later.

meta-linguistic awareness: Being able to reflect on language and evaluate it.

metamemory: Knowledge about one's own memory system and its functioning.

method of loci: A classic mnemonic device in which the to-be-remembered items are mentally placed, one by one, into a set of prememorized locations, with retrieval consisting of a mental walk through the locations.

Mind wandering: The situation in which a person's attention and thoughts wander from the current task to some other, inappropriate line of thought.

mirror neurons: Neurons in the cortex specialised for planning and executing one's own movement, as well as simulating the movement of others that are being observed.

misinformation effect: Incorrectly claiming to remember information that was not part of some original experience.

misleading information: Incorrect information about an event that is encountered after the event. This may then distort a person's memory for what actually happened in the original event.

mnemonic device: Any mental device or strategy that provides a useful rehearsal strategy for storing and remembering difficult material; see *method of loci*, for instance.

modality effect: In sensory memory research, the advantage in recall of the last few items in a list when those items have been presented orally rather than visually.

modal model of memory: The standard model of memory derived by Atkinson and Shiffrin (1968, 1971), which is made of three primary components: the sensory registers, the short-term store and the long-term store.

morphemes: The smallest unit of meaning in language.

motor cortex: The band of cortex at the back of the frontal lobe responsible for processing information about voluntary muscle movements throughout the body.

motor theory of speech perception: The idea that people perceive language, at least in part, by comparing the sounds that they are hearing with how they themselves would move their own vocal apparatus to make those sounds.

multiconstraint theory: Holyoak and Thargard's (1997) multiconstraint theory of analogical problem-solving claims that we work under three constraints as we develop analogies: constraints related to the similarity of the source and target domains (superficial similarities can be misleading), the structure of the problems (must be parallel mappings), and our purposes or goals in developing the analogies.

myelin sheath: The fatty coating on a neuron's axon that can facilitate neural communication.

N

N400: ERP function associated with semantic anomaly.

naming: The characteristic that human languages have names or labels for all the objects and concepts encountered by the speakers of the language (e.g. as opposed to most animal communication systems).

negative afterimages: Perceiving that an image of an object remains after gazing away from it, which complements in brightness and colour the original stimulus (e.g. gazing away from a yellow image produces a blue negative afterimage).

negative priming: Slower to respond to the target trials when they were preceded by irrelevant distractor primes compared to control trials where the ignored object on the prime trial was an unrelated item.

negative set: In problem-solving, a tendency to become accustomed to a single approach or way of thinking about a problem, making it difficult to recognise or generate alternative approaches.

neocortex (or cerebral cortex): The top layer of the brain, newest (*neo-*) in terms of the evolution of the species, divided into left and right hemispheres; the locus of most higher-level mental processes.

networks: Brain networks involve several regions that form part of circuits that are activated during a mental process.

neural imaging: Quantitative non-invasive techniques that measure neural activity in the brain. Examples of such methods include MRI, fMRI, PET, EEG.

neurogenesis: The creation of new neurons. This may be involved in memory formation even into adulthood.

neuron: A specialised cell that conducts neural information through the nervous system; the basic building block of the nervous system.

nodes of Ranvier: Gaps along the myelin sheath that allow the action potential to jump from one point to another, thereby speeding neural communication.

norepinephrine: A neurotransmitter that is involved in the creation of new memories.

null findings: Findings that disprove the hypothesis.

O

occipital lobe: The lobe at the back of the brain that is most heavily involved in vision.

online comprehension task: Task in which measurements of performance are obtained as comprehension takes place; *online* means happening and being measured right now.

onomatopoeia: When a name is based on its referent sound (*buzz, hum, zoom*, etc.). Onomatopoeia is an exception to *arbitrariness*.

operator: In problem-solving, a legal move or operation that can occur during solution of a problem; the set of legal moves within some problem space (e.g. in algebra, one operator is 'multiply both sides of the equation by the same number').

organisation: Especially in studies of episodic long-term memory, the tendency to recall related words together, or the tendency to impose some form of grouping or clustering on information being stored in/retrieved from memory; related to chunking or grouping in short-term memory.

orienting reflex: The reflexive redirection of attention that orients you toward the unexpected stimulus.

output units: The level at which the input pattern has been categorised in a simple three-level connectionist model.

overlearning: The improved memory that results when a person continues to study material after it has already been memorised.

P

P600: ERP function associated with syntactic anomaly.

parallel processing: Any mental processing in which two or more processes or operations occur simultaneously.

parietal lobe: Portion of the cortex on the top, behind the frontal lobe and in front of the occipital lobe. This part of the cortex is important for sensory processing, spatial processing and working memory.

parse: To divide or separate the words in a sentence into logical or meaningful groupings.

partial report condition: An experimental condition in Sperling's (1960) research in which only a randomly selected portion of the entire stimulus display was to be reported (contrast with *whole report condition*).

part-set cueing effect: The finding that if you cue people with part of a list of words, they will have more difficulty recalling the rest of the set than if they had not been cued at all.

Performance: Any observable behaviour; in the context of linguistics, any behaviour related to language (e.g. speech), influenced not only by linguistic factors but also by factors related to lapses in attention, memory and so on (contrast with *competence*).

phi phenomenon: Illusory movement that occurs when two images are viewed in rapid succession in different points in space, as in the signage above a theatre entrance or chasing Christmas lights.

phonological loop: In Baddeley and Hitch's (1974) working memory model, the articulatory loop is the component responsible for recycling verbal material via rehearsal.

phonological similarity effect: The finding that memory is poorer when people need to remember a set of words that are phonologically similar, compared to a set of words that are phonologically dissimilar.

phonological store: The passive store component of the phonological loop.

Phonology: The study of the sounds of language, including how they are produced and how they are perceived.

phrase structure: The underlying structure of a sentence in terms of the groupings of words into meaningful phrases, such as '[The young man] [ran quickly]'.

polysemy: When a word in a language has multiple meanings.

pop-out effect: In visual search, when a target item is highly discriminable from the distractor items, the attentional search mechanisms use a basic rapid attentional mechanism that operates in parallel across the visual field in a highly automatic fashion. The resulting visual search function is quite fast overall and shows only a minimal, if any, effect of set size.

Positron emission tomography (PET): Non-invasive imaging method that enables researchers and clinicians to observe the metabolic or biochemical function of organs in the body.

power: The ability of a study to show true effects when they are present.

pragmatics: The aspects of language that are 'above and beyond' the words, so-called extra-linguistic factors. For instance, part of our pragmatic knowledge of language rules includes the knowledge that the sentence, 'Do you happen to know what time it is?' is actually an indirect request rather than a sentence to be taken literally.

Primacy effect: In recall performance, the elevated recall of the early positions of the list (contrast with *recency effect*).

prime: The first stimulus in a prime–target pair, intended to exert some influence on the second stimulus (see also *priming*) (Ch. 7).

priming: Mental activation of a concept by some means, or the spread of that activation from one concept to another; also, the activation of some target information by action of a previously presented prime; sometimes loosely synonymous with the notion of accessing information in memory.

proactive interference (PI): Interference or difficulty, especially during recall, because of some previous activity, often the stimuli learned on some earlier list; any interference in which material presented at one time interferes with material presented later.

probability heuristics theory: A theory of human mental reasoning in which people reach conclusions using subjective estimates of the probability of events.

Probes: The target stimuli that need to be detected among study items (memory sets).

problem-solving: Involves a goal-directed sequence of steps, the cognitive operations used to work through a problem, and can be broken down into a collection of subgoals.

problem space: The initial, intermediate and goal states of a problem, along with the problem-solver's knowledge and any external resources that can be used to solve the problem.

Processing fluency: The ease with which something is processed or comes to mind.

process model: A stage model designed to explain the several mental steps involved in performance of some task, usually implying that the stages occur sequentially and that they operate independently of one another.

production: A simple *if–then* rule in models of memory processing, stating the conditions (*if*) necessary for some action (*then*) to be taken, whether that action is a physical response or a mental step or operation.

production system: A large-scale model of some kind of performance or mental activity based on productions.

productivity (*also called generativity*): One of Hockett's linguistic universals, referring to the rule-based nature of language, such that an infinite number of sentences can be generated or produced by applying the rules of the language.

propagated: The movement of an action potential from the dendrites, through the soma, and down the axon.

proposition: A simple idea unit.

propositional textbase: An intermediate level of representation that captures the basic idea units present in a text.

prosody: The up and down pitch of an utterance that can convey emotional information.

prosody: Prosody refers to the patterns of rhythm, stress and intonation in speech and plays a crucial role in how language is perceived and understood.

Prospective memory: Remembering to do something in the future; e.g. remembering to make a phone call tomorrow.

proximity: A Gestalt grouping principle that is used by visual perception to structure and organise together elements of a visual input that are nearer to one another in space.

psycholinguistics: The study of language from the perspective of psychology; the study of language behaviour and processes.

psychological refractory period: A delay in a second decision or response cycle if it is required immediately after a preceding decision.

psychophysics: Field of psychology concerned with the relationship between the properties of physical stimuli and sensory processes.

pure word deafness: Disruption of the perceptual or semantic processing in auditory word comprehension.

R

Recency effect: In recall performance, the elevated recall of the last few items in a list, presumably because the items are stored in and retrieved from short-term memory (contrast with *primacy effect*).

recoding: Mentally transforming or translating a stimulus into another code or format; grouping items into larger units, as when recoding a written word into an acoustic–articulatory code.

recognition task: Any yes/no task in which subjects are asked to judge whether they have seen the stimulus before; more generally, any task asking for a simple yes/no (alternatively, true/false, same/different) response, often including a reaction time measurement.

reconsolidation: When a memory is retrieved, this puts it in a plastic, malleable state where it can be changed before it is stored in memory again.

reductionism: The scientific approach in which a complex event or behaviour is broken down into its constituents; the individual constituents are then studied individually.

reductionistic approach: Understanding complex behaviours or processes by breaking them down into their various components.

reference: In language, the allusion to or indirect mention of an element from elsewhere in the sentence or passage, as by using a pronoun or synonym.

rehearsal: The mental repetition or practising of some to-be-learned material.

reminiscence bump: Superior memory than would otherwise be expected for life events around the age of 20, between the ages of 15 and 25.

repeated name penalty: An increase in reading times when a direct reference is used again (e.g. the person's name) compared to when a pronoun is used.

repetition blindness: The tendency to not perceive a pattern, whether a word, a picture or any other visual stimulus, when it is quickly repeated.

repetition priming: A priming effect caused by the exact repetition of a stimulus; often used in implicit memory tests.

representation (or representation of knowledge): A general term referring to the way information is stored in memory. The term always carries the connotation that we are interested in the format or organisation of the information as it is stored. (Is the information stored in a semantic representation? A sound-based representation?)

representativeness heuristic: A reasoning heuristic in which we judge the likelihood of some event by deciding how representative that event seems to be of the larger group or population from which it was drawn.

response time (RT): The elapsed time, usually measured in milliseconds, between some stimulus event and the subject's response to that event; a particularly common measure of performance in cognitive psychology.

retina–geniculate–striate pathway: A group of neurons connecting the thalamus (lateral geniculate nucleus) to the striate cortex.

retrieval: Accessing information stored in memory, whether or not that access involves conscious awareness.

retrieval cues: Any cue, hint or piece of information used to prompt retrieval of some target information.

retrieval-induced forgetting: Forgetting that occurs when one portion of a set of information is retrieved, but the remainder is not. The remaining part of the set is thought to be inhibited or suppressed, and so is less likely to be remembered.

retroactive interference (RI): The interference from a recent event or experience that influences memory for an earlier event, such as trying to recall the items from list 1 but instead recalling the items from list 2.

Ribot's law: For retrograde amnesia, memory is more affected the closer it is to the time of the injury, with older memories being more likely to be preserved.

S

saccade: Saccades Fast, automatic movement of both eyes

salami: Using the same set of data for more than one publication without declaring it to the publisher.

Sapir–Whorf hypothesis: See *linguistic relativity hypothesis.*

savings score: In a relearning task, the score showing how much was saved on second learning compared with original learning. For instance, if original learning took 10 trials and relearning took only 6 trials, then savings would be 40% (10 – 6)/10.

second-order theory of mind: In conversation, the informal theory we develop that expresses our knowledge of what the other participant knows about us, summarised by the phrase 'what he/she thinks I know' (contrast with *direct theory*).

selective attention: The ability to attend to one source of information while ignoring or excluding other ongoing messages.

self-reference effect: The finding that memory is generally better for information that is related to the self in some way.

self-reflection: Our ability to introspect, interpret and evaluate our thoughts, emotions and actions.

self-understanding: Knowledge about our own characteristics, traits, motivations, strengths and weaknesses.

semanticity: One of Hockett's (1960a, 1960b) linguistic universals, expressing the fact that the elements of language convey meaning.

semantic memory: The long-term memory component in which general world knowledge is stored (contrast with *episodic memory*).

semantics: The study of meaning.

Sensory memory: The initial mental storage system for sensory stimuli. There are presumably as many modalities of sensory memory as there are kinds of stimulation that we can sense.

sequential stages of processing: An assumption in most process models that the separate stages of processing occur in a fixed sequence, with no overlap of the stages.

serial position curve: The display of accuracy in recall across the original positions in the to-be-learned list, often found to have a bowed shape, indicating lower recall in the middle of the list than in the initial or final position.

serial processing: Mental processing in which only one process or operation occurs at a time.

serial recall: A recall task in which subjects must recall the list items in their original order of presentation (contrast with *free recall*).

Short-term memory: The component of the human memory system that holds information for up to 20 seconds; the memory component where current and recently attended information is held; sometimes loosely equated with attention and consciousness.

similarity: A Gestalt grouping principle that is used by visual perception to structure and organise together elements of a visual input that have similar visual features, such as texture, colour, brightness, and so forth.

situation model: A memory representation of a real- or possible-world situation; for example, of a situation described in a passage of text.

social cognition: Social cognition is concerned with how we perceive, store, process and use information from our social environment, and how we plan and execute actions based on such information.

social cues: Verbal or non-verbal indicators (e.g. body language, tone of voice) that convey information about the intentions of another person and guide conversations.

social neuroscience: Interdisciplinary field of research that investigates the roles of neural and biological processes in social behaviour.

sodium–potassium pump: An enzyme in the membrane of cells that regulates the concentration of ions (such as sodium and potassium) in the cell.

soma: The cell body of a neuron.

Source memory: Memory of the exact source of information.

span of apprehension: The number of simple elements (e.g. digits, letters) that can be heard and immediately reported in their correct order; a standard short-term memory task, common on standardised intelligence tests.

span of attention: See *span of apprehension.*

speech act: The intended consequence of an utterance. That is, what you are trying to accomplish when you say something.

split-brain: Refers to patients in whom the corpus callosum has been severed surgically and the resultant changes in their performance because of the surgery or, more generally, to research showing various specialisations of the two cerebral hemispheres.

spotlight attention: A rapid attentional mechanism operating in parallel and automatically across the visual field, especially for detecting simple visual features.

spotlight metaphor: Attention is thought to resemble a spotlight that is directed to a specific location and preferentially attended.

spreading activation: The commonly assumed theoretical process by which long-term memory knowledge is accessed and retrieved. Some form of mental excitation or activation is believed to be passed or spread along the pathways that connect concepts in a memory network. When a concept has been activated, it has been retrieved or accessed within the memory representation. The process is loosely analogous to the spread of neural excitation in the brain.

stereotype threat: Unconscious activation of negative stereotypes leads a person to perform worse on a task than he or she would otherwise.

stereotyping: A generalised belief of judgement about other people based on their social group membership.

Sternberg task: The short-term memory scanning task devised by Saul Sternberg.

story mnemonic: A memory aid in which people construct a narrative story containing the material to help them remember it later.

structuralism: The approach, most closely identified with Titchener, in which the structure of the conscious mind – that is, the sensations, images and feelings that are the elements of consciousness – was studied; the first major school of psychological thought, beginning in the late 1800s (contrast with *functionalism*).

structure-building: The process of comprehension in Gernsbacher's theory of building a mental representation of the meaning of sentences.

subgoal: In problem-solving, an intermediate goal that must be achieved to reach a final goal.

subjective organisations: The grouping or organising of items that are to be learned according to some scheme or basis devised by the subject.

subject-object-verb (SOV): Place the subject and the object before the verb, as in 'Tom, the ice cream, ate'.

subject-verb-object (SVO) language: Place the subject before the verb and followed by the object, as in 'Tom ate the ice cream'.

suppression: In Gernsbacher's theory, the active process of reducing the activation level of concepts no longer relevant to the meaning of a sentence.

surface form: The level of representation in language comprehension that corresponds to a verbatim mental representation of the exact words and syntax used in a passage of text.

surface structure: In linguistics and psycholinguistics, the actual form of a sentence, whether written or spoken (contrast with *deep structure*); the literal string of words or sounds present in a sentence.

synapse: The junction of two neurons; the small gap between the terminal buttons of one neuron and the dendrites of another; as a verb, to form a junction with another neuron.

synaptic cleft: The synaptic cleft (or synaptic gap) is the gap between the pre-synaptic axon terminals and the post-synaptic dendrites.

syntax: The arrangement of words as elements in a sentence to show their relationship to one another; grammatical structure; the rules governing the order of words in a sentence.

T

tabula rasa: Latin term meaning 'blank slate'. The term refers to a standard assumption of behaviourists that learning and experience write a record on the blank slate; in other words, the assumption that learning, as opposed to innate factors, is the most important factor in determining behaviour.

task effects: A second major difficulty for depth of processing (along with *circularity*). Different memory tasks revealed differences in memory performance that were incompatible with a unitary processing mechanism.

temporal lobe: The lobe of the cortex on the sides, below the frontal and parietal lobes. This lobe is important for audition and memory.

thalamus: ('inner room'; 'inner chamber'; 'gateway to the cortex') Major relay station from the sensory systems of the body to the cortex; almost all messages entering the cortex come through the thalamus.

Theory of Mind/ToM: Our ability to infer that other people have desires, emotions, intentions and beliefs that are possibly different from our own.

theory of mind (ToM): Theories we develop of our conversational partners.

top-down or conceptually driven processing: Mental processing is said to be conceptually driven when it is guided and assisted by the knowledge already stored in memory (contrast with *data-driven processing*).

topic maintenance: Making conversational contributions relevant to the topic, sticking to the topic.

transformational rules: In Chomsky's transformational grammar, the syntactic rules that transform an idea (a deep structure sentence) into its surface structure; for instance, rules that form a passive sentence or a negative sentence.

trichromacy theory: A theory that posits that human colour vision results from the activation of specialised nerve cells in the retina that respond to blue, green and red.

U

unauthorised inferences: Not intended, especially said of inferences drawn during a conversation (contrast with *authorised inference*).

unconscious processes: Mental processing outside of awareness.

undoing: See counterfactual reasoning.

units: A unit or grouping of information held in short-term memory.

V

valence: Whether an emotion is positive (e.g. happy) or negative (e.g. angry).

ventral pathway: The neural pathway along the bottom of the cortex, stemming from visual processing areas in the occipital lobe, primarily responsible for processing information about what things are in the world.

ventromedial prefrontal cortex (vmPFC): Part of the cortex involved in the identification and interpretation of emotional stimuli and responses.

Verbal learning: The branch of human experimental psychology, largely replaced by cognitive psychology in the late 1950s and early 1960s, investigating the learning and retention of 'verbal', that is, language-based, stimuli; influenced directly by Ebbinghaus's methods and interests.

verbal protocol: In studies of problem-solving, a word-for-word transcription of what the subject said aloud during the problem-solving attempt.

vigilance or sustained attention: The maintenance of attention for infrequent events over long periods of time.

visual attention: Input attention (specifically as it relates to vision; typically associated with the spotlight metaphor).

visual imagery: The mental representation of visual information; the skill or ability to remember visual information.

visual search: Search a spatial display of items (e.g. a set of 4, 8, or 16 letters) for the presence of a target (e.g. the letter K).

visuo-spatial sketch pad: The visual and perceptual component of Baddeley's working memory model.

voicing: Refers to whether the vocal cords begin to vibrate immediately with the obstruction of airflow (/b/ in *bat*) or whether the vibration is delayed until after the release of air (/p/ in *pat*).

von Restorff effect: In a recall task, the elevated accuracy for an item that was noticeably different during list presentation; for instance, because it was written in a different colour of ink.

W

Well-defined problems: A problem in which the initial and final states and the legal operators are clearly specified.

Wernicke's aphasia: One of two common forms of aphasia in which the language disorder is characterised by a serious disruption of comprehension and the use of invented words as well as semantically inappropriate substitutions (contrast with *Broca's aphasia*). The aphasia is caused by damage in the region of the neocortex called *Wernicke's area.*

whole report condition: Especially in Sperling's (1960) research, the condition in which the entire visual display was to be reported (contrast with *partial report condition*).

word frequency effect: Finding that frequent words in the language are processed more rapidly than infrequent words.

word stem completion task: Complete word stems (e.g. *D___ K*) to form a complete word, generally the first word that comes to mind (e.g. *DUCK*).

working memory: The component, similar to short-term memory, in Baddeley and Hitch's (1974) theory in which verbal rehearsal and other conscious processing take place; also, the component that contains the executive controller in charge of devoting conscious processing resources to the various other components in the memory system.

working memory spans: The amount of information a person can actively maintain in working memory at one time.

Y

Yerkes-Dodson law: The inverted u-shaped function that shows that memory is best at moderate levels of emotional arousal, but poorer at low and high levels of arousal.

Index

Note: Page numbers in **bold** refer to glossary entries

A

absent-mindedness (memory) 214–15
absolute thresholds 57
acalculia 290
Accelerated Long-term Forgetting (ALF) 243
accessibility 213–14, 223, **555**
accessibility heuristic *see* availability heuristic
accuracy
  defined 214, **555**
  measures 21–3
  overconfidence 233–4
  of response **555**
acetylcholine 40, **555**
action potential 38–9, **555**
adaptive memory 193
adaptive thinking 367–72
Adolphs, Ralph 445
advantage of clause recency 312–14, **555**
advantage of first mention 312–13, **555**
affective neuroscience 426
affectivism 445, **555**
agnosia 90–2, 290
  apperceptive 92, **555**
  associative 92, **555–6**
  implications for cognitive science 92–3
agraphia 290, 292, **555**
AI (artificial intelligence) 381–2
alcohol
  attention 116, 132
  empathy 436
  state-dependent learning 195
alertness 105–7, **555**
alexia 290, 292, **555**
algorithmic bias 381–2, **555**
Allen, Rick 5
all-or-none principle 39, **555**
Alzheimer's disease 237
  forgetting 222, 243
  functional magnetic resonance imaging 479
ambiguity **555**
  conceptual knowledge 298
  conversation 325
  polysemy 279
  reference 311
  semantics and syntax 284
  transformational grammar 271–2
American Optical Hardy-Rand-Rittler test 65
American Psychological Association 102
amnesia 213, 236–43, **555**
  anterograde 237–41
    problem-solving 422
    social cognition 430
    syntactic priming 273
  feigned 233
  infantile 199–200, **561**
  related to dissociations 237
  related to focal brain lesions 237
  related to time of injury 236–7
  retrograde 236–8, 430
  social cognition 430–1
AMPA receptors 40
amygdala 44, **555**
  emotion 448
    and language 456
    and memory 452–3
    and perception 449
  social cognition 427–9, 433, 437
analogy 401–2, 405–8, **555**
  neurocognition 408–12
anaphoric reference 311, **555**
animals
  communication 254–6, **555**
  perception 396
  problem-solving 396–7
anomia/anomic aphasia 290, 292, **555**
  dissociation 35
antecedent 345–49, 351, **555**
  reference 311–13
anterior cingulate cortex (ACC)
  emotion and perception 449
  social cognition 427, 436–7, 439
anterior insula (AI) 427, 429, 436, 440
anterior prefrontal cortex (aPFC) 427, 432
anterograde amnesia 237–41
  problem-solving 422
  social cognition 430
  syntactic priming 273
anxiety
  and decision-making 457–8
  emotion regulation 461
  emotional Stroop task 450
  research
    dot-probe paradigm 466
    emotional Stroop task 469–70
    eye-tracking methodology 472

apes
  communication 255–6
  problem-solving 396–7
aphasia 289–94, **555**
  anomic 35, 290, 292, **555**
  Broca's 290–1, 293–4, **556**
  conduction 290–2, **557**
  Wernicke's 290–1, 293–4, **568**
apperceptive agnosia 92, **555**
apraxia 290
arbitrariness 251–2, 255–6, **555**
Archimedes 395–6, 401
arcuate fasciculus 292
Aristotle 3–5, 8, 10, 12
arithmetic 6
arousal *see* intensity
articulation 259
articulatory loop 151, 160, **555**
articulatory suppression effect 151, **555**
articulatory suppression task 152–3, 155
artificial intelligence (AI) 381–2
association
  cognitive neuropsychology 51
  memory 237
  association neurons 38
  associative agnosia 92, **555–6**
  associative interference 218–20, **556**
  associative visual cortex 442
  associators (synaesthetes) 69
  Atkinson, Richard 172, 181
  atmosphere heuristic 342, **556**
  attack dispersion problem 406–10
  attention 101–2, 132–3
    automatic processes 104–18
    capture 107–9, **556**
      hemineglect 118
    controlled *see* controlled attention
    defined 102
    divided 102, 119
    and emotion 448–51
    empathy 436
    history of cognitive psychology 15, 18
    information processing 18
    input 114–15
      hemineglect 118
    lack of 102, 104
      inattention blindness 119–20, 123
    as limited mental resource 103–4
    loneliness 441–2
    memory
      short-term 134
      working 102, 122, 132, 134
    as mental process 103
    as mental resource 125–32
    to places and objects 103–4
    selective (focused) 102, 118–24
    spotlight 110–11, **566**
    sustained *see* vigilance
    visual/focal 74
      eye-tracking methodology 472–3
attentional biases **556**
  emotional stimuli 450–1
  loneliness 441
  dot-probe paradigm 465–6
attentional blink **556** *see also* psychological refractory period
audience design 327
auditory nerve 93–5
auditory sensation and perception 56–7, 93–100
  from ear to brain 94–5
  selective attention 120
auditory sensory (echoic) memory 95–8
Augustine, St. 10
authorised inferences 316, **556**
autism spectrum disorders (ASD) 436
autobiographical memory (ABM) 198–202, **556**
  emotion 455
automatic 273, **556**
automaticity 103–5, 125–7, **556**
  alertness and vigilance 105–7
  contrasting input and controlled attention 114–15
  disadvantages 129–32
  hemineglect 116–18
  integration with conceptually driven processes 127
  memory 127–8
  orienting reflex and attention capture 107–9
  practice 127–9
  video games as mechanisms for improving attention 115
  visual search 109–14
availability 213–14, **556**
  retrieval cues 223
availability heuristic 358–9
axon 36, **556**

B

Baddeley, Alan 134, 169–70
balanced bilingual 331, **556**
Banaji, Mahzarin 464, 470
Bannerman, David 213
Bartlett, Frederic 13
base-rate information 377, **556**
  ignoring 356–8
  and stereotypes 357–8
Bates, Elizabeth 246, 248
Bayes' theorem 358, 373–4
Beeman, Mark 386
Beethoven, Ludwig van 5
behavioural research methods 465, **556**
behaviourism 388, **556**
  history 12–13, 15–17
  language 247–8, 275

belief bias 339
beliefs 255, **556**
  language comprehension 298
benefit 110, **556**
beta movement 80, **556**
bias 214–15
  algorithmic 381–2, 555
  attentional *see* attentional biases
  confirmation 348–50, 374
  familiarity 359, **559**
  heuristics 354, 371
    availability heuristic 359
    representativeness heuristic 356–8
  hindsight 363–5, **561**
  historical 382
  optimism 364–5
  publication 481
  representational 382
  syllogistic reasoning 339–42
  undoing heuristic 361–5
bilingualism 286–7
  cognitive correlates 331
binomial distribution 384
bipolar cells 61–3
blocking (memory) 214–15
bottom-up, data-driven processing 30, **556**
  object recognition 87, 90
  pattern recognition 81–7
  phonology 264–6
  visual search 114
boundary extension 155–6
brain 33–4, 53–5
  cognitive neuropsychology 51
  connectionism 52–3
  cortical specialisation 47–50
  cortical structures 44–5
  damage *see* brain damage/injury
  dissociations and double dissociations 34–5
  emotion 447–8, 456
    and perception 449
  language 287–9
    aphasia 289–94
    comprehension 304
    development 248
    and emotion 456
    lexical memory 278
    polysemy 279–80
    reference 312–14
    situation models 322
  lesions *see* brain damage/injury
  levels of explanation 50
  neurons 35–42
  principles of functioning 45–7
  problem-solving 395
    creativity in teamwork 412
    insight 410–11
  research methods 465, 474, 478–80
  social cognition 427–9
    responding to adverse social signals 438–43
    understanding others 433–8
    understanding the self 429–32
  social neuroscience 50
  split-brain research and lateralisation 47
  subcortical structures 42–4 *see also specific components*
brain damage/injury 34
  agnosia 91–2
  amnesia 236–41, 243
  aphasia 289–94
  cognitive neuropsychology 51
  dissociations and double dissociations 34–5
  emotion 455–6
  hemineglect 116–18
  language 312, 456
  lateralised processes 47
  processing resource limitations 379–80
  social cognition
    self-referential processing 429
    understanding others 434, 436
    understanding the self 430–1
brainstem 437
Brexit referendum 336–7
bridging inference 316–17, **556**
British Psychological Society (BPS) 480, 482
Broadbent, Donald 18, 121–2
broadcast transmission (linguistic universal) 249
Broca, Pierre 10, 291
Broca's aphasia 290–1, 293–4, **556**
Broca's area 289, 291, 456, **556**
  animal communication 255
  working memory 149
Brodmann's areas 45, **556**
Brown–Peterson task 137–40, 168
  learning and remembering 182
  levels of processing framework 187
Buddhist monk problem 393, 421
Byrne, Ruth 336

C

Cacioppo, John 426
cancer risk 481–2
candle problem 399
cardiovascular disease risk 481–2
case grammar 280–2, **556**
  garden path sentences 286
case role 281, **556**
categorical perception 261, **556**
categorical syllogisms 338
caudate 431
central executive 148–51, 157, 169, **557**
  dual-task method 159–60

central nervous system (CNS) *see also* brain
  attention 102
  neurons 36–8
cerebellum 347, 431–2
cerebral cortex *see* neocortex
cerebral hemispheres 46, 557
  specialisation 45–7
cerebral lateralisation 46–7, 557
channel capacity 19, 557
Cherry, E. Colin 18, 120–1, 123
chess masters 418
chimpanzees
  communication 255
  problem-solving 396–7
choking under pressure 458, 557
Chomsky, Noam 16–17, 247–8, 253–4, 256, 269–73, 275
Cicero, *De oratore* 174–5
circularity 185, 557
closure 76, 557
clustering 189, 557
coarticulation 557
  embodiment in speech perception 266
  problem of invariance 263–4
cocktail party effect 120–1, 166, 557
cognition
  defined 6–8, 557
  themes 29–31
cognitive bias 442–3
cognitive neuropsychology 51
cognitive neuroscience 30, 33–5, 53–5, 557
  aphasia 290
  brain structures and their function 42–50
  cognitive neuropsychology 51
  connectionism 52–3
  neurons 35–42
  parallel processing 27
  social cognition 426, 443
cognitive psychology 1–3, 31–2
  cognition 3–8
  defined 1, 557
  history 8–17
  information processing 17–18
    measurement 18–23
  memory 6–8
  standard theory and cognitive science 24–9
  themes 29–31
cognitive reappraisal 461
cognitive revolution 2, 13–14, 557
  language development 247
cognitive science 2–3, 29, 557
  agnosia 92–3
  attention 101
    automaticity 125
    hemineglect 116
  connectionist modelling 83
  history 8–9
  parallel processing 27
  problem-solving 395
coin tosses 355–6, 384
Collins, Allan 203–5
colour blindness 65
colour opponent theory 65, 557
colour perception 64–6
common fate 77, 557
competence 256, 557
comprehension 297, 333–5
  conversation 323–33
  events 322–3
  gesture 323, 332–3
  levels of 299
  as mental structure building 302–4
  metacomprehension 299–300
  online tasks 301–2, 563
  reading 305–11
  reference 311–14
  research 299
  situation models 314–22
comprehension aphasia *see* Wernicke's aphasia
computation 6
computational cognitive modelling *see* connectionism
computational theory of object recognition 87–8
computers
  cognition analogy 18–19, 26
  connectionism 52–3
  general problem-solver 416
  serial processing 26
conceptual knowledge 255, 298–304, 557
conceptual semantics 281–2
conceptually driven processing *see* top-down (conceptually driven) processing
conditional reasoning 338, 345
  evidence on 347–8
  form errors 348
  hypothesis testing 350–1
  memory-related errors 350
  search errors 348–50
  valid arguments 345–6
conduction aphasia 290–2, 557
cone cells 61–3, 66
  colour perception 64–5
confirmation bias 348–50, 374
conjunction fallacy 371–2
conjunction search 111–14
connectionism 52–3, 557
  language processing 275
  semantic memory 208–10
connectionist models 52–3, 557
  pattern recognition 83–7
consent, informed 482
consequent 345–51, 557
consolidation 174, 179, 557
  neurons 41–2
constructive episodic simulation hypothesis 203

constructive search 395, **557**
context 27–8, **557**
  emotion 453
  language
    phonology 262–5
    polysemy 278–80
    processing 284
  memory 193–5, 453
  pattern recognition 81
contradictions, search for 420
contralaterality 45–6, **557–8**
  auditory sensation and perception 94
  hemineglect 116
  visual sensation and perception 62–3
control processes 24, **558**
controlled attention 114–15, 125–8, **558**
  hemineglect 118
  short-term and working memory 163–4, 170
    Engle and Kane's controlled attention theory of working memory 157–8, 169, 170, **559**
controlled mental processes 103
convergence (neuronal communication) 40
conversation 323–4
  cognitive characteristics 325–8
  empirical effects 329–31
  metaphors and idioms 331–2
  structure 324–5
conversational rules 325–6, **558**
cooperative principles 325–6, **558**
coping 461
coronary heart disease risk 481–2
corpus callosum 43, 46, **558**
cortex 44–5, 47–50, **558**
cost 110, **558**
counterfactual reasoning 361, **558**
COVID-19 pandemic 469–70
cramming *see* massed practice
critical lure 225–6, **558**
cue overload 217
cued recall 223, **558**
culture
  emotional fit hypothesis 461
  language 47
    Sapir–Whorf hypothesis 257
    transmission (linguistic universal) 250
  pattern recognition 82–3
  reminiscence bump 201
  stereotype threat 458–9

D

Darwin, Charles 12
data-driven processing *see* bottom-up, data-driven processing
daydreaming *see* mind wandering
D.B. (amnesia patient) 430
decay 215, **558**
  forgetting from short-term memory 137–9
deception, temporary 481
decision-making 336–8, 351–2, 372–4, 382–3
  Bayesian theories 373–4
  and emotion 457–60
  heuristics 352–61
    adaptive thinking and 'fast and frugal' heuristics 367–72
    biases 361–5
  history of cognitive psychology 15
  limitations in 381–2
  quantum theory 374–6
  risky 365–7
declarative memory *see* explicit (declarative) memory
deep structure 271–2, **558**
default mode network (DMN) 104, **558**
  social cognition 428, 432
déjà vu 243
dementia
  forgetting 222, 237, 243
  functional magnetic resonance imaging 479
  social cognition 434
demyelinating diseases 37
dendrites 36, **558**
depression
  attentional inhibition 124
  emotion regulation 461
  emotional Stroop task 450
  transcranial magnetic stimulation 480
depth of processing *see* levels of processing
derivational complexity hypothesis 272
Descartes, René 8–9, 37
descriptive model 353–4
detection 57
  signal detection theory 15, 59–60, 375
developmental psychology 124, 327
dichromacy 65
direct theory 327–8, **558**
directional reception 249
discreteness 250
discrimination 57–9
displacement 250, 252, 255, **558**
dissociation **558**
  amnesia 237–8
  cognitive neuropsychology 51
  cognitive neuroscience 34–5
  episodic and semantic memory 237–8
distributed practice 179, 183, **558**
  metacomprehension 300
divergence (neuronal communication) 40
divided attention 102, 119
domain knowledge **558**
  increasing your 418
  limited 376–8
domain-general 435–6, **558**

dominance
defined 46, 558
hemispheric specialisation 46–7
DONALD + GERALD problem 394, 419
Donders, F. C. 144
dopamine 40
dorsal anterior cingulate cortex (dACC) 433, 436, 440
dorsal mid-anterior cingulate cortex (dmACC) 427
dorsal pathway 49, 95, 558
dorsal posterior insula (dpINS) 440
dorsal striatum 433
dorsal temporal pole (dTP) 433
dorsolateral prefrontal cortex (DLPFC)
defined 149, 558
loneliness 442
reasoning 380
social cognition 427, 433–5
working memory 149–50, 169
dorsomedial prefrontal cortex (DMPFC)
conditional reasoning 347
social cognition 427, 431–3, 435
dot-probe paradigm 465–6
double dissociation 558
amnesia 237
aphasia 291, 294
cognitive neuropsychology 51
cognitive neuroscience 35
downhill change 362, 558
drinking glasses problem 393
dual-coding hypothesis 193
duality of patterning 250
dual-process model of mental processes 103
dual-task method 120, 558
Dvorine Colour Blind test 65
dysexecutive syndrome 151, 558
dysfluencies 256, 558
dyslexia *see* alexia

E

ear 93–4
early selection 121, 559
theory 121–3
Easterbrook hypothesis 454–5, 559
Ebbinghaus, Hermann von
accuracy measures 22
episodic memory 174, 177–80, 185, 199
history of cognitive psychology 11–12, 15
problem-solving 422
echoic (auditory sensory) memory 95–8
ecological validity 7–8, 559
effect sizes 23, 559
reproducibility crisis 481
Eisenberger, Naomi 425
elaborative rehearsal 184–5, 189
electrodermal activity (EDA) 475–6
electroencephalography (EEG) 467, 474
electromyography (EMG) 467, 477–8
elimination by aspects 360–1, 559
embodied cognition
language 317–18
learning and remembering 197
speech perception 266
visuo-spatial sketch pad 154
embodied perception 90
embodiment 30, 559
emergent properties 50, 559
emotion 445–6, 462–3, 559
amygdala 44
and decision-making 457–60
and language 455–7
and memory 202, 452–5
nature of 446–7
neurology 447–8
and perception 448–52
regulation 460–1
social cognition 435–7
emotional fit hypothesis 461
emotional Stroop task 450–1, 469–70, 559
empathy 50, 427–8, 433–8
empiricism 10, 559
enactment effect 189–90, 559
encoding 559
modal model 24
process model 25–6
encoding specificity 194–5, 559
Engle and Kane's controlled attention theory of working memory 157–8, 169–70, 559
enhancement 303–4, 559
Enstellung *see* negative set
epilepsy 243
episodic buffer 148–9, 156–7, 173–4, 559
episodic future thinking 202–3
episodic memory 173–4, 559
brain 34, 43
consolidation 174, 179
context 193–5
dissociation of semantic memory and 237–8
Ebbinghaus tradition 174, 177–9
emotion 455
and episodic future thinking, comparison between 203
improving 175, 177, 187–93
infantile amnesia 199
information storage 180–7
language 314
metamemory 174, 180
mnemonics 174–7
quantum theory 375
research methods 479
social cognition 429–31
erasure 73, 559
ethics, research 480, 482

Euler circles 341, 344
Event Indexing Model 320–2
event-related potentials (ERPs) 474–5
  language 287–9
    comprehension 302
    polysemy 279–80
events 322–3
evolution
  adaptive thinking 368
  language 294
    development 247
    and writing 287
  loneliness 440
  social brain hypothesis 435
excitatory post-synaptic potentials (EPSPs) 36
executive control *see* central executive
executive functions
  and frontal lobe tasks 165
  social cognition 434, 436
experience, and decision-making 377
explicit (declarative) memory 30, 173, **559**
explicit processing 105–7, **559**
expressive aphasia *see* Broca's aphasia
extinction 118
extramission 64
extrinsic self-processing 432, **559**
eye 61–6
eye tracker 472–3, **559**
  loneliness 441
  problem-solving 401
  reading 305–8
eye–mind assumption 305–6, **559**
Eysenck, Hans 481–2

F

facial recognition
  disrupted (prosopagnosia) 45, 51, 91, 290
  double dissociation 51
  fusiform face area 68
facilitation of return 113
false alarms 227, **559**
false fame effect 231
false memories 185, 225–6, **559**
  inferences 315
  integration 226–8
familiarity 185
familiarity bias 359, **559**
fan effect 219–22
fast and frugal heuristics 367–72
feature detection approach to pattern recognition 78–80
feature integration theory 473
feature search 111–14
Fechner, Gustav 10
fifteen pennies problem 419–20
figural effect bias 339
figure-ground principle 75
filtering/selecting 120–2, **559**
fishing 481, **560**
fissure of Rolando 45
fixation 73–4, 305–8, **560**
  eye-tracking methodology 472
flanker task 467–8
flashbulb memories 454, **560**
flexibility 251, **560**
fluid intelligence 166, 169
focal attention *see* visual attention
forgetting 179, 213–14, 244–5
  Accelerated Long-term Forgetting (ALF) 243
  amnesia and implicit memory 236–43
  causes 137–9
  curve 178
  decay 215
  interference 215–24
  proactive interference 139–41
  retrieval-induced 222, **565**
  retroactive interference 139
  seven sins of memory 214–15
  from short-term memory 137–41
fovea 63, **560**
Fox, Elaine 101
framing 365–7
free recall 141–2, **560**
  retrieval cues 223
  tasks 186
Freud, Sigmund 199
Friedman, Naomi 134
frontal lobe 45, **560**
  language 291, 304
  memory 238
    working 150–1, 158, 165
  problem-solving 404, 410
  reasoning 379–80
functional fixedness 398–9, 401, **560**
functional magnetic resonance imaging (fMRI) 35, 478–9
functional near-infrared spectroscopy (fNIRS) 412
functionalism 12, **560**
fusiform face area (FFA) 45, 68
fusiform gyrus 91
future focus, visuo-spatial sketch pad 154
future orientation 31

G

GABA 40, **560**
Gage, Phineas 448
Galton-Crovitz technique 201
galvanic skin response (GSR) 475–6
gambler's fallacy 357
gamma-aminobutyric acid *see* GABA
ganglion cells 61–3
garden path sentences 285–6, **560**
gaze duration 305–6, 308, **560**

Gazzaniga, Michael 33
general problem-solver (GPS) 416–17, **560**
generate and test 395, **560**
generation effect 188, 190, **560**
Gestalt 388, 395–6, **560**
  agnosia 92
  early research 396–7
  history 13
  pattern recognition 75–7
  problem-solving difficulties 398–401
gesture 323, 332–3, **560**
Gigerenzer, Gerd 336, 368–70, 372, 381
given-new strategy 275–6, **560**
glial cells 37
global–local distinction 104, **560**
glutamate 40, **560**
goal 389–90, 392, **560**
go/no-go task (GNGT) 466–7
good continuation 77, **560**
grammar 256, **560** *see also* phonology; semantics; syntax
  case 280–2, **556**
  garden path sentences 286
  phrase structure 269–72, **564**
  semantic grammar theory 285–7
  situation models 316
  transformational 269–73, 275
grapheme–colour synaesthesia 69
graphs, interpreting 19–21
Greenwald, Anthony 470
Gross, James 445
Grossarth-Maticek, Ronald 481–2

H

habituation 108–9, **560**
handedness 47
hearing *see* auditory sensation and perception
heart rate variability (HRV) 474–6
heat maps 472
Helmholtz, Hermann von 10
hemineglect 116–18, **560**
hemispheric specialisation 45–7, **560**
HERA model 238, **560**
heuristic theory 342, **561**
heuristics **560–1**
  availability 358–9
  decision-making 352–65
  elimination by aspects 360–1, **559**
  fast and frugal 367–72
  improving your problem-solving 420–2
  1/N 368
  prototypicality 378
  recognition 369–70
  representativeness 355–8, **565**
  risk aversion and seeking 365–7
  satisficing 369–70, 419
  simulation 359–60, 364
  subgoal 419
  syllogistic reasoning 342–3
  'take the best' 369–70
  undoing 361–5
hidden units 84–6, **561**
hindsight bias 363–5, **561**
hippocampus 43–4, **561**
  emotion 448, 452, 455
  language 312
  memory 241, 452, 455
    amnesia 200, 239, 241
    consolidation 216
  neurogenesis 42
historical bias 382
history of cognitive psychology 8–17
H.M. (amnesia patient) 239–41, 243, 422
Homer 8
hospital births 356, 384–5
Hubel, David 63, 80
hypermnesia 178
hyperscanning 412
hypothalamus 43, 66, 437
hypothesis testing 350–1

I

iconic memory *see* visual sensory (iconic) memory
iconic systems 251
identity reference 311–12
idioms 332
ill-defined problems 387–8, 392, **561**
illicit conversion 342, **561**
imagery *see* visual imagery
imitative behaviour 427–8
immediacy assumption 305–6, **561**
impetus theory 377
implanted memories 232–3, **561**
Implicit Association Test (IAT) 470
implicit memory (non-declarative memory) 30, 173, **561**
  amnesia 240–3
  attention 106–7
implicit priming tasks 471
implicit processing 105–7, **561**
inattention *see* attention: lack of
incubation 404, **561**
independent and non-overlapping stages 27, **561**
indirect replies 329–31
indirect requests 329, **561**
infantile amnesia 199–200, **561**
inferences
  authorised 316, **556**
  bridging 316–17, **556**
  individual differences 319
  language comprehension 315
  problem-solving, improving your 419–20
  speech acts 318–19
  unauthorised 316, **567**

inferior frontal gyrus (IFG) 365, 412, 427, 436
inferior parietal lobe (IPL) 347, 427, 434
inferior temporal gyrus 442
inferolateral frontal cortex (IFLC) 433
information processing 17–18
 measurement 18–23
informed consent 482
inhibition **561**
 empathy 436
 frontal lobe tasks 165
 go/no-go task 466–7
 selective attention 124
 stop signal task 467
inhibition of return 112–14, **561**
inhibitory post-synaptic potentials (IPSPs) 36
input attention 114–15
 hemineglect 118
input units 84, **561**
insight 401–4, **561**
 Gestalt 396–7
 neurocognition 408–12
insula 437
integration 226–8, **561**
integrity, research 480–2
intelligence tests
 bilingualism 287
 short-term memory span 135
 working memory 166, 169
intensity 446–7, **561**
 and memory 452, 454–5
intentionality 257–8
interchangeability 250
interference 215–17, **561**
 associative 218–20, **556**
 auditory sensation and perception 96–8
 overcoming forgetting from 222
 paired-associate learning 217–18
 part-set cueing effect 224
 proactive 139–41, 168, 218, **564**
 retrieval cues 222–4
 retroactive 139, 218, **565**
 short-term memory 137–41, 143, 145
 situation models 220–2
 visual sensation and perception 72–3
 working memory 169
  dual-task method 158–62
  and long-term memory 168
interference tasks 467–70
interneurons 38
intraparietal sulcus 432
intrinsic self-processing 432, **561**
introspection 11, 16, **561**
intrusions 23
invariance, problem of 263–4
invisible gorilla experiment 119
involuntary memory 202, **561**
irrelevant speech effect 151
Ishihara test 65
Island, The (film) 431
isolation effect 180–1
Izquierdo, Ivan 172

J

James, William 10, 12, 125
Johnson-Laird, Philip 336
judgements of learning (JOLs) 300
just noticeable difference (JND) 57, **561**

K

Kahneman, Daniel 3, 354, 356–8, 360, 363, 366–7, 370–2, 379
K.C. (amnesia patient) 237–8, 143
knowledge, and decision-making 377
knowledge-lean 388, **561**
knowledge-rich 388, **561**
Köhler, Wolfgang 396
Kounios, John 386

L

labour-in-vain effect 300
language 246, 294–7, 333–5 *see also* linguistics
 animal communication 254–6
 brain 287–94
 compression 254
 conceptual and rule knowledge 298–304
 conversation 323–32
 cultural influences 47
 defined 248–9, **561**
 development 247–8, 254
 and emotion 455–7
 events 322–3
 gesture 323, 332–3
 hemispheric specialisation 47
 infantile amnesia 200
 learning 288
 levels of analysis 255–8
 lexical factors 276–80
 phonology 258–67
 reading 305–11
 reference 311–14
 semantics 280–7
 situation models 314–22
 syntax 267–76
 universals 248–53
language acquisition device (LAD) 247
Lashley, Karl 238
late selection 124, **561**
 theory 122–4
lateral cortex 429
lateral geniculate nucleus (LGN) 66–7, **561**
lateralisation
 cerebral 46–7, **557**
 language 287

law of disuse 215
law of large numbers 355–7
law of small numbers 355–7
leading questions 228–30
learning 172–4, 210–12
  autobiographical memory 198–202
  context 193–5
  Ebbinghaus tradition 177–9
  facts 195–7
  gesture 333
  improving episodic memory 187–93
  information storage 180–7
  metamemory 180
  mnemonics 174–7
  neurons 40–2
  paired-associate 186, 217–18
  situation models 197–8
  state-dependent 195
  thalamus 43
  verbal 15–16, 567–8
  see also comprehension
levels of analysis, language 255–8
levels of processing **561**
  explicit memory 242
  learning and remembering 184–7, 190
levels of representation theory 299
lexical decision task 276–7, **562**
  aphasia 293
  context effects 27–8
  process model 25–7
lexical factors 276–7
  lexical representation 277–8
  morphemes 277
  polysemy 278–80
lexical memory 277–8
lie detectors 476
life scripts 201
lightning 69–70
limbic system 447, **562**
Linda Problem 371–2, 375–6
linguistic relativity (Sapir–Whorf)
    hypothesis 256, **562**
  strong version 257
  weak version 257–8
linguistic universals 248–53, **562**
linguistics 248, **562**
  history 16–17
  see also psycholinguistics
lobes (of the brain) 44, 45, **562**
Loftus, Elizabeth 213, 228–31
logic, formal
  conditional reasoning 345–50
  hypothesis testing 350–1
  syllogistic reasoning 338–45
loneliness 438, 440–3
  emotional fit hypothesis 461
long-term memory (LTM) 172–4,
    210–12, **562**
  attention 106
  context effects 27
  facts 195–7
  future orientation 202–3
  hypothalamus 43
  improving 202
  metacomprehension 300
  modal model 24–5
  process model 25, 26
  reading 308
  and short-term memory 181
  situation models 197–8
  taxonomy 173
  and working memory 155–7, 167–8, 173–4
  see also amnesia; autobiographical
    memory; episodic memory;
    forgetting; semantic memory
long-term potentiation (LTP) 40–2, **562**
lookahead 392, **562**

M

MacLeod, Colin 464
magnetic resonance imaging (MRI) 478, **562**
maintenance rehearsal 184, 185, 187, 189
mapping 302, **562**
marking *see* erasure
Markowitz, Harry 368
Marks, David 482
Marr, David 87, 88
masking 73, **562**
massed practice 179, 183
  metacomprehension 300
matching heuristic 342
means–end analysis 412–13, 419, **562**
  general problem-solver 416–17
  Tower of Hanoi 413–16
medial frontal cortex (MFC) 431
medial orbitofrontal cortex (mOFC) 427
medial prefrontal cortex (mPFC) 427,
    432, 456
medial temporal lobe 455
meditation 105
memory
  adaptive 193
  attention 127–9
  auditory sensory (echoic) 95–8
  autobiographical 198–202, 455, 556
  availability and accessibility 213–14
  availability heuristic 358–9
  consolidation 41–2, 216
  defined 6–8, **562**
  and emotion 202, 452–5
  episodic *see* episodic memory
  explicit (declarative) 30, 173, 559
  false *see* false memories
  gesture 333
  hindsight bias 364
  history of cognitive psychology 10–12, 15

implanted 232–3, **561**
implicit *see* implicit memory (non-declarative memory)
incorrect 224–36
information processing 18, 22–3
involuntary 202, **561**
irony of 235–6
language 312, 319–22
lexical 277–8
long-term *see* long-term memory (LTM)
modal model 24–5, **562**
neurons 41–2
neurotransmitters 40
non-declarative *see* implicit memory (non-declarative memory)
overconfidence in the accuracy of 233–4
procedural 242–3
process model 25–6
prospective 202, **564**
recovered 235
repressed 235
semantic *see* semantic memory
sensory 24, 28, **566**
short-term *see* short-term memory (STM)
source 234, **566**
speech planning and production 274, 275
synaesthesia 69
trans-saccadic 74–5
visual sensory (iconic) 69–74, 98
working *see* working memory
see also forgetting
memory impairment 230, **562**
mental imagery *see* visual imagery
mental model 343, 344, 350, **562**
mental model theory 343, **562**
mental ontology 479, **562**
mental rotation 154–5
mental rules theory 343–4, **562**
mentalising *see* theory of mind (ToM)
Mesquita, Batja 461
metacognitions 30, 180, **562**
neural basis 426, 427
metacomprehension 299–300, **562**
meta-linguistic awareness 331, **562**
metamemory 174, 180, **562**
metaphors 331–2
method of loci 174–5, **562**
migraine 480
mild cognitive impairment 479
mind wandering 131–2, **562**
mirror drawing 240, 243
mirror neurons 48–9, 412, **562**
misattribution (memory) 214–15
misinformation acceptance 231–3
misinformation effect 229, **562**
misleading information 228–30, **562**
misreading effect 82
missionary–cannibal problem 416, 417
Miyake, Akira 134, 165
mnemonic device 174–7, 189–90, **562**
modal model of memory 24–5, **562**
modality effect 97, **562**
modus ponens 345–50
modus tollens 346–51
monkey communication 254–5
moral reasoning 168
morphemes 277, **563**
combining phonemes into 262
motor cortex 48, **563**
motor theory of speech perception 266, **563**
multiconstraint theory 408, 409, **563**
multidimensional self 431–2
multiple sclerosis 37
myelin sheath 36, **563**

N

N400 279, 287, **563**
naive physics 376–7
naming 252, 255, **563**
negative afterimages 65, **563**
negative priming 124, **563**
negative set 399–401, **563**
Neisser, Ulrich 7
neobehaviourism 13, 14
neocortex 42, **563**
principles of functioning 45–7
structures 44–5
networks 479, **563**
neural basis of cognition *see* cognitive neuroscience
neural imaging 465, 478–9, **563**
neural net models *see* connectionist models
neurocognition 408–12
neurogenesis 42, **563**
neuron 35–6, **563**
communication 38–40
connectionist models 209
learning 40–2
levels of explanation 50
structure 36–8
neurophysiological research methods 474–8
neuroscience *see* cognitive neuroscience
neurotransmitters 40, 42
Newell, Allen 18
Newton, Isaac 401
nine-dot problem 419
NMDA receptors 40–1
nociceptors 37
nodes of Ranvier 36, **563**
non-declarative memory *see* implicit memory
norepinephrine 40, **563**
normative model 353–4, 358
null findings 481, **563**
number agreement 268

O

object recognition 87–93
obsessive-compulsive disorder (OCD) 480
occipital lobe 45, **563**
 agnosia 91, 92
 emotion and perception 449
 visual attention 108
 visual sensation and perception 79
 working memory 149
occipital-parietal cortex 449
ocular dominance 63
odours
 and emotion 448
 and involuntary memory 202
 see also smell
1/N heuristic 368
online comprehension tasks 301–2, **563**
onomatopoeia 251, **563**
open science 480–1
operator 392, **563**
optic nerve 62, 63, 66
optimism bias 364–5
orbitofrontal cortex (OFC) 427, 428
organisation 189–91, **563**
orienting reflex 107–10, **563**
output units 85, 86, **563**
overlearning 178–9, **563**
overload 120

P

P600 287–8, **563**
paired-associate learning 186, 217–18
Paivio, Alan 192, 193
Pandemonium model 78–81
panic disorder 469
parade problem 405–7, 409–10
parallel distributed processing (PDP) models *see* connectionism
parallel processing 27, 80, **563**
parietal lobe 45, **563**
 agnosia 92
 attention 107
 problem-solving 410
 working memory 149, 169
parse **563**
 object recognition 88
 semantics 281
  and syntax 286
 transformational grammar 271–2
partial report condition **563**
 auditory sensation and perception 95–6
 visual sensation and reception 72
participant information sheet (PIS) 482
part-set cueing effect 224, **564**
pattern recognition 75
 conceptually driven 81–3
 Gestalt grouping principles 75–7
 template approach 77–8
 top-down processing 81–7
 visual feature detection 78–80
 see also object recognition
Pavlov, Ivan 13
Pelosi, Anthony 482
perception 56–7, 64, 98–100
 animals 396
 embodied 90
 and emotion 448–52
 psychophysics 57–60
 speech 262–4
 see also auditory sensation and perception; visual sensation and perception
performance 256, **564**
peripheral nervous system (PNS) 42
persistence 96–7, 214
personality, and risk for cancer and cardiovascular disease 481–2
p-hacking 481
phi phenomenon 80, **564**
phonemes 258–61
 boundaries 261
 combining into morphemes 262
 combining into words 262–3
phonemic competence 262–3
phonological loop 148, 150–3, **564**
 dual-task method 160–2
phonological similarity effect 152, **564**
phonological store 151, **564**
phonology 256, 258, **564**
 apparent segments in speech 267
 combining phonemes into morphemes 262
 effect of context 264–5
 embodiment in speech perception 266
 level of language analysis 255
 reading comprehension 311
 sounds in isolation 258–9
  articulation 259
  phonemes 259–61
 speech perception and context 262–4
 top-down and bottom-up processes 265–6
phrase order 268
phrase structure 269–72, **564**
physics, naive 376–7
Pinker, Steven 246–8, 251
Pitt, Brad 91
Plato 9
polygraphs 476
polysemy 278–80, **564**
pop-out effect 111, 114, **564**
positive evidence, search for 348–50
positron emission tomography (PET) 35, 149, 238, 239, **564**
Posner, Michael 101, 109–11, 118, 126
posterior cingulate cortex 433, 435
posterior insula (PI) 427
posterior superior temporal sulcus (pSTS) 427

post-traumatic stress disorder (PTSD) 450
power 481, **564**
practice
  attention 127–9
  distributed 179, 183, 300, **558**
    metacomprehension 300
  episodic memory 177
  massed 179, 183, 300
  working memory 164
pragmatics 299, **564**
precuneus 428, 433
prefrontal cortex (PFC)
  emotion 447, 448
    and language 456
    and perception 449
  problem-solving 410
  reasoning 379–80
  social cognition 427, 428, 431, 434, 442
premotor cortex 431, 432
primacy effect 142–3, 183, **564**
primary auditory cortex 94
primates
  communication 254–6
  Social Brain Hypothesis 435
prime 124, **564**
priming 28, **564**
  attention 126, 127
  comprehension 303
  emotion and decision-making 459
  implicit priming tasks 471
  negative 124, **563**
  polysemy 278–9
  repetition 173, 241–3, **565**
  semantic 286–7, 411
  semantic memory 209
  syntactic 273
proactive interference (PI) 139–41, 168, 218, **564**
probability heuristics theory 342, **564**
probes 145, **564**
problem of invariance 263–4
problem space 391–2, **564**
problem-solving 352, 386–9, 423–4, **564**
  in action 386–7
  analogy 401, 405–12
  automating components of the solution 418–19
  characteristics 389–90
  and emotion 457
  Gestalt 395–401
  gesture 333
  improving your 418–23
  insight 401–4, 408–12
  levels of explanation 50
  means–end analysis 412–17
  pattern recognition 80
  types of problems 387–8
  vocabulary 390–5
  and working memory 168, 169
procedural memory 242–3
process model 25–6, **564**
  assumptions 26–9
  connectionism 52
  emotion regulation 460–1
  Sternberg task 145–6
processing fluency 234, **564**
processing resources, limitations in 378–80
production 416, **564**
production aphasia *see* Broca's aphasia
production effect, episodic memory 188
production system 416–17, **564**
productivity 250, 253, 255, **564**
projectors (synaesthetes) 68–9
pronoun reference 312
propagated 39, **564**
proposition 195–7, **564**
propositional network theories 221
propositional textbase 299, **564**
prosody 276, **564**
  conversation 325
  and emotion 456
prosopagnosia 45, 51, 91, 290
prospective memory 202, **564**
protanopia 65
prototypicality heuristic 378
proximity 76, **565**
pseudo-events 232–3
pseudoisochromatic plate tests 65
psycholinguistics 248, **565**
  language development 247
  lexical factors 277
  metaphors 332
  phonemes 260
  productivity 253
  syntax 267, 269, 273, 284
psychological refractory period 125, **565**
psychology
  history 8–17
  linguistic theory 272
  performance-related aspects of language 256
psychophysics 57–59, **565**
publication bias 481
pure word deafness 290, 292, **565**
p-value 23

Q

quantum theory 374–6
Quillian, M. R. 203–5

R

radiation problem 406–9, 421
rapid fading (linguistic universal) 249
Raven's standard progressive matrices test 379–80
R.B. (amnesia patient) 431

reaction time 465–71
reading 305
  factors affecting 309–11
  gaze duration 305–6
  online
    basic effects 306–8
    benefits 308–9
reasoning 336–8, 372, 382–3
  conditional *see* conditional reasoning
  and decision-making 352–5
  hypothesis testing 350–1
  limitations in 376–82
  relational 338
  syllogistic *see* syllogistic reasoning
  types of 377–8
  and working memory 168, 347, 350
    limitations in processing resources 379, 380
recall
  forgetting versus 235
  retrieval cues 222, 223
  tasks 185, 186, 187
recency effect 142–3, 183, 565
receptive aphasia *see* Wernicke's aphasia
receptor cells 37
recoding 137, 565
recognition 222
recognition by components (RBC) theory 87–90
recognition heuristic 369–70
recognition task 144, 185, 187, 565
recollection 185
reconsolidation 231–2, 565
recovered memories 235
recreational problems 393–5
  goals 392
reductionism 8, 565
reductionistic approach 388–9, 565
Rees, Geraint 33
reference 311, 565
  anaphoric 311, 555
  antecedents, influence of 312–13
  clause recency, advantage of 313–14
  identity 311, 312
  individual differences 319
  pronoun 312
  simple 311–12
  synonym 312
reflexes 37–8
region of proximal learning 300
rehearsal 565
  elaborative 184, 185, 189
  improving episodic memory 190, 191
  learning and remembering 181–4
  maintenance 184, 185, 187, 189
  short-term memory 135, 137, 141–3
  working memory 150, 151
relational reasoning 338
relearning task 177–9, 185, 186
relocation bump 201
remembered self 429–31
remembering *see* memory
reminiscence 178
reminiscence bump 200–1, 565
repeated name penalty 311, 565
repeated transcranial magnetic stimulation (rTMS) 479, 480
repetition blindness 82, 565
repetition priming 173, 241–3, 565
replication 23
representation (of knowledge) 30, 565
representational bias 382
representational momentum 156
representativeness heuristic 355–6, 565
  biases 356–8
repressed memories 235
reproducibility crisis 481–2
research
  ecological validity 7–8
  history 9–10, 15
  language processing 288–9
  methods 464, 483
    ethics 480, 482
    integrity 480–2
    neuroimaging and brain stimulation methods 478–80
    neurophysiological methods 474–8
    purpose 464–5
    reaction time-based tasks 465–71
    visual search tasks 472–3
  psychologists as subjects 199
resource theories of attention 125
respect for research participants 482
response time (RT) 21–2, 214, 565
  short-term memory scanning 144, 146
retina 61–3, 66
retina–geniculate–striate pathway 66–7, 565
retrieval 7, 565
retrieval cues 177, 222–4, 565
  encoding specificity 194
  involuntary memory 202
retrieval-induced forgetting 222, 565
retroactive interference (RI) 139, 218, 565
retrograde amnesia 236–7, 430
rhetoric 174
Ribot's law 237, 565
right shift theory 47
right ventral prefrontal cortex (rvPFC) 439
risk aversion and seeking 365–7
robins 6
rod cells 61–3, 66
Rosch, Eleanor 257
rostrolateral prefrontal cortex (RLPFC) 347
rule knowledge 298–304

S

saccades 305–7, **565**
eye-tracking methodology 472
fixation 73
psychological refractory period 125
trans-saccadic memory 74–5
visual attention 74
salami 481, **565**
salience network (SN) 429
sample stimuli and test words (online comprehension tasks) 301
Sapir, Edward 258
Sapir–Whorf (linguistic relativity) hypothesis 256, **562**
strong version 257
weak version 257–8
satisficing heuristic 369, 370
subgoals 419
savings score 177–8, **565**
schizophrenia 124
Scoville, William 239
second-order theory of mind 327–8, **566**
selecting *see* filtering/selecting
selective attention 102, 118–20, **566**
cocktail party effect 120–1
models 121–4
short-term and working memory 134
self-control 451–2
self-reference effect 188, **566**
self-referential processing 427–31, 443
self-reflection 429, **566**
self-understanding 429, **566**
semantic congruity effect 59
semantic grammar theory 285–7
semantic knowledge 255
semantic memory 173, **566**
brain 34
connectionism 208–10
dissociation of episodic memory and 237–8
feature comparison models 205–7
impairment 209–10
infantile amnesia 199
language 314, 317
network model 203–5
problem-solving 398, 410
quantum theory 375
relatedness 207–8
social cognition 429–31
semantic networks 203–5
semantic priming 286–7, 411
semanticity 250, 251, **566**
animal communication 255
semantics 256, 273, 280–1, **566**
brain 287, 289, 291, 292, 294
case grammar 281–2
comprehension 304
semantic grammar theory 285–7
and syntax 282–4
sensation 56–7, 64, 98–100
psychophysics 57–60
see also auditory sensation and perception; visual sensation and perception
sensorimotor network (SMN) 428–9, 432
sensory cortex 48, 429
sensory effector cells 37
sensory memory 24, 28, **566**
sensory neurons 37
sensory thresholds 57
sequential stages of processing 26, **566**
serial position
learning and remembering 182–3
reading times 310
serial position curve 22–3, 141–3, 183, **566**
serial processing 26, **566**
serial recall 141–2, **566**
tasks 186
set effects *see* negative set
shadowing experiments 120–1, 127–8
Shamay-Tsoory, Simone 425, 433, 437–8
Shiffrin, Richard 172, 181
shifting 165
short-term memory (STM) 134–5, 170–1
attention, lack of 102
capacity 136–7
defined 135, **566**
forgetting from 137–41
limited-capacity bottleneck 135–41
and long-term memory 181, 182
modal model 24
retrieval 141–7
scanning 143–7
and working memory 148, 163–4
sight *see* visual sensation and perception
signal detection theory 15, 59–60, 375
similarity 76–7, **566**
Simon, Herb 18
Simon task 467
simulation heuristic 359–60, 364
situation model 299, 314–15, **566**
creation 316–17
embodied cognition, influences of 317–18
events 323
grammatical aspect 316
individual differences 319
inference making 315
interference 220–2
learning and remembering 197–8
reading 308
speech act 318–19
updating 319–22
six pennies problem 393
sixteen dots problem 421

skin conductance response (SCR) 475–6
Skinner, B. F. 13, 16–17
sleep
 emotion and memory 453, 455
 and interference 215–16
 memory consolidation 41, 179
 problem-solving 405
 prospective memory 202
 thalamus 43
smell
 amygdala 44
 thalamus 43
 see also odours
Smith's feature comparison model 205–7
social anxiety 461, 466
social brain hypothesis 435, 443
social cognition 425–6, 443–4
 defined 426, **566**
 neural basis 427–9
 responding to adverse social signals 438
  loneliness 440–3
  social rejection 438–40
 understanding others 426–8, 433–8
  empathising 435–7
  integrating social understanding networks and processes 437–8
  mentalising 433–5
 understanding the self 427–9
  multidimensional self 431–2
  remembered self 429–31
social cognitive neuroscience 426
social cues 107–8, **566**
social exclusion 461
social neuroscience 50, 426, **566**
social rejection 438–40
social rewards 427, 428
social roles 325
social working memory 435
Socrates 9
sodium–potassium pump 39, **566**
soma 36, **566**
somatosensory cortex 437, 440
source memory 234, **566**
source misattribution 231
source monitoring 195
span of apprehension/attention 71, **566**
spatial cuing task 109–10
 hemineglect 118
specialisation 250
speech
 apparent segments 267
 context 262–5
 perception 262–4
  embodiment 266
  motor theory 266, **563**
 planning and production 274–6
 top-down and bottom-up processes 265–6
speech act 318–19, **566**
Sperling, George 18
split-brain 47, **566**
spotlight attention 110–11, **566**
spotlight metaphor 111, 157, **566**
spreading activation 205, **566**
Squid Game (drama series) 425–6
standard theory *see* modal model of memory
state-dependent learning 195
statistical significance 23
stereotype threat 458–9, **566**
stereotyping 357–8, **566**
Sternberg, Saul 144–7
Sternberg task 144–7, 149, **566**
stop signal task (SST) 467
story mnemonic 176–7, **566**
stress
 cancer and cardiovascular disease risk 481–2
 decision-making
  impaired 457–9
  improved 459–60
 forgetting 215
 hypothalamus 43
striatum 437
stroke
 aphasia 290
 empathising 436
 forgetting 222
 hemineglect 116
Stroop effect 126
Stroop interference 468
Stroop task 276–7, 469
 emotion and self-control 451
 emotional 450–1, 469–70, 559
 social cognition 434, 442
 working memory 166–7
structuralism 11, 13, **567**
structure-building 302–5, **567**
subcortical brain structures 42–4
subgoal 390, 419–20, **567**
subjective organisations 191, **567**
subject-object-verb (SOV) language 249, **567**
subject-verb-object (SVO) language 248, 268, **567**
subliminal stimuli 57
suggestibility (memory) 214–15
superior temporal gyrus 411, 412, 442
superior temporal sulcus (STS) 427, 433
supplementary motor area (SMA) 347, 432, 435, 436
suppression 303–4, **567**
supraliminal stimuli 57
surface form 299, **567**
surface structure 271–2, **567**
survival motivation 193
sustained attention *see* vigilance
syllogistic reasoning 339
 biases 339–42
 categorical syllogisms 338–42
 theories 342–5
symbolic distance effect 58, 59

synaesthesia 68–9
synapse 36, 39–40, **567**
  parallel distributed processing models 83
synaptic cleft 39, **567**
synonym reference 312
syntactic priming 273
syntax 256, 267–9, **567**
  brain 287–9, 291, 294
  cognitive role 273–6
  comprehension 304
  conversation 325
  elements 267–8
  level of language analysis 255
  reading 308
  and semantics 282–4
  transformational grammar 269–73

T

tabula rasa 10, **567**
'take the best' heuristic 369–70
task effects 185–7, **567**
teamwork, creativity in 412
template approach to pattern recognition 77–8
temporal lobe 45, **567**
  agnosia 91, 92
  language 291, 292, 304
  problem-solving 411
  reasoning 379–80
  social cognition 436
Temporal Lobe Epilepsy 243
temporary deception 481
temporo-parietal junction (TPJ) 427, 433, 435
testing effect 183–4
textbase models 308
thalamus 43, 66, **567**
theory of mind (ToM) 327–8, 433–5, **567**
  empathising 435
  integrating social understanding networks and processes 437–8
think-aloud verbal protocol (online comprehension tasks) 301–2
three-eared man procedure 95–6
time measures 21–2
tip-of-the-tongue (TOT) phenomenon 215, 223, 292
Titchener, Edward 11, 12
Tolman, Edward C. 13
tongue twisters 266
top-down (conceptually driven) processing 28, 30, **567**
  attention 127
  object recognition 87
  phonology 264–6
  visual search 114
  visual sensation and perception 81–7
topic maintenance 326–7, **567**
total feedback (linguistic universal) 250
Tower of Hanoi problem 395, 413, 421–2
  anterograde amnesia 241
  automating components of the solution 418–19
  four-disc version 414–16, 420
  general problem-solver 416
  three-disc version 413–14
traditional transmission (linguistic universal) 250
transcranial magnetic stimulation (TMS) 35, 479–80
transformation problems 392
transformational grammar 269–73, 275
transformational rules 272, **567**
transience (memory) 214–15
transitive inference problems 379, 380
trans-saccadic memory 74–5
treatment-resistant depression 480
Treiman, Rebecca 297
Treisman, Anne 473
trichromacy theory 64–5, **567**
triple associations 51
Tsao, Doris 56
turn taking (conversations) 324–5
Tversky, Amos 354, 356–8, 360, 363, 366–7, 370–2, 379
two string problem 398, 399
two trains problem 419–20
typing 27

U

Ueberwasser, Ferdinand 9
unauthorised inferences 316, **567**
unconscious processes 30, **567**
undoing *see* counterfactual reasoning
undoing heuristic 361
  hindsight bias 363–5
  typical bias 361–3
units 136, **567**
unrealistic optimism 364–5
updating 165

V

valence 446–7, **567**
  and memory 452
ventral medial prefrontal cortex (vmPFC) 433, 434
ventral pathway 49–50, 95, 107, **567**
ventral striatum (VS) 427, 429, 433
ventral tegmental area 437
ventral temporal pole (vTP) 433
ventromedial cortex 436
ventromedial prefrontal cortex (vmPFC) **567**
  emotion 448
  social cognition 427, 432, 434, 437

verbal learning 15–16, 567–8
verbal protocol 28–9, 568
think-aloud 301–2
video games, as attention improvement mechanisms 115
vigilance 105–7, 568
history of cognitive psychology 15, 18
visible light spectrum 64
visual attention 74, 568
capture 108
hemineglect 118
visual cortex 62, 66–8
visual imagery 568
improving episodic memory 191–3
mnemonics 175
visual mental imagery theory 344
visual pathways 49–50
visual perception *see* visual sensation and perception
visual persistence 70, 72
visual search 109–14, 568
conceptually driven processing 81
emotion and perception 450
visual search tasks 472–3
visual selection tasks 473
visual sensation and perception 56–7, 61–4, 98–100
agnosia 90–2
attention 74
early parts of a fixation 73–4
from eye to visual cortex 66–8
gathering visual information 64–6
object recognition 87–93
pattern recognition 75–80
selective attention 120
synaesthesia 68–9
top-down processing 81–7
trans-saccadic memory 74–5
visual sensory memory 69–73
visual sensory (iconic) memory 69–74, 98
visuo-spatial sketch pad 148–9, 153–7, 568
dual-task method 160–2
vocal–auditory channel (linguistic universal) 249
voicing 259, 568
von Restorff effect 180–1, 453, 568

W

wakefulness 43
Wason card problem 348–9
water jug problem 400–1
Watson, John B. 12, 13, 16
Weber, Max 10
well-defined problems 387–8, 392, 568
Wernicke, Carl 10, 291
Wernicke's aphasia 290–4, 568
Wernicke's area 290, 291, 456
animal communication 255
whispered speech 264
whole report condition 568
auditory sensation and perception 95–6
visual sensation and perception 71, 72
Whorf, Benjamin 258
Wiesel, Torsten 63, 80
withdrawal from research 482
W.J. (amnesia patient) 430
word frequency effect 25–6, 568
word order 267–8
word stem completion task 106, 568
working backwards 420
working memory 134–5, 148, 169–71
assessment 158
dual-task method 158–62
span tests 158, 163–5
and attention 166–7
Engle's controlled attention model 157–8
lack of 102, 132
selective 122
components 148–50, 157
central executive 150–1
episodic buffer 156
phonological loop 151–3
visuo-spatial sketch pad 153–6
defined 135, 568
emotion and decision-making 457–9
fan effect 219–20
improving 164–6
language 300, 319
and long-term memory 155–7, 167–8, 173–4
problem-solving 410, 418–19
process model 25
reading 308
and reasoning 168
conditional 347, 350
limitations in processing resources 379, 380
and short-term memory 148, 163–4
social cognition 434–6
speech planning and production 274, 275
working memory spans 169, 568
assessment 158, 163–5
and attention 166–7
language 319
and long-term memory 167–8
and reasoning 168
small as better 168–9
Sternberg task 145
World War II 14, 18, 105
wrap-up 309
Wundt, Wilhelm 9–11

Y

Yerkes–Dodson law 454, 568

Z

Zeki, Semir 56
Zipf's law 254